W9-DFD-827

AN INTRODUCTION TO THE FINITE ELEMENT METHOD

AN INTRODUCTION TO THE FINITE ELEMENT METHOD

J. N. Reddy

Professor of Engineering Science and Mechanics
Virginia Polytechnic Institute and State University

McGraw-Hill Book Company

New York St. Louis San Francisco Auckland Bogotá Hamburg
Johannesburg London Madrid Mexico Montreal New Delhi
Panama Paris São Paulo Singapore Sydney Tokyo Toronto

This book was set in Times Roman by Science Typographers, Inc.
The editors were T. Michael Slaughter and Susan Hazlett;
the production supervisor was Leroy A. Young.
The drawings were done by J & R Services, Inc.
The cover was designed by Scott Chelius.
Halliday Lithograph Corporation was printer and binder.

AN INTRODUCTION TO THE FINITE ELEMENT METHOD

234567890 HALHAL 8987654

ISBN 0-07-051346-5

Library of Congress Cataloging in Publication Data
Reddy, J. N.
An introduction to the finite element method.
Includes index.
1. Finite element method. I. Title.
TA347.F5R4 1984 620'.001'515353 83-7923
ISBN 0-07-051346-5

To my dear friend
Surendar

CONTENTS

*These topics can be omitted, without discontinuity, in the first course.

PREFACE

The motivation which led to the writing of the present book has come from my many years of teaching finite-element courses to students from various fields of engineering, meteorology, geology and geophysics, physics, and mathematics. The experience gained as a supervisor and consultant to students and colleagues in universities and industry, who have asked for explanations of the various mathematical concepts related to the finite-element method, helped me introduce the method as a variationally based technique of solving differential equations that arise in various fields of science and engineering. The many discussions I have had with students who had no background in solids and structural mechanics gave rise to my writing a book that should fill the rather unfortunate gap in the literature.

The book is designed for senior undergraduate and first-year graduate students who have had a course in linear algebra as well as in differential equations. However, additional courses (or exposure to the topics covered) in mechanics of materials, fluid flow, and heat transfer should make the student feel more comfortable with the physical examples discussed in the book.

In the present book, the finite-element method is introduced as a variationally based technique of solving differential equations. A continuous problem described by a differential equation is put into an equivalent variational form, and the approximate solution is assumed to be a linear combination, $\Sigma c_j\phi_j$, of approximation functions ϕ_j. The parameters c_j are determined using the associated variational form. The finite-element method provides a systematic technique for deriving the approximation functions for simple subregions by which a geometrically complex region can be represented. In the finite-element method, the approximation functions are piecewise polynomials (i.e., polynomials that are defined only on a subregion, called an element).

The approach taken in the present book falls somewhere in the middle of the approaches taken in books that are completely mathematical and those approaches that are more structural-mechanics-oriented. From my own experience as an engineer and self-taught applied mathematician, I know how unfortunate outcomes may be arrived at if one follows a "formula" without deeper insight into the problem and its approximation. Even the best theories lead ultimately to some sort of guidelines (e.g., which variational formulation is suitable, what kind of element is desirable, what is the quality of the approximation, etc.). However, without a certain theoretical knowledge of variational methods one cannot fully understand various formulations, finite-element models, and their limitations.

In the study of variational and finite-element methods, advanced mathematics are intentionally avoided in the interest of simplicity. However, a minimum of mathematical machinery that seemed necessary is included in Chapters 1 and 2. In Chapter 2, considerable attention is devoted to the construction of variational forms since this exercise is repeatedly encountered in the finite-element formulation of differential equations. The chapter is concerned with two aspects: first, the selection of the approximation functions that meet the specified boundary conditions; second, the technique of obtaining algebraic equations in terms of the undetermined parameters. Thus, Chapter 2 not only equips readers with certain concepts and tools that are needed in Chapters 3 and 4, but it also motivates them to consider systematic methods of constructing the approximation functions, which is the main feature of the finite-element method.

In introducing the finite-element method in Chapters 3 and 4, the traditional solid mechanics approach is avoided in favor of the "differential equation" approach, which has broader interpretations than a single special case. However, when specific examples are considered, the physical background of the problem is stated. Since a large number of physical problems are described by second- and fourth-order ordinary differential equations (Chapter 3), and by the Laplace operator in two dimensions (Chapter 4), considerable attention is devoted to the finite-element formulation, the derivation of the interpolation functions, and the solution of problems described by these equations. Representative examples are drawn from various fields of engineering, especially from heat transfer, fluid mechanics, and solid mechanics. Since this book is intended to serve as a textbook for a first course in the finite-element method, advanced topics such as nonlinear problems, shells, and three-dimensional analyses are omitted.

Since the practice of the finite-element method ultimately depends on one's ability to implement the technique on a digital computer, examples and exercises are designed to let the reader actually compute the solutions of various problems using computers. Ample discussion of the computer implementation of the finite-element method is given in Chapters 3 and 4. Three model programs (FEM1D, FEM2D, and PLATE) are described, and their application is illustrated via several examples. The computer programs are very easy to understand because they are designed along the same lines as the theory presented in the book.

Numerous examples, most of which are applications of the concepts to specific problems in various fields of engineering and applied science, are provided throughout the book. The conclusions of the examples are indicated by the symbol ■. At appropriate intervals in the book an extensive number of exercise problems is included to test and extend the understanding of the concepts discussed. For those who wish to gain additional knowledge of the topics covered in the book, many reference books and research papers are listed at the end of each chapter.

There are several sections that can be skipped in a first reading of the book (such sections are marked with an asterisk); these can be filled in wherever needed later. The material is intended for a quarter or a semester course, although it is better suited for a semester or a two-quarter course.

The following schedule of topics is suggested for a first course using the present textbook:

Undergraduate		Graduate	
Chapter 1	Self-study	Chapter 1	Self-study
Chapter 2	Section 2.1 (self)	Chapter 2	Section 2.1 (self)
	Section 2.2		Section 2.2
	Sections 2.3.1–2.3.3		Section 2.3
Chapter 3	Sections 3.1–3.4	Chapter 3	Sections 3.1–3.7
	Sections 3.6–3.7		
Chapter 4	Sections 4.1–4.4	Chapter 4	Sections 4.1–4.8
	Section 4.7		
	Sections 4.8.1–4.8.4	Chapter 5	Term paper

Due to the intimate relationship between Sections 3.5 and 4.6, 3.6 and 4.7, and 3.7 and 4.8, they can be covered simultaneously. Also, it is suggested that Sections 3.6 and 3.7 (hence, 4.7 and 4.8) be covered after Section 3.2.

The author wishes to thank all those students and colleagues who have contributed by their advice and criticism to the improvement of this work. The author is also thankful to Vanessa McCoy for skillful typing of the manuscript, to Mr. N. S. Putcha and Mr. K. Chandrashekhara for proofreading the pages, and to the editors Michael Slaughter and Susan Hazlett for their help and cooperation in publishing the manuscript.

J. N. Reddy

Tejashwina vadheetamasthu
(May what we study be well studied)

AN INTRODUCTION TO THE FINITE ELEMENT METHOD

CHAPTER

ONE

INTRODUCTION

1.1 GENERAL COMMENTS

Virtually every phenomenon in nature, whether biological, geological, or mechanical, can be described with the aid of the laws of physics, in terms of algebraic, differential, or integral equations relating various quantities of interest. Determining the stress distribution in a pressure vessel with oddly shaped holes and numerous stiffeners and subjected to mechanical, thermal, and/or aerodynamic loads, finding the concentration of pollutants in seawater or in the atmosphere, and simulating weather in an attempt to understand and predict the mechanics of formation of tornadoes and thunderstorms are a few examples of many important practical problems. While the derivation of the governing equations for these problems is not unduly difficult, their solution by exact methods of analysis is a formidable task. In such cases approximate methods of analysis provide alternative means of finding solutions. Among these the finite-difference method and the variational methods such as the Ritz and Galerkin methods are more frequently used in the literature.

In the finite-difference approximation of a differential equation, the derivatives in the equations are replaced by difference quotients which involve the values of the solution at discrete mesh points of the domain. The resulting discrete equations are solved, after imposing the boundary conditions, for the values of the solution at the mesh points. Although the finite-difference method is simple in concept, it suffers from several disadvantages. The most notable are the inaccuracy of the derivatives of the approximated solution, the difficulty in imposing the boundary conditions along nonstraight boundaries, the difficulty in accurately representing geometrically complex domains, and the inability to employ nonuniform and nonrectangular meshes.

In the variational solution of differential equations, the differential equation is put into an equivalent variational form, and then the approximate solution is assumed to be a combination ($\Sigma c_j \phi_j$) of given approximation functions ϕ_j. The parameters c_j are determined from the variational form. The variational methods suffer from the disadvantage that the approximation functions for problems with arbitrary domains are difficult to construct.

The finite-element method overcomes the difficulty of the variational methods because it provides a systematic procedure for the derivation of the approximation functions. The method is endowed with two basic features which account for its superiority over other competing methods. First, a geometrically complex domain of the problem is represented as a collection of geometrically simple subdomains, called *finite elements*. Second, over each finite element the approximation functions are derived using the basic idea that any continuous function can be represented by a linear combination of algebraic polynomials. The approximation functions are derived using concepts from interpolation theory, and are therefore called *interpolation functions*. Thus, the finite-element method can be interpreted as a piecewise application of the variational methods (e.g., Ritz and weighted-residual methods), in which the approximation functions are algebraic polynomials and the undetermined parameters represent the values of the solution at a finite number of preselected points, called *nodes*, on the boundary and in the interior of the element. From interpolation theory one finds that the order (or degree) of the interpolation function depends on the number of nodes in the element.

1.2 HISTORICAL COMMENTS

The idea of representing a given domain as a collection of discrete elements is not novel with the finite-element method. It was recorded that ancient mathematicians estimated the value of π by noting that the perimeter of a polygon inscribed in a circle approximates the circumference of a circle. They predicted the value of π to accuracies of almost 40 significant digits by representing the circle as a polygon of a finitely large number of sides. In modern times the idea found a home in aircraft structural analysis, where, for example, wings and fuselages are treated as assemblages of stringers, skins, and shear panels. In 1941 Hrenikoff introduced the so-called framework method, in which a plane elastic medium was represented as a collection of bars and beams. The use of piecewise continuous functions defined over a subdomain to approximate the unknown function dates back to the work of Courant (1943), who used an assemblage of triangular elements and the principle of minimum potential energy to study the St. Venant torsion problem. Although certain key features of the finite-element method can be found in the works of Hrenikoff (1941) and Courant (1943), the formal presentation of the finite-element method is attributed to Argyris and Kelsey (1960) and to Turner, Clough, Martin, and Topp (1956). However, the term "finite element" was first used by Clough in 1960. Since its inception, the literature on finite-element applications has grown exponentially, and today there are numerous journals which are primarily devoted to the theory and application of the finite-element method. A review of the historical developments and the basic theory of the method can be found in more than two dozen textbooks that are exclusively devoted to the introduction and application of the finite-element method (see the references at the end of the chapter). Several reference papers on literature surveys and computer software of the finite-element method are also listed at the end of this chapter.

1.3 THE BASIC CONCEPT OF THE FINITE-ELEMENT METHOD

Here the basic ideas underlying the finite-element method are introduced via two simple examples:

1. Determination of the area of a circle using the equation of a triangle
2. Determination of the center of mass (or gravity) of an irregular body

1.3.1 Example 1

Consider the problem of determining the area of a circle of radius R by representing it as a collection of triangles. It is assumed that the area of a triangle can be computed (i.e., we know the equation for the area of a triangle; see Fig. 1.1). The approximate area of the circle is the sum of the areas of the triangles used to represent the circle. Although this is a trivial example, nevertheless it illustrates several (not all) ideas of (and steps involved in) the finite-element method. We outline the steps involved in computing an approximate area. In doing so, we introduce certain terms that are used in the finite-element analysis of any problem.

1. *Finite-element discretization.* First, the continuous region (i.e., the circle) is represented as a collection of a finite number n of subregions, say triangles. This is called *discretization of the domain* by triangles. Each subregion is called an *element*. The collection of elements is called the *finite-element mesh*. In the present case we discretize the circle into a mesh of five ($n = 5$) triangles, and two such discretizations are shown in Fig. 1.1*a*. Since all of the elements are of the same size, the mesh is said to be *uniform*.
2. *Element equations.* A typical element (i.e., triangle T_e) is isolated and its properties (i.e., area) are computed. It is here that we bring in the governing equation (i.e., the equation for computing the area) of the element to calculate

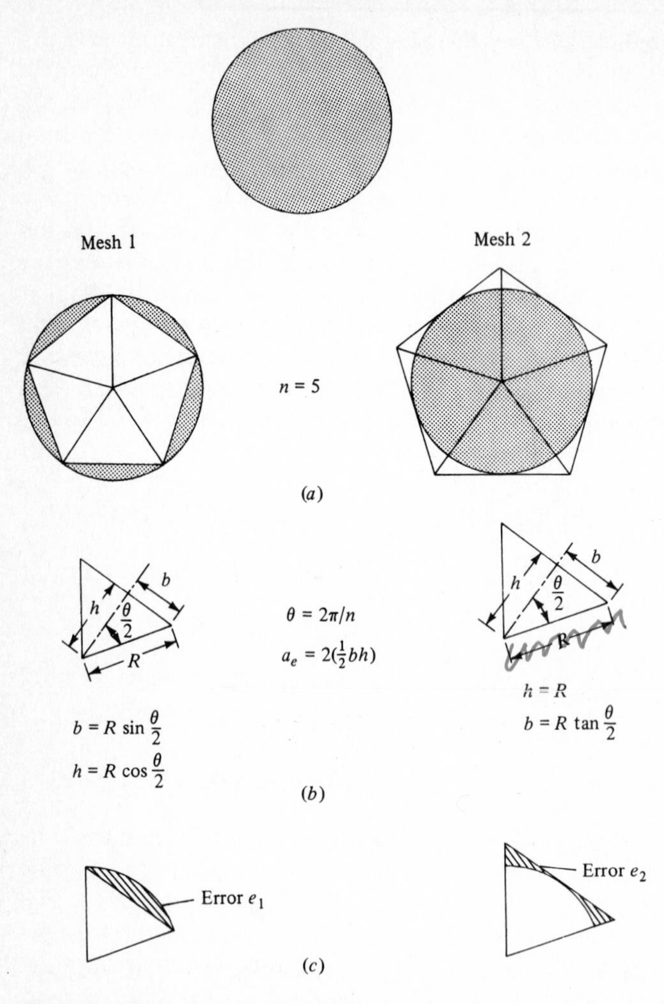

Figure 1.1 Finite-element representation of a circle. (a) Finite-element discretization. (b) Typical elements. (c) Boundary approximation errors.

the required property. Let a_e be the area of element e in mesh 1 and let $\bar{a}_e$ be the area of element e in mesh 2. We have for element e,

$$a_e = \frac{R^2}{2} \sin \frac{2\pi}{n} \qquad \bar{a}_e = R^2 \tan \frac{\pi}{n} \tag{1.1}$$

where R is the radius of the circle. The above equations are called *element equations*.

3. *Assembly of element equations and solution*. The approximate area of the circle is obtained by putting together the element property; this process is called *assembly* of the element equations. The assembly is based, in the present case, on the simple idea that the total area of the assembled elements is equal to the

sum of the areas of individual elements:

$$A_1 = \sum_{e=1}^{n} a_e \qquad A_2 = \sum_{e=1}^{n} \bar{a}_e \tag{1.2}$$

Since the mesh is uniform, a_e or $\bar{a}_e$ is the same for each of the elements in the mesh, and we have

$$A_1^{(n)} = n\frac{R^2}{2}\sin\frac{2\pi}{n} \qquad A_2^{(n)} = nR^2\tan\frac{\pi}{n} \tag{1.3}$$

4. *Convergence and error estimate.* For this simple problem we know the exact solution to the problem: $A_0 = \pi R^2$. We can estimate the error in the approximation and show that the approximate solution converges to the exact in the limit as $n \to \infty$. Consider the typical element e. The error in the approximation is equal to the difference between the area of the sector and that of the triangle (see Fig. 1.1c):

$$e_1 = |S_e - a_e| \qquad e_2 = |S_e - \bar{a}_e| \tag{1.4}$$

where $S_e = \frac{1}{2}R^2\theta$ is the area of the sector. Thus the error estimate for an element in meshes 1 and 2 is given by

$$\begin{aligned} e_1 &= R^2\left(\frac{\pi}{n} - \frac{1}{2}\sin\frac{2\pi}{n}\right) \\ e_2 &= R^2\left(\tan\left(\frac{\pi}{n}\right) - \frac{\pi}{n}\right) \end{aligned} \tag{1.5}$$

The total error (called *global* error) is given by multiplying the e_i's by n:

$$\begin{aligned} E_1^{(n)} &= R^2\left(\pi - \frac{n}{2}\sin\frac{2\pi}{n}\right) = \pi R^2 - A_1^{(n)} \\ E_2^{(n)} &= R^2\left(n\tan\left(\frac{\pi}{n}\right) - \pi\right) = A_2^{(n)} - \pi R^2 \end{aligned} \tag{1.6}$$

We now show that E_1 and E_2 go to zero as $n \to \infty$. Letting $x = 2/n$, we have

$$A_1^{(n)} = R^2\frac{n}{2}\sin\frac{2\pi}{n} = R^2\frac{\sin \pi x}{x}$$

$$\begin{aligned} \lim_{n\to\infty} A_1^{(n)} &= \lim_{x\to 0} R^2\frac{\sin \pi x}{x} \\ &= \lim_{x\to 0} \pi R^2\frac{\cos \pi x}{1} = \pi R^2 \end{aligned}$$

Similarly, letting $y = 1/n$, we have

$$\begin{aligned} \lim_{n\to\infty} A_2^{(n)} &= \lim_{y\to 0} R^2\frac{\tan \pi y}{y} \\ &= \lim_{y\to 0} \pi R^2\frac{\sec^2 \pi y}{1} = \pi R^2 \end{aligned}$$

Hence, $E_1^{(n)}$ and $E_2^{(n)}$ are zero as $n \to \infty$ (see Fig. 1.2c). This completes the proof of the convergence.

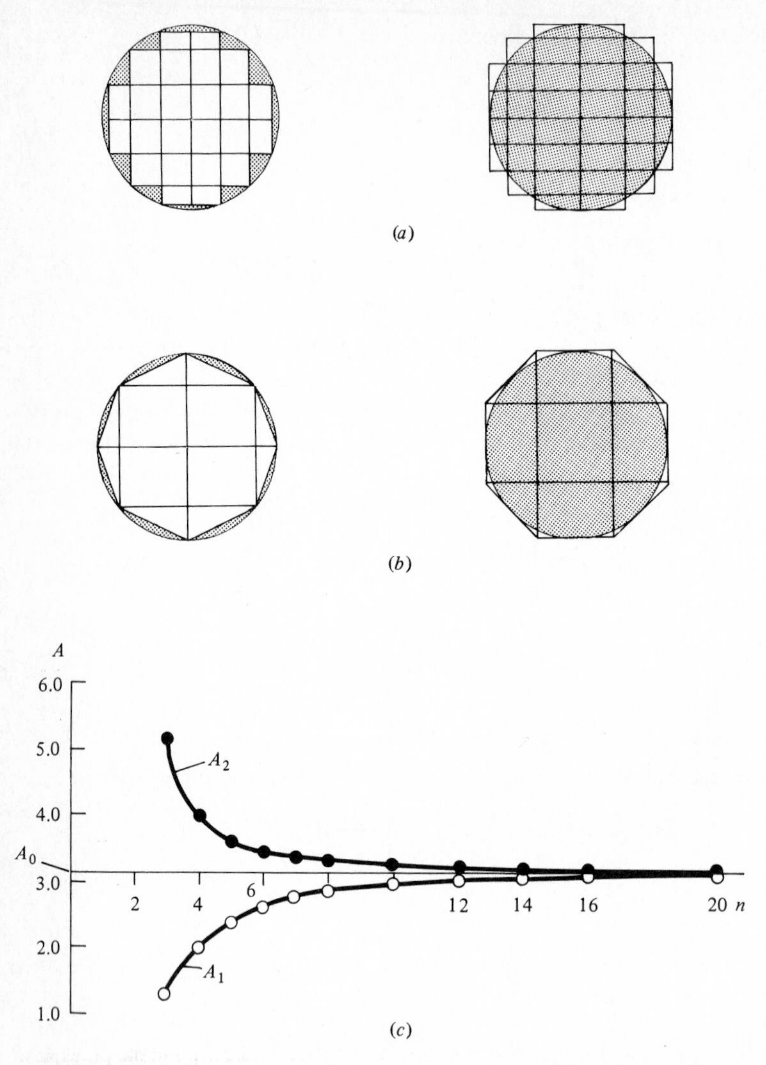

Figure 1.2 Finite-element discretization by rectangular and triangular elements, and convergence of solutions. (*a*) Mesh of rectangular elements. (*b*) Mesh of rectangular and triangular elements. (*c*) Convergence of the finite-element solution.

It should be noted that one can also use either rectangles only or a combination of triangles and rectangles, as shown in Fig. 1.2*a* and *b*. The approximation error in each case is different and therefore so is the solution. Also, note that the equation for the element area is used in its exact form. Therefore, no approximation error in the solution of the equations is introduced. See Sec. 1.3.3 for additional comments on the example.

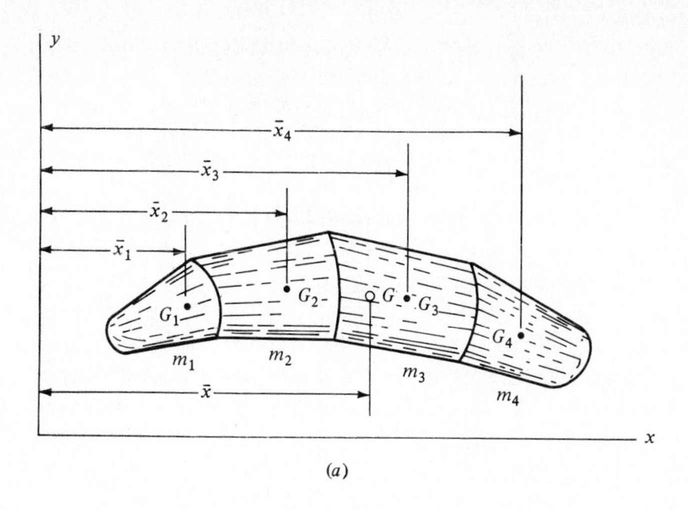

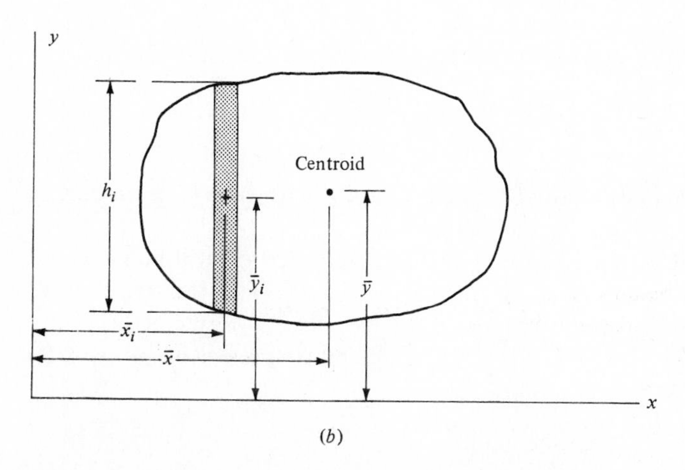

Figure 1.3 Calculation of the center of mass by the method of composite bodies and the principle of Varignon. (*a*) Center of mass of an irregular body. (*b*) Centroid of an irregular area.

1.3.2 Example 2

Another elementary example of the finite-element concept is provided by the calculation of the center of mass of a continuous body. One should recall, from a first course on statics of rigid bodies, that the calculation of the center of an irregular mass (or the centroid of an irregular volume) makes use of the so-called method of composite bodies, in which a body is conveniently divided (mesh discretization) into several parts (elements) of simple shape for which the mass and the center of mass (element properties) can be computed readily. The center of mass of the whole body is then obtained using the moment *principle of Varignon* (a basis for the assembly of element properties; see Fig. 1.3*a*):

$$(m_1 + m_2 + \cdots + m_n)\bar{X} = m_1\bar{x} + m_2\bar{x}_2 + \cdots + m_n\bar{x}_n$$

where $\bar{X}$ is the x coordinate of the center of mass of the whole body, m_i is the mass of the ith part, and $\bar{x}_i$ is the x coordinate of the center of mass of the ith part. Similar expressions hold for the y and z coordinates of the center of mass of the whole body. Analogous relations hold for composite lines, areas, and volumes, wherein the masses are replaced by lengths, areas, and volumes, respectively.

When a given body is not expressible in terms of simple geometric shapes (elements) for which the mass and the center of mass can be represented mathematically, it is necessary to use a method of approximation to represent the properties of an element. As an example consider the problem of finding the centroid $(\bar{x}, \bar{y})$ of the irregular area (region) shown in Fig. 1.3*b*. The region can be divided into a finite number of strips of width Δx and height h. The area of the ith strip is given by $A_i = \Delta x_i h_i$. The area A_i is an approximation of the true area of the strip because h_i is an estimated average height of the strip. The coordinates of the centroid of the region are obtained by applying the moment principle,

$$\bar{x} = \frac{\sum_i A_i \bar{x}_i}{\sum_i A_i} \qquad \bar{y} = \frac{\sum_i A_i \bar{y}_i}{\sum_i A_i}$$

where $\bar{x}_i$ and $\bar{y}_i$ are the coordinates of the centroid of the ith strip (with respect to the coordinate system used for the whole body).

It should be noted that the accuracy of the approximation will be increased by increasing the number of strips (or decreasing the width of the strips) used. Rectangular elements are used in the present discussion for the sake of simplicity only; one may choose to use elements of any size and shape which approximate the given area to satisfactory accuracy.

1.3.3 Some Remarks on the Examples

Although the above examples illustrate the basic idea of the finite-element method, there are several other features that are either not present or not apparent from the discussion of the examples. Some remarks are in order.

1. One can discretize the domain, depending on the shape of the domain, into a mesh of more than one type of element. For example, in the approximation of a circle, one can use either rectangles only or a combination of triangles and rectangles, as shown in Fig. 1.2. Note that a triangular mesh or a mesh of combined triangles and rectangles represents the circle more closely than a corresponding number of rectangular elements. Of course, sector elements represent a circle exactly.

2. If more than one type of element is used in the representation of the domain, one of each kind should be isolated and its properties developed.
3. The governing equations are generally more complex than those considered in the above examples. They are usually differential equations. In many cases, the equations cannot be solved over an element for two reasons. First, the equations do not permit the exact solution. It is here that the variational methods come into play. Second, the discrete equations obtained in the variational methods cannot be solved independent of the remaining elements because the assemblage of the elements is subjected to certain boundary and/or initial conditions.
4. There are two main differences in the form of the approximate solution used in the finite-element method and that used in the classical variational methods (i.e., variational methods applied to the whole domain). First, instead of representing the solution u as a linear combination ($u = \sum c_j \phi_j$) in terms of arbitrary parameters c_j in the variational methods, in the finite-element method the solution is represented as a linear combination ($u = \sum u_j \psi_j$) in terms of the values u_j of u (and possibly its derivatives as well) at the nodal points. Second, the approximate functions in the finite-element method are polynomials that are derived using interpolation theory.
5. The number and the location of the nodes in an element depend on (a) the geometry of the element, (b) the degree of the approximation (i.e., the degree of the polynomials), and (c) the variational form of the equation or equations. By representing the required solution in terms of its values at the nodes, one obtains directly the approximate solution at the nodes.
6. The assembly of elements, in a general case, is based on the idea that the solution (and possibly its derivatives) is continuous at the interelement boundaries. In Example 1, the continuity conditions were not present because the equations used were algebraic equations.
7. In general, the assemblage of finite elements is subjected to boundary and/or initial conditions. The discrete equations associated with the finite-element mesh are solved only after the boundary and/or the initial conditions have been imposed.
8. There are three sources of error in a finite-element solution: (a) errors due to the approximation of the domain (this was the only error present in Example 1), (b) errors due to the approximation of the solution, and (c) errors due to numerical computation (e.g., numerical integration and round-off errors in a computer). The estimation of these errors, in general, is not a simple matter. However, under certain conditions they can be estimated for a given element and problem (see Sec. 3.4).
9. The accuracy and convergence of the finite-element solution depends on the differential equation solved (or the variational form used) and the element used. The word "accuracy" refers to the difference between the exact solution and the finite-element solution, and the word "convergence" refers to the accuracy as the number of elements in the mesh is increased. The nature of the convergence (e.g., monotonically decreasing error or convergence to the

exact solution from above) depends on the formulation of the governing equations. In Example 1, area A_1 converges from below and area A_2 from above to the exact solution A_0 (see Fig. 1.2c).

10. For time-dependent problems, usually a two-stage formulation is followed. In the first stage, the differential equations are approximated by the finite-element method to obtain a set of ordinary differential equations in time. In the second stage, the differential equations in time are solved exactly or further approximated by either variational methods or finite-difference methods to obtain algebraic equations, which are then solved for the nodal values.

1.4 THE PRESENT STUDY

Most introductory finite-element textbooks written for use in engineering schools are intended for students of solid and structural mechanics, and these texts introduce the method as an offspring of matrix methods of structural analysis. A few texts that treat the method as a variationally based technique leave the variational calculus and the associated methods of approximation either to an appendix at the end of the book or to the self-study of the student. This book is written to introduce the finite-element method as a numerical method that employs the philosophy of constructing piecewise variational approximations of solutions to problems described by differential equations. This viewpoint makes the student aware of the generality of the finite-element concept, irrespective of the student's background (engineering or applied science). The viewpoint also enables the student to see the mathematical structure common to various physical theories and thereby to gain additional insight into various engineering problems.

It is clear from the discussion of the basic concept of the finite-element method that one must know two things before he can develop a finite-element formulation of a given problem: first, one should know how to formulate the problem as a variational problem, and second, she should know how to derive an algebraic system of equations associated with the variational problem. Both of these steps are crucial to the development of a finite-element model of any given problem. Therefore, it is necessary to study variational formulations and associated approximations before we proceed to study the finite-element method and its applications. Chapter 2 is devoted to the study of the variational formulation of differential equations and their solution by variational methods of approximation.

Chapter 3 is devoted to the study of the finite-element method as applied to one-dimensional second- and fourth-order differential equations. Most of the finite-element terminology is introduced in this chapter. The finite-element analysis of two-dimensional boundary- and initial-value problems is presented in Chap. 4. The study is limited to problems described by a single second-order differential equation or a system of second-order differential equations. Chapter 5 contains a prelude to advanced topics.

REFERENCES FOR ADDITIONAL READING

Historic papers

Argyris, J. H., and S. Kelsey: *Energy Theorems and Structural Analysis*, Butterworth Scientific Publications, London, 1960.

Clough, R. W.: "The Finite Element Method in Plane Stress Analysis," *J. Struct. Div., ASCE, Proc. 2d Conf. Electronic Computation*, pp. 345–378, 1960.

Courant, R.: "Variational Methods for the Solution of Problems of Equilibrium and Vibration," *Bull. Am. Math. Soc.*, vol. 49, pp. 1–43, 1943.

Hrenikoff, A.: "Solution of Problems in Elasticity by the Framework Method," *J. Appl. Mech., Trans. ASME*, vol. 8, pp. 169–175, 1941.

Turner, M., R. Clough, H. Martin, and L. Topp: "Stiffness and Deflection Analysis of Complex Structures," *J. Aero. Sci.*, vol. 23, pp. 805–823, 1956.

Books

Bathe, K. J.: *Finite Element Procedures in Engineering Analysis*, Prentice-Hall, Englewood Cliffs, N.J. (1982).

Bathe, K. J., and E. L. Wilson: *Numerical Methods in Finite Element Analysis*, Prentice-Hall, Englewood Cliffs, N.J. (1976).

Becker, E. B., G. F. Carey, and J. T. Oden: *Finite Elements, an Introduction*, Vol. I, Prentice-Hall, Englewood Cliffs, N.J. (1981).

Brebbia, C. A., and J. J. Connor: *Fundamentals of Finite Element Techniques for Structural Engineers*, Butterworths, London (1975).

Cheung, Y. K., and M. F. Yeo: *A Practical Introduction to Finite Element Analysis*, Pitman, London, (1979).

Chung, T. J.: *Finite Element Analysis in Fluid Dynamics*, McGraw-Hill, New York (1978).

Ciarlet, P. G.: *The Finite Element Method for Elliptic Problems*, North-Holland, Amsterdam (1978).

Connor, J. C., and C. A. Brebbia: *Finite Element Techniques for Fluid Flow*, Butterworths, London (1976).

Cook, R. D.: *Concepts and Applications of Finite Element Analysis*, John Wiley, New York (1974); 2d ed. (1981).

Desai, C. S.: *Elementary Finite Element Method*, Prentice-Hall, Englewood Cliffs, N.J. (1979).

Desai, C. S., and J. F. Abel: *Introduction to the Finite Element Method*, Van Nostrand Reinhold, New York (1972).

Fenner, R. T.: *Finite Element Methods for Engineers*, Macmillan, London (1975).

Gallagher, R. H.: *Finite Element Analysis Fundamentals*, Prentice-Hall, Englewood Cliffs, N.J. (1975).

Hinton, E., and D. R. J. Owen: *Finite Element Programming*, Academic Press, London (1977).

Hinton, E., and D. R. J. Owen: *An Introduction to Finite Element Computations*, Pineridge Press, Swansea, U.K. (1979).

Huebner, K. H.: *The Finite Element Method for Engineers*, Wiley-Interscience, New York (1975).

Irons, B. M., and S. Ahmad: *Techniques of Finite Elements*, Ellis Horwood, Chichester, U.K. (1979).

Martin, H. C., and G. F. Carey: *Introduction to Finite Element Analysis—Theory and Application*, McGraw-Hill, New York (1973).

Mitchell, A. R., and R. Wait: *The Finite Element Method in Partial Differential Equations*, John Wiley, London (1977).

Nath, B.: *Fundamentals of Finite Elements for Engineers*, Athlone Press, London (1974).

Norrie, D. H., and G. de Vries: *The Finite Element Method: Fundamentals and Applications*, Academic Press, New York (1973).

Norrie, D. H., and G. de Vries: *An Introduction to Finite Element Analysis*, Academic Press, New York (1978).

Oden, J. T.: *Finite Elements of Nonlinear Continua*, McGraw-Hill, New York (1972).

Oden, J. T., and J. N. Reddy: *An Introduction to the Mathematical Theory of Finite Elements*, Wiley-Interscience, New York (1976).

Owen, D. R. J., and E. Hinton: *Finite Elements in Plasticity: Theory and Practice*, Pineridge Press, Swansea, U.K. (1980).

Pinder, G. F., and W. G. Gray: *Finite Elements in Subsurface Hydrology*, Academic Press, New York (1977).

Rao, S. S.: *The Finite Element Method in Engineering*, Pergamon Press, Oxford (1982).

Robinson, J.: *Integrated Theory of Finite Element Methods*, John Wiley, London (1973).

Rockey, K. C., H. R. Evans, D. W. Griffiths, and D. A. Nethercot: *Finite Element Method—A Basic Introduction*, Crosby Lockwood, London (1975).

Segerlind, L. J.: *Applied Finite Element Analysis*, John Wiley, New York (1976).

Strang, G., and G. Fix: *An Analysis of the Finite Element Method*, Prentice-Hall, Englewood Cliffs, N.J. (1973).

Taylor, C., and T. J. Hughes: *Finite Element Programming of the Navier Stokes Equation*, Pineridge Press, Swansea, U.K. (1980).

Tong, P., and J. N. Rossettos: *Finite Element Method: Basic Technique and Implementation*, MIT Press, Cambridge, Mass. (1977).

Wachspress, E. L.: *A Rational Finite Element Basis*, Academic Press, New York (1975).

Zienkiewicz, O. C.: *The Finite Element Method*, 3d expanded and revised ed., McGraw-Hill, London (1977).

Zienkiewicz, O. C., and Y. K. Cheung: *The Finite Element Method in Structural and Continuum Mechanics*, McGraw-Hill, London (1967).

Literature surveys

Akin, J. E., D. L. Fenton, and W. C. T. Stoddart: "The Finite Element Method—A Bibliography of Its Theory and Applications," EM 72-1, Department of Engineering Mechanics, University of Tennessee, Knoxville, Tenn. (1972).

Norrie, D. and G. de Vries: *Finite Element Bibliography*, IFI/Plenum, New York (1976).

Singhal, A. C.: "775 Selected References on the Finite Element Method and Matrix Methods of Structural Analysis," Report S-12, Civil Engineering Department, Laval University, Quebec, Canada (1969).

Whiteman, J. R.: "A Bibliography for Finite Element Methods," TR/9, Department of Mathematics, Brunel University, Oxbridge (1972).

Surveys of finite-element software

Belytschko, T.: "A Survey of Numerical Methods and Computer Programs for Dynamic Structural Analysis," *Nuclear Engineering and Design*, **37**, pp. 23–24 (1976).

Fredriksson, B., and J. Mackerle: "Structural Mechanics Finite Element Computer Programs, Surveys and Availability," LiTH-IKP-R-054, Linkoping Institute of Technology, Department of Mechanical Engineering, Division of Solid Mechanics, Linkoping, Sweden (1975, revised in 1976).

Marcal, P. V., ed.: *On General Purpose Finite Element Computer Programs*, ASME Special Publication, American Society of Mechanical Engineers, New York (1970).

Marcal, P. V.: "Survey of General Purpose Programs for Finite Element Analysis" in *Advances in Computational Methods in Structural Mechanics and Design*, J. T. Oden, R. W. Clough, and Y. Yamamoto, eds., UAH Press, Huntsville, Ala. (1972), pp. 517–528.

Noor, A. K.: "Survey of Computer Programs for Solution of Nonlinear Structural and Solid Mechanics Problems," *Computers and Structures*, **13**, pp. 425–465 (1981).

Pilkey, W., K. Saczalski, and H. Sehaeffer, eds.: *Structural Mechanics Computer Programs, Surveys, Assessments and Availability*, The University Press of Virginia, Charlottesville (1974).

Zukas, J. A., G. H. Jonas, K. D. Kimsey, J. J. Misey, and T. M. Scherrick: "Three-Dimensional Impact Simulations: Resources and Results," in *Computer Analysis of Large-Scale Structures*, K. C. Park and R. F. Jones, Jr., eds., AMD Vol. 49, American Society of Mechanical Engineers New York (1981), pp. 35–68.

CHAPTER

TWO

VARIATIONAL FORMULATION AND APPROXIMATION

2.1 SOME ANCILLARY CONCEPTS AND FORMULAS

2.1.1 Introduction

The finite-element method is a piecewise application of a variational method. Therefore, we begin with a study of the variational methods. There are two basic steps in the variational solution of differential equations:

1. To cast a given differential equation in variational form
2. To determine the approximate solution using a variational method, such as the Ritz method, the Galerkin method, or other methods

The term "variational formulation" is used in the present study to mean the *weak* formulation in which a given differential equation is recast in an equivalent integral form by trading the differentiation between a test function and the dependent variable. For most linear problems the weak formulation is equivalent to the minimization of a quadratic functional $I(u)$, called the *total potential energy* in solid mechanics problems. Analogous to the necessary condition for the minimum of an ordinary function, the necessary condition for a quadratic functional is that its first derivative (or first variation) with respect to the dependent variable be zero. From calculus of variations one knows that the minimizing function is the true solution of the differential equation. This fact provides us the motivation to study the variational formulation of a given differential equation.

In a variational method the dependent variable of a given problem is approximated by a linear combination of appropriately chosen functions: $u = \sum c_j \phi_j$. The parameters c_j are determined such that the function u minimizes the functional $I(u)$ (or, u satisfies the weak formulation of the problem).

The variational form of a given differential equation has several interesting features that facilitate the approximate solution. We illustrate them via an example.

Consider the problem of finding the solution w to the differential equation

$$\frac{d^2}{dx^2}\left[b(x)\frac{d^2w}{dx^2}\right] + f(x) = 0 \qquad \text{for } 0 < x < L \tag{2.1}$$

subject to the end conditions (or boundary conditions)

$$w(0) = \frac{dw}{dx}(0) = 0 \qquad \left(b\frac{d^2w}{dx^2}\right)\bigg|_{x=L} = M_0 \qquad \left[\frac{d}{dx}\left(b\frac{d^2w}{dx^2}\right)\right]_{x=L} = 0 \tag{2.2}$$

This equation arises, for example, in the study of the elastic bending of beams. In this case w denotes the transverse deflection of the beam, L is the total length of the beam, $b(x) > 0$ is the flexural rigidity (i.e., the product of modulus of elasticity and moment of inertia) of the beam, $f(x)$ is the transverse distributed load, and M_0 is the bending moment (see Fig. 2.1a). The solution w is called the *dependent variable* of the problem, and all other quantities (L, b, f, M_0) which are known in advance are called the *data* of the problem.

When $b(x)$ and $f(x)$ are continuous functions of x in (0, L), the data are said to be *smooth*, for which case the solution w to the problem exists and satisfies the differential equation (2.1) at every point x in (0, L) as well as the boundary conditions (2.2) at the boundary points. As an example consider the case in which b and f are nonzero constants. Then the exact solution of Eqs. (2.1) and (2.2) is given by

$$w(x) = \frac{2M_0 - fL^2}{4b}x^2 + \frac{fL}{6b}x^3 - \frac{f}{24b}x^4 \tag{2.3}$$

Thus the solution w and its derivatives up to fourth order are well defined (i.e., they exist and are single-valued) at every point of the domain (0, L).

In most practical situations the data given in a problem are not smooth (i.e., not continuous everywhere in the domain). For example, the flexural rigidity may be discontinuous (e.g., in the case of a composite beam made of dissimilar materials or in the case of a stepped beam), or the transverse loading f may be discontinuous. Suppose that (see Fig. 2.1b)

$$f(x) = f_0 H(a - x) \tag{2.4}$$

and $b(x)$ is continuous. Here $H(a - x)$ denotes the *Heaviside step function*,

$$H(a - x) = \begin{cases} 1 & \text{for } x < a \\ 0 & \text{for } x > a_0 \end{cases} \tag{2.5}$$

In this case, the fourth derivative of the solution (i.e., w) does not exist (i.e., is not single-valued) at $x = a$. Therefore, the exact solution w to Eqs. (2.1) and (2.2) does not exist in the classical sense [i.e., w must satisfy the differential equation (2.1) at all points of the domain]. Similar difficulties are encountered when w and/or its derivatives are specified at points between the endpoints $x = 0$ and $x = L$.

Multiplying Eq. (2.1) with a function v, called the *test function*, that is twice differentiable and satisfies the conditions

$$v(0) = \frac{dv}{dx}(0) = 0$$

integrating the first term twice by parts, and using the boundary conditions (2.2),

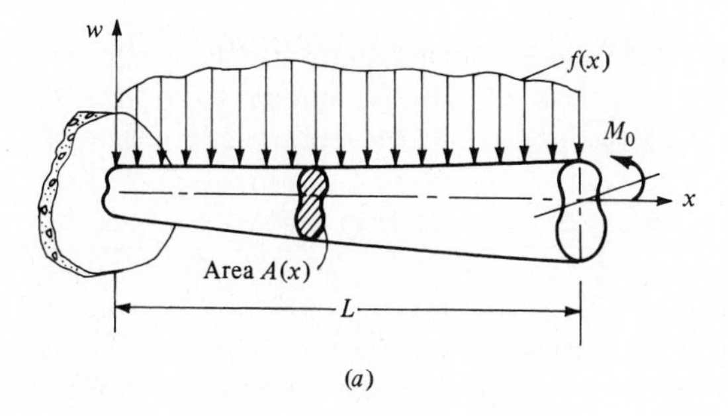

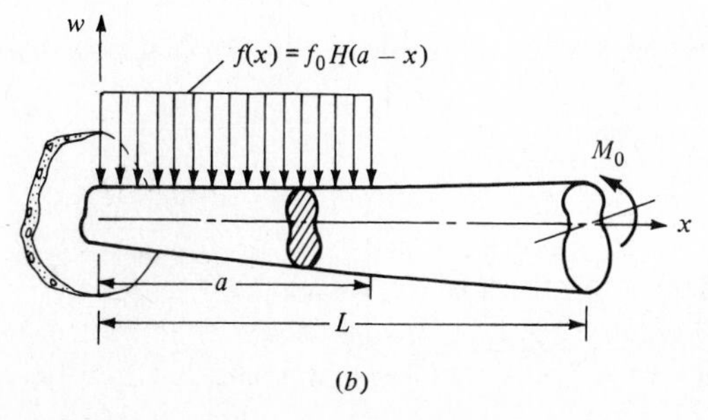

Figure 2.1 Problems with continuous and discontinuous data (H is the Heaviside step function). (a) A cantilevered beam with continuous load. (b) A cantilevered beam with discontinuous load.

we obtain the variational form (see Example 2.3)

$$0 = \int_0^L b \frac{d^2v}{dx^2} \frac{d^2w}{dx^2} dx + \int_0^L vf\, dx - \frac{dv}{dx}(L) M_0 \tag{2.6}$$

The test function v can be viewed as a *variation* in w, consistent with the boundary conditions (2.2). Equation (2.6) is called the *weak*, *generalized*, or *variational equation* associated with Eqs. (2.1) and (2.2), and w is called the *weak*, *generalized*, or *variational solution* of Eq. (2.1). Whenever the classical solution exists, it coincides with the weak solution of the problem. In other words, the variational equation (2.6) is equivalent to Eq. (2.1). By formulating the problem variationally, the continuity requirement on the solution is weakened (now w need be differentiable only twice), and the second pair of boundary conditions in Eq. (2.2) are included in the variational problem (2.6).

The quadratic functional for the problem is obtained by multiplying the expression involving both v and w by one-half and setting $v = w$ in Eq. (2.6):

$$I(w) = \int_0^L \left[\frac{1}{2} b \left(\frac{d^2w}{dx^2} \right)^2 + wf \right] dx - \frac{dw}{dx}(L) M_0 \tag{2.7}$$

The functional represents the total potential energy of the beam. It should be noted that not all differential equations admit functional formulation. On the other hand, one does not need a quadratic functional to develop a finite-element model.

Our goal in this chapter is to illustrate the basic steps in the variational formulation and the associated approximation of various boundary- and initial-value problems. Toward this goal we introduce necessary terminology and notation.

2.1.2 Notation

Derivatives of functions of one variable will be denoted by accents (or primes) in the usual manner:

$$f'(x) \equiv \frac{df(x)}{dx} \qquad f''(x) \equiv \frac{d^2f(x)}{dx^2} \tag{2.8}$$

and so on.

Partial derivatives of functions of several variables will be denoted by appropriate subscripts:

$$\begin{aligned} F_x(x, y, z) &\equiv \frac{\partial F(x, y, z)}{\partial x} \\ F_{xy}(x, y, z) &\equiv \frac{\partial}{\partial x} \left[\frac{\partial F(x, y, z)}{\partial y} \right] \end{aligned} \tag{2.9}$$

and so on.

A function of several variables is said to be of class $C^m(\Omega)$ in a domain Ω if all its partial derivatives up to the mth order inclusive exist and are continuous in the domain Ω. The letters x, y will always be used for rectangular coordinates.

2.1.3 Boundary- and Initial-Value Problems

Domain and boundary The objective of most analyses is to determine unknown functions, called dependent variables, which satisfy a given set of differential equations in a given domain or region and some boundary conditions on the boundary of the domain. A domain is a collection of points in space with the property that if P is a point in the domain, then all points sufficiently close to P belong to the domain. The property implies that a domain consists only of internal points. If any two points of the domain can be joined by a line lying entirely within the domain, then the domain is said to be *convex* and *simply connected*. The boundary of a domain is the set of points such that in any neighborhood of each of these points there are points that belong to the domain as well as points that do not. Note from the definition of a domain that the points on the boundary do not belong to the domain. In the present study we shall also consider domains that are multiply connected (see Fig. 2.2). We shall use the symbol Ω to denote an arbitrary domain and Γ to denote its boundary.

When the dependent variables are functions of one independent variable (say, x), the domain is a line segment (i.e., one-dimensional), and the endpoints of the

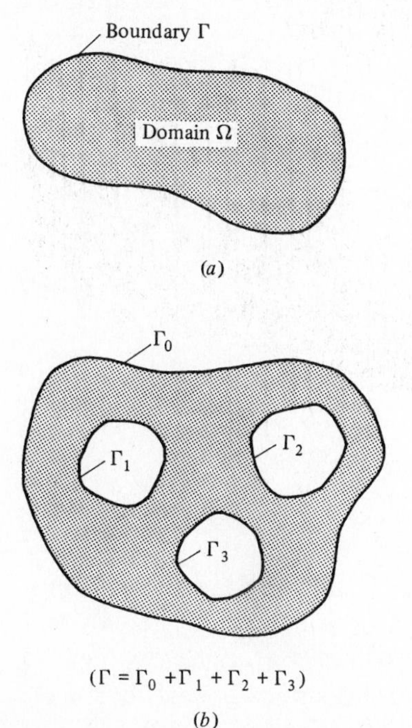

Figure 2.2 Simply connected and multiply connected domains with boundaries in two dimensions. (*a*) Simply connected domain. (*b*) Multiply connected domain ($\Gamma = \Gamma_0 + \Gamma_1 + \Gamma_2 + \Gamma_3$).

domain are called boundary points. It is not uncommon to find problems in one dimension in which the dependent variable and possibly its derivatives are specified at points intermediate to the endpoints (e.g., bending of continuous beams). When the dependent variables are functions of two independent variables (say, x and y), the domain (two-dimensional) is a surface (most often a plane) and the boundary is the closed curve enclosing the domain. If the number of independent variables equals three, the domain is three-dimensional (i.e., a volume) and the boundary is the surface enclosing the volume. Thus, if we are solving the problem of transverse deflection of a string or heat transfer in a fin, the deflection or temperature must be defined in an interval, say, $(0, L)$, where L is the length of the domain. If we are solving the problem of plane elasticity, torsion of a cylindrical member, or flow through an axisymmetric channel, then the displacements, stress function, or velocities must be defined in a plane domain.

A differential equation is said to describe a *boundary-value problem* if the dependent variable and possibly its derivatives are required to take on specified values on the boundary. An *initial-value problem* is one in which the dependent variable and possibly its derivatives are specified initially (i.e., at $t = 0$). Initial-value problems are generally time-dependent problems. Examples of boundary- and initial-value problems are given below.

Boundary-value problem

$$-\frac{d}{dx}\left(a\frac{du}{dx}\right) = f \qquad \text{for } \begin{cases} 0 < x < 1 \\ \Omega \equiv (0, 1) \end{cases} \tag{2.10}$$

$$u(0) = d_0 \qquad \left.\left(a\frac{du}{dx}\right)\right|_{x=1} = g_0 \tag{2.11}$$

Initial-value problem

$$\rho\frac{d^2u}{dt^2} + \alpha u = f \qquad \text{for } 0 < t \leqslant T_0 \tag{2.12}$$

$$u(0) = u_0 \qquad \frac{du}{dt}(0) = v_0 \tag{2.13}$$

Boundary- and initial-value problem

$$-\frac{\partial}{\partial x}\left(a\frac{\partial u}{\partial x}\right) + \rho\frac{\partial u}{\partial t} = f(x, t) \qquad \text{for } \begin{cases} 0 < x < 1 \\ 0 < t \leqslant T_0 \end{cases} \tag{2.14}$$

$$u(0, t) = d_0(t) \qquad \left.\left(a\frac{\partial u}{\partial x}\right)\right|_{x=1} = g_0(t) \qquad u(x, 0) = u_0(x) \tag{2.15}$$

Conditions in Eqs. (2.11) are called *boundary conditions*, and those in Eqs. (2.13) are called *initial conditions*. When the specified values (i.e., d_0, g_0, u_0, and v_0) are nonzero, the conditions are said to be *nonhomogeneous*; otherwise they are said to be *homogeneous*. For example, $u(0) = d_0$ is a nonhomogeneous boundary condition, and the associated homogeneous boundary condition is $u(0) = 0$.

The set of specified quantities (e.g., a, g_0, d_0, ρ, u_0, and v_0) is called the *data* of the problem. Differential equations in which the right-hand side f is zero are called *homogeneous differential equations*. The problem of determining the values of the constant λ such that

$$-\frac{d}{dx}\left(a\frac{du}{dx}\right) = \lambda u \qquad \text{for } 0 < x < 1$$
$$u(0) = 0 \qquad \left.\left(a\frac{du}{dx}\right)\right|_{x=1} = 0 \tag{2.16}$$

is called the *eigenvalue problem* associated with the differential equation (2.10). The values of λ are called the *eigenvalues*.

By *classical* (or exact) *solution* of a differential equation we mean the function that identically satisfies the differential equation (i.e., the classical solution is sufficiently differentiable as required by the equation) and the specified boundary and/or initial conditions. By *variational solution* of a differential equation we mean the solution of an associated variational problem. In other words, the variational solution is not differentiable enough to satisfy the differential equation, but differentiable enough to satisfy a variational equation equivalent to the differential equation.

2.1.4 Gradient and Divergence Theorems

Integration by parts is frequently used in the variational formulation of differential equations. In two-dimensional cases, the integration by parts is better known as the gradient and divergence theorems. In this section, we derive some useful identities for future use.

Let f and g be sufficiently differentiable functions of class $C^4(a, b)$ in one variable. Then we have by integration by parts

$$\int_a^b \frac{df}{dx} g\,dx = -\int_a^b f\frac{dg}{dx}\,dx + f(b)g(b) - f(a)g(a)$$

$$\int_a^b \frac{d^2f}{dx^2} g\,dx = -\int_a^b \frac{df}{dx}\frac{dg}{dx}\,dx + \frac{df}{dx}(b)\,g(b) - \frac{df}{dx}(a)\,g(a)$$

$$\begin{aligned}\int_a^b \frac{d^2}{dx^2}\left[c(x)\frac{d^2f}{dx^2}\right] g\,dx &= -\int_a^b \frac{d}{dx}\left[c(x)\frac{d^2f}{dx^2}\right]\frac{dg}{dx}\,dx \\ &\quad + \left.\frac{d}{dx}\left[c(x)\frac{d^2f}{dx^2}\right] g\right|_{x=a}^{x=b} \\ &= \int_a^b c(x)\frac{d^2f}{dx^2}\frac{d^2g}{dx^2}\,dx + \left[\frac{d}{dx}\left[c(x)\frac{d^2f}{dx^2}\right] g \right. \\ &\quad \left. - c(x)\frac{d^2f}{dx^2}\frac{dg}{dx}\right]_{x=a}^{x=b}\end{aligned} \tag{2.17}$$

where $c(x)$ is a function of class $C^2(a, b)$.

Let ∇ and ∇^2 denote, respectively, the gradient operator and the laplacian operator in three-dimensional space:

$$\begin{aligned} \nabla &= \hat{\mathbf{i}}\frac{\partial}{\partial x} + \hat{\mathbf{j}}\frac{\partial}{\partial y} + \hat{\mathbf{k}}\frac{\partial}{\partial z} \\ \nabla^2 &= \frac{\partial^2}{\partial x^2} + \frac{\partial^2}{\partial y^2} + \frac{\partial^2}{\partial z^2} \end{aligned} \tag{2.18}$$

where $\hat{\mathbf{i}}$, $\hat{\mathbf{j}}$, and $\hat{\mathbf{k}}$ denote the unit basis vectors along x, y, and z (rectangular) coordinates, respectively. If $F(x, y, z)$ and $\mathbf{G}(x, y, z)$ are functions of class $C^1(\Omega)$, the following gradient and divergence theorems hold.

Gradient theorem

$$\begin{gathered} \int_\Omega \operatorname{grad}(F)\, dx\, dy\, dz \equiv \int_\Omega \nabla F\, dx\, dy\, dz = \oint_\Gamma \hat{\mathbf{n}} F\, ds \\ \int_\Omega \left(\hat{\mathbf{i}}\frac{\partial F}{\partial x} + \hat{\mathbf{j}}\frac{\partial F}{\partial y} + \hat{\mathbf{k}}\frac{\partial F}{\partial z} \right) dx\, dy\, dz = \oint_\Gamma \left(\hat{\mathbf{i}} n_x + \hat{\mathbf{j}} n_y + \hat{\mathbf{k}} n_z \right) F\, ds \end{gathered} \tag{2.19a}$$

Divergence theorem

$$\begin{gathered} \int_\Omega \operatorname{div}(\mathbf{G})\, dx\, dy\, dz \equiv \int_\Omega \nabla \cdot \mathbf{G}\, dx\, dy\, dz = \oint_\Gamma \hat{\mathbf{n}} \cdot \mathbf{G}\, ds \\ \int_\Omega \left(\frac{\partial G_x}{\partial x} + \frac{\partial G_y}{\partial y} + \frac{\partial G_z}{\partial z} \right) dx\, dy\, dz = \oint_\Gamma \left(n_x G_x + n_y G_y + n_z G_z \right) ds \end{gathered} \tag{2.19b}$$

Here the dot denotes the scalar product of vectors, $\hat{\mathbf{n}}$ denotes the unit vector normal to the surface Γ of the domain Ω, n_x, n_y, and n_z(G_x, G_y, and G_z) are the rectangular components of $\hat{\mathbf{n}}(\mathbf{G})$, and the circle on the surface integral indicates that the integration is taken on the entire surface of the boundary. The direction cosines n_x, n_y, and n_z of the unit vector $\hat{\mathbf{n}}$ can be written as

$$n_x = \cos(x, \hat{\mathbf{n}}) \qquad n_y = \cos(y, \hat{\mathbf{n}}) \qquad n_z = \cos(z, \hat{\mathbf{n}})$$

where $\cos(x, \hat{\mathbf{n}})$ means the cosine of the angle between the positive x direction and the unit vector $\hat{\mathbf{n}}$.

The following identities, which can be derived using the gradient and divergence theorems, will be useful in the sequel. Let F and G be scalar functions defined in a three-dimensional domain Ω. Then we have

$$\begin{aligned} \int_\Omega (\nabla F) G\, dx\, dy\, dz &= -\int_\Omega (\nabla G) F\, dx\, dy\, dz + \oint_\Gamma \hat{\mathbf{n}}\, FG\, ds \\ \int_\Omega \left(\hat{\mathbf{i}}\frac{\partial F}{\partial x} + \hat{\mathbf{j}}\frac{\partial F}{\partial y} + \hat{\mathbf{k}}\frac{\partial F}{\partial z} \right) G\, dx\, dy\, dz &= -\int_\Omega \left(\hat{\mathbf{i}}\frac{\partial G}{\partial x} + \hat{\mathbf{j}}\frac{\partial G}{\partial y} + \hat{\mathbf{k}}\frac{\partial G}{\partial z} \right) F\, dx\, dy\, dz \\ &\quad + \oint_\Gamma \left(\hat{\mathbf{i}} n_x + \hat{\mathbf{j}} n_y + \hat{\mathbf{k}} n_z \right) FG\, ds \end{aligned} \tag{2.20a}$$

$$-\int_\Omega (\nabla^2 F) G\, dx\, dy\, dz = \int_\Omega \nabla F \cdot \nabla G\, dx\, dy\, dz - \oint_\Gamma \frac{\partial F}{\partial n} G\, ds \tag{2.20b}$$

where $\partial/\partial n$ denotes the normal derivative operator,

$$\frac{\partial}{\partial n} = \hat{\mathbf{n}} \cdot \nabla = n_x \frac{\partial}{\partial x} + n_y \frac{\partial}{\partial y} + n_z \frac{\partial}{\partial z} \tag{2.21}$$

2.1.5 Functionals

In the sequel we shall consider integral expressions of the form [see, for example, $I(w)$ in Eq. (2.7)]

$$I(u) = \int_a^b F(x, u, u')\, dx$$

where the integrand $F(x, u, u')$ is a given function of the arguments x, u, and du/dx. The value $I(u)$ of the integral depends on u, hence the notation $I(u)$ is appropriate. However, for a given u, $I(u)$ represents a scalar value. We shall use the term *functional* to describe functions defined by integrals whose arguments themselves are functions. Loosely speaking, a functional is a "function of functions." Mathematically, a functional is an operator I mapping u into a scalar $I(u)$. The set of all functions $u(x)$ for which $I(u)$ makes sense is called the *domain space* of the functional. The set of images of all functions u under the mapping I is called the *range* of the functional $I(u)$. By definition, the range of a functional is a subset of the real number field.

A functional $l(u)$ is said to be *linear* in u if and only if it satisfies the relation

$$l(\alpha u + \beta v) = \alpha l(u) + \beta l(v)$$

for any scalars α and β and the dependent variables u and v. A functional $B(u, v)$ is said to be *bilinear* if it is linear in each of its arguments u and v.

An example of the linear form is given by the last two terms in Eq. (2.6),

$$l(v) = -\int_0^L vf\, dx + \frac{dv}{dx}(L)\, M_0$$

An example of the bilinear form is given by the first term in Eq. (2.6),

$$B(v, w) = \int_0^L b \frac{dv}{dx} \frac{dw}{dx}\, dx$$

In view of the above definitions, we can rewrite the weak form in Eq. (2.6) in the form

$$B(v, w) = l(v) \qquad \text{for any } v$$

*2.1.6 The Variational Symbol

Consider the function $F = F(x, u, u')$. For an arbitrarily fixed value of the independent variable x, F depends on u and u'. The change αv in u is called the *variation* of u and is denoted by δu,

$$\delta u = \alpha v \qquad \alpha = \text{constant} \tag{2.22}$$

The operator δ is called the *variational symbol*. The variation δu of a function u represents an admissible change in the function $u(x)$ at a *fixed* value of the independent variable. If u is specified at a point (usually on the boundary), the variation of u is zero there because the specified value cannot be varied. Thus the variation of a function u should satisfy the homogeneous form of the boundary conditions on u. The variation δu in u vanishes at points where u is specified, and it is arbitrary elsewhere, that is, δu is a *virtual* change. Associated with this change in u (i.e., u is changed to $u + \alpha v$), there is a change in F. In analogy with the total differential of a function of two variables, the *first variation* of F at u is defined by

$$\delta F = \frac{\partial F}{\partial u}\delta u + \frac{\partial F}{\partial u'}\delta u' \tag{2.23}$$

which can be obtained from

$$\frac{\delta F}{\alpha} = \lim_{\alpha \to 0}\left[\frac{F(x, u + \alpha v, u' + \alpha v') - F(x, u, u')}{\alpha}\right]$$

by expanding $F(x, u + \alpha v, u' + \alpha v')$ in powers of α. Note the analogy between the first variation, Eq. (2.23), and the total differential of F,

$$dF = \frac{\partial F}{\partial x}dx + \frac{\partial F}{\partial u}du + \frac{\partial F}{\partial u'}du' \tag{2.24}$$

Since x is not varied during the variation of u to $u + \delta u$, $dx = 0$ and the analogy between δF and dF becomes apparent. That is, δ acts as a differential operator with respect to dependent variables. It can be easily verified that the laws of variation of sums, products, ratios, powers, and so forth are completely analogous to the corresponding laws of differentiation. For example, if $F_1 = F_1(u)$ and $F_2 = F_2(u)$, then

$$\begin{aligned}
&1.\ \delta(F_1 \pm F_2) = \delta F_1 \pm \delta F_2\\
&2.\ \delta(F_1F_2) = F_2\,\delta F_1 + F_1\,\delta F_2\\
&3.\ \delta\left(\frac{F_1}{F_2}\right) = \frac{F_2\,\delta F_1 - F_1\,\delta F_2}{F_2^2}\\
&4.\ \delta(F_1)^n = n(F_1)^{n-1}\,\delta F_1
\end{aligned} \tag{2.25}$$

Furthermore, the variational operator can commute with differential and integration operators:

$$\frac{d}{dx}(\delta u) = \frac{d}{dx}(\alpha v) = \alpha\frac{dv}{dx} = \alpha v' = \delta u' = \delta\left(\frac{du}{dx}\right) \tag{2.26a}$$

$$\delta\int_a^b u(x)\,dx = \int_a^b \delta u(x)\,dx \tag{2.26b}$$

As an example of the variation of a functional, we consider the quadratic form in Eq. (2.7). We have

$$\delta I(w) = \delta \int_0^L \left[\frac{b}{2} \left(\frac{d^2 w}{dx^2} \right)^2 + wf \right] dx - \delta \left[\frac{dw}{dx}(L) \right] M_0$$

$$= \int_0^L \left[b \left(\frac{d^2 w}{dx^2} \right) \delta \left(\frac{d^2 w}{dx^2} \right) + \delta w f \right] dx - \frac{d\delta w}{dx}(L)\, M_0$$

$$= \left\{ \int_0^L \left(b \frac{d^2 v}{dx^2} \frac{d^2 w}{dx^2} + vf \right) dx - \frac{dv}{dx}(L)\, M_0 \right\} \alpha$$

where $\delta w = \alpha v$ is substituted in the last step. The last equation is precisely the same, except for a constant multiple of α, as that of Eq. (2.6). Thus, the first variation of the quadratic functional in Eq. (2.7) is equivalent to the weak form in Eq. (2.6).

2.2 VARIATIONAL FORMULATION OF BOUNDARY-VALUE PROBLEMS

2.2.1 Motivation

The motivation for the consideration of variational formulations of boundary-value problems comes from the fact that variational methods of approximation (e.g., Ritz, Galerkin, least-squares, collocation, or in general weighted-residual methods) are based on variational (or weak) statements of physical problems. Since the finite-element method is a technique of constructing approximation functions required in a piecewise application of any variational method, it is necessary to study the variational formulation of differential equations. In addition to the above reason, variational formulations also facilitate, in a natural way, the classification of boundary conditions into *natural* and *essential* boundary conditions, which play a crucial role in the derivation of the approximation functions.

In this section, our primary objectives will be to construct the variational formulation of a given differential equation and to classify the boundary conditions associated with the equation.

2.2.2 Variational (or Weak) Formulation

Suppose that we are required to find the variational form of the following partial differential equation in two dimensions:

$$\frac{\partial F}{\partial u} - \frac{\partial}{\partial x}\left(\frac{\partial F}{\partial u_x} \right) - \frac{\partial}{\partial y}\left(\frac{\partial F}{\partial u_y} \right) = 0 \text{ in } \Omega \qquad (2.27a)$$

subjected to the boundary conditions

$$\frac{\partial F}{\partial u_x} n_x + \frac{\partial F}{\partial u_y} n_y = \hat{q} \text{ on } \Gamma_1 \qquad u = \hat{u} \text{ on } \Gamma_2 \tag{2.27b}$$

where $F = F(x, y, u, u_x, u_y)$ and Γ_1 and Γ_2 are disjoint portions whose sum is the total boundary, $\Gamma = \Gamma_1 + \Gamma_2$. Here $u_x = \partial u/\partial x$, and so on, and n_x and n_y are the direction cosines of the unit vector normal to the boundary.

Equations (2.27) describe a general form of most second-order boundary-value problems in two dimensions. Several examples of Eqs. (2.27) will be given later (see Table 2.2). Before we proceed with the variational formulation of Eqs. (2.27), we consider an example of the function F.

Example 2.1 As a special case of Eq. (2.27) consider the case in which F is given by

$$F = \frac{1}{2}\left[k_1\left(\frac{\partial u}{\partial x}\right)^2 + k_2\left(\frac{\partial u}{\partial y}\right)^2\right] - Qu$$

which arises in the study of heat conduction in a two-dimensional medium with conductivities along the x and y directions given by k_1 and k_2, respectively, and heat generation Q. We have

$$\frac{\partial F}{\partial u} = -Q \qquad \frac{\partial F}{\partial u_x} = k_1 \frac{\partial u}{\partial x} \qquad \frac{\partial F}{\partial u_y} = k_2 \frac{\partial u}{\partial y}$$

Equations (2.27) become the Fourier heat conduction equation and the associated boundary conditions, respectively

$$-Q - \frac{\partial}{\partial x}\left(k_1 \frac{\partial u}{\partial x}\right) - \frac{\partial}{\partial y}\left(k_2 \frac{\partial u}{\partial y}\right) = 0 \text{ in } \Omega$$

$$\left(k_1 \frac{\partial u}{\partial x}\right) n_x + \left(k_2 \frac{\partial u}{\partial y}\right) n_y = \hat{q} \text{ on } \Gamma_1 \qquad u = \hat{u} \text{ on } \Gamma_2$$

Here $\hat{q}$ denotes the prescribed heat flux on portion Γ_1 of the boundary, and $\hat{u}$ is the specified temperature on portion Γ_2 of the boundary. ■

Returning to the variational formulation of Eqs. (2.27), we outline three basic steps in the formulation. The first step involves multiplying the equation (with all of the terms on one side of the equality) with a test function v and integrating the product over the domain of the problem:

$$0 = \int_\Omega v\left[\frac{\partial F}{\partial u} - \frac{\partial}{\partial x}\left(\frac{\partial F}{\partial u_x}\right) - \frac{\partial}{\partial y}\left(\frac{\partial F}{\partial u_y}\right)\right] dx\, dy \tag{2.28}$$

The test function v, which can be thought of as a variation in u, is assumed to satisfy the homogeneous form of the second boundary condition in Eq. (2.27b). Other than this requirement, v is an arbitrary continuous function. Note that the integral form in Eq. (2.28) contains the same order of derivatives as the differential equation with which we started. Although the Galerkin method was proposed originally to employ the integral form to seek an approximate solution, in most applications of the Galerkin method one uses a weak form.

The second step involves the transfer of differentiation from the dependent variable u to the test function v, and the identification of the type of boundary conditions the variational form can admit. It is clear that $\partial F/\partial u_x$ contains $u_x = \partial u/\partial x$ and $\partial F/\partial u_y$ contains $u_y = \partial u/\partial y$ (and possibly products of u, u_x, and u_y). Hence, it is desirable to transfer the partial differentiation with respect to x and y to v so that the resulting expression contains derivatives of only first order in u and v. It is important to note that the purpose of the transfer of differentiation from u onto v is to equalize the continuity requirements on u and v. This results in weaker continuity requirements on the solution u in the variational problem than in the original equation. In the process of transferring the differentiation we obtain boundary terms which determine the nature of the (natural or essential) boundary conditions in the solution.

Use of identities of the type of Eq. (2.20) (in component form) on the second and third terms in the square brackets of Eq. (2.28) gives

$$0 = \int_\Omega \left[v \frac{\partial F}{\partial u} + \frac{\partial v}{\partial x}\frac{\partial F}{\partial u_x} + \frac{\partial v}{\partial y}\frac{\partial F}{\partial u_y} \right] dx\, dy - \oint_\Gamma v \left(\frac{\partial F}{\partial u_x} n_x + \frac{\partial F}{\partial u_y} n_y \right) ds \tag{2.29}$$

It is here that we identify the natural and essential boundary conditions. As a general rule, specifying coefficients of v and its derivatives in the boundary integral constitute the *natural boundary condition*. Thus,

$$\frac{\partial F}{\partial u_x} n_x + \frac{\partial F}{\partial u_y} n_y = \hat{q} \text{ on } \Gamma_1$$

is the natural boundary condition in Eq. (2.27*b*). Specification of the dependent variable in the same form as the arbitrary function v in the boundary integral constitutes the *essential boundary condition*. In the present case only v (not its derivative) appears in the boundary integral. Hence, specifying u is the essential boundary condition.

In view of the above classification of boundary conditions, we can now restate the conditions on the test function. The test function should be differentiable as required by Eqs. (2.27) and satisfy the homogeneous form of the specified essential boundary conditions.

The classification of boundary conditions is useful in both the classical variational methods and the finite-element method. As will be seen later, the variables involved in the essential boundary conditions of the problem will be identified as the *primary variables* and those in the natural boundary conditions as the *secondary variables* of the formulation. The primary variables are continuous, whereas the secondary variables can be discontinuous in a problem.

The third and last step in the formulation consists of simplifying the boundary terms with the aid of the specified boundary conditions, and the identification of the associated quadratic functional, if it exists:

$$0 = \int_{\Omega}\left(v\frac{\partial F}{\partial u} + \frac{\partial v}{\partial x}\frac{\partial F}{\partial u_x} + \frac{\partial v}{\partial y}\frac{\partial F}{\partial u_y}\right) dx\,dy - \int_{\Gamma_1} v\hat{q}\,ds$$
$$- \int_{\Gamma_2} v\left(\frac{\partial F}{\partial u_x}n_x + \frac{\partial F}{\partial u_y}n_y\right) ds \tag{2.30}$$

Equation (2.30) is obtained from Eq. (2.29) by splitting the boundary term into two terms, one on Γ_1 and the other on Γ_2, and substituting the natural boundary condition into the first term. The second term vanishes because of the fact that $v = 0$ on Γ_2 (equivalent to $\delta u \equiv \delta\hat{u} = 0$). We have

$$0 = \int_{\Omega}\left(v\frac{\partial F}{\partial u} + \frac{\partial v}{\partial x}\frac{\partial F}{\partial u_x} + \frac{\partial v}{\partial y}\frac{\partial F}{\partial u_y}\right) dx\,dy - \int_{\Gamma_1} v\hat{q}\,ds \tag{2.31}$$

with $v = 0$ on Γ_2. If F is linear in u, say, $\partial F/\partial u = f(x, y)$, the bilinear and linear forms associated with the weak form in Eq. (2.31) are given by expressions involving both u and v, and v alone, respectively:

$$B(v, u) = \int_{\Omega}\left(\frac{\partial v}{\partial x}\frac{\partial F}{\partial u_x} + \frac{\partial v}{\partial y}\frac{\partial F}{\partial u_y}\right) dx\,dy$$
$$l(v) = -\int_{\Omega} vf\,dx\,dy + \int_{\Gamma_1} v\hat{q}\,ds \tag{2.32}$$

The name "weak form" for Eq. (2.31) is appropriate since the continuity required of u is reduced from C^2 in the governing equation to C^1 in the variational equation. Equations of the form of Eq. (2.31) form the basis of variational methods of approximation and, hence, the finite-element method.

Whenever the functional $B(\cdot, \cdot)$ is bilinear and *symmetric*, $B(u, v) = B(v, u)$, and $l(v)$ is linear, the quadratic functional associated with the variational form in Eq. (2.31) is obtained from (see Reddy and Rasmussen, 1982)

$$I(u) = \tfrac{1}{2}B(u, u) - l(u) \tag{2.33}$$

Now we consider some representative examples of differential equations in one and two dimensions, and formulate their variational equations. These examples are of primary interest in the study of the finite-element method.

Example 2.2 Consider the differential equation

$$-\frac{d}{dx}\left(a\frac{du}{dx}\right) - cu + x^2 = 0 \qquad \text{for } 0 < x < 1 \tag{2.34}$$

subjected to the boundary conditions

$$u(0) = 0 \qquad a\frac{du}{dx}(1) = 1 \tag{2.35}$$

The data are $f = -x^2$, $\hat{q} = 1$, and $\hat{u} = 0$.

Following the steps outlined above for the construction of variational statements, we obtain (for test function v)

$$0 = \int_0^1 v\left[-\frac{d}{dx}\left(a\frac{du}{dx}\right) - cu + x^2\right] dx$$
$$= \int_0^1 \left(a\frac{dv}{dx}\frac{du}{dx} - cvu + vx^2\right) dx - \left(va\frac{du}{dx}\right)\Big|_0^1 \tag{2.36}$$

From the boundary term it is clear that the specification of u is an essential boundary condition, and the specification of $a\,du/dx$ is a natural boundary condition. Since $a\,du/dx = 1$ at $x = 1$ and $v = 0$ at $x = 0$ (because u is specified there), we obtain the variational form

$$0 = \int_0^1 \left(a\frac{dv}{dx}\frac{du}{dx} - cvu\right) dx + \int_0^1 vx^2\,dx - v(1) \tag{2.37a}$$

$$0 = B(v, u) - l(v) \tag{2.37b}$$

where

$$B(v, u) = \int_0^1 \left(a\frac{dv}{dx}\frac{du}{dx} - cvu\right) dx$$
$$l(v) = -\int_0^1 vx^2\,dx + v(1) \tag{2.38}$$

Since $B(\cdot, \cdot)$ is bilinear and symmetric, and $l(\cdot)$ is linear, we can compute the quadratic functional from Eq. (2.33),

$$I(u) = \frac{1}{2}\int_0^1 \left[a\left(\frac{du}{dx}\right)^2 - cu^2 + 2ux^2\right] dx - u(1) \tag{2.39}$$

Equations of the type of Eq. (2.34) arise in the study of the deflection of a cable on an elastic foundation ($c \neq 0$) or of heat transfer ($c = 0$) in a fin. In the former case, u denotes the transverse deflection, c the modulus of the foundation, and a the tension in the cable. The first two terms in the quadratic functional represent the elastic strain energy and the last term represents the work done by the distributed force in moving through the displacement u. ■

Additional examples of physical problems governed by Eq. (2.10) are given in Table 2.1.

The next example illustrates the variational formulation of fourth-order differential equations in one dimension.

Example 2.3 Consider the fourth-order differential equation and the boundary conditions in Eqs. (2.1) and (2.2). Since the equation contains a fourth-order derivative, we should integrate it twice by parts to distribute the derivatives

Table 2.1 Some examples of second-order equation in one dimension

$$-\frac{d}{dx}\left(a\frac{du}{dx}\right) = f \qquad \textbf{for } 0 < x < L$$

Essential boundary condition: $u|_{x=0} = d_0$; natural boundary condition: $\left.\left(a\frac{du}{dx}\right)\right|_{x=L} = g_0$

Field	Primary variable u	a	Source term f	Secondary variable g_0
1. Transverse deflection of a cable	Transverse deflection	Tension in cable	Distributed transverse load	Axial force (generally unknown)
2. Axial deformation of a bar	Longitudinal displacement	EA, E = modulus, A = area of cross section	Friction or contact force on surface of bar	Axial force
3. Heat transfer along a fin in heat exchanger	Temperature	Thermal conductivity	Heat generation	Heat flux
4. Flow through pipes	Hydrostatic pressure	$\pi D^4/128\mu$, D = diameter μ = viscosity	Flow source (generally zero)	Flow rate
5. Laminar incompressible flow through a channel under constant pressure gradient	Velocity	Viscosity	Pressure gradient	Axial stress
6. Flow through porous media	Fluid head	Coefficient of permeability	Fluid flux	Flow (seepage)
7. Electro statics	Electrostatic potential	Dielectric constant	Charge density	Electric flux

equally between the dependent variable w and the test function v:

$$
\begin{aligned}
0 &= \int_0^L v\left[\frac{d^2}{dx^2}\left(b\frac{d^2w}{dx^2}\right)+f\right]dx \\
&= \int_0^L\left[\left(-\frac{dv}{dx}\right)\frac{d}{dx}\left(b\frac{d^2w}{dx^2}\right)+vf\right]dx+\left[v\frac{d}{dx}\left(b\frac{d^2w}{dx^2}\right)\right]\Bigg|_0^L \\
&= \int_0^L\left[b\frac{d^2v}{dx^2}\frac{d^2w}{dx^2}+vf\right]dx+\left[v\frac{d}{dx}\left(b\frac{d^2w}{dx^2}\right)-\frac{dv}{dx}b\frac{d^2w}{dx^2}\right]\Bigg|_0^L
\end{aligned} \tag{2.40}
$$

From the last line it follows that the specification of w and dw/dx constitutes the essential (geometric or static) boundary conditions and the specification of $(d/dx)(b\,d^2w/dx^2)$ (*shear force*) and $b\,d^2w/dx^2$ (*bending moment*) constitutes the natural (or dynamic) boundary conditions. In the present case the specified essential boundary conditions are

$$w(0)=\frac{dw}{dx}(0)=0$$

Hence, we require

$$v(0)=\frac{dv}{dx}(0)=0$$

The natural boundary conditions are

$$\left[\frac{d}{dx}\left(b\frac{d^2w}{dx^2}\right)\right]_{x=L}=0 \qquad \text{and} \qquad \left(b\frac{d^2w}{dx^2}\right)\Bigg|_{x=L}=M_0$$

We obtain

$$0=\int_0^L\left[b\frac{d^2v}{dx^2}\frac{d^2w}{dx^2}+vf\right]dx-\frac{dv}{dx}\bigg|_{x=L}M_0 \tag{2.41}$$

or

$$B(v,w)=l(v) \tag{2.42}$$

where

$$
\begin{aligned}
B(v,w) &= \int_0^L b\frac{d^2v}{dx^2}\frac{d^2w}{dx^2}dx \\
l(v) &= -\int_0^L vf\,dx+\frac{dv}{dx}\bigg|_{x=L}M_0
\end{aligned} \tag{2.43}
$$

The quadratic form, commonly known as the *total potential energy* of the beam, is given by

$$I(w)=\int_0^L\left[\frac{b}{2}\left(\frac{d^2w}{dx^2}\right)^2+wf\right]dx-\frac{dw}{dx}\bigg|_{x=L}M_0 \tag{2.44}$$

Note that for the fourth-order problem, the essential boundary conditions involve not only the dependent variable but also its first derivative. As pointed out earlier, at any boundary point only one of the two boundary conditions (essential or natural) can be specified. For example, if w is specified at a boundary point, then one cannot specify $(d/dx)(b\,d^2w/dx^2)$ (shear force) at the same point, and vice versa. Similar comments apply to dw/dx and $b\,d^2w/dx^2$. Note that in the present case, w and dw/dx are the primary variables and $(d/dx)(b\,d^2w/dx^2)$ and $b\,d^2w/dx^2$ are the secondary variables. ∎

The next example is concerned with a second-order differential equation in two dimensions. One should note the special form of the natural boundary conditions used in the example.

Example 2.4 Consider the steady heat conduction in a two-dimensional domain Ω, enclosed by lines AB, BC, CD, DE, EF, FG, GH, and HA (see Fig. 2.3). The governing equation is given by

$$-k\left(\frac{\partial^2 T}{\partial x^2}+\frac{\partial^2 T}{\partial y^2}\right)=0 \text{ in } \Omega \tag{2.45}$$

where k is the conductivity of the material of the domain. We wish to construct the weak or variational formulation of the equation.

Proceeding as described earlier, we have

$$0=\int_\Omega\left[-vk\left(\frac{\partial^2 T}{\partial x^2}+\frac{\partial^2 T}{\partial y^2}\right)\right]dx\,dy$$

$$=\int_\Omega k\left(\frac{\partial v}{\partial x}\frac{\partial T}{\partial x}+\frac{\partial v}{\partial y}\frac{\partial T}{\partial y}\right)dx\,dy-\oint_\Gamma vk\left(\frac{\partial T}{\partial x}n_x+\frac{\partial T}{\partial y}n_y\right)ds \tag{2.46}$$

The boundary Γ can be divided into four segments according to the type of

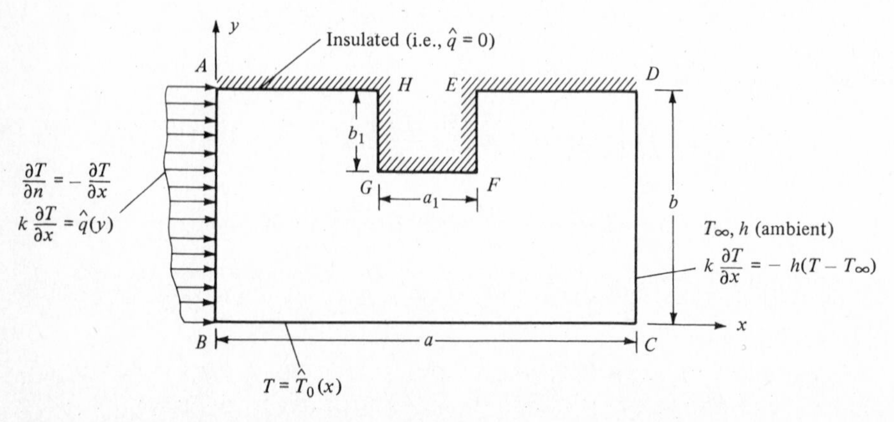

Figure 2.3 Conduction and convection heat transfer in two-dimensional domains.

boundary condition (see Fig. 2.3):

$$\begin{aligned}
&\Gamma_1 = AB \text{ (specified conduction, } \hat{q}(y)) && n_x = -1, n_y = 0 \\
&\Gamma_2 = BC \text{ (specified temperature, } \hat{T}_0(x)) && n_x = 0, n_y = -1 \\
&\Gamma_3 = CD \text{ (convective boundary, } T_\infty) && n_x = 1, n_y = 0 \\
&\Gamma_4 = DEFGHA \left(\text{insulated boundary, } \frac{\partial T}{\partial n} = 0\right)
\end{aligned} \tag{2.47}$$

Using the boundary information, the boundary integral in Eq. (2.46) can be simplified as follows (note that $v = 0$ on Γ_2):

$$\begin{aligned}
\oint_\Gamma v\left(k\frac{\partial T}{\partial n}\right) ds &= -\int_{\Gamma_1} v \cdot \hat{q}\, ds + \int_{\Gamma_2} 0 \cdot \left(k\frac{\partial T}{\partial n}\right) ds \\
&\quad - \int_{\Gamma_3} v[h(T - T_\infty)]\, ds + \int_{\Gamma_4} v \cdot 0\, ds \\
&= -\int_0^b v(0, y)\hat{q}(y)\, dy - h\int_0^b v(a, y)[T(a, y) - T_\infty]\, dy
\end{aligned} \tag{2.48}$$

We obtain the weak form,

$$\begin{aligned}
0 &= \int_\Omega k\left(\frac{\partial v}{\partial x}\frac{\partial T}{\partial x} + \frac{\partial v}{\partial y}\frac{\partial T}{\partial y}\right) dx\, dy + \int_0^b v(0, y)\hat{q}(y)\, dy \\
&\quad + h\int_0^b v(a, y)[T(a, y) - T_\infty]\, dy
\end{aligned} \tag{2.49}$$

Collecting terms involving both v and T into $B(\cdot, \cdot)$ and those involving only v into $l(\cdot)$, we can write Eq. (2.49) in the form

$$B(v, T) = l(v) \tag{2.50}$$

where

$$\begin{aligned}
B(v, T) &= \int_\Omega k\left(\frac{\partial v}{\partial x}\frac{\partial T}{\partial x} + \frac{\partial v}{\partial y}\frac{\partial T}{\partial y}\right) dx\, dy + h\int_0^b v(a, y)T(a, y)\, dy \\
l(v) &= -\int_0^b v(0, y)\hat{q}(y)\, dy + h\int_0^b v(a, y)T_\infty\, dy
\end{aligned} \tag{2.51}$$

The quadratic functional is given by

$$\begin{aligned}
I(T) &= \frac{k}{2}\int_\Omega \left[\left(\frac{\partial T}{\partial x}\right)^2 + \left(\frac{\partial T}{\partial y}\right)^2\right] dx\, dy + \int_0^b T(0, y)\hat{q}(y)\, dy \\
&\quad + h\int_0^b \frac{1}{2}\left[T^2(a, y) + 2T(a, y)T_\infty\right] dy
\end{aligned} \tag{2.52}$$

Note that the boundary integrals in the above example are defined along the y and x axes, respectively. This is because the boundaries are parallel to either the x axis or the y axis.

Equations of the type of Eq. (2.45) also arise in other fields of engineering. Table 2.2 contains some examples of physical problems in which Eq. (2.45) occurs. ■

Table 2.2 Some examples of the Poisson equation $-\nabla \cdot (k\nabla u) = Q$

Natural boundary condition: $k\dfrac{\partial u}{\partial n} + h(u - u_{\infty}) = q$; essential boundary condition: $u = \hat{u}$

Field of application	Primary variable u	Material constant k	Source variable Q	Secondary variables $q, \dfrac{\partial u}{\partial x}, \dfrac{\partial u}{\partial y}$
1. Heat transfer	Temperature T	Conductivity k	Heat source Q	Heat flow q [comes from conduction $k(\partial T/\partial n)$ and convection $h(T - T_{\infty})$] velocities:
2. Irrotational flow of an ideal fluid	Stream function ψ	Density ρ	Mass production σ (normally zero)	$\dfrac{\partial \psi}{\partial x} = -v$ $\dfrac{\partial \psi}{\partial y} = u$
	Velocity potential ϕ	Density ρ	Mass production σ (normally zero)	$\dfrac{\partial \phi}{\partial x} = u$, $\dfrac{\partial \phi}{\partial y} = v$
3. Ground-water flow	Piezometric head ϕ	Permeability K	Recharge Q (or pumping, $-Q$)	Seepage q $q = K\dfrac{\partial \phi}{\partial n}$ velocities: $u = -K\dfrac{\partial \phi}{\partial x}$ $v = -K\dfrac{\partial \phi}{\partial y}$
4. Torsion of constant cross-section members	Stress function ϕ	$k = \dfrac{1}{G}$, G = shear modulus	$Q = 2\theta$, θ = angle of twist per unit length	$\dfrac{\partial \phi}{\partial x} = -\tau_{zy}$ $\dfrac{\partial \phi}{\partial y} = \tau_{zx}$ τ_{zx}, τ_{zy} are shear stresses
5. Electrostatics	Scalar potential ϕ	Dielectric constant ε	Charge density ρ	Displacement flux density D_n
6. Magnetostatics	Magnetic potential ϕ	Permeability μ	Charge density ρ	Magnetic flux density B_n
7. Transverse deflection of elastic membranes	Transverse deflection u	$k = T$, T = tension in membrane	Transversely distributed load	Normal force q

The last example of this section is concerned with a pair of coupled partial differential equations in two dimensions.

Example 2.5 Consider the following pair of partial differential equations:

$$\left.\begin{aligned} C_{11}\frac{\partial^2 u}{\partial x^2} + C_{12}\frac{\partial}{\partial y}\left(\frac{\partial u}{\partial y} + \frac{\partial v}{\partial x}\right) + f = 0 \\ C_{12}\frac{\partial}{\partial x}\left(\frac{\partial u}{\partial y} + \frac{\partial v}{\partial x}\right) + C_{22}\frac{\partial^2 v}{\partial y^2} + g = 0 \end{aligned}\right\} \text{ in } \Omega \tag{2.53}$$

subjected to the boundary conditions

$$\left.\begin{aligned} C_{11}\frac{\partial u}{\partial x}n_x + C_{12}\left(\frac{\partial u}{\partial y} + \frac{\partial v}{\partial x}\right)n_y = \hat{s} \\ C_{12}\left(\frac{\partial u}{\partial y} + \frac{\partial v}{\partial x}\right)n_x + C_{22}\frac{\partial v}{\partial y}n_y = \hat{t} \end{aligned}\right\} \text{ on } \Gamma_1$$

$$\left.\begin{aligned} u = \hat{u} \\ v = \hat{v} \end{aligned}\right\} \text{ on } \Gamma_2 \tag{2.54}$$

Equations such as these might arise in the study of plane elasticity problems.

In the present example we have two dependent variables and two independent variables. We shall construct the variational formulation of these equations. We begin with

$$\begin{aligned} 0 &= \int_\Omega \xi\left[C_{11}\frac{\partial^2 u}{\partial x^2} + C_{12}\frac{\partial}{\partial y}\left(\frac{\partial u}{\partial y} + \frac{\partial v}{\partial x}\right) + f\right] dx\,dy \\ 0 &= \int_\Omega \zeta\left[C_{12}\frac{\partial}{\partial x}\left(\frac{\partial u}{\partial y} + \frac{\partial v}{\partial x}\right) + C_{22}\frac{\partial^2 v}{\partial y^2} + g\right] dx\,dy \end{aligned} \tag{2.55}$$

where ξ and ζ are test functions which satisfy the homogeneous essential boundary conditions on u and v, respectively. Using the integral identities (2.20), we arrive at

$$\begin{aligned} 0 = &-\int_\Omega\left[C_{11}\frac{\partial \xi}{\partial x}\frac{\partial u}{\partial x} + C_{12}\frac{\partial \xi}{\partial y}\left(\frac{\partial u}{\partial y} + \frac{\partial v}{\partial x}\right) - \xi f\right] dx\,dy \\ &+\oint_\Gamma \xi\left[C_{11}\frac{\partial u}{\partial x}n_x + C_{12}\left(\frac{\partial u}{\partial y} + \frac{\partial v}{\partial x}\right)n_y\right] ds \end{aligned} \tag{2.56a}$$

$$\begin{aligned} 0 = &-\int_\Omega\left[C_{12}\frac{\partial \zeta}{\partial x}\left(\frac{\partial u}{\partial y} + \frac{\partial v}{\partial x}\right) + C_{22}\frac{\partial \zeta}{\partial y}\frac{\partial v}{\partial y} - \zeta g\right] dx\,dy \\ &+\oint_\Gamma \zeta\left[C_{12}\left(\frac{\partial u}{\partial y} + \frac{\partial v}{\partial x}\right)n_x + C_{22}\frac{\partial v}{\partial y}n_y\right] ds \end{aligned} \tag{2.56b}$$

From the boundary integrals it follows that the specification of u and v constitutes the essential boundary conditions and the specification of the expression in the square brackets under the boundary integrals constitutes the natural boundary conditions of the problem. Using the natural boundary conditions on u and v on Γ_1, and the homogeneous essential boundary conditions on ξ and ζ on Γ_2, we arrive at the variational forms ($\xi = \zeta = 0$ on Γ_2)

$$\int_\Omega \left[C_{11}\frac{\partial \xi}{\partial x}\frac{\partial u}{\partial x} + C_{12}\frac{\partial \xi}{\partial y}\left(\frac{\partial u}{\partial y} + \frac{\partial v}{\partial x}\right)\right] dx\,dy = \int_{\Gamma_1} \xi\hat{s}\,ds + \int_\Omega \xi f\,dx\,dy$$
$$\int_\Omega \left[C_{12}\frac{\partial \zeta}{\partial x}\left(\frac{\partial u}{\partial y} + \frac{\partial v}{\partial x}\right) + C_{22}\frac{\partial \zeta}{\partial y}\frac{\partial v}{\partial y}\right] dx\,dy = \int_{\Gamma_1} \zeta\hat{t}\,ds + \int_\Omega \zeta g\,dx\,dy \tag{2.57}$$

The bilinear and linear functionals associated with the problem are given by adding the expressions in Eqs. (2.56a and b) and collecting the coefficients:

$$B((\xi,\zeta),(u,v)) = \int_\Omega \left[C_{11}\frac{\partial \xi}{\partial x}\frac{\partial u}{\partial x} + C_{12}\left(\frac{\partial \xi}{\partial y} + \frac{\partial \zeta}{\partial x}\right)\left(\frac{\partial u}{\partial y} + \frac{\partial v}{\partial x}\right) + C_{22}\frac{\partial \zeta}{\partial y}\frac{\partial v}{\partial y}\right] dx\,dy$$

$$l((\xi,\zeta)) = \int_\Omega (\xi f + \zeta g)\,dx\,dy + \int_{\Gamma_1} (\xi\hat{s} + \zeta\hat{t})\,ds \tag{2.58}$$

■

PROBLEMS

For the following differential equations and associated boundary conditions, construct the variational statements (i.e., weak forms and, whenever possible, quadratic functionals):

2.1 *One-dimensional heat conduction/convection*

$$-\frac{d}{dx}\left(a\frac{du}{dx}\right) + f = 0 \qquad \text{for } 0 < x < 1$$

$$u(0) = 0 \qquad a\frac{du}{dx} + h(u - u_\infty) = q \text{ at } x = 1$$

where a and f are functions of x, and h, u_∞, and q are constants.

2.2 *Beam on elastic foundation*

$$\frac{d^2}{dx^2}\left(b\frac{d^2 w}{dx^2}\right) + kw + f = 0 \qquad \text{for } 0 < x < L$$

$$w = b\frac{d^2 w}{dx^2} = 0 \text{ at } x = 0, L$$

where b and f are functions of x, and k is a constant.

2.3 *A nonlinear equation*

$$-\frac{d}{dx}\left(u\frac{du}{dx}\right)+f=0 \qquad \text{for } 0<x<1$$

$$\frac{du}{dx}(0)=0 \qquad u(1)=\sqrt{2}$$

2.4 *Longitudinal deformation of a bar with an end spring* (an eigenvalue problem)

$$-\frac{d}{dx}\left(a\frac{du}{dx}\right)+\lambda u=0 \qquad \text{for } 0<x<L$$

$$u(0)=0 \qquad \left(a\frac{du}{dx}+ku\right)\bigg|_{x=L}=0$$

2.5 *Large-deflection bending of beams*

$$-\frac{d}{dx}\left\{a\left[\frac{du}{dx}+\frac{1}{2}\left(\frac{dv}{dx}\right)^2\right]\right\}+P=0 \qquad \text{for } 0<x<L$$

$$\frac{d^2}{dx^2}\left(b\frac{d^2v}{dx^2}\right)-\frac{d}{dx}\left\{a\frac{dv}{dx}\left[\frac{du}{dx}+\frac{1}{2}\left(\frac{dv}{dx}\right)^2\right]\right\}+f=0$$

$$u=v=0 \text{ at } x=0, L \qquad \frac{dv}{dx}\bigg|_{x=0}=0 \qquad \left(b\frac{d^2v}{dx^2}\right)\bigg|_{x=L}=M_0$$

where a, b, P, and f are functions of x, and M_0 is a constant.

2.6 *A second-order equation*

$$-\frac{\partial}{\partial x}\left(a_{11}\frac{\partial u}{\partial x}+a_{12}\frac{\partial u}{\partial y}\right)-\frac{\partial}{\partial y}\left(a_{21}\frac{\partial u}{\partial x}+a_{22}\frac{\partial u}{\partial y}\right)+f=0 \text{ in } \Omega$$

$$u=u_0 \text{ on } \Gamma_1$$

$$\left(a_{11}\frac{\partial u}{\partial x}+a_{12}\frac{\partial u}{\partial y}\right)n_x+\left(a_{21}\frac{\partial u}{\partial x}+a_{22}\frac{\partial u}{\partial y}\right)n_y=t_0 \text{ on } \Gamma_2$$

where $a_{ij}=a_{ji}$ $(i, j=1,2)$ and f are given functions of position (x, y) in a two-dimensional domain Ω, and u_0 and t_0 are known functions on portions Γ_1 and Γ_2 of the boundary Γ.

2.7 *Navier-Stokes equations for two-dimensional flow of viscous, incompressible fluids* (pressure-velocity formulation)

$$\left.\begin{aligned} u\frac{\partial u}{\partial x}+v\frac{\partial u}{\partial y} &= -\frac{1}{\rho}\frac{\partial P}{\partial x}+\nu\left(\frac{\partial^2 u}{\partial x^2}+\frac{\partial^2 u}{\partial y^2}\right)\\ u\frac{\partial v}{\partial x}+v\frac{\partial v}{\partial y} &= -\frac{1}{\rho}\frac{\partial P}{\partial y}+\nu\left(\frac{\partial^2 v}{\partial x^2}+\frac{\partial^2 v}{\partial y^2}\right)\\ \frac{\partial u}{\partial x}+\frac{\partial v}{\partial y} &= 0 \end{aligned}\right\} \text{ in } \Omega$$

$$\left.\begin{aligned} u&=u_0\\ v&=v_0\end{aligned}\right\} \text{ on } \Gamma_1$$

$$\left.\begin{aligned} \nu\left(\frac{\partial u}{\partial x}n_x+\frac{\partial u}{\partial y}n_y\right)-\frac{1}{\rho}Pn_x &= \hat{t}_x\\ \nu\left(\frac{\partial v}{\partial x}n_x+\frac{\partial v}{\partial y}n_y\right)-\frac{1}{\rho}Pn_y &= \hat{t}_y\end{aligned}\right\} \text{ on } \Gamma_2$$

2.8 *Two-dimensional flow of viscous, incompressible fluids* (stream function-vorticity formulation)

$$\left.\begin{aligned} -\nabla^2\psi - \zeta = 0 \\ -\nabla^2\zeta + \frac{\partial\psi}{\partial x}\frac{\partial\zeta}{\partial y} - \frac{\partial\psi}{\partial y}\frac{\partial\zeta}{\partial x} = 0 \end{aligned}\right\} \text{ in } \Omega$$

Assume that the essential boundary conditions are specified to be zero.

In Probs. 2.9 to 2.11, construct variational formulations of the following time-dependent problems. (*Hint*: Treat the time derivatives in the equation as part of the source term in the differential equation.)

2.9

$$-\frac{\partial}{\partial x}\left(a\frac{\partial u}{\partial x}\right) + b\frac{\partial u}{\partial t} = f \qquad \text{for } 0 < x < 1$$

$$u(0,t) = 0 \qquad \left(a\frac{\partial u}{\partial x}\right)\Bigg|_{x=1} = T_0 \qquad u(x,0) = u_0$$

Here a, b, and f are functions of position x and time t.

2.10

$$\frac{\partial^2}{\partial x^2}\left(a\frac{\partial^2 u}{\partial x^2}\right) + b\frac{\partial^2 u}{\partial t^2} = f \qquad \text{for } 0 < x < L$$

$$u(0,t) = 0 \qquad \frac{\partial u}{\partial x}(0,t) = 0 \qquad \left(a\frac{\partial^2 u}{\partial x^2}\right)\Bigg|_{x=L} = M_0$$

$$\frac{\partial}{\partial x}\left(a\frac{\partial^2 u}{\partial x^2}\right)\Bigg|_{x=L} = P \qquad u(x,0) = u_0 \qquad \frac{\partial u}{\partial t}(x,0) = v_0$$

Here a, b, and f are functions of position x and time t, and M_0, P, u_0, and v_0 are constants.

2.11

$$-k\nabla^2 u + \rho\frac{\partial u}{\partial t} = f \text{ in } \Omega$$

$$u = 0 \text{ on } \Gamma_1 \qquad k\frac{\partial u}{\partial n} = \hat{t} \text{ on } \Gamma_2 \qquad u = u_0 \text{ at } t = 0$$

where k, ρ, and $\hat{t}$ are constants, and f is a given function of position (x, y) in Ω.

2.3 VARIATIONAL METHODS OF APPROXIMATION

2.3.1 Introduction

Our objective in this section is to study the variational methods of approximation. These include the Ritz method, the Galerkin method, the Petrov-Galerkin method, the least-squares method, and the collocation method. All of these methods seek an approximate solution in the form of a linear combination of suitable approximation functions. The parameters in the linear combination are determined such that the approximate solution satisfies the weak form or minimizes the quadratic functional of the equation under study. Various methods differ from each other in the choice of the approximate functions.

The primary objective of this section is to present a number of classical variational methods. The finite-element method makes use of the variational methods to formulate the discrete equations for a subdomain, called element. As we shall see in Chaps. 3 and 4, the choice of the approximation functions in the finite-element method is different from that in the classical variational methods.

2.3.2 The Ritz Method

Consider the variational problem of finding the solution u such that

$$B(v, u) = l(v) \tag{2.59}$$

for all sufficiently differentiable v that satisfy the homogeneous form of essential boundary conditions on u. When the functional B is bilinear and symmetric and l is linear, then the problem in Eq. (2.59) is equivalent to minimizing the quadratic functional

$$I(u) = \tfrac{1}{2}B(u, u) - l(u) \tag{2.60}$$

The Ritz method seeks an approximate solution to Eq. (2.59) in the form of a finite series

$$u_N = \sum_{j=1}^{N} c_j\phi_j + \phi_0 \tag{2.61}$$

where the constants c_j, called the *Ritz coefficients*, are chosen such that Eq. (2.59) holds for $v = \phi_i$ $(i = 1, 2, \ldots, N)$:

$$B\left(\phi_i, \sum_{j=1}^{N} c_j\phi_j + \phi_0\right) = l(\phi_i) \qquad i = 1, 2, \ldots, N \tag{2.62}$$

If B is bilinear, we have

$$\sum_{j=1}^{N} B(\phi_i, \phi_j)c_j = l(\phi_i) - B(\phi_i, \phi_0) \tag{2.63}$$

which represents a system of N linear algebraic equations in N constants c_j. The columns (and rows) of matrix coefficients $b_{ij} = B(\phi_i, \phi_j)$ must be linearly independent in order that the coefficient matrix in Eq. (2.63) can be inverted.

For symmetric bilinear forms, the Ritz method can be viewed as one that seeks solution of the form in Eq. (2.61) in which the parameters are determined by minimizing the quadratic functional corresponding to the symmetric bilinear form, that is, functional $I(u)$ in Eq. (2.60). After substituting u_N from Eq. (2.61) for u into Eq. (2.60) and integrating, the functional $I(u)$ becomes an ordinary (quadratic) function of the parameters $c_1, c_2, \ldots$. Then the necessary condition for the minimum of $I(c_1, c_2, \ldots, c_N)$ is that its partial derivatives with respect to each of the parameters be zero:

$$\frac{\partial I(c_j)}{\partial c_1} = 0, \quad \frac{\partial I(c_j)}{\partial c_2} = 0, \ldots, \quad \frac{\partial I(c_j)}{\partial c_N} = 0 \tag{2.64}$$

Thus there are N linear algebraic equations in N unknowns, c_j $(j = 1, 2, \ldots, N)$.

These equations are exactly the same as those in Eq. (2.63) for linear symmetric bilinear forms. Of course, when $B(\cdot, \cdot)$ is not symmetric, we do not have a quadratic functional. In other words, Eq. (2.63) is more general than Eq. (2.64), and they are the same when $B(\cdot, \cdot)$ is bilinear and symmetric. In most problems of interest in the present study, we will have a symmetric bilinear form.

Returning to the Ritz approximation in Eq. (2.61), we list the properties required of the approximation functions ϕ_i $(i = 1, 2, \ldots, N)$ and ϕ_0. The function ϕ_0 is selected to satisfy the specified essential boundary conditions of the problem. If the specified essential boundary conditions are all homogeneous, then $\phi_0 = 0$. Since ϕ_0 satisfies the specified essential boundary conditions, we require ϕ_i $(i = 1, 2, \ldots, N)$ to satisfy the homogeneous form of the essential boundary conditions so that $u_N = \phi_0$ at the points at which the essential boundary conditions are specified. Since ϕ_i satisfy the homogeneous essential boundary conditions, the choice $v = \phi_i$ is consistent with the requirements of a test function. In addition to the above requirements, we require ϕ_i to satisfy the following conditions:

1. *a.* ϕ_i should be such that $B(\phi_i, \phi_j)$ is well defined and nonzero [i.e., sufficiently differentiable as required by the bilinear form $B(\cdot, \cdot)$].
   b. ϕ_i must satisfy at least the homogeneous form of the essential boundary conditions of the problem.
2. For any N, the set $\{\phi_i\}_{i=1}^N$ along with the columns (and rows) of $B(\phi_i, \phi_j)$ are linearly independent.
3. $\{\phi_i\}$ is complete. (2.65)

The above requirements on the approximation functions guarantee, for linear problems, convergence of the Ritz solution to the exact solution as the value of N is increased. The convergence is understood to be in the following sense:

$$I(u_N) \geq I(u_M) \qquad \text{for } N \leq M \tag{2.66}$$

For any value of N, the previously computed elements of the matrix coefficients b_{ij} and the column vector $F_i = l(\phi_i) - B(\phi_i, \phi_0)$ remain unchanged, and one must add to the existing coefficients newly computed rows and columns. Next, we consider a few examples of the application of the Ritz method.

Example 2.6 Consider the differential equation of Example 2.2 (with $a = c = 1$):

$$-\frac{d^2u}{dx^2} - u + x^2 = 0 \qquad \text{for } 0 < x < 1 \tag{2.67}$$

We shall consider two sets of boundary conditions:

$$\text{Set 1:} \qquad u(0) = 0 \qquad u(1) = 0$$

$$\text{Set 2:} \qquad u(0) = 0 \qquad u'(1) = 1 \tag{2.68}$$

Set 1 The bilinear functional and the linear functional are

$$B(v,u) = \int_0^1 \left(\frac{dv}{dx}\frac{du}{dx} - vu \right) dx \qquad l(v) = -\int_0^1 vx^2\, dx \tag{2.69}$$

Since both boundary conditions $[u(0) = u(1) = 0]$ are of the essential type, we must select ϕ_i in the N-parameter Ritz approximation to satisfy the conditions $\phi_i(0) = \phi_i(1) = 0$. We select the following functions: $\phi_0 = 0$, and

$$\phi_1 = x(1-x),\ \phi_2 = x^2(1-x), \ldots,\ \phi_N = x^N(1-x) \tag{2.70}$$

It should be pointed out that if one selects, for example, the functions $\phi_1 = x^2(1 - x)$, $\phi_2 = x^3(1-x)$, etc., requirement 3 in conditions (2.65) is violated, because the set cannot be used to generate the linear term x of the exact solution. As a rule, one must start with the *lowest order* admissible functions.

The N-parameter Ritz solution for the problem at hand is given by

$$u_N = c_1x(1-x) + c_2x^2(1-x) + \cdots + c_Nx^N(1-x) \tag{2.71}$$

Substituting this expression into the variational problem $B(v,u) = l(v)$, we obtain

$$\int_0^1 \left[\frac{d\phi_i}{dx}\left(\sum_{j=1}^{N} c_j \frac{d\phi_j}{dx} \right) - \phi_i \left(\sum_{j=1}^{N} c_j\phi_j \right) \right] dx = -\int_0^1 \phi_i x^2\, dx$$

or

$$\sum_{j=1}^{N} B(\phi_i, \phi_j)c_j = l(\phi_i) \qquad i = 1, 2, \ldots, N \tag{2.72}$$

Computing the coefficients $b_{ij} \equiv B(\phi_i, \phi_j)$ and $l_i \equiv l(\phi_i)$, we obtain

$$\begin{aligned} b_{ij} &= \int_0^1 \{[ix^{i-1} - (i+1)x^i][jx^{j-1} - (j+1)x^j] \\ &\quad - (x^i - x^{i+1})(x^j - x^{j+1})\}\, dx \\ &= \frac{2ij}{(i+j)[(i+j)^2 - 1]} - \frac{2}{(i+j+1)(i+j+2)(i+j+3)} \\ l_i &= -\int_0^1 x^2(x^i - x^{i+1})\, dx \\ &= -\frac{1}{(3+i)(4+i)} \end{aligned} \tag{2.73}$$

The exact solution for this case is given by

$$u(x) = \frac{\sin x + 2\sin(1-x)}{\sin 1} + x^2 - 2 \tag{2.74}$$

The values of the Ritz coefficients along with a comparison of the Ritz solution with the exact solution are shown in Table 2.3 and Fig. 2.4.

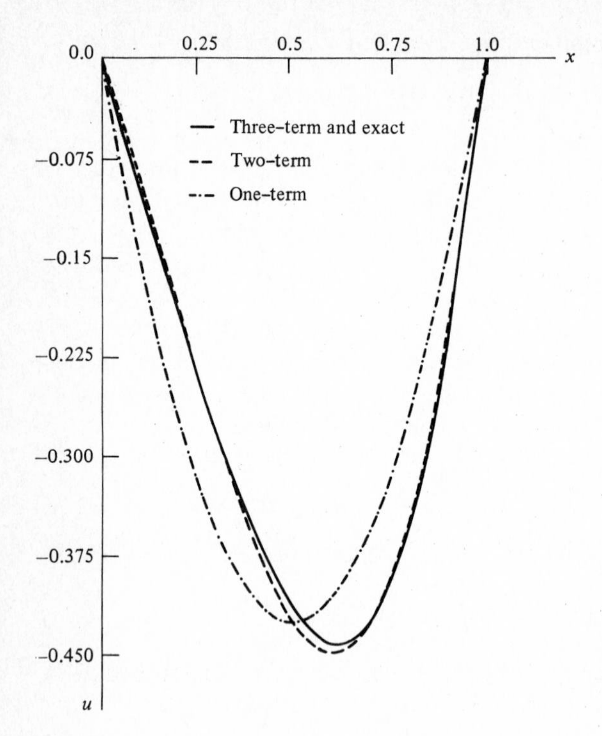

Figure 2.4 Comparison of the exact solution in Eq. (2.74) with the Ritz solution (2.71) for $N = 1$, 2, and 3 of Eq. (2.67) with $u(0) = u(1) = 0$.

Table 2.3 Comparison of the Ritz solution with the exact solution of the equation

$$-\frac{d^2u}{dx^2} - u + x^2 = 0 \qquad \text{for } 0 < x < 1 \qquad u(0) = u(1) = 0$$

Ritz coefficients†		x	Ritz solution ($-10u$) $N = 1$	$N = 2$	$N = 3$	Exact solution
$N = 1$:		0.0	0.0	0.0	0.0	0.0
	$C_1 = -0.1667$	0.1	0.1500	0.0885	0.0954	0.0955
$N = 2$:		0.2	0.2667	0.1847	0.1890	0.1890
	$C_1 = -0.0813$	0.3	0.3500	0.2783	0.2766	0.2763
	$C_2 = -0.1707$	0.4	0.4000	0.3590	0.3520	0.3518
$N = 3$:		0.5	0.4167	0.4167	0.4076	0.4076
	$C_1 = -0.0952$	0.6	0.4000	0.4410	0.4340	0.4342
	$C_2 = -0.1005$	0.7	0.3500	0.4217	0.4200	0.4203
	$C_3 = -0.0702$	0.8	0.2667	0.3486	0.3529	0.3530
		0.9	0.1500	0.2115	0.2183	0.2182
		1.0	0.0	0.0	0.0	0.0

†The four-parameter Ritz solution coincides with the exact solution up to four decimal points.

The same result can be obtained using Eq. (2.64) [instead of Eq. (2.63)]. We have

$$I(u) = \frac{1}{2}\int_0^1 \left[\left(\frac{du}{dx}\right)^2 - u^2 + 2x^2 u\right] dx$$

$$I(c_j) = \frac{1}{2}\int_0^1 \left[\left(\sum_{j=1}^{N} c_j \frac{d\phi_j}{dx}\right)^2 - \left(\sum_{j=1}^{N} c_j\phi_j\right)^2 + 2x^2\left(\sum_{j=1}^{N} c_j\phi_j\right)\right] dx \tag{2.75}$$

$$\frac{\partial I}{\partial c_i} = 0 = \int_0^1 \left[\frac{d\phi_i}{dx}\left(\sum_{j=1}^{N} c_j \frac{d\phi_j}{dx}\right) - \phi_i\left(\sum_{j=1}^{N} c_j\phi_j\right) + \phi_i x^2\right] dx \equiv \sum_{j=1}^{N} b_{ij}c_j - l(\phi_i)$$

where

$$b_{ij} = \int_0^1 \left(\frac{d\phi_i}{dx}\frac{d\phi_j}{dx} - \phi_i\phi_j\right) dx \qquad l(\phi_i) = -\int_0^1 x^2\phi_i\, dx \tag{2.76}$$

Set 2 For the second set of boundary conditions, the bilinear form is the same as that given in Eq. (2.69). The linear form is given by

$$l(v) = -\int_0^1 vx^2\, dx + v(1) \tag{2.77}$$

In this case, the ϕ_i should be selected to satisfy the only essential boundary condition $\phi_i(0) = 0$. The following choice of ϕ_i meets the requirements:

$$\phi_i = x^i \tag{2.78}$$

The coefficients b_{ij} and l_i are given by

$$b_{ij} = \int_0^1 (ijx^{i+j-2} - x^{i+j})\, dx = \frac{ij}{i+j-1} - \frac{1}{i+j+1}$$

$$l_i = -\int_0^1 x^{i+2}\, dx + 1 = -\frac{1}{i+3} + 1 \tag{2.79}$$

The exact solution in the present case is given by

$$u(x) = \frac{2\cos(1-x) - \sin x}{\cos 1} + x^2 - 2 \tag{2.80}$$

A comparison of the Ritz solution with the exact solution is presented in Table 2.4. ■

Example 2.7 Consider the problem of finding the transverse deflection of a cantilevered beam under uniform transverse load of intensity f per unit length, and end moment M_0 (see Example 2.3),

$$\frac{d^2}{dx^2}\left(EI\frac{d^2w}{dx^2}\right) + f = 0 \qquad \text{for } \begin{cases} 0 < x < L \\ EI > 0 \end{cases} \tag{2.81}$$

$$w(0) = \frac{dw}{dx}(0) = 0 \qquad \left(EI\frac{d^2w}{dx^2}\right)\bigg|_{x=L} = M_0 \qquad \left[\frac{d}{dx}\left(EI\frac{d^2w}{dx^2}\right)\right]_{x=L} = 0 \tag{2.82}$$

Table 2.4 Comparison of the Ritz solution with the exact solution of the equation

$$-\frac{d^2u}{dx^2} - u + x^2 = 0 \qquad \text{for } 0 < x < 1 \qquad u(0) = 0 \qquad \frac{du}{dx}(1) = 1$$

Ritz coefficients[†]	x	Ritz solution u $N = 1$	$N = 2$	$N = 3$	Exact solution
$N = 1$:	0	0.0	0.0	0.0	0.0
$C_1 = 1.125$	0.1	0.1125	0.1280	0.1271	0.1262
$N = 2$:	0.2	0.2250	0.2529	0.2518	0.2513
$C_1 = 1.295$	0.3	0.3375	0.3749	0.3740	0.3742
$C_2 = -0.1511$	0.4	0.4500	0.4938	0.4934	0.4943
$N = 3$:	0.5	0.5625	0.6097	0.6099	0.6112
$C_1 = 1.283$	0.6	0.6750	0.7226	0.7234	0.7244
$C_2 = -0.1142$	0.7	0.7875	0.8324	0.8337	0.8340
$C_3 = -0.02462$	0.8	0.9000	0.9393	0.9407	0.9402
	0.9	1.0120	1.043	1.044	1.043
	1.0	1.125	1.144	1.144	1.144

[†] The four-parameter Ritz solution coincides with the exact solution up to four decimal points.

We now construct an N-parameter Ritz solution using the variational form, Eq. (2.42) or (2.44): $B(v, w) = l(v)$, where

$$B(v,w) = \int_0^L EI \frac{d^2v}{dx^2}\frac{d^2w}{dx^2}\,dx \qquad l(v) = \int_0^L - fv\,dx + M_0 \left.\frac{dv}{dx}\right|_{x=L} \tag{2.83}$$

We select algebraic coordinate functions ϕ_i that satisfy the essential boundary conditions $\phi_i(0) = \phi_i'(0) = 0$. We represent w and v by

$$w_N(x) = \sum_{j=1}^{N} c_j\phi_j \qquad v = \phi_i = x^{i+1} \tag{2.84}$$

Substituting Eq. (2.84) into Eq. (2.83), we obtain, for f = constant,

$$b_{ij} = \int_0^L EI(i+1)i x^{i-1} j(j+1)x^{j-1}\,dx$$

$$= \frac{EIij(i+1)(j+1)(L)^{i+j-1}}{i+j-1}$$

$$l(\phi_i) = \frac{-f(L)^{i+2}}{i+2} + M_0(i+1)(L)^i \tag{2.85}$$

For $N = 2$ (two-parameter solution), we have

$$EI(4Lc_1 + 6L^2c_2) = -\frac{fL^3}{3} + 2M_0L$$

$$EI(6L^2c_1 + 12L^3c_2) = -\frac{fL^4}{4} + 3M_0L^2 \tag{2.86}$$

Solving for c_1 and c_2, we obtain

$$c_1 = -\frac{5fL^2 - 12M_0}{24EI} \qquad c_2 = \frac{fL}{12EI} \tag{2.87}$$

and

$$w_2(x) = \frac{12M_0 - 5fL^2}{24EI}x^2 + \frac{fL}{12EI}x^3 \tag{2.88}$$

For the three-parameter ($N = 3$) solution, we get the matrix equations

$$EI\begin{bmatrix} 4 & 6L & 8L^2 \\ 6L & 12L^2 & 18L^3 \\ 8L^2 & 18L^3 & \frac{144}{5}L^4 \end{bmatrix}\begin{Bmatrix} c_1 \\ c_2 \\ c_3 \end{Bmatrix} = \begin{Bmatrix} -fL^2/3 + 2M_0 \\ -fL^3/4 + 3M_0L \\ -fL^4/5 + 4M_0L^2 \end{Bmatrix} \tag{2.89}$$

the solution of which coincides with the exact solution

$$w(x) = \frac{2M_0 - fL^2}{4EI}x^2 + \frac{fL}{6EI}x^3 - \frac{f}{24EI}x^4 \tag{2.90}$$

■

Example 2.8 Consider the Poisson equation in a two-dimensional region:

$$\begin{aligned} -\nabla^2 T &= 1 \text{ in } \Omega = \{(x, y): 0 < (x, y) < 1\} \\ T &= 0 \text{ on sides } x = 1 \text{ and } y = 1 \\ \frac{\partial T}{\partial n} &= 0 \text{ on sides } x = 0 \text{ and } y = 0 \end{aligned} \tag{2.91}$$

The bilinear and linear functionals are given by

$$\begin{aligned} B(v, T) &= \int_0^1 \int_0^1 \left(\frac{\partial v}{\partial x}\frac{\partial T}{\partial x} + \frac{\partial v}{\partial y}\frac{\partial T}{\partial y} \right) dx\, dy \\ l(v) &= \int_0^1 \int_0^1 v\, dx\, dy \end{aligned} \tag{2.92}$$

For an n-parameter Ritz approximation we choose the following set of approximation functions (of course, there are many other choices):

$$\phi_i = \cos\frac{(2i-1)\pi x}{2} \cdot \cos\frac{(2i-1)\pi y}{2} \qquad i = 1, 2, \ldots, n \tag{2.93}$$

Incidentally, ϕ_i also satisfies the natural boundary conditions of the problem. While the choice $\phi_i = \sin(i\pi x) \cdot \sin(i\pi y)$ meets the essential boundary conditions, it is not complete because it cannot be used to generate the solution that does not vanish on the sides $x = 0$ and $y = 0$.

The coefficient matrix b_{ij} and vector l_i can be computed using ϕ_i of Eq. (2.93) in Eq. (2.92). We obtain

$$b_{ij} = \begin{cases} \dfrac{\alpha_i^2}{2} & \text{if } i = j \\ \dfrac{\alpha_i^2 \alpha_j^2}{2\left(\alpha_i^2 - \alpha_j^2\right)^2} & \text{if } i \neq j \end{cases} \tag{2.94}$$

$$l(\phi_i) = \frac{1}{\alpha_i^2} \qquad \alpha_i = (2i - 1)\frac{\pi}{2}$$

The exact solution to this problem is given by

$$T(x, y) = \frac{1}{2}\left\{(1 - y^2) + \frac{32}{\pi^3} \sum_{n=1}^{\infty} \frac{(-1)^n \cos\left[(2n-1)\pi y/2\right] \cosh\left[(2n-1)\pi x/2\right]}{(2n-1)^3 \cosh(2n-1)\pi/2}\right\} \tag{2.95}$$

The one-parameter and two-parameter Ritz solutions are

$$T_1 = \frac{32}{\pi^4} \cos\frac{\pi x}{2} \cos\frac{\pi y}{2} \tag{2.96}$$

$$T_2 = 0.3283988 \cos\frac{\pi x}{2} \cos\frac{\pi y}{2} + 0.001976 \cos\frac{3\pi x}{2} \cos\frac{3\pi x}{2} \tag{2.97}$$

which differ from the exact solution at the center of the region by 9.3 and 0.54 percent, respectively.

If algebraic polynomials are to be used, one can choose $\phi_1 = (1 - x)(1 - y)$ or $\phi_1 = (1 - x^2)(1 - y^2)$, both of which satisfy the (homogeneous) essential boundary conditions. However, the choice $\phi_1 = (1 - x)(1 - y)$ does not meet the natural boundary conditions of the problem. The one-parameter Ritz solution for the choice $\phi_1 = (1 - x^2)(1 - y^2)$ is given by

$$T_1(x, y) = \frac{5}{16}(1 - x^2)(1 - y^2) \tag{2.98}$$

This gives a value of 0.17578 for T at the center of the region. This is an error of 2.94 percent when compared to the exact solution. ■

2.3.3 The Method of Weighted Residuals

Recall the integral form, Eq. (2.28), associated with the differential equation (2.27). Note that one can always write the integral form of a differential equation, whether the equation is linear or nonlinear (in the dependent variables). However, it is not always possible to construct a symmetric variational form and the associated functional when the equation under consideration is nonlinear. While the Ritz method can also be applied to nonlinear problems, it restricts the choice of test functions to those used for the approximation. The weighted residual method is a generalization of the Ritz method in which the test functions can be chosen from an independent set of functions. Further, when it is not possible to construct a weak form, the method of weighted residuals can be used to approximate the integral form of the equation. Since the integral form does not include the natural boundary conditions of the problem, the approximating functions should be selected such that the approximated solution satisfies the boundary conditions of the problem. On the other hand, the test functions can be selected independent of the approximation functions. This flexibility is advantageous in certain nonlinear problems.

In this section, we discuss the general method of weighted residuals first, and then consider certain special cases that are known by specific names (e.g., Galerkin, least squares, etc.). Although a limited use of the weighted-residual method is made in the present work, it is informative to have a knowledge of this method for use in the formulation of certain nonlinear problems readers might encounter in their work.

Consider the operator equation

$$Au = f \text{ in } \Omega \tag{2.99}$$

where A is an operator (linear or nonlinear), often a differential operator, acting on the unknown dependent variable u, and f is a known function of position. Some examples of such operators are provided by

$$
\begin{aligned}
&1.\ Au = -\frac{d}{dx}\left(a\frac{du}{dx}\right)\\
&2.\ Aw = \frac{d^2}{dx^2}\left(b\frac{d^2w}{dx^2}\right)\\
&3.\ Au = -\frac{d}{dx}\left(u\frac{du}{dx}\right)\\
&4.\ Au = -\left[\frac{\partial}{\partial x}\left(k_x\frac{\partial u}{\partial x}\right) + \frac{\partial}{\partial y}\left(k_y\frac{\partial u}{\partial y}\right)\right]\\
&5.\ A(u, v) = u\frac{\partial u}{\partial x} + v\frac{\partial u}{\partial y} + \frac{\partial^2 u}{\partial x^2} + \frac{\partial}{\partial y}\left(\frac{\partial u}{\partial y} + \frac{\partial v}{\partial x}\right)
\end{aligned} \tag{2.100}
$$

An operator A is said to be *linear* if and only if it satisfies the relation

$$A(\alpha u + \beta v) = \alpha A u + \beta A v \tag{2.101}$$

for any scalars α and β and dependent variables u and v. It can be easily verified that all these operators, with the exception of those defined in examples 3 and 5 of Eq. (2.100) are linear. When an operator does not satisfy the condition (2.101), it is said to be *nonlinear*.

The function u (i.e., solution) is not only required to satisfy the operator equation (2.99), it is also required to satisfy the boundary conditions associated with the operator. From the examples considered so far, the boundary conditions associated with operators defined in examples 1, 2, and 4 of Eq. (2.100) are obvious.

In the weighted-residual method the solution u is approximated, in much the same way as in the Ritz method, by expressions of the form

$$u_N = \phi_0 + \sum_{j=1}^{N} c_j \phi_j \tag{2.102}$$

where ϕ_0 must satisfy all the specified boundary conditions ($\phi_0 = 0$ if all the specified boundary conditions are homogeneous) of the problem, and ϕ_j must satisfy the conditions in Eq. (2.65), except that the continuity requirement imposed on ϕ_j is the same as that required by Eq. (2.99) and ϕ_j must satisfy the homogeneous form of the specified boundary conditions. As will be noted shortly, the continuity requirement can be relaxed if the operator permits a weak formulation (as in the Ritz method).

Substitution of the approximation (2.102) into the operator equation (2.99) results in a *residual* (i.e., an error in the equation)

$$E \equiv A(u_N) - f \neq 0 \tag{2.103}$$

Once ϕ_0 and ϕ_j are selected, E is merely a function of the independent variables and the parameters c_j. In the weighted-residual method the parameters are determined by setting the integral (over the domain) of a *weighted* residual of the approximation to zero:

$$\int_\Omega \psi_i(x, y) E(x, y, c_j)\, dx\, dy = 0 \qquad i = 1, 2, \ldots, N \tag{2.104}$$

where ψ_i are *weight functions* (which, in general, are not the same as the approximation functions ϕ_i). Obviously, $\{\psi_i\}$ must be a linearly independent set [otherwise the equations provided by Eq. (2.104) will not be linearly independent and hence are not solvable].

A careful examination of Eq. (2.104) shows that Eqs. (2.28) and (2.104) are alike with $v = \psi_i$. In other words, the weighted-residual method employs a weighted-integral form of the equation to be solved. If the operator permits, one can transfer the differentiation from the solution to the weight functions and thereby relax the continuity requirements on the approximation functions.

When the operator A is linear, Eq. (2.104) can be simplified to the form

$$\sum_{j=1}^{N} \left(\int_\Omega \psi_i A(\phi_j)\, dx\, dy \right) c_j = \int_\Omega \psi_i [f - A(\phi_0)]\, dx\, dy$$

or

$$\sum_{j=1}^{N} A_{ij} c_j = f_i \tag{2.105}$$

Note that the coefficient matrix $[A]$ is not symmetric,

$$A_{ij} = \int_\Omega \psi_i A(\phi_j)\, dx\, dy \neq A_{ji} \tag{2.106}$$

The weighted-residual method (when $\psi_i \neq \phi_i$) is also sometimes referred to as the *Petrov-Galerkin method*. For different choices of ψ_i the method is known by different names. We outline below the most frequently used methods.

The Galerkin method For $\psi_i = \phi_i$, the weighted-residual method is better known as the Galerkin method. When the operator is a linear differential operator of even order, the Galerkin method reduces to the Ritz method. (Because half of the differentiation can be transferred to the weight functions, the resulting coefficient matrix will be symmetric.)

The least-squares method The least-squares method seeks a solution in the form (2.102) and determines the constants c_j by minimizing the integral of the square of the residual (2.103):

$$\frac{\partial}{\partial c_i} \int_\Omega E^2(x, y, c_j)\, dx\, dy = 0$$

or

$$\int_\Omega \frac{\partial E}{\partial c_i} E\, dx\, dy = 0 \tag{2.107}$$

A comparison of Eq. (2.107) with Eq. (2.104) shows that $\psi_i = \partial E/\partial c_i$. If A is a linear operator, Eq. (2.107) becomes

$$\sum_{j=1}^{N} \left(\int_\Omega A(\phi_i) A(\phi_j)\, dx\, dy \right) c_j = \int_\Omega A(\phi_i)[f - A(\phi_0)]\, dx\, dy \tag{2.108}$$

which yields a symmetric coefficient matrix (but requires the same order of differentiation as the operator equation).

The collocation method The collocation method seeks approximate solution u_N to Eq. (2.99) in the form of Eq. (2.102) by requiring the residual in the equation to be identically zero at N selected points $\mathbf{x}^i \equiv (x^i, y^i)$ $(i = 1, 2, \ldots, N)$ in the domain Ω:

$$E\left(x^i, y^i, c_j\right) = 0 \qquad i = 1, 2, \ldots, N \tag{2.109}$$

The selection of the points $\mathbf{x}^i$ is crucial in obtaining a well-conditioned system of equations and ultimately in obtaining an accurate solution. The collocation method can be shown to be a special case of Eq. (2.104) for $\psi_i = \delta(\mathbf{x} - \mathbf{x}^i)$, where $\delta(\mathbf{x})$ is the *Dirac delta function*,

$$\int_\Omega f(\mathbf{x}) \delta(\mathbf{x} - \boldsymbol{\xi})\, dx\, dy = f(\boldsymbol{\xi}) \tag{2.110}$$

The Courant method The so-called Courant method combines the basic concepts of the Ritz method and the least-squares method (for linear operator equations).

The method seeks approximate solution u_N in the form of Eq. (2.102) by minimizing the modified quadratic functional

$$I_p(u_N) = I(u_N) + \frac{\gamma}{2}\|Au_N - f\|^2 \tag{2.111}$$

where $I(u)$ is the quadratic functional associated with $Au = f$, when A is linear, and γ is the *penalty parameter* (preassigned). Obviously the statement makes sense only for operator equations that admit functional formulation. While the method requires a laborious calculation, it improves convergence [i.e., as $N \to \infty$ we have $I_p(u_N) = I(u)$].

One must immediately note that the Courant method is none other than the penalty method now being used in fluid flow problems. In the present case, the condition $Au - f = 0$ can be viewed as the constraint [see Reddy and Rasmussen (1982)].

Example 2.9 We wish to solve the differential equation in Example 2.6 with set 2 boundary conditions. The approximation functions selected there do not meet the condition that

$$\frac{d\phi_i}{dx}(1) = 0 \qquad i = 1,2,\ldots$$

For the weighted-residual method, we select ϕ_0 and $\phi_i (i = 1,2)$:

$$\phi_0 = x \qquad \phi_1 = x(2-x) \qquad \phi_2 = x^2(1 - \tfrac{2}{3}x) \tag{2.112}$$

The residual is given by

$$E = -\frac{d^2}{dx^2}\left(\phi_0 + \sum_{i=1}^{N} c_i\phi_i\right) - \left(\phi_0 + \sum_{i=1}^{N} c_i\phi_i\right) + x^2 \tag{2.113}$$

The Petrov-Galerkin method We choose the weight functions ψ_i to be

$$\psi_1 = x \qquad \psi_2 = x^2 \tag{2.114}$$

Then,

$$\int_0^1 xE\,dx = 0 \qquad \int_0^1 x^2E\,dx = 0$$

or

$$\frac{7}{12}c_1 + \frac{13}{60}c_2 - \frac{1}{12} = 0 \qquad \frac{11}{30}c_1 + \frac{1}{15}c_2 - \frac{1}{20} = 0 \tag{2.115}$$

Solving for $c_i (c_1 = 103/682,\ c_2 = -15/682)$, we have the solution

$$u_{PG} = 1.302053x - 0.173021x^2 - 0.014663x^3 \tag{2.116}$$

The Galerkin method Taking $\psi_i = \phi_i$, we have

$$\int_0^1 x(2-x)E\,dx = 0 \qquad \int_0^1 x^2\left(1 - \frac{2}{3}x\right)E\,dx = 0$$

or

$$\frac{4}{5}c_1 + \frac{11}{30}c_2 - \frac{1}{30} = 0 \qquad \frac{17}{90}c_1 - \frac{3}{35}c_2 - \frac{19}{180} = 0 \tag{2.117}$$

Hence, the solution becomes (with $c_1 = 1571/5210$, $c_2 = -2954/5210$),

$$u_G = 1.60307x - 0.86853x^2 + 0.37799x^3 \tag{2.118}$$

The least-squares method Taking $\psi_i = \partial E/\partial c_i$, we have

$$\int_0^1 (2 - 2x + x^2)E\,dx = 0 \qquad -\int_0^1 \left(2 - 4x + x^2 - \frac{2}{3}x^3\right)E\,dx = 0$$

or

$$\frac{28}{15}c_1 - \frac{7}{10}c_2 - \frac{13}{60} = 0 \qquad -\frac{47}{90}c_1 + \frac{49}{35}c_2 + \frac{19}{180} = 0 \tag{2.119}$$

The least-squares approximation of Eq. (2.67) is given by (with $c_1 = 17346/169932$, $c_2 = -6342/169932$)

$$u_L = 1.204152x - 0.139396x^2 + 0.02488x^3 \tag{2.120}$$

The collocation method Choosing the points $x = \frac{1}{3}$ and $x = \frac{2}{3}$ as the collocation points, we evaluate the residuals at these points and set them to zero:

$$E(\tfrac{1}{3}) = 0: \qquad 117c_1 - 61c_2 = 18$$

$$E(\tfrac{2}{3}) = 0: \qquad 90c_1 + 34c_2 = 18 \tag{2.121}$$

The solution is given by ($c_1 = -1710/9468$, $c_2 = 486/9468$)

$$u_C = 1.3612x - 0.12927x^2 - 0.03422x^3 \tag{2.122}$$

The four approximate solutions are compared in Table 2.5 with the exact solution in Eq. (2.80). For this example the Petrov-Galerkin method gives a more accurate solution. ■

Table 2.5 Comparison of the variational solutions with the exact solution of Eq. (2.67) with $u(0) = 0 \quad u'(1) = 1$

x	u_{exact}	u_{PG}	u_{G}	u_{L}	u_{C}
0.1	0.1262	0.1285	0.1512	0.1191	0.1348
0.2	0.2513	0.2536	0.2889	0.2355	0.2668
0.3	0.3742	0.3754	0.4130	0.3494	0.3958
0.4	0.4943	0.4941	0.5264	0.4610	0.5216
0.5	0.6112	0.6096	0.6316	0.5703	0.6440
0.6	0.7244	0.7221	0.7308	0.6777	0.7628
0.7	0.8340	0.8317	0.8262	0.7831	0.8777
0.8	0.9402	0.9384	0.9201	0.8876	0.9887
0.9	1.0433	1.0424	1.0148	0.9890	1.0954
1.0	1.1442	1.1437	1.1125	1.0533	1.1977

Although it is difficult, without mathematical analysis, to conclude that one method is more accurate than others, the following general observations can be made concerning various methods. When a given differential equation admits a symmetric weak formulation, the Ritz method (then the Galerkin method can be reduced to the Ritz method) offers the simplicity of the selection of ϕ_i. For nonlinear problems in which the differentiability on the dependent variable can be weakened, even though quadratic forms do not exist, the Ritz-Galerkin methods are suitable. Considerably more work may be entailed in the least-squares approach to nonlinear problems than is involved in the other methods. Moreover, higher-order equations result for the parameters in the least-squares method. Approaches that employ a combination of two or more methods (for example, least-squares and collocation) are also being used in problems where such approaches are believed to result in more accurate solutions. In this introductory course, no attempt will be made to discuss these advanced ideas.

*2.3.4 Time-Dependent Problems

In time-dependent (unsteady) problems, the undetermined parameters c_i in eq. (2.61) are assumed to be functions of time, while ϕ_i are assumed to depend on spatial coordinates. This leads to two stages of solution, both of which employ approximate methods. In the solution of time-dependent problems, the spatial approximation is considered first and the time (or timelike) approximation next. Such a procedure is commonly known as *semidiscrete approximation* (in space). Semidiscrete variational approximation in space, as discussed in the preceding sections, results in a set of ordinary differential equations in time, which must be further approximated to obtain a set of algebraic equations. The spatial approximation of time-dependent problems leads to a matrix differential equation (in time) of the form (irrespective of the space dimension)

$$[A]\left\{\frac{\partial c}{\partial t}\right\} + [B]\{c\} = \{P\} \tag{2.123}$$

for equations involving first-order time derivatives, and

$$[A]\left\{\frac{\partial^2 c}{\partial t^2}\right\} + [B]\{c\} = \{P\} \tag{2.124}$$

for equations containing second-order time derivatives, where

$$A_{ij} = \int_\Omega \phi_i \phi_j \, dx \, dy \tag{2.125}$$

In the following paragraphs we discuss approximation schemes for first- and second-order time derivatives.

Approximations of first-order (time) derivatives Consider the matrix (differential) equation of the form

$$[A]\{\dot{c}\} + [B]\{c\} = \{P\} \qquad \text{for } 0 < t \leqslant T_0 \tag{2.126}$$

where $[A]$, $[B]$, and $\{P\}$ are known matrices and $\{c\}$ is the column matrix of the undetermined parameters. A superposed dot on $\{c\}$ indicates differentiation with respect to time t. Equation (2.126) is valid for any time $t > 0$.

We introduce a *θ family of approximation* which approximates a weighted average of the time derivative of a dependent variable at two consecutive time steps by linear interpolation of the values of the variable at the two time steps:

$$\theta\{\dot{c}\}_{n+1} + (1-\theta)\{\dot{c}\}_n = \frac{\{c\}_{n+1} - \{c\}_n}{\Delta t_{n+1}} \qquad \text{for } 0 \leqslant \theta \leqslant 1 \tag{2.127}$$

where $\{\cdot\}_n$ refers to the value of the enclosed quantity at time $t = t_n = \Sigma_{i=1}^{n} \Delta t_i$, and $\Delta t_n = t_n - t_{n-1}$ is the nth time step. If the time interval $[0, T_0]$ is divided into equal time steps, then $t_n = n\,\Delta t$. From Eq. (2.127) we can obtain a number of well-known difference schemes by choosing the value of θ:

$$\theta = \begin{cases} 0 & \text{forward difference (Euler) scheme} \quad \text{conditionally stable} \\ \frac{1}{2} & \text{Crank-Nicolson scheme} \\ \frac{2}{3} & \text{Galerkin method} \quad \text{unconditionally stable} \\ 1 & \text{backward-difference scheme} \quad \text{conditionally stable} \end{cases} \tag{2.128}$$

Using the approximation (2.127) for time t_n and t_{n+1} in Eq. (2.126), we obtain

$$\begin{aligned} [A]\{c\}_{n+1} = [A]\{c\}_n &+ \theta\Delta t_{n+1}(\{P\}_{n+1} - [B]\{c\}_{n+1}) \\ &+ (1-\theta)\,\Delta t_{n+1}(\{P\}_n - [B]\{c\}_n) \end{aligned} \tag{2.129}$$

Rearranging the terms to write $\{c\}_{n+1}$ in terms of $\{c\}_n$, we obtain

$$\begin{aligned} ([A] + \theta\Delta t_{n+1}[B])\{c\}_{n+1} = [[A] &- (1-\theta)\,\Delta t_{n+1}[B]]\{c\}_n \\ &+ \Delta t_{n+1}\left[\theta\{P\}_{n+1} + (1-\theta)\{P\}_n\right] \end{aligned}$$

or

$$[\hat{A}]\{c\}_{n+1} = [\hat{B}]\{c\}_n + \{P\}_{n,n+1} \equiv \{\hat{P}\}_{n,n+1} \tag{2.130}$$

where

$$\begin{aligned} [\hat{A}] &= [A] + \theta\Delta t_{n+1}[B] \\ [\hat{B}] &= [A] - (1-\theta)\,\Delta t_{n+1}[B] \end{aligned} \tag{2.131}$$

The solution at time $t = t_{n+1}$ is obtained in terms of the solution (known) at time t_n by inverting the matrix $[\hat{A}]$. At $t = 0$, the solution is known from the *initial* conditions of the problem, and therefore, Eq. (2.130) can be used to obtain the solution at $t = \Delta t_1$. Since the column vector $\{P\}$ is known at all times, $\{P\}_{n+1}$ is known in advance.

It must be pointed out that one can expect better results if smaller time steps are used. In practice, however, one wishes to take as large a time step as possible to cut down the computational expense. Larger time steps, in addition to decreasing the accuracy of the solution, can introduce some unwanted, numerically induced oscillations into the solution. Thus an estimate of an upper bound on the time step proves to be very useful. A stability analysis shows that the numerical scheme (2.130) is stable (i.e., has no unbounded oscillations) if the minimum eigenvalue λ of the equation

$$\det\left([\hat{B}] - \lambda[\hat{A}]\right) = 0 \tag{2.132}$$

is nonnegative. More specifically, we have

$$\begin{aligned} &\text{Stable without oscillations} \qquad && 0 < \lambda < 1 \\ &\text{Stable with oscillations} && -1 < \lambda < 0 \\ &\text{Unstable} && \lambda < -1 \end{aligned} \tag{2.133}$$

The Crank-Nicolson and Galerkin methods can be shown to be stable methods (numerical oscillations may occur, but they never become unstable).

Example 2.10 Consider the one-dimensional partial differential equation (which might arise in one-dimensional heat-conduction problems)

$$\frac{\partial u}{\partial t} - \frac{\partial^2 u}{\partial x^2} = 0 \qquad 0 < x < 1 \tag{2.134}$$

with the boundary conditions

$$u(0, t) = 0 \qquad \frac{\partial u}{\partial x}(1, t) = 0 \tag{2.135}$$

and the initial condition

$$u(x,0) = 1.0 \tag{2.136}$$

where u is the dependent variable, t is the time, and x is the independent coordinate.

First, we construct the *quasi-variational statement* in which the time is fixed (at arbitrary t). We have, following the usual procedure,

$$\begin{aligned} 0 &= \int_0^1 v\left(\frac{\partial u}{\partial t} - \frac{\partial^2 u}{\partial x^2}\right) dx \\ &= \int_0^1 \left(v\frac{\partial u}{\partial t} + \frac{\partial v}{\partial x}\frac{\partial u}{\partial x}\right) dx \end{aligned} \tag{2.137}$$

The boundary term vanished in view of the homogeneous natural boundary conditions at $x = 1$ and specified essential boundary condition at $x = 0$.

Consider a two-parameter (semidiscrete) Ritz approximation of the form

$$u(x, t) = \phi_0(x) + \sum_{j=1}^{2} c_j(t)\phi_j(x) \tag{2.138}$$

with $\phi_0 = 0$, $\phi_1 = x$, and $\phi_2 = x^2$. Substituting Eq. (2.138) into Eq. (2.137), we obtain

$$\int_0^1 \left[\phi_i \left(\sum_{j=1}^{2} \frac{dc_j}{dt} \phi_j \right) + \sum_{j=1}^{2} c_j \frac{d\phi_i}{dx} \frac{d\phi_j}{dx} \right] dx = 0$$
$$[A]\{\dot{c}\} + [B]\{c\} = \{0\} \tag{2.139}$$

where

$$A_{ij} = \int_0^1 \phi_i \phi_j \, dx \qquad B_{ij} = \int_0^1 \frac{d\phi_i}{dx} \frac{d\phi_j}{dx} dx \qquad i, j = 1, 2 \tag{2.140}$$

Carrying out the indicated integration, we get

$$\frac{1}{60} \begin{bmatrix} 20 & 15 \\ 15 & 12 \end{bmatrix} \begin{Bmatrix} \dot{c}_1 \\ \dot{c}_2 \end{Bmatrix} + \frac{1}{3} \begin{bmatrix} 3 & 3 \\ 3 & 4 \end{bmatrix} \begin{Bmatrix} c_1 \\ c_2 \end{Bmatrix} = \begin{Bmatrix} 0 \\ 0 \end{Bmatrix} \tag{2.141}$$

This completes the semidiscretization in space.

Equation (2.141) can be solved either exactly (e.g., using Laplace transform methods) or approximately (e.g., using the temporal approximations described above or the variational methods of approximation). In any case, one should know the initial conditions on c_i. Since the initial condition $u(0, x) = 1$ cannot be satisfied exactly by the present approximation [because $1 = c_1(0)x + c_2(0)x^2$ implies $c_1(0) = 0$, $c_2(0) = 0$ and yet the condition is not satisfied], we find the initial values $c_i(0)$ by an approximate method, say, the Galerkin method.

The residual of the approximation in the initial condition is

$$E = u(x,0) - 1 = xc_1(0) + x^2c_2(0) - 1 \tag{2.142}$$

Using the Galerkin method (the same result can be obtained using the least-squares method),

$$\int_0^1 \left[xc_1(0) + x^2c_2(0) - 1 \right] x \, dx = 0$$
$$\int_0^1 \left[xc_1(0) + x^2c_2(0) - 1 \right] x^2 \, dx = 0 \tag{2.143}$$

we obtain the (approximate) initial conditions

$$c_1(0) = 4 \qquad c_2(0) = -\frac{10}{3}. \tag{2.144}$$

We can solve the ordinary differential equation (2.141), subjected to the initial conditions (2.144), by exact means. Using the Laplace transform method, we obtain

$$s[A]\{\bar{c}\} - [A]\{c(0)\} + [B]\{\bar{c}\} = \{0\} \tag{2.145}$$

where s is the variable in the Laplace transform and $\bar{c}_i$ denotes the Laplace transform of c_i:

$$\bar{c}_i = \int_0^\infty e^{-st} c_i(t) \, dt \tag{2.146}$$

We obtain

$$\begin{bmatrix} 20s + 60 & 15s + 60 \\ 15s + 60 & 12s + 80 \end{bmatrix} \begin{Bmatrix} \bar{c}_1 \\ \bar{c}_2 \end{Bmatrix} = \begin{Bmatrix} 30 \\ 20 \end{Bmatrix} \tag{2.147a}$$

and

$$\bar{c}_1 = \frac{12s + 240}{3s^2 + 104s + 240} \qquad \bar{c}_2 = -\frac{10s + 120}{3s^2 + 104s + 240} \tag{2.147b}$$

Inverting, we get

$$\begin{aligned} c_1(t) &= 1.6408e^{-32.1807t} + 2.3592e^{-2.486t} \\ c_2(t) &= -(2.265e^{-32.1807t} + 1.068e^{-2.486t}) \end{aligned} \tag{2.148}$$

If we use the θ family of approximation, we obtain

$$\begin{bmatrix} \frac{1}{3} + \Delta t\,\theta & \frac{1}{4} + \Delta t\,\theta \\ \frac{1}{4} + \Delta t\,\theta & \frac{1}{5} + \dfrac{4\Delta t\,\theta}{3} \end{bmatrix} \begin{Bmatrix} c_1 \\ c_2 \end{Bmatrix}_{n+1} = \begin{bmatrix} \frac{1}{3} - (1-\theta)\,\Delta t & \frac{1}{4} - (1-\theta)\,\Delta t \\ \frac{1}{4} - (1-\theta)\,\Delta t & \frac{1}{5} - \dfrac{4(1-\theta)\,\Delta t}{3} \end{bmatrix} \begin{Bmatrix} c_1 \\ c_2 \end{Bmatrix}_{n} \tag{2.149}$$

Upon selecting the values of θ and Δt, solution c_i at $t = (n+1)\,\Delta t$ can be obtained in terms of c_i at $t = n\,\Delta t$; the initial conditions are given by Eqs. (2.144).

Table 2.6 shows a comparison of the approximate solution, Eq. (2.138), with c_i given by Eqs. (2.148) and (2.149), with the analytical solution

$$\theta(x, t) = 2 \sum_{n=0}^{\infty} \frac{e^{-\lambda_n^2 t} \sin \lambda_n x}{\lambda_n} \qquad \lambda_n = \frac{(2n+1)\pi}{2} \tag{2.150}$$

■

Temporal approximations for second-order time derivatives In structural dynamics problems the equations of motion involve the second-order time derivatives of the dependent variable(s). The semidiscrete (spatial) approximation of the equations results in a matrix differential equation of the form

$$[A]\{\ddot{c}\} + [B]\{c\} = \{F\} \qquad 0 < t < t_0 \tag{2.151}$$

There are several approximation schemes available for time derivatives. The most commonly used one is the *Newmark direct integration method*. In the Newmark direct integration method the first time derivative $\{\dot{c}\}$ and the function (of time) $\{c\}$ itself are approximated at the $(n+1)$th time step ($\Delta t_1 = \Delta t_2 = \cdots = \Delta t$) by the following expressions:

$$\begin{aligned} \{\dot{c}\}_{n+1} &= \{\dot{c}\}_n + \left[(1-\alpha)\{\ddot{c}\}_n + \alpha\{\ddot{c}\}_{n+1}\right] \Delta t \\ \{c\}_{n+1} &= \{c\}_n + \{\dot{c}\}_n \Delta t + \left[(\tfrac{1}{2} - \beta)\{\ddot{c}\}_n + \beta\{\ddot{c}\}_{n+1}\right](\Delta t)^2 \end{aligned} \tag{2.152}$$

Table 2.6 Comparison of the approximate solutions with the analytical solution ($\Delta t = 0.05$) of Example 2.10

t	x	Analytical	Ritz-Laplace	Ritz-Galerkin $\theta = \frac{2}{3}$	Ritz-Galerkin $\theta = \frac{1}{2}$
0.2	0.5	0.5532	0.5555	0.5610	0.5548
	1.0	0.7723	0.7844	0.7913	0.7848
0.4	0.5	0.3356	0.3376	0.3441	0.3372
	1.0	0.4745	0.4777	0.4868	0.4771
0.6	0.5	0.2049	0.2054	0.2113	0.2045
	1.0	0.2897	0.2906	0.2989	0.2900
0.8	0.5	0.1251	0.1249	0.1297	0.1246
	1.0	0.1769	0.1767	0.1836	0.1763
1.0	0.5	0.0764	0.0760	0.0797	0.0757
	1.0	0.1080	0.1075	0.1127	0.1071
1.2	0.5	0.0466	0.04626	0.0489	0.0460
	1.0	0.0659	0.00654	0.0692	0.0652
1.4	0.5	0.0285	0.0281	0.0300	0.0406
	1.0	0.0402	0.0398	0.0425	0.0396

where α and β are parameters that control the accuracy and the stability of the scheme, and the subscript n indicates that the solution is evaluated at the nth time step (i.e., at time $t = t_n$). The choice $\alpha = \frac{1}{2}$ and $\beta = \frac{1}{4}$ is known to give an unconditionally stable (in linear problems) scheme, which corresponds to the *constant-average-acceleration method*. The case $\alpha = \frac{1}{2}$ and $\beta = \frac{1}{6}$ corresponds to the *linear acceleration method*.

Rearranging Eqs. (2.151) and (2.152), we arrive at

$$[\hat{A}]\{c\}_{n+1} = \{\hat{F}\}_{n,n+1} \tag{2.153}$$

where

$$[\hat{A}] = [B] + a_0[A] \qquad \{\hat{F}\} = \{F\}_{n+1} + [A]\big(a_0\{c\}_n + a_1\{\dot{c}\}_n + a_2\{\ddot{c}\}_n\big) \tag{2.154}$$

Once the solution $\{c\}$ is known at $t_{n+1} = (n+1)\,\Delta t$, the first and second derivatives (velocity and acceleration) of $\{c\}$ at t_{n+1} can be computed from [rearranging the expressions in Eqs. (2.152)]

$$\begin{aligned}
\{\ddot{c}\}_{n+1} &= a_0\big(\{c\}_{n+1} - \{c\}_n\big) - a_1\{\dot{c}\}_n - a_2\{\ddot{c}\}_n \\
\{\dot{c}\}_{n+1} &= \{\dot{c}\}_n + a_3\{\ddot{c}\}_n + a_4\{\ddot{c}\}_{n+1}
\end{aligned} \tag{2.155}$$

$$a_0 = \frac{1}{\beta \Delta t^2} \qquad a_1 = a_0 \Delta t \qquad a_2 = \frac{1}{2\beta} - 1 \qquad a_3 = (1-\alpha)\,\Delta t \qquad a_4 = \alpha\,\Delta t$$

For a given set of initial conditions $\{c\}_0$, $\{\dot{c}\}_0$, and $\{\ddot{c}\}_0$, we can solve Eq. (2.153) repeatedly, marching forward in time, for the column vector $\{c\}$ and its time derivatives at any time $t > 0$.

A couple of comments are in order on the selection of the time step and the computation of the initial conditions. Although the Newmark method is unconditionally stable (i.e., the solution is stable for any value of Δt; however, it may be inaccurate), it is helpful to have a means to determine the value of Δt for which the solution is also accurate. The following formula gives an estimate for the time increment:

$$\Delta t = \frac{T_{\min}}{\pi} \tag{2.156}$$

where $T_{\min}$ is the smallest period of natural vibration associated with the approximate problem. An estimate for Δt can also be obtained from the condition that the smallest eigenvalue of the eigenvalue problem

$$\left(a_0[A] - \lambda[\hat{A}]\right)\{u\} = 0 \tag{2.157}$$

is less than 1.

Initial values of $\{\ddot{c}\}$ are generally not known from the problem description. In that case one can make use of Eq. (2.151) at time $t = 0$ to compute $\{\ddot{c}\}$.

Example 2.11 Consider the dynamics (i.e., unsteady transverse motion) of a uniform beam clamped at the ends. The equation of motion, in nondimensional form, can be written as

$$\frac{\partial^4 w}{\partial x^4} + \frac{\partial^2 w}{\partial t^2} = 0 \qquad \text{for } 0 < x < 1 \text{ and } t > 0 \tag{2.158}$$

subject to the boundary conditions

$$w = \frac{\partial w}{\partial x} = 0 \qquad x = 0, 1;\ t > 0 \tag{2.159}$$

and the initial conditions

$$w = \sin \pi x - \pi x(1 - x) \qquad \frac{\partial w}{\partial t} = 0 \qquad 0 < x < 1, t = 0 \tag{2.160}$$

Note that all the boundary conditions in the present problem are of the essential type. Hence, the selection criteria for the approximation functions in the Galerkin method and the Ritz method coincide. The following approximation functions meet the boundary conditions:

$$\phi_1 = 1 - \cos 2\pi x, \phi_2 = 1 - \cos 4\pi x, \ldots, \phi_n = 1 - \cos 2n\pi x \tag{2.161}$$

The semidiscrete approximation results in

$$\sum_{j=1}^{N} \int_0^1 \left(\frac{d^2 c_j}{dt^2} \phi_i \phi_j + c_j \frac{d^2\phi_i}{dx^2} \frac{d^2\phi_j}{dx^2} \right) dx = 0$$

or

$$[A]\{\ddot{c}\} + [B]\{c\} = 0 \tag{2.162}$$

where

$$A_{ij} = \int_0^1 \phi_i \phi_j \, dx \qquad B_{ij} = \int_0^1 \frac{d^2\phi_i}{dx^2} \frac{d^2\phi_j}{dx^2} dx \tag{2.163}$$

We consider the special case of a one-parameter ($N = 1$) approximation. Then we have

$$A_{11} = \frac{3}{2} \qquad B_{11} = \frac{(2\pi)^4}{2} \tag{2.164}$$

and Eq. (2.162) becomes (with $k^2 = B_{11}/A_{11}$)

$$\frac{d^2 c_1}{dt^2} + k^2 c_1 = 0 \qquad k = \frac{(2\pi)^2}{\sqrt{3}} = 22.7929 \tag{2.165}$$

The exact solution of this equation is given by

$$c_1 = a \sin kt + b \cos kt \tag{2.166}$$

where a and b are constants to be determined using the initial conditions. The residuals in the initial values of w and $\partial w/\partial t$ are

$$\begin{aligned} R_1^0 &= w(x,0) - \sin \pi x + \pi x(1 - x) \\ &= c_1(0)(1 - \cos 2\pi x) - \sin \pi x + \pi x(1 - x) \\ R_2^0 &= \frac{\partial w}{\partial t}(x,0) - 0 = \frac{dc_1}{dt}(0)(1 - \cos 2\pi x) \end{aligned} \tag{2.167}$$

Using the Galerkin method on these residuals with $\phi_1 = 1 - \cos 2\pi x$, we get

$$c_1(0) = 0.1107 \qquad \frac{dc_1}{dt}(0) = 0 \tag{2.168}$$

which give $a = 0$, $b = 0.1107$. Hence the solution is given by

$$w(x, t) = 0.1107(1 - \cos 2\pi x) \cos 22.7929t \tag{2.169}$$

If the Newmark direct integration is used to solve the differential equation (2.165) subject to the initial conditions in Eq. (2.168), we obtain

$$\hat{A}_{11} = \frac{(2\pi)^4}{2} + \frac{3a_0}{2} \qquad \hat{F}_1(0) = 0.1661a_0 - 57.5105a_2 \tag{2.170}$$

and for $\alpha = \frac{1}{2}$ and $\beta = \frac{1}{4}$ we have $a_0 = 4/(\Delta t)^2$. Then we must solve Eq. (2.153) repeatedly, marching forward in time.

The numerical results for the center deflection (as a function of time) given by Eq. (2.169) and the Newmark method are shown in Table 2.7 for various values of the time step (see also Fig. 2.5). ■

Table 2.7 Comparison of solutions of the time-dependent problem of Example 2.11

t	Galerkin in space, Newmark's integration in time			Galerkin in space, exact in time
	$\Delta t = 0.01$	$\Delta t = 0.005$	$\Delta t = 0.0025$	
0.02	0.19898	0.19884	0.19877	0.19879
0.04	0.13626	0.13574	0.13550	0.13558
0.06	0.04595	0.04498	0.04468	0.04455
0.08	−0.05367	−0.05495	−0.05550	−0.05534
0.10	−0.14242	−0.14368	−0.14421	−0.14406
0.12	−0.20233	−0.20312	−0.20347	−0.20336
0.14	−0.22126	−0.22116	−0.22113	−0.22113
0.16	−0.19537	−0.19411	−0.19355	−0.19374
0.18	−0.12992	−0.12750	−0.12640	−0.12677
0.20	−0.03816	−0.03490	−0.03343	−0.03392
0.22	0.06133	0.06482	0.06637	0.06586
0.24	0.14840	0.15132	0.15261	0.15219
0.26	0.20542	0.20699	0.20767	0.20744
0.28	0.22083	0.22046	0.22028	0.22032

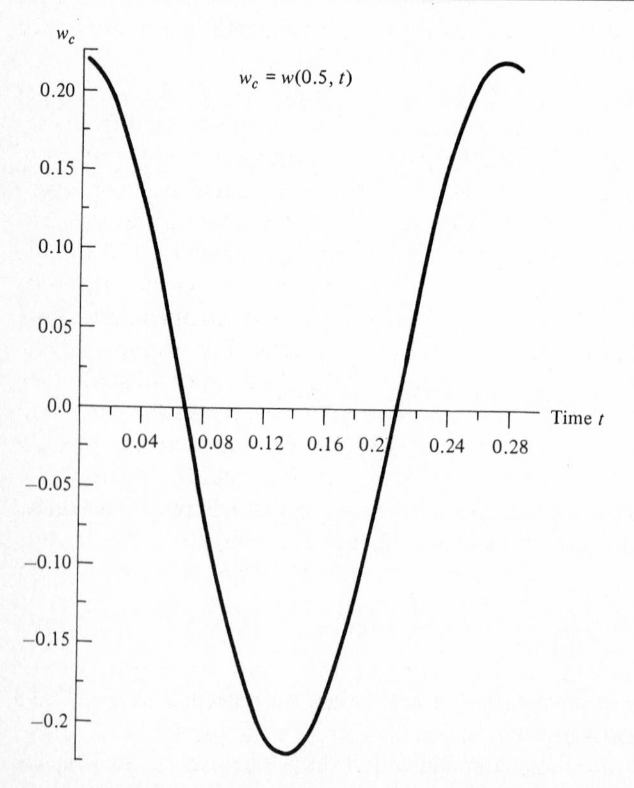

Figure 2.5 Center deflection versus time for a clamped beam subjected to nonzero initial conditions (Example 2.11).

We close this section with a couple of comments on the time approximations discussed in the preceding. First we note that the semidiscrete approximations discussed here are not applicable to wave propagation problems, because in such problems one cannot assume separability of spatial and temporal parts of the solution. Next we point out that the stability criteria for a given discrete model depend on the variational formulation used. For additional details on this topic the reader is asked to consult the references at the end of this chapter.

We close this chapter with some remarks on the classical variational methods. Most of the comments made here are applicable to the methods reviewed in the present study.

2.3.5 Some Remarks

The classical variational methods (i.e., Ritz, Galerkin, least-squares, etc.) presented in the preceding sections provide simple means of finding approximate solutions to physical problems. The formulative and computational efforts involved are less compared to most other methods, such as the finite-difference and the finite-element methods. Furthermore, the approximate solutions obtained are continuous functions of position (as opposed to piecewise continuous functions or functions known only at discrete points) in the domain.

The main disadvantage, from the practical point of view, of the variational methods that prevented them from being competitive with traditional finite-difference methods is the difficulty encountered in selecting the approximation functions. Apart from the properties the functions are required to satisfy, there exists no systematic procedure of constructing them. The selection process becomes more difficult when the domain is geometrically complex and/or the boundary conditions are complicated. If the functions are not selected from the domain space of the operator of the equation being solved (completeness property), the resulting solution could be either zero or wrong. For example, if one selects $\phi_1 = 1 - \cos(2\pi x/L)$ for a one-parameter Ritz solution of a cantilevered beam of length L, the resulting solution is wrong because it vanishes at $x = L$ (where the deflection should be maximum). While the function satisfies the essential boundary conditions $[w(0) = w'(0) = 0]$ of the problem, it cannot be used to represent a solution that is nonzero at $x = L$. An appropriate choice would be $1 - \cos(\pi x/L)$. Thus, without a judicious choice of the approximation functions, the resulting solution can be unacceptable. One cannot automatize the procedure for a given equation because the choice of approximation functions differs with the boundary conditions.

From the preceding discussion it appears that the variational methods could provide powerful means of finding approximate solutions, provided we can find a way to construct systematically the approximation functions, for almost any geometry, that depend only on the differential equation being solved and not on the boundary conditions of the problem. The latter property enables one to develop a computer program for a particular class of problems (each problem in the class differs from others only in the data), that is, a *general-purpose* computer program. Since the functions must be constructed for a geometrically complex domain, it seems that (recalling the discussion of the method of composites for the determination of the center of mass of an irregular shape) the region must be represented (or approximated if required) as an assemblage of simple geometric shapes for which the construction of approximation functions becomes simpler.

The finite-element method to be discussed in the forthcoming chapters is based on these ideas. In the finite-element method a given domain is represented (discretized) by a collection of geometrically simple shapes (elements), and on each element of the collection, the governing equation is *formulated* using any one of the variational methods. The approximation functions are systematically generated for each (typical) element using the essential boundary conditions (the *specified values* are the unknown values of the dependent variable and/or its derivatives) of the element. The elements are *connected* together by imposing the continuity of the dependent variables across the interelement boundaries. With this brief qualitative description (or preview) of the finite-element method, we proceed to Chap. 3.

PROBLEMS

2.12 Compute the coefficient matrix and the right-hand side of the N-parameter Ritz system of the equation

$$-\frac{d}{dx}\left[(1+x)\frac{du}{dx}\right]=0 \qquad 0<x<1$$

$$u(0)=0 \qquad u(1)=1$$

Use algebraic polynomials for the approximation functions. Specialize your result for $N=2$ and compute the Ritz coefficients.

2.13 Give trigonometric functions for the two-parameter Ritz approximation of the equation in Prob. 2.12.

2.14 Set up the equations for the N-parameter Ritz approximation of the equation (associated with a simply supported beam subjected to uniform loading, $f=f_0$):

$$\frac{d^2}{dx^2}\left(EI\frac{d^2w}{dx^2}\right)+f_0=0 \qquad 0<x<L$$

$$w=EI\frac{d^2w}{dx^2}=0 \qquad \text{at } x=0, L$$

(a) Use algebraic polynomials.
(b) Use trigonometric functions.

Compare the two-parameter Ritz solutions with the exact solution.

2.15 Repeat Prob. 2.14 for $f=f_0\sin(\pi x/L)$.

2.16 Repeat Prob. 2.14 for $f=f_0\,\delta(x-L/2)$, where $\delta(x)$ is the Dirac delta function.

2.17 Find the first two eigenvalues of Prob. 2.12 using the Ritz method with algebraic polynomials.

2.18 Set up the N-parameter Ritz equations for the eigenvalue problem

$$-\frac{d}{dx}\left(a\frac{du}{dx}\right)-\lambda u=0 \qquad 0<x<L$$

$$u(0)=0 \qquad a\frac{du}{dx}+ku=0 \text{ at } x=L$$

where a and k are constants and λ is the eigenvalue. Use trigonometric functions. Specialize the results for $N=3$ and compute the three eigenvalues.

2.19 Find a one-parameter Galerkin solution to the following pair of equations:

$$\left.\begin{aligned}\frac{d^2u}{dx^2}-\frac{v}{a}-f=0\\ \frac{d^2v}{dx^2}+g=0\end{aligned}\right\} \qquad 0<x<1$$

$$u(0)=\frac{du}{dx}(0)=0 \qquad v(1)=\frac{dv}{dx}(1)=1$$

where a, f, and g are constants.

2.20 Find a one-parameter approximate solution to the nonlinear equation

$$-2u\frac{d^2u}{dx^2}+\left(\frac{du}{dx}\right)^2=4 \qquad 0<x<1$$

$$u(0)=1 \qquad u(1)=0$$

and compare with the exact solution $u_0=1-x^2$. Use (a) the Galerkin method, (b) the least-squares method, and (c) the Petrov-Galerkin method with $v=1$.

2.21 Give a one-parameter Galerkin solution to the equation

$$-\nabla^2u=1 \text{ in } \Omega=\text{unit square}$$

$$u=0 \text{ on } \Gamma$$

2.22 Repeat Prob. 2.21 for an equilateral triangular domain. (*Hint*: Use the product of equations of the lines representing the sides of the triangle for the approximation function.)

2.23 Compute the critical buckling load N_{cr} of a simply-supported square plate of length a subjected to uniformly distributed edge load in the x direction. Use the approximation function

$$\phi_1=\sin\frac{\pi x}{a}\sin\frac{\pi y}{a}$$

and the integral form of the governing equation

$$D\int_0^a\int_0^a v\nabla^4 w\,dx\,dy+N_{cr}\int_0^a\int_0^a v\frac{\partial^2 w}{\partial x^2}\,dx\,dy=0$$

2.24 Repeat Prob. 2.18 with the collocation method, with collocation points $x=L/4, L/2, 3L/4$.

2.25 Find the first three eigenvalues of the equation

$$-\frac{d^2u}{dx^2}-\lambda u=0 \qquad 0<x<1$$

$$u=0 \text{ at } x=0,1$$

using collocation at $x=\frac{1}{4},\frac{1}{2},\frac{3}{4}$.

2.26 Consider the differential equation

$$-\frac{d^2u}{dx^2}=\cos\pi x \qquad 0<x<1$$

subjected to the following two sets of boundary conditions:

1. $u(0) = 0 \qquad u(1) = 0$
2. $u(0) = 0 \qquad \dfrac{du}{dx}(1) = 0$

Determine a three-parameter solution using (*a*) the Ritz method, (*b*) the least-squares method, and (*c*) collocation at $x = \frac{1}{4}, \frac{1}{2}, \frac{3}{4}$, and compare with the exact solutions:

1. $u_0 = \dfrac{1}{\pi^2}(\cos \pi x + 2x - 1)$
2. $u_0 = \dfrac{1}{\pi^2}(\cos \pi x - 1)$

2.27 Consider the nondimensionalized heat-conduction equation

$$\frac{\partial u}{\partial t} - \frac{\partial^2 u}{\partial x^2} = 0 \qquad 0 < x < 1$$

subjected to the boundary conditions

$$u(0, t) - \frac{\partial u}{\partial x}(0, t) = 0 \qquad \frac{\partial u}{\partial x}(1, t) = 0$$

and the initial condition

$$u(x, 0) = 1$$

Find a one-parameter (*a*) Ritz, (*b*) Galerkin, and (*c*) Petrov-Galerkin (with $v = 1$) solutions.

2.28 Consider the partial differential equation

$$\frac{\partial^2 u}{\partial x^2} - \frac{\partial^2 u}{\partial t^2} = 0 \qquad 0 < x < 1$$

subjected to the boundary conditions

$$u(0, t) = u(1, t) = 0$$

and the initial conditions

$$u(x, 0) = x(1 - x) \qquad \frac{\partial u}{\partial t}(x, 0) = 0$$

Find the one-parameter solution using (*a*) the Galerkin method and (*b*) collocation at $x = \frac{1}{2}$. Choose approximation in the form

$$u(x, t) = c_1(t)x(1 - x)$$

Solve the resulting ordinary differential equation for c_1 exactly.

2.29 Repeat Prob. 2.28 for a two-parameter Galerkin solution.

2.30 Show that Eqs. (2.151) and (2.152) can be expressed in the alternative form [to eq. (2.153)]

$$[H]\{\ddot{c}\}_{n+1} = \{F\}_{n+1} - [B]\{b\}_n$$

where

$$[H] = \beta(\Delta t)^2[B] + [A]$$

$$\{b\}_n = \{c\}_n + \Delta t\{\dot{c}\}_n + \left(\tfrac{1}{2} - \beta\right)(\Delta t)^2\{\ddot{c}\}_n$$

2.31 Using the Newmark integration scheme (2.152), express the equation

$$[M]\{\ddot{c}\} + [C]\{\dot{c}\} + [K]\{c\} = \{F\} \qquad (a)$$

in the form

$$[\bar{A}]\{c\}_{n+1} = \{P\}_{n+1}$$

where

$$[\bar{A}] = [K] + a_0[M] + a_5[C] \qquad a_5 = \frac{\alpha}{\beta \Delta t}$$

$$\{P\}_{n+1} = \{F\}_{n+1} + [M]\big(a_0\{c\}_n + a_1\{\dot{c}\}_n + a_2\{\ddot{c}\}_n\big) + [C]\big(a_5\{c\}_n + a_6\{\dot{c}\}_n + a_7\{\ddot{c}\}_n\big)$$

$$a_6 = \frac{\alpha}{\beta} - 1 \qquad a_7 = \frac{\Delta t}{2}\left(\frac{\alpha}{\beta} - 2\right)$$

Hint: Express Eq. (a) at time $t + \Delta t$, and write the second equation in (2.152) for $\{\ddot{c}\}_{n+1}$ in terms of $\{c\}_{n+1}$. Then substitute $\{\ddot{c}\}_{n+1}$ into the first equation in (2.152) to obtain equations for $\{\dot{c}\}_{n+1}$ and $\{\ddot{c}\}_{n+1}$ each in terms of $\{c\}_{n+1}$ only. Finally, substitute the expressions thus obtained into Eq. (a) and collect coefficients of $\{c\}_{n+1}$ and known quantities on the right-hand side of the equality sign to obtain the desired equation.

REFERENCES FOR ADDITIONAL READING

Becker, M.: *The Principles and Applications of Variational Methods*, MIT Press, Cambridge, Mass. (1964).

Biot, M. A.: *Variational Principles in Heat Transfer*, Clarendon, London (1972).

Finlayson, B. A.: *The Method of Weighted Residuals and Variational Principles*, Academic Press, New York (1972).

Forray, M. J.: *Variational Calculus in Science and Engineering*, McGraw-Hill, New York (1968).

Hildebrand, F. B.: *Methods of Applied Mathematics*, 2d ed., Prentice-Hall, New York (1965).

Lanczos, C.: *The Variational Principles of Mechanics*, The University of Toronto Press, Toronto (1964).

Langhaar, H. L.: *Energy Methods in Applied Mechanics*, John Wiley, New York (1962).

Leipholz, H.: *Direct Variational Methods and Eigenvalue Problems in Engineering*, Noordhoff, Leyden, The Netherlands (1977).

Lippmann, H.: *Extremum and Variational Principles in Mechanics*, Springer-Verlag, New York (1972).

Mikhlin, S. G.: *Variational Methods in Mathematical Physics*, Pergamon Press (distributed by Macmillan), New York (1964).

Mikhlin, S. G.: *The Numerical Performance of Variational Methods*, Wolters-Noordhoff, Groningen, The Netherlands (1971).

Oden, J. T., and J. N. Reddy: *Variational Methods in Theoretical Mechanics*, Springer-Verlag, New York (1976); 2d ed. (1983).

Reddy, J. N., and M. L. Rasmussen: *Advanced Engineering Analysis*, John Wiley, New York (1982).

Rektorys, K.: *Variational Methods in Mathematics, Science and Engineering*, Reidel Publishing, Boston (1977).

Schechter, R. S.: *The Variational Methods in Engineering*, McGraw-Hill, New York (1967).

Washizu, K.: *Variational Methods in Elasticity and Plasticity*, 2d ed., Pergamon Press, New York (1975); 3d ed. (1982).

Weinstock, R.: *Calculus of Variations with Applications to Physics and Engineering*, McGraw-Hill, New York (1952).

CHAPTER

THREE

FINITE-ELEMENT ANALYSIS OF ONE-DIMENSIONAL PROBLEMS

3.1 INTRODUCTORY COMMENTS

As pointed out in the closing of Chap. 2, the traditional variational methods (i.e., Ritz, Galerkin, least-squares, etc.) described there cease to be effective due to a serious shortcoming, namely, the difficulty in choosing the approximation functions. The approximation functions, apart from satisfying the continuity, linear independence, completeness, and essential boundary conditions, are arbitrary; the selection becomes even more difficult when the given domain is geometrically complex. Since the quality of the approximation is directly affected by the choice of the approximation functions, it is discomforting to know that there exists no systematic procedure to construct them. Due to this shortcoming, despite the simplicity in obtaining approximate solutions (once the approximating functions are chosen), the variational methods of approximation were never regarded as competitive computationally when compared with traditional finite-difference schemes.

Ideally speaking, an effective computational method should have the following features:

1. The method should have a sound mathematical as well as physical basis.
2. The method should not have limitations with regard to the geometry and the physical composition of the domain as well as the nature of "loading."
3. The formulative procedure should be independent of the shape of the domain and the boundary conditions.
4. The method should be flexible enough to allow choosing a desired degree of approximation without reformulating the entire problem.
5. The method should involve a systematic procedure that can be automated for use on digital computers.

The finite-element method not only overcomes the shortcoming of the traditional variational methods, it is also endowed with the features of an effective computational technique. The method consists of representing a given domain, however complex it may be, by geometrically simple shapes over which the approximation functions can be systematically derived. Then the Ritz-Galerkin type approximations of the governing equations are developed over each element. Finally, the equations over all elements of the collection are connected by the continuity of the primary variable or variables, the boundary conditions of the problem are imposed, and then the connected set of equations is solved. A

Table 3.1 Steps involved in the finite-element analysis of a typical problem

1. Discretization (or representation) of the given domain into a collection of preselected finite elements. (This step can be postponed until after the finite-element formulation of the equation is completed.)
   a. Construct the finite-element mesh of preselected elements.
   b. Number the nodes and elements.
   c. Generate the geometric properties (e.g., coordinates, cross-sectional areas, etc.) needed for the problem.

2. Derivation of element equations for all typical elements in the mesh.
   a. Construct the variational formulation of the given differential equation over the typical element.
   b. Assume that a typical dependent variable u is of the form

   $$u = \sum_{i=1}^{n} u_i \psi_i$$

   and substitute it into step 2*a* to obtain element equations in the form

   $$[K^{(e)}]\{u^{(e)}\} = \{F^{(e)}\}$$

   c. Derive or select, if already available in the literature, element interpolation functions ψ_i and compute the element matrices.

3. Assembly of element equations to obtain the equations of the whole problem.
   a. Identify the interelement continuity conditions among the primary variables (relationship between the local degrees of freedom and the global degrees of freedom—connectivity of elements) by relating element nodes to global nodes.
   b. Identify the "equilibrium" conditions among the secondary variables (relationship between the local source or force components and the globally specified source components).
   c. Assemble element equations using steps 3*a* and 3*b* and the superposition property described in the example of Sec. 3.2.3.

4. Imposition of the boundary conditions of the problem.
   a. Identify the specified global primary degrees of freedom.
   b. Identify the specified global secondary degrees of freedom (if not already done in step 3*b*).
5. Solution of the assembled equations.
6. Postprocessing of the results.
   a. Compute the gradient of the solution or other desired quantities from the primary degrees of freedom computed in step 5.
   b. Represent the results in tabular and/or graphic form.

flowchart of the basic steps involved in the finite-element analysis of a problem is given in Table 3.1.

In the sections that follow, our objective will be to introduce many fundamental ideas that form the basis of the finite-element method. In doing so, we postpone some issues of practical and theoretical complexity to later sections of this chapter and to Chaps. 4 and 5.

3.2 ONE-DIMENSIONAL SECOND-ORDER EQUATIONS

Consider the problem of finding the function $u(x)$ that satisfies the differential equation

$$-\frac{d}{dx}\left(a\frac{du}{dx}\right) - f = 0 \qquad 0 < x < L \tag{3.1}$$

and the boundary conditions

$$u(0) = 0 \qquad \left(a\frac{du}{dx}\right)\bigg|_{x=L} = P \tag{3.2}$$

where $a = a(x)$, $f = f(x)$, and P are the data of the problem. Examples of physical problems in which Eq. (3.1) arises are given in Table 2.1. A step-by-step procedure of the finite-element analysis of the problem is given below.

3.2.1 Discretization of the Domain into Elements

The domain $\Omega \equiv (0, L)$ of the problem is divided into a set of line elements, called the *finite-element mesh*. The mesh shown in Fig. 3.1 is a *nonuniform mesh* because the elements are not of equal length. The intersection of any two elements is termed the *interelement boundary*. The intersection points, and possibly some intermediate points, are called the *global nodes*. The coordinate of the eth global node is x_e.

The number of elements used in a problem depends mainly on the element type and accuracy desired. Whenever a problem is solved by the finite-element method for the first time, one is required to investigate the convergence characteristics of the finite-element approximation by gradually *refining the mesh* (i.e., increasing the number of elements) and by comparing the solution with those obtained by *higher-order* elements. The *order* of an element will be discussed shortly.

3.2.2 Derivation of Element Equations

A typical element $\Omega^e = (x_A, x_B)$ is isolated from the mesh. In a one-dimensional problem, the element is a line element. Suppose that the typical element is the eth element. We wish to solve the given equation over the element, using any one of the variational methods of approximation discussed in Chap. 2. The equations

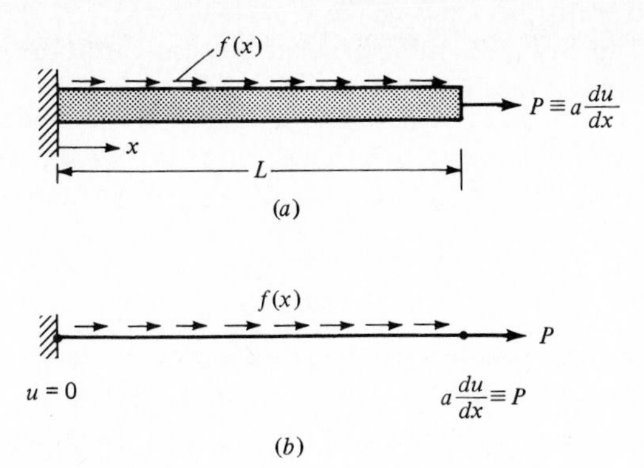

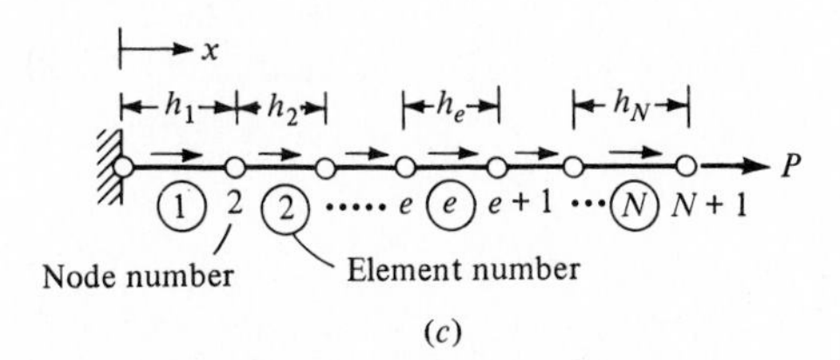

Figure 3.1 Finite-element discretization of a one-dimensional second-order differential equation (3.1). (a) Physical problem. (b) Mathematical idealization. (c) Finite-element discretization.

resulting from the application of a variational method are generally relations between the *primary* variables (i.e., variables involved in the specification of the essential boundary conditions) and the *secondary* variables (i.e., variables involved in the specification of the natural boundary conditions). The following three steps describe the finite-element formulation of the equation over the typical element.

Variational formulation of the equation over an element Since Eq. (3.1) is valid over the domain $(0, L)$, it is valid, in particular, over the element $\Omega^e = (x_A, x_B)$.

Figure 3.2 Finite-element discretization of a one-dimensional domain of Eq. (3.1). (a) A typical finite element from the finite-element mesh shown in Fig. 3.1c; x = global coordinate, $\bar{x}$ = local (element) coordinate $= x - x_A$. (b) A typical element with the definition of the primary variable u and secondary variable $(a\, du/dx)$ at the nodes.

Following the procedure described in Sec. 2.2, we can construct the variational formulation of Eq. (3.1) over the element (see Fig. 3.2):

$$0 = \int_{x_A}^{x_B} v\left[-\frac{d}{dx}\left(a\frac{du}{dx}\right) - f\right]dx$$

$$= \int_{x_A}^{x_B}\left(a\frac{dv}{dx}\frac{du}{dx} - vf\right)dx + \left[v\left(-a\frac{du}{dx}\right)\right]_{x_A}^{x_B}$$

where v denotes an arbitrary continuous function (test function or weight function). From the boundary term in the above expression, we immediately note that the specification of u at $x = x_A$ and x_B constitute the essential boundary conditions, and the specification of $(-a\,du/dx)$ at $x = x_A$ and x_B constitute the natural boundary conditions for the element. Thus the basic unknowns at the element nodes are the primary variable u and the secondary variable $(a\,du/dx)$. Let

$$u(x_A) \equiv u_1^{(e)} \qquad u(x_B) \equiv u_2^{(e)} \tag{3.3a}$$

$$\left(-a\frac{du}{dx}\right)\bigg|_{x_A} \equiv P_1^{(e)} \qquad \left(a\frac{du}{dx}\right)\bigg|_{x_B} \equiv P_2^{(e)} \tag{3.3b}$$

These boundary conditions are shown on the typical element. Students of engineering recognize that Fig. 3.2b is the so-called *free-body diagram* of the typical element. With the notation in Eqs. (3.3), the variational form becomes

$$0 = \int_{x_A}^{x_B}\left(a\frac{dv}{dx}\frac{du}{dx} - vf\right)dx - P_1^{(e)}v(x_A) - P_2^{(e)}v(x_B) \tag{3.4a}$$

or

$$0 = B(v, u) - l(v) \tag{3.4b}$$

where the bilinear and linear forms are given by

$$\begin{aligned} B(v, u) &= \int_{x_A}^{x_B} a\frac{dv}{dx}\frac{du}{dx}dx \\ l(v) &= \int_{x_A}^{x_B} vf\,dx + v(x_A)P_1^{(e)} + v(x_B)P_2^{(e)} \end{aligned} \tag{3.4c}$$

Note that the natural boundary conditions (3.3b) are included in the variational form. The associated quadratic form for the element is given by [from Eq. (2.33)]:

$$I_e(u) = \int_{x_A}^{x_B}\left[\frac{a}{2}\left(\frac{du}{dx}\right)^2 - uf\right]dx - P_1^{(e)}u(x_A) - P_2^{(e)}u(x_B) \tag{3.5}$$

Variational approximation of the equation over an element Now suppose that we wish to find an approximate solution of the variational problem (3.4) with the boundary conditions (3.3a) using the Ritz method. Let the Ritz approximation of

u on the element be given by

$$u_e(x) = \sum_{j=1}^{n} \alpha_j^{(e)} \psi_j^{(e)}(x) \tag{3.6}$$

where α_j are the parameters to be determined, and $\psi_j(x)$ are the approximation functions to be constructed shortly. Substituting Eq. (3.6) for u and $v = \psi_i$ into Eq. (3.4a), we obtain

$$\begin{aligned} 0 &= \sum_{j=1}^{n} \left(\int_{x_A}^{x_B} a \frac{d\psi_i}{dx} \frac{d\psi_j}{dx} dx \right) \alpha_j - \int_{x_A}^{x_B} \psi_i f \, dx - P_1^{(e)} \psi_i(x_A) - P_2^{(e)} \psi_i(x_B) \\ &= \sum_{j=1}^{n} K_{ij}^{(e)} \alpha_j^{(e)} - F_i^{(e)} \qquad i = 1, 2, \ldots, n \end{aligned}$$

or

$$[K^{(e)}]\{\alpha^{(e)}\} = \{F^{(e)}\} \tag{3.7}$$

where the coefficient matrix $K_{ij}^{(e)}$, called *stiffness matrix*, and the column vector $F_i^{(e)}$, called *force vector*, are given by

$$K_{ij} = B(\psi_i, \psi_j) \qquad F_i = l(\psi_i) \tag{3.8a}$$

or, more specifically,

$$\begin{aligned} K_{ij}^{(e)} &= \int_{x_A}^{x_B} a \frac{d\psi_i}{dx} \frac{d\psi_j}{dx} dx \\ F_i^{(e)} &= \int_{x_A}^{x_B} \psi_i f \, dx + P_1^{(e)} \psi_i(x_A) + P_2^{(e)} \psi_i(x_B) \end{aligned} \tag{3.8b}$$

Now we turn to the question of constructing the approximation functions $\psi_i^{(e)}$ for the element. It is here that the finite-element method differs from the Ritz method.

Derivation of the approximation functions for an element The approximation functions $\psi_i^{(e)}$ are constructed using precisely the same conditions as those in Eq. (2.65). However, instead of using them as guidelines as in the Ritz method, we use them to construct $\psi_i^{(e)}$. To satisfy condition 1 in Eq. (2.65), we must select ψ_i such that Eq. (3.6) is differentiable at least once with respect to x and satisfies the essential boundary conditions (3.3a). Conditions 2 and 3 require $\{\psi_i\}$ to be linearly independent and complete. Conditions 1a, 2, and 3 are met if we choose at least linear approximation of the form

$$u(x) = c_1 + c_2 x = c_1 \phi_1 + c_2 \phi_2 \tag{3.9}$$

The continuity is obviously satisfied, $\phi_1 = 1$ and $\phi_2 = x$ are linearly independent and the set $\{1, x\}$ is complete. To satisfy the remaining requirement, condition 1b in Eq. (2.65), we require u to satisfy the essential boundary conditions of the element, Eq. (3.3a):

$$\begin{aligned} u(x_e) &\equiv u_1^{(e)} = c_1 + c_2 x_e \\ u(x_{e+1}) &\equiv u_2^{(e)} = c_1 + c_2 x_{e+1} \end{aligned} \tag{3.10a}$$

In matrix form we have

$$\begin{Bmatrix} u_1^{(e)} \\ u_2^{(e)} \end{Bmatrix} = \begin{bmatrix} 1 & x_e \\ 1 & x_{e+1} \end{bmatrix} \begin{Bmatrix} c_1 \\ c_2 \end{Bmatrix} \tag{3.10b}$$

Solving for c_1 and c_2 in terms of $u_1^{(e)}$ and $u_2^{(e)}$, we obtain

$$c_1 = \frac{u_1^{(e)} x_{e+1} - u_2^{(e)} x_e}{x_{e+1} - x_e} \qquad c_2 = \frac{u_2^{(e)} - u_1^{(e)}}{x_{e+1} - x_e} \tag{3.10c}$$

Substituting Eq. (3.10c) for c_i in Eq. (3.9) and collecting the coefficients of $u_i^{(e)}$, we get

$$\begin{aligned} u(x) &= \frac{u_1^{(e)} x_{e+1} - u_2^{(e)} x_e}{x_{e+1} - x_e} + \frac{u_2^{(e)} - u_1^{(e)}}{x_{e+1} - x_e} x \\ &= \frac{x_{e+1} - x}{x_{e+1} - x_e} u_1^{(e)} + \frac{x - x_e}{x_{e+1} - x_e} u_2^{(e)} \\ &= \sum_{i=1}^{2} u_i^{(e)} \psi_i^{(e)} \end{aligned} \tag{3.11}$$

where

$$\psi_1^{(e)} = \frac{x_{e+1} - x}{x_{e+1} - x_e} \qquad \psi_2^{(e)} = \frac{x - x_e}{x_{e+1} - x_e} \qquad x_e \leqslant x \leqslant x_{e+1} \tag{3.12}$$

Expression (3.11) satisfies the essential boundary conditions of the element, and the $\{\psi_i\}$ are continuous, linearly independent, and complete over the element. Comparing Eq. (3.11) with Eq. (3.6), we immediately note that $n = 2$ and $\alpha_j^{(e)} = u_j^{(e)}$.

Since the approximation functions are derived from Eq. (3.9) in such a way that $u(x)$ is equal to $u_1^{(e)}$ at node 1 ($x_A = x_e$) and to $u_2^{(e)}$ at node 2 ($x_B = x_{e+1}$), that is, interpolated, they are also called *Lagrange family of interpolation functions.* The interpolation functions have the following properties, in addition to the property that $\psi_i^{(e)} = 0$ outside the element Ω^e (see Fig. 3.3a):

$$\begin{aligned} &1.\ \psi_i^{(e)}(x_j) = \begin{cases} 0 & \text{if } i \neq j \\ 1 & \text{if } i = j \end{cases} \qquad x_1 = x_e \qquad x_2 = x_{e+1} \\ &2.\ \sum_{i=1}^{2} \psi_i^{(e)}(x) = 1 \end{aligned} \tag{3.13}$$

The first property is the result of the interpolation, and the second property is the result of including a constant term in the approximation (3.9). In other words, property 1 implies that Eq. (3.11) satisfies the conditions (3.3a) and that ψ_i are linearly independent. Conversely, if Eq. (3.11) satisfies conditions (3.3a) and the ψ_i are linearly independent, then it follows that ψ_i has property 1. Property 2 of ψ_i allows us to model problems in which the dependent variable is constant over an

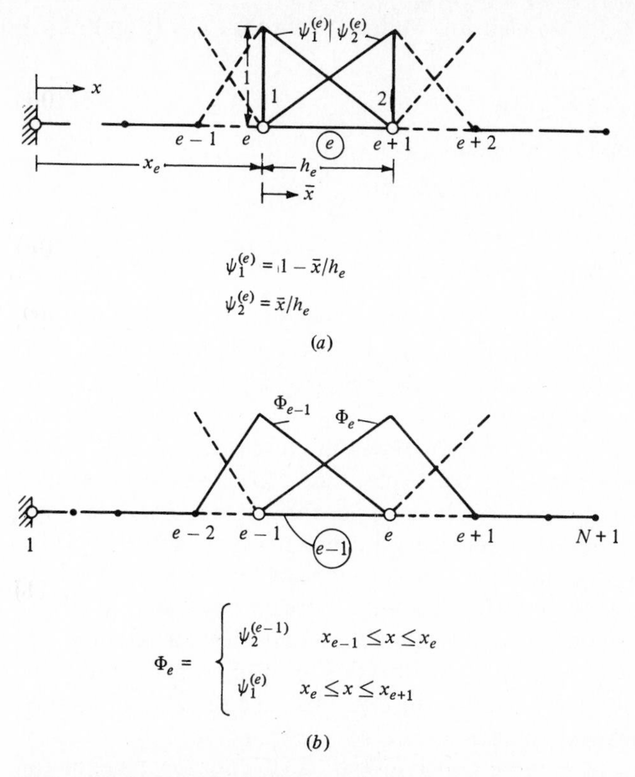

Figure 3.3 Local and global interpolation functions for the two-node linear element ($x_A = x_e$, $x_B = x_{e+1}$). (*a*) Local interpolation functions for a linear element. (*b*) Global interpolation functions for the two-node element mesh.

interval (say, an element), $u_1 = u_2 = u$. Then, we have from Eq. (3.11) that

$$u = u_1\psi_1 + u_2\psi_2 = u(\psi_1 + \psi_2)$$

or

$$1 = \psi_1 + \psi_2$$

One should note that the derivation of interpolation functions does not depend on the problem. The interpolation functions depend on the type of element (geometry, number of nodes, and number of primary unknowns per node). Thus one can derive the finite element equations of the form of Eq. (3.7) using an approximation of the form of Eq. (3.6). When the integrals in Eq. (3.8*b*) are to be evaluated, one must choose an appropriate set of functions ψ_i from a list (or library) of functions available. Thus, the linear interpolation functions derived above are not only useful in the finite-element approximation of the problem at hand, they are also useful in all problems that admit linear interpolation of the variables. A discussion of higher-order one-dimensional Lagrange interpolation functions is given in Sec. 3.6.2.

Returning to Eq. (3.7) for an element, we have from Eqs. (3.11) and (3.13) that

$$[K^{(e)}]\{u^{(e)}\} = \{F^{(e)}\} \tag{3.14}$$

where, for the linear element ($x_A = x_e$ and $x_B = x_{e+1}$),

$$\begin{aligned} K_{ij}^{(e)} &= \int_{x_e}^{x_{e+1}} a \frac{d\psi_i}{dx} \frac{d\psi_j}{dx} dx \\ F_i^{(e)} &= \int_{x_e}^{x_{e+1}} \psi_i f \, dx + P_i^{(e)} \end{aligned} \tag{3.15}$$

Property 1 of Eqs. (3.13) was used to simplify $F_i^{(e)}$ of Eq. (3.8*b*) to yield $F_i^{(e)}$ above. Equations (3.14) and (3.15) represent the *finite-element model* of Eqs. (3.4) [equivalently, Eq. (3.1)] over an element.

It should be pointed out that the approximation chosen for u in Eq. (3.9) satisfies the minimum requirements in Eq. (2.65). One can choose a quadratic, cubic, and higher-order approximation by adding various powers of x (not leaving out any lower-order terms, otherwise the completeness condition will be violated). For higher-order (i.e., quadratic, cubic, etc.) elements, the steps involved in the derivation of the approximation functions remain the same as those described in the preceding pages. Note that an nth-order element requires the inversion of an $(n+1) \times (n+1)$ matrix [see Eqs. (3.10)], which limits the application of the procedure outlined between Eqs. (3.9) and (3.12). An alternative derivation that avoids the inversion will be presented in Sec. 3.6. For higher-order elements, Eq. (3.9) requires identification of additional nodes (in addition to nodes $x = x_A$ and $x = x_B$) in the element. In other words, a quadratic approximation (which has three constants c_1, c_2, and c_3) requires identification of three nodes (the third one, say, at the center of the element). Thus, there is a relationship between the order of the approximation used for the dependent variable u and the number of nodes in the element.

A local coordinate system (i.e., a coordinate system in the element) proves to be more convenient in the derivation of the interpolation functions. Let $\bar{x}$ denote the local coordinate whose origin is at node 1 (left end) of an element (see Fig. 3.3). The local coordinate $\bar{x}$ is related to the global coordinate x [which is used to describe Eq. (3.1)] by the linear "translation" transformation

$$x = \bar{x} + x_e \tag{3.16}$$

In the local coordinate system, Eqs. (3.9) and (3.10) take the form

$$\left.\begin{aligned} & u(\bar{x}) = \hat{c}_1 + \hat{c}_2 \bar{x} \\ & u_1^{(e)} \equiv u(0) = \hat{c}_1 \qquad u_2^{(e)} \equiv u(h_e) = \hat{c}_1 + \hat{c}_2 h_e \\ & \begin{Bmatrix} u_1^{(e)} \\ u_2^{(e)} \end{Bmatrix} = \begin{bmatrix} 1 & 0 \\ 1 & h_e \end{bmatrix} \begin{Bmatrix} \hat{c}_1 \\ \hat{c}_2 \end{Bmatrix} \qquad h_e = x_{e+1} - x_e \\ & \hat{c}_1 = u_1^{(e)} \qquad \hat{c}_2 = \frac{u_2^{(e)} - u_1^{(e)}}{h_e} \end{aligned}\right\} \tag{3.17a}$$

and finally we have

$$u = \sum_{j=1}^{2} u_j^{(e)} \psi_j^{(e)}(\bar{x})$$
$$\psi_1^{(e)}(\bar{x}) = 1 - \frac{\bar{x}}{h_e} \qquad \psi_2^{(e)}(\bar{x}) = \frac{\bar{x}}{h_e} \qquad 0 \leq \bar{x} \leq h_e \tag{3.17b}$$

Note that mathematically and geometrically (see Fig. 3.3*a*), Eqs. (3.12) and (3.17*b*) are the same; one can obtain Eqs. (3.17*b*) from Eqs. (3.12) by means of the transformation (3.16). Local coordinate systems are also convenient to use in the numerical evaluation of integrals. For example, the coefficient matrix $K_{ij}^{(e)}$ and column vector $F_i^{(e)}$ in Eqs. (3.15) can be written in the local coordinate system as

$$K_{ij}^{(e)} = \int_0^{h_e} \hat{a} \frac{d\psi_i}{d\bar{x}} \frac{d\psi_j}{d\bar{x}} d\bar{x}$$
$$F_i^{(e)} = \int_0^{h_e} \psi_i \hat{f}\, d\bar{x} + P_i^{(e)} \tag{3.18a}$$

where

$$\hat{a} \equiv a \text{ evaluated at } x_e + \bar{x} = a(x_e + \bar{x})$$
$$\hat{f} \equiv f(x_e + \bar{x}) \tag{3.18b}$$

Equations (3.15) and (3.18) (the form of which is problem-dependent) are valid for elements of any order; one should only make note of the range of the subscripts i and j. In other words, for the two-node (linear) element, i and j take on values of 1 and 2, and for the three-node (quadratic) element, i and j take on values of 1, 2, and 3.

When the linear interpolation functions are used to approximate the dependent variable of the present problem, we obtain (for elementwise constant values of a and f)

$$[K^{(e)}] = \frac{a_e}{h_e}\begin{bmatrix} 1 & -1 \\ -1 & 1 \end{bmatrix} \qquad \{F^{(e)}\} = \frac{f_e h_e}{2}\begin{Bmatrix} 1 \\ 1 \end{Bmatrix} + \begin{Bmatrix} P_1^{(e)} \\ P_2^{(e)} \end{Bmatrix} \tag{3.19}$$

3.2.3 Assembly (or Connectivity) of Element Equations

Since Eq. (3.14) is derived for an arbitrarily typical element, it holds for any element from the finite-element mesh. For the sake of discussion, suppose that the domain of the problem, $\Omega = (0, L)$, is divided into three elements of possibly unequal lengths. Since these elements are connected at nodes 2 and 3 and u is continuous, u_2 of element Ω^e should be the same as u_1 of element Ω^{e+1} for $e = 1, 2$. To express this correspondence mathematically, we label the values of u at the global nodes with U_i $(i = 1, 2, \ldots, N)$ where N is the total number of global nodes. Then we have the following correspondence between the local (element) nodal values and the global nodal values (see Fig. 3.4):

$$u_1^{(1)} = U_1 \qquad u_2^{(1)} = U_2 = u_1^{(2)} \qquad u_2^{(2)} = U_3 = u_1^{(3)} \qquad u_2^{(3)} = U_4 \tag{3.20}$$

We shall call these relations *interelement continuity conditions*.

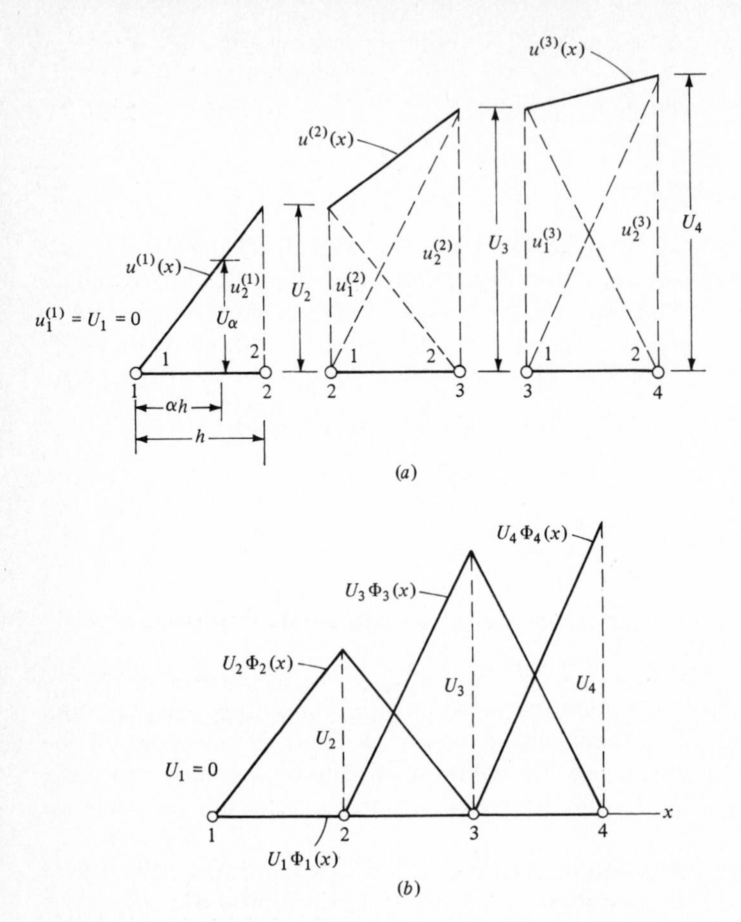

Figure 3.4 Correspondence of local and global nodal values and representation of the finite-element solution by global interpolation functions for the model problem. (*a*) Correspondence of element and global nodal values. (*b*) Representation of the solution by global interpolation functions.

The correspondence between the local nodes and the global nodes can be conveniently expressed in the form of an array, called the *boolean connectivity matrix*,

$$b_{ij} = \text{the global node number corresponding to the } j\text{th node of element } i \qquad (3.21)$$

For the mesh at hand, the matrix $[B] = [b_{ij}]$ is given by

$$[B] = \begin{bmatrix} 1 & 2 \\ 2 & 3 \\ 3 & 4 \end{bmatrix}$$

Repetition of a number in $[B]$ indicates that the coefficients of $[K^{(e)}]$ associated with the number add up; for example, (2, 2) location of the assembled matrix has both $K_{22}^{(1)}$ and $K_{11}^{(2)}$. To illustrate the use of the connectivity matrix further, we consider the network of linear elastic springs in Prob. 3.7. The connectivity matrix for the problem is given by

$$[B] = \begin{bmatrix} 1 & 2 \\ 2 & 4 \\ 2 & 3 \\ 2 & 5 \\ 3 & 4 \\ 4 & 5 \end{bmatrix} \tag{3.22}$$

The repetition of number 2 in locations (1, 2), (2, 1), (3, 1), and (4, 1) of $[B]$ indicates that the (2, 2) location of the assembled matrix is given by

$$K_{22} = K_{22}^{(1)} + K_{11}^{(2)} + K_{11}^{(3)} + K_{11}^{(4)}$$

Similarly,

$$K_{44} = K_{22}^{(2)} + K_{22}^{(5)} + K_{11}^{(6)}$$

We shall make use of the connectivity matrix in the computer implementation of the finite-element method.

Returning to the problem at hand, we express the element equations (3.14), with $[K]$ and $\{F\}$ given by Eqs. (3.19), in terms of the global nodal values U_i for each element:

Element 1

$$\frac{a_1}{h_1}\begin{bmatrix} 1 & -1 & 0 & 0 \\ -1 & 1 & 0 & 0 \\ 0 & 0 & 0 & 0 \\ 0 & 0 & 0 & 0 \end{bmatrix}\begin{Bmatrix} U_1 \\ U_2 \\ U_3 \\ U_4 \end{Bmatrix} = \frac{f_1 h_1}{2}\begin{Bmatrix} 1 \\ 1 \\ 0 \\ 0 \end{Bmatrix} + \begin{Bmatrix} P_1^{(1)} \\ P_2^{(1)} \\ 0 \\ 0 \end{Bmatrix} \tag{3.23a}$$

Element 2

$$\frac{a_2}{h_2}\begin{bmatrix} 0 & 0 & 0 & 0 \\ 0 & 1 & -1 & 0 \\ 0 & -1 & 1 & 0 \\ 0 & 0 & 0 & 0 \end{bmatrix}\begin{Bmatrix} U_1 \\ U_2 \\ U_3 \\ U_4 \end{Bmatrix} = \frac{f_2 h_2}{2}\begin{Bmatrix} 0 \\ 1 \\ 1 \\ 0 \end{Bmatrix} + \begin{Bmatrix} 0 \\ P_1^{(2)} \\ P_2^{(2)} \\ 0 \end{Bmatrix} \tag{3.23b}$$

Element 3

$$\frac{a_3}{h_3}\begin{bmatrix} 0 & 0 & 0 & 0 \\ 0 & 0 & 0 & 0 \\ 0 & 0 & 1 & -1 \\ 0 & 0 & -1 & 1 \end{bmatrix}\begin{Bmatrix} U_1 \\ U_2 \\ U_3 \\ U_4 \end{Bmatrix} = \frac{f_3 h_3}{2}\begin{Bmatrix} 0 \\ 0 \\ 1 \\ 1 \end{Bmatrix} + \begin{Bmatrix} 0 \\ 0 \\ P_1^{(3)} \\ P_2^{(3)} \end{Bmatrix} \tag{3.23c}$$

Equations (3.23) indicate the contribution of each element to the overall problem. Note that none of the element equations is solvable because the element matrices are singular and because parts of the right-hand sides are not known. By superimposing (i.e., adding) Eqs. (3.23) we obtain the equations of the system:

$$\begin{bmatrix} \frac{a_1}{h_1} & -\frac{a_1}{h_1} & 0 & 0 \\ -\frac{a_1}{h_1} & \frac{a_1}{h_1}+\frac{a_2}{h_2} & -\frac{a_2}{h_2} & 0 \\ 0 & -\frac{a_2}{h_2} & \frac{a_2}{h_2}+\frac{a_3}{h_3} & -\frac{a_3}{h_3} \\ 0 & 0 & -\frac{a_3}{h_3} & \frac{a_3}{h_3} \end{bmatrix} \begin{Bmatrix} U_1 \\ U_2 \\ U_3 \\ U_4 \end{Bmatrix} = \frac{1}{2} \begin{Bmatrix} f_1 h_1 \\ f_1 h_1 + f_2 h_2 \\ f_2 h_2 + f_3 h_3 \\ f_3 h_3 \end{Bmatrix} + \begin{Bmatrix} P_1^{(1)} \\ P_2^{(1)} + P_1^{(2)} \\ P_2^{(2)} + P_1^{(3)} \\ P_2^{(3)} \end{Bmatrix} \tag{3.24}$$

Equation (3.24) represents the *global* finite-element model of Eq. (3.1).

An alternative derivation of the assembly of element equations is also presented here. The procedure to be described is based on the idea that the quadratic functional (or the variational formulation) associated with the problem is equal to the sum of the quadratic functionals I_e of Eq. (3.5), of all elements:

$$I(U_i) = \sum_{e=1}^{N} I_e\left(u_i^{(e)}\right) \tag{3.25a}$$

or, in matrix form,

$$0 = \left[\delta I(U_i)\right]\{\delta U_i\} = \sum_{e=1}^{N} \left[\delta I_e\left(u_i^{(e)}\right)\right]\{\delta u_i^{(e)}\} \tag{3.25b}$$

where $\delta u_i^{(e)}$ can be viewed as the test function values $v_i^{(e)}$,

$$I_e\left(u_i^{(e)}\right) = \tfrac{1}{2}\{u^{(e)}\}^T [K^{(e)}]\{u^{(e)}\} - \{u^{(e)}\}^T\{F^{(e)}\} \tag{3.25c}$$

N denotes the number of elements in the mesh and δ the variational operator. Since Eq. (3.25b) is more general than Eq. (3.25a), we shall work with the former.

Carrying out the algebra in Eq. (3.25*b*), we obtain

$$\begin{aligned}
0 &= \sum_{e=1}^{N} \sum_{i=1}^{2} \left[\sum_{j=1}^{2} K_{ij}^{(e)} u_j^{(e)} - F_i^{(e)} \right] \delta u_i^{(e)} \\
&= \left(K_{11}^{(1)} u_1^{(1)} + K_{12}^{(1)} u_2^{(1)} - F_1^{(1)} \right) \delta u_1^{(1)} + \left(K_{21}^{(1)} u_1^{(1)} + K_{22}^{(1)} u_2^{(1)} - F_2^{(1)} \right) \delta u_2^{(1)} \\
&\quad + \left(K_{11}^{(2)} u_1^{(2)} + K_{12}^{(2)} u_2^{(2)} - F_1^{(2)} \right) \delta u_1^{(2)} + \left(K_{21}^{(2)} u_1^{(2)} + K_{22}^{(2)} u_2^{(2)} - F_2^{(2)} \right) \delta u_2^{(2)} \\
&\quad + \left(K_{11}^{(3)} u_1^{(3)} + K_{12}^{(3)} u_2^{(3)} - F_1^{(3)} \right) \delta u_1^{(3)} + \left(K_{21}^{(3)} u_1^{(3)} + K_{22}^{(3)} u_2^{(3)} - F_2^{(3)} \right) \delta u_2^{(3)}
\end{aligned} \tag{3.26}$$

Using the correspondence in Eq. (3.20) in the above equation and collecting the coefficients of δU_i $(i = 1, 2, \ldots, N + 1)$, we obtain

$$\begin{aligned}
0 &= \delta U_1 \left[K_{11}^{(1)} U_1 + K_{12}^{(1)} U_2 - F_1^{(1)} \right] \\
&\quad + \delta U_2 \left[K_{21}^{(1)} U_1 + \left(K_{22}^{(1)} + K_{11}^{(2)} \right) U_2 + K_{12}^{(2)} U_3 - \left(F_2^{(1)} + F_1^{(2)} \right) \right] \\
&\quad + \delta U_3 \left[K_{21}^{(2)} U_2 + \left(K_{22}^{(2)} + K_{11}^{(3)} \right) U_3 + K_{12}^{(3)} U_4 - \left(F_2^{(2)} + F_1^{(3)} \right) \right] \\
&\quad + \delta U_4 \left[K_{21}^{(3)} U_3 + K_{22}^{(3)} U_4 - F_2^{(3)} \right]
\end{aligned} \tag{3.27}$$

Since the variations in U_i are arbitrary, the equation above implies that the coefficient of each δU_i $(i = 1, 2, \ldots, N + 1)$, where $N + 1$ is the total number of global nodes, should be equal to zero separately. The result, expressed in matrix form, is

$$\begin{bmatrix}
K_{11}^{(1)} & K_{12}^{(1)} & 0 & 0 \\
K_{21}^{(1)} & K_{22}^{(1)} + K_{11}^{(2)} & K_{12}^{(2)} & 0 \\
0 & K_{21}^{(2)} & K_{22}^{(2)} + K_{11}^{(3)} & K_{12}^{(3)} \\
0 & 0 & K_{21}^{(3)} & K_{22}^{(3)}
\end{bmatrix}
\begin{Bmatrix} U_1 \\ U_2 \\ U_3 \\ U_4 \end{Bmatrix}
=
\begin{Bmatrix} F_1^{(1)} \\ F_2^{(1)} + F_1^{(2)} \\ F_2^{(2)} + F_1^{(3)} \\ F_2^{(3)} \end{Bmatrix} \tag{3.28}$$

which is the same as Eq. (3.24).

3.2.4 Imposition of Boundary Conditions

Equations (3.24) and (3.28) are valid for any problem described by the differential equation (3.1), irrespective of the boundary conditions. The coefficient matrix (3.24) is singular prior to the imposition of the essential boundary conditions (on the primary variables). Upon the imposition of suitable boundary conditions of the problem, we obtain a nonsingular matrix that can be inverted. To this end we note first that the natural boundary conditions are included in the column vector $\{F^{(e)}\}$ through $P_j^{(e)}$. At all global nodes between the boundary nodes, the sum of the contributions of the natural boundary condition from node 2 of element e and node 1 of element $e + 1$ should equal the specified value of the secondary variable $(a\,du/dx)$:

$$P_2^{(e)} + P_1^{(e+1)} = \text{specified value of } \left(a\frac{du}{dx}\right)\Big|_{x=x_{e+1}} \tag{3.29}$$

For the problem at hand, $(a\,du/dx)$ at $x = h_1$ and $x = h_1 + h_2$ are specified to be zero. [Since u is not specified at global nodes 2 and 3, it is understood that the secondary variable $(a\,du/dx)$ is specified to be zero at these nodes.] We have (see remark 10 in Sec. 3.2.7)

$$P_2^{(1)} + P_1^{(2)} = 0 \qquad P_2^{(2)} + P_1^{(3)} = 0 \tag{3.30}$$

For the longitudinal deformation of a bar, these equations can be interpreted as the *equilibrium* conditions for the internal forces. Mathematically, this amounts to assuming that the difference of the values of $(a\,du/dx)$ from elements e and $e+1$ at global node $e+1$ equals the specified value of $(a\,du/dx)$ at node $e+1$ (which is equal to zero in the present example):

$$\left(a\frac{du}{dx}\right)\Big|_{x=x_{e+1}}^{(e)} - \left(a\frac{du}{dx}\right)\Big|_{x=x_{e+1}}^{(e+1)} \equiv P_2^{(e)} + P_1^{(e+1)} = 0 \tag{3.31}$$

The difference accounts for any point source applied at the point $x = x_{e+1}$. If no point source is applied, the difference is *assumed* to be zero. Note that $(a\,du/dx)$ is unknown at $x = 0$ and it is equal to P at $x = L$.

Next, we discuss the imposition of the essential boundary conditions of the problem. In the present case, the only known essential boundary condition is

$$U_1 = u_1^{(1)} = 0 \tag{3.32}$$

The modified columns of the unknowns and $P_i^{(e)}$'s are given by

$$\{\Delta\} = \begin{Bmatrix} 0 \\ U_2 \\ U_3 \\ U_4 \end{Bmatrix} \qquad \{P\} = \begin{Bmatrix} \left(-a\dfrac{du}{dx}\right)\Big|_{x=0} \equiv P_1^{(1)} \\ 0 \\ 0 \\ P \end{Bmatrix} \tag{3.33}$$

3.2.5 Solution of Equations

The global finite-element equations (3.28) can be partitioned conveniently into the following form:

$$\begin{bmatrix} [K^{11}] & [K^{12}] \\ [K^{21}] & [K^{22}] \end{bmatrix} \begin{Bmatrix} \{\Delta^1\} \\ \{\Delta^2\} \end{Bmatrix} = \begin{Bmatrix} \{F^1\} \\ \{F^2\} \end{Bmatrix} \tag{3.34}$$

where $\{\Delta^1\}$ is the column of the known displacements (U_1), $\{\Delta^2\}$ is the column of the unknown displacements (U_2, U_3, U_4), $\{F^1\}$ is the column of the unknown forces $(P_1^{(1)})$, and $\{F^2\}$ is the column of the known forces $(0, 0, P)$. Writing Eq.

(3.34) as two matrix equations, we obtain

$$[K^{11}]\{\Delta^1\} + [K^{12}]\{\Delta^2\} = \{F^1\} \tag{3.35}$$

$$[K^{21}]\{\Delta^1\} + [K^{22}]\{\Delta^2\} = \{F^2\} \tag{3.36}$$

From Eq. (3.36) we have

$$\{\Delta^2\} = [K^{22}]^{-1}(\{F^2\} - [K^{21}]\{\Delta^1\}) \tag{3.37}$$

Once $\{\Delta^2\}$ is known, $\{F^1\}$ can be computed from Eq. (3.35). In the present case, we have

$$[K^{11}] = K_{11}^{(1)} \qquad [K^{12}] = \{K_{12}^{(1)} \quad 0 \quad 0\}$$

$$[K^{21}] = \begin{Bmatrix} K_{21}^{(1)} \\ 0 \\ 0 \end{Bmatrix} \qquad [K^{22}] = \begin{bmatrix} K_{22}^{(1)} + K_{11}^{(2)} & K_{12}^{(2)} & 0 \\ K_{21}^{(2)} & K_{22}^{(2)} + K_{11}^{(3)} & K_{12}^{(3)} \\ 0 & K_{21}^{(3)} & K_{22}^{(3)} \end{bmatrix} \tag{3.38}$$

$$\{F^1\} = F_1^{(1)} \qquad \{F^2\} = \begin{Bmatrix} F_2^{(1)} + F_1^{(2)} \\ F_2^{(2)} + F_1^{(3)} \\ F_2^{(3)} \end{Bmatrix}$$

For a = constant, $h_1 = h_2 = h_3 = L/3$, and f = constant, we obtain

$$[K^{22}]^{-1} = \frac{L}{3a}\begin{bmatrix} 1 & 1 & 1 \\ 1 & 2 & 2 \\ 1 & 2 & 3 \end{bmatrix} \tag{3.39a}$$

$$\{\Delta^2\} = \begin{Bmatrix} U_2 \\ U_3 \\ U_4 \end{Bmatrix} = \frac{fL^2}{18a}\begin{Bmatrix} 5 \\ 8 \\ 9 \end{Bmatrix} + \frac{PL}{3a}\begin{Bmatrix} 1 \\ 2 \\ 3 \end{Bmatrix} \tag{3.39b}$$

and the unknown natural boundary condition at $x = 0$ (corresponding to the reaction force in the problem of the deformation of a bar)

$$P_1^{(1)} \equiv \left(-a\frac{du}{dx}\right)\bigg|_{x=0} = -(fL + P) \tag{3.40}$$

Since $\psi_i^{(e)}$ ($e = 1, 2, 3$) is zero in any element Ω^f for $e \neq f$ (see Fig. 3.4), the (global) finite-element solution for the entire domain is given by

$$u(x) = \sum_{e=1}^{3}\left(\sum_{i=1}^{2} u_i^{(e)}\psi_i^{(e)}\right) \equiv \sum_{I=1}^{4} U_I\Phi_I(x) \tag{3.41}$$

where $\Phi_I(x)$, $I = 1, 2, \ldots, N + 1$, are the piecewise continuous *global interpolation functions*,

$$\Phi_I(x) = \begin{cases} \psi_2^{(I-1)}(x) & x_{I-1} \leqslant x \leqslant x_I \\ \psi_1^{(I)}(x) & x_I \leqslant x \leqslant x_{I+1} \end{cases} \tag{3.42}$$

For computational purposes, Eq. (3.41) can be written in the form

$$u(x) = \begin{cases} U_1\psi_1^{(1)}(x) + U_2\psi_2^{(1)}(x) & 0 \leqslant x \leqslant \frac{L}{3} \\ U_2\psi_1^{(2)}(x) + U_3\psi_2^{(2)}(x) & \frac{L}{3} \leqslant x \leqslant \frac{2L}{3} \\ U_3\psi_1^{(3)}(x) + U_4\psi_2^{(3)}(x) & \frac{2L}{3} \leqslant x \leqslant L \end{cases} \tag{3.43}$$

For example, to compute u at $x = x_0$, $L/3 \leqslant x_0 \leqslant 2L/3$, one should use the second line of Eq. (3.43).

3.2.6 Postprocessing of the Solution

The solution of the finite-element equations gives the nodal values of the primary unknown or unknowns u (e.g., displacement, velocity, temperature, etc.). Post-processing of the results includes one or more of the following:

1. Computation of any secondary variables (e.g., gradient of the solution)
2. Interpretation of the results to check whether the solution makes sense (intuition and experience are the guides when other results are not available for comparison)
3. Tabular and/or graphic presentation of the results

Since the calculation of the secondary variables involves the derivative of the solution, we compute the derivative of the solution for the linear element noting that $\psi_i^{(e)} = 0$ in element Ω^f for $e \neq f$,

$$\frac{du}{dx} = \sum_{e=1}^{3} \sum_{i=1}^{2} u_i^{(e)} \frac{d\psi_i^{(e)}}{dx} \tag{3.44a}$$

$$= \frac{u_2^{(e)} - u_1^{(e)}}{h_e} \qquad e = 1, 2, 3 \tag{3.44b}$$

Note that du/dx is a constant in each element. This is expected since u is assumed to be linear in each element. Also, note that this constant is different in different elements. Thus, the derivative of the solution is piecewise constant.

In the present problem, the exact solution can be obtained very easily:

$$u_{\text{exact}} = \frac{f}{2a}(2xL - x^2) + \frac{P}{a}x \tag{3.45}$$

Therefore, we can compare the finite-element solution with the exact solution. Evaluating the exact solution at $x = h$, $2h$, and $3h$, we obtain the same result as in Eqs. (3.39*b*). Hence, the finite-element solution is exact at the nodes. If we need to compute $u(x)$ at any intermediate point (i.e., between any two nodes), we can employ Eq. (3.43). For example, at $x = \alpha h$, $0 < \alpha < 1$, we have

(see Fig. 3.4a)

$$u(\alpha h) = \frac{h - \alpha h}{h} u_1^{(1)} + \frac{\alpha h - 0}{h} u_2^{(1)}$$
$$= (1 - \alpha) u_1^{(1)} + \alpha u_2^{(1)} \equiv U_\alpha \tag{3.46}$$

That is, u at $x = \alpha h$ is a weighted average of its values at the nodes on either side of the point $x = \alpha h$. Similarly, for $x_0 = h + \alpha h$, we obtain from the second line of Eq. (3.43) the value $u(x_0) = (1 - \alpha) u_1^{(2)} + \alpha u_2^{(2)}$.

In Fig. 3.5 the exact solution and its derivative are compared with the finite-element solution. Note that at interelement boundaries du/dx is discontinuous (has different values in going from one element to the next). However, the average of du/dx at nodes e and $e + 1$ coincides with the exact value of du/dx at $x = x_e + h/2$.

This completes the finite-element approximation of Eqs. (3.1) and (3.2). There are several observations we can make from the procedure described above for the example problem.

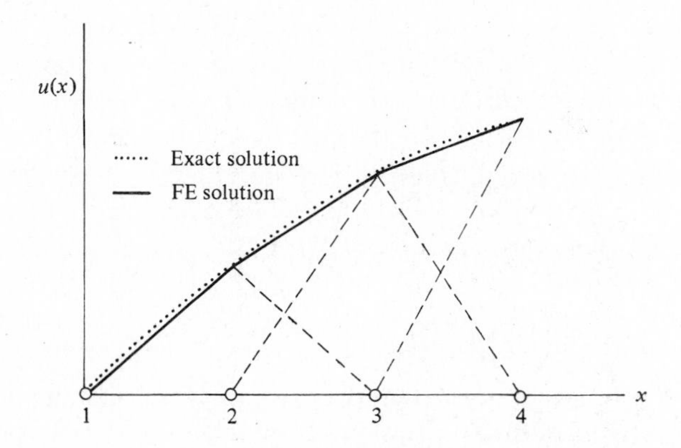

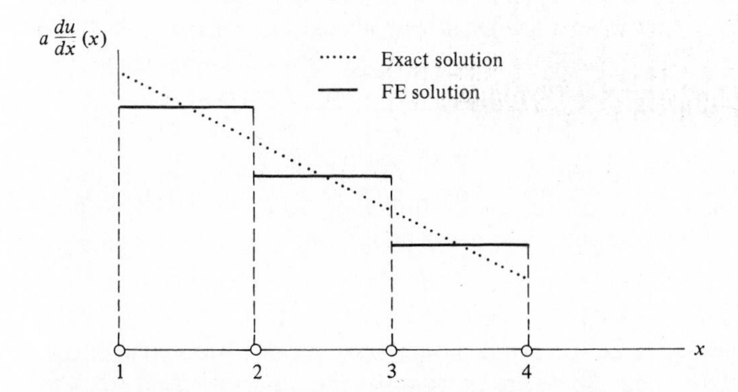

Figure 3.5 Comparison of the finite-element solution with the exact solution of the differential equation (3.1) with boundary conditions (3.2).

3.2.7 Some Remarks on the Finite-Element Method

Remark 1 Although the Ritz method was used to set up the element equations, any other variational method, such as the least-squares method, could be used (see Sec. 2.3.3 and Sec. 5.2.1).

Remark 2 Steps 1 through 6 (Table 3.1) are common for any problem. Element geometry and the characteristics of the problem enter the formulation in steps 1 and 2. The derivation of interpolation functions depends only on the element geometry, the number and position of nodes in the element, and the number of independent degrees of freedom per node. Of course, the degree of approximation used for a variable depends on the order of the derivative appearing in the variational formulation of the equation; the number of nodes in the element depends on the degree of approximation.

Remark 3 The interpolation functions ψ_i derived in step 2 for an element (for any degree of approximation) satisfy the properties in Eqs. (3.13). Property 1 is the direct result of the (interpolation) requirement that $u(x_j) = \Sigma u_i^{(e)}\psi_i^{(e)}(x_j) = u_j^{(e)}$. Property 2 is the result of including the constant term c_1 in the approximation. This permits approximating the constant state of the primary variable of the problem (e.g., temperature, displacement, etc.) in the element and therefore in the entire domain if needed.

Remark 4 The element equations are derived for the linear operator

$$A \equiv -\frac{d}{dx}\left(a\frac{d}{dx}\right)$$

Hence, they are valid for any physical problem that is described by the operator A (one needs only to interpret the quantities). Examples of problems described by this operator are listed in Table 2.1. Thus, a computer program written for the finite-element analysis of the bar can be used to analyze any of the problems in Table 2.1. Also, note that the property $a = a(x)$ (which depends on the geometry as well as on the material) can be different in each of the elements.

Remark 5 Integration of the element matrices in Eqs. (3.15) can be implemented on the computer employing numerical quadrature (see Sec. 3.6.4). When these integrals are algebraically complicated, one has no other choice but to use numerical integration to evaluate the matrix elements in Eqs. (3.15).

Remark 6 Since the geometry of the problem is exactly represented, the integration is carried out exactly, and the linear approximation is able to represent the exact solution at the nodes (for a = constant, f = constant), all of the three errors mentioned in Sec. 1.3.3 are zero in the present problem.

***Remark** 7 Physical (or direct) approach.* The conventional (or direct) approach (when the finite-element method was not around) to the solution of the problem we discussed above is not much different. The difference lies only in the derivation of the element equations (3.14). The conventional method uses the laws of physics to obtain the force-displacement relations in Eq. (3.14). "Breaking the problem into pieces" is not common in areas other than structural mechanics, but the derivation of element equations by direct means, more or less, follows the same procedure. For example, consider the free-body diagram of a portion of an elastic bar. From elementary mechanics, we have

$$\text{Force} = (\text{area of cross section})\,(\text{stress}) = A\sigma$$
$$\text{Stress} = (\text{modulus of elasticity})\,(\text{strain}) = E\varepsilon$$
$$\text{Strain} = (\text{elongation}/\text{original length}) = e/h$$

where h is the original length of the portion (or element) under consideration. The strain defined above is the average strain. (Mathematically, $\varepsilon = du/dx$, u being the displacement, which may include rigid body motion and elongation of the bar.) Using this relation, we write ($a_e = A_e E_e$)

$$P_1^{(e)} = \frac{a_e}{h_e}\left(u_1^{(e)} - u_2^{(e)}\right) \qquad P_2^{(e)} = \frac{a_e}{h_e}\left(u_2^{(e)} - u_1^{(e)}\right) \tag{3.47}$$

or

$$\frac{a_e}{h_e}\begin{bmatrix} 1 & -1 \\ -1 & 1 \end{bmatrix}\begin{Bmatrix} u_1^{(e)} \\ u_2^{(e)} \end{Bmatrix} = \begin{Bmatrix} P_1^{(e)} \\ P_2^{(e)} \end{Bmatrix} \tag{3.48}$$

which is the same as Eq. (3.19) for $f = 0$. Note that in deriving the element equations we have used the knowledge of mechanics of materials and the assumption that the strain is constant over the length of the element. Equations of the type of Eq. (3.48) can also be derived for a spring element, a pipe-flow element, an electrical resistor element, and so on (see Problems at the end of this section). If a more accurate, say, linear, representation of the strain is required, the direct procedure is difficult to use. In such cases, the variational method used in constructing the finite-element equations, which does not require the knowledge of the laws of physics but uses the governing equation in deriving the element equations, offers advantage over the physical approach.

***Remark** 8* Another interpretation of Eqs. (3.19) can be given in terms of the finite-difference approximation. The axial force at any point x is given by $P(x) = EA\,du/dx$. Using the forward-difference approximation, we have

$$-P_1^{(e)} \equiv P(x)|_{x=x_e} = EA\frac{u(x_{e+1}) - u(x_e)}{h_e} \tag{3.49}$$

and using the backward-difference approximation,

$$P_2^{(e)} \equiv P(x)|_{x=x_{e+1}} = EA\frac{u(x_{e+1}) - u(x_e)}{h_e} \tag{3.50}$$

Remark 9 For the model problem considered, the element matrices $[K^e]$ in Eqs. (3.15) are symmetric, $K_{ij}^{(e)} = K_{ji}^{(e)}$. In most linear problems of interest, we have *symmetry* of the coefficient matrix. This enables one to compute $K_{ij}^{(e)}$ ($i = 1, 2, \ldots, n$) for $j \leqslant i$ only. In other words, one needs to compute only the diagonal terms and the upper (or lower) diagonal terms. Because of the symmetry of the element matrices, the assembled global matrix will also be symmetric. Thus, one needs to store only the upper triangle (including the diagonal) of the assembled matrix in a finite-element program. Another property characteristic of the finite-element method is the *sparseness* of the assembled matrix. Since the global interpolation function Φ_e is nonzero only on two neighboring elements (see Fig. 3.3) at node e, its contribution to nodes $e - 2$, $e - 3$, etc., and to nodes $e + 2$, $e + 3$, etc. is zero; that is, entries K_{13}, K_{24}, etc. in the global matrix are zero. In general, $K_{IJ} = 0$ if nodes I and J do not belong to the same element. Thus, if the global nodes are numbered sequentially, the contribution to entries two or more places away, on either side of the main diagonal of the global coefficient matrix, is zero, and the matrix is *banded*. When a matrix is banded and symmetric, one needs to store only the entries in the upper (or lower) band of the matrix. Equation solvers which are written for the solution of banded symmetric equations are available for use in such cases. While symmetry of the coefficient matrix is a property transferred from variational formulation of the equation being modeled, the sparseness of the matrix is a result of the finite-element interpolation functions, which have *local support* (i.e., have nonzero values over a small portion of the domain).

Remark 10 The compatibility (or "equilibrium") of the secondary (dual) variables (or "forces") P_i at the interelement boundaries is expressed by Eqs. (3.30). This amounts to imposing the condition that the secondary variable $(a\,du/dx)$ at the node be continuous. Although Eqs. (3.30) are imposed in solving the global equations, the continuity of $(a\,du/dx)$ will not be satisfied because the finite-element solution is defined only elementwise. Thus, in general, we have at the nodes

$$P_2^{(1)} + P_1^{(2)} \neq 0 \qquad (a)$$

In most finite-element books, this point is not made clear to the reader. These books consider the quadratic form (3.25a) of the total problem and omit the sum of the interelement contributions,

$$\sum_{e=1}^{N} \left(\sum_{i=1}^{2} P_i^{(e)} u_i^{(e)} \right) \qquad (b)$$

in the quadratic form of the problem. However, this amounts to imposing equilibrium conditions of the form of Eqs. (3.30). When the secondary variable is specified to be nonzero (say, P_0) at an interelement boundary (say, at global node 2), then we have

$$P_2^{(1)} + P_1^{(2)} = P_0 \qquad (c)$$

In other finite-element books, P_0 is included in the functional as P_0U_2, where U_2 is the value of u at global node 2.

To fix the ideas, consider Eq. (3.1). The variational form of Eq. (3.1) over the entire domain is given by

$$0 = \int_0^L \left(a \frac{dv}{dx} \frac{du}{dx} - vf \right) dx - v(L)P \tag{d}$$

When u is approximated by functions that are defined only on a local interval (which is the case in the finite-element method), use of the above variational form implies the omission of the sum of the interelement contributions of Eq. (b). Substituting Eq. (3.41) for u and $v = \Phi_I$ into Eq. (d), we obtain

$$0 = \int_0^L \left[a \frac{d\Phi_I}{dx} \left(\sum_{J=1}^{4} U_J \frac{d\Phi_J}{dx} \right) - \Phi_I f \right] dx - \Phi_I(L)P \tag{e}$$

Since Φ_I is nonzero only between x_{I-1} and x_{I+1}, the integral becomes

$$0 = \int_{x_{I-1}}^{x_{I+1}} \left[a \frac{d\Phi_I}{dx} \left(U_{I-1} \frac{d\Phi_{I-1}}{dx} + U_I \frac{d\Phi_I}{dx} + U_{I+1} \frac{d\Phi_{I+1}}{dx} \right) - \Phi_I f \right] dx$$
$$- \Phi_I(L)P \tag{f}$$

and we have (for a three-element mesh)

$$I = 1: \quad 0 = \int_{x_1=0}^{x_2} \left[a \frac{d\Phi_1}{dx} \left(U_1 \frac{d\Phi_1}{dx} + U_2 \frac{d\Phi_2}{dx} \right) - \Phi_1 f \right] dx - \Phi_1(L)P$$

$$I = 2: \quad 0 = \int_{x_1=0}^{x_3} \left[a \frac{d\Phi_2}{dx} \left(U_1 \frac{d\Phi_1}{dx} + U_2 \frac{d\Phi_2}{dx} + U_3 \frac{d\Phi_3}{dx} \right) - \Phi_2 f \right] dx$$
$$- \Phi_2(L)P \tag{g}$$

$$I = 3: \quad 0 = \int_{x_2}^{x_4=L} \left[a \frac{d\Phi_3}{dx} \left(U_2 \frac{d\Phi_2}{dx} + U_3 \frac{d\Phi_3}{dx} + U_4 \frac{d\Phi_4}{dx} \right) - \Phi_3 f \right] dx$$
$$- \Phi_3(L)P$$

$$I = 4: \quad 0 = \int_{x_3}^{x_4=L} \left[a \frac{d\Phi_4}{dx} \left(U_3 \frac{d\Phi_3}{dx} + U_4 \frac{d\Phi_4}{dx} \right) - \Phi_4 f \right] dx - \Phi_4(L)P$$

These equations, upon performing the integration, yield Eq. (3.24), with the last column (containing P's) in Eq. (3.24) replaced by [cf. Eq. (3.33)]

$$\begin{Bmatrix} 0 \\ 0 \\ 0 \\ P \end{Bmatrix} \tag{h}$$

Although this procedure gives the assembled equations directly, it is algebraically complicated (especially for two-dimensional problems) and not amenable to simple computer implementation.

The following two examples illustrate the steps involved in the finite-element analysis of one-dimensional second-order equations.

Example 3.1 Consider the differential equation (3.1) (corresponding to the transverse deflection of a nonuniform cable fixed at both ends and subjected to distributed transverse force, or to heat transfer in a fin of nonhomogeneous material or a rod of variable cross section subjected to a distributed heat source along the length) with a, f, and the boundary conditions given by

$$a(x)=1+x \qquad f(x)=1+4x \qquad \text{for } 0<x<1$$
$$u(0)=u(1)=0 \tag{3.51}$$

Step 1 We wish to determine the finite-element solution to Eqs. (3.1) and (3.51) using four linear elements (of equal size).

Step 2 The variational formulation (3.4) of the equation and the algebraic expressions (3.15) for $K_{ij}^{(e)}$ and $F_i^{(e)}$ do not change because the differential equation is the same as that considered before. We have

$$0=\int_{x_e}^{x_{e+1}}\left[(1+x)\frac{dv}{dx}\frac{du}{dx}-v(1+4x)\right]dx-v(x_e)P_1^{(e)}-v(x_{e+1})P_2^{(e)} \tag{3.52}$$

where

$$P_1^{(e)}=\left[-(1+x)\frac{du}{dx}\right]_{x=x_e} \qquad P_2^{(e)}=\left[(1+x)\frac{du}{dx}\right]_{x=x_{e+1}} \tag{3.53}$$

and

$$K_{ij}^{(e)}=\begin{cases}\displaystyle\int_{x_e}^{x_{e+1}}(1+x)\frac{d\psi_i^{(e)}}{dx}\frac{d\psi_j^{(e)}}{dx}\,dx & \text{in global coordinate system} \quad (3.54a)\\[2ex] \displaystyle\int_0^{h_e}(1+x_e+\bar{x})\frac{d\psi_i^{(e)}}{d\bar{x}}\frac{d\psi_j^{(e)}}{d\bar{x}}\,d\bar{x} & \text{in local coordinate system} \quad (3.54b)\end{cases}$$

$$F_i^{(e)}=\begin{cases}\displaystyle\int_{x_e}^{x_{e+1}}(1+4x)\psi_i^{(e)}\,dx+P_i^{(e)} & \text{global} \quad (3.55a)\\[2ex] \displaystyle\int_0^{h_e}(1+4x_e+4\bar{x})\psi_i^{(e)}\,d\bar{x}+P_i^{(e)} & \text{local} \quad (3.55b)\end{cases}$$

We shall illustrate the computation of $K_{ij}^{(e)}$ and $F_i^{(e)}$ using the linear interpolation functions in Eqs. (3.12) and (3.17b).

(*a*) *Global coordinate system.* Using $\psi_i^{(e)}$ from Eqs. (3.12) in Eqs. (3.54*a*) and (3.55*a*) and carrying out the integration with respect to x, we obtain

$$\frac{d\psi_1^{(e)}}{dx} = \frac{-1}{h_e} \qquad \frac{d\psi_2^{(e)}}{dx} = +\frac{1}{h_e}$$

$$K_{11}^{(e)} = \int_{x_e}^{x_{e+1}} (1+x)\left(-\frac{1}{h_e}\right)^2 dx = \frac{1}{h_e^2}\left(x + \frac{x^2}{2}\right)\Bigg|_{x_e}^{x_{e+1}}$$

$$= \frac{1}{h_e^2}\left[x_{e+1} - x_e + \frac{1}{2}(x_{e+1}^2 - x_e^2)\right] = \frac{1}{h_e}\left[1 + \frac{1}{2}(x_{e+1} + x_e)\right]$$

$$K_{12}^{(e)} = \int_{x_e}^{x_{e+1}} (1+x)\left(-\frac{1}{h_e}\right)\frac{1}{h_e}\, dx = -\frac{1}{h_e}\left[1 + \frac{1}{2}(x_{e+1} + x_e)\right]$$

$$K_{22}^{(e)} = \int_{x_e}^{x_{e+1}} (1+x)\left(\frac{1}{h_e}\right)^2 dx = \frac{1}{h_e}\left[1 + \frac{1}{2}(x_{e+1} + x_e)\right]$$

$$F_1^{(e)} = \int_{x_e}^{x_{e+1}} (1+4x)\left(\frac{x_{e+1} - x}{h_e}\right) dx + P_1^{(e)} \tag{3.56}$$

$$= \frac{1}{h_e}\left(x_{e+1}x - \frac{x^2}{2}\right)\Bigg|_{x_e}^{x_{e+1}} + \frac{4}{h_e}\left(\frac{x_{e+1}x^2}{2} - \frac{x^3}{3}\right)\Bigg|_{x_e}^{x_{e+1}} + P_1^{(e)}$$

$$= \frac{h_e}{2} + \frac{2h_e}{3}(x_{e+1} + 2x_e) + P_1^{(e)}$$

$$F_2^{(e)} = \int_{x_e}^{x_{e+1}} (1+4x)\left(\frac{x - x_e}{h_e}\right) dx + P_2^{(e)}$$

$$= \frac{1}{h_e}\left(\frac{x^2}{2} - x_e x\right)\Bigg|_{x_e}^{x_{e+1}} + \frac{4}{h_e}\left(\frac{x^3}{3} - \frac{x^2 x_e}{2}\right)\Bigg|_{x_e}^{x_{e+1}} + P_2^{(e)}$$

$$= \frac{h_e}{2} + \frac{2h_e}{3}(x_e + 2x_{e+1}) + P_2^{(e)}$$

(*b*) *Local coordinates.* To show that we get the same results using the local coordinate system, we compute $K_{11}^{(e)}$ and $F_1^{(e)}$:

$$K_{11}^{(e)} = \int_0^{h_e} (1 + x_e + \bar{x})\left(\frac{-1}{h_e}\right)^2 d\bar{x}$$

$$= \frac{1}{h_e}\left(1 + \frac{h_e}{2} + x_e\right)$$

$$= \frac{1}{h_e}\left(1 + \frac{x_{e+1} + x_e}{2}\right)$$

$$F_1^{(e)} = \int_0^{h_e} (1 + 4x_e + 4\bar{x})\left(1 - \frac{\bar{x}}{h_e}\right) d\bar{x} + P_1^{(e)}$$

$$= \left(h_e + 4h_e x_e + 2h_e^2\right) - \frac{1}{h_e}\left(\frac{h_e^2}{2} + 2h_e^2 x_e + \frac{4}{3}h_e^3\right) + P_1^{(e)}$$

$$= \frac{h_e}{2} + h_e\left(2x_e + \frac{2}{3}h_e\right) + P_1^{(e)}$$

$$= \frac{h_e}{2} + \frac{2h_e}{3}(x_{e+1} + 2x_e) + P_1^{(e)}$$

and so on. Thus we obtain the same results as in Eqs. (3.56). We have

$$[K^e] = \frac{1}{h_e}\left(\begin{bmatrix} 1 & -1 \\ -1 & 1 \end{bmatrix} + \frac{x_{e+1} + x_e}{2}\begin{bmatrix} 1 & -1 \\ -1 & 1 \end{bmatrix}\right) \tag{3.57a}$$

$$\{F^e\} = \begin{Bmatrix} P_1^e \\ P_2^e \end{Bmatrix} + \frac{h_e}{2}\begin{Bmatrix} 1 \\ 1 \end{Bmatrix} + \frac{2h_e}{3}\begin{Bmatrix} x_{e+1} + 2x_e \\ 2x_{e+1} + x_e \end{Bmatrix} \tag{3.57b}$$

Note that the elements of the matrices $[K^{(e)}]$ and $\{F^{(e)}\}$ depend on the coordinates x_e and x_{e+1} of the element e, and therefore they are not the same for any two elements.

Step 3 Suppose that the interval (0, 1) is divided into *four* equal elements, $h_e = \frac{1}{4}$, $e = 1, 2, 3, 4$. Then Eqs. (3.57) become:

Element 1

$$e = 1,\ x_e = 0,\ x_{e+1} = h_e$$

$$[K^{(1)}] = 4\begin{bmatrix} 1 + 0.125 & -1 - 0.125 \\ -1 - 0.125 & 1 + 0.125 \end{bmatrix}$$

$$\{F^{(1)}\} = \begin{Bmatrix} P_1^{(1)} \\ P_2^{(1)} \end{Bmatrix} + \frac{1}{8}\begin{Bmatrix} 1 \\ 1 \end{Bmatrix} + \frac{1}{24}\begin{Bmatrix} 1 \\ 2 \end{Bmatrix}$$

Element 2

$$e = 2,\ x_e = h_e,\ x_{e+1} = 2h_e$$

$$[K^{(2)}] = 4\begin{bmatrix} 1 + 0.375 & -1 - 0.375 \\ -1 - 0.375 & 1 + 0.375 \end{bmatrix}$$

$$\{F^{(2)}\} = \begin{Bmatrix} P_1^{(2)} \\ P_2^{(2)} \end{Bmatrix} + \frac{1}{8}\begin{Bmatrix} 1 \\ 1 \end{Bmatrix} + \frac{1}{24}\begin{Bmatrix} 4 \\ 5 \end{Bmatrix}$$

Element 3

$$e = 3,\ x_e = 2h_e,\ x_{e+1} = 3h_e$$

$$[K^{(3)}] = 4\begin{bmatrix} 1 + 0.625 & -1 - 0.625 \\ -1 - 0.625 & 1 + 0.625 \end{bmatrix}$$

$$\{F^{(3)}\} = \begin{Bmatrix} P_1^{(3)} \\ P_2^{(3)} \end{Bmatrix} + \frac{1}{8}\begin{Bmatrix} 1 \\ 1 \end{Bmatrix} + \frac{1}{24}\begin{Bmatrix} 7 \\ 8 \end{Bmatrix}$$

Element 4

$$e = 4,\ x_e = 3h_e,\ x_{e+1} = 4h_e$$

$$[K^{(4)}] = 4\begin{bmatrix} 1 + 0.875 & -1 - 0.875 \\ -1 - 0.875 & 1 + 0.875 \end{bmatrix}$$

$$\{F^{(4)}\} = \begin{Bmatrix} P_1^{(4)} \\ P_2^{(4)} \end{Bmatrix} + \frac{1}{8}\begin{Bmatrix} 1 \\ 1 \end{Bmatrix} + \frac{1}{24}\begin{Bmatrix} 10 \\ 11 \end{Bmatrix}$$

The assembled matrix equation is given by

$$[K]\{U\} = \{F\} \tag{3.58a}$$

where

$$[K] = \begin{bmatrix} 4.5 & -4.5 & 0 & 0 & 0 \\ -4.5 & 4.5 + 5.5 & -5.5 & 0 & 0 \\ 0 & -5.5 & 5.5 + 6.5 & -6.5 & 0 \\ 0 & 0 & -6.5 & 6.5 + 7.5 & -7.5 \\ 0 & 0 & 0 & -7.5 & 7.5 \end{bmatrix} \tag{3.58b}$$

$$\{F\} = \begin{Bmatrix} P_1^{(1)} \\ P_2^{(1)} + P_1^{(2)} \\ P_2^{(2)} + P_1^{(3)} \\ P_2^{(3)} + P_1^{(4)} \\ P_2^{(4)} \end{Bmatrix} + \begin{Bmatrix} 0.125 \\ 0.250 \\ 0.250 \\ 0.250 \\ 0.125 \end{Bmatrix} + \begin{Bmatrix} 0.04167 \\ 0.250 \\ 0.500 \\ 0.750 \\ 0.4583 \end{Bmatrix} \tag{3.58c}$$

Step 4 The boundary conditions (i.e., known or specified U_i and P_i) are

$$U_1 = U_5 = 0$$
$$P_2^{(1)} + P_1^{(2)} = 0 \qquad P_2^{(2)} + P_1^{(3)} = 0 \qquad P_2^{(3)} + P_1^{(4)} = 0 \tag{3.59}$$

The unknown secondary variables (corresponding to tensions for the cable problem) are

$$P_1^{(1)} = \left[-(1+x)\frac{du}{dx}\right]_{x=0} = -\frac{du}{dx}(0)$$

$$P_2^{(4)} = \left[(1+x)\frac{du}{dx}\right]_{x=1} = 2\frac{du}{dx}(1) \tag{3.60}$$

Note that at any given boundary point in the problem, we know either the primary variable U or the secondary variable P; at no point will we know in advance (i.e., before solving the problem) both variables.

Step 5 The solution of equations consists of two steps (the second step can be considered as a postcomputation): (1) solve for the unknown primary variables first and then (2) solve for the unknown secondary variables from the information gained in step (1). In the present problem, we solve for U_2, U_3, and U_4 by inverting the submatrix obtained by deleting the first row and column and the last row and column:

$$\begin{bmatrix} 10.0 & -5.5 & 0 \\ -5.5 & 12.0 & -6.5 \\ 0 & -6.5 & 14.0 \end{bmatrix} \begin{Bmatrix} U_2 \\ U_3 \\ U_4 \end{Bmatrix} = \begin{Bmatrix} 0.50 \\ 0.75 \\ 1.00 \end{Bmatrix} \tag{3.61}$$

Equations of the type (3.61) can be solved using any standard solution routines available on a computer. In the present case, it is not too tedious (algebraically) to solve by Cramer's rule:

$$\begin{aligned} U_2 &= \frac{0.5(12 \times 14 - 6.5 \times 6.5) + 5.5(0.75 \times 14 + 6.5 \times 1)}{10(12 \times 14 - 6.5 \times 6.5) + 5.5(-5.5 \times 14)} \\ &= \frac{156.375}{834} = 0.1875 \\ U_3 &= \frac{10(0.75 \times 14 + 6.5 \times 1) + 5.5(0.5 \times 14 - 0)}{834} \\ &= \frac{208.5}{834} = 0.25 \\ U_4 &= \frac{10(12 \times 1 + 6.5 \times 0.75) + 5.5(-5.5 \times 1 + 6.5 \times 0.5)}{834} \\ &= \frac{156.375}{834} = 0.1875 \end{aligned} \tag{3.62}$$

The unknown secondary variables can be computed using the first and the last equations of Eq. (3.58):

$$4.5U_1 - 4.5U_2 = P_1^{(1)} + 0.125 + 0.04167$$

$$-7.5U_4 + 7.5U_5 = P_2^{(4)} + 0.125 + 0.4583\,.$$

or

$$\begin{aligned} P_1^{(1)} &= -4.5U_2 - 0.16667 = -1.0104 \\ P_2^{(4)} &= -7.5U_4 - 0.5833 = -1.9896 \end{aligned} \tag{3.63}$$

For the cable problem, $P_1^{(1)}$ and $P_2^{(4)}$ are the reaction forces at the fixed ends (i.e., $x = 0, 1$) of the cable. It is easy to verify that the force equilibrium is satisfied:

$$\begin{aligned} P_1^{(1)} + P_2^{(4)} + \int_0^1 f(x)\,dx &= 0 \\ -1.0104 - 1.9896 + (1 + 2) &= 0 \end{aligned} \tag{3.64}$$

Step 6 If we wish to find the value of $u(x)$ at $x = 0.4$, we simply evaluate the finite-element approximation for u at $x = 0.4$:

$$u(x) = \begin{cases} 0.1875(4x) & \text{for } 0 \leqslant x \leqslant 0.25 \\ 0.1875(2 - 4x) + 0.25(4x - 1) & \text{for } 0.25 \leqslant x \leqslant 0.5 \\ 0.25(3 - 4x) + 0.1875(4x - 2) & \text{for } 0.5 \leqslant x \leqslant 0.75 \\ 0.1875(4 - 4x) & \text{for } 0.75 \leqslant x \leqslant 1 \end{cases} \tag{3.65}$$

$$\begin{aligned} u(0.4) &= U_2(2 - 4x)|_{x=0.4} + U_3(4x - 1)|_{x=0.4} \\ &= 0.4U_2 + 0.6U_3 = 0.225 \end{aligned}$$

The derivative du/dx at the same point is given by

$$\begin{aligned} \frac{du}{dx}(0.4) &= U_2(-4) + U_3(4) = (U_3 - U_2)4 \\ &= 0.25 \end{aligned}$$

The exact solution of Eq. (3.1), with the data and boundary conditions given in Eqs. (3.51), is given by

$$u_0(x) = x(1 - x) \qquad \frac{du_0}{dx}(x) = 1 - 2x \tag{3.66}$$

Note that the finite-element solution and the exact solution agree with each other at the finite-element nodes. However, at any point between nodes they differ because the exact solution varies quadratically, whereas the finite-element solution varies linearly between nodes (see Fig. 3.6). The derivative of the solution is elementwise constant and matches with the derivative of the exact solution only at element midpoints. The finite-element solution at $x = 0.4$ differs from the exact solution by 6.25 percent. However, as the number of elements is increased, the finite-element solution converges to the exact solution. The finite-element solution of two, four, eight, and ten elements is compared with the exact solution in Table 3.2 (page 93). ■

Example 3.2 (One-dimensional conductive/convective heat transfer) Consider the one-dimensional heat transfer in an insulated (i.e., no transfer of heat through

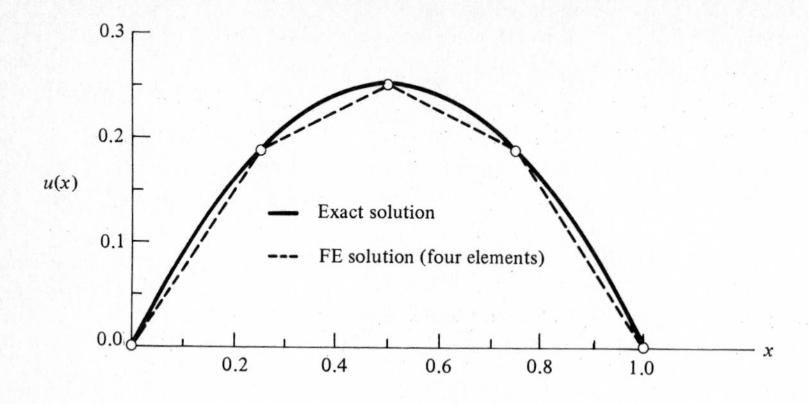

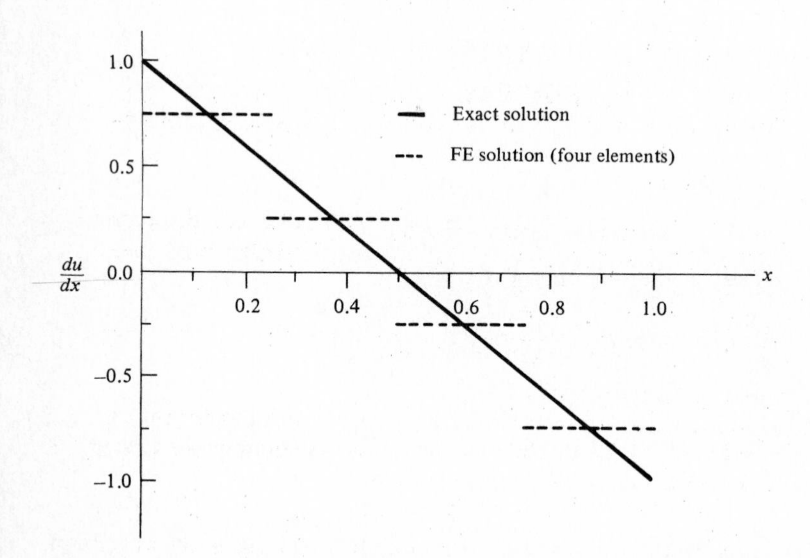

Figure 3.6 Comparison of the finite-element solution with the exact solution of Eq. (3.1) and Eqs. (3.51).

the surface) rod of cross-sectional area A, length L, conductivity k, convection coefficient β, and surrounding media (ambient) temperature T_∞. The governing differential equation is given by

$$-\frac{d}{dx}\left(kA\frac{dT}{dx}\right) = f \qquad 0 < x < L \tag{3.67}$$

where f is the distributed heat source (in most problems, it is zero). The boundary conditions are

$$T(0) = T_0 \qquad \left[kA\frac{dT}{dx} + \beta(T - T_\infty) + \hat{q}\right]_{x=L} = 0 \tag{3.68}$$

Table 3.2 Comparison of the finite-element solution with the exact solution of Eqs. (3.1) and (3.51)

x	Finite-element solution u				Exact solution
	2 elements	4 elements	8 elements	10 elements†	u_0
0.0	0.00	0.000	0.0	0.00	0.00
0.1	0.05	0.075	0.0875	0.09	0.09
0.2	0.10	0.150	0.15625	0.16	0.16
0.3	0.15	0.200	0.20625	0.21	0.21
0.4	0.20	0.225	0.2375	0.24	0.24
0.5	0.25	0.250	0.250	0.25	0.25
0.6	0.20	0.225	0.2375	0.24	0.24
0.7	0.15	0.200	0.20625	0.21	0.21
0.8	0.10	0.150	0.15625	0.16	0.16
0.9	0.05	0.075	0.0875	0.09	0.09
1.0	0.00	0.000	0.0	0.00	0.00

† The 10-element solution differs from u_0 at points between x_e and x_{e+1} ($e = 1, 2, \ldots, 10$).

The second boundary condition in Eqs. (3.68) accounts for the convective heat transfer through the end, $x = L$, and $\hat{q}$ is the specified heat flux there.

Suppose that the data are discontinuous:

$$a \equiv kA = \begin{cases} a_1 & 0 \leqslant x \leqslant L_1 \\ a_2 & L_1 \leqslant x \leqslant L \end{cases} \tag{3.69}$$

In other words, either the cross-sectional area (see Fig. 3.7*b*) and/or the material is different (see Fig. 3.7*a*) in two portions of the rod. We now proceed with the finite-element analysis of Eqs. (3.67) through (3.69).

Step 1 The data in Eq. (3.69) require us to divide the length of the rod into at least two elements. Of course, one can divide the rod into more than two elements.

Step 2 The variational formulation of Eq. (3.67), with boundary conditions (3.68), over an element $\Omega^e = (x_A, x_B)$ is given by

$$0 = \int_{x_A}^{x_B} \left(a \frac{dv}{dx} \frac{dT}{dx} - vf \right) dx + \left[v\left(-a \frac{dT}{dx} \right) \right]_{x_A}^{x_B} \tag{3.70a}$$

$$= \int_{x_A}^{x_B} \left(a \frac{dv}{dx} \frac{dT}{dx} - vf \right) dx + \{ v[\beta(T - T_\infty) + \hat{q}] \}_{x_A}^{x_B}$$

$$= \int_{x_A}^{x_B} \left(a \frac{dv}{dx} \frac{dT}{dx} - vf \right) dx - \beta v(x_A) T(x_A) + \beta v(x_B) T(x_B)$$

$$- T_\infty \beta [v(x_B) - v(x_A)] - v(x_A)\hat{q}(x_A) + v(x_B)\hat{q}(x_B) \tag{3.70b}$$

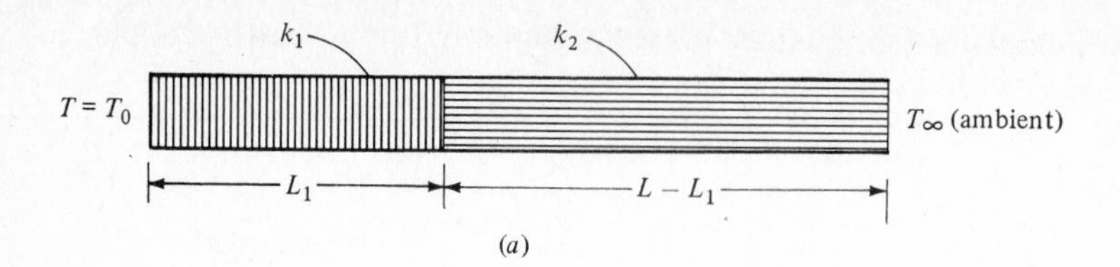

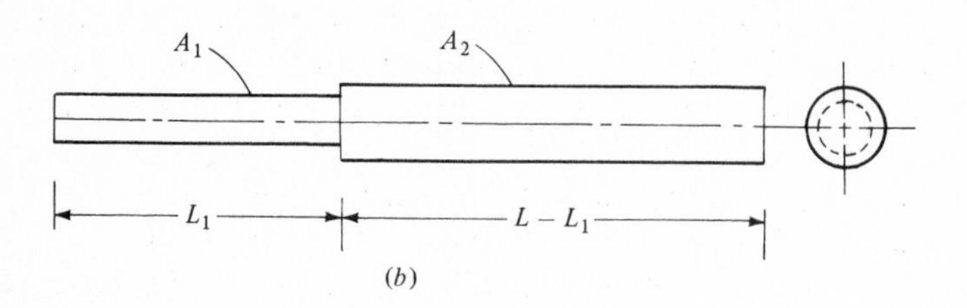

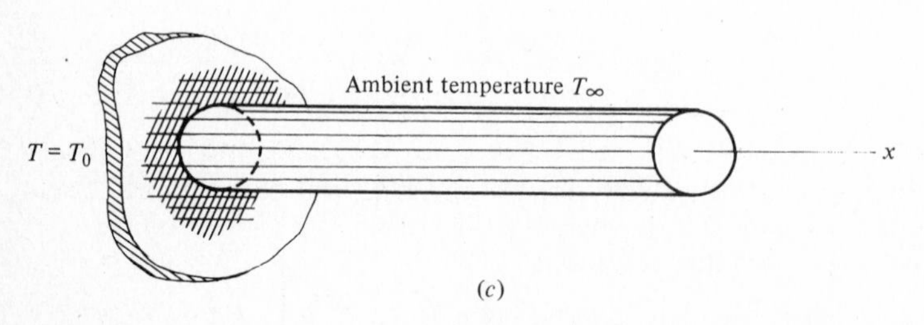

Figure 3.7 One-dimensional heat flow (conduction and convection). (*a*) Heat transfer in a dissimilar-material (composite) insulated rod. (*b*) Heat transfer in a stepped insulated rod. (*c*) Heat transfer in an uninsulated rod.

where v is a test function. Note that the $P_i^{(e)}$ in Eq. (3.53) are replaced by

$$
\begin{aligned}
P_1^{(e)} &= -\left(a\frac{dT}{dx}\right)\bigg|_{x=x_A} = [\beta(T - T_\infty) + q]_{x=x_A} \\
P_2^{(e)} &= \left(a\frac{dT}{dx}\right)\bigg|_{x=x_B} = -[\beta(T - T_\infty) + q]_{x=x_B}
\end{aligned}
\tag{3.71}
$$

Of course, one can keep $P_i^{(e)}$ till the assembly is completed and the boundary conditions are imposed on the global system of equations. Then the $P_i^{(e)}$ can be replaced by the expressions in Eqs. (3.71) (see step 3 below).

From an examination of the boundary term in Eq. (3.70*a*), we note that the essential boundary conditions involve the specification of T at the boundary points; hence, T is the primary variable (or the *degree of freedom*) at the finite-element nodes. Since the variational form (3.70*b*) involves only the first derivative of T, the linear interpolation functions in Eqs. (3.17*b*) are acceptable. Substituting Eq. (3.17*b*) for T and $v = \psi_i$ into the variational form (3.70*b*), we obtain the following element equations (for $f = 0$, $x_A = x_e$, and $x_B = x_{e+1}$):

$$0 = \sum_{j=1}^{2} \left\{ \int_{x_e}^{x_{e+1}} a \frac{d\psi_i}{dx} \frac{d\psi_j}{dx} dx + \beta\left[\psi_i(x_{e+1})\psi_j(x_{e+1}) - \psi_i(x_e)\psi_j(x_e)\right] \right\} T_j^{(e)}$$

$$- \{\beta T_\infty[\psi_i(x_{e+1}) - \psi_i(x_e)] + \psi_i(x_e)q(x_e) - \psi_i(x_{e+1})q(x_{e+1})\}$$

$$= \sum_{j=1}^{2} K_{ij}^{(e)} T_j^{(e)} - F_i^{(e)}$$

or

$$[K^{(e)}]\{T^{(e)}\} = \{F^{(e)}\} \tag{3.72}$$

where

$$K_{ij}^{(e)} = \int_{x_e}^{x_{e+1}} a \frac{d\psi_i}{dx} \frac{d\psi_j}{dx} dx + \beta\left[\psi_i(x_{e+1})\psi_j(x_{e+1}) - \psi_i(x_e)\psi_j(x_e)\right]$$

$$F_i^{(e)} = \beta T_\infty\left[\psi_i(x_{e+1}) - \psi_i(x_e)\right] + \psi_i(x_e)q(x_e) - \psi_i(x_{e+1})q(x_{e+1}) \tag{3.73}$$

Equations (3.72) and (3.73) are valid for a typical element; one can use any admissible ψ_i for evaluating K_{ij} and F_i. When the linear interpolation functions are used, we obtain

$$[K^{(e)}] = \frac{a_e}{h_e}\begin{bmatrix} 1 & -1 \\ -1 & 1 \end{bmatrix} + \begin{bmatrix} -\beta_1^{(e)} & 0 \\ 0 & \beta_2^{(e)} \end{bmatrix}$$

$$\{F^{(e)}\} = T_\infty \begin{Bmatrix} -\beta_1^{(e)} \\ \beta_2^{(e)} \end{Bmatrix} + \begin{Bmatrix} -q_1^{(e)} \\ q_2^{(e)} \end{Bmatrix} \tag{3.74}$$

where $\beta_i^{(e)}$ and $q_i^{(e)}$ denote the film coefficient and the heat flux, respectively, at node i of element e.

Step 3 Before we assemble the element equations, we choose specific values of the parameters:

$$a_1 = 76\ (\text{W} \cdot \text{cm})/°\text{C} \qquad a_2 = 96\ (\text{W} \cdot \text{cm})/°\text{C}$$

$$L_1 = 4\ \text{cm} \qquad L = 10\ \text{cm} \qquad \beta_2^{(2)} = 10\ \text{W/cm}^2 \qquad T_\infty = 30°\text{C} \tag{3.75}$$

$$T_0 = 100°\text{C} \qquad \beta_1^{(1)} = 0.0 \qquad \beta_2^{(1)} = \beta_1^{(2)}\ (= 10\ \text{W/cm}^2)$$

Note that the film coefficients between any two elements need not be given because the convective contribution at any interelement node is assumed to be zero:

$$T_\infty \beta_2^{(e)} - T_\infty \beta_1^{(e+1)} = 0$$

Of course, this is only a mathematical manipulation that is employed to handle the convective terms at the nodes between the boundary nodes. In reality, there is no contribution of the convective terms to nodes other than the boundary nodes.

Returning to the calculation of the element matrices, we have

$$[K^{(1)}] = 19\begin{bmatrix} 1 & -1 \\ -1 & 1 \end{bmatrix} + 10\begin{bmatrix} 0 & 0 \\ 0 & 1 \end{bmatrix} = \begin{bmatrix} 19 & -19 \\ -19 & 29 \end{bmatrix}$$

$$[K^{(2)}] = 16\begin{bmatrix} 1 & -1 \\ -1 & 1 \end{bmatrix} + 10\begin{bmatrix} -1 & 0 \\ 0 & 1 \end{bmatrix} = \begin{bmatrix} 6 & -16 \\ -16 & 26 \end{bmatrix}$$

$$\{F^{(1)}\} = \begin{Bmatrix} -\beta_1^{(1)}T_\infty \\ \beta_2^{(1)}T_\infty \end{Bmatrix} + \begin{Bmatrix} -q_1^{(1)} \\ q_2^{(1)} \end{Bmatrix} \qquad \{F^{(2)}\} = \begin{Bmatrix} -\beta_1^{(2)}T_\infty \\ 300 \end{Bmatrix} + \begin{Bmatrix} -q_1^{(2)} \\ q_2^{(2)} \end{Bmatrix}$$

and the assembled matrix equations are given by

$$\begin{bmatrix} 19 & -19 & 0 \\ -19 & 29+6 & -16 \\ 0 & -16 & 26 \end{bmatrix}\begin{Bmatrix} U_1 \\ U_2 \\ U_3 \end{Bmatrix} = \begin{Bmatrix} -\beta_1^{(1)}T_\infty \\ 0 \\ 300 \end{Bmatrix} + \begin{Bmatrix} -q_1^{(1)} \\ q_2^{(1)} - q_1^{(2)} \\ q_2^{(2)} \end{Bmatrix} \tag{3.76a}$$

Alternatively, if we carry the $P_i^{(e)}$ in the element equations, the assembled equations take the following equivalent form:

$$\begin{bmatrix} 19 & -19 & 0 \\ -19 & 19+16 & -16 \\ 0 & -16 & 16 \end{bmatrix}\begin{Bmatrix} U_1 \\ U_2 \\ U_3 \end{Bmatrix} = \begin{Bmatrix} P_1^{(1)} \\ P_2^{(1)} + P_1^{(2)} \\ P_2^{(2)} \end{Bmatrix} = \begin{Bmatrix} P_1^{(1)} \\ 0 \\ -10U_3 + 300 \end{Bmatrix} \tag{3.76b}$$

where $P_2^{(2)} = -[\beta(T(x) - T_\infty)]|_{x=10}$ is substituted into the last equation. Note that Eq. (3.76b) is the same as Eq. (3.76a) with $P_1^{(1)} = -q_1^{(1)}$ and $q_2^{(2)} \equiv \hat{q} = 0$.

Step 4 The specified boundary conditions and the equilibrium of internal heat flux give

$$U_1 = 100°\text{C} \qquad q_2^{(1)} - q_1^{(2)} = 0 \qquad q_2^{(2)} = \hat{q} = 0 \tag{3.77}$$

The heat flux at $x = 0$ is unknown:

$$q_1^{(1)} = \left[-a\frac{dT}{dx}\right]_{x=0} \tag{3.78}$$

Step 5 After imposing the boundary conditions (3.77) on Eqs. (3.76), we get

$$\begin{bmatrix} 35 & -16 \\ -16 & 26 \end{bmatrix}\begin{Bmatrix} U_2 \\ U_3 \end{Bmatrix} = \begin{Bmatrix} 19U_1 \\ 300 \end{Bmatrix} = \begin{Bmatrix} 1900 \\ 300 \end{Bmatrix}$$

$$q_1^{(1)} = 19U_2 - 19U_1$$

Solving the matrix equation for U_2 and U_3, we obtain, using Cramer's rule,

$$U_2 = \frac{1900 \times 26 + 300 \times 16}{35 \times 26 - 16 \times 16} = 82.88°\text{C}$$
$$U_3 = \frac{35 \times 300 + 16 \times 1900}{654} = 62.54°\text{C} \tag{3.79}$$

The heat flux at $x = 0$ is given by

$$q_1^{(1)} = -325.28 \text{ W/cm}^2$$

The finite-element solution (in °C) is given by

$$T(x) = \begin{cases} 100\left(1 - \dfrac{x}{4}\right) + 82.88\left(\dfrac{x}{4}\right) & 0 \leqslant x \leqslant 4 \\ \quad = 100 - 4.28x & \\ 82.88\left(\dfrac{10 - x}{6}\right) + 62.54\left(\dfrac{x - 4}{6}\right) & 4 \leqslant x \leqslant 10 \\ \quad = 96.44 - 3.39x & \end{cases} \tag{3.80}$$

Step 6 The exact solution of Eqs. (3.67) through (3.69) coincides with the finite-element solution. ■

It should be pointed out that if the rod is *not* insulated, then the second boundary condition (3.68) also applies on the surface of the rod (see Fig. 3.7*c*). Consequently, the variational form (3.70) must be modified accordingly. We have

$$\begin{aligned} 0 &= \int_{x_A}^{x_B} \int_{\text{area}} \left(k \frac{dv}{dx} \frac{dT}{dx} - vQ\right) dA\, dx - \int_S v\left(a \frac{dT}{dx}\right) dS \\ &= \int_{x_A}^{x_B} \left(kA \frac{dv}{dx} \frac{dT}{dx} - vf\right) dx + \int_S v[\beta(T - T_\infty) + q]\, dS \end{aligned} \tag{3.81}$$

where $A = A(x)$ is the area of cross section of the member, $f = AQ$ is the heat generation per unit length, and S denotes the surface of the rod. In order to simplify the surface integral, it is necessary (without the knowledge of surface elements) to assume that the area of cross section is constant within each element. (This assumption is not necessary when the rod is insulated.) Then we have

$$0 = \int_{x_A}^{x_B} \left(a \frac{dv}{dx} \frac{dT}{dx} - vf\right) dx + \int_{x_A}^{x_B} cv[\beta(T - T_\infty) + q]\, dx \tag{3.82}$$

where c is the perimeter.

The finite-element model of Eq. (3.82) is given by

$$[K^{(e)} + H^{(e)}]\{T\} = \{F^{(e)} + Q^{(e)}\} \tag{3.83a}$$

where

$$K_{ij}^{(e)} = \int_{x_e}^{x_{e+1}} a \frac{d\psi_i^{(e)}}{dx} \frac{d\psi_j^{(e)}}{dx} dx \qquad H_{ij}^{(e)} = c\beta \int_{x_e}^{x_{e+1}} \psi_i^{(e)} \psi_j^{(e)} dx$$
$$F_i^{(e)} = \int_{x_e}^{x_{e+1}} \psi_i^{(e)} f^{(e)} dx \qquad Q_i^{(e)} = \int_{x_e}^{x_{e+1}} c(\beta T_\infty - q)\psi_i^{(e)} dx \tag{3.83b}$$

Using the linear interpolation functions, we obtain (for constant values of a, f, and q)

$$[K^{(e)}] = \frac{a_e}{h_e}\begin{bmatrix} 1 & -1 \\ -1 & 1 \end{bmatrix} \qquad [H^{(e)}] = \frac{c\beta h_e}{6}\begin{bmatrix} 2 & 1 \\ 1 & 2 \end{bmatrix}$$
$$\{F^{(e)}\} = \frac{fh_e}{2}\begin{Bmatrix} 1 \\ 1 \end{Bmatrix} \qquad \{Q^{(e)}\} = c(\beta T_\infty - q)\frac{h_e}{2}\begin{Bmatrix} 1 \\ 1 \end{Bmatrix} \tag{3.84}$$

Some comments are in order on higher-order elements for the solution of one-dimensional second-order equations of the type of Eq. (3.1). Note that Eqs. (3.4) through (3.8) are valid for any order element. One needs to substitute the proper integration limits x_A and x_B and interpolation functions to evaluate $[K^{(e)}]$ and $\{F^{(e)}\}$. For higher-order (than the linear) elements, additional nodes are identified intermediate to the end nodes. The variational formulation, hence the column vector $\{F^{(e)}\}$, should be modified to account for the possible specification of the secondary variable ($a\, du/dx$) at the intermediate nodes. For an element with n nodes (hence the interpolation polynomials are of degree $n - 1$), the variational formulation (3.4a) takes the form

$$0 = \int_{x_A}^{x_B} \left(a \frac{dv}{dx}\frac{du}{dx} - vf \right) dx - P_1^{(e)} v(x_A) - P_2^{(e)} v(x_A + \bar{x}_2) - \cdots - P_n^{(e)} v(x_B) \tag{3.85a}$$

where $\bar{x}_i$ is the local coordinate of the ith node in the element ($\bar{x}_1 = 0$). The element column vector $\{F^{(e)}\}$ takes the form

$$F_i^{(e)} = \int_{x_A}^{x_B} \psi_i^{(e)} f\, dx + P_1^{(e)} \psi_i^{(e)}(\bar{x}_1 + x_A) + \cdots + P_n^{(e)} \psi_i^{(e)}(\bar{x}_n + x_A)$$

Due to the interpolation property of $\psi_i^{(e)}$ ($i = 1, 2, \ldots, n$)

$$\psi_i^{(e)}(\bar{x}_j + x_A) = \delta_{ij}$$

the force vector can be simplified to the form

$$F_i^{(e)} = \int_{x_A}^{x_B} \psi_i^{(e)} f\, dx + P_i^{(e)} \tag{3.85b}$$

We will make use of Eq. (3.85b) in Prob. 3.4.

The variational formulation used in the preceding sections was based on the original differential equation of second order. These formulations involve only the dependent variable of the problem as the primary variable. Such formulations are known in the finite-element literature as the *displacement formulations*, and the associated finite-element models are called the *displacement models*. Other formulations of the original second-order equations are possible. These alternative formulations involve decomposing the given second-order equation into a pair of first-order equations, and the use of other variational methods, such as the least-squares method, to construct the finite-element model. Some discussion of these alternative models is given in Chap. 5.

A reader (or instructor) who is eager to implement the method on a digital computer/calculator can do so at this time without going through Secs. 3.3 through 3.5. In fact, it is suggested that Secs. 3.6 and 3.7 be studied to get a feel for the power of the finite-element method. Some of the following problems can be solved both by hand and by calculator/computer.

PROBLEMS

Many of the following problems are designed for hand calculation. This is believed to give the student deeper understanding of what is involved in the formulation and solution of a problem by the finite-element method. The hand calculations can be verified, in most cases, by solving the same problems with the computer program FEM1D, which is described in Sec. 3.7.

3.1 Derive the Lagrange linear interpolation functions (3.17*b*) by the following alternative procedure. Start with ψ_1 and express it as a function that vanishes at $\bar{x} = h_e$ ($\bar{x}$ is the local coordinate):

$$\psi_1(\bar{x}) = c(h_e - \bar{x}) \qquad c = \text{constant}$$

Then find the constant c such that $\psi_1(0) = 1$. The function ψ_1 thus derived satisfies condition (1) of Eq. (3.13). Repeat the same procedure for ψ_2. Verify that condition (2) of Eq. (3.13) is also satisfied by the ψ_i (see Sec. 3.6).

3.2 Derive the Lagrange quadratic interpolation functions for a three-node (one-dimensional) element (with equally spaced nodes) using the procedure described in Prob. 3.1, and verify that they satisfy condition 2 in Eq. (3.13). Note that the Lagrange interpolation function at node i vanishes at the other two nodes and takes the value of unity at node i. Use the local coordinate $\bar{x}$ for simplicity.

3.3 Give the finite-element formulation of the following equation over an element:

$$-\frac{d}{dx}\left(a\frac{du}{dx}\right) + b\frac{du}{dx} + cu = f \qquad \text{for } x_A < x < x_B$$

3.4 Use the quadratic interpolation functions derived in Prob. 3.2 to solve Eqs. (3.1) and (3.2). Use two elements. Compare the solution with that obtained using four linear elements and with the exact solution. Use $a = 1, f = x$, and $P = 0$.

3.5 Solve the differential equation

$$-\frac{d^2u}{dx^2} - cu + x^2 = 0 \qquad \text{for } 0 < x < 1$$

for the following three sets of boundary conditions:

(*a*)	$u(0) = u(1) = 0$	Dirichlet boundary conditions
(*b*)	$u(0) = 0, \dfrac{du}{dx}(1) = 1$	mixed boundary conditions
(*c*)	$\dfrac{du}{dx}(0) = 1, \dfrac{du}{dx}(1) = \dfrac{4}{3}$	Neumann boundary conditions

Note that for Neumann boundary conditions, none of the primary dependent variables is specified, and therefore the coefficient matrix remains unaltered. When $c = 0$, the coefficient matrix is singular and cannot be inverted. In such cases, one of the U_i (say, U_2) should be set to a constant (e.g., zero) to remove the "rigid-body" mode (i.e., to determine the arbitrary constant in the solution). Use three linear finite elements, and compare your answer with that given in Example 2.8 for $c = 1$. Also, determine du/dx at $x = 0$.

3.6 Consider the composite structure of axially loaded members shown in Fig. P3.6. Write the continuity conditions (i.e., the correspondence of element nodal values to global nodal values) and the equilibrium conditions (i.e., the relationship between $P_i^{(e)}$ at the interelement nodes) for the structure. Derive the assembled coefficient (stiffness) matrix for the structure, and set up the equations for the unknown displacements and forces.

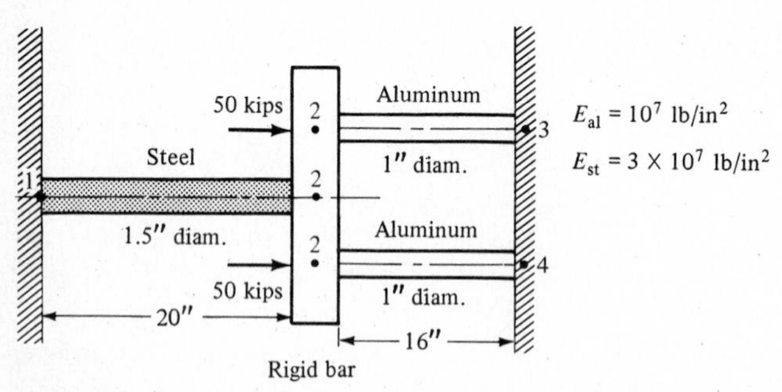

Figure P3.6

3.7 Consider the system of linear elastic springs shown in Fig. P3.7. Calculate the displacements U_2, U_3, U_4, and U_5 and the forces in the springs. [*Hint*: Use the free-body diagrams of the blocks to write the force equilibrium equations analogous to Eq. (3.30)].

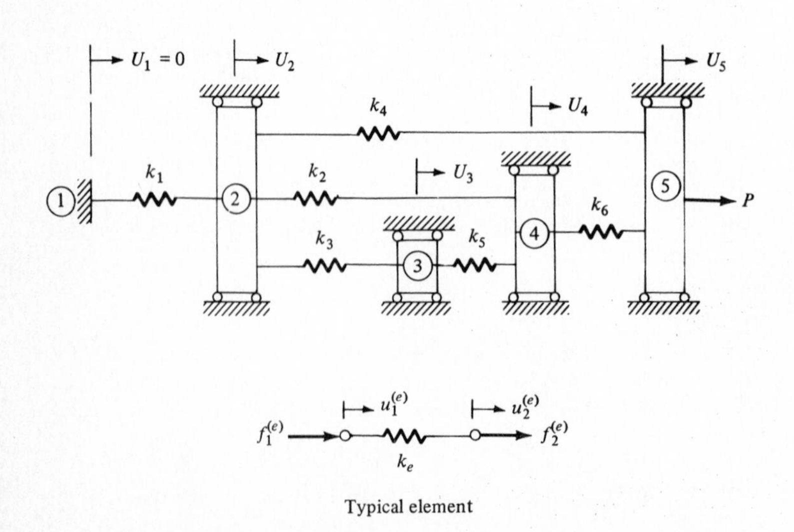

Figure P3.7

3.8 Consider the hydraulic network (the flow is assumed to be laminar) shown in Fig. P3.8. A typical element (which is a circular pipe of constant cross-sectional area) with two nodes is also shown in the figure. The unknown primary degree of freedom at each node is the pressure P, and the secondary degree of freedom is the flow (or discharge) Q. The element equations relating the primary variables to the secondary variables is given by

$$\frac{c(d_e)^4}{h_e}\begin{bmatrix} 1 & -1 \\ -1 & 1 \end{bmatrix}\begin{Bmatrix} P_1^{(e)} \\ P_2^{(e)} \end{Bmatrix} = \begin{Bmatrix} Q_1^{(e)} \\ Q_2^{(e)} \end{Bmatrix} \qquad c = \frac{\pi}{128\mu}$$

where d_e is the diameter of the pipe, h_e is the length of the pipe, and μ is the viscosity of the fluid. Determine the unknown pressures and flows using the minimum number of elements.

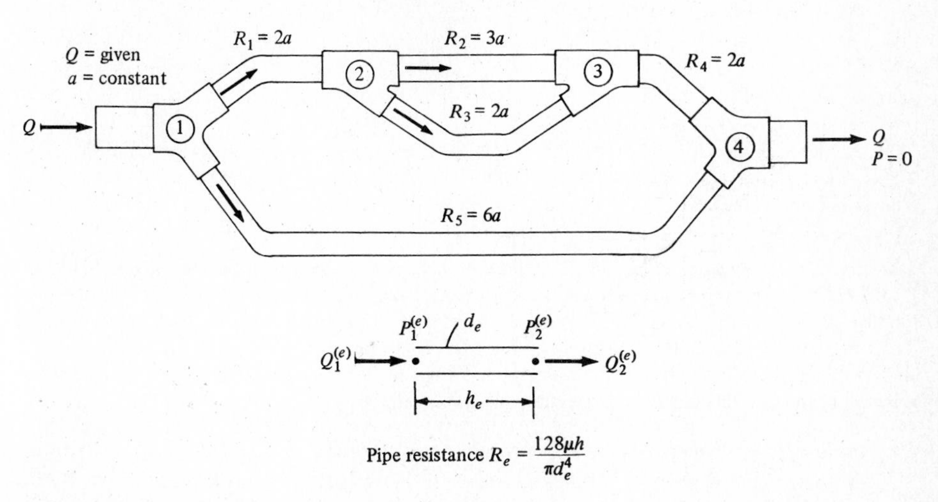

Figure P3.8

3.9 Consider the direct-current electric network shown in Fig. P3.9. We wish to determine the voltages V and currents I in the network using the finite-element method. A typical finite element in this case

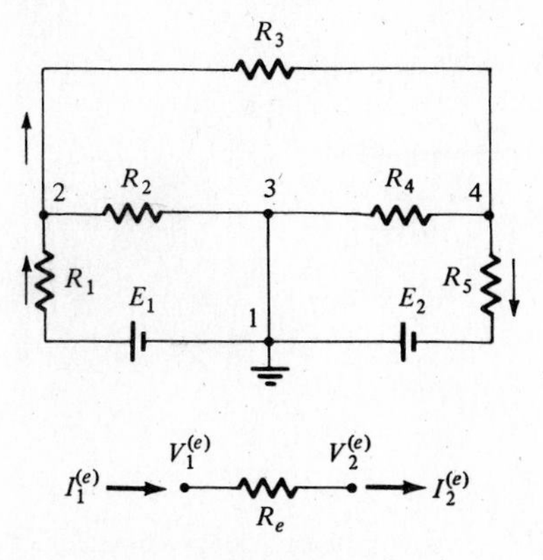

Figure P3.9

consists of a resistor R, with the primary degree of freedom being the voltage and the secondary degree of freedom being the current. The element equations are provided by Ohm's law:

$$\frac{1}{R_e}\begin{bmatrix} 1 & -1 \\ -1 & 1 \end{bmatrix}\begin{Bmatrix} V_1^{(e)} \\ V_2^{(e)} \end{Bmatrix} = \begin{Bmatrix} I_1^{(e)} \\ I_2^{(e)} \end{Bmatrix}$$

The continuity conditions at the interelement nodes require that the net current flow into any junction (node) be always zero in a closed loop. Determine the unknown voltages and currents.

3.10 The following differential equation arises in connection with the heat transfer in an insulated rod:

$$-\frac{d}{dx}\left(kA\frac{dT}{dx}\right) = Q \qquad \text{for } 0 < x < L$$

$$T(0) = T_0 \qquad \left[kA\frac{dT}{dx} + \beta(T - T_\infty) + \hat{q}\right]_{x=L} = 0$$

The variables have the same meaning as in Example 3.2. Solve the equation for temperature values at $x = L/2$ and L. Take the following values for the data, and use linear finite elements: $\hat{q} = 0$, $L = 10$ cm, $kA = 1.0$ (W · cm)/K, $\beta = 25$ W/(cm^2 · K), $T_0 = 50$°C, $Q = 0$, $T_\infty = 5$°C.

3.11 Repeat Prob. 3.10 when the rod is not insulated.

3.12 An insulating wall is constructed of three homogeneous layers in intimate contact (see Fig. P3.12). In the steady-state condition the temperatures at the boundaries of the layers are characterized by the external surface temperatures T_1 and T_4 and the interface temperatures T_2 and T_3. Formulate the problem to determine the temperatures T_i ($i = 1, 2, 3, 4$) when the ambient temperatures T_0 and T_5 and the (surface) film coefficients β_0 and β_i are known. Assume that there is no internal heat generation, and that the heat flow is one-dimensional ($\partial T/\partial y = 0$).

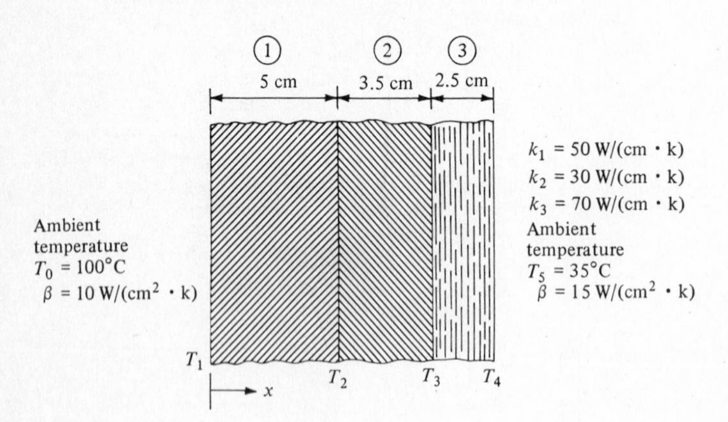

Figure P3.12

3.13 Repeat Prob. 3.10 for the case in which the area A is given by

$$A = A_0\left(1 + \frac{x}{L}\right)$$

3.14 Find the three-element finite-element solution to the stepped-bar problem (axial deformation of a bar). See Fig. P3.14 for the geometry and data.

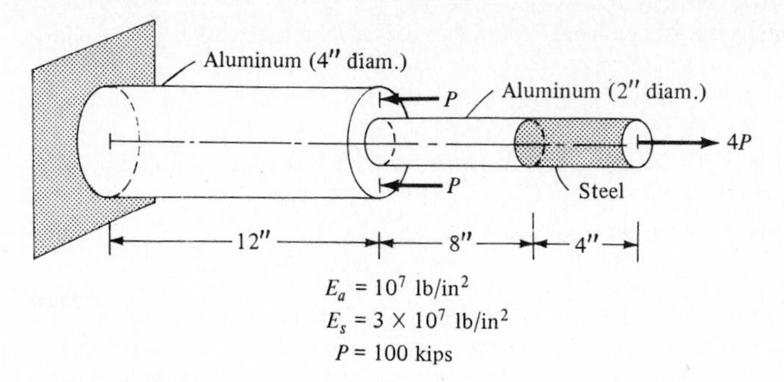

Figure P3.14

3.15 The differential equation governing (radially) symmetric two-dimensional field problems is given by

$$-\frac{1}{r}\frac{d}{dr}\left(rk\frac{du}{dr}\right) = f(r) \qquad \text{for } 0 < r < L$$

where r is the radial coordinate. Equations of this type arise in all fields listed in Table 2.1 when the domain is circular (or annular) and radial symmetry exists. Derive the finite-element model of the above equation such that the following two types of boundary conditions are accounted for:

$$u = u_0 \qquad k\frac{du}{dr}n_r + \beta(u - u_\infty) + \hat{q} = 0$$

Note that the integration is with respect to the polar coordinate r. (Hence, multiply the equation with $r\,dr$ and integrate from 0 to L.)

3.16 The governing equation for an unconfined aquifer with flow in the radial direction is given by the differential equation in Prob. 3.15, where k denotes the coefficient of permeability, f the recharge, and u is the piezometric head. Pumping is considered to be a negative recharge. Now consider the following problem. A well penetrates an aquifer and pumping is done at a rate of $Q = 150$ m^3/h. The permeability of the aquifer is $k = 25$ m^3/(h $\cdot$ m^2). The aquifer is unconfined, and radial symmetry exists in the flow field (with the origin of the radial coordinate being at the pump). A constant head of $u_0 = 50$ m exists at a radial distance of $L = 200$ m. Determine the piezometric head at radial distances (nonuniform mesh of linear elements) of 0, 10, 20, 40, 80, and 140 m (see Fig. P3.16). You are required only to set up the matrix equations for the unknowns (so that the equations can be solved on a digital computer).

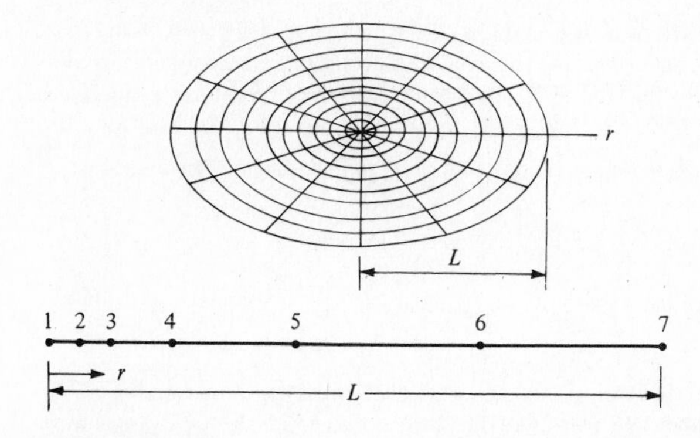

Figure P3.16

3.17 The equation governing the one-dimensional flow of an inviscid fluid is given by

$$-\frac{d}{dx}\left(\rho A \frac{d\phi}{dx}\right) = 0 \qquad 0 < x < L$$

where ϕ is the velocity potential, ρ the density, and A the cross-sectional area of the pipe. The velocity u is related to the velocity potential ϕ by

$$u = \frac{d\phi}{dx}$$

Note that constant terms in ϕ do not contribute to the velocity field. Determine the velocity field in a variable-cross-section pipe of length L using two quadratic elements. Take the inlet velocity to be u_0 and assume that the velocity potential is zero at the exit (to eliminate the constant term in the velocity potential). The area of cross section is

$$A = A_0\left[2 - \left(\frac{x}{L}\right)\right]$$

3.18 Repeat Prob. 3.17 with the following approximation and three linear elements (see Fig. P3.18):

$$A^{(e)} = \tfrac{1}{2}[A(x_e) + A(x_{e+1})] \qquad e = 1,2,3,4$$

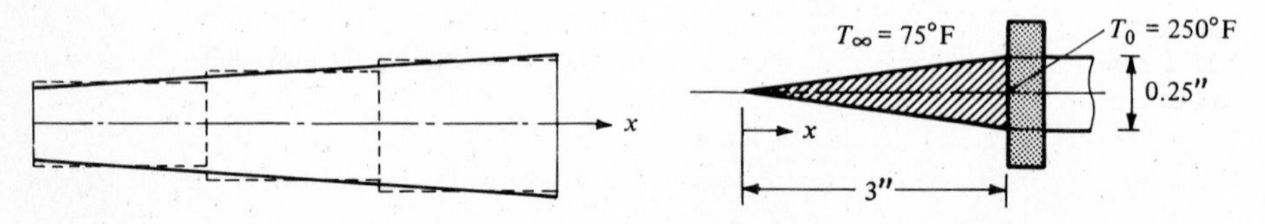

Figure P3.18 **Figure P3.19**

3.19 Find the temperature distribution in the tapered fin shown in Fig. P3.19. Assume that the temperature at the root of the fin is 250°F, the conductivity $k = 120$ Btu/(h · ft · °F), and the film coefficient $\beta = 15$ Btu/(h · ft^2 · °F), and use three linear elements. (The ambient temperature at the top and bottom of the fin is $T_\infty = 75$°F.)

3.20 The equation governing the axial deformation of an elastic bar in the presence of applied mechanical loads f and P and the temperature change T is given by

$$-\frac{d}{dx}\left[EA\left(\frac{du}{dx} - \alpha T\right)\right] = f \qquad 0 < x < L$$

where α is the thermal expansion coefficient, E the modulus of elasticity, and A the area of the cross section. Using three linear finite elements, determine the axial displacements in a nonuniform rod of length 30 in, fixed at the left end and subjected to an axial force of $P = 400$ lb and a temperature change of 60°F. Take $A(x) = 6 - x/10$ in^2, $E = 30 \times 10^6$ lb/in^2, and $\alpha = 12 \times 10^{-6}$ in/(in · °F).

3.21 Consider the following pair of differential equations governing the bending of elastic beams:

$$-\frac{d^2M}{dx^2} = f \qquad 0 < x < L$$

$$-\frac{d^2w}{dx^2} + \frac{M}{EI} = 0 \qquad 0 < x < L$$

where w is the transverse deflection, M the bending moment, f the distributed transverse loading, E the modulus of elasticity, and I the moment of inertia of the beam. Construct the variational form associated with the above pair of equations (do not eliminate the bending moment), and obtain the associated finite-element model for an element in the form

$$\begin{bmatrix} [K^{11}] & [K^{12}] \\ [K^{21}] & [K^{22}] \end{bmatrix} \begin{Bmatrix} \{w\} \\ \{M\} \end{Bmatrix} = \begin{Bmatrix} \{F^1\} \\ \{F^2\} \end{Bmatrix}$$

3.22 Solve the problem of the bending of a simply supported beam under uniformly distributed load. Assume that EI is constant and use two linear elements in the half beam.

3.23 Find the first three eigenvalues λ associated with the equation (use linear elements)

$$-\frac{d^2u}{dx^2} = \lambda u \qquad 0 < x < L$$

$$u(0) = u(L) = 0$$

3.24 Repeat Prob. 3.17 for an assemblage of three pipes, with $A_1 = A_0$, $A_2 = 1.5A_0$, and $A_3 = 2A_0$. Use linear elements.

3.25 Find the first two eigenvalues λ of the equation

$$-\frac{d^2u}{dx^2} = \lambda u \qquad 0 < x < 1$$

$$u(0) = 0 \qquad u(1) + \frac{du}{dx}(1) = 0$$

using linear elements.

3.26 Consider the partial differential equation arising in connection with the unsteady heat transfer in an insulated rod

$$\frac{\partial u}{\partial t} - \frac{\partial}{\partial x}\left(a\frac{\partial u}{\partial x}\right) = f \qquad 0 < x < L$$

$$u(0) = u_0 \qquad \left[a\frac{\partial u}{\partial x} + \beta(u - u_\infty) + \hat{q}\right]_{x=L} = 0$$

Following the procedure outlined in Chap. 2, derive the semidiscrete variational formulation, the semidiscrete finite-element formulation, and the fully discretized finite-element equations for a typical element.

3.27 Using a two-element (linear) model and the semidiscrete finite-element equations derived in Prob. 3.26, determine the nodal temperatures as functions of time for the case in which $a = 1, f = 0$, $u_0 = 1$, and $\hat{q} = 0$. (Use the Laplace transform technique to solve the ordinary differential equations in time.)

3.28 Consider the problem of steady flow in a channel with two parallel flat walls (see Fig. P3.28). Let the distance between the walls be denoted by $2b$. Then the equations governing Stoke's flow (see Prob.

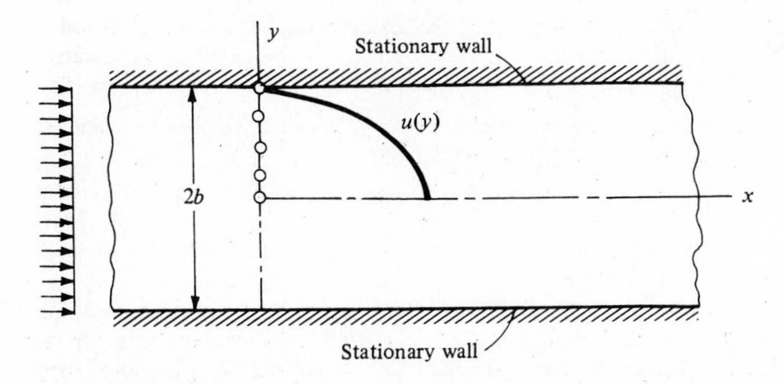

Figure P3.28

2.7) take the simple form [because $u = u(y)$, $v = 0$]

$$\frac{dP}{dx} = \mu \frac{d^2u}{dy^2} \tag{a}$$

with the boundary conditions

$$u = 0 \qquad \text{for } y = \pm b \tag{b}$$

Because $\partial P/\partial y = 0$, Eq. (a) implies that $\mu\, d^2u/dy^2$ is constant, and hence dP/dx is constant, say, equal to f_0. Solve Eqs. (a) and (b) using a four-element (linear) finite-element model. Compare the finite-element solution with the exact solution

$$u = -\frac{1}{2\mu} f_0 (b^2 - y^2)$$

3.29 Consider the problem of so-called Couette flow between two parallel flat walls, one of which is at rest, the other moving in its own plane with a velocity u_0 (see Fig. P3.29). The governing equation of the flow is given by Eq. (a) in Prob. 3.28. The boundary conditions of the problem are

$$u(0) = 0 \qquad u(2b) = u_0 = 1$$

Solve the problem for $f_0 = -3/b$ and $\mu = 1$ using a nonuniform mesh of four linear elements ($h_1 = 0.8b$, $h_2 = 0.5b$, $h_3 = 0.3b$, and $h_4 = 0.4b$), and compare the finite-element solution with the exact solution

$$u = \frac{u_0 y}{2b} - \frac{2b^2}{\mu} f_0 \frac{y}{2b}\left(1 - \frac{y}{2b}\right)$$

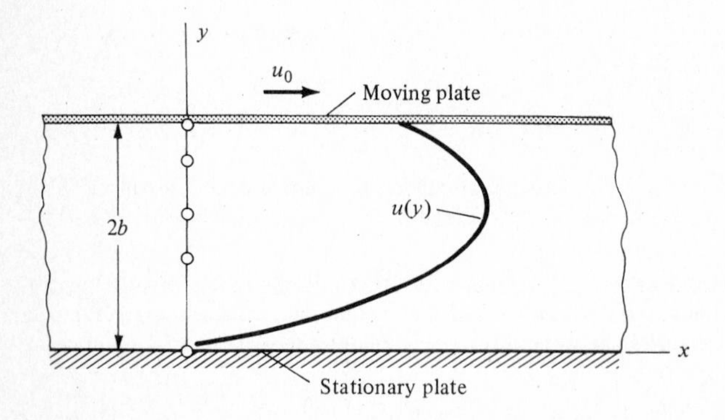

Figure P3.29

3.30 Consider a set of four linear (algebraic) simultaneous equations of the form

$$\sum_{j=1}^{4} K_{ij} U_j = F_i \qquad i = 1, 2, 3, 4$$

Modify and solve the equations for the following two cases:

(a) $U_1 = \alpha$.

(b) $U_1 = \alpha$, $U_2 = \beta$.

3.31 (Axisymmetric radial heat flow through a circular cylinder) Consider the axisymmetric radial heat flow by conduction through a homogeneous hollow circular cylinder of inside radius r_i and outside radius r_o. If the cylinder is sufficiently long, the end effects can be neglected. The governing equation is given by

$$-\frac{d}{dr}\left(kA\frac{dT}{dr}\right) = AQ \qquad \text{for } r_i < r < r_o$$

where k is the thermal conductivity, $A = 2\pi rl$ the surface area, l the length of the cylinder, Q the heat generation per unit area, r the radial coordinate, and T the temperature. The heat flux is given by

$$q = -kA\frac{dT}{dr}$$

Using the data $r_o = 3.25$ in, $r_i = 1.75$ in, $l = 10$ ft, $Q = 0$, $k = 0.04$ Btu/(h · ft · °F), $T_i = 400$°F, $T_o = 80$°F, and three linear elements, determine the temperature distribution in the cylinder. Compare the finite-element solution with the exact solution

$$T = T_i - (T_i - T_o)\frac{\ln(r/r_i)}{\ln(r_o/r_i)}$$

where T_i and T_o are the specified temperatures at the inner and outer walls, respectively.

3.32 (Elastic, axisymmetric deformation of cylinders and disks) Consider the axisymmetric deformation of a long hollow cylinder (plane strain) or a thin annular disk (plane stress). The equilibrium in terms of the normal (radial) stress σ_r and the tangential (hoop) stress σ_θ is given by

$$\frac{d}{dr}[t\sigma_r] + \frac{t}{r}(\sigma_r - \sigma_\theta) + tf_r = 0 \qquad \text{for } r_i < r < r_o \tag{i}$$

where $t = t(r)$ is the thickness of the disk and f_r is the body force. The constitutive (i.e., stress-strain) relations are given by

$$\sigma_r = C_{11}\varepsilon_r + C_{12}\varepsilon_\theta \qquad \sigma_\theta = C_{12}\varepsilon_r + C_{22}\varepsilon_\theta \tag{ii}$$

where ε_r and ε_θ are the strains and C_{ij} are the elastic constants. For the axisymmetric case, the strain-displacement relations are given by

$$\varepsilon_r = \frac{du}{dr} \qquad \varepsilon_\theta = \frac{u}{r} \tag{iii}$$

For an isotropic body, the elastic constants are given in terms of the modulus of elasticity E and Poisson's ratio ν by

	C_{11}	C_{12}	C_{22}
Plane strain	$\dfrac{(1-\nu)E}{(1+\nu)(1-2\nu)}$	$\dfrac{\nu E}{(1+\nu)(1-2\nu)}$	C_{11}
Plane stress	$\dfrac{E}{1-\nu^2}$	$\dfrac{\nu E}{1-\nu^2}$	C_{11}

Using (ii) and (iii) in (i), multiplying the result with rv (v is a test function), and integrating once with respect to r, obtain the variational formulation and the associated finite-element model of Eqs. (i) to (iii) in terms of the radial displacement. Use the finite-element model to find a three-element solution to the problem of a constant-thickness ring clamped (i.e., $u = 0$) along the inner ($r = r_i$) boundary and free of traction (i.e., $\sigma_r = 0$) on the outer boundary ($r = r_o$), and subjected to the body force $f_r = c_0 r$. Compare the finite-element solution with the exact solution

$$u = \frac{c_0 r}{8C_{11}}\left[1 - \left(\frac{r_i}{r}\right)^2\right]\left[c_1 - (r^2 - r_i^2)\right]$$

$$c_1 = (r_o^2 - r_i^2)\frac{(3+\nu)r_o^2 + (1-\nu)r_i^2}{(1+\nu)r_o^2 + (1-\nu)r_i^2}$$

3.33 Consider the second-order equation (3.1). Rewrite it as a pair of first-order equations,

$$-\left(\frac{d\sigma}{dx}+f\right)=0 \qquad a\frac{du}{dx}-\sigma=0$$

Note that σ is defined by the second equation. Now give a variational formulation of the pair over an element, and construct the associated finite-element model using the Ritz procedure.

3.34 (Continuation of Prob. 3.33) Repeat Prob. 3.33 using the penalty function method. Consider the functional in Eq. (3.5) and treat the equation $du/dx - \sigma/a = 0$ as a constraint. The modified functional for the penalty formulation becomes

$$I_P(u,\sigma)=I(u)+\frac{\gamma}{2}\int_{x_e}^{x_{e+1}}\left(\frac{du}{dx}-\frac{\sigma}{a}\right)^2 dx$$

where $I(u)$ is the quadratic functional in Eq. (3.5) and γ is the penaly parameter. Use I_P to develop the penalty finite-element model.

3.3 ONE-DIMENSIONAL FOURTH-ORDER EQUATIONS

Here we consider the (displacement) finite-element formulation of the one-dimensional fourth-order differential equation that arises in the elastic bending of beams. The equation can be decomposed into a pair of second-order equations involving the bending moment $(b\,d^2w/dx^2)$ and the transverse deflection w, and they can be formulated using the ideas presented in Sec. 3.2. Such a formulation is called the *mixed formulation*, and we discuss it in Chap. 5. The displacement formulation of the fourth-order equation involves the same steps as described in Sec. 3.2 for second-order equations, but the mathematical details are somewhat different, especially in the finite-element formulation of the equation.

Consider the fourth-order differential equation

$$\frac{d^2}{dx^2}\left(b\frac{d^2w}{dx^2}\right)+f=0 \qquad 0<x<L \tag{3.86}$$

where $b = b(x)$ and $f = f(x)$ are given functions of x (i.e., data), and w is the dependent variable. In the case of bending of beams, b denotes the product of the modulus of elasticity E and the moment of inertia I of the beam, f denotes the transversely distributed load (acting downward), and w is the transverse deflection (positive upward) of the beam. In addition to satisfying the differential equation (3.86), w must also satisfy appropriate boundary conditions; since the equation is fourth-order, four boundary conditions are needed to solve it. Although the first three steps of the finite-element analysis do not require the knowledge of the boundary conditions, for the sake of completing the description of the problem, we choose the following boundary conditions (which correspond to a cantilevered beam subjected to a point load F_0 and a moment M_0 at the free end; see Fig. 3.8a):

$$\begin{gathered} w(0)=0 \qquad \frac{dw}{dx}(0)=0 \\ \frac{d}{dx}\left(b\frac{d^2w}{dx^2}\right)\bigg|_{x=L}=F_0 \qquad \left(b\frac{d^2w}{dx^2}\right)\bigg|_{x=L}=M_0 \end{gathered} \tag{3.87}$$

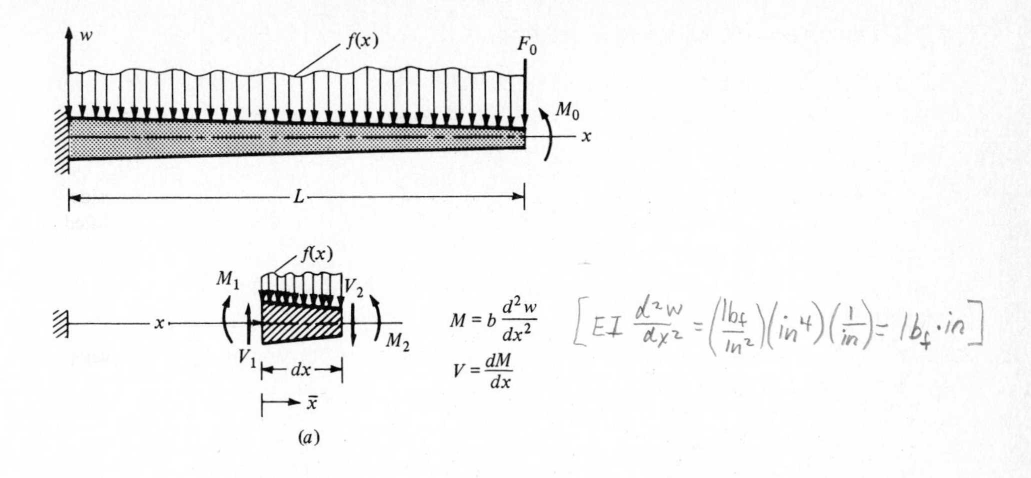

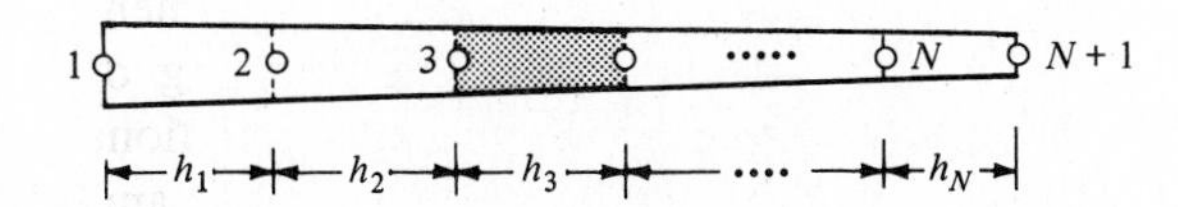

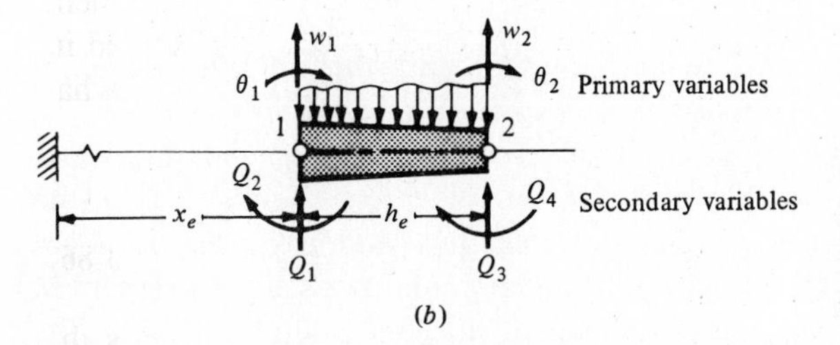

Figure 3.8 Finite element for the fourth-order equation (3.86). (*a*) Bending of elastic beams: sign convention. (*b*) Finite-element discretization and a typical element.

The variational formulation of Eq. (3.86) should verify that the form of the boundary conditions (3.87) is right. We now give a step-by-step procedure for the finite-element analysis of Eq. (3.86).

3.3.1 Discretization of the Domain into Elements

The domain is divided into a set (say, N) of line elements, each element having at least the two end nodes. Although the element, geometrically, is the same as that used in the second-order equations, the number and form of the primary unknowns at each node are dictated by the variational formulation of the differential equation (3.86). Due to the physical background of Eq. (3.86), the element is referred to as the *beam element*.

3.3.2 Derivation of Element Equations

In this step we isolate a typical element $\Omega^e = (x_e, x_{e+1})$ (see Fig. 3.8*b*) and construct the variational form of Eq. (3.86) over the element. The variational formulation will provide us the primary and secondary variables of the problem; the variables will be labeled at each node. Then suitable approximations for the primary variables are developed, and the element equations are derived.

Variational formulation of the equation over a typical element Following the procedure used earlier, we write

$$\begin{aligned}
0 &= \int_{x_e}^{x_{e+1}} v\left[\frac{d^2}{dx^2}\left(b\frac{d^2w}{dx^2}\right) + f\right] dx \\
&= \int_{x_e}^{x_{e+1}} \left[-\frac{dv}{dx}\frac{d}{dx}\left(b\frac{d^2w}{dx^2}\right) + vf\right] dx + \left[v\frac{d}{dx}\left(b\frac{d^2w}{dx^2}\right)\right]_{x=x_e}^{x_{e+1}} \\
&= \int_{x_e}^{x_{e+1}} \left(b\frac{d^2v}{dx^2}\frac{d^2w}{dx^2} + vf\right) dx + \left[v\frac{d}{dx}\left(b\frac{d^2w}{dx^2}\right) - \frac{dv}{dx}b\frac{d^2w}{dx^2}\right]_{x=x_e}^{x_{e+1}}
\end{aligned} \tag{3.88}$$

where v is a test function that is differentiable twice with respect to x. An examination of the boundary terms indicates that the essential boundary conditions involve the specification of w and dw/dx, and the natural boundary conditions involve the specification of $b\, d^2w/dx^2$ and $(d/dx)(b\, d^2w/dx^2)$ at the endpoints of the element. Thus there are two essential boundary conditions and two natural boundary conditions; therefore, we must identify w and dw/dx as the primary variables at each node (so that the essential boundary conditions are included in the interpolation). The natural boundary conditions always remain in the variational form and end up on the right-hand side (i.e., source vector) of the matrix equation. For the sake of mathematical convenience, we introduce the following notation:

$$\begin{aligned}
\frac{dw}{dx} &\equiv -\theta \\
Q_1^{(e)} &\equiv \left[\frac{d}{dx}\left(b\frac{d^2w}{dx^2}\right)\right]_{x=x_e} \\
Q_2^{(e)} &\equiv \left[b\frac{d^2w}{dx^2}\right]_{x=x_e} \\
Q_3^{(e)} &\equiv -\left[\frac{d}{dx}\left(b\frac{d^2w}{dx^2}\right)\right]_{x=x_{e+1}} \\
Q_4^{(e)} &\equiv -\left[b\frac{d^2w}{dx^2}\right]_{x=x_{e+1}}
\end{aligned} \tag{3.89}$$

For the bending of beams, $Q_1^{(e)}$ and $Q_3^{(e)}$ denote the shear forces, and $Q_2^{(e)}$ and $Q_4^{(e)}$ denote the bending moments (see Fig. 3.8a). With this notation, the variational form (3.88) becomes

$$\begin{aligned}0 &= \int_{x_e}^{x_{e+1}} \left(b \frac{d^2 v}{dx^2} \frac{d^2 w}{dx^2} + vf \right) dx - v(x_e) Q_1^{(e)} - \left[-\frac{dv}{dx}(x_e) \right] Q_2^{(e)} \\ &\quad - v(x_{e+1}) Q_3^{(e)} - \left[-\frac{dv}{dx}(x_{e+1}) \right] Q_4^{(e)} \\ &\equiv B(v, w) - l(v) \end{aligned} \tag{3.90a}$$

where the bilinear and linear forms are given by

$$\begin{aligned} B(v, w) &= \int_{x_e}^{x_{e+1}} b \frac{d^2 v}{dx^2} \frac{d^2 w}{dx^2} dx \\ l(v) &= -\int_{x_e}^{x_{e+1}} vf\, dx + v(x_e) Q_1^{(e)} \\ &\quad + \left(-\frac{dv}{dx} \right)\Big|_{x_e} Q_2^{(e)} + v(x_{e+1}) Q_3^{(e)} + \left(-\frac{dv}{dx} \right)\Big|_{x_{e+1}} Q_4^{(e)} \end{aligned} \tag{3.90b}$$

The quadratic functional, known in solid mechanics as the *total potential energy* of the beam element, is given by [from Eq. (2.33)]

$$\begin{aligned} I_e(w) &= \int_{x_e}^{x_{e+1}} \left[\frac{b}{2} \left(\frac{d^2 w}{dx^2} \right)^2 + wf \right] dx - w(x_e) Q_1^{(e)} \\ &\quad - \left(-\frac{dw}{dx} \right)\Big|_{x_e} Q_2^{(e)} - w(x_{e+1}) Q_3^{(e)} - \left(-\frac{dw}{dx} \right)\Big|_{x_{e+1}} Q_4^{(e)} \end{aligned} \tag{3.91}$$

The first term in the square brackets is known as the elastic strain energy due to bending, the second term is the work done by the distributed load, and the remaining terms account for the work done by the end (generalized) forces Q_i in moving through the (generalized) displacements of the element.

Derivation of the interpolation functions The variational form (3.90a) requires that the interpolation functions be continuous with continuous derivatives up to order 3 (so that Q_1 and Q_3 are nonzero), and that they satisfy certain (interpolation) properties so that the approximation for w satisfies the end conditions (of an element)

$$\begin{aligned} w(x_e) &= w_1 \qquad w(x_{e+1}) = w_2 \\ \theta(x_e) &= \theta_1 \qquad \theta(x_{e+1}) = \theta_2 \end{aligned} \tag{3.92}$$

In an effort to satisfy the end conditions, we automatically satisfy the continuity conditions. Hence, we pay attention to the satisfaction of the end conditions (3.92), which form the basis for the interpolation procedure.

Since there is a total of four conditions in an element, a four-parameter polynomial must be selected for w:

$$w(x) = c_1 + c_2 x + c_3 x^2 + c_4 x^3 \tag{3.93}$$

Note that the continuity conditions are automatically met (for $c_4 \neq 0$). The next step involves expressing c_i in terms of w_1, w_2, θ_1, and θ_2 such that conditions (3.92) are satisfied:

$$\begin{aligned}
w_1 &= w(x_e) = c_1 + c_2 x_e + c_3 x_e^2 + c_4 x_e^3 \\
\theta_1 &= \left(-\frac{dw}{dx}\right)\bigg|_{x=x_e} = -c_2 - 2c_3 x_e - 3c_4 x_e^2 \\
w_2 &= w(x_{e+1}) = c_1 + c_2 x_{e+1} + c_3 x_{e+1}^2 + c_4 x_{e+1}^3 \\
\theta_2 &= \left(-\frac{dw}{dx}\right)\bigg|_{x=x_{e+1}} = -c_2 - 2c_3 x_{e+1} - 3c_4 x_{e+1}^2
\end{aligned} \tag{3.94a}$$

or

$$\begin{Bmatrix} w_1 \\ \theta_1 \\ w_2 \\ \theta_2 \end{Bmatrix} = \begin{bmatrix} 1 & x_e & x_e^2 & x_e^3 \\ 0 & -1 & -2x_e & -3x_e^2 \\ 1 & x_{e+1} & x_{e+1}^2 & x_{e+1}^3 \\ 0 & -1 & -2x_{e+1} & -3x_{e+1}^2 \end{bmatrix} \begin{Bmatrix} c_1 \\ c_2 \\ c_3 \\ c_4 \end{Bmatrix} \tag{3.94b}$$

Inverting the matrix equation (3.94*b*) to write c_i in terms of w_1, w_2, θ_1, and θ_2, and substituting the result into Eq. (3.93), we obtain

$$\begin{aligned}
w(x) &= \phi_1^{(e)} w_1^{(e)} + \phi_2^{(e)} \theta_1^{(e)} + \phi_3^{(e)} w_2^{(e)} + \phi_4^{(e)} \theta_2^{(e)} \\
&= \sum_{j=1}^{4} u_j \phi_j^{(e)}
\end{aligned} \tag{3.95a}$$

where

$$u_1 = w_1 \qquad u_2 = \theta_1 \qquad u_3 = w_2 \qquad u_4 = \theta_2 \tag{3.95b}$$

$$\begin{aligned}
\phi_1^{(e)} &= 1 - 3\left(\frac{x - x_e}{h_e}\right)^2 + 2\left(\frac{x - x_e}{h_e}\right)^3 \\
\phi_2^{(e)} &= -(x - x_e)\left(1 - \frac{x - x_e}{h_e}\right)^2 \\
\phi_3^{(e)} &= 3\left(\frac{x - x_e}{h_e}\right)^2 - 2\left(\frac{x - x_e}{h_e}\right)^3 \\
\phi_4^{(e)} &= -(x - x_e)\left[\left(\frac{x - x_e}{h_e}\right)^2 - \frac{x - x_e}{h_e}\right]
\end{aligned} \tag{3.96}$$

In terms of the local (or element) coordinate $\bar{x}$, the interpolation functions $\phi_i^{(e)}$

take the simple form

$$
\begin{aligned}
\phi_1^{(e)} &= 1 - 3\left(\frac{\bar{x}}{h_e}\right)^2 + 2\left(\frac{\bar{x}}{h_e}\right)^3 \\
\phi_2^{(e)} &= -\bar{x}\left(1 - \frac{\bar{x}}{h_e}\right)^2 \\
\phi_3^{(e)} &= 3\left(\frac{\bar{x}}{h_e}\right)^2 - 2\left(\frac{\bar{x}}{h_e}\right)^3 \\
\phi_4^{(e)} &= -\bar{x}\left[\left(\frac{\bar{x}}{h_e}\right)^2 - \frac{\bar{x}}{h_e}\right]
\end{aligned}
\tag{3.97}
$$

The first two conditions of Eq. (3.92) imply that

$$
\begin{aligned}
\phi_1^{(e)}(x_e) &= 1 \qquad \phi_i^{(e)}(x_e) = 0 \qquad i \neq 1 \\
\phi_3^{(e)}(x_{e+1}) &= 1 \qquad \phi_i^{(e)}(x_{e+1}) = 0 \qquad i \neq 3.
\end{aligned}
\tag{3.98a}
$$

The last two conditions of Eq. (3.92) imply that

$$
\begin{aligned}
-\frac{d\phi_2^{(e)}}{dx}\bigg|_{x=x_e} &= 1 \qquad \frac{d\phi_i^{(e)}}{dx}\bigg|_{x=x_e} = 0 \qquad i \neq 2 \\
-\frac{d\phi_4^{(e)}}{dx}\bigg|_{x=x_{e+1}} &= 1 \qquad \frac{d\phi_i^{(e)}}{dx}\bigg|_{x=x_{e+1}} = 0 \qquad i \neq 4
\end{aligned}
\tag{3.98b}
$$

It is easy to verify that the $\phi_i^{(e)}$ do satisfy conditions (3.98). We note the following properties of $\phi_i^{(e)}$ (see Fig. 3.9):

$$
\begin{aligned}
&\phi_{2i-1}^{(e)}(\bar{x}_j) = \delta_{ij} \qquad \phi_{2i}^{(e)}(\bar{x}_j) = 0 \qquad \sum_{i=1}^{2} \phi_{2i-1}^{(e)} = 1 \\
&\frac{d\phi_{2i-1}^{(e)}}{dx}(\bar{x}_j) = 0 \qquad -\frac{d\phi_{2i}^{(e)}}{dx}(\bar{x}_j) = \delta_{ij}
\end{aligned}
\tag{3.99}
$$

where $x_1 = 0$ and $x_2 = h_e$ are the local coordinates of nodes 1 and 2 of element e.

It should be pointed out that the order of the interpolation functions derived above is the minimum required for the variational formulation (3.90). If higher-order (i.e., higher than cubic) approximation of w is desired, one must identify additional dependent (primary) unknowns at each of the two nodes. For example, if we add d^2w/dx^2 as the primary unknown at each of the two nodes, there will be a total of six conditions, and a fifth-order polynomial is required to interpolate the end conditions. It should also be noted that a cubic polynomial that interpolates w at four nodal points (two internal and two end nodes) of the

or, preferably, a higher-order element will be used.

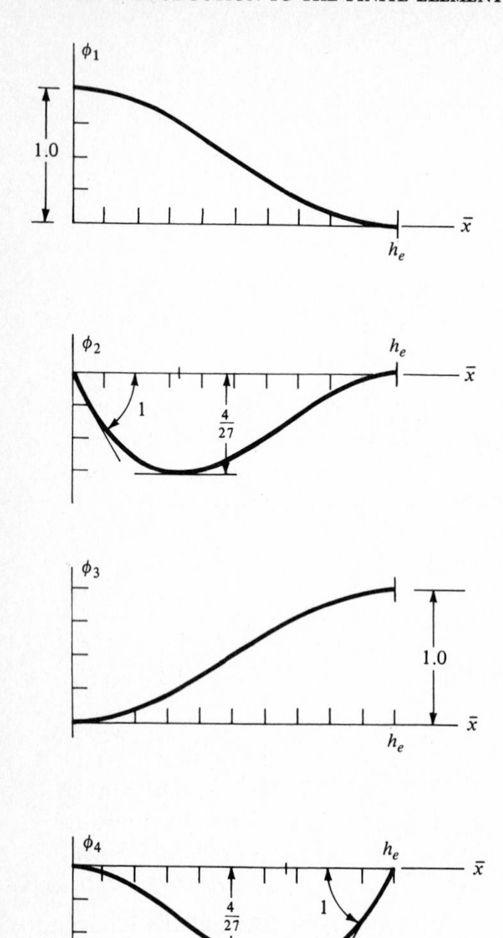

Figure 3.9 Hermite cubic interpolation functions defined in Eqs. (3.97).

element is not admissible, although the continuity conditions are met, because the polynomial does not satisfy the second set of end conditions (3.92). In the case of the bending of beams this amounts to violating the slope continuity at the interelement boundaries.

Finite-element model of the equation Using the approximation of the form of Eqs. (3.95) for w, and $v = \phi_i$ in the variational form (3.90), and noting the properties (3.99), we obtain

$$0 = \sum_{j=1}^{4} \left(\int_{x_e}^{x_{e+1}} b \frac{d^2\phi_i}{dx^2} \frac{d^2\phi_j}{dx^2} dx \right) u_j^{(e)} + \int_{x_e}^{x_{e+1}} \phi_i f \, dx - Q_i^{(e)} \qquad (3.100a)$$

or

$$\sum_{j=1}^{4} K_{ij}^{(e)} u_j^{(e)} - F_i^{(e)} = 0 \tag{3.100b}$$

where

$$\begin{aligned} K_{ij}^{(e)} &= \int_{x_e}^{x_{e+1}} b \frac{d^2\phi_i}{dx^2} \frac{d^2\phi_j}{dx^2} dx \\ F_i^{(e)} &= -\int_{x_e}^{x_{e+1}} \phi_i f \, dx + Q_i^{(e)} \end{aligned} \tag{3.101}$$

Equations (3.100) represent the finite-element model of Eq. (3.86); any admissible interpolation functions ϕ_i can be used in Eq. (3.101).

For the case in which b and f are constant over an element, the element matrices $[K^{(e)}]$ (stiffness) and $\{F^{(e)}\}$ (generalized forces) are given by

$$[K^{(e)}] = \frac{2b}{h^3} \begin{bmatrix} 6 & -3h & -6 & -3h \\ -3h & 2h^2 & 3h & h^2 \\ -6 & 3h & 6 & 3h \\ -3h & h^2 & 3h & 2h^2 \end{bmatrix} \qquad \{F^{(e)}\} = -\frac{fh}{12} \begin{Bmatrix} 6 \\ -h \\ 6 \\ h \end{Bmatrix} + \begin{Bmatrix} Q_1 \\ Q_2 \\ Q_3 \\ Q_4 \end{Bmatrix} \tag{3.102}$$

It can easily be verified that the term $-(fh/12)\{\vdots\}$ in $\{F^{(e)}\}$ represents the "statically equivalent" forces and moments at nodes 1 and 2 due to the uniformly distributed load over the element. When f is an algebraically complicated function of x, the mechanics of a materials-type approach become less appealing, whereas Eqs. (3.101) provides a straightforward way of computing the components.

3.3.3 Assembly of Element Equations

To show the assembly procedure, we select a two-element model and assume that f and $b = EI$ are constant. Then there are three global nodes and a total of six global primary and six secondary variables in the problem. We note the following correspondence between the local (primary) degrees of freedom u_i and the global (primary) degrees of freedom U_i (see Fig. 3.10):

$$\begin{aligned} &u_1^{(1)} = w_1^{(1)} = U_1 \qquad u_2^{(1)} = \theta_1^{(1)} = U_2 \qquad u_3^{(1)} = w_2^{(1)} = w_1^{(2)} = U_3 \\ &u_4^{(1)} = \theta_2^{(1)} = \theta_1^{(2)} = U_4 \qquad u_3^{(2)} = w_2^{(2)} = U_5 \qquad u_4^{(2)} = \theta_2^{(2)} = U_6 \end{aligned} \tag{3.103}$$

The connectivity matrix $[B]$ (which will be used in computer implementation) is given by

$$[B] = \begin{bmatrix} 1 & 2 \\ 2 & 3 \end{bmatrix} \tag{3.104}$$

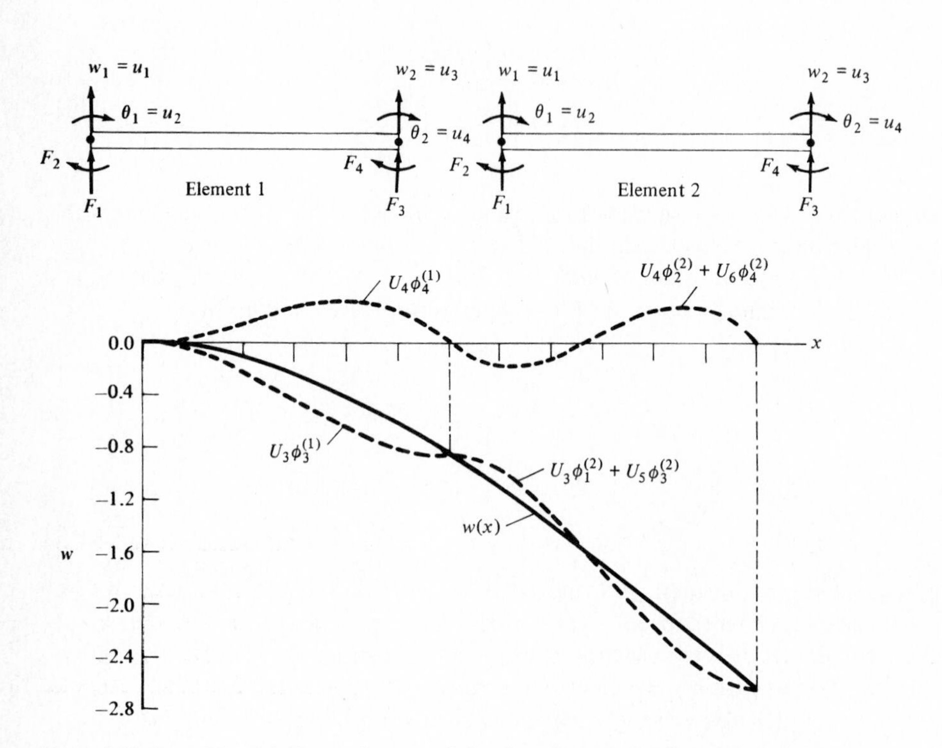

Figure 3.10 Assembly of the beam elements; finite-element representation of the solution $w(x)$.

Since there are two primary unknowns (or degrees of freedom) per node, repetition of a number in $[B]$ indicates that the coefficients associated with both degrees of freedom add up. The repetition of the number 2 (which corresponds to global degrees of freedom 3 and 4) in $[B]$ indicates that the global K_{33}, K_{34}, K_{43}, and K_{44} have contributions from both elements 1 and 2:

$$\begin{aligned} K_{33} &= K_{33}^{(1)} + K_{11}^{(2)} \qquad & K_{34} &= K_{34}^{(1)} + K_{12}^{(2)} \\ K_{43} &= K_{43}^{(1)} + K_{21}^{(2)} \qquad & K_{44} &= K_{44}^{(1)} + K_{22}^{(2)} \end{aligned} \tag{3.105}$$

In general, the assembled (stiffness and force) matrices for the assembly of beam elements have the form shown in Eqs. (3.106) and (3.107):

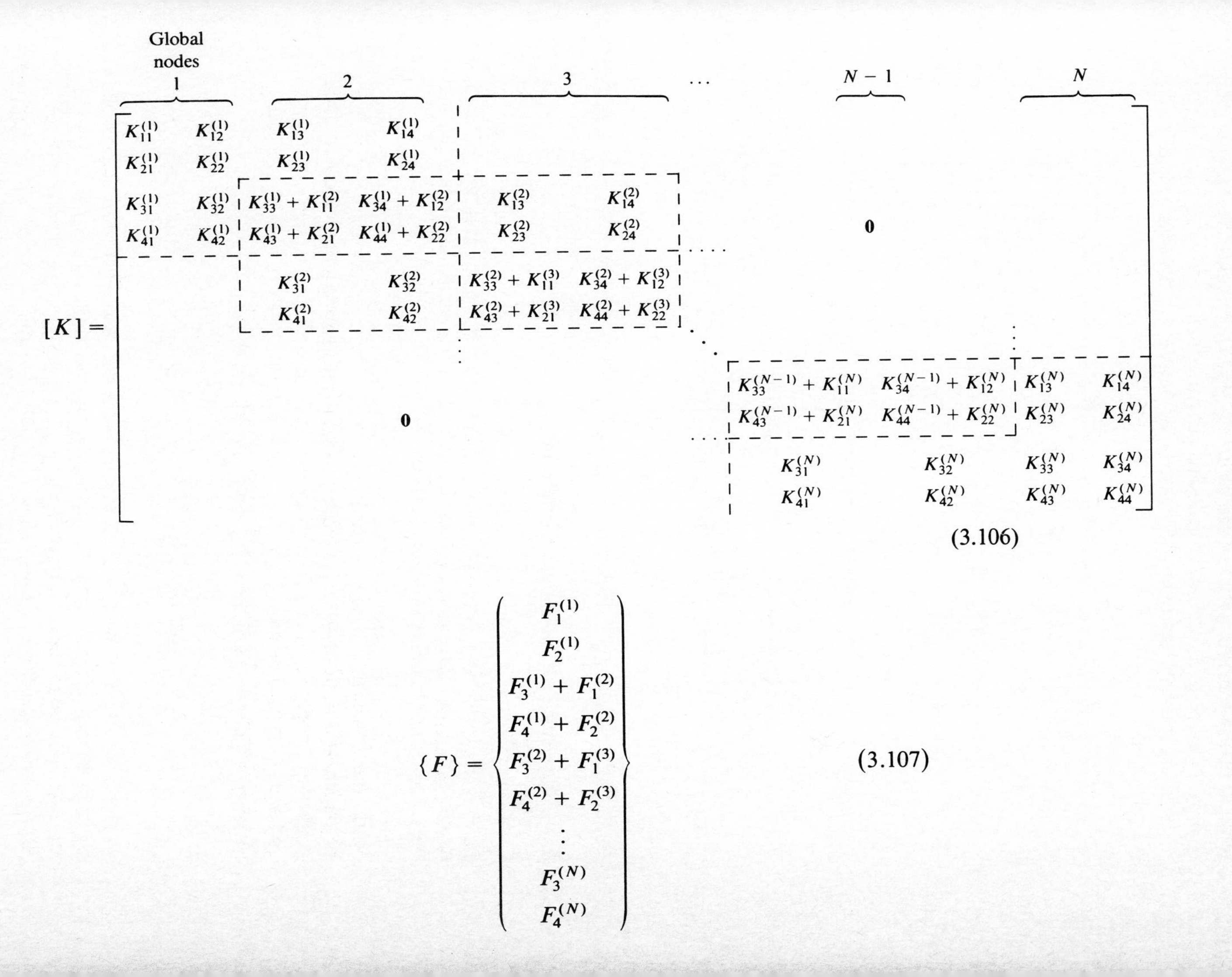

$$
[K] = \begin{bmatrix}
K_{11}^{(1)} & K_{12}^{(1)} & K_{13}^{(1)} & K_{14}^{(1)} & & & & & & \\
K_{21}^{(1)} & K_{22}^{(1)} & K_{23}^{(1)} & K_{24}^{(1)} & & & & \mathbf{0} & & \\
K_{31}^{(1)} & K_{32}^{(1)} & K_{33}^{(1)} + K_{11}^{(2)} & K_{34}^{(1)} + K_{12}^{(2)} & K_{13}^{(2)} & K_{14}^{(2)} & & & & \\
K_{41}^{(1)} & K_{42}^{(1)} & K_{43}^{(1)} + K_{21}^{(2)} & K_{44}^{(1)} + K_{22}^{(2)} & K_{23}^{(2)} & K_{24}^{(2)} & \cdots & & & \\
 & & K_{31}^{(2)} & K_{32}^{(2)} & K_{33}^{(2)} + K_{11}^{(3)} & K_{34}^{(2)} + K_{12}^{(3)} & & & & \\
 & & K_{41}^{(2)} & K_{42}^{(2)} & K_{43}^{(2)} + K_{21}^{(3)} & K_{44}^{(2)} + K_{22}^{(3)} & & & & \\
 & & & & \vdots & \ddots & & & \vdots & \\
 & & & \mathbf{0} & & & K_{33}^{(N-1)} + K_{11}^{(N)} & K_{34}^{(N-1)} + K_{12}^{(N)} & K_{13}^{(N)} & K_{14}^{(N)} \\
 & & & & & & K_{43}^{(N-1)} + K_{21}^{(N)} & K_{44}^{(N-1)} + K_{22}^{(N)} & K_{23}^{(N)} & K_{24}^{(N)} \\
 & & & & & \cdots & K_{31}^{(N)} & K_{32}^{(N)} & K_{33}^{(N)} & K_{34}^{(N)} \\
 & & & & & & K_{41}^{(N)} & K_{42}^{(N)} & K_{43}^{(N)} & K_{44}^{(N)}
\end{bmatrix} \tag{3.106}
$$

$$
\{F\} = \begin{Bmatrix}
F_1^{(1)} \\
F_2^{(1)} \\
F_3^{(1)} + F_1^{(2)} \\
F_4^{(1)} + F_2^{(2)} \\
F_3^{(2)} + F_1^{(3)} \\
F_4^{(2)} + F_2^{(3)} \\
\vdots \\
F_3^{(N)} \\
F_4^{(N)}
\end{Bmatrix} \tag{3.107}
$$

Of course, one can obtain the same result by writing out Eqs. (3.100) for each element in terms of U_i and adding all element equations.

For the problem at hand, the assembled system of equations is given by $(h = L/2)$

$$\frac{2EI}{h^3}\begin{bmatrix} 6 & -3h & -6 & -3h & 0 & 0 \\ -3h & 2h^2 & 3h & h^2 & 0 & 0 \\ -6 & 3h & 6+6 & 3h-3h & -6 & -3h \\ -3h & h^2 & 3h-3h & 2h^2+2h^2 & 3h & h^2 \\ 0 & 0 & -6 & 3h & 6 & 3h \\ 0 & 0 & -3h & h^2 & 3h & 2h^2 \end{bmatrix}\begin{Bmatrix} U_1 \\ U_2 \\ U_3 \\ U_4 \\ U_5 \\ U_6 \end{Bmatrix}$$

$$= -\frac{fh}{12}\begin{Bmatrix} 6 \\ -h \\ 6+6 \\ h-h \\ 6 \\ h \end{Bmatrix} + \begin{Bmatrix} Q_1^{(1)} \\ Q_2^{(1)} \\ Q_3^{(1)} + Q_1^{(2)} \\ Q_4^{(1)} + Q_2^{(2)} \\ Q_3^{(2)} \\ Q_4^{(2)} \end{Bmatrix} \tag{3.108}$$

3.3.4 Imposition of Boundary Conditions

First, we write the equilibrium conditions for the secondary variables. At global node 1, $Q_1^{(1)}$ and $Q_2^{(1)}$ (corresponding to the shear force F_L and the bending moment M_L) are not known. At global node 2, there are no externally applied shear forces and bending moments. Hence,

$$Q_3^{(1)} + Q_1^{(2)} = 0 \qquad Q_4^{(1)} + Q_2^{(2)} = 0 \tag{3.109a}$$

At global node 3, the shear force is given as F_0, and the bending moment is given as M_0:

$$\begin{aligned} Q_3^{(2)} &\equiv -\left[\frac{d}{dx}\left(EI\frac{d^2w}{dx^2}\right)\right]_{x=L} = -F_0 \\ Q_4^{(2)} &\equiv -\left[b\frac{d^2w}{dx^2}\right]_{x=L} = -M_0 \end{aligned} \tag{3.109b}$$

Next, we identify the specified primary variables. Since the beam is clamped at global node 1, it follows that the deflection and the slope are zero at that node:

$$\begin{aligned} u_1^{(1)} &\equiv w_1^{(1)} = U_1 = 0 \\ u_2^{(1)} &\equiv \theta_1^{(1)} = U_2 = 0 \end{aligned} \tag{3.110}$$

Using Eqs. (3.109) and (3.110) in eq. (3.108), we obtain

$$\frac{2EI}{h^3}\left[\begin{array}{cc|cccc} 6 & -3h & -6 & -3h & 0 & 0 \\ -3h & 2h^2 & 3h & h^2 & 0 & 0 \\ \hline -6 & 3h & 12 & 0 & -6 & -3h \\ -3h & h^2 & 0 & 4h^2 & 3h & h^2 \\ 0 & 0 & -6 & 3h & 6 & 3h \\ 0 & 0 & -3h & h^2 & 3h & 2h^2 \end{array}\right]\left\{\begin{array}{c} U_1 = 0 \\ U_2 = 0 \\ \hline U_3 \\ U_4 \\ U_5 \\ U_6 \end{array}\right\}$$

$$= -\frac{fh}{12}\left\{\begin{array}{c} 6 \\ -h \\ \hline 12 \\ 0 \\ 6 \\ h \end{array}\right\} + \left\{\begin{array}{c} F_L \\ M_L \\ \hline 0 \\ 0 \\ -F_0 \\ -M_0 \end{array}\right\} \tag{3.111}$$

3.3.5 Solution of Equations

The fact that the first two degrees of freedom U_1 and U_2 are known and the remaining four degrees of freedom U_3 through U_6 are to be determined provides us the motivation to partition (shown by dashed lines) the matrix equation (3.111). Equation (3.111) is in the form

$$\begin{bmatrix} [K^{11}] & [K^{12}] \\ [K^{21}] & [K^{22}] \end{bmatrix}\begin{Bmatrix} \{\Delta^1\} \\ \{\Delta^2\} \end{Bmatrix} = \begin{Bmatrix} \{F^1\} \\ \{F^2\} \end{Bmatrix} \tag{3.112}$$

where the meaning of $[K^{11}]$, etc. is obvious from Eq. (3.111). Equation (3.112) can be written, after carrying out the matrix multiplication, in the form

$$[K^{11}]\{\Delta^1\} + [K^{12}]\{\Delta^2\} = \{F^1\} \tag{3.113a}$$

$$[K^{21}]\{\Delta^1\} + [K^{22}]\{\Delta^2\} = \{F^2\} \tag{3.113b}$$

Since $\{\Delta^1\}$ and $\{F^2\}$ are known $(\{\Delta^1\} = \{0\})$, we can use Eq. (3.113*b*) to solve for $\{\Delta^2\}$, and then use Eq. (3.113*a*) to compute the unknown reactions $\{F^1\}$:

$$\{\Delta^2\} \equiv \begin{Bmatrix} U_3 \\ U_4 \\ U_5 \\ U_6 \end{Bmatrix} = \frac{h^3}{2EI}\begin{bmatrix} 12 & 0 & -6 & -3h \\ 0 & 4h^2 & 3h & h^2 \\ -6 & 3h & 6 & 3h \\ -3h & h^2 & 3h & 2h^2 \end{bmatrix}^{-1}\begin{Bmatrix} -fh \\ 0 \\ -F_0 - fh/2 \\ -M_0 - fh^2/12 \end{Bmatrix}$$

Note that $\{\Delta^1\} = \{0\}$ in the present problem, and therefore we do not take the term $[K^{21}]\{\Delta^1\}$ to the right-hand side of Eq. (3.113*b*) to modify vector $\{F^2\}$. Inverting the matrix and performing the matrix multiplication, we obtain

$$\{\Delta^2\} = \frac{h}{6EI}\begin{bmatrix} 2h^2 & -3h & 5h^2 & -3h \\ -3h & 6 & -9h & 6 \\ 5h^2 & -9h & 16h^2 & -12h \\ -3h & 6 & -12h & 12 \end{bmatrix}\begin{Bmatrix} -fh \\ 0 \\ -F_0 - fh/2 \\ -M_0 - fh^2/12 \end{Bmatrix}$$

$$= \frac{h}{6EI}\begin{Bmatrix} -5h^2F_0 + 3hM_0 - \frac{17}{4}fh^3 \\ 9hF_0 - 6M_0 + 7fh^2 \\ -16h^2F_0 + 12hM_0 - 12fh^3 \\ 12hF_0 - 12M_0 + 8fh^2 \end{Bmatrix} \tag{3.114}$$

$$\{F^1\} = \begin{Bmatrix} F_L \\ M_L \end{Bmatrix} = \frac{2EI}{h^3}\begin{bmatrix} -6 & -3h & 0 & 0 \\ 3h & h^2 & 0 & 0 \end{bmatrix}\begin{Bmatrix} U_3 \\ U_4 \\ U_5 \\ U_6 \end{Bmatrix} + \frac{fh}{12}\begin{Bmatrix} 6 \\ -h \end{Bmatrix}$$

$$= \begin{Bmatrix} F_0 + 2fh \\ -2h(F_0 + fh) + M_0 \end{Bmatrix} \tag{3.115}$$

It can be easily verified that the reactions F_L and M_L satisfy the static equilibrium equations of the beam:

$$F_L - F_0 - 2fh = 0 \qquad M_L + 2fh^2 + 2hF_0 - M_0 = 0 \tag{3.116}$$

For $EI = 58 \times 10^8$ lb · in², $L = 150$ in ($h = 75$ in), $F_0 = 8000$ lb, $f = 100$ lb/in, and $M_0 = 0$, the finite-element solution for the nodal values becomes

$$\begin{aligned} U_3 &= -0.8713 \text{ in} \qquad & U_4 &= 0.02012 \text{ rad} \\ U_5 &= -2.643 \text{ in} \qquad & U_6 &= 0.02522 \text{ rad} \end{aligned} \tag{3.117}$$

The reactions F_L and M_L are given by

$$F_L = 23{,}000 \text{ lb} \qquad M_L = 2.325 \times 10^6 \text{ lb} \cdot \text{in} \tag{3.118}$$

3.3.6 Postprocessing of the Solution

The finite-element solution $w(x)$ is given by (see Fig. 3.10)

$$w(x) = \begin{cases} U_3\phi_3^{(1)} + U_4\phi_4^{(1)} & \text{for } 0 \leqslant x \leqslant h \\ U_3\phi_1^{(2)} + U_4\phi_2^{(2)} + U_5\phi_3^{(2)} + U_6\phi_4^{(2)} & \text{for } h \leqslant x \leqslant 2h \end{cases} \tag{3.119}$$

The exact solution of Eqs. (3.86) and (3.87) can be obtained by direct integration, and is given by

$$EIw_0(x) = -\frac{f}{24}(L-x)^4 + \frac{F_0}{6}(x-L)^3 + \frac{M_0}{2}x^2 - \left(\frac{fL^3}{6} + \frac{F_0L^2}{2}\right)x$$
$$+ \frac{fL^4}{24} + \frac{F_0L^3}{6} \qquad \text{for } 0 \leqslant x \leqslant L$$
$$EI\theta(x) = -\frac{f}{6}(L-x)^3 - \frac{F_0}{2}(x-L)^2 - M_0x \tag{3.120}$$
$$+ \left(\frac{fL^3}{6} + \frac{F_0L^2}{2}\right) \qquad \text{for } 0 \leqslant x \leqslant L$$
$$M(x) = -\frac{f}{2}(L-x)^2 + F_0(x-L) + M_0 \qquad \text{for } 0 \leqslant x \leqslant L$$

The finite-element solution (3.119) and the exact solution (3.120) are compared in Table 3.3. Note that the finite-element solution coincides with the exact solution at the nodes. At other points, the difference between the finite-element solution and the exact solution is less than 2 percent; the difference cannot be seen on the graph (see Fig. 3.11).

This completes the finite-element formulation of the fourth-order differential equation (3.86).

In most beam problems, the flexural rigidity $b \equiv EI$ is a constant in each element (i.e., a beam can be represented by a finite-element mesh that consists of constant-cross-section, homogeneous beam elements); therefore, the element stiffness matrix (3.102) can be used. We consider two examples below.

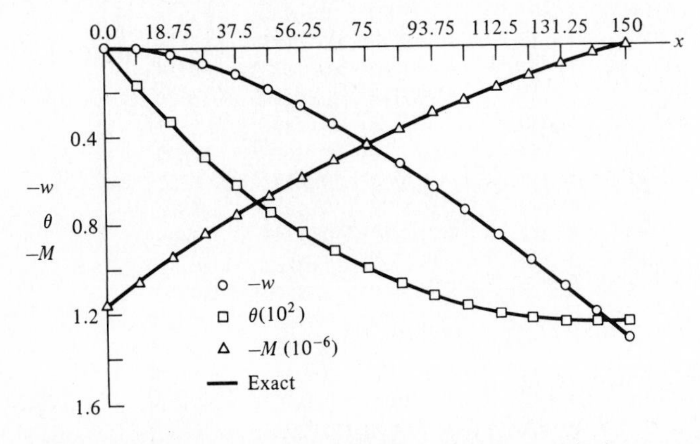

Figure 3.11 Comparison of the finite-element solution with the exact solution of Eqs. (3.86) and (3.87) for the case in which $M_0 = 0$.

Example 3.3 Consider the discontinuously loaded beam shown in Fig. 3.12. The differential equation (3.86) is valid with the following discontinuous data:

$$b \equiv EI = \begin{cases} 2 \times 10^7 \text{ (lb} \cdot \text{ft}^2) & 0 \leqslant x \leqslant 10 \text{ ft} \\ 10^7 \text{ (lb} \cdot \text{ft}^2) & 10 \text{ ft} \leqslant x \leqslant 28 \text{ ft} \end{cases} \tag{3.121a}$$

$$f = \begin{cases} 2400 \text{ (lb/ft)} & 0 \leqslant x \leqslant 10 \text{ ft} \\ 10{,}000\delta(x - 22) \text{ (lb)} & 10 \text{ ft} \leqslant x \leqslant 28 \text{ ft} \end{cases} \tag{3.121b}$$

where $\delta(x - a)$ is the Dirac delta function, defined by

$$\int_{-\infty}^{\infty} g(x)\delta(x - a)\, dx \equiv g(a) \qquad -\infty < x, a < \infty \tag{3.122}$$

Note that the Dirac delta function enables us to include a point load in the description of the source term f. As will be seen later, the point load at a global node goes into the corresponding location in the global force vector.

The geometry and loading in the present case require us to use at least three elements (see Fig. 3.12b) to represent the domain $[0, L] = 0 \leqslant x \leqslant 28$ ft. Let us use the minimum number of elements to analyze the problem. Then there are four nodes and eight global degrees of freedom (before imposing the boundary conditions).

Table 3.3 Comparison of the exact solutions and the finite-element solutions of Eqs. (3.86) and (3.87) with $M_0 = 0$

	Finite-element solution† (2 elements)			Exact solution		
x	$-w$	θ	$-M \times 10^{-6}$	$-w$	θ	$-M \times 10^{-6}$
0.0	0.0	0.0	2.2781	0.0	0.0	2.325
9.375	0.0168	0.003536	2.0977	0.0171	0.003586	2.1138
18.75	0.0654	0.006781	1.9172	0.0662	0.006838	1.9113
28.125	0.1430	0.009734	1.7367	0.1443	0.009770	1.7177
37.50	0.2470	0.012396	1.5563	0.2484	0.012396	1.5328
46.875	0.3746	0.014765	1.3758	0.3758	0.01473	1.3567
56.25	0.5229	0.016843	1.1953	0.5237	0.01679	1.1895
65.625	0.6895	0.018629	1.0148	0.6897	0.0858	1.0310
75.00	0.8713	0.02012	0.8344	0.8713	0.02012	0.8813
84.375	1.0660	0.021384	0.7242	1.0663	0.02143	0.74033
93.75	1.2717	0.022465	0.61406	1.2725	0.02252	0.60820
103.125	1.4867	0.023369	0.50391	1.4879	0.023404	0.48486
112.50	1.7093	0.024094	0.39375	1.7107	0.024094	0.37031
121.875	1.9379	0.024642	0.28359	1.9392	0.024606	0.26455
131.25	2.1708	0.035011	0.17344	2.1716	0.024954	0.16758
140.625	2.4063	0.025202	0.06328	2.4066	0.025153	0.07939
150.00	2.6428	0.02522	0.04688	2.6428	0.02522	0.0

† $M = EI\, d^2w/dx^2$

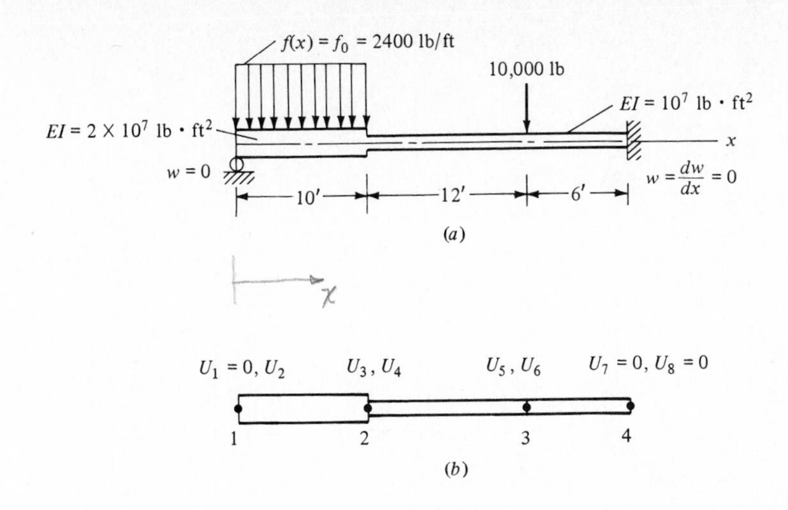

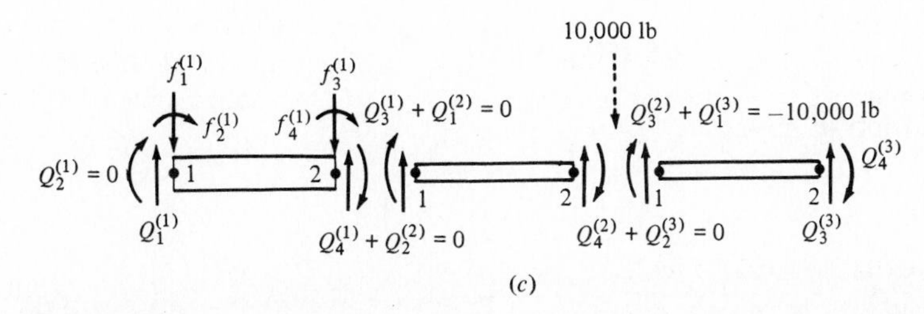

10,000 lb

$f_1^{(1)}$ $f_3^{(1)}$ $f_2^{(1)}$ $f_4^{(1)}$ $Q_3^{(1)} + Q_1^{(2)} = 0$ $Q_3^{(2)} + Q_1^{(3)} = -10{,}000$ lb

$Q_2^{(1)} = 0$ $Q_4^{(3)}$

$Q_1^{(1)}$ $Q_4^{(1)} + Q_2^{(2)} = 0$ $Q_4^{(2)} + Q_2^{(3)} = 0$ $Q_3^{(3)}$

(c)

Figure 3.12 Finite-element mesh and equilibrium conditions for the beam-bending problem considered in Example 3.3. (*a*) Physical problem. (*b*) Finite-element mesh of three elements. (*c*) Equilibrium conditions among the generalized forces (i.e., secondary variables).

The element stiffness matrix and the force vector are given by Eqs. (3.102), with $f = 2400$ in element 1 and $f = 0$ in both elements 2 and 3. The point load will be directly included in the global force vector (through equilibrium of forces). The element stiffness matrix and the force vector for each element are given by

$$[K^{(1)}] = \frac{2(10^7)}{10^3}\begin{bmatrix} 12 & -60 & -12 & -60 \\ -60 & 400 & 60 & 200 \\ -12 & 60 & 12 & 60 \\ -60 & 200 & 60 & 400 \end{bmatrix}$$

$$\{F^{(1)}\} = \begin{Bmatrix} -12{,}000 \\ 20{,}000 \\ -12{,}000 \\ -20{,}000 \end{Bmatrix} + \begin{Bmatrix} Q_1^{(1)} \\ Q_2^{(1)} \\ Q_3^{(1)} \\ Q_4^{(1)} \end{Bmatrix}$$

$$[K^{(2)}] = \frac{10^7}{12^3}\begin{bmatrix} 12 & -72 & -12 & -72 \\ -72 & 576 & 72 & 288 \\ -12 & 72 & 12 & 72 \\ -72 & 288 & 72 & 576 \end{bmatrix} \qquad \{F^{(2)}\} = \begin{Bmatrix} 0 \\ 0 \\ 0 \\ 0 \end{Bmatrix} + \begin{Bmatrix} Q_1^{(2)} \\ Q_2^{(2)} \\ Q_3^{(2)} \\ Q_4^{(2)} \end{Bmatrix} \tag{3.123}$$

$$[K^{(3)}] = \frac{10^7}{6^3}\begin{bmatrix} 12 & -36 & -12 & -36 \\ -36 & 144 & 36 & 72 \\ -12 & 36 & 12 & 36 \\ -36 & 72 & 36 & 144 \end{bmatrix} \qquad \{F^{(3)}\} = \begin{Bmatrix} 0 \\ 0 \\ 0 \\ 0 \end{Bmatrix} + \begin{Bmatrix} Q_1^{(3)} \\ Q_2^{(3)} \\ Q_3^{(3)} \\ Q_4^{(3)} \end{Bmatrix}$$

The assembled system of equations is given by

$$[K]\{U\} = \{F\} \tag{3.124a}$$

or, in explicit form, we have

$$10^7 \left[\begin{array}{c|cccccc|cc} 0.024 & -0.12 & -0.024 & -0.12 & & & & & \\ \hline & 0.80 & 0.12 & 0.40 & & & & & \\ & & 0.0309 & 0.0783 & -0.00694 & -0.04167 & & & \\ & & & 1.133 & 0.0417 & 0.167 & & & \\ & & & & 0.0625 & -0.125 & & -0.0556 & -0.167 \\ & & & & & 1.0 & & 0.1667 & 0.333 \\ \hline \text{Symmetric} & & & & & & & 0.0556 & 0.1667 \\ & & & & & & & & 0.6667 \end{array}\right]$$

$$\begin{Bmatrix} U_1 \\ \hline U_2 \\ U_3 \\ U_4 \\ U_5 \\ U_6 \\ \hline U_7 \\ U_8 \end{Bmatrix} = \begin{Bmatrix} -12{,}000 \\ \hline 20{,}000 \\ -12{,}000 \\ -20{,}000 \\ 0 \\ 0 \\ \hline 0 \\ 0 \end{Bmatrix} + \begin{Bmatrix} Q_1^{(1)} \\ \hline Q_2^{(1)} \\ Q_3^{(1)} + Q_1^{(2)} \\ Q_4^{(1)} + Q_2^{(2)} \\ Q_3^{(2)} + Q_1^{(3)} \\ Q_4^{(2)} + Q_2^{(3)} \\ \hline Q_3^{(3)} \\ Q_4^{(3)} \end{Bmatrix} \tag{3.124b}$$

where the interelement continuity conditions were already used. The equilibrium of internal forces is given by (it is here that we include the known forces into the assembled equations)

$$Q_2^{(1)} = 0 \qquad Q_3^{(1)} + Q_1^{(2)} = 0 \qquad Q_4^{(1)} + Q_2^{(2)} = 0$$
$$Q_4^{(2)} + Q_2^{(3)} = 0 \qquad Q_3^{(2)} + Q_1^{(3)} = -10{,}000 \tag{3.125a}$$

Note that the forces $Q_1^{(1)}$, $Q_3^{(3)}$, and $Q_4^{(3)}$ are not known (the reactions at the supports).

The boundary conditions on the primary degrees of freedom are given by

$$U_1 = U_7 = U_8 = 0 \tag{3.125b}$$

Using the known forces and displacements in Eqs. (3.125), we can partition the global system of equations (which is obtained by deleting the columns and rows corresponding to the known U_i) and condense out the known degrees of freedom to obtain the final equation [see the submatrix enclosed by the dashed lines in Eq. (3.124*b*)]

$$\begin{bmatrix} K_{22} & K_{23} & K_{24} & 0 & 0 \\ & K_{33} & K_{34} & K_{35} & K_{36} \\ & & K_{44} & K_{45} & K_{46} \\ & \text{Symmetric} & & K_{55} & K_{56} \\ & & & & K_{66} \end{bmatrix} \begin{Bmatrix} U_2 \\ U_3 \\ U_4 \\ U_5 \\ U_6 \end{Bmatrix} = \begin{Bmatrix} 20{,}000 \\ -12{,}000 \\ -20{,}000 \\ -10{,}000 \\ 0 \end{Bmatrix} \tag{3.126}$$

The unknown reactions can be computed from the equation (which is obtained from the remaining portions of the assembled matrix)

$$\begin{Bmatrix} Q_1^{(1)} \\ Q_3^{(3)} \\ Q_4^{(3)} \end{Bmatrix} = \begin{Bmatrix} 12{,}000 \\ 0 \\ 0 \end{Bmatrix} + \begin{bmatrix} K_{12} & K_{13} & K_{14} & 0 & 0 \\ K_{72} & K_{73} & K_{74} & K_{75} & K_{76} \\ 0 & 0 & 0 & K_{85} & K_{86} \end{bmatrix} \begin{Bmatrix} U_2 \\ U_3 \\ U_4 \\ U_5 \\ U_6 \end{Bmatrix} \tag{3.127}$$

Equations (3.126) and (3.127) can be used to obtain all of the unknown displacements and forces of the problem. The solution U_i ($i = 2, \ldots, 6$) is given by

$$U_2 = 0.03856 \qquad U_3 = -0.2808 \qquad U_4 = 0.01214$$
$$U_5 = -0.1103 \qquad U_6 = -0.02752 \tag{3.128}$$

Plots of the finite-element solution for w, θ, and M are shown in Fig. 3.13. One should note that the exact solution of the problem is also defined element-wise (because of the discontinuity in the loading and the flexural rigidity). The reader is asked to compare the exact solution with the finite-element solution. ■

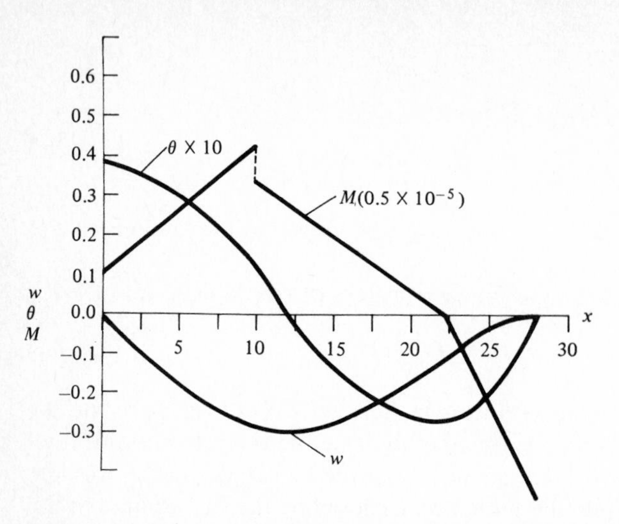

Figure 3.13 Finite-element approximation of the solution of the fourth-order equations (3.86 and 3.87) (see Fig. 3.12 for the boundary conditions).

Example 3.4 Consider the cantilevered beam shown in Fig. 3.14. We shall use two elements, as dictated by the loading. The element matrices are given by

$$[K^1] = EI\begin{bmatrix} 0.015625 & -0.375 & -0.015625 & -0.375 \\ -0.375 & 12.0 & 0.375 & 6.0 \\ -0.015625 & 0.375 & 0.015625 & 0.375 \\ -0.375 & 6.0 & 0.375 & 12.0 \end{bmatrix} \qquad \{F^1\} = \begin{Bmatrix} Q_1^{(1)} \\ Q_2^{(1)} \\ Q_3^{(1)} \\ Q_4^{(1)} \end{Bmatrix}$$

$$[K^2] = EI\begin{bmatrix} 0.00463 & -0.1667 & -0.00463 & -0.1667 \\ -0.1667 & 8.0 & 0.1667 & 4.0 \\ -0.00463 & 0.1667 & 0.00463 & 0.1667 \\ -0.1667 & 4.0 & 0.1667 & 8.0 \end{bmatrix}$$

$$\{F^2\} = \{f^{(2)}\} + \begin{Bmatrix} Q_1^{(2)} \\ Q_2^{(2)} \\ Q_3^{(2)} \\ Q_4^{(2)} \end{Bmatrix}$$

where $f_i^{(2)}$ are given by

$$f_i^{(2)} = \int_0^{72} f(\bar{x})\phi_i(\bar{x})\,d\bar{x} \qquad f(\bar{x}) = \frac{100}{6}\bar{x}$$

and ϕ_i are the Hermite interpolation functions given in Eqs. (3.97). Carrying out the integration, we obtain

$$\{f^{(2)}\} = \{-90.0, 1440.0, -210.0, -2160.0\}^T$$

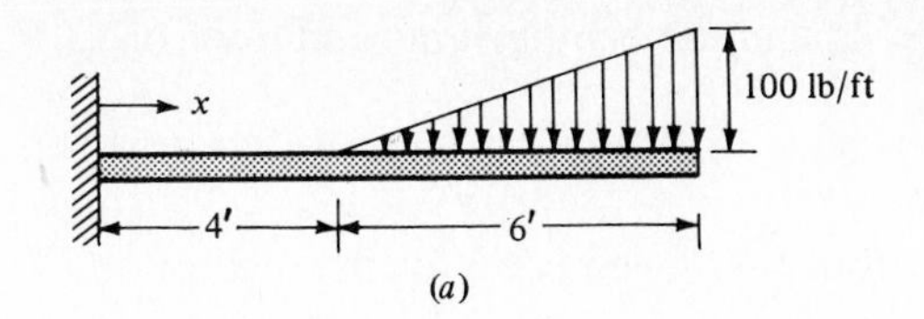

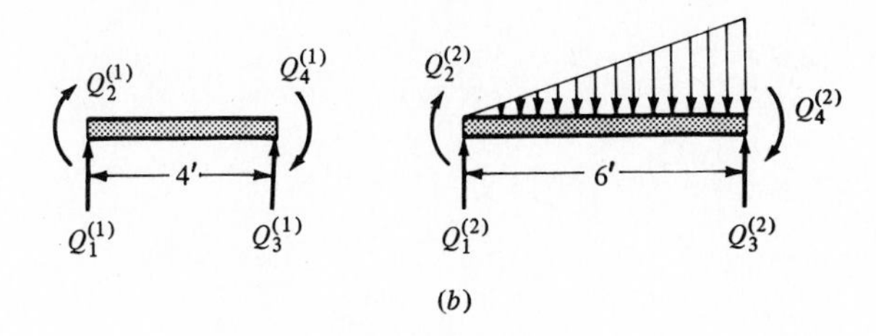

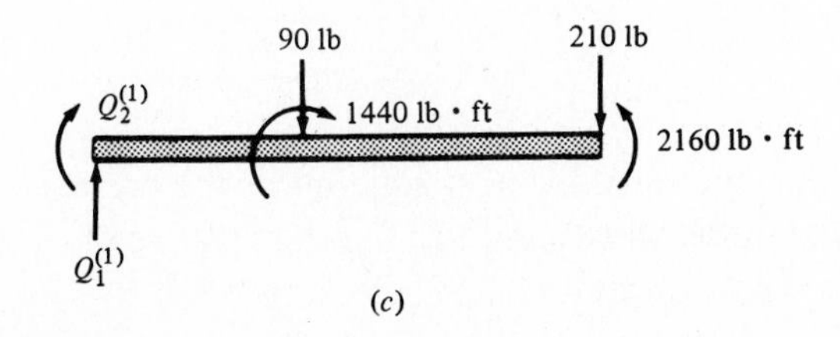

Figure 3.14 Finite-element analysis of a cantilevered beam with discontinuous loading. (*a*) Physical problem. (*b*) Finite-element discretization and generalized element forces. (*c*) Assembly of finite elements and generalized global forces.

The assembled system of equations is given by

$$EI\begin{bmatrix} 0.01563 & -0.375 & -0.01563 & -0.375 & 0.0 & 0.0 \\ & 12.0 & 0.375 & 6.0 & 0.0 & 0.0 \\ \hline & & 0.02025 & 0.2083 & -0.00463 & -0.1667 \\ \text{Symmetric} & & & 20.0 & 0.1667 & 4.0 \\ & & & & 0.00463 & 0.1667 \\ & & & & & 8.0 \end{bmatrix}\begin{Bmatrix} U_1 \\ U_2 \\ \hline U_3 \\ U_4 \\ U_5 \\ U_6 \end{Bmatrix}$$

$$= \begin{Bmatrix} 0 \\ 0 \\ \hline -90.0 \\ 1440.0 \\ -210.0 \\ -2160.0 \end{Bmatrix} + \begin{Bmatrix} Q_1^{(1)} \\ Q_2^{(1)} \\ \hline Q_3^{(1)} + Q_1^{(2)} \\ Q_4^{(1)} + Q_2^{(2)} \\ Q_3^{(2)} \\ Q_4^{(2)} \end{Bmatrix}$$

Using the boundary conditions on U_i and the equilibrium conditions on $Q_i^{(e)}$,

$$U_1 = U_2 = 0 \qquad Q_3^{(1)} + Q_1^{(2)} = 0 \qquad Q_4^{(1)} + Q_2^{(2)} = 0 \qquad Q_3^{(2)} = Q_4^{(2)} = 0$$

we obtain a system of 4 × 4 equations (not given here) for the unknowns U_3, U_4, U_5, and U_6. The solution is given by

$$U_3 = -0.192\frac{10^6}{EI} \qquad U_4 = 0.72\frac{10^4}{EI} \qquad U_5 = -0.853\frac{10^6}{EI} \qquad U_6 = 0.99\frac{10^4}{EI}$$

The exact solution of the problem is given (in feet) by

$$w(x) = \begin{cases} \dfrac{1}{EI}[-1200x^2 + 50x^3] & 0 \leqslant x \leqslant 4 \text{ ft} \\ \dfrac{1}{EI}\left[-1200x^2 + 50x^3 - \dfrac{5}{36}(x-4)^5\right] & 4 \text{ ft} \leqslant x \leqslant 10 \text{ ft} \end{cases}$$

The reader is asked to verify that the finite-element solution in the present case coincides with the exact solution at the nodes. ■

We close this section with some comments concerning the computation of element matrices and the superposition of the bar element and the beam element to obtain the *frame element*.

1. The computation of element matrices, in any formulation, can be simplified by writing the integrals in the element coordinate. In most physical problems, the loading (on beams) is given in a graphic (i.e., sketch) form rather than as a function expressed in terms of the global coordinate. When the load is discontinuous, it is most convenient to express it in the element coordinates and compute its contribution to the force vector (see Example 3.4 and Fig. 3.14). As will be seen later, two-dimensional problems involve the specification of the secondary variables on discrete portions of the boundary of the domain. In such cases, one can use appropriate one-dimensional interpolation functions to compute the contributions of the specified secondary variables to the nodal force.
2. The two line elements discussed in Secs. 3.2 and 3.3 are identical with regard to the geometry and the number of nodes. Therefore, these two elements can be combined into one line element that would have two nodes with three degrees of freedom (u, w, θ) per node. Such an element is useful in structural mechanics, in the analysis of members undergoing bending as well as extensional deformation. The element stiffness matrix for this element is given by superposing the element matrices associated with the two-node elements of the second-order and fourth-order equations. When the coefficients a and b are

constant in the element, the superposed element matrix is given by

$$[K] = \frac{2b}{h^3}\begin{bmatrix} \mu & 0 & 0 & -\mu & 0 & 0 \\ 0 & 6 & -3h & 0 & -6 & -3h \\ 0 & -3h & 2h^2 & 0 & 3h & h^2 \\ -\mu & 0 & 0 & \mu & 0 & 0 \\ 0 & -6 & 3h & 0 & 6 & 3h \\ 0 & -3h & h^2 & 0 & 3h & 2h^2 \end{bmatrix} \qquad \mu = \frac{ah^2}{2b} \tag{3.129}$$

In most problems of practical interest, for example, frames, such elements are encountered in many arbitrary orientations (with respect to a coordinate system of the frame). Therefore, the transformation of element matrices to a common (global) coordinate system is necessary before assembly is carried out. For a frame element oriented at angle α from the positive x axis in the counterclockwise direction, the element stiffness matrix referred to global coordinates is given by

$$[K^{(e)}] = \frac{2EI}{h^3}\begin{bmatrix} \mu\cos^2\alpha + 6\sin^2\alpha & & & & & \\ (\mu-6)\cos\alpha\sin\alpha & \mu\sin^2\alpha + 6\cos^2\alpha & & & \text{Symmetric} & \\ 3h\sin\alpha & -3h\cos\alpha & 2h^2 & & & \\ -(\mu\cos^2\alpha + 6\sin^2\alpha) & -(\mu-6)\sin\alpha\cos\alpha & -3h\sin\alpha & \mu\cos^2\alpha + 6\sin^2\alpha & & \\ -(\mu-6)\cos\alpha\sin\alpha & -(\mu\sin^2\alpha + 6\cos^2\alpha) & 3h\cos\alpha & (\mu-6)\cos\alpha\sin\alpha & \mu\sin^2\alpha + 6\cos^2\alpha & \\ 3h\sin\alpha & -3h\cos\alpha & h^2 & -3h\sin\alpha & 3h\cos\alpha & 2h^2 \end{bmatrix} \tag{3.130}$$

The reader is asked to verify this result and then obtain the corresponding force vector.

PROBLEMS

3.35–3.46 For the beam problems shown in Figs. P3.35 through P3.46 give:

(*a*) The assembled stiffness matrix and force vector

(*b*) The specified global displacements and forces, and the equilibrium conditions

(*c*) The condensed matrix equations for the primary unknowns (i.e., generalized displacements) and the secondary unknowns (i.e., generalized forces) separately

Solve for the unknown displacements if the number of unknown displacements is < 4 (use Cramer's rule), and compute the bending moment at point C. Use the minimum number of elements required in each problem.

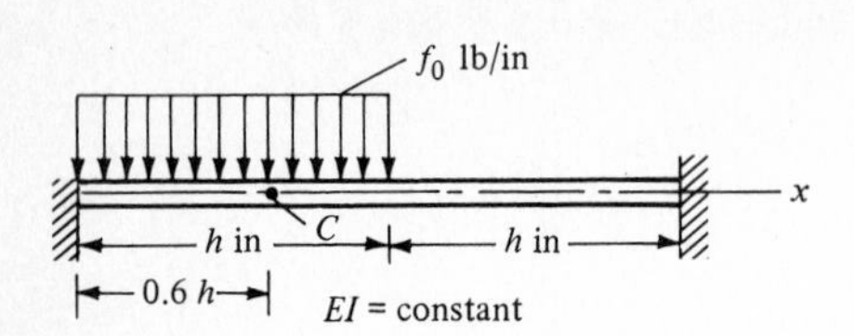

Figure P3.35

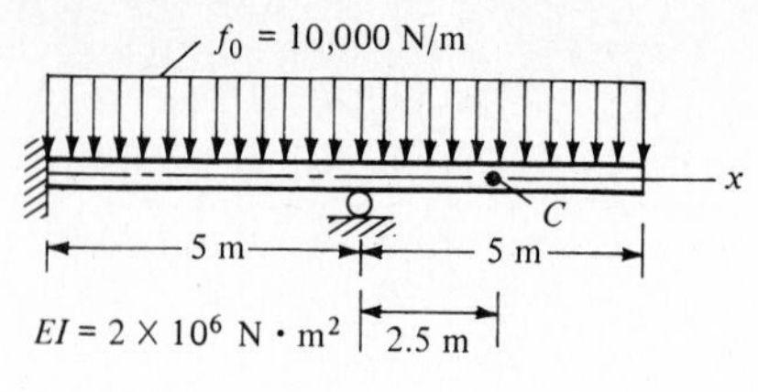

Figure P3.36

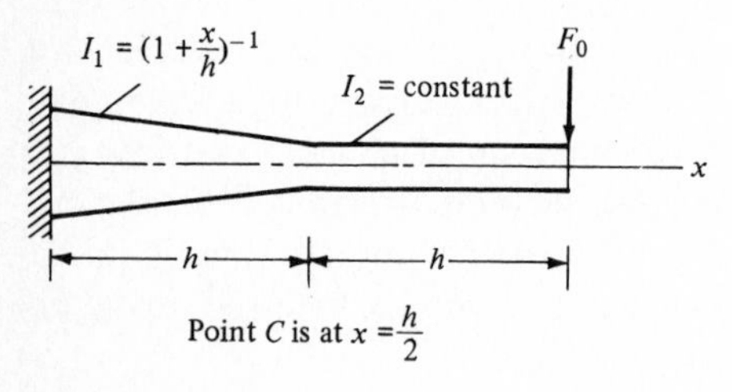

Figure P3.37

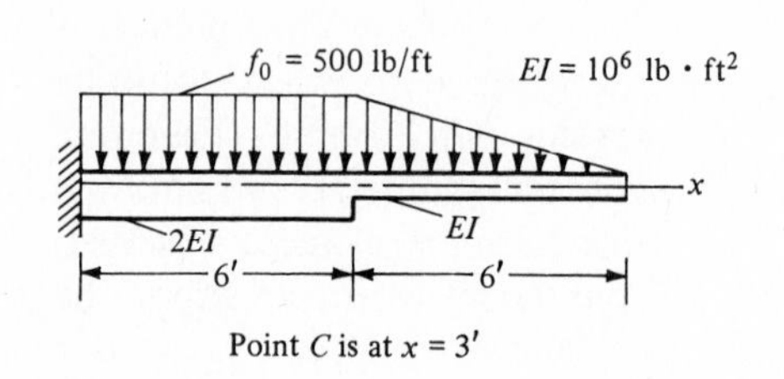

Figure P3.38

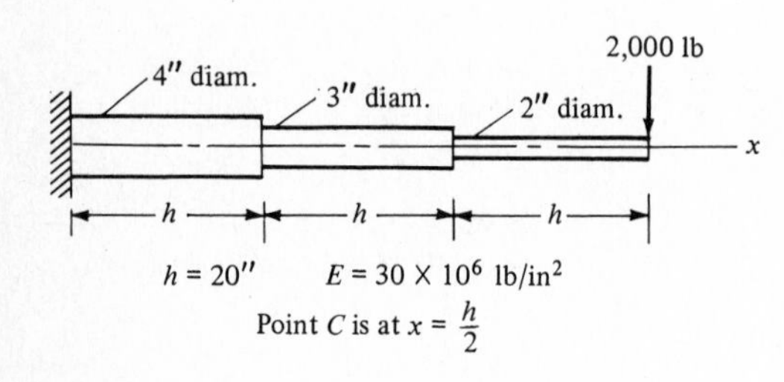

Figure P3.39

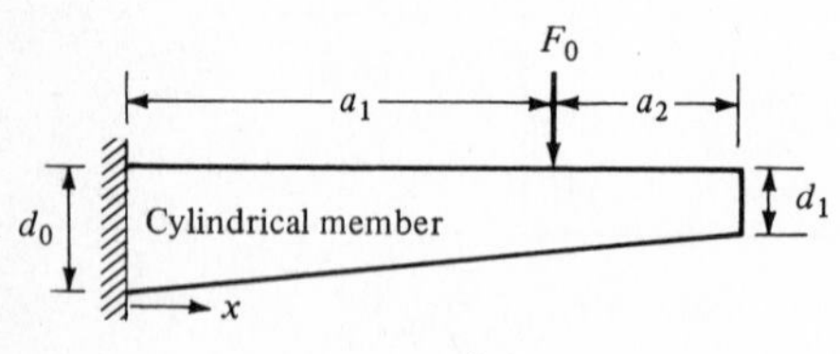

Figure P3.40

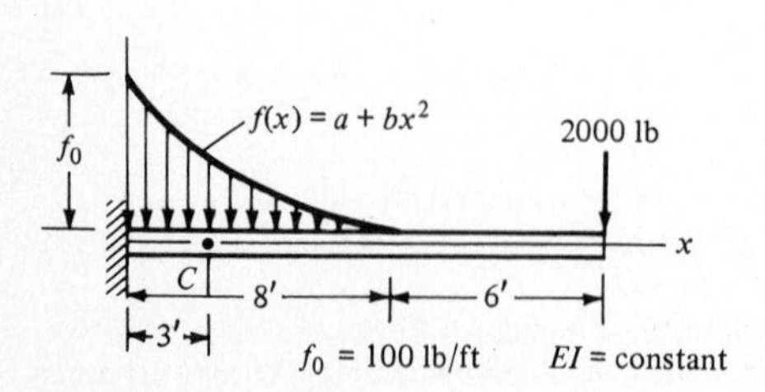

Figure P3.41

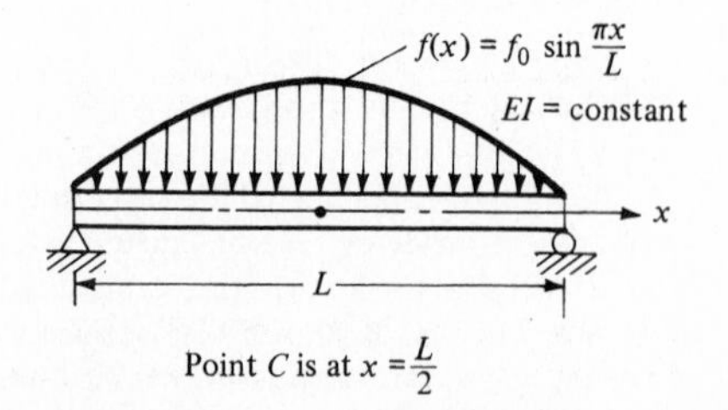

Figure P3.42

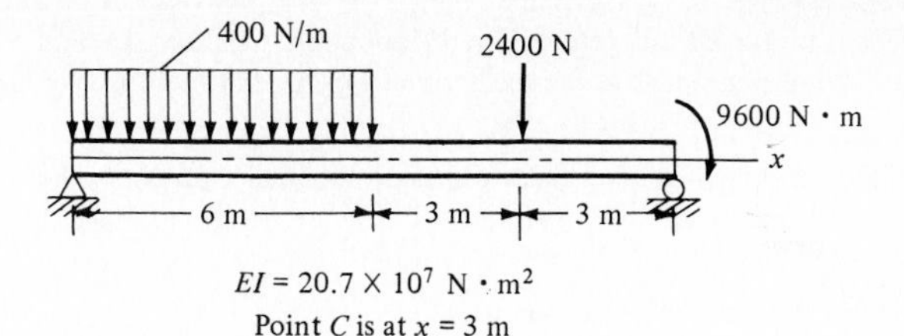

Figure P3.43

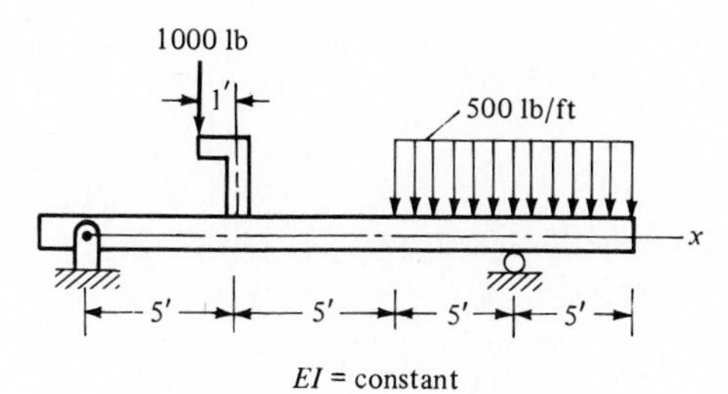

Figure P3.44

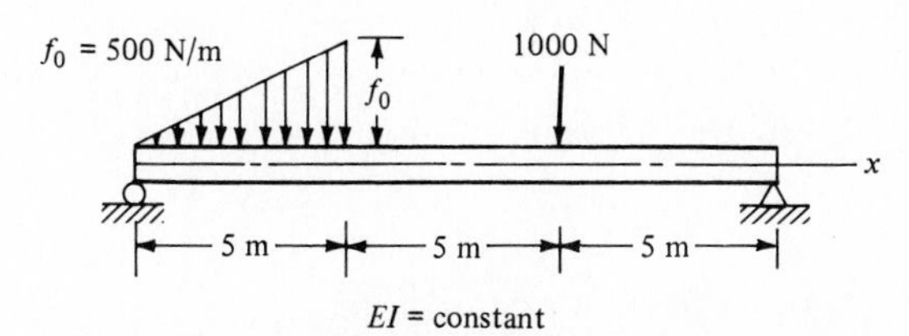

Figure P3.45

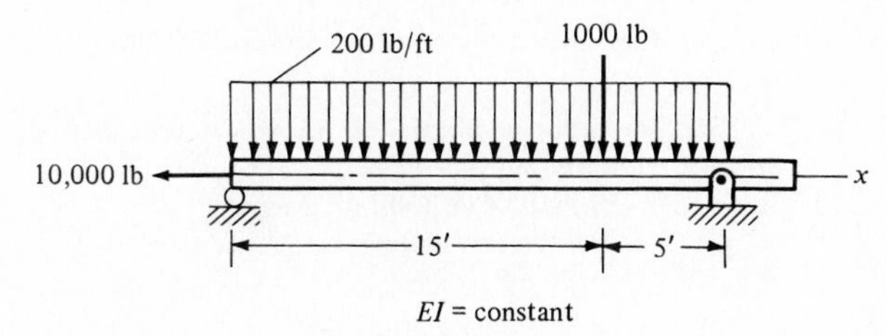

Figure P3.46

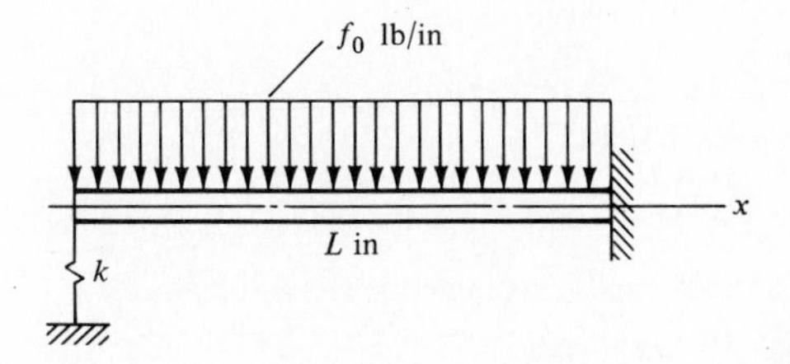

Figure P3.47

3.47 The free end of the linear elastic cantilever beam shown in Fig. P3.47 is supported by a linear elastic spring with a spring constant of k (lb/in). Determine the reaction and the compression in the spring using the one-element model. Assume that $k = EI/L^3$ (lb/in). [*Hint*: Replace the spring by the spring force S acting upward at the free end, and note that $S = -kw(0) = -kU_1$.]

3.48 Consider a simply supported beam on an elastic foundation (foundation modulus k) and subjected to uniform loading. Determine the displacement at the midspan using a one-element model. Take $L = 1$, $EI = 1$, and $f = f_0 = 1$.

3.49 Consider the axisymmetric bending of linearly elastic circular plates of constant thickness. The differential equation is given by

$$\frac{1}{r}\frac{d}{dr}\left[-D_{11}\left(r\frac{d^3w}{dr^3}+\frac{d^2w}{dr^2}\right)+\frac{D_{22}}{r}\frac{dw}{dr}\right]+f=0 \qquad a<r<b$$

where D_{11} and D_{22} are the material stiffnesses (constant). Give:

(a) The variational formulation of the equation over a typical element

(b) The finite-element formulation in the form

$$[K^{(e)}]\{u^{(e)}\} = \{F^{(e)}\}$$

Comment on the interpolation functions that are admissible in the problem.

3.50 By rewriting the differential equation in Prob. 3.49 in the form

$$\frac{1}{r}\frac{d}{dr}\left[r\frac{dM_r}{dr}+(M_r-M_\theta)\right]+f=0$$

$$\frac{d^2w}{dr^2}-\bar{D}_{22}M_r+\bar{D}_{12}M_\theta=0$$

$$\frac{1}{r}\frac{dw}{dr}-\bar{D}_{11}M_\theta+\bar{D}_{12}M_r=0$$

where

$$\bar{D}_{11}=\frac{D_{11}}{D_{12}^2-D_{11}D_{22}} \qquad \bar{D}_{22}=\frac{D_{22}}{D_{11}}\bar{D}_{11} \qquad \bar{D}_{12}=\frac{D_{12}}{D_{11}}\bar{D}_{11}$$

derive a mixed finite-element model that involves w, M_r and M_θ as primary degrees of freedom.

3.51 Solve the problem of a clamped isotropic ($\upsilon = 0.3$) circular plate with uniformly distributed load using the element developed in Prob. 3.49.

3.52 Consider the fourth-order equation (3.86) and its variational formulation (3.90). Suppose that we employ a two-node element with *three* primary variables at each node (w, θ, κ), $\theta = -dw/dx$, $\kappa = d^2w/dx^2$. Show that the associated interpolation (Hermite) functions are given by

$$\phi_1 = (2h^5 - 20x^3h^2 + 30x^4h - 12x^5)/2h^5$$

$$\phi_2 = (2h^5x - 12x^3h^3 + 16x^4h^2 - 6x^5h)/2h^5$$

$$\phi_3 = (x^2h^5 - 3x^3h^4 + 3x^4h^3 - x^5h^2)/2h^5$$

$$\phi_4 = (-20x^3h^2 + 30x^4h - 12x^5)/2h^5$$

$$\phi_5 = (8x^3h^3 - 14x^4h^2 + 6x^5h)/2h^5$$

$$\phi_6 = (x^3h^4 - 2x^4h^3 + x^5h^2)/2h^5$$

where x is the element coordinate with the origin at node 1. Also compute the element matrices.

3.53–3.58 For the truss problems illustrated in Figs. P3.53 through P3.58 give:

(a) The transformed element matrices (the transformed force vector can be calculated from $\{F^{(e)}\} = [T^{(e)}]^T\{\bar{F}^{(e)}\}$; see below)

(b) The assembled element matrices

(c) The condensed matrix equations for the unknown generalized displacements and forces

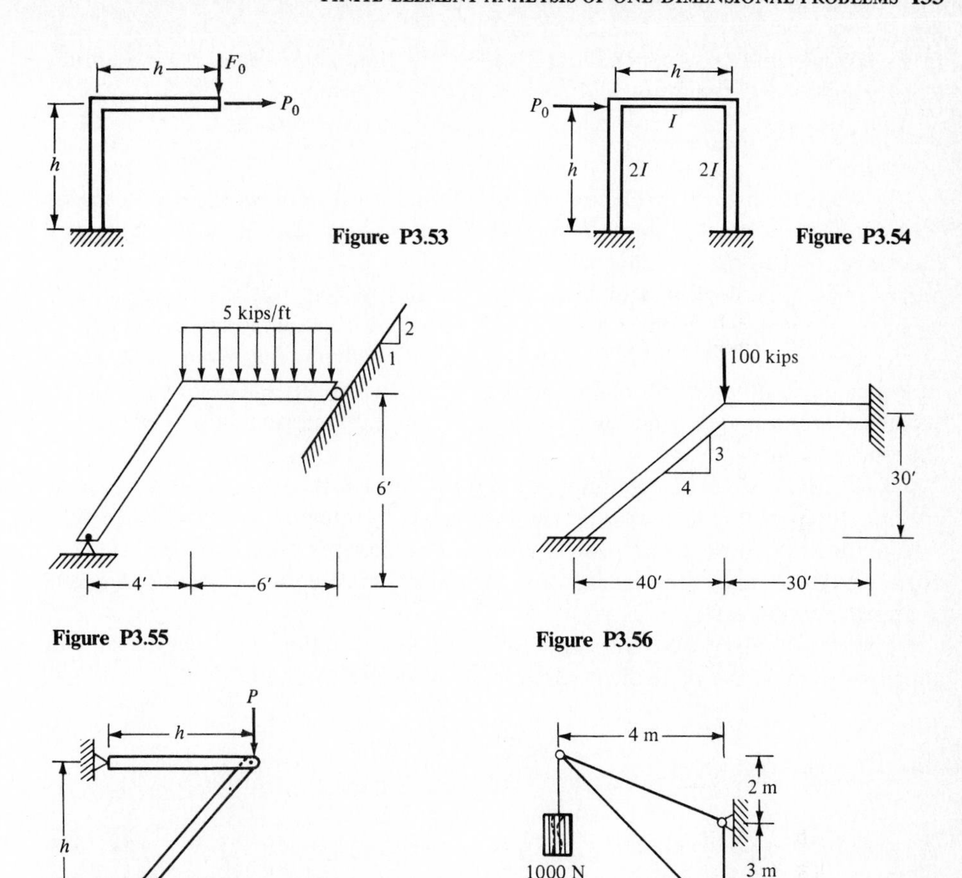

Figure P3.53

Figure P3.54

Figure P3.55

Figure P3.56

Figure P3.57

Figure P3.58

The transformation matrix $[T^{(e)}]$ for the eth element can be calculated from

$$[T^{(e)}] = \begin{bmatrix} [\lambda] & [0] \\ [0] & [\lambda] \end{bmatrix} \qquad [\lambda] = \begin{bmatrix} \cos\alpha & \sin\alpha & 0 \\ -\sin\alpha & \cos\alpha & 0 \\ 0 & 0 & 1 \end{bmatrix}$$

where α is the orientation of the element, measured counterclockwise from the positive x axis.

3.4 APPROXIMATION ERRORS IN THE FINITE-ELEMENT METHOD

3.4.1 Introduction

The errors introduced into the finite-element solution of a given differential equation can be attributed to three basic sources:

1. *Boundary error*. Error due to the approximation of the domain

2. *Quadrature and finite arithmetic errors*. Errors due to the numerical evaluation of integrals and the numerical computation on a computer
3. *Approximation error*. Error due to the approximation of the solution

In one-dimensional problems discussed in the present work, the domains considered were straight lines. Therefore, no approximation of the domain was necessary. In two-dimensional problems posed on nonrectangular domains, domain (or boundary) approximation errors are introduced into the finite-element problems. In general, these errors can be interpreted as errors in the specification of the data of the problem because we are now solving the given differential equation on a modified domain. As we refine the mesh, the domain is more accurately represented, and therefore, the boundary approximation errors are expected to approach zero.

When finite-element computations are performed on a computer, round-off errors and errors due to the numerical evaluation of integrals are introduced into the solution. In most linear problems with a reasonably small number of total degrees of freedom in the system, these errors are expected to be small compared to approximation errors.

The error introduced into the finite-element solution u_h because of the approximation of the dependent variable u is inherent to any problem

$$\begin{aligned} u \approx u_h &= \sum_{e=1}^{N} \sum_{i=1}^{n} u_i^{(e)} \psi_i^{(e)} \\ &= \sum_{I=1}^{M} U_I \Phi_I \end{aligned} \tag{3.131}$$

where N is the number of elements in the mesh, M is the total number of global nodes, and n is the number of nodes in an element. We wish to know how the error $E = u - u_h$, measured in a meaningful way, behaves as the number of elements in the mesh is increased.

3.4.2 Various Measures of Errors

There are several ways one can measure the "difference" (or distance) between any two functions u and u_h. The *pointwise error* is the difference of u and u_h at each point of the domain. We can also define the difference of u and u_h to be the maximum of all absolute values of the differences of u and u_h in the domain $\Omega = (a, b)$:

$$\|u - u_h\|_\infty \equiv \max_{a \leqslant x \leqslant b} |u(x) - u_h(x)| \tag{3.132}$$

This measure of difference is called *supmetric*. Note that the supmetric is a real number, whereas the pointwise error is a function, and does not qualify for a distance or *norm* in a strict mathematical sense.

More generally used measures (or norms) of the difference of two functions are the L_2 norm (pronounced "L two norm"), and the energy norm. For any square-integrable functions u and u_h defined on the domain $\Omega = (a, b)$, the two norms are defined by

$$L_2 \text{ norm:} \qquad \|u - u_h\|_0 = \left\{ \int_a^b |u - u_h|^2 \, dx \right\}^{1/2} \tag{3.133}$$

$$\text{Energy norm:} \qquad \|u - u_h\|_m = \left\{ \int_a^b \sum_{i=0}^{m} \left| \frac{d^i u}{dx^i} - \frac{d^i u_h}{dx^i} \right|^2 dx \right\}^{1/2} \tag{3.134}$$

where $2m$ is the order of the differential equation being solved. The word "energy norm" is used to indicate that it contains the same order derivatives as the quadratic functional (which, for most solid mechanics problems, denotes the energy) associated with the equation. Various measures of the distance between two functions are illustrated in Fig. 3.15a. These definitions can be easily modified for two-dimensional domains.

The finite-element solution u_h is said to *converge in the energy norm* to the true solution u if

$$\|u - u_h\|_m \leqslant ch^p \qquad \text{for } p > 0 \tag{3.135}$$

where c is a constant independent of u and u_h and h is the characteristic length of an element. The constant p is called the *rate of convergence*.

3.4.3 Accuracy of the Solution

Returning to the question of estimating the approximation error, we consider a $2m$th-order differential equation in one dimension ($m = 1$, second-order equations; $m = 2$, fourth-order equations):

$$\sum_{i=1}^{m} (-1)^i \frac{d^i}{dx^i}\left(a_i \frac{d^i u}{dx^i} \right) = f \qquad \text{for } 0 < x < L \tag{3.136}$$

where the coefficients $a_1(x)$ and $a_2(x)$ are assumed to be positive. Suppose that the boundary conditions of the problem are given by

$$u(0) = u(L) = 0 \qquad m = 1, 2 \tag{3.137}$$

$$\frac{du}{dx}(0) = \frac{du}{dx}(L) = 0 \qquad m = 2 \tag{3.138}$$

The variational formulation of Eqs. (3.136) to (3.138) is given by

$$0 = \int_0^L \left(\sum_{i=1}^{m} a_i \frac{d^i v}{dx^i} \frac{d^i u}{dx^i} - vf \right) dx \tag{3.139}$$

The quadratic functional corresponding to the variational form is

$$I(u) = \int_0^L \frac{1}{2} \left[\sum_{i=1}^{m} a_i \left(\frac{d^i u}{dx^i} \right)^2 \right] dx - \int_0^L uf \, dx \tag{3.140}$$

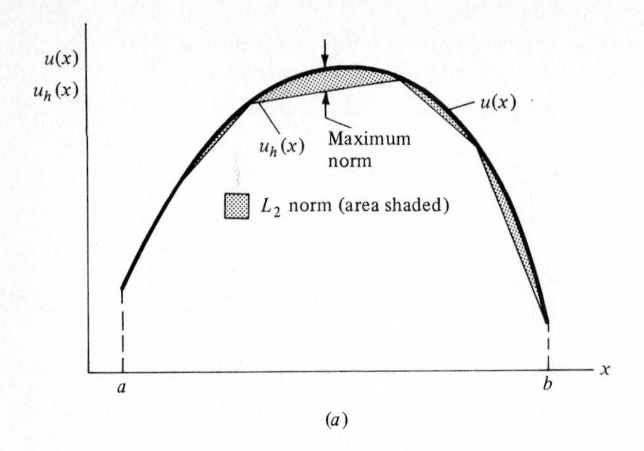

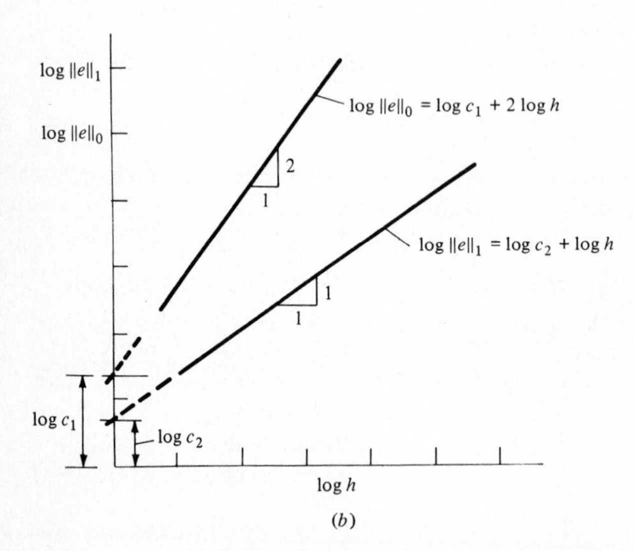

Figure 3.15 Various measures of the difference between two functions. (a) Measures of the difference of two functions u and u_h. (b) Plots of the logarithm of the L_2 error and the energy error versus the logarithm of h.

Now consider a finite-element discretization of the domain by N elements of equal length h. If u_h denotes the finite-element solution in Eq. (3.131), we have from Eq. (3.140)

$$I(u_h) = \int_0^L \frac{1}{2}\left[\sum_{i=1}^{m} a_i \left(\frac{d^i u_h}{dx^i}\right)^2\right] dx - \int_0^L u_h f\, dx \tag{3.141}$$

In the following paragraphs, we show that the energy I associated with the finite-element solution approaches the true energy from above, and then give an error estimate. We confine our discussion, for the sake of simplicity, to the second-order equation ($m = 1$).

From Eqs. (3.140) and (3.141), and

$$f = -\frac{d}{dx}\left(a_1 \frac{du}{dx}\right)$$

we have

$$\begin{aligned}
I(u_h) - I(u) &= \int_0^L \frac{1}{2}\left[a_1\left(\frac{du_h}{dx}\right)^2 - a_1\left(\frac{du}{dx}\right)^2 + 2f(u - u_h)\right] dx \\
&= \int_0^L \left[\frac{a_1}{2}\left(\frac{du_h}{dx}\right)^2 - \frac{a_1}{2}\left(\frac{du}{dx}\right)^2 - \frac{d}{dx}\left(a_1 \frac{du}{dx}\right)(u - u_h)\right] dx \\
&= \int_0^L \left\{\frac{a_1}{2}\left[\left(\frac{du_h}{dx}\right)^2 - \left(\frac{du}{dx}\right)^2\right] + a_1 \frac{du}{dx}\frac{d}{dx}(u - u_h)\right\} dx \\
&= \int_0^L \frac{a_1}{2}\left[\left(\frac{du_h}{dx}\right)^2 + \left(\frac{du}{dx}\right)^2 - 2\frac{du}{dx}\frac{du_h}{dx}\right] dx \\
&= \int_0^L \frac{a_1}{2}\left(\frac{du_h}{dx} - \frac{du}{dx}\right)^2 dx \geqslant 0
\end{aligned} \tag{3.142}$$

Thus, we have

$$I(u_h) \geqslant I(u) \tag{3.143}$$

The equality holds only for $u = u_h$. Equation (3.143) implies that the convergence, if at all it converges, of the energy of the finite-element solution to the true energy is from above. Since the relation in Eq. (3.143) holds for any u_h, the inequality also indicates that the true solution u minimizes the energy. A similar relation can be established for the fourth-order equation ($m = 2$).

Now suppose that the finite-element interpolation functions Φ_I ($I = 1, 2, \ldots, M$) are complete polynomials of degree k. Then the error in the energy norm can be shown to satisfy the inequality

$$\|e\|_m \equiv \|u - u_h\|_{k+1} \leqslant ch^p \qquad p = k + 1 - m > 0 \tag{3.144}$$

where c is a constant. This estimate implies that the error goes to zero at the rate of p as h is decreased (or, the number of elements is increased). In other words, the logarithm of the error in the energy norm versus the logarithm of h is a straight line whose slope is $k + 1 - m$. The greater the degree of the interpolation functions, the faster the rate of convergence. Note also that the error in the energy goes to zero at the rate of $k + 1 - m$; the error in the L_2 norm will be even faster, namely, $k + 1$ (in other words, derivatives converge slower than the solution itself).

Error estimates of the type in Eq. (3.144) are very useful because this gives an idea of the accuracy of the approximate solution, whether we know the true solution or not. While the estimate gives an idea of how rapidly the finite-element solution converges to the true solution, it does not tell us when to stop refining the mesh. This decision rests with the analysts because only they know what a reasonable tolerance is for the problems they are solving.

As an example of estimating the error in the approximation, that is, Eq. (3.144), consider the linear (two-node) element for a second-order equation ($m = 1$). We have for an element

$$u_h = u_1(1 - s) + u_2 s \tag{3.145}$$

where $s = \bar{x}/h$ and $\bar{x}$ is the local coordinate. Since u_2 can be viewed as a function of u_1 via Eq. (3.145), one can expand u_2 in the Taylor series around node 1 to obtain

$$u_2 = u_1 + u_1' + \tfrac{1}{2}u_1'' + \cdots \tag{3.146}$$

where $u' \equiv du/ds$. Substituting this expression into Eq. (3.145), we get

$$u_h = u_1 + u_1's + \tfrac{1}{2}u_1''s + \cdots \tag{3.147}$$

Expanding the true solution by the Taylor series about node 1, we get

$$u = u_1 + u_1's + \tfrac{1}{2}u_1''s^2 + \cdots \tag{3.148}$$

Therefore, we have from Eqs. (3.147) and (3.148)

$$\begin{aligned} |u_h - u| &\leqslant \frac{1}{2}(s - s^2) \max_{0 \leqslant s \leqslant 1} \left| \frac{d^2u}{ds^2} \right| \\ &= \frac{1}{2}(s - s^2)h^2 \max_{0 \leqslant \bar{x} \leqslant h} \left| \frac{d^2u}{d\bar{x}^2} \right| \end{aligned} \tag{3.149}$$

and

$$\left| \frac{d}{d\bar{x}}(u_h - u) \right| \leqslant \frac{1}{2}h \max_{0 \leqslant \bar{x} \leqslant h} \left| \frac{d^2u}{d\bar{x}^2} \right| \tag{3.150}$$

Equations (3.149) and (3.150) lead to

$$\|u - u_h\|_0 \leqslant c_1 h^2 \qquad \text{and} \qquad \|u - u_h\|_1 \leqslant c_2 h \tag{3.151}$$

where the constants c_1 and c_2 depend only on the length L of the domain.

The reader is asked to carry a similar error analysis for the fourth-order equation.

Example 3.5 Here we consider a computational example to verify the error estimates in Eq. (3.151). Consider the differential equation

$$-\frac{d^2u}{dx^2} = 2 \qquad 0 < x < 1$$

$$u(0) = u(1) = 0$$

The exact and the finite-element solutions are given by

$$u(x) = x(1 - x) \tag{3.152}$$

$$N = 2: \qquad u_h = \begin{cases} h^2\left(\dfrac{x}{h}\right) & 0 \leqslant x \leqslant h \\ h^2\left(2 - \dfrac{x}{h}\right) & \text{for } h \leqslant x \leqslant 2h \end{cases} \tag{3.153a}$$

$$N = 3: \qquad u_h = \begin{cases} 2h^2\left(\dfrac{x}{h}\right) & 0 \leqslant x \leqslant h \\ 2h^2\left(2 - \dfrac{x}{h}\right) + 2h^2\left(\dfrac{x}{h} - 1\right) & h \leqslant x \leqslant 2h \\ 2h^2\left(3 - \dfrac{x}{h}\right) & 2h \leqslant x \leqslant 3h \end{cases} \tag{3.153b}$$

$$N = 4: \qquad u_h = \begin{cases} 3h^2\left(\dfrac{x}{h}\right) & 0 \leqslant x \leqslant h \\ 3h^2\left(2 - \dfrac{x}{h}\right) + 4h^2\left(\dfrac{x}{h} - 1\right) & h \leqslant x \leqslant 2h \\ 4h^2\left(3 - \dfrac{x}{h}\right) + 3h^2\left(\dfrac{x}{h} - 2\right) & 2h \leqslant x \leqslant 3h \\ 3h^2\left(4 - \dfrac{x}{h}\right) & 3h \leqslant x \leqslant 4h \end{cases} \tag{3.153c}$$

For the two-element case, the errors are given by ($h = 0.5$)

$$\|u - u_h\|_0^2 = \int_0^h (x - x^2 - hx)^2\, dx + \int_h^{2h} (x - x^2 - 2h^2 + xh)^2\, dx$$
$$= 0.002083$$

$$\left\| \frac{du}{dx} - \frac{du_h}{dx} \right\|_0^2 = \int_0^h (1 - 2x - h)^2\, dx + \int_h^{2h} (1 - 2x + h)^2\, dx \tag{3.154}$$
$$= 0.08333$$

Similar calculations can be made for $N = 3$ and $N = 4$. The table below contains the errors for $N = 2$, 3, and 4:

h	$\log_{10} h$	$\|e\|_0$	$\log_{10}\|e\|_0$	$\|e\|_1$	$\log_{10}\|e\|_1$
$\frac{1}{2}$	−0.301	0.04564	−1.341	0.2887	−0.5396
$\frac{1}{3}$	−0.477	0.02028	−1.693	0.1925	−0.7157
$\frac{1}{4}$	−0.601	0.01141	−1.943	0.1443	−0.8406

A plot of $\log\|e\|_0$ and $\log\|e\|_1$ versus $\log h$ will show that (see Fig. 3.15b)

$$\log\|e\|_0 = 2\log h + \log c_1$$
$$\log\|e\|_1 = \log h + \log c_2 \tag{3.155}$$

In other words, the rate of convergence of the finite-element solution is 2 in the L_2 norm and 1 in the energy norm, verifying the estimates in Eqs. (3.151). ■

Much of the discussion presented in this section can be carried to curved elements and two-dimensional elements. When curved elements (i.e., elements with nonstraight sides) are involved, the error estimate also depends on the Jacobian of the transformation. Due to the introductory nature of the present study, these topics are not discussed here. Interested readers can consult Ciarlet (1978), Mitchell and Wait (1977), Oden and Reddy (1976), and Strang and Fix (1973) (see References for Additional Reading, Chap. 1).

*3.5 TIME-DEPENDENT PROBLEMS

3.5.1 Introduction

This section is devoted to the finite-element formulation of one-dimensional time-dependent problems. For a model problem, we consider the differential equation

$$c_1 \frac{\partial u}{\partial t} + c_2 \frac{\partial^2 u}{\partial t^2} - \frac{\partial}{\partial x}\left(a \frac{\partial u}{\partial x}\right) + \frac{\partial^2}{\partial x^2}\left(b \frac{\partial^2 u}{\partial x^2}\right) + f = 0 \qquad 0 < x < L \tag{3.156}$$

which contains as special cases the time-dependent second-order (for $c_2 = b = 0$ or $c_1 = b = 0$) and fourth-order (for $c_1 = a = 0$) problems. The second-order problems, for example, involve finding the transverse motion of a cable, the longitudinal motion of a rod, and the temperature transients in a fin. The fourth-order problem involves finding the transverse motion of a beam. In the sections following this introduction, we discuss the semidiscrete variational and finite-element formulations of Eq. (3.156) and the fully discrete formulation of the form discussed in Sec. 2.3.4.

3.5.2 Semidiscrete Finite-Element Models

The semidiscrete formulation involves the approximation of the spatial variation of the dependent variable, which follows essentially the same steps as described in the previous sections. The first step involves the construction of the semidiscrete variational formulation of the equation over a typical element. For Eq. (3.156), the typical element is either the bar element (when $b = 0$) of Sec. 3.2 or the beam element (when $a = 0$) of Sec. 3.3.

We begin with the semidiscrete variational formulation, where we have

$$0 = \int_{x_e}^{x_{e+1}} v\left[c_1 \frac{\partial u}{\partial t} + c_2 \frac{\partial^2 u}{\partial t^2} - \frac{\partial}{\partial x}\left(a \frac{\partial u}{\partial x}\right) + \frac{\partial^2}{\partial x^2}\left(b \frac{\partial^2 u}{\partial x^2}\right) + f\right] dx$$

$$= \int_{x_e}^{x_{e+1}} \left(c_1 v \frac{\partial u}{\partial t} + c_2 v \frac{\partial^2 u}{\partial t^2} + a \frac{\partial v}{\partial x}\frac{\partial u}{\partial x} + b \frac{\partial^2 v}{\partial x^2}\frac{\partial^2 u}{\partial x^2} + vf\right) dx$$

$$+ \left\{ v\left[-a \frac{\partial u}{\partial x} + \frac{\partial}{\partial x}\left(b \frac{\partial^2 u}{\partial x^2}\right)\right]\right\}_{x=x_e}^{x=x_{e+1}} - \left[b \frac{\partial v}{\partial x}\frac{\partial^2 u}{\partial x^2}\right]_{x=x_e}^{x=x_{e+1}}$$

$$= \int_{x_e}^{x_{e+1}} \left(c_1 v \frac{\partial u}{\partial t} + c_2 v \frac{\partial^2 u}{\partial t^2} + a \frac{\partial v}{\partial x} \frac{\partial u}{\partial x} + b \frac{\partial^2 v}{\partial x^2} \frac{\partial^2 u}{\partial x^2} + vf \right) dx$$
$$- \hat{Q}_1 v(x_e) - \hat{Q}_3 v(x_{e+1}) - \hat{Q}_2 \left[-\frac{dv}{dx}(x_e) \right] - \hat{Q}_4 \left[-\frac{dv}{dx}(x_{e+1}) \right] \tag{3.157}$$

where $\hat{Q}_2 = Q_2$ and $\hat{Q}_4 = Q_4$ are given by Eqs. (3.89), and $\hat{Q}_1$ and $\hat{Q}_3$ are defined by

$$\hat{Q}_1 = \left[-a \frac{\partial u}{\partial x} + \frac{\partial}{\partial x} \left(b \frac{\partial^2 u}{\partial x^2} \right) \right]_{x=x_e}$$
$$\hat{Q}_3 = \left[a \frac{\partial u}{\partial x} - \frac{\partial}{\partial x} \left(b \frac{\partial^2 u}{\partial x^2} \right) \right]_{x=x_{e+1}} \tag{3.158}$$

Next, we assume that u is interpolated by an expression of the form

$$u = \sum_{j=1}^{r} u_j(t) \psi_j(x) \tag{3.159}$$

where $r \geq 2$ if $b = 0$ and c_1 or c_2 is equal to zero, and $r = 4$ if $a = c_1 = 0$; ψ_j are the corresponding interpolation functions. Equation (3.159) implies that at any arbitrarily fixed time $t > 0$, the function u can be approximated by a linear combination of ψ_j, with u_j being the value of u, at time t, at the jth node of the element. In other words, the time and spatial variations of u are separable. Obviously, one cannot use such approximations for wave propagation problems in which the time and the spatial variations of u cannot be separated. Substituting for $v = \psi_i(x)$ and Eq. (3.159) into Eq. (3.157), we obtain

$$0 = \int_{x_e}^{x_{e+1}} \left(c_1 \psi_i \sum_{j=1}^{r} \frac{du_j}{dt} \psi_j + c_2 \psi_i \sum_{j=1}^{r} \frac{d^2 u_j}{dt^2} \psi_j + a \frac{d\psi_i}{dx} \sum_{j=1}^{r} u_j \frac{d\psi_j}{dx} \right.$$
$$\left. + b \frac{d^2 \psi_i}{dx^2} \sum_{j=1}^{r} u_j \frac{d^2 \psi_j}{dx^2} + \psi_i f \right) dx - \hat{Q}_i$$

or

$$[M^1]\{\dot{u}\} + [M^2]\{\ddot{u}\} + ([K^1] + [K^2])\{u\} = \{F\} \tag{3.160}$$

where

$$M_{ij}^1 = \int_{x_e}^{x_{e+1}} c_1 \psi_i \psi_j \, dx \qquad M_{ij}^2 = \int_{x_e}^{x_{e+1}} c_2 \psi_i \psi_j \, dx$$
$$K_{ij}^1 = \int_{x_e}^{x_{e+1}} a \frac{d\psi_i}{dx} \frac{d\psi_j}{dx} dx \qquad K_{ij}^2 = \int_{x_e}^{x_{e+1}} b \frac{d^2\psi_i}{dx^2} \frac{d^2\psi_j}{dx^2} dx \tag{3.161}$$
$$F_i = -\int_{x_e}^{x_{e+1}} \psi_i f \, dx + \hat{Q}_i$$

This completes the semidiscrete finite-element formulation of Eq. (3.156) over an element.

3.5.3. Time Approximations

As special cases, Eq. (3.160) contains equations of the form in Eqs. (2.123) and (2.124). Therefore, the time approximation of Eq. (3.160) for the two cases, Case 1: $c_2 = 0$, and Case 2: $c_1 = 0$, proceeds along the same lines as described in Sec. 2.3.4. We repeat the main steps of the procedure for the sake of completeness.

Case 1 We have from Eq. (3.160)

$$[M^1]\{\dot{u}\} + [K]\{u\} = \{F\} \tag{3.162}$$

Using the θ family of approximation in Eq. (2.127), we write Eq. (3.162) [see Eq. (2.130)] as a set of linear algebraic equations in a matrix form

$$[\hat{K}^1]\{u\}_{n+1} = \{\hat{F}^1\} \tag{3.163a}$$

where

$$\begin{aligned} [\hat{K}^1] &= [M^1] + \theta\Delta t[K] \\ \{\hat{F}^1\} &= ([M^1] - (1-\theta)\,\Delta t[K])\{u\}_n \\ &\quad + \Delta t(\theta\{F\}_{n+1} + (1-\theta)\{F\}_n) \end{aligned} \tag{3.163b}$$

and Δt is the time step. Equations (3.163) are valid for a typical element. The assembly, the imposition of boundary conditions, and the solution of the assembled equations are the same as described before. The calculation of $[\hat{K}]$ and $\{\hat{F}\}$ requires the knowledge of the initial conditions $\{u\}_0$ and the time variation of $\{F\}$. Note that for $\theta = 0$ we obtain $[\hat{K}^1] = [M^1]$. When the mass matrix $[M^1]$ is diagonal (i.e., the mass of the element is equally lumped at the nodes of the element), Eqs. (3.163) become *explicit*, and one can solve for $\{u\}_{n+1}$ directly.

Case 2 Equation (3.160) takes the form

$$[M^2]\{\ddot{u}\} + [K]\{u\} = \{F\}, \tag{3.164}$$

which resembles Eq. (2.124). Using the Newmark direct integration method [i.e., Eqs. (2.152)] in Eq. (3.164), we obtain

$$[\hat{K}^2]\{u\}_{n+1} = \{\hat{F}^2\} \tag{3.165a}$$

where

$$\begin{aligned} [\hat{K}^2] &= [K] + a_0[M^2] \\ \{\hat{F}^2\} &= \{F\}_{n+1} + [M^2](a_0\{u\}_n + a_1\{\dot{u}\}_n + a_2\{\ddot{u}\}_n) \end{aligned} \tag{3.165b}$$

and $a_0, a_1, \ldots$ are the parameters defined in Eqs. (2.155). Again, Eqs. (3.165) are valid over a typical element. Note that the calculation of $[\hat{K}^2]$ and $\{\hat{F}^2\}$ requires knowledge of the initial conditions $\{u\}_0$, $\{\dot{u}\}_0$, and $\{\ddot{u}\}_0$. In practice, one does not know $\{\ddot{u}\}_0$; it must be calculated from Eq. (3.164):

$$\{\ddot{u}\}_0 = [M^2]^{-1}(\{F\} - [K]\{u\}_0) \tag{3.166}$$

The remaining procedure stays the same as the static (i.e., non-time-dependent) problems.

Example 3.6 Consider the transient heat-conduction problem of Example 2.10. We have $a = 1$, $b = 0$, $c_1 = 1$, $c_2 = 0$, and $f = 0$. For the linear element ($r = 2$), Eq. (3.162) takes the form

$$\frac{h}{6}\begin{bmatrix} 2 & 1 \\ 1 & 2 \end{bmatrix}\begin{Bmatrix} \dot{u}_1 \\ \dot{u}_2 \end{Bmatrix} + \frac{1}{h}\begin{bmatrix} 1 & -1 \\ -1 & 1 \end{bmatrix}\begin{Bmatrix} u_1 \\ u_2 \end{Bmatrix} = \begin{Bmatrix} \hat{Q}_1 = P_1 \\ \hat{Q}_3 = P_2 \end{Bmatrix} \tag{3.167}$$

where h is the length of the element.

For a one-element model, we have $h = 1$ and

$$\begin{bmatrix} \frac{h}{3} + \frac{\theta\,\Delta t}{h} & \frac{h}{6} - \frac{\theta\,\Delta t}{h} \\ \frac{h}{6} - \frac{\theta\,\Delta t}{h} & \frac{h}{3} + \frac{\theta\,\Delta t}{h} \end{bmatrix}\begin{Bmatrix} u_1^{n+1} \\ u_2^{n+1} \end{Bmatrix} = \begin{Bmatrix} \left(\frac{h}{3} - \frac{\theta\,\Delta t}{h}\right)u_1^n + \left(\frac{h}{6} + \frac{\theta\,\Delta t}{h}\right)u_2^n + \Delta t\, P_1^n \\ \left(\frac{h}{6} + \frac{\theta\,\Delta t}{h}\right)u_1^n + \left(\frac{h}{3} - \frac{\theta\,\Delta t}{h}\right)u_2^n + \Delta t\, P_2^n \end{Bmatrix} \tag{3.168}$$

The boundary and initial conditions are

$$\begin{aligned} &u_1^n = 0 \qquad P_2^n = 0 \qquad \text{for any } n \text{ (i.e., } t > 0) \\ &u_1^0 = u_2^0 = 1 \qquad \text{for any } x \end{aligned} \tag{3.169}$$

Let $\theta = 0.5$ and $\Delta t = 0.05$. Then we have

$$\begin{bmatrix} 0.3583 & 0.1416 \\ 0.1416 & 0.3583 \end{bmatrix}\begin{Bmatrix} u_1^{n+1} \\ u_2^{n+1} \end{Bmatrix} = \begin{Bmatrix} 0.3083u_1^n + 0.1913u_2^n + 0.05P_1^n \\ 0.1913u_1^n + 0.3083u_2^n \end{Bmatrix}$$

or

$$u_2^{n+1} = \frac{0.3083}{0.3583}u_2^n = 0.86045u_2^n \tag{3.170}$$

The equation can be evaluated repeatedly for u_2 at different values of time. Clearly, the one-element model is not accurate. Note that the initial condition $u_1^0 = 1$ and the boundary condition $u_1^n = 0$ are contradicting (a kind of singularity). Thus a nonuniform mesh with small elements in the vicinity of the singularity is desirable for an accurate prediction of the solution. The following nonuniform meshes of both linear and quadratic elements were used:

Mesh	x_1	x_2	x_3	x_4	x_5	x_6	x_7	x_8	x_9
2(1)	0.0	0.2	1.0						
4(2)	0.0	0.2	0.5	0.75	1.0				
6(3)	0.0	0.1	0.2	0.35	0.5	0.75	1.0		
8(4)	0.0	0.1	0.2	0.35	0.5	0.6	0.75	0.9	1.0

The first column indicates the number of linear (quadratic) elements in the mesh. The results for $u(1, t)$ for two different time steps and parameters θ are presented in Table 3.4. The best results, when compared to the analytical solution [Eq. (2.150)], are given when $\theta = 0.5$ and $\Delta t = 0.025$. ■

Table 3.4 Effect of time step, mesh, degree of approximation, and parameter θ on the accuracy of the solution of Example 3.6

		Analytic	Linear element				Quadratic element			
Case†	t	$u(1, t)$	2	4	6	8	1	2	3	4
	0.2	0.7723	0.8128	0.7614	0.7738	0.7393	0.7679	0.7632	0.7734	0.7732
	0.4	0.4745	0.4620	0.4648	0.4736	0.4771	0.4668	0.4710	0.4746	0.4741
1	0.6	0.2897	0.2626	0.2822	0.2867	0.3242	0.2837	0.2883	0.2873	0.2890
	0.8	0.1769	0.1493	0.1711	0.1735	0.2360	0.1725	0.1761	0.1743	0.1764
	1.0	0.1080	0.0848	0.1037	0.1051	0.1860	0.1048	0.1074	0.1060	0.1076
	0.2	0.7723	0.8129	0.7674	0.7753	0.7410	0.7680	0.7729	0.7710	0.7728
	0.4	0.4745	0.4627	0.4665	0.4727	0.4775	0.4672	0.4739	0.4740	0.4741
2	0.6	0.2897	0.2632	0.2828	0.2870	0.3244	0.2841	0.2892	0.2895	0.2894
	0.8	0.1769	0.1497	0.1715	0.1742	0.2363	0.1728	0.1765	0.1767	0.1767
	1.0	0.1080	0.0851	0.1040	0.1057	0.1863	0.1051	0.1077	0.1079	0.1079
	0.2	0.7723	0.8214	0.7723	0.7805	0.7466	0.7744	0.7768	0.7779	0.7773
	0.4	0.4745	0.4730	0.4747	0.4811	0.4851	0.4754	0.4820	0.4825	0.4825
3	0.6	0.2897	0.2720	0.2904	0.2946	0.3311	0.2916	0.2967	0.2971	0.2971
	0.8	0.1769	0.1564	0.1776	0.1804	0.2416	0.1789	0.1826	0.1829	0.1829
	1.0	0.1080	0.0899	0.1086	0.1104	0.1902	0.1097	0.1124	0.1126	0.1126
	0.2	0.7723	0.8170	0.7697	0.7775	0.7436	0.7712	0.7747	0.7749	0.7744
	0.4	0.4745	0.4682	0.4708	0.4772	0.4816	0.4715	0.4782	0.4786	0.4786
4	0.6	0.2897	0.2677	0.2868	0.2910	0.3279	0.2881	0.2932	0.2935	0.2935
	0.8	0.1769	0.1533	0.1747	0.1775	0.2390	0.1760	0.1797	0.1800	0.1800
	1.0	0.1080	0.0877	0.1064	0.1082	0.1883	0.1075	0.1102	0.1103	0.1104

†Case 1: $\theta = 0.5$, $\Delta t = 0.05$.
Case 2: $\theta = 0.5$, $\Delta t = 0.025$.
Case 3: $\theta = 0.65$, $\Delta t = 0.05$.
Case 4: $\theta = 0.65$, $\Delta t = 0.025$.

Example 3.7 Consider the transverse motion of the clamped-clamped beam of Example 2.11. We have $a = 0$, $b = 1$, $c_1 = 0$, $c_2 = 1$ and $f = 0$. For a typical beam element we have

$$\frac{h}{420}\begin{bmatrix} 156 & -22h & 54 & 13h \\ -22h & 4h^2 & -13h & -3h^2 \\ 54 & -13h & 156 & 22h \\ 13h & -3h^2 & 22h & 4h^2 \end{bmatrix}\begin{Bmatrix} \ddot{u}_1 \\ \ddot{u}_2 \\ \ddot{u}_3 \\ \ddot{u}_4 \end{Bmatrix}$$

$$+\frac{2}{h^3}\begin{bmatrix} 6 & -3h & -6 & -3h \\ -3h & 2h^2 & 3h & h^2 \\ -6 & 3h & 6 & 3h \\ -3h & h^2 & 3h & 2h^2 \end{bmatrix}\begin{Bmatrix} u_1 \\ u_2 \\ u_3 \\ u_4 \end{Bmatrix} = \begin{Bmatrix} Q_1 \\ Q_2 \\ Q_3 \\ Q_4 \end{Bmatrix} \tag{3.171}$$

For a one-element model in the half-beam ($h = 0.5$), the boundary and initial conditions are

$$u_1 = u_2 = u_4 = 0 \qquad \text{for all} \quad t \geqslant 0 \tag{3.172}$$

$$\left.\begin{aligned} u_3 &= 1 - \frac{\pi}{4} = 0.2146 \qquad \dot{u}_3 = 0 \\ \ddot{u}_3 &= \frac{-K_{33}u_3}{M_{33}} = -110.932 \end{aligned}\right\} \quad \text{for } t = 0 \tag{3.173}$$

Table 3.5 Effect of time step and mesh on the accuracy of the transient response of a clamped beam (see Example 2.11)

	$\Delta t = 0.005$		$\Delta t = 0.0025$			Galerkin in space and exact in time [see Eq. (2.169)]
t	2†	4	2	4	6	
0.01	0.2098	0.2097	0.2098	0.2097	0.2097	0.2157
0.02	0.1950	0.1951	0.1950	0.1951	0.1951	0.1988
0.03	0.1698	0.1698	0.1695	0.1696	0.1696	0.1716
0.04	0.1347	0.1349	0.1345	0.1348	0.1348	0.1356
0.05	0.0931	0.0935	0.0930	0.0932	0.0933	0.0925
0.06	0.0482	0.0483	0.0480	0.0483	0.0483	0.0447
0.07	0.0014	0.0017	0.0014	0.0016	0.0016	−0.0055
0.08	−0.0460	−0.0455	−0.0464	−0.0458	−0.0458	−0.0553
0.09	−0.0923	−0.0916	−0.0928	−0.0921	−0.0920	−0.1023
0.10	−0.1341	−0.1335	−0.1346	−0.1341	−0.1341	−0.1441
0.11	−0.1684	−0.1682	−0.1685	−0.1682	−0.1681	−0.1783
0.12	−0.1932	−0.1931	−0.1932	−0.1932	−0.1932	−0.2034
0.13	−0.2088	−0.2087	−0.2089	−0.2086	−0.2086	−0.2179
0.14	−0.2150	−0.2148	−0.2154	−0.2152	−0.2151	−0.2211
0.15	−0.2112	−0.2110	−0.2113	−0.2112	−0.2112	−0.2129

† Number of elements in half-beam.

Let $\Delta t = 0.0025$, $\alpha = 0.5$, and $\beta = 0.25$. We have

$$a_0 = \frac{4}{(\Delta t)^2} \qquad a_1 = \frac{4}{\Delta t} \qquad a_2 = 1 \qquad a_3 = a_4 = 0.5\,\Delta t$$

In view of the boundary conditions (3.172), we have from Eqs. (3.165)

$$(K_{33} + a_0 M_{33}) u_3^{n+1} = F_3^n \equiv M_{33}(a_0 u_3^n + a_2 \ddot{u}_3^n) \tag{3.174}$$

which can be solved for u_3 at different times. At $t = \Delta t$ (i.e., $n = 0$), we have

$$\left[\frac{156h}{420} + \frac{4}{(\Delta t)^2}\frac{12}{h^3}\right] u_3^1 = \frac{12}{h^3}\left[\frac{4}{(\Delta t)^2} u_3^0 - 110.932\right] \tag{3.175}$$

$$u_3^1 \approx u_3^0 = 0.2146$$

A comparison of the finite-element solution obtained using two different time steps and various number of elements (in the half-beam) with the Galerkin solution of Eq. (2.169) is presented in Table 3.5. The solutions agree very closely. The two-element solution ($\Delta t = 0.0025$) and the Galerkin solution are compared in graphic form in Fig. 3.16 (see Fig. 2.5).

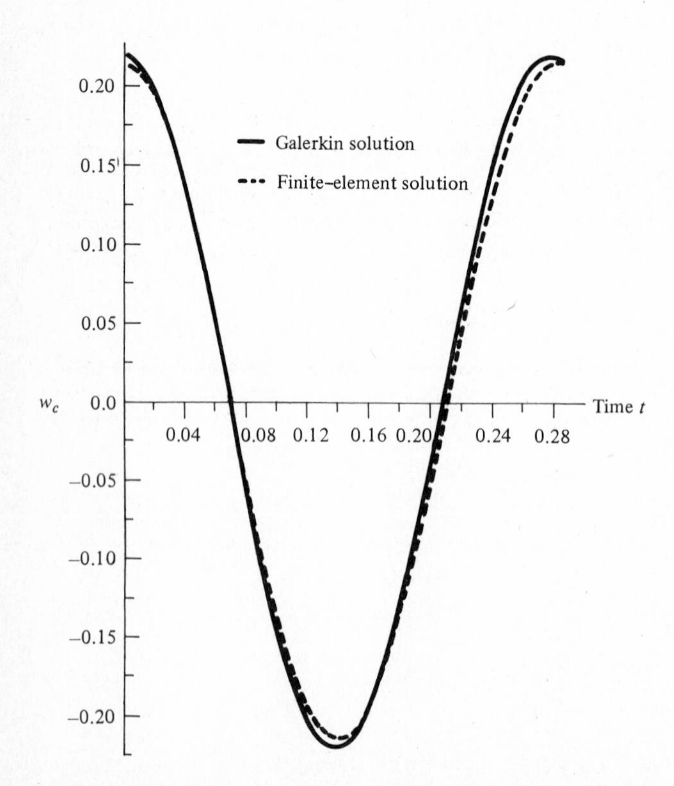

Figure 3.16 Comparison of the Galerkin solution in Eq. (2.169) with the finite-element solution for the center deflection of a clamped beam subjected to nonzero initial conditions.

PROBLEMS

3.59 Show that the error estimate for the fourth-order equation (3.86) is given by

$$\|w - w_h\|_2 \leqslant ch^2$$

where c is a constant, w_h is the finite-element solution obtained by using the Hermite cubic interpolation, and w is the exact solution of the problem.

3.60 Consider a Lagrange quadratic element extending between $-h \leqslant x \leqslant h$ and having the three nodes at $x = -h$, $x = \alpha h$, and $x = h$. The transformation between x and a normalized coordinate ξ is given by

$$x = h\left[\xi + \alpha(1 - \xi^2)\right]$$

If the dependent variable u is interpolated by a quadratic polynomial in ξ, show that the error $e = u - u_h$ is given by

$$\|e\|_1 \leqslant \frac{c\alpha}{(1 - 2\alpha)}h$$

Hint: First show that

$$|u - u_h| \leqslant c \max_{-1 \leqslant \xi \leqslant 1} \left|\frac{d^3u}{d\xi^3}\right|$$

and then

$$\frac{d^3u}{d\xi^3} = -6h^2\alpha(1 - 2\alpha\xi)\frac{d^2u}{dx^2} + h^3(1 - 2\alpha\xi)^3\frac{d^3u}{dx^3}$$

$$\frac{d}{dx} = \frac{1}{h(1 - 2\alpha\xi)}\frac{d}{d\xi}$$

3.61 Consider a two-linear-element approximation of the problem in Example 3.6. Determine the eigenvalues λ by finding the roots of the polynomial $|[B] - \lambda[A]| = 0$, where $[A]$ and $[B]$ are the matrices of the form in Eq. (2.131) (after imposing the boundary conditions of the problem). Determine the condition on the ratio $\Delta t/h^2$ by requiring the smallest eigenvalue to be $\lambda > -1$ (take $\theta = 0.5$).

Use computer program FEM1D discussed in Section 3.7.8 to solve Probs. 3.62-3.68.

3.62 Consider the axial motion of an elastic bar (governed by the second-order equation $EA\ \partial^2 u/\partial x^2 = \rho\ \partial^2 u/\partial t^2$, $0 < x < L$) with the following data: length of bar $L = 500$ mm, area of cross section $A = 1$ mm^2, modulus of elasticity $E = 20{,}000$ N/mm^2, density $\rho = 0.008$ (N $\cdot$ μs^2)/mm^4, boundary conditions

$$u(0, t) = 0 \qquad EA\frac{\partial u}{\partial x}(L, t) = 1$$

and zero initial conditions. Using 50 linear elements and $\Delta t = 0.002$ μs, determine the axial displacement and plot the displacement as a function of position along the bar for $t = 0.08$ μs.

3.63 Consider the following nondimensionalized differential equation governing the plane-wall transient [see Myers (1971), p. 101]:

$$-\frac{\partial^2 T}{\partial x^2} + \frac{\partial T}{\partial t} = 0 \qquad 0 < x < 1$$

with the boundary conditions

$$T(0, t) = 1 \qquad T(1, t) = 0$$

and the initial condition

$$T(x, 0) = 0$$

Solve the problem using (*a*) four, six, and eight linear elements, and (*b*) two, three, and four quadratic elements. Compare the finite-element solutions with the exact solution.

3.64 Consider a simply supported beam of length L subjected to a point load

$$P(t) = \begin{cases} P_0 \sin \dfrac{\pi t}{\tau} & \text{for } 0 \leqslant t \leqslant \tau \\ 0 & \text{for } t \geqslant \tau \end{cases}$$

at a distance c from the left end of the beam (assumed to be at rest at $t = 0$). The transverse deflection $w(x, t)$ is given by [see Harris and Crede (1961), pp. 8–53]

$$w(x,t) = \frac{2P_0L^3}{\pi^4 EI} \sum_{i=1}^{\infty} \frac{1}{i^4} \sin \frac{i\pi c}{L} \sin \frac{i\pi x}{L} \times \left[\frac{1}{1 - T_i^2/4\tau^2} \left(\sin \frac{\pi t}{\tau} - \frac{T_i}{2\tau} \sin \omega_i t \right) \right] \qquad \text{for } 0 \leqslant t \leqslant \tau$$

$$w(x,t) = \frac{2P_0L^3}{\pi^4 EI} \sum_{i=1}^{\infty} \frac{1}{i^4} \sin \frac{i\pi c}{L} \sin \frac{i\pi x}{L} \times \left[\frac{(T_i/\tau)\cos(\pi\tau/T_i)}{T_i^2/4\tau^2 - 1} \sin \omega_i \left(t - \frac{\tau}{2} \right) \right] \qquad \text{for } t \geqslant \tau$$

where

$$T_i = \frac{2\pi}{\omega_i} = \frac{2L^2}{i^2\pi} \frac{A\rho}{EI} = \frac{T_1}{i^2}$$

Use the following data:

$$P_0 = 1000 \text{ lb}, \tau = 20 \times 10^{-6} \text{ s}, L = 30 \text{ in},$$

$$E = 30 \times 10^6 \text{ lb/in}^2, \rho = 733 \times 10^{-6} \text{ lb/in}^3, \Delta t = 10^{-6} \text{ s},$$

and assume that the beam is of square cross section $\frac{1}{2}$ in by $\frac{1}{2}$ in. Using five elements in the half-beam, obtain the finite-element solution and compare with the series solution at the midspan for the case $c = L/2$.

3.65 Repeat Prob. 3.64 for $c = L/4$ and eight elements in the full span.

3.66 Repeat Prob. 3.64 for $c = L/8$ and eight elements in the full span.

3.67 Consider a cantilevered beam with a point load at the free end. Using the load and data of Prob. 3.64, find the finite-element solution for the transverse deflection.

3.68 Repeat Prob. 3.64 for a clamped beam with the load at the midspan.

3.69 Repeat Prob. 3.61 for the mixed formulation of the heat conduction equation in Example 3.6 (use $\theta = 0.5$).

3.70 Use the stability criteria described in Prob. 3.61 ($\lambda_{\min} > -1$) for the fourth-order equation in Example 3.7. (Use the standard values of the Newmark parameters.)

3.6 ISOPARAMETRIC ELEMENTS AND NUMERICAL INTEGRATION

3.6.1 Summary of One-Dimensional Elements

Before we proceed to discuss the isoparametric element concept and the numerical integration techniques, we summarize the basic elements developed in Secs. 3.2 and 3.3. The preceding sections of this chapter were devoted to the finite-element formulations of two *classes* of boundary-value problems in one dimension:

1. Second-order differential equations (bar element)
2. Fourth-order differential equations (beam element)

The frame element, obtained by superposing the bar element and the beam element, was discussed in Sec. 3.3. It should be noted that the variational form as well as the algebraic form of the finite-element model of a given equation in an interval (x_A, x_B) does not depend on the type (linear, quadratic, etc.) of the element. Therefore, one can construct the finite-element model of a given equation without selecting the element type.

The variational formulation of a given differential equation helps us to identify the primary and secondary variables and the type and the minimum degree of interpolation functions that are admissible. The finite-element interpolation functions are constructed such that the finite-element approximation over an element satisfies the end conditions on the primary variables (i.e., it satisfies the essential boundary conditions of the problem) and includes elementwise constant states of the primary variables as well as of the secondary variables. The latter requirement ensures completeness of the set of finite-element interpolation functions and hence the convergence.

Table 3.6 contains a summary of one-dimensional Ritz-finite elements for general second-order and fourth-order equations.

3.6.2 Natural (or Normal) Coordinates

In this section, we discuss the transformation from the global (or problem) coordinate system x to a local coordinate system ξ which has the origin at the center of the element and is scaled such that $\xi = -1$ at the left end node and $\xi = 1$ at the right end node. The transformation is achieved by the linear "stretch" transformation given by

$$\xi = \frac{2x - (x_e + x_{e+1})}{h_e} \tag{3.176}$$

where x_e and x_{e+1} denote the global coordinates of the left and right end nodes, respectively, of element e, and h_e is the element length (see Fig. 3.17). The coordinate ξ is called *normal coordinate* (or *natural coordinate*) to imply that it is a normalized (nondimensional) coordinate whose values are always between -1 and 1. The transformation (3.176) transforms points x $(x_e \leqslant x \leqslant x_{e+1})$ to points $\xi\,(-1 \leqslant \xi \leqslant 1)$.

The normal coordinate system is convenient in two ways. (1) It is convenient in constructing the interpolation functions. (2) It is convenient in the numerical integration (by the Gauss-Legendre quadrature) of the coefficient matrices. First, we discuss the derivation of the Lagrange family of interpolation functions in terms of the natural coordinate system. Recall from Eqs. (3.13) that the interpola-

Table 3.6 Summary of the finite-element equations for one-dimensional problems

Type of equation	Second-order equations	Fourth-order equations
1. Differential equation	$-\frac{d}{dx}\left(a\frac{du}{dx}\right) + cu = f$	$\frac{d^2}{dx^2}\left(b\frac{d^2w}{dx^2}\right) - \frac{d}{dx}\left(a\frac{dw}{dx}\right) + cw = f$
2. Variational form	$0 = \int_{x_A}^{x_B}\left[a\frac{dv}{dx}\frac{du}{dx} + cvu - vf\right]dx - P_A v(x_A) - P_B v(x_B)$ $P_A = -\left(a\frac{du}{dx}\right)_A, \quad P_B = \left(a\frac{du}{dx}\right)_B$	$0 = \int_{x_A}^{x_B}\left[b\frac{d^2v}{dx^2}\frac{d^2w}{dx^2} + a\frac{dv}{dx}\frac{dw}{dx} + cvw - vf\right]dx$ $-Q_1 v(x_A) - Q_2\left(-\frac{dv}{dx}\right)_A - Q_3 v(x_B) - Q_4\left(-\frac{dv}{dx}\right)_B$ $Q_1 = \left[-a\frac{dw}{dx} + \frac{d}{dx}\left(b\frac{d^2w}{dx^2}\right)\right]_A, \quad Q_2 = \left[b\frac{d^2w}{dx^2}\right]_A$ $Q_3 = \left[a\frac{dw}{dx} - \frac{d}{dx}\left(b\frac{d^2w}{dx^2}\right)\right]_B, \quad Q_4 = -\left[b\frac{d^2w}{dx^2}\right]_B$
3. Primary and secondary variables	Primary variable: u Secondary variable: $a\frac{du}{dx}$	Primary variables: $w, \theta \equiv -\frac{dw}{dx}$ Secondary variables: $\frac{d}{dx}\left[b\frac{d^2w}{dx^2}\right] - a\frac{dw}{dx}, b\frac{d^2w}{dx^2}$
4. Finite-element equations	$[K^{(e)}]\{u\} = \{f\} + \{P\}$ $K_{ij}^{(e)} = \int_{x_A}^{x_B}\left(a\frac{d\psi_i}{dx}\frac{d\psi_j}{dx} + c\psi_i\psi_j\right)dx, \quad i, j = 1, 2, \ldots, n.$ $f_i^{(e)} = \int_{x_A}^{x_B}\psi_i f\,dx, \quad P_1 = P_A, \quad P_n = P_B$ $P_i = \text{specified}\ (i = 2, 3, \ldots, n-1).$	$[K^{(e)}]\{\Delta\} = \{f\} + \{Q\}$ $K_{ij}^{(e)} = \int_{x_A}^{x_B}\left(b\frac{d^2\psi_i}{dx^2}\frac{d^2\psi_j}{dx^2} + a\frac{d\psi_i}{dx}\frac{d\psi_j}{dx} + c\psi_i\psi_j\right)dx,$ $f_i^{(e)} = \int_{x_A}^{x_B}\psi_i f\,dx, \quad i, j = 1, 2, 3, 4.$

5. A typical element

Lagrange linear element

(a) Generalized displacements: u_i

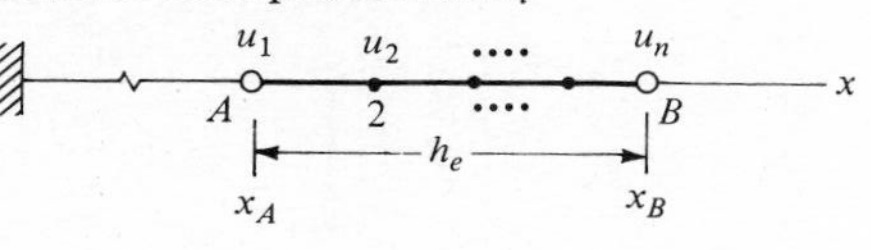

(b) Generalized forces: F_i

Hermite cubic element

(a) Generalized displacements: $w_1, \theta_1, w_2, \theta_2$

(b) Generalized forces: F_i

6. Explicit form of the coefficient matrices

For *Lagrange linear element* (& a, c, and f are constants),

$$[K^{(e)}] = \frac{a}{h}\begin{bmatrix} 1 & -1 \\ -1 & 1 \end{bmatrix} + \frac{ch}{6}\begin{bmatrix} 2 & 1 \\ 1 & 2 \end{bmatrix},$$

$$\{f^{(e)}\} = \frac{fh}{2}\begin{Bmatrix} 1 \\ 1 \end{Bmatrix}.$$

For *Hermite cubic element* (& a, b, c, and f are constants):

$$[K^{(e)}] = \frac{2b}{h^3}\begin{bmatrix} 6 & -3h & -6 & -3h \\ -3h & 2h^2 & 3h & h^2 \\ -6 & 3h & 6 & 3h \\ -3h & h^2 & 3h & 2h^2 \end{bmatrix} + \frac{a}{30h}\begin{bmatrix} 36 & -3h & -36 & -3h \\ -3h & 4h^2 & 3h & -h^2 \\ -36 & 3h & 36 & 3h \\ -3h & -h^2 & 3h & 4h^2 \end{bmatrix} + \frac{ch}{420}\begin{bmatrix} 156 & -22h & 54 & 13h \\ -22h & 4h^2 & -13h & -3h^2 \\ 54 & -13h & 156 & 22h \\ 13h & -3h^2 & 22h & 4h^2 \end{bmatrix}$$

$$\{f^{(e)}\} = \frac{fh}{12}\{-6, h, -6, -h\}^T$$

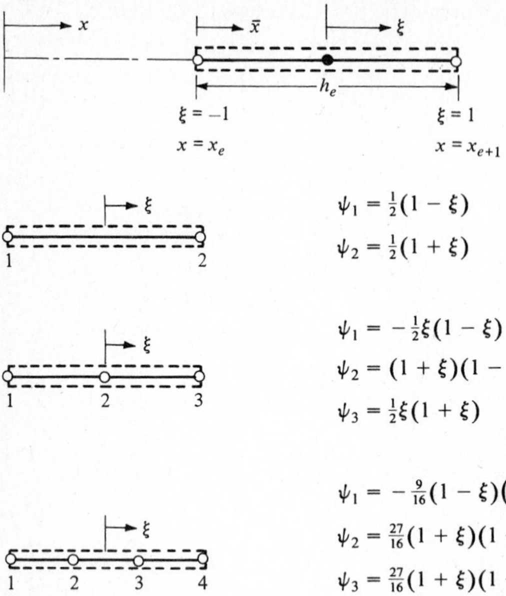

$$\psi_1 = \tfrac{1}{2}(1-\xi)$$
$$\psi_2 = \tfrac{1}{2}(1+\xi)$$

$$\psi_1 = -\tfrac{1}{2}\xi(1-\xi)$$
$$\psi_2 = (1+\xi)(1-\xi)$$
$$\psi_3 = \tfrac{1}{2}\xi(1+\xi)$$

$$\psi_1 = -\tfrac{9}{16}(1-\xi)\left(\tfrac{1}{3}+\xi\right)\left(\tfrac{1}{3}-\xi\right)$$
$$\psi_2 = \tfrac{27}{16}(1+\xi)(1-\xi)\left(\tfrac{1}{3}-\xi\right)$$
$$\psi_3 = \tfrac{27}{16}(1+\xi)(1-\xi)\left(\tfrac{1}{3}+\xi\right)$$
$$\psi_4 = -\tfrac{9}{16}\left(\tfrac{1}{3}+\xi\right)\left(\tfrac{1}{3}-\xi\right)(1+\xi)$$

Figure 3.17 Natural (or normal) coordinate transformation and the Lagrange family of interpolation functions.

tion functions satisfy the property

$$\psi_i(\xi_j) = \begin{cases} 1 & \text{if } i = j \\ 0 & \text{if } i \neq j \end{cases} \tag{3.177}$$

where ξ_j denotes the ξ coordinate of the jth node in the element. For an element with n nodes, ψ_i $(i = 1, 2, \ldots, n)$ are functions of degree $n - 1$. To construct ψ_i which satisfy property (3.177), we proceed as follows (an alternative method). For each ψ_i, we form the product of $n - 1$ linear functions $\xi - \xi_j$ $(j = 1, 2, \ldots, i - 1, i + 1, \ldots, n, j \neq i)$:

$$\psi_i = c_i(\xi - \xi_1)(\xi - \xi_2) \cdots (\xi - \xi_{i-1})(\xi - \xi_{i+1}) \cdots (\xi - \xi_n) \tag{3.178a}$$

Note that ψ_i is zero at all nodes except i. Next we determine the constant c_i such that $\psi_i = 1$ at $\xi = \xi_i$:

$$c_i = \left[(\xi_i - \xi_1)(\xi_i - \xi_2) \cdots (\xi_i - \xi_{i-1})(\xi_i - \xi_{i+1}) \cdots (\xi_i - \xi_n)\right]^{-1} \tag{3.178b}$$

Thus the interpolation function associated with node i is given by

$$\psi_i(\xi) = \frac{(\xi - \xi_1)(\xi - \xi_2) \cdots (\xi - \xi_{i-1})(\xi - \xi_{i+1}) \cdots (\xi - \xi_n)}{(\xi_i - \xi_1)(\xi_i - \xi_2) \cdots (\xi_i - \xi_{i-1})(\xi_i - \xi_{i+1}) \cdots (\xi_i - \xi_n)} \tag{3.179}$$

Interpolation functions which satisfy property (3.177) are said to belong to the *Lagrange family of interpolation functions*, and the associated finite elements belong to the *Lagrange family of finite elements*. The interpolation functions ψ_i in Eqs. (3.17*b*) provide an example of the Lagrange interpolation functions ($n = 2$). The interpolation functions ϕ_i in Eqs. (3.97) do not belong to a Lagrange family of interpolation functions because not all ϕ_i satisfy property (3.177). Figure 3.17 contains the linear, quadratic, and cubic (Lagrange) interpolation functions expressed in terms of the normal coordinate (for equally spaced nodes).

Recall from Eqs. (3.15) and (3.101) that the element matrices involve the derivatives of the interpolation functions with respect to the global coordinate x. Since ψ_i is derived in the natural coordinate system, a transformation of the form

$$x = f(\xi), \qquad \xi = g(x) \tag{3.180}$$

is required in order to rewrite the integrals in terms of ξ. The functions f and g are assumed to be one-to-one transformations. An example of $f(\xi)$ and $g(x)$ is provided by Eq. (3.176):

$$\begin{aligned} f(\xi) &= \tfrac{1}{2}(\xi h_e + x_e + x_{e+1}) \\ g(x) &= \frac{2x - (x_e + x_{e+1})}{h_e} \end{aligned} \tag{3.181}$$

In this case $f(\xi)$ and $g(x)$ are linear functions of ξ and x, respectively. It should be pointed out that when f and g are nonlinear functions, a straight line is transformed (or mapped) into a curve of the same degree as the transformation.

The transformations (3.181) can be chosen in terms of the interpolation functions

$$x = \sum_{i=1}^{r} x_i \psi_i(\xi) \tag{3.182}$$

where ψ_i are the Lagrange interpolation functions of degree $r - 1$. Equation (3.182) describes the shape of an element, and therefore, the ψ_i are sometimes called *shape functions*. Then we have

$$dx = \left(\sum_{i=1}^{r} x_i \frac{d\psi_i}{d\xi} \right) d\xi = J\, d\xi \tag{3.183}$$

where J is the *jacobian* of the transformation $J = dx/d\xi$,

$$J = \sum_{i=1}^{r} x_i \frac{d\psi_i}{d\xi} \tag{3.184}$$

The derivative of ψ_i with respect to x is given by

$$\frac{d\psi_i}{dx} = \frac{d\psi_i}{d\xi}\frac{d\xi}{dx} = \frac{1}{dx/d\xi}\frac{d\psi_i}{d\xi} = (J)^{-1}\frac{d\psi_i}{d\xi} \tag{3.185}$$

3.6.3 Isoparametric Elements

Suppose that a dependent variable u is approximated by expressions of the form

$$u = \sum_{i=1}^{s} u_i \psi_i \tag{3.186}$$

Note that u is approximated by interpolation functions of degree $s - 1$. In general, the degree of approximation used to describe the coordinate transformation (3.182) is not equal to the degree of approximation (3.186) used to represent a dependent variable. In other words, two independent sets of nodes can exist for a region: one set of nodes for the coordinate transformation (3.182), which describes the shape of the element, and the second set of nodes for the interpolation (3.186) of the dependent variable. Depending on the relationship between the degree of approximation r used for the coordinate transformation and the degree of approximation s used for the interpolation, elements are classified into three categories:

$$\begin{array}{lll} \text{1. } \textit{Subparametric elements:} & r < s & \\ \text{2. } \textit{Isoparametric elements:} & r = s & \qquad (3.187) \\ \text{3. } \textit{Superparametric elements:} & r > s & \end{array}$$

Of these three types of elements, isoparametric elements are more commonly used due to the ease and efficiency of calculation in the finite-element implementation.

3.6.4 Numerical Integration

The evaluation of integrals of the form [see Eqs. (3.15) and (3.101)]

$$\int_a^b F(x)\, dx \tag{3.188}$$

by classical (i.e., exact integration) methods is either difficult or impossible due to the complicated form of the integrand F. For example, when the isoparametric elements are used, calculation of the integrals in Eqs. (3.15) is made difficult (except for some simple cases) by the presence of the jacobian in the integrand. Numerical integration is also required when the integrand is to be evaluated inexactly (as in the penalty method) and when the integrand depends on a quantity that is known only at discrete points (e.g., in nonlinear problems). The basic idea behind all numerical integration (also called *quadrature*) techniques is to find a function $P(x)$ that is both a suitable approximation of $F(x)$ and simple to integrate. The interpolating polynomials of degree n, denoted by P_n, which interpolate the integrand at $n + 1$ points of the interval $[a, b]$ often produce suitable approximation and possess the desired property of simple integrability. An illustration of the approximation of the function $F(x)$ by the polynomial $P_4(x)$ that exactly matches the function $F(x)$ at the indicated base points is given in Fig. 3.18*a*. The exact value of Eq. (3.188) is given by the area under the solid curve, and the approximate value

$$\int_a^b P_4(x)\,dx \tag{3.189}$$

is given by the area under the dashed curve. It should be noted that the difference (i.e., the error in the approximation) $E = F(x) - P_4(x)$ is not always of the same sign, and therefore the overall integration error may be small (because positive errors in one part cancel negative errors in other parts), even when P_4 is not a good approximation of F.

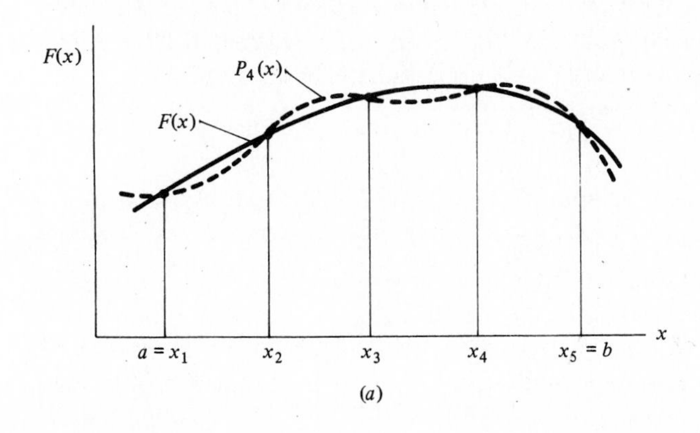

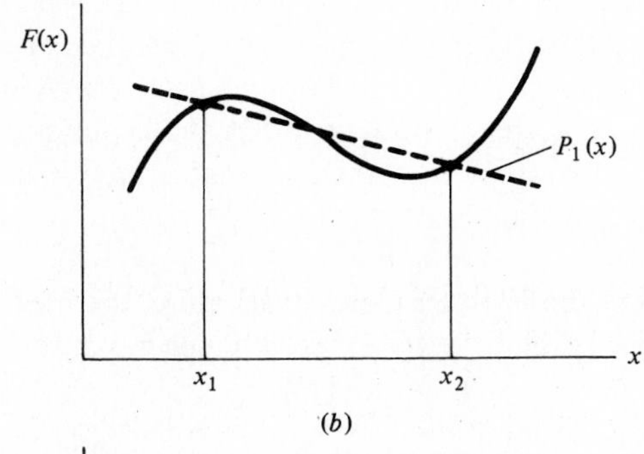

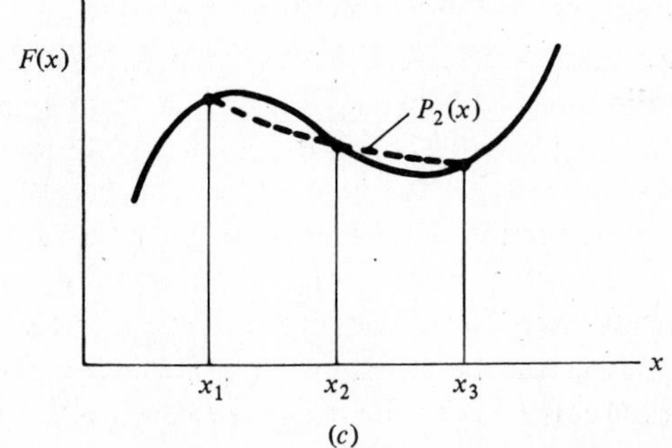

Figure 3.18 Numerical integration by the Newton-Cotes quadrature. (*a*) Approximation of a function by $P_4(x)$. (*b*) The trapezoidal rule. (*c*) Simpson's rule.

The commonly used integration methods can be classified into two basic groups: (1) The Newton-Cotes formula that employs values of the function at equally spaced base or sample points and (2) the Gauss quadrature formula that employs unequally spaced sample points.

The Newton-Cotes quadrature For $n + 1$ equally spaced base points, the Newton-Cotes integration formula is given by

$$\int_a^b F(x)\, dx = (b - a) \sum_{i=1}^{n+1} w_i F(x_i) \tag{3.190}$$

where w_i are the weighting coefficients and x_i the base points, which are equally spaced. For $n = 1$, Eq. (3.190) gives the familiar *trapezoidal rule*, in which the required area under the solid curve in Fig. 3.18*b* is approximated by the area under the dotted straight line [i.e., $F(x)$ is approximated by $P_1(x)$]:

$$\int_{a=x_1}^{b=x_2} F(x)\, dx = \frac{h}{2}[F(x_1) + F(x_2)] \qquad E = 0(h^3) \tag{3.191}$$

where E denotes the error in the approximation and h is the spacing between two base points. For $n = 2$, Eq. (3.190) gives the familiar *Simpson's rule* (see Fig. 3.18*c*):

$$\int_{a=x_1}^{b=x_3} F(x)\, dx = \frac{h}{3}[F(x_1) + 4F(x_2) + F(x_3)] \qquad E = 0(h^4) \tag{3.192}$$

The weighting coefficients for $n = 1, 2, \ldots, 6$ are given in Table 3.7. Note that $\sum_{i=1}^{n+1} w_i = 1$.

A comment is in order on the use of the Newton-Cotes integration formula (3.190). When n is even (i.e., when there is an even number of intervals or an odd number of base points), the formula is exact for $F(x)$ being a polynomial of degree $n + 1$ or less; when n is odd, the formula is exact for a polynomial of degree n or less. Conversely, an nth-order polynomial is integrated exactly by choosing $n + 1$ base points.

The Gauss-Legendre quadrature In the Newton-Cotes quadrature, the base points have been specified (to be equally spaced). If the x_i are not specified, then there

Table 3.7 Weighting coefficients for the Newton-Cotes formula (3.190)†

n	w_1	w_2	w_3	w_4	w_5	w_6	w_7
1	1/2	1/2					
2	1/6	4/6	1/6				
3	1/8	3/8	3/8	1/8			
4	7/90	32/90	12/90	32/90	7/90		
5	19/288	75/288	50/288	50/288	75/288	19/288	
6	41/840	216/840	27/840	272/840	27/840	216/840	41/840

†For $n = 0$, we have $h = b - a$, $w_1 = 1$, $x_1 = (b - a)/2$.

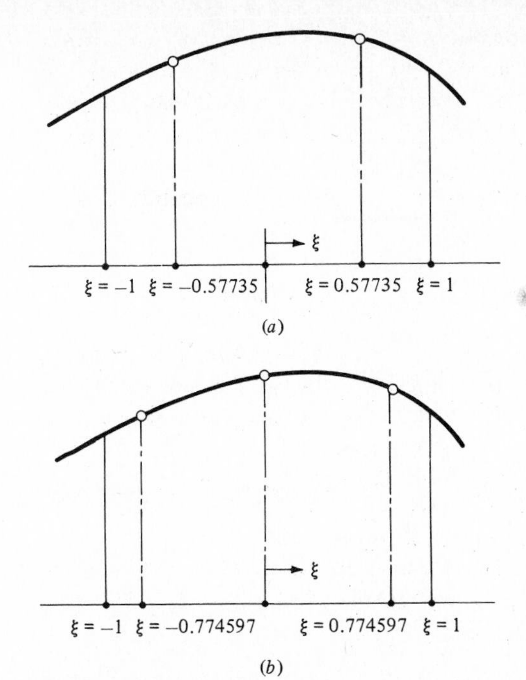

Figure 3.19 Gauss-Legendre quadrature. (*a*) The two-point Gauss-Legendre quadrature. (*b*) The three-point Gauss-Legendre quadrature.

will be $2n + 2$ undetermined parameters (the weights w_i and base points x_i) which define a polynomial of degree $2n + 1$. The Gauss-Legendre quadrature is based on this idea. The base points x_i and the weights w_i are chosen so that the sum of the $n + 1$ appropriately weighted values of the function yields the integral exactly when $F(x)$ is a polynomial of degree $2n + 1$ or less. The Gauss-Legendre quadrature formula is given by (see Fig. 3.19)

$$\int_a^b F(x)\,dx = \int_{-1}^{1} \hat{F}(\xi)\,d\xi = \sum_{i=1}^{n} w_i \hat{F}(\xi_i) \tag{3.193}$$

where w_i are the weight factors, ξ_i the base points [roots of the Legendre polynomial $P_{n+1}(\xi)$], and $\hat{F}$ is the transformed (to the natural coordinate system) integrand

$$\hat{F}(\xi) = F(x(\xi)) \cdot J(\xi) \tag{3.194}$$

The weight factors and Gauss points for the Gauss-Legendre quadrature (3.193) are given, for $n = 1, \ldots, 6$, in Table 3.8. The Gauss-Legendre quadrature is more frequently used than the Newton-Cotes quadrature because the Gauss-Legendre quadrature requires fewer base points (hence a saving in computation) to achieve the same accuracy. The error in the approximation is zero if the $(2n + 2)$th derivative of the integrand vanishes. In other words, a polynomial of degree n is integrated exactly by employing $(n + 1)/2$ Gauss points. When $n + 1$ is odd, one should pick the nearest larger integer.

Table 3.8 Weights and Gauss points for the Gauss-Legendre quadrature $\int_{-1}^{1} F(\xi)\, d\xi = \sum_{i=1}^{n} F(\xi_i) w_i$

Points ξ_i	n	Weights w_i
0.0000000000	One-point formula	2.0000000000
$\pm$0.5773502692	Two-point formula	1.0000000000
0.0000000000	Three-point formula	0.8888888889
$\pm$0.7745966692		0.5555555555
$\pm$0.3399810435	Four-point formula	0.6521451548
$\pm$0.8611363116		0.3478548451
0.0000000000	Five-point formula	0.5688888889
$\pm$0.5384693101		0.4786286705
$\pm$0.9061798459		0.2369268850
$\pm$0.2386191861	Six-point formula	0.4679139346
$\pm$0.6612093865		0.3607615730
$\pm$0.9324695142		0.1713244924

Example 3.8 Consider the numerical evaluation of the following two integrals:

$$K_{11} = \int_{x_e}^{x_{e+1}} \left(\frac{d\psi_1}{dx}\right)^2 dx, \qquad G_{11} = \int_{x_e}^{x_{e+1}} (\psi_1)^2\, dx \tag{3.195}$$

where ψ_1 is the quadratic Lagrange polynomial associated with node 1 of the three-node Lagrange element (see Fig. 3.17):

$$\psi_1(\xi) = -\frac{\xi}{2}(1 - \xi)$$

If we choose the linear transformation (3.176), we have ($J_e = h_e/2$)

$$dx = \frac{h_e}{2} d\xi \qquad \frac{d\psi_1}{dx} = \frac{d\psi_1}{d\xi}\frac{d\xi}{dx} = \frac{2}{h_e}\frac{d\psi_1}{d\xi} \tag{3.196}$$

and expressions K_{11} and G_{11} become

$$K_{11} = \frac{2}{h_e}\int_{-1}^{1}\left(\frac{d\psi_1}{d\xi}\right)^2 d\xi \qquad G_{11} = \frac{h_e}{2}\int_{-1}^{1}(\psi_1)^2\, d\xi \tag{3.197}$$

where $h_e = x_{e+1} - x_e$. Note that the integrand in K_{11} is a polynomial of degree 2, and that in G_{11} is a polynomial of degree 4:

$$\frac{d\psi_1}{d\xi} = -\tfrac{1}{2} + \xi \qquad \left(\frac{d\psi_1}{d\xi}\right)^2 = \tfrac{1}{4} + (\xi)^2 - \xi$$

$$(\psi_1)^2 = \tfrac{1}{4}(\xi^2 - 2\xi^3 + \xi^4) \tag{3.198}$$

We use both kinds of quadratures to evaluate the integrals.

The Newton-Cotes quadrature Let $K_{11}^{(m)}$ denote the result obtained by using the m-point Newton-Cotes formula. We have $b - a = 2$, and

$$K_{11}^{(2)} = \frac{4}{2h_e}\left[\left(\frac{1}{4} + 1 + 1\right) + \left(\frac{1}{4} + 1 - 1\right)\right] = \frac{5}{h_e}$$

$$K_{11}^{(3)} = \frac{4}{6h_e}\left(\frac{5}{2} + 4 \times \frac{1}{4}\right) = \frac{7}{3h_e}$$

$$K_{11}^{(4)} = \frac{4}{8h_e}\left[\frac{5}{2} + 3\left(\frac{1}{4} + \frac{1}{9}\right)2\right] = \frac{7}{3h_e}$$

$$G_{11}^{(2)} = \frac{h_e}{4}\frac{1}{2}[(1 + 2 + 1) + (1 - 2 + 1)] = \frac{h_e}{2} \tag{3.199}$$

$$G_{11}^{(3)} = \frac{h_e}{4}\frac{1}{6}(4 + 4 \times 0) = \frac{h_e}{6}$$

$$G_{11}^{(4)} = \frac{h_e}{4}\frac{1}{8}\left[4 + 3\left(\frac{1}{9} + \frac{1}{81}\right)2\right] = \frac{4h_e}{27}$$

$$G_{11}^{(5)} = \frac{h_e}{4}\frac{1}{90}\left[7(4) + 32\left(\frac{1}{4} + \frac{1}{16}\right)2\right] = \frac{2h_e}{15}$$

$$G_{11}^{(6)} = \frac{h_e}{4}\frac{1}{288}\left[19(4) + 75\left(\frac{9}{25} + \frac{81}{625}\right)2 + 50\left(\frac{1}{25} + \frac{1}{625}\right)2\right] = \frac{2h_e}{15}.$$

Thus K_{11} is integrated exactly by the three-point formula, whereas G_{11} is integrated exactly by the five-point formula, confirming the statement made earlier.

The Gauss-Legendre quadrature Here we have

$$K_{11}^{(1)} = \frac{2}{h_e}\left(\frac{1}{4}\right)2 = \frac{1}{h_e}$$

$$K_{11}^{(2)} = \frac{2}{h_e}\left[\frac{1}{4} + (-0.57735)^2\right]2 = \frac{2.33333}{h_e}\left(= \frac{7}{3h_e}\right)$$

$$K_{11}^{(3)} = \frac{2}{h_e}\left[\frac{1}{4}(0.88889) + 2\left(\frac{1}{4} + 0.77459^2\right)(0.55555)\right] = \frac{2.33333}{h_e}\left(= \frac{7}{3h_e}\right)$$

$$G_{11}^{(2)} = \frac{h_e}{2}\frac{1}{4}\left[(0.57735)^2 + (0.57735)^4\right]2 = 0.11111h_e\left(= \frac{h_e}{9}\right) \tag{3.200}$$

$$G_{11}^{(3)} = \frac{h_e}{2}\frac{1}{4}\left[(0.77459)^2 + (0.77459)^4\right](0.55555)2 = 0.13333h_e\left(= \frac{2h_e}{15}\right)$$

Note that the Gauss-Legendre quadrature gives the same accuracy for fewer integration points. As noted earlier, the second-degree polynomial ($n = 2$) is integrated exactly by $\left[(n + 1)/2 = \frac{3}{2}\right]$ two-point Gauss quadrature and the fourth-degree polynomial by three-point Gauss quadrature.

When an isoparametric element is used, evaluation of K_{11} and G_{11} by hand becomes more involved. However, for a digital computer such computations are trivial. It should be pointed out that the jacobian matrix will be the same ($J_e = h_e/2$) when the element is straight, even if the coordinate transformation involves the quadratic or cubic elements. However, when the element is curved, the jacobian is a function of ξ when the transformation involves other than linear interpolation functions. For example, when the quadratic isoparametric element is used, we have from Eq. (3.184)

$$J_e = \sum_{i=1}^{3} x_i^e \frac{d\psi_i^{(e)}}{d\xi}$$

where x_i^e is the global coordinate of node i. For a straight element we have $x_1^e = x_e$, $x_2^e = x_e + h_e/2$, and $x_3^e = x_{e+1} (= x_e + h_e)$ and

$$J_e = x_e\left(-\frac{1}{2} + \xi\right) + \left(x_e + \frac{h_e}{2}\right)(-2\xi) + (x_e + h_e)\left(\frac{1}{2} + \xi\right) = \frac{h_e}{2} \tag{3.201}$$

Thus J_e does not depend on ξ.

The Gauss-Legendre quadrature of the finite-element matrices

$$K_{ij} = \int_{x_e}^{x_{e+1}} \frac{d\psi_i}{dx}\frac{d\psi_j}{dx}\,dx \qquad G_{ij} = \int_{x_e}^{x_{e+1}} \psi_i \psi_j \, dx \qquad F_i = \int_{x_e}^{x_{e+1}} \psi_i \, dx \tag{3.202}$$

for the quadratic and cubic interpolation functions (see Fig. 3.17) yields the following values (up to the fifth decimal point):

Quadratic (three-point formula)

$$[K] = \frac{1}{h_e}\begin{bmatrix} 2.33333 & -2.66667 & 0.33333 \\ -2.66667 & 5.33333 & -2.66667 \\ 0.33333 & -2.66667 & 2.33333 \end{bmatrix}$$

$$[G] = \frac{h_e}{10}\begin{bmatrix} 1.33333 & 0.66667 & -0.33333 \\ 0.66667 & 5.33333 & 0.66667 \\ -0.33333 & 0.66667 & 1.33333 \end{bmatrix} \tag{3.203}$$

$$\{F\} = h_e\begin{Bmatrix} 0.166667 \\ 0.666667 \\ 0.166667 \end{Bmatrix}$$

Cubic (four-point formula)

$$[K] = \frac{1}{h_e}\begin{bmatrix} 3.7 & -4.725 & 1.35 & -0.325 \\ -4.725 & 10.8 & -7.425 & 1.35 \\ 1.35 & -7.425 & 10.8 & -4.725 \\ -0.325 & 1.35 & -4.725 & 3.7 \end{bmatrix}$$

$$[G] = \frac{h_e}{10}\begin{bmatrix} 0.761905 & 0.589286 & -0.214286 & 0.113095 \\ 0.589286 & 3.85714 & -0.482143 & -0.214286 \\ -0.214286 & -0.482143 & 3.85714 & 0.589286 \\ 0.113095 & -0.214286 & 0.589286 & 0.761905 \end{bmatrix}$$

$$\{F\} = h_e\begin{Bmatrix} 0.125 \\ 0.375 \\ 0.375 \\ 0.125 \end{Bmatrix} \tag{3.204}$$

One can use these results to analyze the second-order differential equation (3.1), for constant coefficient a and source term f, for any set of boundary conditions. ■

Example 3.9 Consider the problem of radial axisymmetric heat flow in a long cylinder (see Prob. 3.31). The governing differential equation is given by

$$\begin{gathered} -\frac{d}{dr}\left(kr\frac{dT}{dr}\right) = Qr \qquad r_i < r < r_o \\ T(r_i) = T_i \qquad T(r_o) = T_o \end{gathered} \tag{3.205}$$

The variational formulation of the equation over an element between points A and B is given by

$$0 = \int_{r_A}^{r_B}\left(kr\frac{dv}{dr}\frac{dT}{dr} - vQr\right)dr - q_A v(r_A) - q_B v(r_B) \tag{3.206}$$

where v is the test function, and

$$q_A = -\left(kr\frac{dT}{dr}\right)_{r=r_A} \qquad q_B = \left(kr\frac{dT}{dr}\right)_{r=r_B} \tag{3.207}$$

We assume a finite-element interpolation of the form

$$T = \sum_{i=1}^{n} T_i\psi_i \tag{3.208}$$

where n is the number of nodes per element and ψ_i are the Lagrange interpolation functions of order $n-1$. The finite-element equations for an element are given by

$$0 = \sum_{j=1}^{n}\left[\int_{r_A}^{r_B}\left(kr\frac{d\psi_i}{dr}\frac{d\psi_j}{dr} - \psi_i Qr\right)dr\right]T_j - q_A\psi_i(r_A) - q_B\psi_i(r_B) \tag{3.209}$$

for $i = 1, 2, \dots, n$.

It should be pointed out that if $n > 2$ and there exists specified heat flux q_0 at a node or nodes between the end nodes, one must append the variational form with the term $-q_0 v(r_0)$, where r_0 is the global coordinate of the node at which q_0 is specified. Alternatively, one can include all specified heat fluxes at internal nodes (of an element) directly in the global force vector. Both ways are equivalent to modifying the differential equation (3.205) to account for the specified point sources. Suppose that there are M such point sources of strength q_j acting at point r_j $(j = 1, 2, \ldots, M)$ of the domain. Then the differential equation (3.205) becomes

$$-\frac{d}{dt}\left(kr\frac{dT}{dr}\right) - Qr - \sum_{j=1}^{M} q_j \delta(r - r_j) = 0 \tag{3.210}$$

where $\delta(r - r_j)$ is the Dirac delta function defined in Eq. (3.122). The variational form becomes

$$0 = \int_{r_A}^{r_B} \left(kr\frac{dv}{dr}\frac{dT}{dr} - vQr\right) dr - q_A v(r_A) - q_B v(r_B) - \sum_{j=1}^{m} q_j v(r_j) \tag{3.211}$$

where m is the number of point sources in the element (not including the point sources at the end nodes A and B).

Equations (3.209) can be expressed in matrix form

$$[K^{(e)}]\{T^{(e)}\} = \{F^{(e)}\} \tag{3.212a}$$

and

$$K_{ij}^{(e)} = \int_{r_A}^{r_B} kr\frac{d\psi_i}{dr}\frac{d\psi_j}{dr}\,dr$$

$$F_i^{(e)} = \int_{r_A}^{r_B} Qr\psi_i\,dr + q_A\psi_i(r_A) + q_B\psi_i(r_B) \tag{3.212b}$$

For any Lagrange element, r_A and r_B denote the global coordinates of the first and last nodes, respectively, of the element. Note that the last two terms in F_i vanish for any intermediate node of an element [because $\psi_i(r_A) = 0$ for $i \neq 1$ and $\psi_i(r_B) = 0$ for $i \neq n$].

The calculation of element matrices using the n-point Gauss quadrature, where n is the number of nodes in an element, is straightforward. Using the transformation (3.176), we write

$$\xi = \frac{2r - (r_A + r_B)}{h_e} \qquad h_e = r_B - r_A$$

or

$$r = \frac{1 - \xi}{2} r_A + \frac{1 + \xi}{2} r_B \tag{3.213}$$

We have for any straight-line element and constant k

$$K_{ij}^{(e)} = \frac{2k}{h_e}\int_{-1}^{1}\left(\frac{1 - \xi}{2} r_A + \frac{1 + \xi}{2} r_B\right)\frac{d\psi_i}{d\xi}\frac{d\psi_j}{d\xi}\,d\xi \tag{3.214}$$

Table 3.9 Comparison of the finite-element solutions with the exact solutions of the radial axisymmetric heat-flow problem of Example 3.9

r	Linear (6 elements)	Quadratic (3 elements)	Cubic (2 elements)	Exact
1.75	400.0000	400.0000	400.0000	400.0000
2.00	331.0090	330.9848	330.9645	330.9736
2.25	270.1346	270.0892	270.7763	270.0879
2.50	215.6680	215.6293	215.6240	215.6242
2.75	166.3887	166.3560	166.3525	166.3553
3.00	121.3946	121.3789	121.3734	121.3767
3.25	80.0000	80.0000	80.0000	80.0000

For linear ($n = 2$), quadratic ($n = 3$), and cubic ($n = 4$) Lagrange elements, the coefficient matrix $[K^{(e)}]$ for the first element in the mesh is given by ($h_e = 0.25$)

$$[K^{(1)}]_{\text{linear}} = k\begin{bmatrix} 7.5 & -7.5 \\ -7.5 & 7.5 \end{bmatrix}$$

$$[K^{(1)}]_{\text{quad.}} = k\begin{bmatrix} 8.6667 & -10.0000 & 1.3333 \\ -10.0000 & 21.3333 & -11.3333 \\ 1.3333 & -11.3333 & 10.0000 \end{bmatrix} \tag{3.215}$$

$$[K^{(1)}]_{\text{cubic}} = k\begin{bmatrix} 9.0583 & -11.6625 & 3.5250 & -0.9208 \\ -11.6625 & 28.5750 & -21.0375 & 4.1250 \\ 3.5250 & -21.0375 & 32.6250 & -15.1125 \\ -0.9208 & 4.1250 & -15.1125 & 11.9083 \end{bmatrix}$$

The remaining steps, that is, assembly of elements, imposition of boundary conditions, and solution of equations (on a computer), should be familiar to the reader by now. The finite-element solutions obtained by using six linear elements, three quadratic elements, and two cubic elements are compared in Table 3.9 with the exact solution (see Prob. 3.31 for the data and the exact solution). Note that for the same number of global nodes in the mesh, the cubic element gives the best results. ■

In Sec. 3.7, we study the computer implementation of the steps involved in the finite-element analysis of one-dimensional problems. As a part of the element calculations there, the computer implementation of the numerical integration ideas presented in this section will be studied. A model finite-element program (FEM1D) for the solution of one-dimensional problems discussed in this chapter is also described, and its application is demonstrated via several examples. The reader will find the program easy to understand because it truly reflects the steps discussed in the theory.

3.7 COMPUTER IMPLEMENTATION

3.7.1 Introductory Comments

It should be clear to the reader by now that the steps involved in the finite-element analysis of a general class of problems (e.g., second-order one-dimensional equations, fourth-order one-dimensional equations, and so on) are systematic and can be implemented on a digital computer. Indeed, the success of the finite-element method is largely due to the ease with which the analysis of a class, without regard to a specific problem, can be implemented on a digital computer. For different geometries, boundary conditions, and problem data, a specific problem from the general class can be solved by simply reading the input data to the program. For example, if we develop a general program to solve equations of the form

$$-\frac{d}{dx}\left(a\frac{du}{dx}\right)+\frac{d^2}{dx^2}\left(b\frac{d^2u}{dx^2}\right)+cu=f \qquad 0<x<L \tag{3.216}$$

then all physical problems described by Eq. (3.1) (see Table 2.1) and beam problems can be solved for any compatible boundary conditions.

The purpose of this section is to discuss some fundamental ideas regarding the development of a computer program for second- and fourth-order one-dimensional differential equations. The ideas presented here are meant to be illustrative of a typical finite-element program development. One can make use of the ideas to develop a program of one's own (for the class of problems one is interested in).

3.7.2 General Outline

A finite-element program consists of three basic parts (see Fig. 3.20):

1. Preprocessor
2. Processor
3. Postprocessor

In the preprocessor part of the program, the input data of the problem are read in and/or generated. This includes the geometry (e.g., length of the domain, boundary conditions, etc.), the data of the problem (e.g., coefficients in the differential equation, source term, etc.), finite-element mesh information (e.g., number of elements, element length, connectivity matrix, etc.), and indicators for various options (print, no print, static analysis, transient analysis, degree of interpolation, etc.). In the processor part, all steps in the finite-element method, except for postprocessing, discussed in the preceding section are performed. These include: generation of the element matrices using numerical integration, assembly of element equations, imposition of the boundary conditions, and solution of the

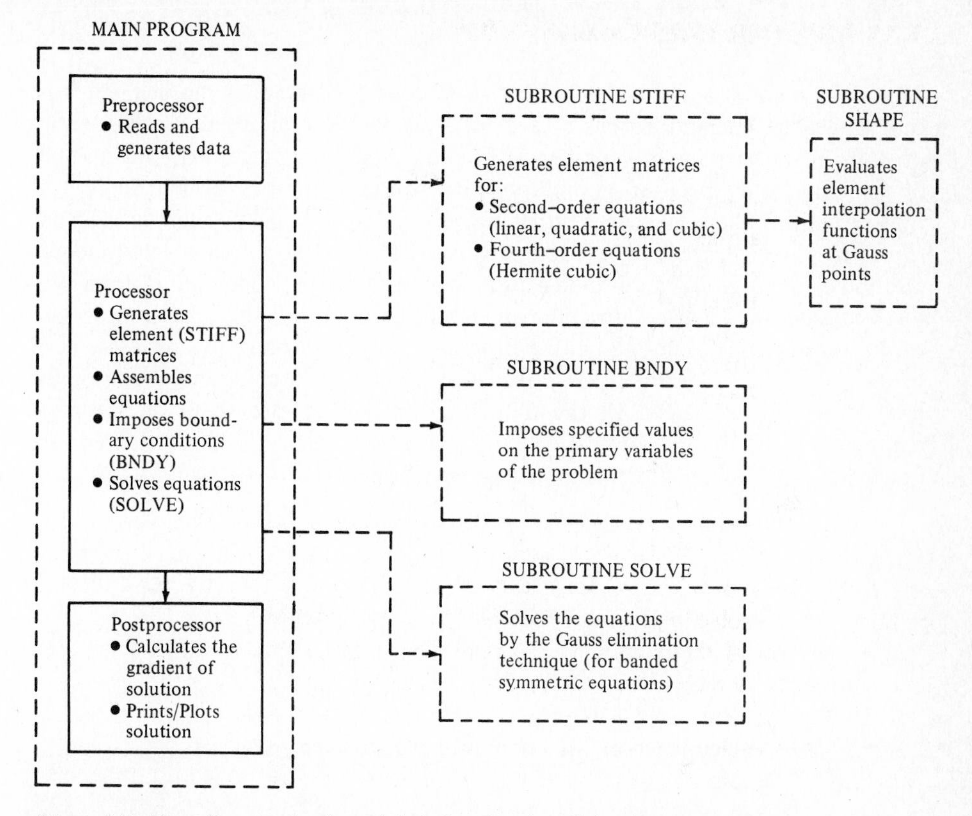

Figure 3.20 Three functional units and their functions in a finite-element program.

equations for the nodal point values of the primary variables. In the postprocessor part of the program, the output data is processed in a desired format for printout and/or plotting, and secondary variables that are derivable from the solution are computed and printed.

The preprocessor and postprocessors can be a few Fortran statements to read and print pertinent information, simple subroutines (e.g., subroutines to generate mesh and compute the gradient of the solution), or complex programs linked to other units via disk and tape files. The processor, where typically large amounts of computing time are spent, can consist of several subroutines, each having a special purpose (e.g., a subroutine for the calculation of element matrices, a subroutine for the imposition of boundary conditions, and a subroutine for the solution of the equations). The degree of sophistication and the complexity of a finite-element program depend on the general class of problems being programmed, the generality of the data in the equation, and the intended user of the program. It is always desirable to describe, through comment statements, all variables used in the computer program.

3.7.3 Input Data (Preprocessor)

The input data to a finite-element program consist of reading the element type (i.e., Lagrange element or beam element), the number of elements (if a series of uniform meshes have to be used, give the minimum and maximum number of elements), the specified boundary point sources (number, global degrees of freedom, and values of the specified forces), the specified boundary conditions (number, global degrees of freedom, and values of the specified displacements), and the element global coordinates and element material properties. If a uniform mesh is used, the length of the domain should be read in, and global coordinates of the nodes can be generated in the program. A description of the variables used in the computer program FEM1D (see Appendix I for a complete listing of the program) is given below. In the following sections, a discussion of the basic components of a typical finite-element program is given, and then the ideas are illustrated via Fortran statements (taken from FEM1D).

```
.....................................................................
.                                                                   .
.     D E S C R I P T I O N   O F   T H E   V A R I A B L E S       .
.                                                                   .
. AL........LENGTH OF THE DOMAIN                                    .
.                                                                   .
. AXO,AX1,..PARAMETERS THAT DEFINE LINEAR VARIATION OF THE          .
. BXO,BX1,..COEFFICIENTS IN THE DIFFERENTIAL EQUATIONS (SEE         .
. CXO,CX1...TABLE 3.6): A=AXO+X*AX1, B=BXO+X*BX1, CX=CXO+X*CX1      .
.                                                                   .
. ALFA,BETA,                                                        .
. AO,A1,A2,.PARAMETERS IN THE TIME APPROXIMATION SCHEMES            .
. A3,A4.....                                                        .
.                                                                   .
. COEF(I,J).ARRAY FOR STORING AXO,AX1,...,FO,F1 (F=FO+X*F1) FOR     .
.           THE I-TH ELEMENT: AXO=COEF(I,1),AX1=COEF(I,2), ETC.     .
.                                                                   .
. DT........TIME INCREMENT FOR TIME-DEPENDENT PROBLEMS              .
. EC(I,J)...ARRAY OF GLOBAL COORDINATES OF NODE J OF ELEMENT I      .
. ELX.......ARRAY OF THE GLOBAL COORDINATES OF ELEMENT NODES        .
. ELSTIF....ELEMENT COEFFICIENT (STIFFNESS) MATRIX                  .
. ELF.......ELEMENT FORCE VECTOR                                    .
.                                                                   .
. GF,GFO,...COLUMN VECTORS FOR DISPLACEMENTS AND THEIR TIME         .
. GF1,GF2   DERIVATIVES                                             .
.                                                                   .
. ICONT.....INDICATOR FOR CONTINUITY OF THE COEFFIECIENTS AND       .
.           AND UNIFORMITY OF THE MESH:                             .
.           ICONT=1, A,B, ETC. ARE CONTINUOUS, AND MESH IS          .
.                    UNIFORM IN THE PROBLEM                         .
.           ICONT=0, OTHERWISE                                      .
.                                                                   .
. IELEM.....INDICATOR FOR THE TYPE OF ELEMENT:                      .
.           IELEM=0, 2-NODE ELEMENT FOR FOURTH-ORDER PROBLEMS       .
.           IELEM=1, 2-NODE ELEMENT FOR SECOND-ORDER PROBLEMS       .
.           IELEM=2, 3-NODE ELEMENT FOR SECOND-ORDER PROBLEMS       .
.           IELEM=3, 4-NODE ELEMENT FOR SECOND-ORDER PROBLEMS       .
.                                                                   .
. ITEM......INDICATOR FOR TRANSIENT (TIME-DEPENDENT) ANALYSIS       .
.           ITEM=0, STATIC (NOT TIME-DEPENDENT) ANALYSIS            .
.           ITEM=1, FIRST-ORDER TIME DERIVATIVES INVOLVED           .
.           ITEM=2, SECOND-ORDER TIME DERIVATIVES INVOLVED          .
.                                                                   .
. GSTIF.....ASSEMBLED COEFFICIENT (STIFFNESS) MATRIX IN             .
.           UPPER-HALF BANDED FORM (NEQ BY NHBW)                    .
.                                                                   .
. GF........ASSEMBLED (GLOBAL) FORCE VECTOR BEFORE SUBROUTINE       .
.           'SOLVE' IS CALLED; AND IT CONTAINS THE SOLUTION OF      .
.           EQUATIONS WHEN COMES OUT OF THE SUBROUTINE 'SOLVE'      .
```

```
.                                                                   .
. NBDY......NUMBER OF SPECIFIED PRIMARY DEGREES OF FREEDOM          .
. IBDY......COLUMN OF SPECIFIED PRIMARY DEGREES OF FREEDOM          .
. VBDY......COLUMN OF SPECIFIED VALUES OF ENTRIES IN IBDY           .
. NBF.......NUMBER OF SPECIFIED SECONDARY DEGREES OF FREEDOM        .
. IBF.......COLUMN OF SPECIFIED SECONDARY DEGREES OF FREEDOM        .
. VBF.......COLUMN OF SPECIFIED VALUES OF ENTRIES IN IBF            .
. NRMAX.....ROW-DIMENSION OF 'GSTIF' IN THE DIMENSION STATEMENT     .
. NCMAX.....COLUMN-DIMENSION OF 'GSTIF' IN DIMENSION STATEMENT      .
. NPE.......NUMBER OF NODES PER ELEMENT                             .
. NDF.......NUMBER OF DEGREES OF FREEDOM PER NODE                   .
. NEM.......NUMBER OF ELEMENTS IN THE MESH                          .
. NEQ.......NUMBER OF EQUATIONS IN THE FINITE ELEMENT MODEL         .
. NHBW......HALF BAND WIDTH OF THE GLOBAL STIFFNESS MATRIX          .
. NGP.......NUMBER OF GAUSS POINTS USED IN THE INTEGRATION          .
. NN .......NUMBER OF TOTAL DEGREES OF FREEDOM IN THE ELEMENT       .
. NNM.......NUMBER OF NODES IN THE FINITE ELEMENT MESH              .
. NOD(I,J)..GLOBAL NODE NUMBER CORRESPONDING TO THE J-TH NODE       .
.           OF THE I-TH ELEMENT (CONNECTIVITY MATRIX)               .
.                                                                   .
. NPRNT.....INDICATOR FOR PRINT (NPRNT=1) OR NOPRINT (NPRNT=0)      .
.           OF ELEMENT MATRICES AND GLOBAL MATRICES                 .
.                                                                   .
. NTIME.....NUMBER OF TIME STEPS                                    .
. X.........VECTOR (COLUMN) OF GLOBAL COORDINATES                   .
.                                                                   .
.....................................................................
```

3.7.4 Calculation of Element Matrices (Processor)

The element matrices are computed in a subroutine, STIFF, by evaluating the integrals of the form of Eqs. (3.15). In order to include the linear, quadratic, and cubic Lagrange elements, as well as the beam element in the program, the corresponding interpolation (or shape) functions SF and their first and second derivatives, GDSF and GDDSF, with respect to the global coordinates are computed in subroutine SHAPE, which is called from subroutine STIFF (see Fig. 3.20). Note that there are two types of elements, the bar element and the beam element, and three different orders of interpolation functions for the bar element. Therefore, an indicator, IELEM, is needed to specify the type and order of the element.

The global node information can be transferred to corresponding element nodes by means of the array NOD. For example, the global node N is the same as local node 2 of element $N-1$ and local node 1 of element N. But this correspondence is already stored in NOD:

$$N = \text{NOD}(N-1,2) = \text{NOD}(N,1)$$

For example, the global coordinate of nodes 1 and 2 of element N are given by

$$\text{ELX}(1) \text{ of element } N = X(N) = X(\text{NOD}(N,1))$$

$$\text{ELX}(2) \text{ of element } N-1 = X(N) = X(\text{NOD}(N-1,2))$$

We need to store the gaussian weights and points associated with two-, three-, and four-point integration in order to be able to evaluate polynomials of degree up to 6 (square of the cubic polynomials). This can be done by defining matrix arrays WT(4, 4) and GAUSS(4, 4). The nth columns of WT and GAUSS contain the weights and zeros, respectively, corresponding to n-point Gauss quadrature. (The undefined entries should be filled with zeros.) For example, the third column of WT contains

$$\mathrm{WT}(1,3) = 0.55555 \qquad \mathrm{WT}(2,3) = 0.88888$$
$$\mathrm{WT}(3,3) = 0.55555 \qquad \mathrm{WT}(4,3) = 0.0$$

Thus,

$$\text{GAUSS(NI, NGP)} = \text{NIth Gauss weight corresponding to the NGP-point Gauss rule}$$

where NGP denotes the number of Gauss points, which must be selected to achieve good accuracy of the element matrices. For the beam element, NGP should be 4, and for the bar element, NGP should be equal to the number $(n + 1)/2$, where n is the degree of the highest polynomial appearing in the element matrices. The highest polynomial is of degree $n = 2r$, where r is the degree of the interpolation polynomial, and appears in the calculation of the "mass" matrix

$$C_{ij} = \int_{x_e}^{x_{e+1}} \psi_i \psi_j \, dx$$

Thus, for a Lagrange interpolation function of degree r (the associated element has $r + 1$ nodes), the integration rule should be one that uses $r + \frac{1}{2}$ points, or (rounding to an integer) $r + 1$ points. In other words, NGP should be equal to the number of nodes per element.

The transformation (3.176) for an element (x_e, x_{e+1}) becomes

$$\xi = \frac{2x - (x_e + x_{e+1})}{x_{e+1} - x_e} = \frac{2(x - x_e) - h_e}{h_e}$$

or

$$x = \frac{h_e}{2}(1 + \xi) + x_e \tag{3.217}$$

Equations (3.185) and (3.217) must be used to rewrite all functions, that are known in the global coordinate system x, in the element coordinates.

The summation in the Gauss quadrature formula (3.193) is replaced by the DOLOOP 70, in which the integrand is evaluated at the Gauss points and multiplied by the Gauss weights. The coefficients AX, BX, and CX define the coefficients in the differential equation (3.216). For example, consider the finite-element matrix corresponding to the first term in Eq. (3.216) [for the case $a = 1 + x = 1 + x(\xi)$]:

$$A_{ij} = \int_{x_e}^{x_{e+1}} (1 + x) \frac{d\psi_i}{dx} \frac{d\psi_j}{dx} dx = \int_{-1}^{1} (1 + x) \left(\frac{1}{\mathrm{GJ}} \frac{d\psi_i}{d\xi} \right) \left(\frac{1}{\mathrm{GJ}} \frac{d\psi_j}{d\xi} \right) J \, d\xi$$

$$= \sum_{I=1}^{n} \left[(1 + x) \left(\frac{1}{\mathrm{GJ}} \frac{d\psi_i}{d\xi} \right) \left(\frac{1}{\mathrm{GJ}} \frac{d\psi_j}{d\xi} \right) \mathrm{GJ} \right]_{\xi = \xi_I} W_I \tag{3.218}$$

where GJ is the jacobian of the transformation. In Fortran form we have

$$\mathtt{A(I,J)} = \sum_{\mathtt{NI=1}}^{\mathtt{NGP}} \mathtt{AX\ GDSF(I)GDSF(J)\ GJ\ W_{NI}} \tag{3.219}$$

A Fortran form of the above discussion is given below:

```
C        ....................................................................
C
C        CALCULATION OF COEFFICIENT MATRICES AND COLUMN VECTOR
C                        FOR THE MODEL EQUATION:
C
C                 -(AX.U')' + (BX.U'')'' + CX.U = F
C
C        HERE THE SYMBOL ' DENOTES DIFFERENTIATION WITH RESPECT TO
C        X, AND AX,BX,CX AND F ARE GIVEN FUNCTIONS OF THE GLOBAL
C        COORDINATE X(SEE TABLE 3.6 FOR ONE-DIMENSIONAL EQUATIONS)
C
C        ....................................................................
C
C          X.........GLOBAL COORDINATE APPEARING IN THE EQUATION
C          XI .......LOCAL COORDINATE
C          ELX(I)....GLOBAL COORDINATE OF THE ELEMENT'S I-TH NODE
C          GAUSS.....MATRIX OF GAUSS POINTS: VALUES IN N-TH COLUMN
C                    CORRESPOND TO GAUSS POINTS FOR N-POINT FORMULA
C          WT........GAUSS WEIGHTS CORRESPONDING TO GAUSS POINTS
C          SF(I).....ELEMENT INTERPOLATION FUNCTION FOR NODE I
C          GDSF(I)...FIRST DERIVATIVE OF SF(I) WITH RESPECT TO X
C          GDDSF(I)..SECOND DERIVATIVE OF SF(I) WITH RESPECT TO X
C          GJ........JACOBIAN OF THE TRANSFORMATION
C          H.........ELEMENT LENGTH
C          NPE.......NODES PER ELEMENT
C
C          NOTE: IT IS ASSUMED THAT ALL NECESSARY VARIABLES, SUCH
C                AS NPE,NN,IELEM,ELX, ETC. ARE TRANSFERRED FROM THE
C                MAIN PROGRAM (SEE PROGRAM FEM1D IN APPENDIX I)
C
C     ......................................................................
C
      IMPLICIT REAL*8(A-H,O-Z)
      DIMENSION ELSTIF(4,4),ELF(4),ELX(4),GAUSS(4,4),WT(4,4)
      DATA GAUSS/4*0.0D0,-0.57735027D0,0.57735027D0,2*0.0D0,
     *-0.77459667D0,0.0D0,0.77459667D0,0.0D0,-0.86113631D0,
     *-0.33998104D0,0.33998104D0,0.86113631D0/
      DATA WT/2.0D0,3*0.0D0,2*1.0D0,2*0.0D0,0.55555555D0,
     *0.88888888D0,0.55555555D0,0.0D0,0.34785485D0,
     *0.02*0.65214515D0,0.34785485D0/
C
      NN=NPE*NDF
C
C     INITIALIZE ALL ARRAYS
C
      DO 10 I=1,NN
      ELF(I) = 0.0
      DO 10 J=1,NN
   10 ELSTIF(I,J)=0.0
C
C     DO-LOOP ON NUMBER OF GAUSS POINTS BEGINS HERE
C
      DO 40 NI=1,NGP
      XI=GAUSS(NI,NGP)
C
C     CALL SUBROUTINE 'SHAPE' TO EVALUATE THE INTERPOLATION
C     FUNCTIONS AND THEIR DERIVATIVES AT THE GAUSS POINTS
C
      CALL SHAPE(XI,H,NPE,NN)
      CNST = GJ*WT(NI,NGP)
      X=0.5*H*(1.0+XI)+ELX(1)
C
```

```
C       DEFINE THE COEFFICIENTS OF THE DIFFERENTIAL EQUATIONS
C          (ONLY UPTO LINEAR VARIATION IS ACCOUNTED HERE)
C
        AX=AX0+AX1*X
        BX=BX0+BX1*X
        CX=CX0+CX1*X
        F=F0+F1*X
C
C       COMPUTE THE COEFFICIENT MATRIX AND COLUMN VECTOR FOR THE
C                                DIFFERENTIAL EQUATION
C
        DO 20 I = 1,NN
        ELF(I) = ELF(I) + CONST*SF(I)*F
        DO 20 J = 1,NN
        ELSTIF(I,J)=ELSTIF(I,J)+AX*CNST*GDSF(I)*GDSF(J)
      *                            +BX*CNST*GDDSF(I)*GDDSF(J)
      *                            +CX*CNST*SF(I)*SF(J)
   20 CONTINUE
   40 CONTINUE
        STOP
        END
```

3.7.5 Assembly in a Banded Matrix Form

The assembly of element matrices should be carried out as soon as the element matrices of each element are computed, rather than waiting till element matrices of all elements are computed. The latter requires storing of the element matrices of each element. In the former case we can perform the assembly in the same loop in which the subroutine STIFF is called to calculate element matrices. Another point that enables us to save storage and computing time is the assembly of element matrices in upper-banded form. When element matrices are symmetric (this is the case in most problems of interest to us in this book), the resulting global (or assembled) matrix is also symmetric, with many zeros away from the main diagonal (see Fig. 3.21). Therefore, it makes sense to store only the upper *half-band* of the assembled matrix. General-purpose equation solvers are available for such banded systems of equations.

The half-bandwidth NHBW of the assembled (i.e., global) finite-element matrix can be determined in the finite-element program itself. The compactness property of the finite-element interpolation functions (i.e., $\psi_i^{(e)}$ are defined to be

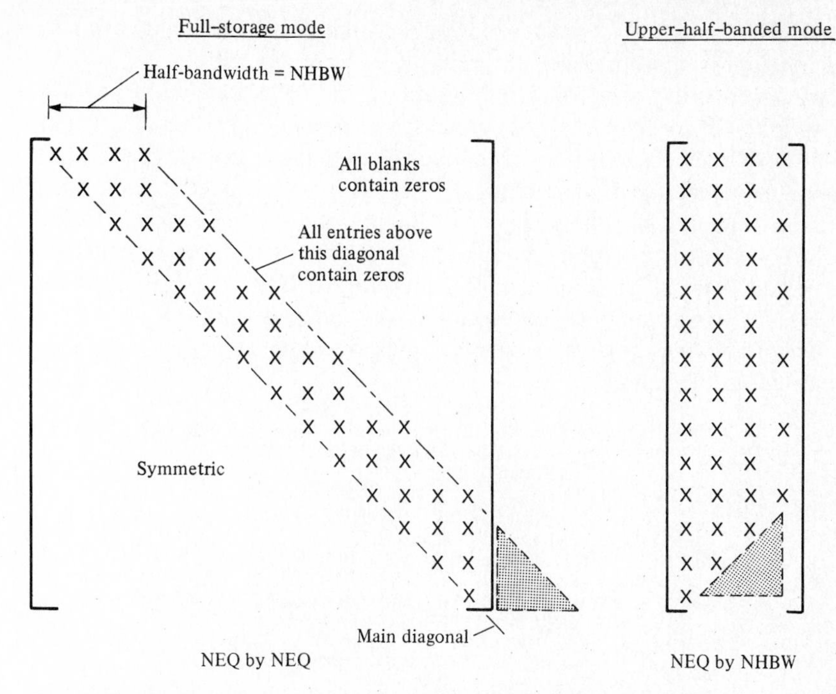

Figure 3.21 Finite-element matrix storage in upper-half-banded form.

nonzero only over the element e) is responsible for the banded character of the assembled matrix. Recall from earlier discussions that if two global nodes do not belong to the same element, then the corresponding entries in the global matrix contain zeros:

$$K_{IJ} = 0 \text{ if } I \text{ and } J \text{ do not belong to the same element}$$

This property enables us to determine the half-bandwidth NHBW of the assembled matrix:

$$\text{NHBW} = \max_{\substack{1 \leq N \leq \text{NEM} \\ 1 \leq I, J \leq \text{NPE}}} [|\text{NOD}(N, I) - \text{NOD}(N, J)| + 1] \times \text{NDF} \tag{3.220}$$

where

$$\text{NDF} = \text{number of degrees of freedom (i.e., primary unknowns) per node}$$
$$\text{NEM} = \text{number of elements in the finite-element mesh}$$
$$\text{NPE} = \text{number of nodes per element}$$

Clearly, for one-dimensional problems, the maximum difference between nodes of an element is equal to NPE − 1. Hence,

$$\text{NHBW} = [(\text{NPE} - 1) + 1] \times \text{NDF} = \text{NPE} \times \text{NDF} \tag{3.221}$$

Of course, NHBW is always less than or equal to the total number of primary degrees of freedom in the finite-element mesh of the problem.

We now describe the banded assembly procedure. The logic should be such that the assembly of element matrices should be skipped whenever $J < I$ and $J >$ NHBW. The main diagonal, $I = J$, of the assembled square matrix (i.e., full storage form) becomes the first column of the assembled banded matrix (i.e., banded storage form), as shown in Fig. 3.21. The upper diagonals (parallel to the main diagonal) take the position of respective columns in the banded matrix. Thus, the banded matrix has the (actual) dimension of NEQ × NHBW, where NEQ denotes the total number of equations (or primary unknowns) in the problem. The following Fortran statements describe the assembly of finite-element matrices in the banded form:

```
C        ELX(I)...GLOBAL COORDINATE OF THE I-TH NODE OF AN ELEMENT
C        ELSTIF...ELEMENT COEFFICIENT (STIFFNESS) MATRIX
C        ELF .....ELEMENT FORCE VECTOR
C        GF.......COLUMN OF GLOBAL (ASSEMBLED) FORCES
C        GSTIF....UPPER-HALF BANDED FORM OF THE COEFFICIENT MATRIX
C        NPE......NUMBER OF NODES PER ELEMENT
C        NDF......NUMBER OF PRIMARY UNKNOWNS PER NODE
C        NEQ......NUMBER OF EQUATIONS IN THE FINITE ELEMENT MODEL
C        NHBW.....HALF BAND WIDTH OF THE GLOBAL COEFFICIENT MATRIX
C        NN.......NUMBER OF UNKNOWNS PER ELEMENT
C        NOD......CONNECTIVITY MATRIX
C        X........GLOBAL COORDINATE OF GLOBAL NODE I
         .........................................................

C       INITIALIZE GLOBAL MATRICES
     60 DO 70 I=1,NEQ
        GF(I)=0.0
        DO 70 J=1,NHBW
     70 GSTIF(I,J)=0.0
C       DO-LOOP ON NUMBER OF ELEMENTS BEGINS HERE
        DO 150 N = 1, NEM
C       TRANSFER THE GLOBAL DATA TO THE ELEMENT
        DO 100 I=1,NPE
        NI=NOD(N,I)
    100 ELX(I)=X(NI)
C       CALL SUBROUTINE 'STIFF' TO CALCULATE ELEMENT EQUATIONS
C       SUBROUTINE STIFF COMPUTES MATRIX ELSTIF AND VECTOR ELF
        CALL STIFF (NN,NPE,ELSTIF,ELF)
C       ASSEMBLY OF ELEMENT MATRICES AND VECTORS BEGIN HERE
        DO 140 I = 1, NPE
        NR = (NOD(N,I) - 1)*NDF
        DO 140 II = 1, NDF
        NR = NR + 1
        L = (I-1)*NDF + II
        GF(NR) = GF(NR) + ELF(L)
        DO 140 J = 1, NPE
        NCL = (NOD(N,J)-1)*NDF
        DO 140 JJ = 1, NDF
        M = (J-1)*NDF + JJ
        NC = NCL-NR+JJ+1
        IF(NC)140,140,130
    130 GSTIF(NR,NC) = GSTIF(NR,NC) + ELSTIF(L,M)
    140 CONTINUE
    150 CONTINUE
```

3.7.6 Imposition of Boundary Conditions

Imposition of boundary conditions (on the primary unknowns) can be done as described earlier (see also Prob. 3.30). The procedure involves modifying the assembled matrix by moving the known products to the right-hand column of the matrix equation, replacing the columns and rows corresponding to the known primary variable by zeros, except on the main diagonal where the variable is set to unity, and replacing the corresponding component of the right-hand column by the specified value of the variable. To fix the ideas, consider the following n algebraic equations in matrix form:

$$\begin{bmatrix} K_{11} & K_{12} & K_{13} & \cdots \\ K_{21} & K_{22} & K_{23} & \cdots \\ K_{31} & K_{32} & K_{33} & \cdots \\ \cdots & \cdots & \cdots & \cdots \end{bmatrix} \begin{bmatrix} U_1 \\ U_2 \\ U_3 \\ \vdots \end{bmatrix} = \begin{bmatrix} F_1 \\ F_2 \\ F_3 \\ \vdots \end{bmatrix} \tag{3.222}$$

Suppose that $U_2 = \alpha$ is specified. (Recall that when the generalized displacement at a node is known, the corresponding generalized force is unknown, and vice versa.) We now proceed to enforce the boundary condition in the following manner. Set $K_{22} = 1$, and $F_2 = \alpha$; further set $K_{2i} = K_{i2} = 0$ for $i \neq 2$ ($i = 1, 3, \ldots, n$). We have

$$\begin{bmatrix} K_{11} & 0 & K_{13} & K_{14} & \cdots & K_{1n} \\ 0 & 1 & 0 & 0 & \cdots & 0 \\ K_{31} & 0 & K_{33} & K_{34} & \cdots & K_{3n} \\ \vdots & & & & & \vdots \\ K_{n1} & 0 & K_{n3} & K_{n4} & \cdots & K_{nn} \end{bmatrix} \begin{Bmatrix} U_1 \\ U_2 \\ U_3 \\ \vdots \\ U_n \end{Bmatrix} = \begin{Bmatrix} \hat{F}_1 \\ \alpha \\ \\ \vdots \\ \hat{F}_n \end{Bmatrix} \tag{3.223}$$

where

$$\hat{F}_i = F_i - K_{i2}\alpha \qquad i = 1, 3, 4, 5, \ldots, n \qquad i \neq 2 \tag{3.224}$$

Thus, in general, if $U_k = \alpha$ is known, we have

$$\begin{aligned} K_{kk} &= 1 \qquad F_k = \alpha \\ F_i &\to F_i - K_{ik}\alpha \\ K_{ki} &= K_{ik} = 0 \end{aligned} \tag{3.225}$$

where

$$i = 1, 2, \ldots, k-1, k+1, \ldots, n, \qquad i \neq k$$

This procedure enables us to retain the original order of the matrix, and the imposed boundary conditions are printed as part of the solution. However, the storage is not reduced (which would be desirable if we can delete the column and row corresponding to the specified degree of freedom). The following Fortran statements describe the same procedure for a banded matrix (see subroutine

BNDY in Appendix I):

```
C     .............................................................
C
C     GSTIF.....GLOBAL (ASSEMBLED) COEFFICIENT MATRIX
C               IN UPPER-HALF BANDED FORM (NEQ BY NHBW)
C     GF........GLOBAL (ASSEMBLED) VECTOR OF FORCES
C     NBDY......NUMBER OF SPECIFIED PRIMARY DEGREES OF FREEDOM
C     IBDY......COLUMN OF SPECIFIED PRIMARY DEGREES OF FREEDOM
C     VBDY......COLUMN OF SPECIFIED VALUES OF ENTRIES IN IBDY
C     NBF.......NUMBER OF SPECIFIED SECONDARY DEGREES OF FREEDOM
C     IBF.......COLUMN OF SPECIFIED SECONDARY DEGREES OF FREEDOM
C     VBF.......COLUMN OF SPECIFIED VALUES OF FORCES IN IBF
C     NRMAX.....ROW-DIMENSION OF GSTIF IN THE DIMENSION STATEMENT
C     NCMAX.....COLUMN-DIMENSION OF GSTIF IN DIMENSION STATEMENT
C     NEQ.......TOTAL NUMBER OF EQUATIONS IN THE FEM MODEL
C     NHBW......HALF BAND WIDTH OF MATRIX GSTIF
C
C     .............................................................
C
C
C     IMPOSE SPECIFIED VALUES OF THE SECONDARY VARIABLES (FORCES)
C
      IF(NBF .EQ. 0)GOTO 170
      DO 160 NF=1, NBF
      NB=IBF(NF)
  160 GF(NB)=GF(NB)+VBF(NF)
C
C     IMPOSE SPECIFIED VALUES OF THE PRIMARY DEGREES OF FREEDOM
C
  170 CALL BNDY(NRMAX,NCMAX,NEQ,NHBW,GSTIF,GF,NBDY,IBDY,VBDY)
C
C     SUBROUTINE BNDY IS GIVEN BELOW
C
      SUBROUTINE BNDY(NRMAX,NCMAX,NEQ,NHBW,S,SL,NBDY,IBDY,VBDY)
C
C     .............................................................
C
C     SUBROUTINE USED TO INCLUDE THE SPECIFIED PRIMARY DEGREES OF
C            FREEDOM INTO THE ASSEMBLED SYSTEM OF EQUATIONS
C
C
C     .............................................................
C
      IMPLICIT REAL*8 (A-H,O-Z)
      DIMENSION S(NRMAX,NCMAX),SL(NRMAX)
      DIMENSION IBDY(NBDY),VBDY(NBDY)
      DO 300 NB = 1, NBDY
      IE = IBDY(NB)
      SVAL = VBDY(NB)
      IT=NHBW-1
      I=IE-NHBW
      DO 100 II=1,IT
      I=I+1
      IF (I .LT. 1)  GO TO 100
      J=IE-I+1
      SL(I)=SL(I)-S(I,J)*SVAL
      S(I,J)=0.0
  100 CONTINUE
      S(IE,1)=1.0
      SL(IE)=SVAL
      I=IE
      DO 200 II=2,NHBW
      I=I+1
      IF (I .GT. NEQ)  GO TO 200
      SL(I)=SL(I)-S(IE,II)*SVAL
      S(IE,II)=0.0
  200 CONTINUE
  300 CONTINUE
      RETURN
      END
```

3.7.7 Solution of Equations and Postprocessing

Subroutine SOLVE, given in Appendix I, solves a banded system of equations and returns the solution in array GF. The routine performs gaussian elimination and back substitution to obtain the solution. On most computing systems, a variety of equation solvers are available, and one can use any of the programs that suits the needs.

Postprocessing involves the computation of the gradient of the solution, and possibly plotting. Subroutine STRESS listed below illustrates the computation of w and dw/dx at points intermediate to element nodes.

```
      SUBROUTINE STRESS(NPE,NDF,IELEM,WO,ELX)
C
C     ..............................................................
C
C     X.........GLOBAL COORDINATE
C     XI .......LOCAL COORDINATE
C     SF........ELEMENT INTERPOLATION FUNCTIONS
C     GDSF......FIRST DERIVATIVE OF SF WITH RESPECT TO X
C     GDDSF.....SECOND DERIVATIVE OF SF WITH RESPECT TO X
C     WO........COLUMN OF NODAL VALUES OF PRIMARY UNKNOWNS
C     W.........PRIMARY DEGREES OF FREEDOM AT GAUSS POINTS
C     DW........FIRST DERIVATIVE OF W: DW/DX
C     DDW.......SECOND DERIVATIVE OF W: D(DW)/DX
C
C     NOTE: W, DW, AND DDW ARE COMPUTED AT NINE POINTS OF EACH
C           ELEMENT (DW AND DDW ARE NOT EXPECTED TO BE ACCURATE
C           AT THE NODAL POINTS OF THE ELEMENT)
C     ..............................................................
C
      IMPLICIT REAL*8 (A-H,O-Z)
      COMMON/SHP/SF(4),GDSF(4),GDDSF(4),GJ
      DIMENSION   GAUSS(9),WO(4),ELX(4)
      DATA GAUSS/-1.0D0,-0.75D0,-0.50D0,-0.25D0,0.0D0,0.25D0,
     *           0.50D0,0.75D0,1.0D0/
C
      NET=NPE
      IF(IELEM .EQ. 0)NET=4
      H = ELX(NPE)-ELX(1)
      DO 70 NI=1,9
      XI = GAUSS(NI)
      CALL SHAPE(XI,H,NPE,NET,IELEM)
      X = 0.5*H*(1.0+XI)+ELX(1)
      W=0.0
      DW=0.0
      DDW=0.0
      DO 65 I=1,NET
      W = W + SF(I)*WO(I)
      DW=DW+GDSF(I)*WO(I)
      IF(IELEM .NE. 0)GOTO 65
      DDW=DDW+GDDSF(I)*WO(I)
   65 CONTINUE
      IF(IELEM.EQ.0) PRINT 10,X,W,DW,DDW
      IF(IELEM.GT.0) PRINT 10, X,W,DW
   70 CONTINUE
   10 FORMAT (10X,8E13.5)
      RETURN
      END
```

3.7.8 Applications of the Computer Program FEM1D

The computer program listed in Appendix I is intended to illustrate certain fundamental steps in the computer implementation of the finite-element method for linear one-dimensional problems. One can modify certain parts of the program to develop a finite-element program that can be used for the solution of more special or general problems. Let us not forget that a fundamental step in the finite-element analysis of a problem is to formulate a finite-element model of the

Table 3.10 Description of the variables in the data input to FEM1D

COLUMNS	VARIABLE	DESCRIPTION OF THE VARIABLE
* PROBLEM DATA CARD 1 (20A4)		PROBLEM HEADING CARD
1-80	TITLE	ENTER THE TITLE OF THE PROBLEM FOR LABELING THE OUTPUT; EACH PROBLEM SHOULD HAVE A TITLE
* PROBLEM DATA CARD 2 (16I5)		RIGHT-JUSTIFY EACH ENTRY
1-5	NPRNT	INDICATOR FOR PRINT OUT OF ELEMENT OR GLOBAL MATRICES: NPRNT=0, NO PRINT NPRNT=1, ELEMENT MATRICES ONLY NPRNT=2, GLOBAL MATRICES ONLY
6-10	IELEM	ELEMENT TYPE: IELEM=0,BEAM; IELEM=1, 2 OR 3 FOR LINEAR, QUADRATIC OR CUBIC BAR ELEMENTS
11-15	NPE	NODES PER ELEMENT: NPE=2 FOR BEAM, AND NPE=IELEM+1 FOR BAR
16-20	NDF	NUMBER OF DEGREES OF FREEDOM (PRIMARY UNKNOWNS) PER NODE: NDF=2 IF IELEM=0; OTHERWISE 1
21-25	NEM	NUMBER OF ELEMENTS IN THE MESH
26-30	NBDY	NUMBER OF SPECIFIED PRIMARY DEGREES OF FREEDOM IN THE MODEL FOR BEAMS, N-TH GLOBAL NODE HAS DEGREES OF FREEDOM N*(NDF+1)+1 AND N*(NDF+1)+2
31-35	NBF	NUMBER OF SPECIFIED SECONDARY DEGREES OF FREEDOM IN THE MODEL INCLUDE ONLY NONZERO FORCES
36-40	ICONT	INDICATOR FOR CONTINUITY OF THE DATA (COEFFICIENTS AX,BX,CX,F, MESH INFORMATION): ICONT=1, IF AX,BX,CX, AND F ARE CONTINUOUS FUNCTIONS AND THE MESH IS UNIFORM THROUGH OUT; ICONT=0, OTHERWISE; IN THIS CASE ELEMENT DATA WILL BE READ LATER
41-45	ITEM	INDICATOR FOR TRANSIENT PROBLEM ITEM=0, STEADY STATE OR STATIC; ITEM=1,TRANSIENT ANALYSIS WITH FIRST-ORDER TIME DERIVATIVES; ITEM=2, TRANSIENT ANALYSIS WITH SECOND-ORDER TIME DERIVATIVES
46-50	NTIME	NUMBER OF TIME STEPS
* PROBLEM DATA CARD 3 (8F10.4)		TWO ENTRIES ON THE SAME CARD
1-10	AL	LENGTH OF THE DOMAIN
11-20	X0	COORDINATE OF THE FIRST NODE

NBMX

```
* PROBLEM DATA CARD 4 (8F10.4)     *** SKIP IF ICONT = 0 ***
                                   READ THE COEFFICIENTS IN THE
                                   DIFFERENTIAL EQUATION,
                                   -(AX.U')'+(BX.U'')''+CX.U=F
                                   UPTO LINEAR FUNCTIONS ARE USED:

    1-10,11-20       AX0, AX1      COEFFICIENTS IN AX:AX=AX0+AX1*X

    21-30,31-40      BX0, BX1      COEFFICIENTS IN BX:BX=BX0+BX1*X

    41-50,51-60      CX0, CX1      COEFFICIENTS IN CX:CX=CX0+CX1*X

    61-70,71-80      F0, F1        COEFFICIENTS IN F: F=F0+F1*X

* PROBLEM DATA CARD 5 (8F10.4)     *** SKIP IF ITEM = 0 ***

    1-10,11-20       ALFA, BETA    PARAMETERS IN TIME-INTEGRATION
                                   SCHEMES: ALFA=THETA WHEN ITEM=1
                                   (BETA IS NOT NEEDED; GIVE ZERO)
                                   FOR ITEM=2, GIVE VALUES OF ALFA
                                   AND BETA FOR THE NEWMARK SCHEME

    21-30            DT            TIME STEP FOR TRANSIENT PROBLEM

    31-40            C0            VALUE OF THE COEFFICIENT OF THE
                                   TIME DERIVATIVE IN THE EQUATION

* PROBLEM DATA CARD 6 (16I5)       *** ENTER ZERO IF NBDY = 0 ***

    1-80             IBDY(I)       ARRAY OF SPECIFIED PRIMARY
                                   DEGREES OF FREEDOM; 16 PER CARD

* PROBLEM DATA CARD 7 (8F10.4)     *** ENTER ZERO IF NBDY = 0 ***

    1-80             VBDY(I)       ARRAY OF THE VALUES OF THE
                                   SPECIFIED DEGREES OF FREEDOM IN
                                   IBDY (SHOULD BE IN THE SAME
                                   ORDER AS THE ENTRIES IN IBDY);
                                   EIGHT ENTRIES PER CARD

* PROBLEM DATA CARD 8 (16I5)       *** SKIP IF NBF = 0 ***

    1-80             IBF(I)        ARRAY OF SPECIFIED NONZERO
                                   SECONDARY DEGREES OF OF FREEDOM

* PROBLEM DATA CARD 9 (8F10.4)     *** SKIP NBF = 0 ***

    1-80             VBF(I)        ARRAY OF THE VALUES OF ENTRIES
                                   IN ARRAY IBF

***NOTE: IF ICONT=0 AND ITEM=0, SKIP THE REMAINING DATA CARDS ***

* PROBLEM DATA CARD 10 (8F10.4)    *** SKIP IF ITEM = 0 ***

    1-80             GF0           INITIAL CONDITIONS ON U

* PROBLEM DATA CARD 11 (8F10.4)    *** SKIP IF ITEM = 0 ***

    1-80             GF1           INITIAL CONDITIONS ON U*=DU/DT

* PROBLEM DATA CARD 12 (8F10.4)    *** SKIP IF ITEM = 0 ***

    1-80             GF2           INITIAL CONDITIONS ON DU*/DT

NOTE: THE FOLLOWING CARDS ARE READ IN A DO-LOOP ON NUMBER OF
      ELEMENTS. HENCE, READ DATA CARDS 13 AND 14 FOR EACH ELEMENT
```

Table 3.10 (*Continued*)

```
* PROBLEM DATA CARD 13 (8F10.4)        *** SKIP IF ICONT = 1 ***

      1-80             ELX(I)          ARRAY OF GLOBAL COORDINATES OF
                                       ELEMENT NODES. NPE ENTRIES

* PROBLEM DATA CARD 14 (8F10.4)        *** SKIP IF ICONT = 1 ***

      1-80             COEF(I,J)       COEFFICIENTS AX, BX, CX, AND FX
                                       FOR EACH ELEMENT; SEE DESCRIP-
                                       TION OF VARIABLES ON PAGE 166
```

equations and select appropriate interpolation functions. The element matrices can be programmed with ease by modifying appropriate statements in STIFF.

Here we illustrate the generation of the data input to the computer program FEM1D by reconsidering the problems in Examples 3.1, 3.3, 3.6, and 3.7. Typical output of the program is also given for each problem. A list of the variables in the data input to FEM1D is given in Table 3.10.

Example 3.10 (Example 3.1: transverse deflection of a cable fixed at both ends) Using the finite-element mesh and the data considered in Example 3.1, we find that

Element type IELEM = 1 NPE = 2 NDF = 1 NEM = 4

Since a steady-state (i.e., equilibrium) solution is desired, we set

ITEM = 0 NTIME = 0

The differential equation governing the problem is given by Eq. (3.1). Hence,

$$a = 1 + x \quad (\text{AX0} = 1, \text{AX1} = 1) \qquad b = c = 0$$

$$f = 1 + 4x \quad (\text{F0} = 1, \text{F1} = 4)$$

for all x in the domain (0, 1). Since the data are the same for all elements, we have ICONT = 1.

The boundary conditions of the problem are $u(0) = u(1) = 0$ ($U_1 = U_5 = 0$), and there are no specified nonzero forces in the problem. Therefore, we have
NBF = 0 NBDY = 2 IBDY(1) = 1 IBDY(2) = 5
VBDY(1) = 0.0 VBDY(2) = 0.0

The Fortran form of the data for the problem is summarized below.

```
 TRANSVERSE DEFLECTION OF A CABLE FIXED AT BOTH ENDS (SEE EXAMPLE 3.1)
   2    1    2    1    4    2    0    1    0    0
1.0       0.0
1.0       1.0       0.0       0.0       0.0       0.0       1.0       4.0
    1    5
0.0       0.0
```

The program output is given in Table 3.11. ■

Example 3.11 (Example 3.3: bending of discontinuously loaded beam) The governing equation of this problem is described by Eq. (3.86) with b and f given by Eqs. (3.87). [Note that f in Eq. (3.86) is equal to $-f$ in the program.] For the finite-element mesh and the data considered in Example 3.3, we have

$$\text{IELEM} = 0 \quad \text{NPE} = 2 \quad \text{NDF} = 2 \quad \text{NEM} = 3 \quad \text{ICONT} = 0$$

$$\text{ITEM} = \text{NTIME} = 0 \quad \text{NBDY} = 3 \quad \text{IBDY}(1) = 1 \quad \text{IBDY}(2) = 7$$

$$\text{IBDY}(3) = 8 \quad \text{VBDY(I)} = 0.0, I = 1, 2, 3 \quad \text{NBF} = 1 \quad \text{IBF}(1) = 5$$

$$\text{VBF}(1) = -10000.0 \quad \text{AL} = 28.0 \quad \text{X0} = 0.0$$

The element coordinates and the data for each element are as follows:

Element 1

$$x_1^{(1)} = 0.0 \quad x_2^{(1)} = 10.0 \quad [\text{ELX}(1) = 0.0, \text{ELX}(2) = 10.0]$$

$$b = 2(10^7) \quad f = -2400 \quad a = c = 0.0$$

Element 2

$$x_1^{(2)} = 10.0 \quad x_2^{(2)} = 22.0$$

$$b = 1(10^7) \quad a = c = f = 0.0$$

Element 3

$$x_1^{(3)} = 22.0 \quad x_2^{(3)} = 28.0$$

$$b = 1(10^7) \quad a = c = f = 0.0$$

The Fortran form of the data is given below.

```
BENDING OF DISCONTINUOUSLY LOADED BEAM (SEE EXAMPLE 3.3)
   2     0     2     2     3     3     1     0     0     0
28.0      0.0
   1     7     8
0.0       0.0       0.0
   5
-10000.0
0.0       10.0
0.0       0.0       2.0       0.0       0.0       0.0       -2400.0   0.0
10.0      22.0
0.0       0.0       1.0       0.0       0.0       0.0       0.0       0.0
22.0      28.0
0.0       0.0       1.0       0.0       0.0       0.0       0.0       0.0
```

The output for the problem is given in Table 3.12. ■

Example 3.12 (Example 3.6: transient heat conduction in a fin) From Example 3.6, we have (for the two-element mesh of quadratic elements)

$$\text{IELEM} = 2 \quad \text{NPE} = 3 \quad \text{NDF} = 1 \quad \text{NEM} = 2 \quad \text{NBDY} = 1$$

$$\text{IBDY}(1) = 1 \quad \text{VBDY}(1) = 0.0 \quad \text{NBF} = 0$$

$$\text{ICONT} = 0 \text{ (because of nonuniform mesh)}$$

Table 3.11 Computer output for example 3.10

```
*** S E C O N D - O R D E R   P R O B L E M S ***

TRANSVERSE DEFLECTION OF A CABLE FIXED AT BOTH ENDS (SEE EXAMPLE 3.1)
      NO. OF ELEMENTS IN THE MESH ......... =   4
      NO. OF NODES IN THE MESH ............ =   5
      NO. OF DEG. OF FREEDOM PER NODE ......=   1
      FEM MESH  AND COEFFICIENTS (AXO,AX1, ETC.) FOLLOW :

          0.0            0.25000D+00    0.50000D+00    0.75000D+00    0.10000D+01
          0.10000D+01    0.10000D+01    0.0            0.0            0.0            0.0
          0.10000D+01    0.40000D+01

      ELEMENT MATRICES :

          0.45000D+01   -0.45000D+01
         -0.45000D+01    0.45000D+01
          0.16667D+00    0.20833D+00
      SPECIFIED DEGREES OF FREEDOM:

          1    5

      VALUES OF SPECIFIED DEGREES OF FREEDOM:

       0.0       0.0

      SPECIFIED NONZERO SECONDARY DEGREES OF FREEDOM :

          0

      VALUES OF THE SPECIFIED SECONDARY DEGREES OF FREEDOM :

       0.0
```

S O L U T I O N :

0.0 0.18750D+00 0.25000D+00 0.18750D+00 0.0

X	DISPL.	1ST DERIV.	2ND DERIV.
0.0	0.0	0.75000D+00	
0.31250D-01	0.23437D-01	0.75000D+00	
0.62500D-01	0.46875D-01	0.75000D+00	
0.93750D-01	0.70312D-01	0.75000D+00	
0.12500D+00	0.93750D-01	0.75000D+00	
0.15625D+00	0.11719D+00	0.75000D+00	
0.18750D+00	0.14062D+00	0.75000D+00	
0.21875D+00	0.16406D+00	0.75000D+00	
0.25000D+00	0.18750D+00	0.75000D+00	
0.25000D+00	0.18750D+00	0.25000D+00	
0.28125D+00	0.19531D+00	0.25000D+00	
0.31250D+00	0.20312D+00	0.25000D+00	
0.34375D+00	0.21094D+00	0.25000D+00	
0.37500D+00	0.21875D+00	0.25000D+00	
0.40625D+00	0.22656D+00	0.25000D+00	
0.43750D+00	0.23437D+00	0.25000D+00	
0.46875D+00	0.24219D+00	0.25000D+00	
0.50000D+00	0.25000D+00	0.25000D+00	
0.50000D+00	0.25000D+00	-0.25000D+00	
0.53125D+00	0.24219D+00	-0.25000D+00	
0.56250D+00	0.23437D+00	-0.25000D+00	
0.59375D+00	0.22656D+00	-0.25000D+00	
0.62500D+00	0.21875D+00	-0.25000D+00	
0.65625D+00	0.21094D+00	-0.25000D+00	
0.68750D+00	0.20312D+00	-0.25000D+00	
0.71875D+00	0.19531D+00	-0.25000D+00	
0.75000D+00	0.18750D+00	-0.25000D+00	
0.75000D+00	0.18750D+00	-0.75000D+00	
0.78125D+00	0.16406D+00	-0.75000D+00	
0.81250D+00	0.14062D+00	-0.75000D+00	
0.84375D+00	0.11719D+00	-0.75000D+00	
0.87500D+00	0.93750D-01	-0.75000D+00	
0.90625D+00	0.70312D-01	-0.75000D+00	
0.93750D+00	0.46875D-01	-0.75000D+00	
0.96875D+00	0.23437D-01	-0.75000D+00	
0.10000D+01	0.0	-0.75000D+00	

Table 3.12 Output for example 3.11

```
 *** F O U R T H - O R D E R   P R O B L E M S ***

BENDING OF DISCONTINUOUSLY LOADED BEAM (SEE EXAMPLE 3.3)
      NO. OF ELEMENTS IN THE MESH ......... =    3
      NO. OF NODES IN THE MESH ............ =    4
      NO. OF DEG. OF FREEDOM PER NODE ......=    2
      FEM MESH  AND COEFFICIENTS (AX0,AX1, ETC.) FOLLOW :


          1   0.0           0.1000D+02
          0.0             0.0             0.20000D+01    0.0            0.0            0.0
         -0.24000D+04     0.0

      ELEMENT MATRICES :

          0.24000D-01    -0.12000D+00    -0.24000D-01    -0.12000D+00
         -0.12000D+00     0.80000D+00     0.12000D+00     0.40000D+00
         -0.24000D-01     0.12000D+00     0.24000D-01     0.12000D+00
         -0.12000D+00     0.40000D+00     0.12000D+00     0.80000D+00
         -0.12000D+05     0.20000D+05    -0.12000D+05    -0.20000D+05

          2   0.1000D+02    0.2200D+02
          0.0             0.0             0.10000D+01    0.0            0.0            0.0
          0.0             0.0

          3   0.2200D+02    0.2800D+02
          0.0             0.0             0.10000D+01    0.0            0.0            0.0
          0.0             0.0
      SPECIFIED DEGREES OF FREEDOM:

          1    7    8

      VALUES OF SPECIFIED DEGREES OF FREEDOM:

         0.0        0.0        0.0

      SPECIFIED NONZERO SECONDARY DEGREES OF FREEDOM :

          5
```

```
VALUES OF THE SPECIFIED SECONDARY DEGREES OF FREEDOM :

-0.100D+05

S O L U T I O N :

   0.0            0.38555D+06   -0.28084D+07    0.12142D+06   -0.11033D+07   -0.27517D+06
   0.0            0.0

     X             DISPL.       1ST DERIV.     2ND DERIV.

  0.0            0.0           -0.38555D+06   0.10000D+05
  0.12500D+01  -0.47306D+06  -0.37049D+06   0.14103D+05
  0.25000D+01  -0.92409D+06  -0.35030D+06   0.18207D+05
  0.37500D+01  -0.13467D+07  -0.32497D+06   0.22310D+05
  0.50000D+01  -0.17344D+07  -0.29452D+06   0.26414D+05
  0.62500D+01  -0.20808D+07  -0.25894D+06   0.30517D+05
  0.75000D+01  -0.23796D+07  -0.21823D+06   0.34621D+05
  0.87500D+01  -0.26242D+07  -0.17239D+06   0.38724D+05
  0.10000D+02  -0.28084D+07  -0.12142D+06   0.42828D+05
  0.10000D+02  -0.28084D+07  -0.12142D+06   0.65655D+05
  0.11500D+02  -0.29197D+07  -0.29046D+05   0.57504D+05
  0.13000D+02  -0.29017D+07   0.51096D+05   0.49352D+05
  0.14500D+02  -0.27726D+07   0.11901D+06   0.41200D+05
  0.16000D+02  -0.25507D+07   0.17470D+06   0.33049D+05
  0.17500D+02  -0.22546D+07   0.21816D+06   0.24897D+05
  0.19000D+02  -0.19024D+07   0.24939D+06   0.16745D+05
  0.20500D+02  -0.15125D+07   0.26839D+06   0.85936D+04
  0.22000D+02  -0.11033D+07   0.27517D+06   0.44192D+03
  0.22000D+02  -0.11033D+07   0.27517D+06   0.44192D+03
  0.22750D+02  -0.89791D+06   0.27116D+06  -0.11134D+05
  0.23500D+02  -0.69876D+06   0.25847D+06  -0.22710D+05
  0.24250D+02  -0.51238D+06   0.23709D+06  -0.34286D+05
  0.25000D+02  -0.34529D+06   0.20704D+06  -0.45861D+05
  0.25750D+02  -0.20399D+06   0.16830D+06  -0.57437D+05
  0.26500D+02  -0.95004D+05   0.12088D+06  -0.69013D+05
  0.27250D+02  -0.24836D+05   0.64783D+05  -0.80589D+05
  0.28000D+02   0.0            0.0           -0.92165D+05
..........................................................................................
```

Table 3.13 Edited computer output for example 3.12

```
 *** S E C O N D - O R D E R   P R O B L E M S ***

TRANSIENT HEAT CONDUCTION IN A FIN (SEE EXAMPLE 3.6)

      TIME-DEPENDENT (TRANSIENT) ANALYSIS

      ALFA =    0.5000D+00  BETA =    0.1000D+01  TIME INCREMENT =   0.2500D-01  CO = 0.1000D+01

      NO. OF ELEMENTS IN THE MESH ......... =   2
      NO. OF NODES IN THE MESH ............ =   5
      NO. OF DEG. OF FREEDOM PER NODE ......=   1
      FEM MESH  AND COEFFICIENTS (AXO,AX1, ETC.) FOLLOW :

      TIME =    0.25000D-01

        1   0.0          0.2000D+00   0.5000D+00
        0.10000D+01    0.0          0.0          0.0          0.0          0.0
        0.0            0.0

        2   0.5000D+00   0.7500D+00   0.1000D+01
        0.10000D+01    0.0          0.0          0.0          0.0          0.0
        0.0            0.0
      SPECIFIED DEGREES OF FREEDOM:

        1

      VALUES OF SPECIFIED DEGREES OF FREEDOM:

       0.0
```

```
  SPECIFIED NONZERO SECONDARY DEGREES OF FREEDOM :

      0

  VALUES OF THE SPECIFIED SECONDARY DEGREES OF FREEDOM :

   0.0

  GLOBAL MATRICES :

      0.12500D+00   -0.33333D-01   -0.83333D-02
      0.40000D+00   -0.33333D-01    0.0
      0.25000D+00   -0.33333D-01   -0.83333D-02
      0.40000D+00   -0.33333D-01    0.0
      0.12500D+00    0.0            0.0

      0.83333D-01    0.33333D+00    0.16667D+00    0.33333D+00    0.83333D-01

 S O L U T I O N :

      0.0            0.91286D+00    0.95436D+00    0.99585D+00    0.99585D+00
     -0.20000D+02   -0.17427D+01   -0.91286D+00   -0.82988D-01   -0.82988D-01

(STRESS CALCULATIONS ARE ELIMINATED IN THIS OUTPUT)


   TIME =    0.50000D-01


 S O L U T I O N :

      0.0            0.62415D+00    0.97528D+00    0.98408D+00    0.99964D+00
     -0.10000D+02   -0.66285D+01   -0.20156D-01   -0.25827D+00    0.52935D-01
```

ITEM = 1 (transient analysis with first-order time derivative)

NTIME = 16 (number of time steps) DT = 0.025 AL = 1.0

X0 = 0.0 ALFA (theta) = 0.5 C0 = 1

GF0(I) = 1.0 I = 1, 2,..., 5 (initial conditions)

GF1(I) = GF2(I) = 0.0 (not required in the problem but should be read)

Element 1

$$x_1^{(1)} = 0.0 \qquad x_2^{(1)} = 0.2 \qquad x_3^{(1)} = 0.5$$

$$a = 1.0 \qquad b = c = f = 0.0$$

Element 2

$$x_1^{(2)} = 0.5 \qquad x_2^{(2)} = 0.75 \qquad x_3^{(2)} = 1.0$$

$$a = 1.0 \qquad b = c = f = 0.0$$

The Fortran form of the input data for the problem is given below.

```
TRANSIENT HEAT CONDUCTION IN A FIN (SEE EXAMPLE 3.6)
    2     2     3     1     2     1     0     0     1    16
1.0        0.0
0.5        1.0        0.025      1.0
    1
0.0
1.0        1.0        1.0        1.0        1.0
0.0        0.0        0.0        0.0        0.0
0.0        0.0        0.0        0.0        0.0
0.0        0.2        0.5
1.0        0.0        0.0        0.0        0.0        0.0        0.0        0.0
0.5        0.75       1.0
1.0        0.0        0.0        0.0        0.0        0.0        0.0        0.0
```

The output of the program is given in Table 3.13. ■

Example 3.13 (Example 3.7: transient response of a clamped beam) From the discussion of Example 3.7, we have (for the half-beam model)

IELEM = 0 NPE = 2 NDF = 2 NEM = 4 NBDY = 3

NBF = 0 ICONT = 1 (uniform mesh and continuous data)

ITEM = 2 (transient analysis with second-order time derivative)

NTIME = 20 AL = 0.5 X0 = 0.0 $a = 0.0$ $b = 1.0$ $c = f = 0.0$

ALFA = 0.5 BETA = 0.25 DT = 0.0025 C0 = 1.0

IBDY(1) = 1 IBDY(2) = 2 IBDY(3) = 10 $(U_1 = U_2 = U_{10} = 0.0)$

VBDY(I) = 0.0, I = 1, 2, 3

GF0(1)($= w(0)$) = 0.0

GF0(2)($= \theta(0)$) = 0.0

GF0(3)($= w(0.125)$) = −0.0391

GF0(4)($= \theta(0.125)$) = 0.5463, etc.

$$\left.\begin{aligned} w &= -\sin \pi x + \pi x(1 - x) \\ \theta &= \pi \cos \pi x - x(1 - 2x) \end{aligned}\right\}$$

GF1(I) = GF2(I) = 0.0, I = 1, 2,..., 10 $(\dot{w} = \dot{\theta} = 0.0)$

Table 3.14 Edited computer output for example 3.13

```
*** F O U R T H - O R D E R   P R O B L E M S ***

TRANSIENT RESPONSE OF A CLAMPED BEAM (SEE EXAMPLE 3.7)

     TIME-DEPENDENT (TRANSIENT) ANALYSIS


     ALFA =    0.5000D+00   BETA =    0.2500D+00   TIME INCREMENT =    0.2500D-02   CO = 0.1000D+01

     NO. OF ELEMENTS IN THE MESH ......... =    4
     NO. OF NODES IN THE MESH ............ =    5
     NO. OF DEG. OF FREEDOM PER NODE ......=    2
     FEM MESH  AND COEFFICIENTS (AXO,AX1, ETC.) FOLLOW :

        0.0            0.12500D+00    0.25000D+00    0.37500D+00    0.50000D+00
        0.0            0.0            0.10000D+01    0.0            0.0            0.0
        0.0            0.0

     ELEMENT MATRICES :

        0.46429D-01   -0.81845D-03    0.16071D-01    0.48363D-03
       -0.81845D-03    0.18601D-04   -0.48363D-03   -0.13951D-04
        0.16071D-01   -0.48363D-03    0.46429D-01    0.81845D-03
        0.48363D-03   -0.13951D-04    0.81845D-03    0.18601D-04
       -0.30451D+02    0.62736D+01    0.30451D+02   -0.24672D+01
     SPECIFIED DEGREES OF FREEDOM:

        1    2    10

     VALUES OF SPECIFIED DEGREES OF FREEDOM:

      0.0         0.0         0.0

     SPECIFIED NONZERO SECONDARY DEGREES OF FREEDOM :

        0
```

Table 3.14 (*Continued*)

```
VALUES OF THE SPECIFIED SECONDARY DEGREES OF FREEDOM :

 0.0

TIME =    0.25000D-02

S O L U T I O N :

    0.0            0.0            -0.38983D-01    0.54565D+00   -0.11788D+00    0.65002D+00
   -0.18731D+00    0.41613D+00    -0.21431D+00    0.0
    0.0            0.0             0.93979D-01   -0.51826D+00    0.17410D+00   -0.46645D+00
    0.22872D+00   -0.53898D+00     0.23514D+00    0.0
    0.0            0.0             0.37750D+02   -0.79060D+03    0.63608D+02   -0.35812D+03
    0.89590D+02    0.20434D+03     0.10134D+03    0.0

   (STRESS CALCULATIONS ARE NOT INCLUDED IN THIS OUTPUT)

TIME =    0.50000D-02

S O L U T I O N :

    0.0            0.0            -0.38648D-01    0.54224D+00   -0.11724D+00    0.64801D+00
   -0.18647D+00    0.41516D+00    -0.21340D+00    0.0
    0.0            0.0             0.17334D+00   -0.22095D+01    0.34236D+00   -0.11391D+01
    0.45044D+00   -0.23778D+00     0.48803D+00    0.0
    0.0            0.0             0.25740D+02   -0.56240D+03    0.70998D+02   -0.17998D+03
    0.87786D+02    0.36619D+02     0.10097D+03    0.0

TIME =    0.10000D-01

S O L U T I O N :

    0.0            0.0            -0.37527D-01    0.52856D+00   -0.11460D+00    0.63935D+00
   -0.18310D+00    0.41106D+00    -0.20976D+00    0.0
    0.0            0.0             0.27197D+00   -0.34308D+01    0.70090D+00   -0.26011D+01
    0.89983D+00   -0.11361D+01     0.96758D+00    0.0
    0.0            0.0             0.20612D+02   -0.83069D+03    0.55704D+02   -0.61909D+03
    0.93177D+02    0.45732D+03     0.10147D+03    0.0
```

X	DISPL.	1ST DERIV.	2ND DERIV.
0.0	0.0	0.0	-0.59534D+01
0.15625D-01	-0.70919D-03	-0.89653D-01	-0.55222D+01
0.31250D-01	-0.27666D-02	-0.17257D+00	-0.50910D+01
0.46875D-01	-0.60669D-02	-0.24875D+00	-0.46597D+01
0.62500D-01	-0.10505D-01	-0.31818D+00	-0.42285D+01
0.78125D-01	-0.15975D-01	-0.38089D+00	-0.37973D+01
0.93750D-01	-0.22372D-01	-0.43685D+00	-0.33660D+01
0.10938D+00	-0.29591D-01	-0.48607D+00	-0.29348D+01
0.12500D+00	-0.37527D-01	-0.52856D+00	-0.25036D+01
0.12500D+00	-0.37527D-01	-0.52856D+00	-0.24543D+01
0.14063D+00	-0.46069D-01	-0.56385D+00	-0.20623D+01
0.15625D+00	-0.55115D-01	-0.59301D+00	-0.16703D+01
0.17188D+00	-0.64569D-01	-0.61604D+00	-0.12783D+01
0.18750D+00	-0.74335D-01	-0.63296D+00	-0.88628D+00
0.20313D+00	-0.84317D-01	-0.64374D+00	-0.49426D+00
0.21875D+00	-0.94420D-01	-0.64840D+00	-0.10224D+00
0.23438D+00	-0.10455D+00	-0.64694D+00	0.28977D+00
0.25000D+00	-0.11460D+00	-0.63935D+00	0.68179D+00
0.25000D+00	-0.11460D+00	-0.63935D+00	0.73471D+00
0.26563D+00	-0.12449D+00	-0.62573D+00	0.10076D+01
0.28125D+00	-0.13414D+00	-0.60786D+00	0.12805D+01
0.29688D+00	-0.14347D+00	-0.58572D+00	0.15534D+01
0.31250D+00	-0.15242D+00	-0.55932D+00	0.18263D+01
0.32813D+00	-0.16092D+00	-0.52865D+00	0.20991D+01
0.34375D+00	-0.16892D+00	-0.49372D+00	0.23720D+01
0.35938D+00	-0.17633D+00	-0.45452D+00	0.26449D+01
0.37500D+00	-0.18310D+00	-0.41106D+00	0.29178D+01
0.37500D+00	-0.18310D+00	-0.41106D+00	0.29143D+01
0.39063D+00	-0.18916D+00	-0.36480D+00	0.30078D+01
0.40625D+00	-0.19449D+00	-0.31707D+00	0.31014D+01
0.42188D+00	-0.19906D+00	-0.26788D+00	0.31950D+01
0.43750D+00	-0.20285D+00	-0.21723D+00	0.32885D+01
0.45313D+00	-0.20584D+00	-0.16511D+00	0.33821D+01
0.46875D+00	-0.20801D+00	-0.11154D+00	0.34756D+01
0.48438D+00	-0.20932D+00	-0.56499D-01	0.35692D+01
0.50000D+00	-0.20976D+00	0.0	0.36627D+01

GF2(I) is not used in the program, but it is generated in the program from given GF1(I) at time equal to zero. The Fortran form of the data input to the program is given below.

```
TRANSIENT RESPONSE OF A CLAMPED BEAM (SEE EXAMPLE 3.7)
    1    0    2    2    4    3    0    1    2   20
0.5       0.0
0.0       0.0       1.0       0.0       0.0       0.0       0.0       0.0
0.5       0.25      0.0025    1.0
    1    2   10
0.0       0.0       0.0
0.0       0.0      -0.0391    0.5463   -0.1181    0.6506   -0.1876    0.4168
-0.2146   0.0
0.0       0.0       0.0       0.0       0.0       0.0       0.0       0.0
0.0       0.0
0.0       0.0       0.0       0.0       0.0       0.0       0.0       0.0
0.0       0.0
```

The program output (edited for the book) is given in Table 3.14. ■

This completes the discussion of the illustrative example problems. The reader can use program FEM1D to solve some of the problems on pages 191–192.

PROBLEMS

3.71 Derive quadratic interpolation functions in the natural coordinate system for a three-node element, with the inside node being at a distance s from the left end node.

3.72 Rewrite the Hermite cubic polynomials in Eqs. (3.97) in terms of the natural coordinate.

3.73 Rewrite the interpolation functions of Prob. 3.52 in terms of the natural coordinate.

3.74 Consider a three-node line element (equally spaced nodes). Assuming interpolation functions of the form

$$\psi_i = \sin\left[\pi\left(a_i + b_i\xi + c_i\xi^2\right)\right]$$

determine the lowest-order polynomials ψ_i such that they satisfy the interpolation requirement

$$\psi_i(\xi_j) = \delta_{ij} \qquad \xi_1 = -1 \qquad \xi_2 = 0 \qquad \xi_3 = 1$$

Discuss the numerical evaluation of element matrices for such elements.

3.75 The *error* in the finite-element approximation is often measured in terms of the *energy* associated with the problem under consideration. For the second-order problem considered in Eq. (3.1), the energy of the problem is given by the functional I,

$$I = \sum_{e=1}^{N} I_e(u_e) \tag{a}$$

where N is the total number of elements in the finite-element mesh, and I_e is the functional given in Eq. (3.5). The error in the solution is defined to be the difference between the true or exact solution u_0 and the finite-element solution u_h,

$$E = u_0 - u_h \tag{b}$$

and the error in the energy is defined by

$$\|E\|_1 \equiv \sum_{e=1}^{N} I_e(E) \tag{c}$$

If the finite-element approximation u_h is an interpolant of the true solution u_0, determine the error in the solution and the energy of Eq. (3.1) when (*a*) u_h is a linear interpolant and (*b*) u_h is a quadratic interpolant. [*Hint*: Use pertinent references at the end of the chapter to determine the interpolation error in terms of the mesh parameter (spacing) h, and use the result in Eq. (*c*) to determine the error in energy in the form $\|E\|_1 = c_0 h^p$, where c_0 and p are constants; p determines the rate of convergence ($p > 0$) or divergence ($p < 0$).]

In Probs. 3.76 through 3.80, compute the matrix coefficients using the linear transformation (3.176) as well as (*a*) the Newton-Cotes integration formula and (*b*) the Gauss-Legendre quadrature. Use the appropriate number of integration points, and verify the results with those obtained by the exact integration.

3.76

$$K_{12} = \int_{x_e}^{x_{e+1}} (1 + x) \frac{d\psi_1}{dx} \frac{d\psi_2}{dx} \, dx$$

$$G_{12} = \int_{x_e}^{x_{e+1}} (1 + x) \psi_1 \psi_2 \, dx$$

where ψ_i are the linear (Lagrange) interpolation functions.

3.77 Repeat Prob. 3.76 for the quadratic interpolation functions.

3.78

$$K_{11} = \int_{x_e}^{x_{e+1}} \left(\frac{d\psi_1}{dx} \right)^2 dx$$

$$G_{11} = \int_{x_e}^{x_{e+1}} (\psi_1)^2 \, dx$$

where ψ_i are the Hermite cubic functions in Eqs. (3.97) (see Prob. 3.72).

3.79 Repeat Prob. 3.78 for the case in which the interpolation functions are the fifth-order Hermite polynomials of Prob. 3.73.

3.80

$$K_{11} = \int_{x_e}^{x_{e+1}} \left(\frac{d\psi_1}{dx} \right)^2 dx$$

where $\psi_1 = \sin[\pi(\xi + 3\xi^2)/4]$. Use three- and five-point Gauss quadrature to compute K_{11}.

3.81 Solve the problem in Example 3.1 using the quadratic interpolation functions. Use one-element and two-element meshes and compare the results with those obtained in Example 3.1.

3.82 Modify the finite-element equations (3.58) for the following boundary conditions:

(*a*) $$u(0) = 0 \qquad \left(u + a \frac{du}{dx} \right) \bigg|_{x=1} = \beta$$

(*b*) $$\left(u + a \frac{du}{dx} \right) \bigg|_{x=0} = \alpha \qquad \left(u + a \frac{du}{dx} \right) \bigg|_{x=1} = \beta$$

Exercises involving computer programming

3.83 Write a subroutine, PREPRO, to read and generate the data required to solve the general second-order equation (3.1). You are required to debug the program to ensure that no mistake was made in the logic.

3.84 Write a subroutine, SHAPE2, to evaluate ψ_i, $d\psi_i/dx$, and $d^2\psi_i/dx^2$, and compute the jacobian of the transformation for a given Gauss point when ψ_i are the Hermite fifth-order polynomials developed in Prob. 3.52.

3.85 Write a (or modify) subroutine, STRESS, to calculate the secondary variables (e.g., heat flux, stresses) of axisymmetric heat flow (Prob. 3.31) and axisymmetric elastic deformation (Prob. 3.32) of a long cylinder.

3.86 Modify (or write your own) program FEM1D given in Appendix I to incorporate subroutines PREPRO and SHAPE1 developed in the previous problems.

3.87 Write a computer program (or modify FEM1D) to analyze beam problems using the higher-order element discussed in Prob. 3.52.

3.88 Develop a computer program, FRAME, to analyze frames using the frame element discussed in Sec. 3.3. Note that program FEM1D can be modified by rewriting subroutine STIFF and adding a subroutine TRANS (for transforming element matrices from the local coordinates to the global coordinates).

3.89 Modify subroutine STIFF to incorporate the penalty finite-element model of the beam-bending problem. Do not forget to use reduced integration for the penalty terms [see Reddy (1982)].

3.90 Modify the main program in FEM1D to account for convection boundary conditions in a heat-transfer problem (see Example 3.2 and Probs. 3.10 and 3.31).

Use program FEM1D or your own computer program to analyze the following problems and Probs. 3.61 to 3.69.

3.91 Solve the problem in Example 3.1 using two, four, and six linear elements.

3.92 Repeat Prob. 3.91 using the quadratic elements.

3.93 Solve the heat-transfer problem, Prob. 3.10, using four linear elements.

3.94 Solve the unconfined-aquifer problem, Prob. 3.16, using six linear elements (nonuniform mesh).

3.95 Solve the one-dimensional flow problem, Prob. 3.17, using four elements.

p. 107 **3.96** Solve the Couette flow problem, Prob. 3.29, using (a) four- and six-element (nonuniform) mesh of linear elements and (b) an equivalent mesh of quadratic elements.

3.97 Solve the problem of radial axisymmetric heat flow in a long cylinder (Prob. 3.31) using six linear elements.

p. 107 **3.98** Solve the problem of axisymmetric deformation of a circular cylinder (Prob. 3.32) using (a) four linear elements and (b) two quadratic elements.

3.99–3.110 Solve the beam problems, 3.35 through 3.46, using the minimum number of Hermite cubic elements.

3.111–3.112 Solve the axisymmetric bending of a circular plate for clamped and simply supported boundary conditions. Use (a) four linear elements and (b) two quadratic elements in the half-domain.

REFERENCES FOR ADDITIONAL READING

References on fluid mechanics

Duncan, W. J., A. S. Thom, and A. D. Young: *Mechanics of Fluids*, 2d ed., American Elsevier, New York (1970).

deVries, G., and D. H. Norrie: "The Application of the Finite-Element Technique to Potential Flow Problems," *Trans ASME, Series E, Journal of Applied Mechanics*, **38**, pp. 798–802 (1971).

Harr, M. E.: *Ground Water and Seepage*, McGraw-Hill, New York (1962).

Shames, I. H.: *Mechanics of Fluids*, McGraw-Hill, New York (1962).

Vallentine, H. R.: *Applied Hydrodynamics*, Butterworths, London (1959).

References on heat transfer

Kreith, F.: *Principles of Heat Transfer*, 3d ed., Harper & Row, New York (1973).
Myers, G. E.: *Analytical Methods in Conduction Heat Transfer*, McGraw-Hill, New York (1971).
Nagotov, E. P.: *Applications of Numerical Methods to Heat Transfer*, McGraw-Hill, New York (1978).

References on elasticity (solid mechanics)

Boresi, A. P., and P. P. Lynn: *Elasticity in Engineering Mechanics*, Prentice-Hall, Englewood Cliffs, N.J. (1974).
Dym, C. L., and I. H. Shames: *Solid Mechanics: A Variational Approach*, McGraw-Hill, New York (1973).
Timoshenko, S. P., and J. N. Goodier: *Theory of Elasticity*, McGraw-Hill, New York (1970).

References on variational methods (also see References for Additional Reading in Chap. 2)

Crandall, S. H.: *Engineering Analysis*, McGraw-Hill, New York (1956).
Finlayson, B. A.: *The Method of Weighted Residuals and Variational Principles*, Academic Press, New York (1972).
Lynn, P. P., and S. K. Arya: "Use of the Least Squares Criterion in the Finite Element Formulation," *International Journal for Numerical Methods in Engineering*, **6**, pp. 75–83 (1973).
Reddy, J. N.: "Some Computational Aspects of Mixed Finite-Element Approximations," *Proceedings of the 12th Annual Meeting of the Society of Engineering Science*, The University of Texas, Austin, pp. 965–980 (1975).
Reddy, J. N., and M. L. Rasmussen: *Advanced Engineering Analysis*, John Wiley, New York (1982).

References on the penalty-function method

Courant, R.: "Variational Methods for the Solution of Equilibrium and Vibrations," *Bulletin of the American Mathematical Society*, **49**, pp. 1–23 (1943).
Fiacco, A. V., and G. P. McCormick: *Nonlinear Programming: Sequential Unconstrained Methods for Solving Constrained Minimization Techniques*, John Wiley, New York (1975) (history of the penalty-function method).
Reddy, J. N. (ed.): *Penalty-Finite Element Methods in Mechanics*, AMD-vol. 51, The American Society of Mechanical Engineers, 345 East 47th St., New York (1982).
Reddy, J. N.: "On Penalty Function Methods in the Finite-Element Analysis of Flow Problems," *International Journal for Numerical Methods in Fluids*, **2**, pp. 110–120 (1982).

Other references

Carnahan, B., H. A. Luther, and J. O. Wilkes: *Applied Numerical Methods*, John Wiley, New York (1969) (interpolation and numerical integration).
Conte, S. D.: *Elementary Numerical Analysis*, McGraw-Hill, New York (1965) (numerical integration).
Harris, C. M., and C. E. Crede: *Shock and Vibration Handbook*, vol. 1, McGraw-Hill, New York (1961).
Martin, H. C.: *Introduction to Matrix Methods of Structural Analysis*, McGraw-Hill, New York (1966) (frame-element applications).

CHAPTER

FOUR

FINITE-ELEMENT ANALYSIS OF TWO-DIMENSIONAL PROBLEMS

4.1 INTRODUCTION

The finite-element analysis of two-dimensional problems involves the same basic steps as that described for one-dimensional problems in Chap. 3. The analysis is somewhat complicated by the fact that two-dimensional problems are described by partial differential equations. The boundary Γ of a two-dimensional domain Ω is, in general, a curve. Therefore, the finite elements are simple two-dimensional geometric shapes that can be used to approximate a given two-dimensional domain. In other words, in two-dimensional problems we not only seek an approximate solution to a given partial differential equation, but we also approximate the given domain by a suitable finite-element mesh. Consequently, we will have approximation errors (due to the approximation of the solution) as well as discretization errors (due to the approximation of the domain) in the finite-element analysis of two-dimensional problems. The finite-element mesh (discretization) consists of simple two-dimensional elements, such as triangles, rectangles, and/or quadrilaterals, that are connected to each other at nodal points on the boundaries of the elements. The ability to represent domains with irregular geometries by a collection of finite elements makes the method a valuable practical tool for the solution of boundary-value problems arising in various fields of engineering.

The objective here is to extend the basic steps discussed earlier for one-dimensional problems to two-dimensional problems. Once again, we begin with the discussion of problems described by second-order partial differential equations. We shall consider two types of second-order equations: (1) equations involving a dependent unknown u that is a scalar-valued function of position in Ω and (2) equations involving a dependent unknown $\mathbf{u} = (u_1, u_2)$ that is a vector-valued function of position in Ω. The components u_1 and u_2 of $\mathbf{u}$ are, in general, coupled by a pair of partial differential equations. Equations of the first type arise in the mathematical description of various physical phenomena, including heat transfer, seepage flow, the transverse deflection of a membrane, and the torsion of a cylindrical member (see Table 2.2). Equations of the second type arise in connection with the description of plane elasticity problems ($\mathbf{u}$ denotes the displacement vector) and incompressible fluid flow problems ($\mathbf{u}$ denotes the velocity vector). The discussion of the fourth-order differential equations, which arise in connection with the bending of elastic plates and in the stream function formulation of incompressible fluid flow, will not be considered in this first course. However, we shall consider an alternative formulation of plate bending that involves the solution of a set of second-order equations (see Sec. 4.5.4).

4.2 SECOND-ORDER EQUATIONS INVOLVING A SCALAR-VALUED FUNCTION

4.2.1 Description of the Model Equation

Consider the problem of finding the solution u of the second-order partial differential equation

$$-\frac{\partial}{\partial x}\left(a_{11}\frac{\partial u}{\partial x} + a_{12}\frac{\partial u}{\partial y}\right) - \frac{\partial}{\partial y}\left(a_{21}\frac{\partial u}{\partial x} + a_{22}\frac{\partial u}{\partial y}\right) + a_{00}u - f = 0 \quad (4.1)$$

for given data a_{ij} $(i, j = 1, 2)$, a_{00} and f and specified boundary conditions. The form of the boundary conditions will be apparent from the variational formulation. As a special case, one can obtain the Poisson equation from Eq. (4.1) by setting $a_{11} = a_{22} = a$ and $a_{12} = a_{21} = a_{00} = 0$:

$$-\nabla \cdot (a\nabla u) = f \text{ in } \Omega \quad (4.2)$$

where ∇ is the gradient operator. If $\hat{\mathbf{i}}$ and $\hat{\mathbf{j}}$ denote the unit vectors directed along the x and y axes, respectively, the gradient operator can be expressed as (see Sec. 2.1.4)

$$\nabla = \hat{\mathbf{i}}\frac{\partial}{\partial x} + \hat{\mathbf{j}}\frac{\partial}{\partial y} \quad (4.3)$$

In the following, we shall discuss the variational formulation and the finite-element formulation of the model equation (4.1).

4.2.2 Variational Formulation

In two dimensions there is more than one geometric shape that can be used as a finite element (see Fig. 4.1). As we shall see shortly, the interpolation functions depend not only on the number of nodes in the element, but also on the shape of

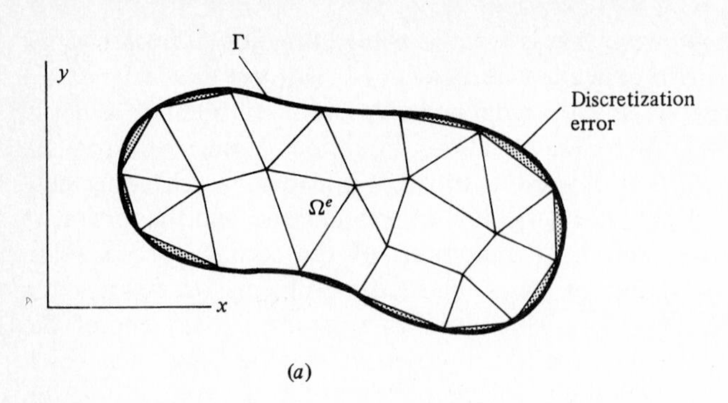

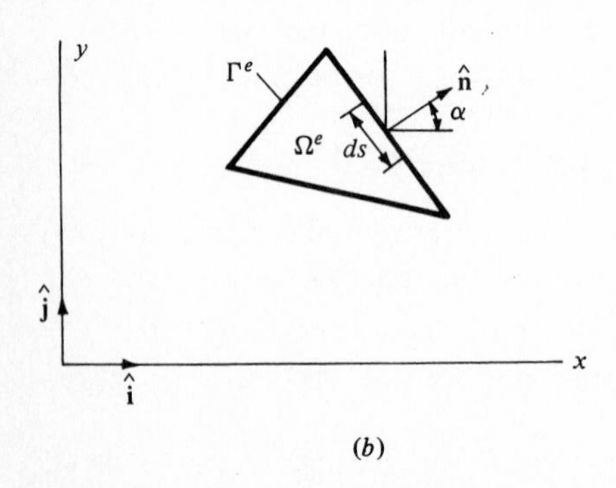

Figure 4.1 Finite-element discretization of an irregular domain. (*a*) Discretization of a domain by triangular and quadrilateral elements. (*b*) A typical triangular element (boundary Γ^e, the unit normal $\hat{\mathbf{n}}$ on the boundary of the element).

the element. At this stage of the development, we do not have to confine our discussion to any particular element. We assume that Ω^e is a typical element, whether triangular or quadrilateral, of the finite-element mesh, and we develop the finite-element model of Eq. (4.1). Various two-dimensional elements will be discussed in the next section.

Following the procedure of Chap. 2, we develop the variational form of Eq. (4.1) over a typical element. We multiply Eq. (4.1) with a test function v (assumed to be differentiable once with respect to x and y) and integrate over the element domain Ω^e:

$$0 = \int_{\Omega^e} v\left[-\frac{\partial}{\partial x}(F_1) - \frac{\partial}{\partial y}(F_2) + a_{00}u - f\right] dx\, dy \tag{4.4}$$

where

$$F_1 = a_{11}\frac{\partial u}{\partial x} + a_{12}\frac{\partial u}{\partial y} \qquad F_2 = a_{21}\frac{\partial u}{\partial x} + a_{22}\frac{\partial u}{\partial y}$$

In order to distribute the differentiation among u and v equally, we integrate the first two terms in Eq. (4.4) by parts. First we note the identities

$$
\begin{aligned}
-v\frac{\partial F_1}{\partial x} &= -\frac{\partial}{\partial x}(vF_1) + \frac{\partial v}{\partial x}F_1 \\
-v\frac{\partial F_2}{\partial y} &= -\frac{\partial}{\partial y}(vF_2) + \frac{\partial v}{\partial y}F_2
\end{aligned}
\tag{4.5}
$$

Next we recall the component form of the gradient (or divergence) theorem in Eqs. (2.19),

$$\int_{\Omega^e}\frac{\partial}{\partial x}(vF_1)\,dx\,dy = \oint_{\Gamma^e} vF_1 n_x\,ds \tag{4.6a}$$

$$\int_{\Omega^e}\frac{\partial}{\partial y}(vF_2)\,dx\,dy = \oint_{\Gamma^e} vF_2 n_y\,ds \tag{4.6b}$$

where n_x and n_y are the components (i.e., the direction cosines) of the unit normal $\hat{\mathbf{n}}$

$$\hat{\mathbf{n}} = n_x\hat{\mathbf{i}} + n_y\hat{\mathbf{j}} = \cos\alpha\,\hat{\mathbf{i}} + \sin\alpha\,\hat{\mathbf{j}} \tag{4.7}$$

on the boundary Γ_e, and ds is the arc length of an infinitesimal element along the boundary (see Fig. 4.1b). Using Eqs. (4.5) and (4.6) in Eq. (4.4), we obtain

$$
\begin{aligned}
0 = &\int_{\Omega^e}\left[\frac{\partial v}{\partial x}\left(a_{11}\frac{\partial u}{\partial x} + a_{12}\frac{\partial u}{\partial y}\right) + \frac{\partial v}{\partial y}\left(a_{21}\frac{\partial u}{\partial x} + a_{22}\frac{\partial u}{\partial y}\right) + a_{00}vu - vf\right]dx\,dy \\
&-\oint_{\Gamma^e} v\left[n_x\left(a_{11}\frac{\partial u}{\partial x} + a_{12}\frac{\partial u}{\partial y}\right) + n_y\left(a_{21}\frac{\partial u}{\partial x} + a_{22}\frac{\partial u}{\partial y}\right)\right]ds
\end{aligned}
\tag{4.8}
$$

From an inspection of the boundary term in Eq. (4.8), we note that the specification of u constitutes the essential boundary condition (hence u is the primary variable), and the specification of

$$q_n \equiv n_x\left(a_{11}\frac{\partial u}{\partial x} + a_{12}\frac{\partial u}{\partial y}\right) + n_y\left(a_{21}\frac{\partial u}{\partial x} + a_{22}\frac{\partial u}{\partial y}\right) \tag{4.9}$$

constitutes the natural boundary condition (hence q_n is the secondary variable) of the formulation. The function $q_n = q_n(s)$ denotes the projection of the vector $\mathbf{F} = F_1\hat{\mathbf{i}} + F_2\hat{\mathbf{j}}$ along the unit normal $\hat{\mathbf{n}}$. When $a_{11} = a_{22} = a$ and $a_{21} = a_{12} = 0$, q_n is precisely equal to the product of a and the directional derivative of u with respect to $\hat{\mathbf{n}}$. The secondary variable q_n is of physical interest in most problems. For example, in the case of the heat transfer through an anisotropic medium (a_{ij} denote the conductivities of the medium), q_n denotes the heat flux across the boundary of the element.

The variational form of Eq. (4.1) is given by

$$
\begin{aligned}
0 = &\int_{\Omega^e}\left[\frac{\partial v}{\partial x}\left(a_{11}\frac{\partial u}{\partial x} + a_{12}\frac{\partial u}{\partial y}\right) + \frac{\partial v}{\partial y}\left(a_{21}\frac{\partial u}{\partial x} + a_{22}\frac{\partial u}{\partial y}\right) + a_{00}vu - vf\right]dx\,dy \\
&-\oint_{\Gamma^e} vq_n\,ds
\end{aligned}
\tag{4.10}
$$

This variational equation (4.10) forms the basis of the finite-element model of Eq. (4.1).

4.2.3 Finite-Element Formulation

The variational form in Eq. (4.10) indicates that the approximation chosen for u should be at least bilinear in x and y so that the first two terms in Eq. (4.10) and q_n in Eq. (4.9) are nonzero. Suppose that u is approximated by the expression

$$u = \sum_{j=1}^{n} u_j \psi_j \tag{4.11}$$

where u_j are the values of u at the point (x_j, y_j) and ψ_j are linear interpolation functions with the property

$$\psi_i(x_j, y_j) = \delta_{ij} \tag{4.12}$$

The specific form of ψ_i will be derived for linear triangular and rectangular elements in Sec. 4.2.4 (also, see Sec. 4.4).

Substituting Eq. (4.11) for u and ψ_i for v into the variational form (4.10), we obtain

$$\begin{aligned} 0 = \sum_{j=1}^{n} \Bigg\{ \int_{\Omega^e} \bigg[& \frac{\partial \psi_i}{\partial x}\left(a_{11}\frac{\partial \psi_j}{\partial x} + a_{12}\frac{\partial \psi_j}{\partial y}\right) \\ & + \frac{\partial \psi_i}{\partial y}\left(a_{21}\frac{\partial \psi_j}{\partial x} + a_{22}\frac{\partial \psi_j}{\partial y}\right) + a_{00}\psi_i\psi_j \bigg] dx\,dy \Bigg\} u_j \\ & - \int_{\Omega^e} f\psi_i \,dx\,dy - \oint_{\Gamma^e} \psi_i q_n \,ds \qquad i = 1, 2, \ldots, n \end{aligned}$$

or

$$\sum_{j=1}^{n} K_{ij}^{(e)} u_j^{(e)} = F_i^{(e)} \tag{4.13}$$

where

$$\begin{aligned} K_{ij}^{(e)} = \int_{\Omega^e} \bigg[& \frac{\partial \psi_i}{\partial x}\left(a_{11}\frac{\partial \psi_j}{\partial x} + a_{12}\frac{\partial \psi_j}{\partial y}\right) \\ & + \frac{\partial \psi_i}{\partial y}\left(a_{21}\frac{\partial \psi_j}{\partial x} + a_{22}\frac{\partial \psi_j}{\partial y}\right) + a_{00}\psi_i\psi_j \bigg] dx\,dy \end{aligned} \tag{4.14}$$

$$F_i^{(e)} = \int_{\Omega^e} f\psi_i \,dx\,dy + \oint_{\Gamma^e} q_n \psi_i \,ds$$

Note that $K_{ij}^{(e)} = K_{ji}^{(e)}$ (i.e., $[K]$ is symmetric) only when $a_{12} = a_{21}$. Equation (4.13) represents the finite-element model of Eq. (4.1). In the next section, we shall discuss the derivation of interpolation functions.

4.2.4 Interpolation Functions

An examination of the variational form (4.10) and the finite-element matrices in Eqs. (4.14) shows that ψ_i should be at least bilinear functions of x and y. As pointed out before, there is a correspondence between both the number and location of nodal points and the number of primary unknowns per node in a finite element and the number of terms used in the polynomial approximations of a dependent variable over an element. In one-dimensional second-order problems, the number of nodes n in an element uniquely defines the degree r of the polynomial, with the correspondence between n and r being $n = r + 1$. In two-dimensional second-order problems, the correspondence between the number of nodes (which is equal to the number of terms in the approximating polynomial) and the degree of the polynomial is not unique. For example, the polynomial

$$u(x, y) = c_1 + c_2 x + c_3 y \tag{4.15}$$

contains three (linearly independent) terms, and it is linear in both x and y. On the other hand, the polynomial

$$u(x, y) = c_1 + c_2 x + c_3 y + c_4 xy \tag{4.16}$$

contains four (linearly independent) terms, but it is also linear in both x and y. The former requires an element with three nodes (with one primary unknown per node), and the latter requires an element with four nodes. A two-dimensional element with three nodes is a triangle with the nodes at the vertices of the triangle. When the number of nodes is equal to 4, one can choose a triangle with the fourth node at the center (or centroid) of the triangle or a rectangle (or quadrilateral) with the nodes at the vertices of the rectangle. A polynomial with five constants is the (incomplete) quadratic polynomial

$$u(x, y) = c_1 + c_2 x + c_3 y + c_4 xy + c_5 (x^2 + y^2) \tag{4.17}$$

which can be used to construct an element with five nodes (e.g., a rectangle with a node at each corner and at the midpoint of the rectangle). Similarly, the quadratic polynomial

$$u(x, y) = c_1 + c_2 x + c_3 y + c_4 xy + c_5 x^2 + c_6 y^2 \tag{4.18}$$

with six constants can be used to construct an element with six nodes (e.g., a triangle with a node at each vertex and at the midpoint of each side). Examples of three-, four-, five-, and six-node elements are shown in Fig. 4.2.

Here we derive the linear interpolation functions for the three-node triangular element and the four-node rectangular element. The procedure is the same as that used for the one-dimensional linear element.

Interpolation functions for the three-node triangular element Consider the linear approximation (4.15). The set $\{1, x, y\}$ is linearly independent and complete. We

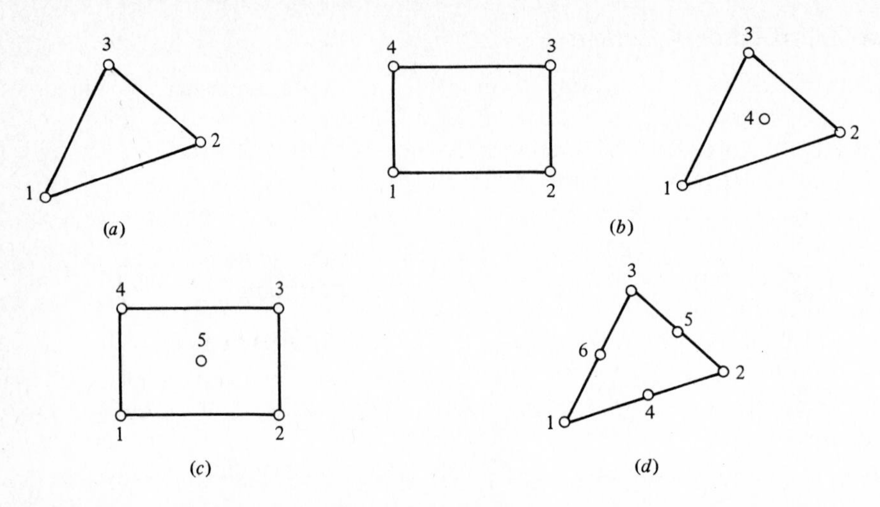

Figure 4.2 Finite elements in two dimensions. (*a*) A three-node element. (*b*) Four-node elements. (*c*) A five-node element. (*d*) A six-node element.

must rewrite the approximation (4.15) such that it satisfies the conditions

$$u(x_i, y_i) = u_i \qquad i = 1, 2, 3 \tag{4.19}$$

where (x_i, y_i) $(i = 1, 2, 3)$ are the (global) coordinates of the three vertices of the triangle. In other words, we determine the three constants c_i in Eq. (4.15) in terms of u_i from Eq. (4.19):

$$\begin{aligned} u_1 &\equiv u(x_1, y_1) = c_1 + c_2 x_1 + c_3 y_1 \\ u_2 &\equiv u(x_2, y_2) = c_1 + c_2 x_2 + c_3 y_2 \\ u_3 &\equiv u(x_3, y_3) = c_1 + c_2 x_3 + c_3 y_3 \end{aligned} \tag{4.20}$$

In matrix form, we have [see Eq. (3.10)]

$$\begin{Bmatrix} u_1 \\ u_2 \\ u_3 \end{Bmatrix} = \begin{bmatrix} 1 & x_1 & y_1 \\ 1 & x_2 & y_2 \\ 1 & x_3 & y_3 \end{bmatrix} \begin{Bmatrix} c_1 \\ c_2 \\ c_3 \end{Bmatrix} \tag{4.21}$$

Note that the nodes are numbered counterclockwise. Solving Eq. (4.21) for c_i $(i = 1, 2, 3)$, we obtain

$$\begin{aligned} c_1 &= \frac{1}{2A_e}\left[u_1(x_2 y_3 - x_3 y_2) + u_2(x_3 y_1 - x_1 y_3) + u_3(x_1 y_2 - x_2 y_1)\right] \\ c_2 &= \frac{1}{2A_e}\left[u_1(y_2 - y_3) + u_2(y_3 - y_1) + u_3(y_1 - y_2)\right] \\ c_3 &= \frac{1}{2A_e}\left[u_1(x_3 - x_2) + u_2(x_1 - x_3) + u_3(x_2 - x_1)\right] \end{aligned} \tag{4.22}$$

where A_e is the area of the triangle,

$$2A_e = \begin{vmatrix} 1 & x_1 & y_1 \\ 1 & x_2 & y_2 \\ 1 & x_3 & y_3 \end{vmatrix}$$

$$= (x_2y_3 - x_3y_2) + (x_3y_1 - x_1y_3) + (x_1y_2 - x_2y_1) \tag{4.23}$$

The determinant will have an opposite sign if the nodes are numbered clockwise. Substituting for c_i from Eqs. (4.22) into Eq. (4.15), we obtain

$$u(x, y) = u_1\psi_1(x, y) + u_2\psi_2(x, y) + u_3\psi_3(x, y)$$

$$= \sum_{i=1}^{3} u_i\psi_i^{(e)} \tag{4.24}$$

where $\psi_i^{(e)}$ are the linear interpolation functions for the triangular element,

$$\psi_i^{(e)} = \frac{1}{2A_e}(\alpha_i + \beta_i x + \gamma_i y) \qquad i = 1, 2, 3 \tag{4.25a}$$

and α_i, β_i, and γ_i are the constants

$$\left.\begin{aligned} \alpha_i &= x_j y_k - x_k y_j \\ \beta_i &= y_j - y_k \\ \gamma_i &= x_k - x_j \end{aligned}\right\} \quad \begin{array}{l} i \neq j \neq k, \text{ and } i,\ j, \text{ and } k \\ \text{permute in a natural order} \end{array} \tag{4.25b}$$

For example, α_2 is given by setting $i = 2$, $j = 3$, and $k = 1$ in Eq. (4.25*b*):

$$\alpha_2 = x_3y_1 - x_1y_3$$

The linear interpolation functions $\psi_i^{(e)}$ are shown in Fig. 4.3. Note that ψ_i has the properties

$$\begin{aligned} \psi_i(x_j, y_j) &= \delta_{ij} \qquad i, j = 1, 2, 3 \\ \sum_{i=1}^{3} \psi_i &= 1 \end{aligned} \tag{4.26}$$

Also note that Eq. (4.24) determines a plane surface passing through u_1, u_2, and u_3. Hence, use of the linear interpolation functions ψ_i of a triangle will result in

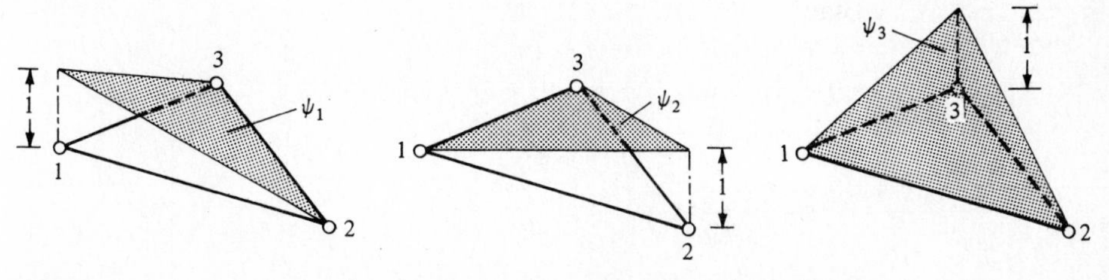

Figure 4.3 Linear interpolation functions for the three-node triangular element.

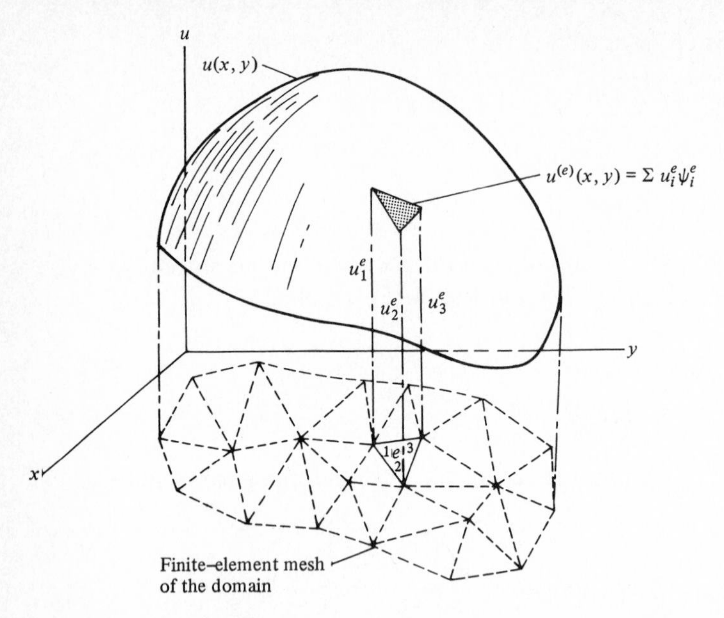

Figure 4.4 Representation of a continuous function $u(x, y)$ by linear interpolation functions of three-node triangular elements.

the approximation of the curved surface $u(x, y)$ by a planar function $\Sigma_{i=1}^{3} u_i \psi_i$ (see Fig. 4.4).

Linear interpolation functions for the four-node rectangular element Here we consider approximation of the form (4.16) and use a rectangular element with sides a and b (see Fig. 4.5). For the sake of convenience, we choose a local coordinate system (ξ, η) to derive the interpolation functions. We assume that

$$u(\xi, \eta) = c_1 + c_2 \xi + c_3 \eta + c_4 \xi \eta \tag{4.27}$$

and require

$$\begin{aligned} u_1 &= u(0,0) = c_1 \\ u_2 &= u(a,0) = c_1 + c_2 a \\ u_3 &= u(a, b) = c_1 + c_2 a + c_3 b + c_4 ab \\ u_4 &= u(0, b) = c_1 + c_3 b \end{aligned} \tag{4.28}$$

Solving for c_i $(i = 1, 2, 3, 4)$, we obtain

$$\begin{Bmatrix} c_1 \\ c_2 \\ c_3 \\ c_4 \end{Bmatrix} = \begin{bmatrix} 1 & 0 & 0 & 0 \\ 1 & a & 0 & 0 \\ 1 & a & b & ab \\ 1 & 0 & b & 0 \end{bmatrix}^{-1} \begin{Bmatrix} u_1 \\ u_2 \\ u_3 \\ u_4 \end{Bmatrix} = \frac{1}{ab} \begin{bmatrix} ab & 0 & 0 & 0 \\ -b & b & 0 & 0 \\ -a & 0 & 0 & a \\ 1 & -1 & 1 & -1 \end{bmatrix} \begin{Bmatrix} u_1 \\ u_2 \\ u_3 \\ u_4 \end{Bmatrix} \tag{4.29}$$

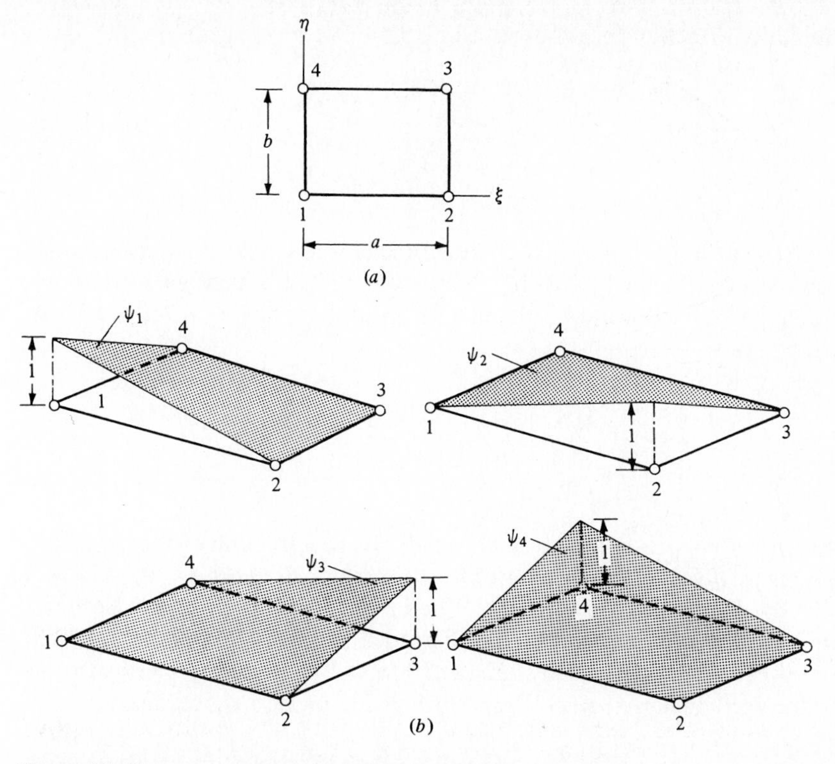

Figure 4.5 Linear rectangular element and associated interpolation functions. (*a*) A four-node rectangular element. (*b*) Linear interpolation functions.

Substituting Eq. (4.29) into Eq. (4.27), we obtain

$$u(\xi, \eta) = \{1 \quad \xi \quad \eta \quad \xi\eta\}\begin{Bmatrix} c_1 \\ c_2 \\ c_3 \\ c_4 \end{Bmatrix} = \{\psi_1 \quad \psi_2 \quad \psi_3 \quad \psi_4\}\begin{Bmatrix} u_1 \\ u_2 \\ u_3 \\ u_4 \end{Bmatrix}$$

$$= \sum_{i=1}^{4} u_i \psi_i(\xi, \eta) \tag{4.30}$$

where

$$\begin{aligned} \psi_1(\xi, \eta) &= \left(1 - \frac{\xi}{a}\right)\left(1 - \frac{\eta}{b}\right) \\ \psi_2(\xi, \eta) &= \frac{\xi}{a}\left(1 - \frac{\eta}{b}\right) \\ \psi_3(\xi, \eta) &= \frac{\xi}{a}\frac{\eta}{b} \\ \psi_4(\xi, \eta) &= \left(1 - \frac{\xi}{a}\right)\frac{\eta}{b} \end{aligned} \tag{4.31}$$

The interpolation functions are shown in Fig. 4.5*b*. We have again

$$\psi_i(\xi_j, \eta_j) = \delta_{ij} \qquad i, j = 1,2,3,4$$
$$\sum_{i=1}^{4} \psi_i = 1 \tag{4.32}$$

Note that the linear interpolation functions for the four-node rectangular element can also be obtained by taking the tensor product of the linear interpolation functions (3.17*b*) associated with sides 1–2 and 2–3 (which can be viewed as line elements):

$$\begin{Bmatrix} 1 - \dfrac{\xi}{a} \\ \dfrac{\xi}{a} \end{Bmatrix} \left\{ 1 - \frac{\eta}{b} \;\middle|\; \frac{\eta}{b} \right\} = \begin{bmatrix} \psi_1 & \psi_4 \\ \psi_2 & \psi_3 \end{bmatrix} \tag{4.33}$$

The conventional procedure given above for the construction of the interpolation functions involves the inversion of an $n \times n$ matrix [see Eq. (4.29)], where n is the number of nodes in the element. When n is large, the inversion becomes very tedious. The alternative procedure discussed in Prob. 3.1 for one-dimensional elements proves to be algebraically simple. Here we illustrate the procedure for the four-node rectangular element. Property 1 of Eq. (4.32) requires that

$$\psi_1(\xi_i, \eta_i) = 0 \qquad i = 2,3,4$$
$$\psi_1(\xi_1, \eta_1) = 1$$

That is, ψ_1 is identically zero on lines $\xi = a$ and $\eta = b$. Hence, $\psi_1(\xi, \eta)$ must be of the form

$$\psi_1(\xi, \eta) = c_1(a - \xi)(b - \eta)$$

Using the condition $\psi_1(\xi_1, \eta_1) = \psi_1(0,0) = 1$, we get $c_1 = 1/ab$. Hence,

$$\psi_1(\xi, \eta) = \frac{1}{ab}(a - \xi)(b - \eta) = \left(1 - \frac{\xi}{a}\right)\left(1 - \frac{\eta}{b}\right)$$

Likewise, one can obtain the remaining three interpolation functions.

4.2.5 Computation of Element Matrices

Computation of the element matrices $[K^{(e)}]$ and $\{F^{(e)}\}$ in Eqs. (4.14) by the conventional methods (i.e., by exact integration) is, in general, not easy. However, when a_{ij}, a_{00}, and f are constants, it is possible to evaluate the integrals exactly over the triangular and rectangular elements discussed in the previous section. The boundary integral in $\{F^{(e)}\}$ of Eqs. (4.14) can be evaluated whenever q_n is known. For an interior element (i.e., an element that does not have any of its sides on the boundary of the problem), the contribution from the boundary integral cancels with similar contributions from adjoining elements of the mesh (analogous to the $P_i^{(e)}$ and $Q_i^{(e)}$ in the one-dimensional problems). A more detailed discussion is given below.

For the sake of brevity, we rewrite $[K^{(e)}]$ in Eqs. (4.14) as the sum of four basic matrices $[S^{\alpha\beta}]$ $(\alpha, \beta = 1, 2)$ and $[S]$,

$$[K^{(e)}] = a_{11}[S^{11}] + a_{12}[S^{12}] + a_{21}[S^{12}]^T + a_{22}[S^{22}] + a_{00}[S] \tag{4.34a}$$

where

$$S_{ij}^{11} = \int_{\Omega^e} \frac{\partial \psi_i}{\partial x} \frac{\partial \psi_j}{\partial x} dx\, dy \qquad S_{ij}^{12} = \int_{\Omega^e} \frac{\partial \psi_i}{\partial x} \frac{\partial \psi_j}{\partial y} dx\, dy$$
$$S_{ij}^{22} = \int_{\Omega^e} \frac{\partial \psi_i}{\partial y} \frac{\partial \psi_j}{\partial y} dx\, dy \qquad S_{ij} = \int_{\Omega^e} \psi_i \psi_j \, dx\, dy \tag{4.34b}$$

Also we let

$$f_i^{(e)} = \int_{\Omega^e} f \psi_i \, dx\, dy \qquad Q_i^{(e)} = \oint_{\Gamma^e} q_n \psi_i \, ds \tag{4.35}$$

We now proceed to compute the matrices in Eqs. (4.34) and (4.35) using the linear interpolation functions derived in the previous section.

Element matrices for a linear triangular element For a triangle, the following exact integral formulas are available for evaluating the integrals. Let

$$I_{mn} \equiv \int_{\Delta} x^m y^n \, dx\, dy \tag{4.36}$$

Then we have

$$I_{00} = A \quad \text{area of the triangle}$$
$$I_{10} = A \cdot \bar{x} \qquad \bar{x} = \frac{1}{3} \sum_{i=1}^{3} x_i$$
$$I_{01} = A \cdot \bar{y} \qquad \bar{y} = \frac{1}{3} \sum_{i=1}^{3} y_i$$
$$I_{11} = \frac{A}{12}\left(\sum_{i=1}^{3} x_i y_i + 9\bar{x}\bar{y} \right) \tag{4.37}$$
$$I_{20} = \frac{A}{12}\left(\sum_{i=1}^{3} x_i^2 + 9\bar{x}^2 \right)$$
$$I_{02} = \frac{A}{12}\left(\sum_{i=1}^{3} y_i^2 + 9\bar{y}^2 \right)$$

Using the linear interpolation functions (4.25) in Eqs. (4.34*b*) and (4.35), and noting that

$$\frac{\partial \psi_i}{\partial x} = \frac{\beta_i}{2A} \qquad \frac{\partial \psi_i}{\partial y} = \frac{\gamma_i}{2A} \tag{4.38}$$

we obtain

$$S_{ij}^{11} = \frac{1}{4A}\beta_i\beta_j \qquad S_{ij}^{12} = \frac{1}{4A}\beta_i\gamma_j \qquad S_{ij}^{22} = \frac{1}{4A}\gamma_i\gamma_j$$

$$S_{ij} = \frac{1}{4A}\Big\{\big[\alpha_i\alpha_j + (\alpha_i\beta_j + \alpha_j\beta_i)\bar{x} + (\alpha_i\gamma_j + \alpha_j\gamma_i)\bar{y}\big] \tag{4.39a}$$

$$+\frac{1}{A}\big[I_{20}\beta_i\beta_j + I_{11}(\gamma_i\beta_j + \gamma_j\beta_i) + I_{02}\gamma_i\gamma_j\big]\Big\}$$

In view of the identity $\alpha_i + \beta_i\bar{x} + \gamma_i\bar{y} = 2A/3$ [which follows from Eqs. (4.25*b*) and (4.37)], we have

$$f_i^{(e)} = \frac{f}{2}(\alpha_i + \beta_i\bar{x} + \gamma_i\bar{y}) = \frac{fA}{3} \tag{4.39b}$$

Once the coordinates of the element nodes are known, one can compute α_i, β_i, and γ_i from Eq. (4.25*b*) and substitute into Eq. (4.39) to obtain the element matrices.

Element matrices for a linear rectangular element Since a_{00}, a_{ij}, and f are constants, we can use the interpolation functions of Eqs. (4.31) with ξ and η replaced by x and y, respectively. We have

$$[S^{11}] = \frac{b}{6a}\begin{bmatrix} 2 & -2 & -1 & 1 \\ -2 & 2 & 1 & -1 \\ -1 & 1 & 2 & -2 \\ 1 & -1 & -2 & 2 \end{bmatrix} \qquad [S^{12}] = \frac{1}{4}\begin{bmatrix} 1 & 1 & -1 & -1 \\ -1 & -1 & 1 & 1 \\ -1 & -1 & 1 & 1 \\ 1 & 1 & -1 & -1 \end{bmatrix}$$

$$[S^{22}] = \frac{a}{6b}\begin{bmatrix} 2 & 1 & -1 & -2 \\ 1 & 2 & -2 & -1 \\ -1 & -2 & 2 & 1 \\ -2 & -1 & 1 & 2 \end{bmatrix} \qquad [S] = \frac{ab}{36}\begin{bmatrix} 4 & 2 & 1 & 2 \\ 2 & 4 & 2 & 1 \\ 1 & 2 & 4 & 2 \\ 2 & 1 & 2 & 4 \end{bmatrix} \tag{4.40}$$

$$\{f\} = \frac{fab}{4}\{1 \quad 1 \quad 1 \quad 1\}^T$$

Evaluation of the boundary integrals Here we consider the evaluation of boundary integrals of the type

$$Q_i^{(e)} = \oint_{\Gamma^e} q_n^{(e)}\psi_i^{(e)}(s)\,ds \tag{4.41}$$

when $q_n^{(e)}$ is a known function of the distance s along the boundary Γ^e. It is not necessary to compute such integrals when a portion of Γ^e does not coincide with the boundary Γ of the domain Ω. On portions of Γ^e that are in the interior of the domain, $q_n^{(e)}$ on side (i, j) of element e cancels with $q_n^{(f)}$ on side (p, q) of element f when sides (i, j) of element e and (p, q) of element f are the same (i.e., interface of the elements e and f). This can be viewed as the equilibrium of internal "forces" (see Fig. 4.6*b*, *c*). When Γ^e falls on the boundary of the domain, $q_n^{(e)}$ is either known (in general, as a function of s) or to be determined in the

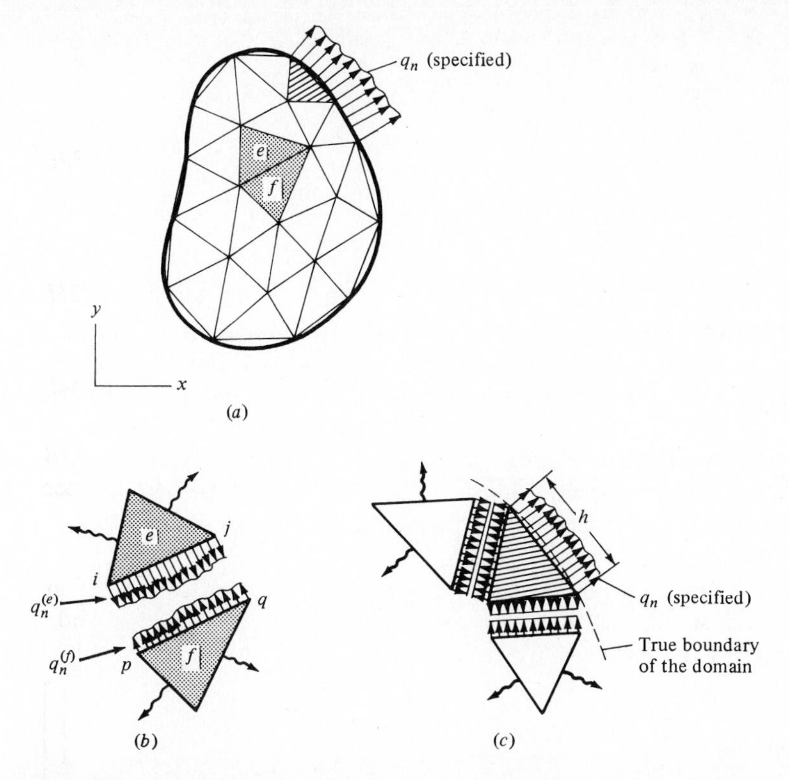

Figure 4.6 Computation of boundary forces and equilibrium of secondary variables at interelement boundaries. (*a*) Finite-element discretization. (*b*) Equilibrium of forces at interfaces. (*c*) Computation of forces on the true boundary.

postcomputation. In the latter case, the primary variable will be specified (on the portion of the boundary where q_n is not specified).

The boundary Γ^e of linear two-dimensional elements is a set of linear one-dimensional elements. Therefore, the evaluation of the boundary integral amounts to evaluating line integrals. It should not be surprising to the reader that when two-dimensional interpolation functions are restricted to (i.e., evaluated on) the boundary, we get the corresponding one-dimensional interpolation functions. To fix the ideas, consider a finite element that has a portion of its boundary on the boundary of the domain (see Fig. 4.6*c*), and assume that q_n is known there. Then

$$\int_0^h q_n(s)\psi_i(s)\,ds \equiv Q_i \tag{4.42}$$

gives the contribution of q_n to node i. Here h denotes the length of the side that is subjected to the force q_n, and $\psi_i(s)$ are the one-dimensional interpolation functions. When $\psi_i(x, y)$ are linear [then $\psi_i(s)$ are linear], i takes the values of 1 and 2, and when $\psi_i(x, y)$ are quadratic [then $\psi_i(s)$ are quadratic], i takes the values of 1, 2, and 3. For example, when ψ_i are linear ($\psi_1 = 1 - s/h$, $\psi_2 = s/h$), we have

$$Q_i = \begin{cases} \int_0^h q_n \psi_i \, ds & \text{for any } q_n \\ \dfrac{q_n h}{2} & \text{for } q_n \text{ constant} \end{cases} \tag{4.43}$$

4.2.6 Assembly of Element Matrices

The assembly of finite-element equations is based on the same principle as that employed in one-dimensional problems. We illustrate the procedure by considering a finite-element mesh consisting of a triangular element and a quadrilateral element (see Fig. 4.7*a*). Let $K_{ij}^{(1)}$ $(i, j = 1, 2, 3)$ denote the coefficient matrix corresponding to the triangular element, and let $K_{ij}^{(2)}$ $(i, j = 1, 2, 3, 4)$ denote that corresponding to the quadrilateral element. From the finite-element mesh shown in Fig. 4.7*a*, we note the following correspondence between the global and element nodal values:

$$U_1 = u_1^{(1)} \qquad U_2 = u_2^{(1)} = u_1^{(2)} \qquad U_3 = u_3^{(1)} = u_4^{(2)} \qquad U_4 = u_2^{(2)} \qquad U_5 = u_3^{(2)} \tag{4.44}$$

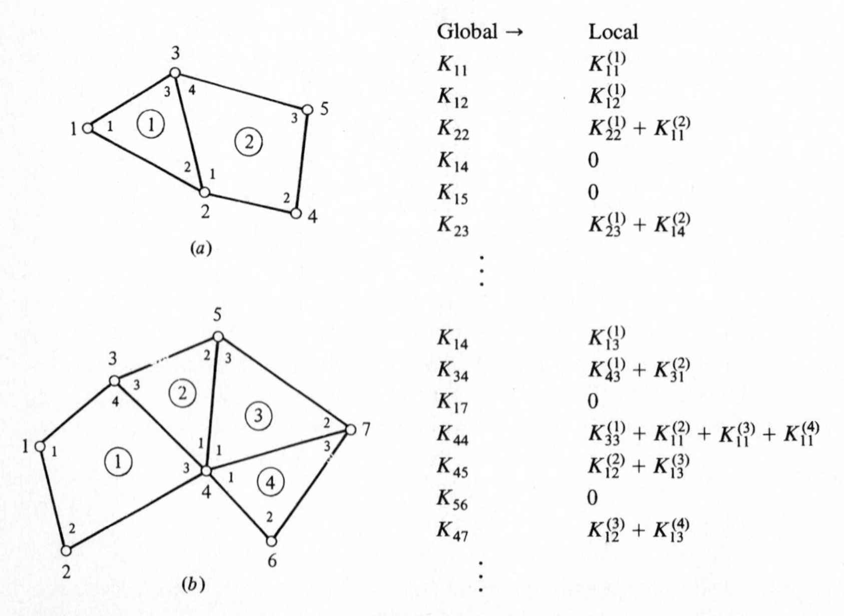

(a)
(b)
Global →
Local
K_{11} $K_{11}^{(1)}$
K_{12} $K_{12}^{(1)}$
K_{22} $K_{22}^{(1)} + K_{11}^{(2)}$
K_{14} 0
K_{15} 0
K_{23} $K_{23}^{(1)} + K_{14}^{(2)}$
K_{14} $K_{13}^{(1)}$
K_{34} $K_{43}^{(1)} + K_{31}^{(2)}$
K_{17} 0
K_{44} $K_{33}^{(1)} + K_{11}^{(2)} + K_{11}^{(3)} + K_{11}^{(4)}$
K_{45} $K_{12}^{(2)} + K_{13}^{(3)}$
K_{56} 0
K_{47} $K_{12}^{(3)} + K_{13}^{(4)}$

Figure 4.7 Assembly of finite-element coefficient matrices using the correspondence between global and element nodes (one unknown per node). (*a*) Assembly of two elements. (*b*) Assembly of several elements.

Note that the continuity of the nodal values at the interelement nodes guarantees the continuity of the primary variable along the entire interelement boundary. To see this, recall that the finite-element approximation is linear along the boundaries of linear triangular and quadrilateral elements. Since a linear polynomial is uniquely defined by two constants (namely, the nodal values on the boundary) and the constants are the same in both elements that share the boundary, it follows that the primary variable is uniquely defined along the interelement boundary.

Next we use the interelement continuity conditions (4.44) and Eqs. (3.25) to assemble the element equations:

$$0 = \sum_{e=1}^{2} \{\delta u^{(e)}\}^T([K^{(e)}]\{u^{(e)}\} - \{F^{(e)}\})$$

$$= \sum_{e=1}^{2} \sum_{i=1}^{n_e} \delta u_i^{(e)} \left[\sum_{j=1}^{n_e} K_{ij}^{(e)} u_j^{(e)} - F_i^{(e)} \right] \qquad n_1 = 3 \qquad n_2 = 4$$

or

$$\begin{aligned}
0 = {} & \delta U_1\left[K_{11}^{(1)}U_1 + K_{12}^{(1)}U_2 + K_{13}^{(1)}U_3 - F_1^{(1)}\right] \\
& + \delta U_2\left[K_{21}^{(1)}U_1 + K_{22}^{(1)}U_2 + K_{23}^{(1)}U_3 - F_2^{(1)}\right] \\
& + \delta U_3\left[K_{31}^{(1)}U_1 + K_{32}^{(1)}U_2 + K_{33}^{(1)}U_3 - F_3^{(1)}\right] \\
& + \delta U_2\left[K_{11}^{(2)}U_2 + K_{12}^{(2)}U_4 + K_{13}^{(2)}U_5 + K_{14}^{(2)}U_3 - F_1^{(2)}\right] \\
& + \delta U_4\left[K_{21}^{(2)}U_2 + K_{22}^{(2)}U_4 + K_{23}^{(2)}U_5 + K_{24}^{(2)}U_3 - F_2^{(2)}\right] \\
& + \delta U_5\left[K_{31}^{(2)}U_2 + K_{32}^{(2)}U_4 + K_{33}^{(2)}U_5 + K_{34}^{(2)}U_3 - F_3^{(2)}\right] \\
& + \delta U_3\left[K_{41}^{(2)}U_2 + K_{42}^{(2)}U_4 + K_{43}^{(2)}U_5 + K_{44}^{(2)}U_3 - F_4^{(2)}\right]
\end{aligned} \tag{4.45}$$

Collecting the coefficients of δU_i $(i = 1, 2, \ldots, 5)$ separately and setting them to zero separately, we obtain

$$\begin{bmatrix}
K_{11}^{(1)} & K_{12}^{(1)} & K_{13}^{(1)} & 0 & 0 \\
K_{21}^{(1)} & K_{22}^{(1)} + K_{11}^{(2)} & K_{23}^{(1)} + K_{14}^{(2)} & K_{12}^{(2)} & K_{13}^{(2)} \\
K_{31}^{(1)} & K_{32}^{(1)} + K_{41}^{(2)} & K_{33}^{(1)} + K_{44}^{(2)} & K_{42}^{(2)} & K_{43}^{(2)} \\
0 & K_{21}^{(2)} & K_{24}^{(2)} & K_{22}^{(2)} & K_{23}^{(2)} \\
0 & K_{31}^{(2)} & K_{34}^{(2)} & K_{32}^{(2)} & K_{33}^{(2)}
\end{bmatrix}
\begin{Bmatrix} U_1 \\ U_2 \\ U_3 \\ U_4 \\ U_5 \end{Bmatrix}
=
\begin{Bmatrix} F_1^{(1)} \\ F_2^{(1)} + F_1^{(2)} \\ F_3^{(1)} + F_4^{(2)} \\ F_2^{(2)} \\ F_3^{(2)} \end{Bmatrix} \tag{4.46}$$

Note that the element matrices overlap in locations corresponding to the global nodes 2 and 3, which are shared by both elements.

The assembly procedure described above can be interpreted in such a way that one can avoid the lengthy algebra in Eq. (4.45). A close examination of the finite-element mesh in Fig. 4.7*a* shows the following correspondence between pairs of global nodes and pairs of element nodes:

Global nodes	Element nodes
(1, 1)	(1, 1) of element 1
(1, 2)	(1, 2) of element 1
(1, 3)	(1, 3) of element 1
(1, 4)	No correspondence
(1, 5)	No correspondence
(2, 2)	(2, 2) of element 1 and (1, 1) of element 2
(2, 3)	(2, 3) of element 1 and (1, 4) of element 2
(2, 4)	(1, 2) of element 2
(2, 5)	(1, 3) of element 2
(3, 3)	(3, 3) of element 1 and (4, 4) of element 2
(3, 4)	(4, 2) of element 2
(3, 5)	(4, 3) of element 2
(4, 4)	(2, 2) of element 2
(4, 5)	(2, 3) of element 2
(5, 5)	(3, 3) of element 2

This correspondence provides us an easy way of assembling the element matrices to obtain the global coefficient matrix with appropriate entries from the element matrices. For example, consider the finite-element mesh shown in Fig. 4.7*b*. The location (3, 4) of the global coefficient matrix contains $K_{43}^{(1)} + K_{31}^{(2)}$, and the location (4, 4) contains $K_{33}^{(1)} + K_{11}^{(2)} + K_{11}^{(3)} + K_{11}^{(4)}$. The location 4 in the assembled column vector contains $F_3^{(1)} + F_1^{(2)} + F_1^{(3)} + F_1^{(4)}$. Locations (1, 5), (1, 6), (1, 7), (2, 5), (2, 6), (2, 7), (3, 6), (3, 7), and (5, 6) of the global matrix contain zero entries [because $K_{ij} = 0$ when (i, j) is not in the same element].

Example 4.1 Consider the Poisson equation (see Example 2.8)

$$-\nabla^2 u = 1 \quad \text{in } \Omega = \{(x, y):\ 0 < (x, y) < 1\}$$

$$\frac{\partial u}{\partial x}(0, y) = \frac{\partial u}{\partial y}(x, 0) = 0 \qquad u(1, y) = u(x, 1) = 0 \tag{4.47}$$

We wish to solve the problem using the linear triangular and rectangular elements.

Solution by linear triangular elements Due to the symmetry along the diagonal $x = y$, one need only model either the lower or the upper triangular domain (see Fig. 4.8). Along the line of symmetry, the normal derivative of the primary variable is zero. Suppose that as a first choice we use a (uniform) mesh of four linear triangular elements to represent (i.e., discretize) the domain. In the present case, there is no discretization error involved in the problem.

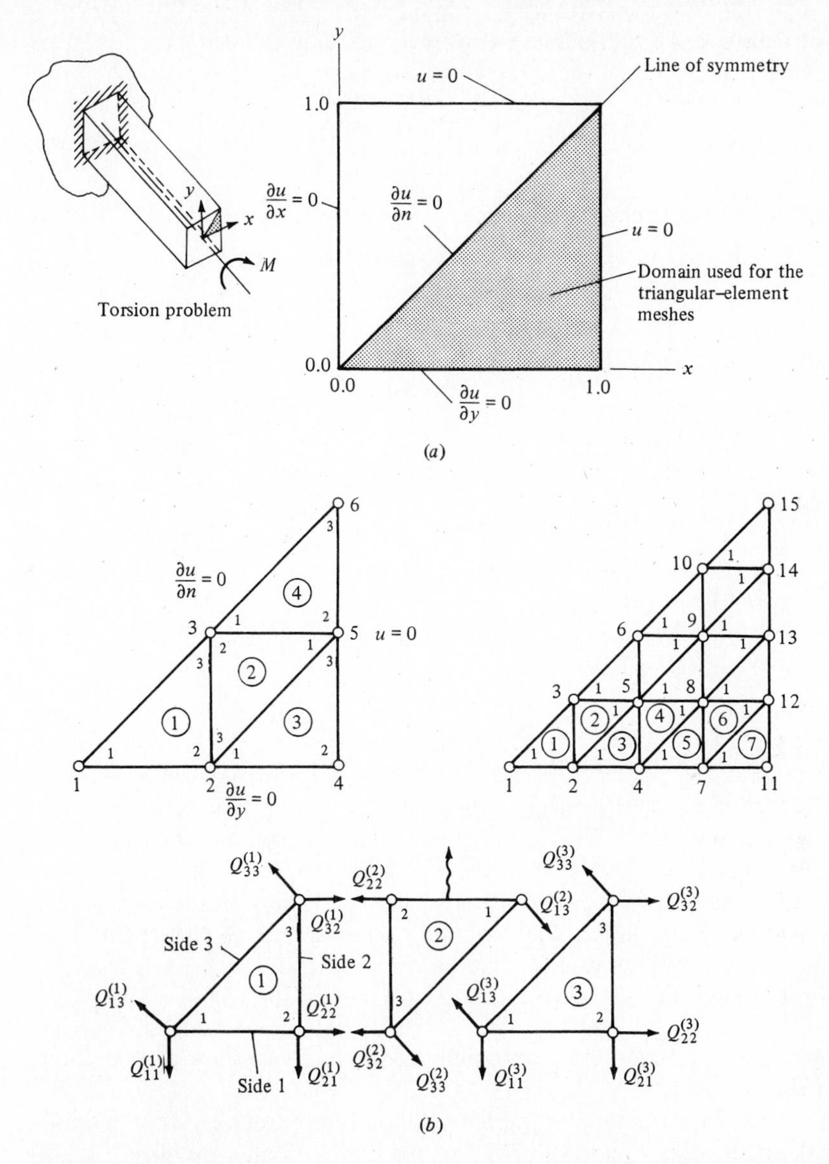

Figure 4.8 Finite-element discretization by triangular elements and equilibrium of secondary variables at the interelement boundaries. (*a*) Domain and boundary conditions for Example 4.1. (*b*) Finite-element discretizations (2×2 and 4×4) by triangular elements.

Note that the elements 1, 3, and 4 are identical (in orientation as well as geometry). Element 2 is geometrically identical to element 1, except that it is oriented differently. If we number the local nodes of element 2 to match those of element 1 (i.e., the diagonal numbered as 3 and 1), then all four elements have the same element matrices, and it is necessary to compute them only for element 1. (When the element matrices are computed on a computer, we do not have to worry about numbering local nodes so as to give the same element matrices.) For a typical element of the mesh, the element equations are given by

$$K_{ij}^{(e)} = \frac{1}{4A}\left(\beta_i^{(e)}\beta_j^{(e)} + \gamma_i^{(e)}\gamma_j^{(e)}\right) \qquad f_i^{(e)} = \frac{fA}{3} \tag{4.48}$$

For example, for element 1 we have

$$\alpha_1 = x_2 y_3 - x_3 y_2 = \tfrac{1}{4} \qquad \alpha_2 = 0 \qquad \alpha_3 = 0$$

$$\beta_1 = y_2 - y_3 = -\tfrac{1}{2} \qquad \beta_2 = \tfrac{1}{2} \qquad \beta_3 = 0$$

$$\gamma_1 = x_3 - x_2 = 0 \qquad \gamma_2 = -\tfrac{1}{2} \qquad \gamma_3 = \tfrac{1}{2}$$

$$\psi_1 = (1 - 2x) \qquad \psi_2 = 2(x - y) \qquad \psi_3 = 2y$$

$$[K^{(1)}] = \frac{1}{2}\begin{bmatrix} 1 & -1 & 0 \\ -1 & 2 & -1 \\ 0 & -1 & 1 \end{bmatrix} \qquad \{F^{(1)}\} = \frac{1}{24}\begin{Bmatrix} 1 \\ 1 \\ 1 \end{Bmatrix} + \begin{Bmatrix} Q_1^{(1)} \\ Q_2^{(1)} \\ Q_3^{(1)} \end{Bmatrix}$$

where $Q_i^{(1)}$ is given by

$$\begin{aligned} Q_i^{(1)} &= \oint_{\Gamma^e} q_n^{(1)} \psi_i^{(1)}(s)\, ds \\ &= \int_0^{0.5} \left[q_n^{(1)} \psi_i^{(1)}(x, y)\right]_{y=0} dx + \int_0^{0.5} \left[q_n^{(1)} \psi_i^{(1)}(x, y)\right]_{x=0.5} dy \\ &\quad + \int_{0.5}^{0} \left[q_n^{(1)} \psi_i^{(1)}(x, y)\right]_{x=y} dx \\ &\equiv Q_{i1}^{(1)} + Q_{i2}^{(1)} + Q_{i3}^{(1)} \end{aligned}$$

Here $Q_{ij}^{(e)}$ denotes the value of $Q_i^{(e)}$ on side j $(j = 1, 2, 3)$ of element e. For example, $Q_1^{(1)}$ is given by (because $\psi_1^{(1)} = 0$ at $x = 0.5$)

$$Q_1^{(1)} = Q_{11}^{(1)} + Q_{13}^{(1)}$$

Note that $Q_{ij}^{(e)}$ are scalars and the arrows indicated in Fig. 4.8*b* show the direction of $q_n^{(e)}$, not the sign of $Q_{ij}^{(e)}$.

It is of interest in the computer implementation of the finite-element method to develop the connectivity matrix (3.21). For the mesh at hand, we have

$$[B] = \begin{bmatrix} 1 & 2 & 3 \\ 5 & 3 & 2 \\ 2 & 4 & 5 \\ 3 & 5 & 6 \end{bmatrix}.$$

The half-bandwidth of the assembled matrix is equal to 4.

The assembled coefficient matrix for the finite-element mesh is 6 × 6 (because there are six global nodes with one unknown per node). The assembled matrix can be obtained directly by using the correspondence between the global nodes and the local nodes. We have

$$
[K] = \begin{array}{c} \\ 1 \\ 2 \\ 3 \\ 4 \\ 5 \\ 6 \end{array}
\left[\begin{array}{cccccc}
1 & 2 & 3 & 4 & 5 & 6 \\
K_{11}^{(1)} & K_{12}^{(1)} & K_{13}^{(1)} & 0 & 0 & 0 \\
 & K_{22}^{(1)} + K_{33}^{(2)} + K_{11}^{(3)} & K_{23}^{(1)} + K_{32}^{(2)} & K_{12}^{(3)} & K_{13}^{(3)} + K_{31}^{(2)} & 0 \\
 & & K_{33}^{(1)} + K_{22}^{(2)} + K_{11}^{(4)} & 0 & K_{21}^{(2)} + K_{12}^{(4)} & K_{13}^{(4)} \\
\text{Symmetric} & & & K_{22}^{(3)} & K_{23}^{(3)} & 0 \\
 & & & & K_{11}^{(2)} + K_{33}^{(3)} + K_{22}^{(4)} & K_{23}^{(4)} \\
 & & & & & K_{33}^{(4)}
\end{array}\right]
\tag{4.49a}
$$

$$
\{F\} = \begin{array}{c} 1 \\ 2 \\ 3 \\ 4 \\ 5 \\ 6 \end{array}
\left\{\begin{array}{c}
F_1^{(1)} \\
F_2^{(1)} + F_3^{(2)} + F_1^{(3)} \\
F_3^{(1)} + F_2^{(2)} + F_1^{(4)} \\
F_2^{(3)} \\
F_1^{(2)} + F_3^{(3)} + F_2^{(4)} \\
F_3^{(4)}
\end{array}\right\}
\tag{4.49b}
$$

Noting that $[K^{(1)}] = [K^{(2)}] = [K^{(3)}] = [K^{(4)}]$ and $\{F^{(1)}\} = \{F^{(2)}\} = \{F^{(3)}\} = \{F^{(4)}\}$, we obtain from Eqs. (4.48) and (4.49) the following assembled system of equations:

$$
\frac{1}{2}\left[\begin{array}{rrr|rrr}
1 & -1 & 0 & 0 & 0 & 0 \\
-1 & 4 & -2 & -1 & 0 & 0 \\
0 & -2 & 4 & 0 & -2 & 0 \\
\hline
0 & -1 & 0 & 2 & -1 & 0 \\
0 & 0 & -2 & -1 & 4 & -1 \\
0 & 0 & 0 & 0 & -1 & 1
\end{array}\right]
\left\{\begin{array}{c} U_1 \\ U_2 \\ U_3 \\ \hline U_4 \\ U_5 \\ U_6 \end{array}\right\}
= \frac{1}{24}\left\{\begin{array}{c} 1 \\ 3 \\ 3 \\ \hline 1 \\ 3 \\ 1 \end{array}\right\}
+ \left\{\begin{array}{c}
Q_1^{(1)} \\
Q_2^{(1)} + Q_3^{(2)} + Q_1^{(3)} \\
Q_3^{(1)} + Q_2^{(2)} + Q_1^{(4)} \\
\hline
Q_2^{(3)} \\
Q_1^{(2)} + Q_3^{(3)} + Q_2^{(4)} \\
Q_3^{(4)}
\end{array}\right\}
\tag{4.50}
$$

The specified boundary conditions on the primary degrees of freedom of the problem are

$$U_4 = U_5 = U_6 = 0 \tag{4.51}$$

The specified secondary degrees of freedom (i.e., q_n) and the internal equilibrium of the secondary variables are given by

$$\begin{aligned}
Q_1^{(1)} &= Q_{11}^{(1)} + Q_{13}^{(1)} = \text{specified to be zero} = 0 \\
Q_2^{(1)} + Q_3^{(2)} + Q_1^{(3)} &= \left(Q_{21}^{(1)} + Q_{11}^{(3)}\right) + \left[\left(Q_{22}^{(1)} + Q_{32}^{(2)}\right) + \left(Q_{13}^{(3)} + Q_{33}^{(2)}\right)\right] \\
&= \text{specified to be zero} + [\text{cancel to give zero}] = 0 \\
Q_3^{(1)} + Q_2^{(2)} + Q_1^{(4)} &= Q_{33}^{(1)} + Q_{13}^{(4)} = 0 \\
Q_2^{(3)} &= Q_{21}^{(3)} + Q_{22}^{(3)} = \text{specified to be zero} + \text{unknown} = Q_{22}^{(3)}, \text{ unknown} \\
Q_1^{(2)} + Q_3^{(3)} + Q_2^{(4)} &= Q_{32}^{(3)} + Q_{22}^{(4)}, \text{ unknown} \\
Q_3^{(4)} &= Q_{32}^{(4)} + Q_{33}^{(4)} = \text{unknown} + \text{zero due to symmetry} = Q_{32}^{(4)}, \text{ unknown}
\end{aligned} \tag{4.52}$$

The "equilibrium" relations given above are illustrated in Fig. 4.8*b*. Thus we have the following equations for the three unknown generalized displacements U_1, U_2, and U_3 and three unknown generalized forces $Q_{22}^{(3)}$, $Q_{32}^{(3)} + Q_{22}^{(4)}$, and $Q_{32}^{(4)}$ [see Eqs. (4.50) and (4.52)]:

$$\begin{bmatrix} 0.5 & -0.5 & 0 \\ -0.5 & 2.0 & -1.0 \\ 0 & -1.0 & 2.0 \end{bmatrix} \begin{Bmatrix} U_1 \\ U_2 \\ U_3 \end{Bmatrix} = \frac{1}{24} \begin{Bmatrix} 1 \\ 3 \\ 3 \end{Bmatrix} \tag{4.53}$$

$$\begin{Bmatrix} Q_{22}^{(3)} \\ Q_{32}^{(3)} + Q_{22}^{(4)} \\ Q_{32}^{(4)} \end{Bmatrix} = -\frac{1}{24} \begin{Bmatrix} 1 \\ 3 \\ 1 \end{Bmatrix} + \begin{bmatrix} 0 & -0.5 & 0 \\ 0 & 0 & -1 \\ 0 & 0 & 0 \end{bmatrix} \begin{Bmatrix} U_1 \\ U_2 \\ U_3 \end{Bmatrix} \tag{4.54}$$

Solving Eq. (4.53) for U_i $(i = 1, 2, 3)$, we get

$$\begin{Bmatrix} U_1 \\ U_2 \\ U_3 \end{Bmatrix} = \frac{1}{24} \begin{bmatrix} 3 & 1 & 0.5 \\ 1 & 1 & 0.5 \\ 0.5 & 0.5 & 0.75 \end{bmatrix} \begin{Bmatrix} 1 \\ 3 \\ 3 \end{Bmatrix} = \frac{1}{24} \begin{Bmatrix} 7.5 \\ 5.5 \\ 4.25 \end{Bmatrix} = \begin{Bmatrix} 0.31250 \\ 0.22917 \\ 0.17708 \end{Bmatrix} \tag{4.55}$$

and from Eq. (4.54) we have

$$\begin{Bmatrix} Q_{22}^{(3)} \\ Q_{32}^{(3)} + Q_{22}^{(4)} \\ Q_{32}^{(4)} \end{Bmatrix} = \begin{Bmatrix} -\frac{1}{24} - 0.5U_2 \\ -\frac{1}{8} - U_3 \\ -\frac{1}{24} \end{Bmatrix} = \begin{Bmatrix} -0.197917 \\ -0.302083 \\ -0.041667 \end{Bmatrix} \tag{4.56}$$

Solution by linear rectangular elements Note that we cannot exploit the symmetry along the diagonal $x = y$ to our advantage when we use rectangular

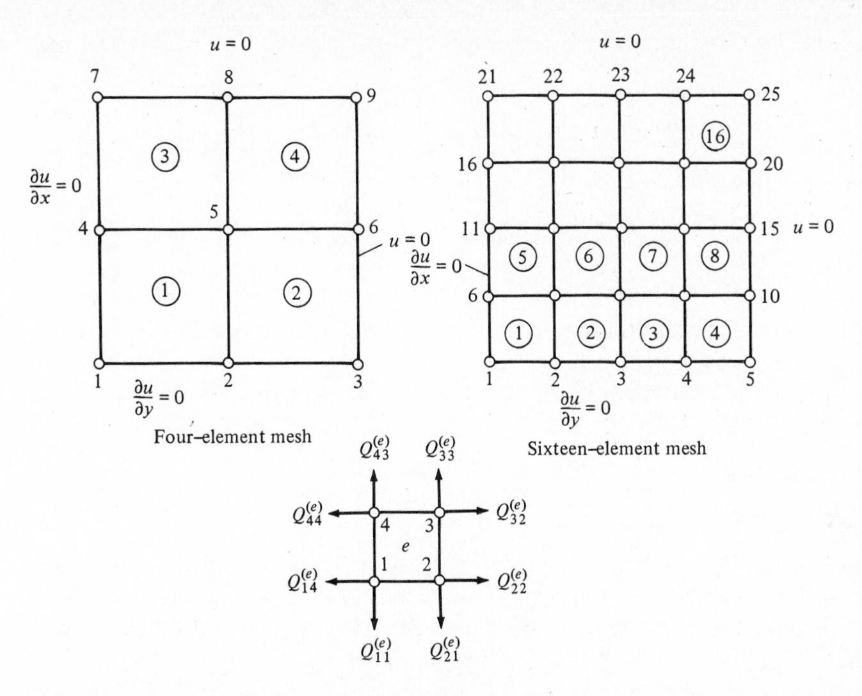

Figure 4.9 Finite-element discretization of the domain of Example 4.1 by linear rectangular elements.

elements. Therefore, we use a 2×2 uniform mesh of four linear rectangular elements (see Fig. 4.9) to discretize a quadrant of the domain. Once again, no discretization error is introduced in the present case.

Since all elements in the mesh are identical, we will compute the element matrices for only one element, say element 1. We have

$$\psi_1 = (1 - 2x)(1 - 2y) \qquad \psi_2 = 2x(1 - 2y) \qquad \psi_3 = 4xy$$

$$\psi_4 = (1 - 2x)2y$$

$$K_{ij}^{(e)} = \int_0^{0.5} \int_0^{0.5} \left(\frac{\partial \psi_i}{\partial x} \frac{\partial \psi_j}{\partial x} + \frac{\partial \psi_i}{\partial y} \frac{\partial \psi_j}{\partial y} \right) dx\, dy \tag{4.57}$$

$$f_i^{(e)} = \int_0^{0.5} \int_0^{0.5} f \psi_i \, dx\, dy$$

Evaluating the integrals in Eqs. (4.57), we obtain [see Eqs. (4.34) and (4.40)]

$$[K^{(e)}] = \frac{1}{6} \begin{bmatrix} 4 & -1 & -2 & -1 \\ -1 & 4 & -1 & -2 \\ -2 & -1 & 4 & -1 \\ -1 & -2 & -1 & 4 \end{bmatrix} \qquad \{F^{(e)}\} = \frac{1}{16} \begin{Bmatrix} 1 \\ 1 \\ 1 \\ 1 \end{Bmatrix} + \begin{Bmatrix} Q_1^{(e)} \\ Q_2^{(e)} \\ Q_3^{(e)} \\ Q_4^{(e)} \end{Bmatrix} \tag{4.58a}$$

where

$$Q_i^{(e)} = \int_{x_1}^{x_2} \left[q_n^{(e)} \psi_i(x, y) \right]_{y=0} dx + \int_{y_2}^{y_3} \left[q_n^{(e)} \psi_i(x, y) \right]_{x=a} dy$$

$$+ \int_{x_3}^{x_4} \left[q_n^{(e)} \psi_i(x, y) \right]_{y=b} dx + \int_{y_4}^{y_1} \left[q_n^{(e)} \psi_i(x, y) \right]_{x=0} dy \quad (4.58b)$$

and (x_i, y_i) denote local coordinates of the element nodes (and $a = x_2 - x_1 = x_3 - x_4$, $b = y_4 - y_1 = y_3 - y_2$).

The assembled coefficient matrix for the four-element mesh is given by

$$[K] = \begin{array}{c} \\ 1 \\ 2 \\ 3 \\ 4 \\ 5 \\ 6 \\ 7 \\ 8 \\ 9 \end{array} \begin{array}{ccccccccc} 1 & 2 & 3 & 4 & 5 & 6 & 7 & 8 & 9 \\ K_{11}^{(1)} & K_{12}^{(1)} & 0 & K_{14}^{(1)} & K_{13}^{(1)} & 0 & 0 & 0 & 0 \\ & K_{22}^{(1)} + K_{11}^{(2)} & K_{12}^{(2)} & K_{24}^{(1)} & K_{23}^{(1)} + K_{14}^{(2)} & K_{13}^{(2)} & 0 & 0 & 0 \\ & & K_{22}^{(2)} & 0 & K_{24}^{(2)} & K_{23}^{(2)} & 0 & 0 & 0 \\ & & & K_{44}^{(1)} + K_{11}^{(3)} & K_{43}^{(1)} + K_{12}^{(3)} & 0 & K_{14}^{(3)} & K_{13}^{(3)} & 0 \\ \text{Symmetric} & & & & K_{33}^{(1)} + K_{44}^{(2)} + K_{11}^{(4)} + K_{22}^{(3)} & K_{43}^{(2)} + K_{12}^{(4)} & K_{24}^{(3)} & K_{23}^{(3)} + K_{14}^{(4)} & K_{13}^{(4)} \\ & & & & & K_{33}^{(2)} + K_{22}^{(4)} & 0 & K_{24}^{(4)} & K_{23}^{(4)} \\ & & & & & & K_{44}^{(3)} & K_{43}^{(3)} & 0 \\ & & & & & & & K_{44}^{(4)} + K_{33}^{(3)} & K_{43}^{(4)} \\ & & & & & & & & K_{33}^{(4)} \end{array} \quad (4.59)$$

The assembled force vector and the connectivity matrix (for future use) are given by

$$\{F\} = \begin{Bmatrix} F_1^{(1)} \\ F_2^{(1)} + F_1^{(2)} \\ F_2^{(2)} \\ F_4^{(1)} + F_1^{(3)} \\ F_3^{(1)} + F_4^{(2)} + F_2^{(3)} + F_1^{(4)} \\ F_3^{(2)} + F_2^{(4)} \\ F_4^{(3)} \\ F_3^{(3)} + F_4^{(4)} \\ F_3^{(4)} \end{Bmatrix} \qquad [B] = \begin{bmatrix} 1 & 2 & 5 & 4 \\ 2 & 3 & 6 & 5 \\ 4 & 5 & 8 & 7 \\ 5 & 6 & 9 & 8 \end{bmatrix} \tag{4.60}$$

The maximum difference between any two nodes of any element in the mesh is 4. Therefore, the half-bandwidth of the assembled matrix is 5.

The assembled equations (4.59) for the problem at hand are given by

$$\frac{1}{6}\begin{bmatrix} 4 & -1 & 0 & -1 & -2 & 0 & 0 & 0 & 0 \\ & 8 & -1 & -2 & -2 & -2 & 0 & 0 & 0 \\ & & 4 & 0 & -2 & -1 & 0 & 0 & 0 \\ & & & 8 & -2 & 0 & -1 & -2 & 0 \\ & & & & 16 & -2 & -2 & -2 & -2 \\ & & & & & 8 & 0 & -2 & -1 \\ & \text{Symmetric} & & & & & 4 & -1 & 0 \\ & & & & & & & 8 & -1 \\ & & & & & & & & 4 \end{bmatrix} \begin{Bmatrix} U_1 \\ U_2 \\ U_3 \\ U_4 \\ U_5 \\ U_6 \\ U_7 \\ U_8 \\ U_9 \end{Bmatrix}$$

$$= \frac{1}{16}\begin{Bmatrix} 1 \\ 2 \\ 1 \\ 2 \\ 4 \\ 2 \\ 1 \\ 2 \\ 1 \end{Bmatrix} + \begin{Bmatrix} Q_1^{(1)} \\ Q_2^{(1)} + Q_1^{(2)} \\ Q_2^{(2)} \\ Q_4^{(1)} + Q_1^{(3)} \\ Q_3^{(1)} + Q_4^{(2)} + Q_2^{(3)} + Q_1^{(4)} \\ Q_3^{(2)} + Q_2^{(4)} \\ Q_4^{(3)} \\ Q_3^{(3)} + Q_4^{(4)} \\ Q_3^{(4)} \end{Bmatrix} \tag{4.61}$$

The specified primary degrees of freedom are

$$U_3 = U_6 = U_7 = U_8 = U_9 = 0 \tag{4.62}$$

and the specified secondary degrees of freedom and the equilibrium of the internal generalized forces give (see Fig. 4.9)

$$\begin{gathered}
Q_1^{(1)} = 0 \qquad Q_2^{(1)} + Q_1^{(2)} = 0 \qquad Q_2^{(2)} = Q_{21}^{(2)} + Q_{22}^{(2)} = Q_{22}^{(2)}, \text{ unknown} \\
Q_4^{(1)} + Q_1^{(3)} = 0 \qquad Q_3^{(1)} + Q_4^{(2)} + Q_2^{(3)} + Q_1^{(4)} = 0
\end{gathered} \tag{4.63}$$

and all other secondary variables in Eq. (4.61) are unknown. The condensed form of the matrix equation for the unknown U_i is given by [see the enclosed quantities in Eq. (4.61)]

$$\frac{1}{6}\begin{bmatrix} 4 & -1 & -1 & -2 \\ -1 & 8 & -2 & -2 \\ -1 & -2 & 8 & -2 \\ -2 & -2 & -2 & 16 \end{bmatrix} \begin{Bmatrix} U_1 \\ U_2 \\ U_4 \\ U_5 \end{Bmatrix} = \frac{1}{16}\begin{Bmatrix} 1 \\ 2 \\ 2 \\ 4 \end{Bmatrix} \tag{4.64}$$

Since $U_2 = U_4$ (because of symmetry), we can write Eq. (4.64) as

$$\frac{1}{6}\begin{bmatrix} 4 & -1-1 & -2 \\ 2(-1) & 2(8-2) & 2(-2) \\ -2 & -2-2 & 16 \end{bmatrix} \begin{Bmatrix} U_1 \\ U_2 \\ U_5 \end{Bmatrix} = \frac{1}{16}\begin{Bmatrix} 1 \\ 2\times 2 \\ 4 \end{Bmatrix} \tag{4.65}$$

where the second row is multiplied by 2 to regain the symmetry of the coefficient matrix. The solution is given by

$$\begin{Bmatrix} U_1 \\ U_2 \\ U_5 \end{Bmatrix} = \frac{6}{16} \times \frac{1}{560}\begin{bmatrix} 176 & 40 & 32 \\ 40 & 60 & 20 \\ 32 & 20 & 44 \end{bmatrix} \begin{Bmatrix} 1 \\ 4 \\ 4 \end{Bmatrix} = \begin{Bmatrix} 0.31071 \\ 0.24107 \\ 0.19286 \end{Bmatrix} \tag{4.66}$$

The unknown secondary variables $Q_{22}^{(2)}$, $Q_{32}^{(2)} + Q_{22}^{(4)}$, and $Q_{32}^{(4)} + Q_{33}^{(4)}$ are given by

$$\begin{aligned}
\begin{Bmatrix} Q_2^{(2)} \\ Q_3^{(2)} + Q_2^{(4)} \\ Q_3^{(4)} \end{Bmatrix} &= \begin{Bmatrix} Q_{22}^{(2)} \\ Q_{32}^{(2)} + Q_{22}^{(4)} \\ Q_{32}^{(4)} + Q_{33}^{(4)} \end{Bmatrix} = -\frac{1}{16}\begin{Bmatrix} 1 \\ 2 \\ 1 \end{Bmatrix} + \frac{1}{6}\begin{bmatrix} 0 & -1 & 0 & -2 \\ 0 & -2 & 0 & -2 \\ 0 & 0 & 0 & -2 \end{bmatrix}\begin{Bmatrix} U_1 \\ U_2 \\ U_4 \\ U_5 \end{Bmatrix} \\
&= \begin{Bmatrix} -0.17857 \\ -0.29286 \\ -0.12679 \end{Bmatrix}
\end{aligned} \tag{4.67}$$

It should be noted that $Q_3^{(4)} = 2Q_{32}^{(4)}$ (because $Q_{32}^{(4)} = Q_{33}^{(4)}$ by symmetry).

The finite-element solutions obtained by two different meshes of triangular elements and two different meshes of rectangular elements are compared in Table 4.1 with the 50-term series solution (at $x = 0$ for varying y) in Eq. (2.95) and the one-parameter Ritz solution in Eq. (2.96); see also Fig. 4.10. The finite-element solution obtained by 16 triangular elements is the most accurate when compared to the series solution.

Table 4.1 Comparison of the finite-element solutions $u(0, y)$ with the series solution and the Ritz solution of Eqs. (4.47)

	Triangular elements		Rectangular elements		Ritz,	Series solution,
y	4 elements	16 elements	4 elements	16 elements	Eq. (2.96)	Eq. (2.95)
0.0	0.3125	0.3013	0.3107	0.2984	0.3125	0.2947
0.25	0.2709†	0.2805	0.2759†	0.2824	0.2930	0.2789
0.50	0.2292	0.2292	0.2411	0.2322	0.2344	0.2293
0.75	0.1146†	0.1393	0.1205†	0.1414	0.1367	0.1397
1.0	0.0000	0.0000	0.0000	0.0000	0.0000	0.0000

† Interpolated values.

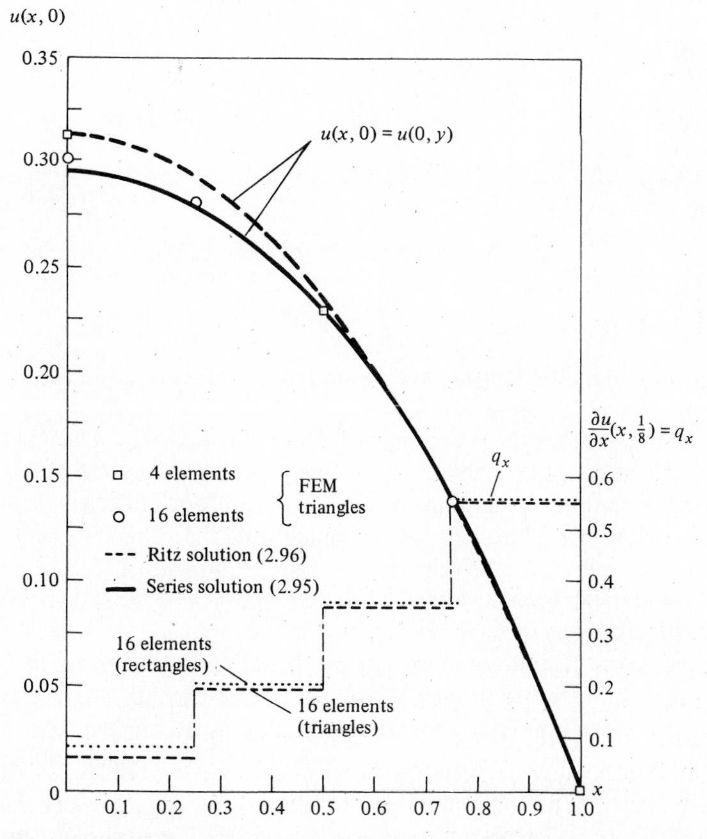

Figure 4.10 Comparison of the finite-element solution with the two-parameter Ritz solution and series solution of Eqs. (4.47).

The components of the gradient of the solution can be computed at any interior point of the domain. For a point (x, y) in element e, we have

$$\left.\begin{aligned} q_x(x, y) &= \frac{\partial u}{\partial x} = \sum_{i=1}^{n} u_i^{(e)} \frac{\partial \psi_i^{(e)}}{\partial x} \\ q_y(x, y) &= \frac{\partial u}{\partial y} = \sum_{i=1}^{n} u_i^{(e)} \frac{\partial \psi_i^{(e)}}{\partial y} \end{aligned}\right\} \quad n = 3 \text{ or } 4 \tag{4.68}$$

Note that for a linear triangular element, q_x and q_y are constants over an entire element, whereas they are linear (q_x is linear in y and q_y is linear in x) in a linear rectangular element. For example, consider element 1:

Triangular element (4 elements)

$$\begin{aligned} q_x^{(1)} &= \frac{1}{2A} \sum_{i=1}^{3} u_i^{(1)} \beta_i^{(1)} = 2(u_2 - u_1) = 2(U_2 - U_1) = -0.16667 \\ q_y^{(1)} &= \frac{1}{2A} \sum_{i=1}^{3} u_i^{(1)} \gamma_i^{(1)} = 2(U_3 - U_2) = -0.10417 \end{aligned} \tag{4.69a}$$

Rectangular element (4 elements)

$$\begin{aligned} q_x^{(1)} &= \sum_{i=1}^{4} u_i^{(1)} \frac{\partial \psi_i^{(1)}}{\partial x} = -2U_1(1 - 2y) + 2U_2(1 - 2y) + 4yU_5 - 4yU_4 \\ q_x^{(1)}(0.25, 0.25) &= -0.11785 \\ q_y^{(1)} &= \sum_{i=1}^{4} u_i^{(1)} \frac{\partial \psi_i^{(1)}}{\partial y} = -2U_1(1 - 2x) + 2U_2(1 - 2x) + 4xU_5 - 4xU_4 \\ q_y^{(1)}(0.25, 0.25) &= -0.11785 \end{aligned} \tag{4.69b}$$

A plot of q_x, obtained by the 16-element mesh, along the x axis (for $y = 0.125$) is shown in Fig. 4.10.

It should be pointed out that the problem considered here [i.e., Eq. (4.47)] has several physical interpretations (see Table 2.2). The problem can be viewed as one of finding the temperature u in a unit square with uniform internal heat generation, the $x = 0$ and $y = 0$ sides being insulated and the other two sides subjected to zero temperature (or, equivalently, a square of dimension 2 with all four sides subjected to zero temperature; the $x = 0$ and $y = 0$ sides can be interpreted as lines of symmetry).

Another interpretation of the equation is that it defines the torsion of a 2-inch-square cross-section member subjected to twist per unit length θ equal to $(2G)^{-1}$, where G is the shear modulus (pounds per square inch). In this case, u denotes the Prandtl stress function, and the components of the gradient of the solution are the stresses (which are of primary interest):

$$\tau_{zx} = \frac{\partial u}{\partial y} \qquad \tau_{zy} = -\frac{\partial u}{\partial x} \tag{4.70}$$

A third interpretation of Eq. (4.47) is provided by the groundwater (seepage) flow problem. In this case, u denotes the piezometric head (measured from the bottom of an aquifer) and f is the charge (pumping is negative). The flow velocities are given by

$$v_1 = -a_{11}\frac{\partial u}{\partial x} \qquad v_2 = -a_{22}\frac{\partial u}{\partial y} \tag{4.71}$$

where a_{11} and a_{22} are the coefficients of permeability. (In the present example, we have $a_{11} = a_{22} = 1$.)

The last example of the problem considered here is one of a 2-inch-square (elastic) membrane fixed on the boundary and subjected to uniformly distributed load of unit intensity. In this case, u denotes the transverse deflection of the membrane. ■

The next two examples are concerned with the irrotational flow of an ideal fluid (i.e., a nonviscous fluid). Examples of physical problems that can be approximated by such flows are provided by flow around bodies such as weirs, airfoils, buildings, and so on, and by flow of water through the earth and dams. The equation governing these flows is a special case of Eq. (4.1), namely, the Laplace equation. Therefore, one can use the finite-element equations developed earlier to model these physical problems. Due to the nonrectangular boundaries involved in the two examples to be discussed, only triangular-element meshes were employed. It is possible to use meshes consisting of both triangular and rectangular elements to obtain the same discretization accuracy. However, in the interest of brevity, we will not use meshes with two different kinds of elements. We will return to these examples in Sec. 4.8 on the computer implementation of the finite-element method.

Example 4.2 (Confined flow about a circular cylinder) The irrotational flow of an ideal fluid (i.e., a nonviscous fluid) about a circular cylinder (placed with its axis perpendicular to the plane of the flow between two long horizontal walls; see Fig. 4.11*a*) is to be analyzed using the finite-element method. The equation governing the flow is given by

$$-\nabla^2 u = 0 \quad \text{in } \Omega \tag{4.72}$$

where u can be one of two functions: (1) u is the *stream function* or (2) u is the *velocity potential*. If u is the stream function ψ, the velocity components of the flow field are given by

$$u_1 = \frac{\partial \psi}{\partial y} \qquad u_2 = -\frac{\partial \psi}{\partial x} \tag{4.73}$$

If u is the velocity potential ϕ, the velocity components can be computed from

$$u_1 = \frac{\partial \phi}{\partial x} \qquad u_2 = \frac{\partial \phi}{\partial y} \tag{4.74}$$

In either case, the velocity field is not affected by a constant term in the solution u.

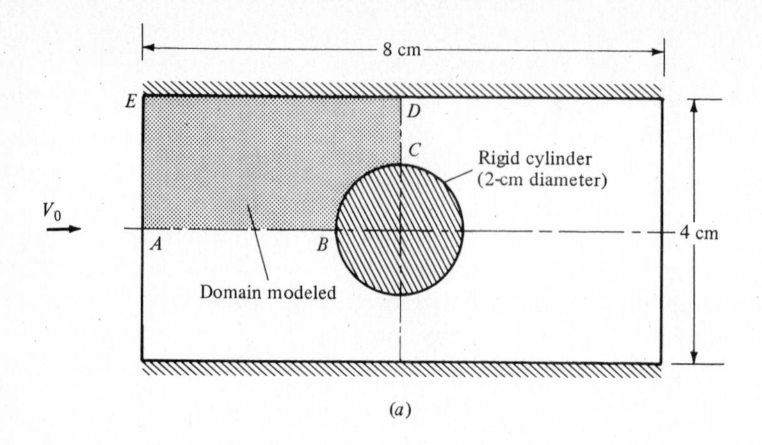

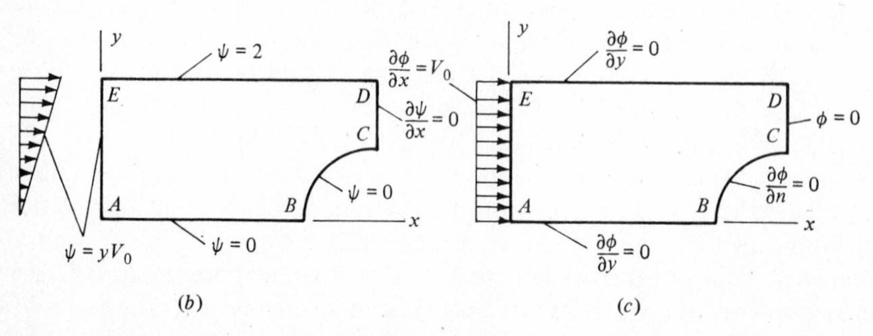

Figure 4.11 Domain and boundary conditions for the stream function and velocity potential formulations of irrotational flow about a cylinder. (a) Domain of the flow about a circular cylinder. (b) Stream function formulation. (c) Velocity potential formulation.

Stream function formulation The boundary conditions on the stream function ψ can be determined as follows (see Fig. 4.11b). Streamlines have the property that flow perpendicular to a streamline is zero. Therefore, the fixed walls correspond to streamlines. Due to the biaxial symmetry about the horizontal and vertical centerlines, only a quadrant (say, $ABCDE$) of the domain need be used in the analysis. The fact that the velocity component perpendicular to the horizontal line of symmetry is equal to zero allows us to use that line as a streamline. Since the velocity field depends on the relative difference of two streamlines, we take the value of the streamline that coincides with the axis of symmetry of the cylinder to be zero ($\psi_A = 0$) and then determine the value of ψ on the upper wall from the condition

$$\frac{\partial \psi}{\partial y} = V_0$$

where V_0 is the velocity of the fluid parallel to the streamline. Since V_0 is given only at the inlet, we determine the value of the streamline at point E by

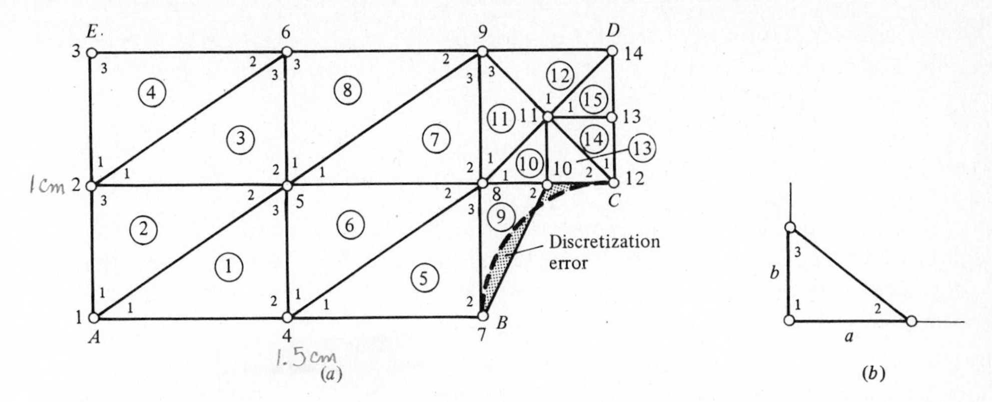

Figure 4.12 Finite-element mesh of triangular elements for irrotational flow about a circular cylinder (mesh: 15 elements and 14 nodes). (*a*) A finite-element mesh of triangular elements (15 elements and 14 nodes). (*b*) A typical right-angle element.

integrating the above equation:

$$\int_{\psi_A}^{\psi_E} d\psi = \psi_E - \psi_A = \int_{y_A}^{y_E} V_0\, dy = V_0(y_E - y_A)$$

or

$$\psi_E = \psi_A + V_0(y_E - y_A) \tag{4.75}$$

Since the velocity at the inlet is uniform (i.e., constant), ψ varies at the inlet linearly from zero at point A to the value $V_0 y_E$ at point E.

In selecting a mesh, we should note that the velocity field is uniform (i.e., streamlines are parallel) at the inlet, and it takes a parabolic profile at the exit (along CD). Therefore, the mesh at the inlet should be uniform, and the mesh close to the cylinder should be relatively more refined (to be able to model the curved boundary and capture the rapid change in ψ). A finite-element mesh of 15 triangular elements and 14 nodes is used (see Fig. 4.12*a*). The coarse mesh is selected for illustrative purposes, and the results of a more refined mesh (see Fig. 4.13) will be discussed subsequently. It should be noted that the discretization error (see the shaded area in Fig. 4.12*a*) is not zero because of the approximation of the circular arc by straight lines.

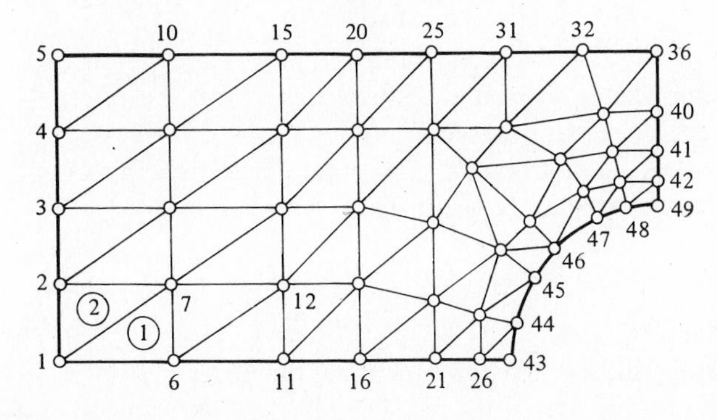

Figure 4.13 Refined finite-element mesh for the flow about a cylinder (69 elements and 49 nodes).

The computation of the element matrices is straightforward. For a typical right-angle element (see Fig. 4.12*b*), the element coefficient matrix is given by

$$[K^{(e)}] = \frac{1}{2}\begin{bmatrix} \alpha+\beta & -\alpha & -\beta \\ -\alpha & \alpha & 0 \\ -\beta & 0 & \beta \end{bmatrix} \qquad \alpha = \frac{a}{b} \qquad \beta = \frac{b}{a} \tag{4.76}$$

Note that the element matrix depends only on the ratio of the element's sides, and not on the element size. For element 1 of the mesh, we have (note the correspondence between the local node numbers of the typical element and element 1)

$$[K^{(1)}] = \begin{bmatrix} \frac{1}{3} & -\frac{1}{3} & 0 \\ -\frac{1}{3} & \frac{13}{12} & -\frac{3}{4} \\ 0 & -\frac{3}{4} & \frac{3}{4} \end{bmatrix}$$

The assembly of element matrices is not presented here in the interest of brevity. The correspondence between the local nodes and the global nodes (i.e., the connectivity matrix) is given by

$$[B] = \begin{array}{c} \\ 1\\2\\3\\4\\5\\6\\7\\8\\9\\10\\11\\12\\13\\14\\15 \end{array} \begin{array}{c} \begin{array}{ccc} 1 & 2 & 3 \end{array} \\ \begin{bmatrix} 1 & 4 & 5 \\ 1 & 5 & 2 \\ 2 & 5 & 6 \\ 2 & 6 & 3 \\ 4 & 7 & 8 \\ 4 & 8 & 5 \\ 5 & 8 & 9 \\ 5 & 9 & 6 \\ 7 & 10 & 8 \\ 8 & 10 & 11 \\ 8 & 11 & 9 \\ 11 & 14 & 9 \\ 10 & 12 & 11 \\ 12 & 13 & 11 \\ 11 & 13 & 14 \end{bmatrix} \end{array} \tag{4.77}$$

Note that the assembled force vector will have contributions from the internal forces $Q_i^{(e)}$ (because $f = 0$ in the present problem) only.

The specified boundary degrees of freedom (i.e., generalized displacements) associated with the primary nodal values are ($V_0 = 1$)

$$\begin{aligned} &U_1 = U_4 = U_7 = U_{10} = U_{12} = 0.0 && \text{zero streamline} \\ &U_3 = U_6 = U_9 = U_{14} = 2.0 && \text{upper boundary} \\ &U_2 = 1.0 \quad \text{due to the linear variation of } \psi \text{ along the line } x = 0 \end{aligned} \tag{4.78}$$

The known generalized global forces are

$$F_5 = F_8 = F_{11} = F_{13} = 0 \tag{4.79}$$

Note that at points C and D, one has boundary conditions on both U and F. However, we choose to impose the essential boundary conditions instead of the natural boundary conditions (like we did in Example 4.1).

The condensed system of equations consists of four equations in the four unknowns U_5, U_8, U_{11}, and U_{13}:

$$\begin{bmatrix} K_{55} & K_{58} & K_{5(11)} & K_{5(13)} \\ & K_{88} & K_{8(11)} & K_{8(13)} \\ & & K_{11(11)} & K_{11(13)} \\ \text{Symmetric} & & & K_{13(13)} \end{bmatrix} \begin{Bmatrix} U_5 \\ U_8 \\ U_{11} \\ U_{13} \end{Bmatrix} = \begin{Bmatrix} \hat{F}_5 \\ \hat{F}_8 \\ \hat{F}_{11} \\ \hat{F}_{13} \end{Bmatrix} \tag{4.80}$$

where the modified force components $\hat{F}_i$ (due to the nonzero specified displacements) are given by

$$\hat{F}_i = F_i - \sum_{j=1}^{14} K_{ij} U_j \qquad i = 5, 8, 11, 13 \qquad j \neq 5, 8, 11, 13 \tag{4.81}$$

For example, $\hat{F}_5$ is given by [in view of Eqs. (4.78) and (4.79)]

$$\hat{F}_5 = -K_{52} - 2\left(K_{53} + K_{56} + K_{59} + K_{5(14)}\right)$$

Solving the system of Eqs. (4.80), we obtain

$$U_5 = 0.93913 \qquad U_8 = 0.60436 \qquad U_{11} = 1.0863 \qquad U_{13} = 1.0432 \tag{4.82}$$

Since the gradients of ψ (i.e., the velocity vector) are of interest, we should compute $\partial\psi/\partial x \equiv -u_2$ and $\partial\psi/\partial y \equiv u_1$ in each element. The results are tabulated in Table 4.2. Note that elements 1, 4, 5, 8, 12, and 13 have zero vertical velocity, consistent with the expected flow field. The maximum horizontal velocity occurs in element 13. Figure 4.14 shows plots of the horizontal velocity obtained using the two meshes along the (horizontal) line of symmetry.

Velocity potential formulation The boundary conditions on the velocity potential ϕ can be derived as follows (see Fig. 4.11c). The fact that $u_2 = \partial\phi/\partial y = 0$ on the upper wall as well as on the horizontal line of symmetry gives the boundary conditions there. Along AE the velocity $u_1 = \partial\phi/\partial x = -\partial\phi/\partial n$ is specified to be $V_0(= 1)$. Along CD we must know either ϕ or $\partial\phi/\partial n = \partial\phi/\partial x$. Since $u_1 = \partial\phi/\partial x$ is not known there, we assume that ϕ is known and set ϕ equal to a constant (say, zero) to complete the boundary conditions. Note that setting $\phi = 0$ on CD eliminates the arbitrary constant (i.e, the rigid-body motion) in the solution. Since the velocity field depends on the gradient of ϕ, one can use any arbitrary constant for ϕ on CD and get the same results for $\nabla\phi$.

Table 4.2 Gradients of solutions (cm / s) of stream function and velocity potential formulations of the flow about a cylinder†

Element	$\frac{\partial \psi}{\partial x} = -u_2$	$\frac{\partial \psi}{\partial y} = u_1$	$\frac{\partial \phi}{\partial x} = u_1$	$\frac{\partial \phi}{\partial y} = u_2$
1	0.0	0.9391	0.9757	0.0570
2	−0.0406	1.0000	1.0029	0.0162
3	−0.0406	1.0609	1.0029	0.0356
4	0.0	1.0000	1.0184	0.0123
5	0.0	0.6044	0.8048	0.4024
6	−0.2232	0.9391	1.0350	0.0570
7	−0.2232	1.3956	1.0350	0.1707
8	0.0	1.0609	1.1251	0.0356
9	−1.2087	0.6044	1.1717	0.4024
10	−1.2087	2.1727	1.1717	0.5509
11	−0.4317	1.3956	1.5519	0.1707
12	0.0	1.8273	1.6217	0.2405
13	0.0	2.1727	2.4132	0.5509
14	−0.0863	2.0863	1.8623	0.0
15	−0.0863	1.9137	1.8623	0.0

† ψ denotes the stream function and ϕ denotes the velocity potential.

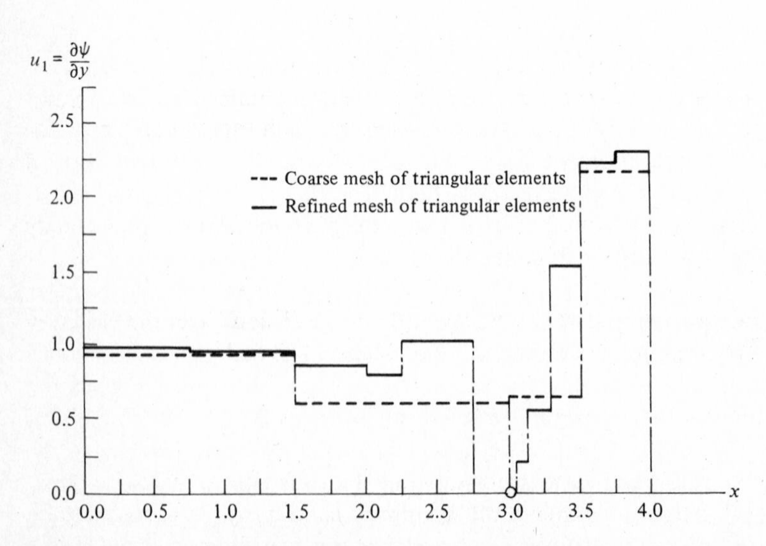

Figure 4.14 Plot of the horizontal velocity along the horizontal line of symmetry and the surface of the cylinder (stream function formulation).

The coarse mesh used in the stream function formulation is used here. The specified primary nodal values are (see Fig. 4.12)

$$U_{12} = U_{13} = U_{14} = 0 \tag{4.83}$$

The generalized forces at all nodes except 1, 2, 3, 12, 13, and 14 are zero. At nodes 5, 8, and 11 internal equilibrium of forces gives $F_5 = F_8 = F_{11} = 0$; we have $q_n \equiv \partial\phi/\partial n = 0$, therefore $F_i = 0$, at nodes 4, 6, 7, 9, and 10. At nodes 12, 13, and 14, $Q_i^{(e)}$ are unknown. At nodes 1, 2, and 3, $\partial\phi/\partial n = q_n$ is specified to be -1. We must compute the integral (4.41) on boundary 1–3 of element 2, and boundary 1–3 of element 4. Since the boundaries are along the y axis, we have $\psi_1^{(2)} = 1 - y$, $\psi_3^{(2)} = y$, and $h = 1$. Therefore, we obtain

$$\begin{aligned} Q_i^{(2)} &= \int_0^1 (-1)\psi_i(y)\,dy = -\tfrac{1}{2} \\ Q_i^{(4)} &= \int_0^1 (-1)\psi_i(y)\,dy = -\tfrac{1}{2} \end{aligned} \tag{4.84a}$$

for $i = 1$ and 3. Substitution of these values into the global force components at global nodes 1, 2, and 3 gives

$$F_1 = -0.5 \qquad F_2 = -(0.5 + 0.5) \qquad F_3 = -0.5 \tag{4.84b}$$

The condensed matrix in the present formulation is 11×11. Solution of the equations (on a digital computer) is given by

$$\begin{aligned} &U_1 = 4.8656 \quad U_2 = 4.8944 \quad U_3 = 4.8371 \quad U_4 = 3.4020 \\ &U_5 = 3.3450 \quad U_6 = 3.3094 \quad U_7 = 2.1948 \quad U_8 = 1.7924 \\ &U_9 = 1.6217 \quad U_{10} = 1.2066 \quad U_{11} = 0.9311 \end{aligned} \tag{4.85}$$

The velocity components of the present formulation $u_1 = \partial\phi/\partial x$ and $u_2 = \partial\phi/\partial y$ are tabulated in Table 4.2, along with those obtained in the stream function formulation. The results are in fair agreement with those obtained by the stream function formulation for u_1. ■

The next example illustrates the use of the finite-element model in the solution of groundwater flow problems.

Example 4.3 (Groundwater flow or seepage) The governing differential equation for a homogeneous aquifer with flow in the xy plane is given by

$$-\left(a_{11}\frac{\partial^2\phi}{\partial x^2} + a_{22}\frac{\partial^2\phi}{\partial y^2}\right) + f = 0 \quad \text{in } \Omega \tag{4.86}$$

where a_{11} and a_{22} (assumed to be constant) are the coefficients of permeability (meters per day) along the x and y directions, respectively, ϕ is the *piezometric head* (meters), measured from a reference level (usually the bottom of the aquifer), and f is the rate of pumping $[\text{m}^3/(\text{day}\cdot\text{m}^3)]$. We know from the previous discussions that the natural and essential boundary conditions associated with Eq. (4.86) are

Natural

$$a_{11}\frac{\partial \phi}{\partial x}n_x + a_{22}\frac{\partial \phi}{\partial y}n_y = \phi_n \quad \text{on } \Gamma_1$$

Essential (4.87)

$$\phi = \phi_0 \quad \text{on } \Gamma_2$$

where Γ_1 and Γ_2 are the portions of the boundary Γ of Ω such that $\Gamma_1 + \Gamma_2 = \Gamma$.

Consider the problem of finding the lines of constant potential ϕ in a 3000-m × 1500-m rectangular aquifer Ω (see Fig. 4.15) bounded on the long sides by an impermeable material (i.e., $\partial\phi/\partial n = 0$) and on the short sides by a constant piezometric head of 200 m ($\phi_0 = 200$ m). Further, suppose that a river is passing through the aquifer, infiltrating the aquifer at a rate of 0.24 m^3/day per unit length (meters), and that two pumps are located at (1000, 670) and (1900, 900), pumping at a rate of $Q_1 = 1200$ $m^3/(\text{day} \cdot m^3)$ and $Q_2 = 2400$ $m^3/(\text{day} \cdot m^3)$, respectively.

A mesh of 64 triangular elements and 45 nodes is used to model the domain (see Fig. 4.16*a*). The river forms the interelement boundary between the sets (33, 35, 37, 39) and (26, 28, 30, 32) of elements. Note that neither pump is located at a node. This is done intentionally for the purpose of illustrating the calculation of the generalized forces due to a point source within an element. We must calculate the generalized force components due to the distributed line source (i.e., the river) and the point sources (i.e., the pumps). Calculation of the element coefficient matrices should be a routine task by now. Let us concentrate on the calculation of the generalized forces from the given information.

First, consider the line source. We can view the river as a source of constant intensity, 0.24 $m^3/(\text{day} \cdot m)$. Since the length of the river is equally divided by nodes 21 through 25 (into four parts), we can compute the contribution of the

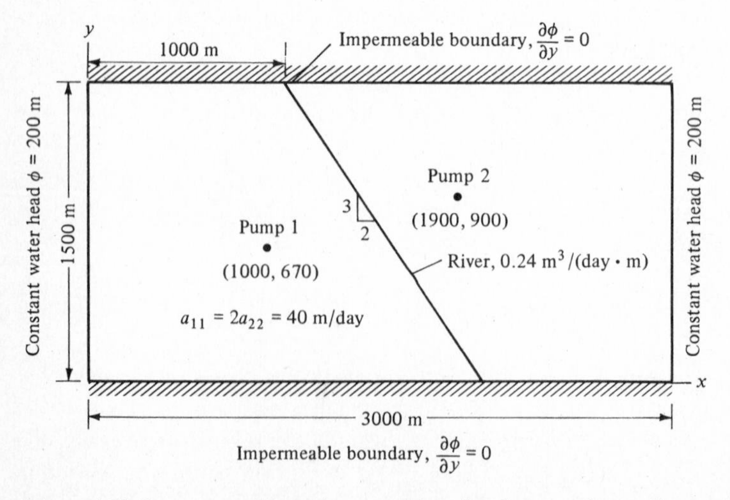

Figure 4.15 Geometry and boundary conditions for the groundwater flow problem of Example 4.3.

infiltration of the river at each of the nodes 21 through 25 by evaluating the integrals (see Fig. 4.16*b*):

$$\begin{aligned}
&\text{Node 21:} \quad && \int_0^h (0.24)\psi_1^{(1)}\, ds \\
&\text{Node 22:} \quad && \int_0^h (0.24)\psi_2^{(1)}\, ds + \int_0^h (0.24)\psi_1^{(2)}\, ds \\
&\text{Node 23:} \quad && \int_0^h (0.24)\psi_2^{(2)}\, ds + \int_0^h (0.24)\psi_1^{(3)}\, ds \\
&\text{Node 24:} \quad && \int_0^h (0.24)\psi_2^{(3)}\, ds + \int_0^h (0.24)\psi_1^{(4)}\, ds \\
&\text{Node 25:} \quad && \int_0^h (0.24)\psi_2^{(4)}\, ds
\end{aligned} \tag{4.88}$$

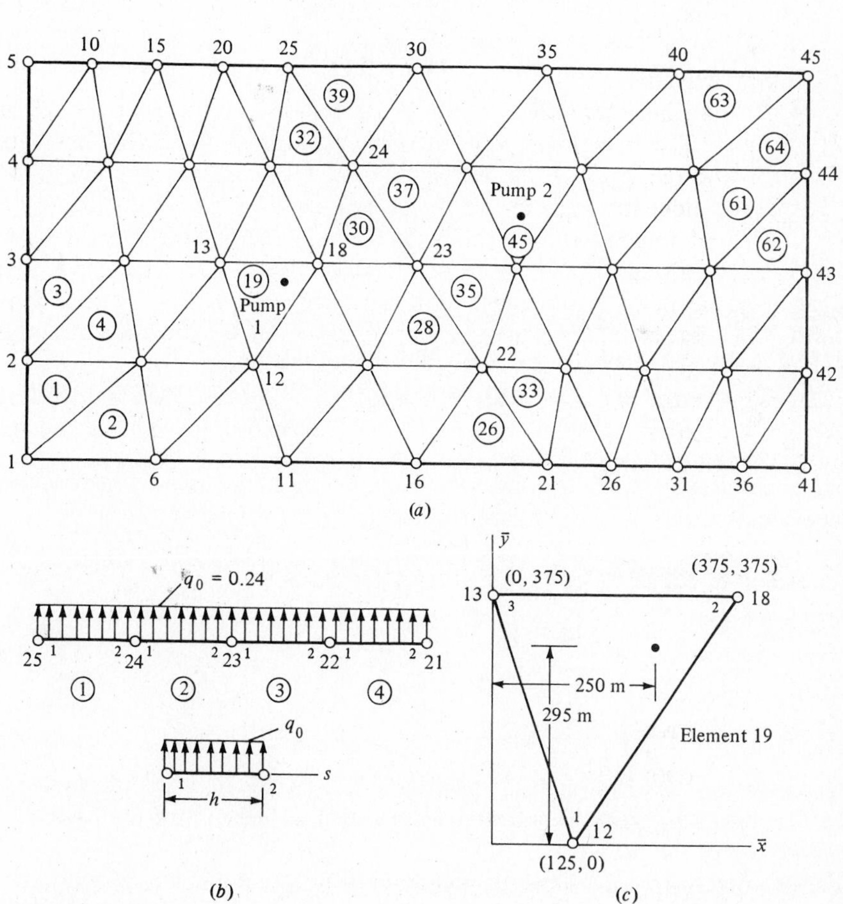

Figure 4.16 Finite-element mesh, and computation of force components for the groundwater (seepage) flow. (*a*) Finite-element mesh of triangular elements (45 nodes and 64 elements). (*b*) Computation of global forces due to the infiltration of the river. (*c*) Computation of global forces for pump 1, located inside element 19.

For constant intensity q_0 and the linear interpolation functions $\psi_1(s) = 1 - s/h$ and $\psi_2(s) = s/h$, the contribution of these integrals is well known:

$$\int_0^h q_0 \psi_i \, ds = \frac{q_0 h}{2} \qquad h = \tfrac{1}{4}\{(1000)^2 + (1500)^2\}^{1/2} \qquad q_0 = 0.24 \quad (4.89)$$

Next, we consider the contribution of the point sources. Since the point sources are located inside an element, we distribute the source to the nodes of the element by interpolation. For example, the source at pump 1 (located in element 19) gives

$$Q_1(x, y) = -1200\delta(x - 1000)\delta(y - 670) \qquad (4.90a)$$

where $\delta(\cdot)$ is the Dirac delta function in Eq. (2.110). We have

$$f_i^{(19)} = \int_{\text{area}} Q_1(x, y)\psi_i \, dx\, dy = -1200\psi_i^{(19)}(1000, 670) \qquad (4.90b)$$

The interpolation functions for element 19 are (in terms of the coordinates $\bar{x}$ and $\bar{y}$; see Fig. 4.16c)

$$\psi_i(\bar{x}, \bar{y}) = \frac{1}{2A}(\alpha_i + \beta_i \bar{x} + \gamma_i \bar{y})$$

$$A = \tfrac{1}{2}(375)^2 \qquad \alpha_1 = (375)^2 \qquad \alpha_2 = -375(125) \qquad \alpha_3 = 375(125) \qquad (4.91a)$$

$$\beta_1 = 0 \qquad \beta_2 = 375 \qquad \beta_3 = -375 \qquad \gamma_1 = -375 \qquad \gamma_2 = 125 \qquad \gamma_3 = 250$$

We have $\bar{x} = x - 750$ and $\bar{y} = y - 375$, and therefore,

$$\psi_1(250, 295) = 0.2133 \qquad \psi_2(250, 295) = 0.595 \qquad \psi_3(250, 295) = 0.1911 \qquad (4.91b)$$

Similar computations can be done for pump 2. Thus, the known generalized displacements (meters) and the nonzero forces [$\text{m}^3/(\text{day} \cdot \text{m}^3)$] are given by

$$U_1 = U_2 = U_3 = U_4 = U_5 = U_{41} = U_{42} = U_{43} = U_{44} = U_{45} = 200.0$$

$$\begin{aligned} &F_{21} = 54.08 \qquad F_{22} = F_{23} = F_{24} = 108.17 \qquad F_{25} = 54.08 \\ &F_{12} = -255.6 \qquad F_{13} = -299.2 \qquad F_{18} = -715.2 \qquad F_{28} = -1440.0 \\ &F_{29} = -410.4 \qquad F_{34} = -549.6 \end{aligned} \qquad (4.92)$$

Global forces at nodes 6 through 11, 14 through 17, 19, 20, 26, 27, 30 through 33, and 35 through 40 are zero. This completes the data generation for the finite-element modeling of the problem.

The solution of the equations (on a computer) for the unknown U_i (piezometric heads at the nodes) is shown in Fig. 4.17a along with the velocity vectors. The lines of constant ϕ are shown in Fig. 4.17b. The greatest drawdown (of water) occurs at node 28, which has the largest portion of the discharge from pump 2. Of course, in a standard finite-element analysis one should see that any point sources

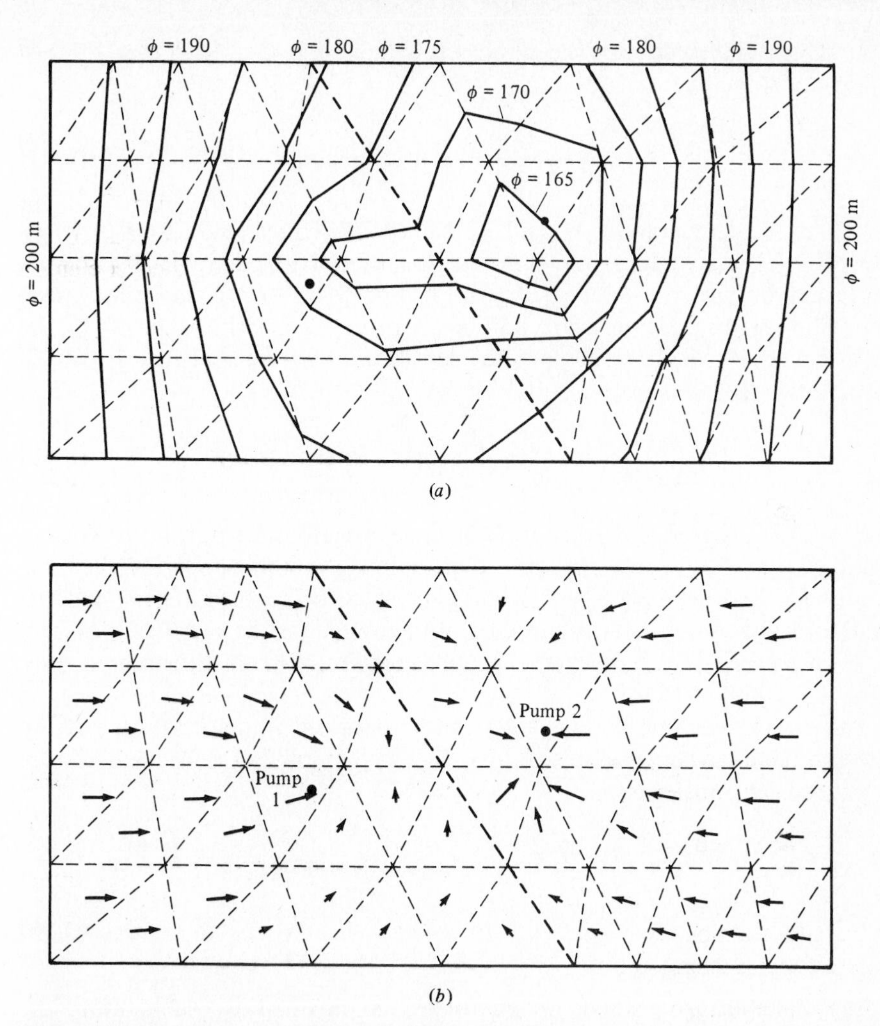

Figure 4.17 Plots of constant piezometric head, and velocity vector for the groundwater flow. (*a*) Lines of constant ϕ. (*b*) Plot of velocity vectors.

in the problem are located at a node point. The solution of the same problem by an alternative mesh that puts pumps 1 and 2 at nodal points is left as an exercise. ■

The last example of this section is concerned with the finite-element-analysis of a heat transfer problem.

Example 4.4 (Convection heat transfer problems) Here we consider a more general heat transfer problem (than the one considered in Example 4.1). More specifically, we consider the finite-element analysis of heat transfer problems that involve convection phenomena (i.e., energy transfer between a solid body and a surrounding fluid medium). The governing equation is a special case of that considered before:

$$-\frac{\partial}{\partial x}\left(k_x\frac{\partial T}{\partial x}\right)-\frac{\partial}{\partial y}\left(k_y\frac{\partial T}{\partial y}\right)=f \tag{4.93}$$

where T is the temperature, k_x and k_y are conductivities [Btu/(h · ft · °F) or W/(h · m · K)] along the x and y directions, respectively, and f is the internal heat generation per unit volume [Btu/(h · ft^3) or W/(h · m^3)]. As we have seen before, the essential boundary condition involves the specification of T; the natural boundary condition involves the specification of the heat flux $\hat{q}_n$ (Btu/h · ft^2 or W/h · m^2) such that

$$\left(k_x\frac{\partial T}{\partial x}n_x+k_y\frac{\partial T}{\partial y}n_y\right)+\beta(T-T_\infty)-\hat{q}=0 \tag{4.94}$$

where β is the convective conductance [or the convective heat transfer coefficient in Btu/(h · ft^2 · °F) or W/(h · m^2 · K)], and T_∞ is the temperature of the surrounding fluid medium. The first term accounts for heat transfer by conduction, the second by convection, and the third accounts for the specified heat flux. It is the presence of the term $\beta(T - T_\infty)$ that requires some modification of Eqs. (4.10), (4.13), and (4.14).

The variational form of Eq. (4.93) can be obtained as a special case of Eq. (4.8). The boundary integral in Eq. (4.8) should be modified to account for the convective heat transfer term:

$$0=\int_{\Omega^e}\left(k_x\frac{\partial v}{\partial x}\frac{\partial T}{\partial x}+k_y\frac{\partial v}{\partial y}\frac{\partial T}{\partial y}-vf\right)dx\,dy+\oint_{\Gamma^e}v\left[\beta^{(e)}(T-T_\infty)-q_n^{(e)}\right]ds \tag{4.95}$$

where v is a test function.

The finite-element model of Eq. (4.95) is obtained by substituting the finite-element approximation of the form (4.11) for T in Eq. (4.95), resulting in

$$\sum_{j=1}^{n}\left[K_{ij}^{(e)}+H_{ij}^{(e)}\right]T_j^{(e)}=F_i^{(e)}+P_i^{(e)} \tag{4.96}$$

where

$$K_{ij}^{(e)}=\int_{\Omega^e}\left(k_x\frac{\partial\psi_i}{\partial x}\frac{\partial\psi_j}{\partial x}+k_y\frac{\partial\psi_i}{\partial y}\frac{\partial\psi_j}{\partial y}\right)dx\,dy$$

$$F_i^{(e)}=\int_{\Omega^e}f\psi_i\,dx\,dy+\oint_{\Gamma^e}q_n^{(e)}\psi_i\,ds\equiv f_i^{(e)}+Q_i^{(e)} \tag{4.97}$$

$$H_{ij}^{(e)}=\beta^{(e)}\oint_{\Gamma^e}\psi_i\psi_j\,ds\qquad P_i^{(e)}=\beta^{(e)}\oint_{\Gamma^e}\psi_iT_\infty\,ds$$

Note that by setting $\beta = 0$, we get the heat transfer model that accounts for no convection.

For linear interpolation functions, the coefficient matrices $H_{ij}^{(e)}$ and $P_i^{(e)}$ are given, for a triangular element, by

$$[H^{(e)}] = \frac{\beta_{12}^{(e)} h_{12}^{(e)}}{6}\begin{bmatrix} 2 & 1 & 0 \\ 1 & 2 & 0 \\ 0 & 0 & 0 \end{bmatrix} + \frac{\beta_{23}^{(e)} h_{23}^{(e)}}{6}\begin{bmatrix} 0 & 0 & 0 \\ 0 & 2 & 1 \\ 0 & 1 & 2 \end{bmatrix} + \frac{\beta_{31}^{(e)} h_{31}^{(e)}}{6}\begin{bmatrix} 2 & 0 & 1 \\ 0 & 0 & 0 \\ 1 & 0 & 2 \end{bmatrix} \tag{4.98a}$$

$$\{P^{(e)}\} = \frac{\beta_{12}^{(e)} T_\infty h_{12}^{(e)}}{2}\begin{Bmatrix} 1 \\ 1 \\ 0 \end{Bmatrix} + \frac{\beta_{23}^{(e)} T_\infty h_{23}^{(e)}}{2}\begin{Bmatrix} 0 \\ 1 \\ 1 \end{Bmatrix} + \frac{\beta_{31}^{(e)} T_\infty h_{31}^{(e)}}{2}\begin{Bmatrix} 1 \\ 0 \\ 1 \end{Bmatrix} \tag{4.98b}$$

where $\beta_{ij}^{(e)}$ are the convective coefficients for side i–j of element e, and $h_{ij}^{(e)}$ is the length of the side i–j. For a rectangular element, similar equations hold [for example, add a column and row of zeros to the first two matrices, and add two more matrices for sides 3–4 and 4–1 in Eq. (4.98a)].

Note that, in actual calculations, one is not required to compute $[H^{(e)}]$ and $\{P^{(e)}\}$ for all elements; the contributions from the convective terms should be calculated only for elements that have their boundaries exposed to ambient temperature. Even in those elements, one should compute the contributions to the nodes on the side that is subjected to convective heat transfer boundary conditions.

As an example, consider the heat transfer in an 8- by 4-cm rectangular plate with a 2-cm circular hole (see Fig. 4.11a for the geometry of the domain). The temperature on the outside boundary (i.e., on long and short sides) is 20°C; the circular hole contains a fluid at 200°C. The thermal conductivity of the plate is 10 W/(cm · K), and the convective heat transfer coefficient between the plate and the fluid medium is 15 W/(cm² · K). We wish to determine the temperature distribution in the plate (for $f = \hat{q} = 0$).

Using the biaxial symmetry, we model only a quadrant of the domain. The two finite-element meshes in Figs. 4.12 and 4.13 are employed in the present calculations. We will discuss the formulation for the coarse mesh, and then present results for the refined one. Sides 1–2 of elements 9 and 13 are exposed to ambient temperature. Therefore the element matrices of elements 9 and 13 should be modified to include the following matrices:

$$[H^{(9)}] = \frac{15h}{6}\begin{bmatrix} 2 & 1 & 0 \\ 1 & 2 & 0 \\ 0 & 0 & 0 \end{bmatrix} \qquad \{P^{(9)}\} = \frac{15(473)h}{2}\begin{Bmatrix} 1 \\ 1 \\ 0 \end{Bmatrix} \qquad h = \sqrt{1 + (0.5)^2}$$

$$[H^{(13)}] = \frac{15(0.5)}{6}\begin{bmatrix} 2 & 1 & 0 \\ 1 & 2 & 0 \\ 0 & 0 & 0 \end{bmatrix} \qquad \{P^{(13)}\} = \frac{15(473)(0.5)}{2}\begin{Bmatrix} 1 \\ 1 \\ 0 \end{Bmatrix} \tag{4.99}$$

For example, for element 13, the element equations are given by

$$\begin{bmatrix} 10+2.5 & -5+1.25 & -5+0 \\ -5+1.25 & 5+2.5 & 0+0 \\ -5+0 & 0+0 & 5+0 \end{bmatrix} \begin{Bmatrix} T_1^{(13)} \\ T_2^{(13)} \\ T_3^{(13)} \end{Bmatrix} = \begin{Bmatrix} Q_1^{(13)} \\ Q_2^{(13)} \\ Q_3^{(13)} \end{Bmatrix} + \begin{Bmatrix} 1773.75 \\ 1773.75 \\ 0.0 \end{Bmatrix} \tag{4.100}$$

The specified boundary conditions on the primary and secondary degrees of freedom are (note that $\partial T/\partial n = 0$ on the lines of symmetry)

$$U_1 = U_2 = U_3 = U_6 = U_9 = U_{14} = 20°\text{C} = 293\text{ K} \tag{4.101a}$$

$$F_4 = F_5 = F_8 = F_{11} = F_{13} = 0 \qquad F_7 = 3966.23$$

$$F_{10} = 3966.23 + 1773.75 \qquad F_{12} = 1773.75 \tag{4.101b}$$

The solution of the equations is given by (kelvins)

$$U_4 = 323.32 \quad U_5 = 313.45 \quad U_7 = 398.02 \quad U_8 = 357.74$$

$$U_{10} = 389.79 \quad U_{11} = 337.19 \quad U_{12} = 395.08 \quad U_{13} = 340.62 \tag{4.102}$$

Plots of the temperature along the lines of symmetry are given in Fig. 4.18 for the two different meshes shown in Figs. 4.12 and 4.13. The magnitude and the direction of the gradient of the temperature in each element (obtained using the refined mesh) are given in Table 4.3. ■

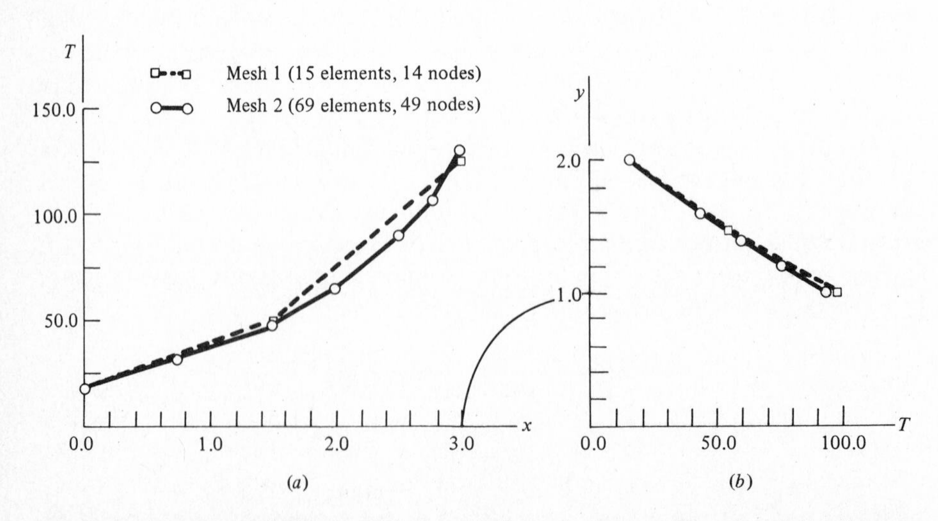

Figure 4.18 Temperature distribution (°C) along the lines of symmetry (Example 4.4). (*a*) Temperature distribution along the horizontal line of symmetry. (*b*) Temperature distribution along the vertical line of symmetry.

Table 4.3 The finite-element solution for the magnitude and direction of the gradient of the temperature of Example 4.4

Element	Magnitude	Direction† (negative)	Element	Magnitude	Direction (negative)
1	164.36	6.7	44	953.47	27.19
2	150.44	0.0	46	945.12	44.39
3	159.81	19.72	48	725.35	46.49
4	114.5	0.0	49	778.85	66.41
5	139.24	34.69	50	732.16	65.33
7	111.18	56.31	51	802.89	79.11
8	0.0	0.0	52	792.37	79.11
16	92.50	90.0	53	658.36	90.0
24	214.85	90.0	54	815.52	90.0
25	512.08	15.19	55	882.31	86.61
26	445.39	10.14	56	1018.10	60.21
27	559.19	38.37	57	831.16	62.22
28	369.12	34.92	58	925.68	75.95
29	528.34	55.05	59	870.03	75.67
30	332.01	61.13	60	927.49	86.77
31	518.67	72.00	61	949.28	85.89
32	332.95	90.0	62	1031.2	59.29
33	697.98	9.25	63	1083.1	75.26
34	723.34	10.69	64	1163.3	72.53
35	666.18	31.88	65	982.68	74.49
37	654.40	55.22	66	1050.1	85.15
39	595.00	73.89	67	1029.3	86.09
41	493.27	90.0	68	1114.7	85.56
42	911.43	9.3369	69	1146.4	84.34

† Measured counterclockwise from the positive x axis.

4.3 SOME COMMENTS ON MESH GENERATION AND IMPOSITION OF BOUNDARY CONDITIONS

4.3.1 Discretization of a Domain

The representation of a given domain (it is assumed that the domain is already decided upon from physical considerations) by a collection of simple geometric shapes, that is, finite elements, requires engineering judgment on the part of the finite-element practitioner. The number, type (i.e., linear, quadratic), shape (i.e., triangular, rectangular), size, and density (i.e., mesh refinement) of elements used in a given problem depend on a number of considerations. The first consideration is to discretize the domain as close to exact as possible with elements that are admissible. (As we shall see later, one can use different elements for the approximation of a domain and the solution.) In discretizing a domain, consideration must be given to an accurate representation of point sources, distributed sources with discontinuities (i.e., sudden change in the intensity of the source), and material and geometric discontinuities, including a reentrant corner. The discretization should include, for example, nodes at point sources (so that the point source is accurately lumped at the node) and reentrant corners, and element interfaces where abrupt changes in geometry and material properties occur. The second consideration, which requires some engineering judgment, is to discretize the body or portions of the body into sufficiently small elements so that steep gradients of the solution are accurately calculated. The engineering judgment should come from both a qualitative understanding (or intuition) of the behavior of the solution and an estimate of the computational costs involved in the mesh refinement (i.e., reducing the size of the elements). For example, consider the irrotational flow about a cylinder of Example 4.2. From the physical laws governing the problem (i.e., continuity of the flow), we know that the flow entering at the left should leave at the right. Since the outlet section is smaller than the inlet section, it is expected that the flow accelerates in the vicinity of the cylinder. On the other hand, the velocity field far from the cylinder (i.e., near the inlet) is essentially uniform. Such knowledge of the qualitative behavior of the flow allows us to employ a coarse mesh (i.e., elements are relatively large in size) at sites sufficiently far from the cylinder, and a fine one at closer distances to the cylinder (see Fig. 4.13). Another purpose for using a refined mesh near the cylinder (at least one layer of elements along the surface of the cylinder) is to accurately represent the curved boundary of the domain there. Concerning the shape of the element, thus far we have discussed only two geometric shapes, triangular and rectangular; since a mesh of rectangular elements would involve large discretization errors (in representing a curved boundary) compared to that of an equivalent mesh of triangular elements, a mesh of triangular elements was used in Example 4.3. Of course, one can use a mesh of both rectangular elements (away from the curved boundary) and triangular elements (near the curved boundary), or a mesh of quadrilateral elements, which will be discussed in the forthcoming sections. In general, a refined mesh is required in places where acute changes in geometry, boundary conditions, loading, material properties, or solution occur.

A mesh refinement should meet three conditions: (1) all previous meshes should be contained in the current (refined) mesh, (2) every point in the body can be included within an arbitrarily small element at any stage of the mesh refinement, (3) the same order of approximation for the solution should be retained through all stages of the refinement process. The last requirement eliminates comparison of two different approximations in two different meshes. The requirement is met because one is unlikely to change the degree of approximation in the midst of a convergence study. When a mesh is refined, care should be taken to avoid elements with very large aspect ratios (i.e., the ratio of one side of an element to the other). Recall from the element matrices (4.40) and (4.76) that the coefficient matrices depend on the ratios of a to b and b to a. If the value of a/b or b/a is very large, the resulting coefficient matrices are ill-conditioned (i.e., numerically not invertible). The lower and upper limits on b/a are believed to be 0.1 and 10, respectively.

The words "coarse" and "fine" are relative. In any given problem, one begins with a finite-element mesh that is believed to be adequate (based on experience and engineering judgment) to solve the problem at hand. Then, as a second choice, one selects a mesh that consists of a larger number of elements (and includes the first one as a subset) to solve the problem once again. If there is a significant difference between the two solutions, one sees the benefit of mesh refinement (and further mesh refinement may be warranted). If the difference is negligibly small, further mesh refinements are not necessary. Such numerical experiments with mesh refinements are not always feasible in practice (mostly due to computational costs involved). In cases where computational cost is the prime concern, one must depend on one's judgment concerning what is a reasonably good mesh. Since most practical problems with geometric complexity are approximated in their engineering formulations (of the governing equations), one cannot be overconcerned with the numerical accuracy of the solution. A feel for the relative proportions and directions of various errors introduced into the analysis helps the finite-element practitioner to make a decision on when to stop further refining a mesh. In summary, scientific (or engineering) knowledge and experience with a given class of problems is an essential part of any analysis.

4.3.2 Generation of Finite-Element Data

An important part of the finite-element modeling is the mesh generation, which involves numbering the nodes and elements, and the generation of nodal coordinates and the boolean connectivity matrix. While the task of generating such data is quite simple, the type of the data has an effect on the computational efficiency as well as on accuracy. More specifically, the numbering of the nodes directly affects the bandwidth of the final (assembled) equations, and therefore the storage requirement and computational cost. For example, consider the finite-element meshes shown in Fig. 4.19. In Fig. 4.19*a*, the nodes are numbered in a rowwise fashion from left to right, starting with the bottom row, moving to the next row above, and repeating the procedure when the numbering along the particular row is done. In Fig. 4.19*b*, the numbering is done columnwise from bottom to top, starting with the leftmost column and proceeding to the next (on the right) when each column is done. The elements can be numbered arbitrarily. The half-bandwidths [see Eq. (3.220)] for the two cases are given, respectively, by 11 * DF and 7 * DF, where DF is the number of primary unknowns per node. Note that the element numbering has no effect on the half-bandwidth. Smaller bandwidths are obtained by numbering the nodes in the direction of the smallest number of subdivisions. In a general-purpose mesh-generating program, such options should be included to minimize the half-bandwidth. The saving of computational cost due to a smaller bandwidth in the solution of equations can be substantial, especially in problems where a large number of nodes is involved. While element numbering does not affect the half-bandwidth, it affects the computer time required to assemble (usually, a small percentage of the time required to solve the equations) the global coefficient matrix.

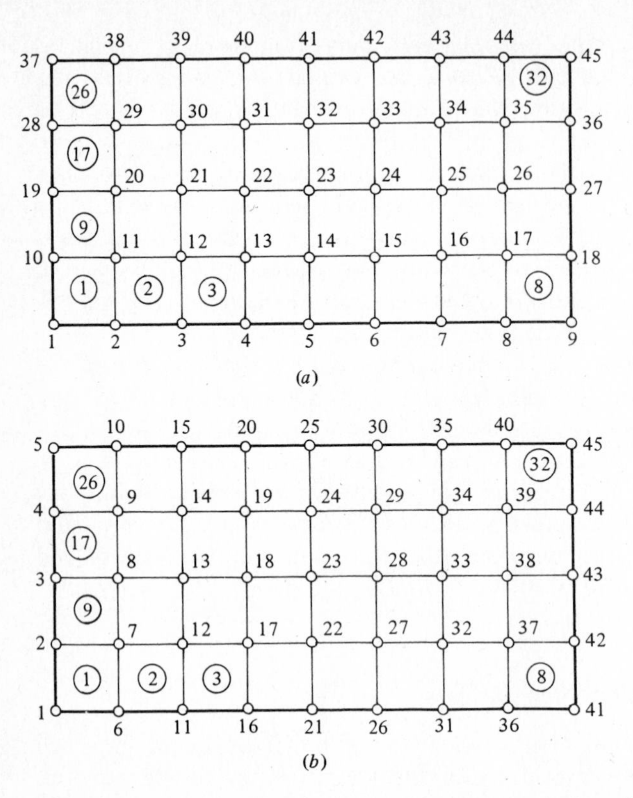

Figure 4.19 Two commonly used finite-element node numbering schemes. (*a*) Scheme I. (*b*) Scheme II.

The accuracy of the finite-element solution can also depend on the choice of the finite-element mesh. For instance, if the mesh selected violates the symmetry present in the problem, the resulting solution will be less accurate than one obtained using a mesh that agrees with the physical symmetry present in the problem. Geometrically, a triangular element has fewer (or no) lines of symmetry when compared to a rectangular element, and therefore one should use meshes of triangular elements with care (e.g., select a mesh that does not contradict the mathematical symmetry present in the problem).

4.3.3 Imposition of Boundary Conditions

In most problems of interest one encounters situations where the portion of the boundary on which natural boundary conditions are specified has points in common with the portion of the boundary on which the essential boundary conditions are specified. In other words, at these points both the primary degrees of freedom and the secondary degrees of freedom are specified. Such points are called *singular points*. Obviously, one cannot impose both boundary conditions at the same point. As a general rule, we should impose the essential boundary condition (i.e., the boundary condition on the primary variables) at the singular points and disregard the natural boundary condition (i.e., the boundary condition on the secondary variables). For instance, in all of the examples considered in this chapter, we followed the above rule and imposed only the essential boundary conditions at the singular points.

Another type of singularity one encounters in the solution of boundary-value problems is the specification of two different values of a primary variable at the same boundary point. An example of such cases is provided by Prob. 4.11, where u is specified to be zero on the boundary defined by the line $x = 0$ (for any y), and it is specified to be unity on the boundary defined by the line $y = 1$ (for any x). Consequently, at $x = 0$ and $y = 1$, u has two different values. In the finite-element analysis, one must make a choice between the two values. In either case, the true boundary condition is replaced by an approximate condition. The closeness of the approximate boundary condition to the true one depends on the size of the element containing the point (see Fig. 4.20). A mesh refinement in

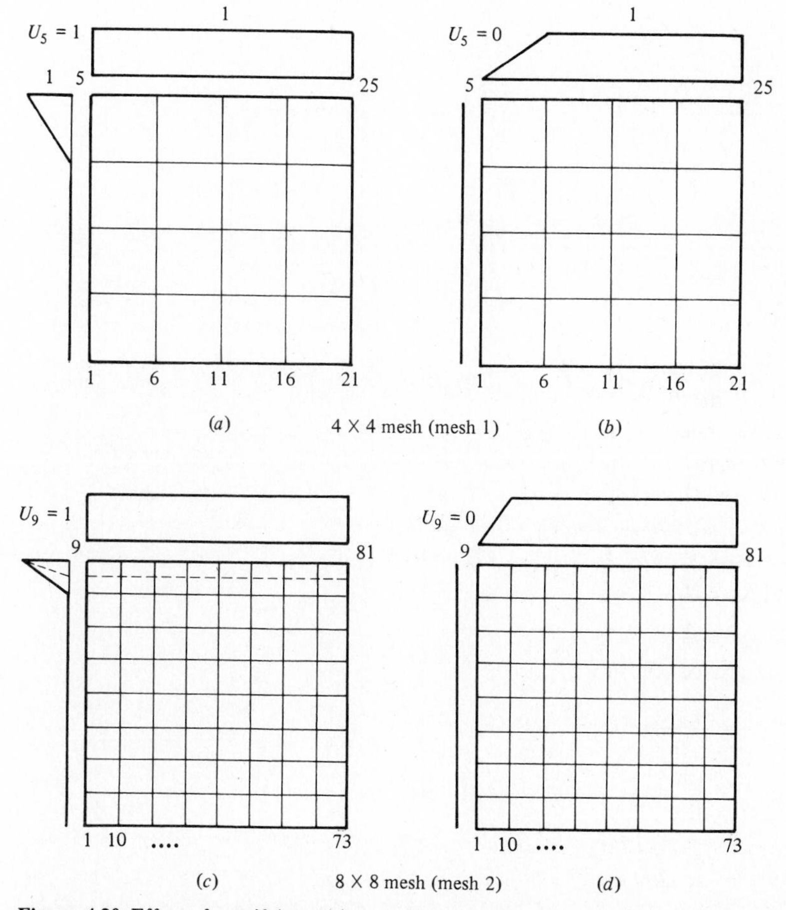

Figure 4.20 Effect of specifying (either of the) two values of a primary variable at a boundary node [node 5 in (a) and (b) and node 9 in (c) and (d)].

the vicinity of the singular point often yields an acceptable solution (at least away from the point; see Prob. 4.13).

Some of the comments made in this section are illustrated using the problem of Example 4.1. The problem has a mathematical symmetry about the $x = y$ line. We investigate the effect of various choices of triangular-element meshes and mesh refinement on the solution.

Example 4.5 First we investigate the effect of the choice of each of the finite-element meshes shown in Fig. 4.21. The finite-element solutions obtained by the three meshes are compared with the exact solution in Table 4.4. Clearly, the solution obtained using mesh 3 is less accurate. This is expected because mesh 3 is symmetric about the diagonal line connecting node 3 to node 7, whereas the mathematical symmetry is about the diagonal line connecting node 1 to node 9 (see Fig. 4.21). Mesh 1 is the most desirable of the three because it does not contradict the mathematical symmetry present in the problem. Note that the sum of the pointwise errors is the least in mesh 1.

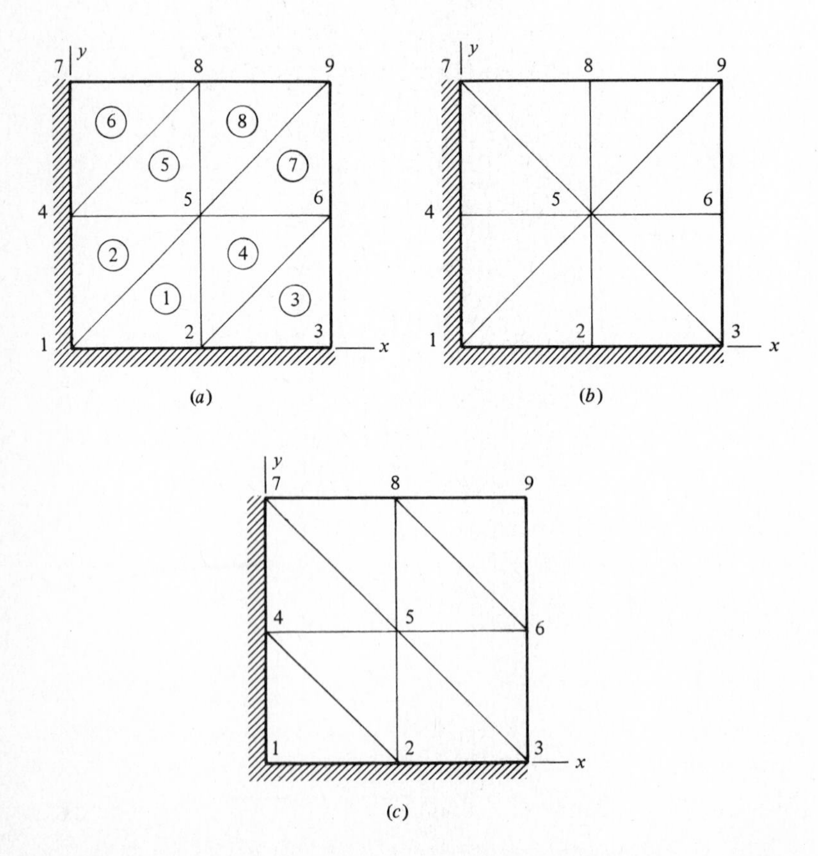

Figure 4.21 Various types of triangular-element meshes for the domain of Example 4.1. (*a*) Mesh 1. (*b*) Mesh 2. (*c*) Mesh 3.

Table 4.4 Comparison of the finite-element solutions obtained by various triangular-element meshes with the series solution of the problem in Example 4.1

Node	Finite-element solution			Series solution
	Mesh 1	Mesh 2	Mesh 3	
1	0.31250	0.29167	0.25000	0.29469
2	0.22917	0.20833	0.20833	0.22934
4	0.22917	0.20833	0.20833	0.22934
5	0.17708	0.18750	0.16667	0.18114

Next, we investigate the effect of mesh refinement with rectangular elements. Four different meshes of rectangular elements are shown in Fig. 4.22. Each of the meshes contains the previous mesh as a subset. The mesh shown in Fig. 4.22*c* is nonuniform; it is obtained by subdividing the first two rows and columns of elements of the mesh shown in Fig. 4.22*b*. The finite-element solutions obtained by these meshes are compared in Table 4.5. The numerical convergence of the finite-element solution of both refined meshes to the series solution is apparent from the results presented. ■

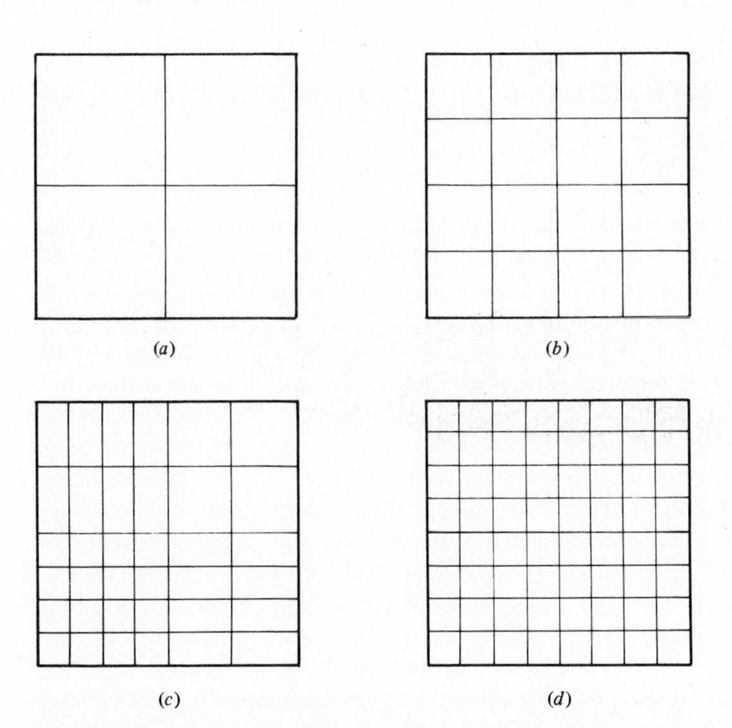

Figure 4.22 Mesh refinement: meshes in (*a*), (*b*), and (*d*) are uniform; mesh in (*c*) is nonuniform. (*a*) 2×2 mesh. (*b*) 4×4 mesh. (*c*) 6×6 mesh. (*d*) 8×8 mesh.

Table 4.5 Convergence of the finite-element solution (with the mesh refinement†) of the problem in Example 4.1

Location		Finite-element solution				Series solution
x	y	2×2	4×4	6×6	8×8	
0.0	0.0	0.31071	0.29839	0.29641	0.29560	0.29469
0.125	0.0	—	—	0.29248	0.29167	0.29077
0.250	0.0	—	0.28239	0.28055	0.27975	0.27888
0.375	0.0	—	—	0.26022	0.24943	0.25863
0.50	0.0	0.24107	0.23220	0.23081	0.23005	0.22934
0.625	0.0	—	—	—	0.19067	0.19009
0.750	0.0	—	0.14137	0.14064	0.14014	0.13973
0.875	0.0	—	—	—	0.07709	0.07687
0.125	0.125	—	—	0.28862	0.28781	0.28692
0.250	0.250	—	0.26752	0.26580	0.26498	0.26415
0.375	0.375	—	—	0.22960	0.22873	0.22799
0.50	0.50	0.19286	0.18381	0.18282	0.18179	0.18114
0.625	0.625	—	—	—	0.12813	0.12757
0.750	0.750	—	0.07506	0.07481	0.07332	0.07282
0.875	0.875	—	—	—	0.02561	0.02510

†See Fig. 4.22 for the finite-element meshes.

4.4 TWO-DIMENSIONAL FINITE ELEMENTS AND INTERPOLATION FUNCTIONS

4.4.1 Introduction

In the previous section, the linear triangular and rectangular elements were discussed, their interpolation functions were derived, and their application to the finite-element solution of second-order two-dimensional problems with one dependent variable was illustrated. The objective of this section is to construct higher-order triangular as well as rectangular elements. The reader should be reminded that the finite-element formulation of a differential equation does not require a priori knowledge of the type of the element. The selection of an appropriate element can be made depending on the admissibility conditions implied by the finite-element formulation, the geometry of the domain being modeled, and the degree of accuracy required. On the other hand, the construction of interpolation functions does not depend on the specific differential equation being solved. The construction procedure depends only on the geometry, the number and position of the nodes, and the number of (primary) dependent unknowns identified at the nodes of the element. Thus, one can develop a library of finite elements, which can be used wherever they are applicable. It is the objective of this section to develop a library of two-dimensional triangular and rectangular elements of the Lagrange family (i.e., elements with only one unknown, namely, the dependent variable itself, per node) and give interpolation functions for quadratic and cubic *serendipity elements*. Once we have a library of elements at our disposal, we can choose an appropriate element (and functions) from the library of available elements.

4.4.2 Triangular Elements

Higher-order triangular elements (i.e., triangular elements with interpolation functions of higher degree) can be systematically developed with the help of the so-called *Pascal's triangle*, which contains the terms of polynomials of various degrees in the two variables x and y, as shown in Fig. 4.23. One can view the position of the terms as the nodes of the triangle, with the constant term and the first and last terms of a given row being the vertices of the triangle. (Of course, the shape of the triangle is arbitrary—not an equilateral triangle, as might appear from the position of the terms in Pascal's triangle.) For example, a triangular element of order 2 (i.e., the degree of the polynomial is 2) contains six nodes, as can be seen from the top three rows of Pascal's triangle. The position of the six nodes in the triangle is at the three vertices and at the midpoints of the three sides. The polynomial involves six constants, which can be expressed in terms of the nodal values of the variable being interpolated:

$$u = \sum_{i=1}^{6} u_i \psi_i(x, y) \tag{4.103}$$

where ψ_i are the quadratic interpolation functions obtained following the same procedure as that used for the linear element in Sec. 4.2. In general, a pth-order triangular element has a number of n nodes

$$n = \frac{(p+1)(p+2)}{2} \tag{4.104}$$

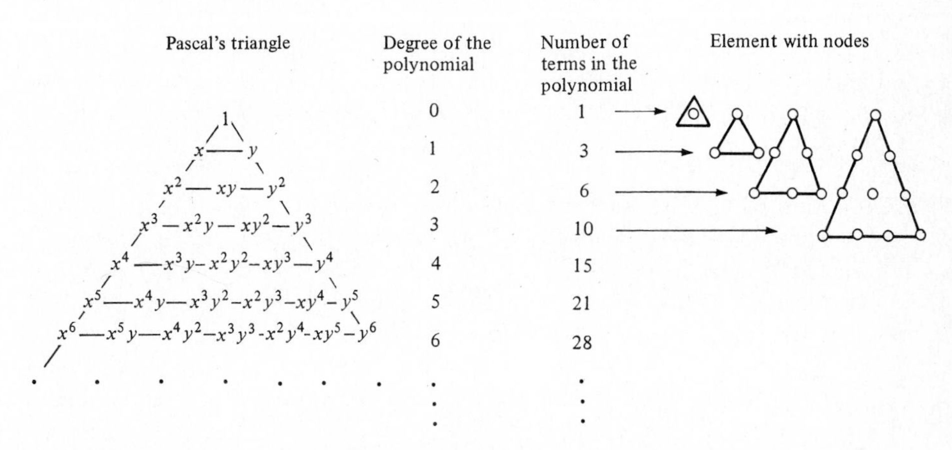

Figure 4.23 Pascal's triangle for the generation of the Lagrange family of triangular elements.

and a complete polynomial of pth degree is given by

$$u(x, y) = \sum_{i=1}^{n} a_i x^r y^s \qquad r + s \leqslant p$$
$$= \sum_{j=1}^{n} u_j \psi_j \tag{4.105}$$

The location of the entries in Pascal's triangle gives a symmetric location of nodal points in elements that will produce exactly the right number of nodes to define a Lagrange interpolation of any degree. It should be noted that the Lagrange family of triangular elements (of order greater than zero) should be used for second-order problems that require only the dependent variable or variables (*not* their derivatives) of the problem to be continuous at interelement boundaries. It can be easily seen that the pth-degree polynomial associated with the pth-order Lagrange element, when evaluated on the boundary of the element, yields a pth-degree polynomial in the boundary coordinate. For example, the quadratic polynomial associated with the quadratic (six-node) triangular element shown in Fig. 4.24 is given by

$$u^e(x, y) = a_1 + a_2 x + a_3 y + a_4 xy + a_5 x^2 + a_6 y^2 \tag{4.106}$$

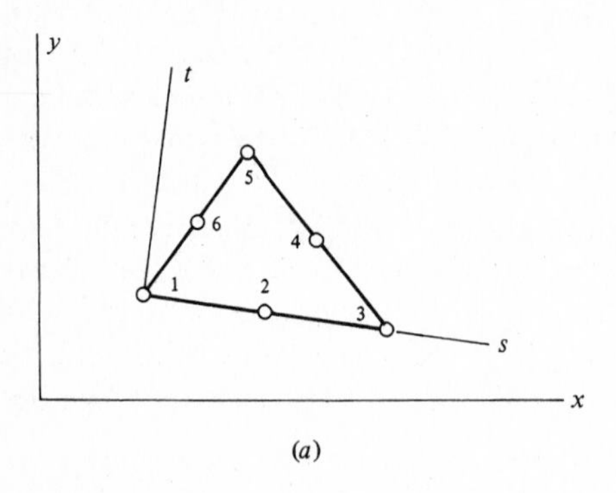

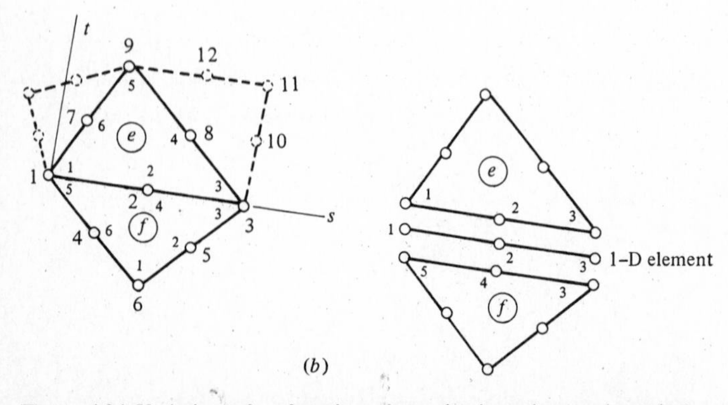

Figure 4.24 Variation of a function along the interelement boundaries of Lagrange (triangular) elements. (a) A typical higher-order element. (b) Interelement continuity of a quadratic function.

The element shown in Fig. 4.24*a* is an arbitrary quadratic triangular element. By rotating and translating the (x, y) coordinate system we obtain the (s, t) coordinate system. Since the transformation from the (x, y) system to the (s, t) system involves only rotation (which is linear) and translation, a kth-degree polynomial in the (x, y) coordinate system is still a kth-degree polynomial in the (s, t) system:

$$u^e(s, t) = \hat{a}_1 + \hat{a}_2 s + \hat{a}_3 t + \hat{a}_4 st + \hat{a}_5 s^2 + \hat{a}_6 t^2 \tag{4.107}$$

where $\hat{a}_i$ $(i = 1, 2, \ldots, 6)$ are constants that depend on a_i and the angle of rotation α. Now by setting $t = 0$, we get the restriction of u to side 1–2–3 of element e,

$$u^e(s, 0) = \hat{a}_1 + \hat{a}_2 s + \hat{a}_5 s^2 \tag{4.108}$$

which is a quadratic polynomial in s. If a neighboring element f has its side 5–4–3 common with side 1–2–3 of element e, then the function u on side 5–4–3 of element f is also a quadratic polynomial,

$$u^f(s, 0) = \hat{b}_1 + \hat{b}_2 s + \hat{b}_5 s^2 \tag{4.109}$$

Since these polynomials are uniquely defined by the same nodal values $U_1 = u_1^e = u_5^f$, $U_2 = u_2^e = u_4^f$, and $U_3 = u_3^e = u_3^f$, we have $u^e(s, 0) = u^f(s, 0)$, and hence the function u is uniquely defined on the interelement boundary of elements e and f.

The ideas described above can be easily extended to three dimensions, in which case Pascal's triangle takes the form of a Christmas tree and the elements are of a pyramid shape, called tetrahedral elements. We shall not elaborate on this any further because the scope of the present study is limited to two-dimensional elements only. (See Chap. 5 for an introduction to 3-D elements.)

Note that the formal procedure of constructing the interpolation functions [see Eqs. (4.19) through (4.24)] involves the inversion of an $n \times n$ matrix, where n is the number of terms in the polynomial used to represent a function. When $n > 3$, this procedure is algebraically very tedious, and therefore one should devise an alternative way of developing the interpolation functions (see Sec. 3.6 for one-dimensional elements).

The alternative derivation of the interpolation functions for the higher-order (Lagrange family of) triangular elements is simplified by the use of the *area coordinates* L_i (see Prob. 4.10). Recall that for triangular elements it is possible to construct three nondimensionalized coordinates L_i $(i = 1, 2, 3)$ (see Fig. P4.10), which relate respectively to the sides directly opposite nodes 1, 2, and 3, such that

$$L_i = \frac{A_i}{\sum_{j=1}^{3} A_j}$$

$$x = \sum_{i=1}^{3} L_i x_i \tag{4.110}$$

$$y = \sum_{i=1}^{3} L_i y_i$$

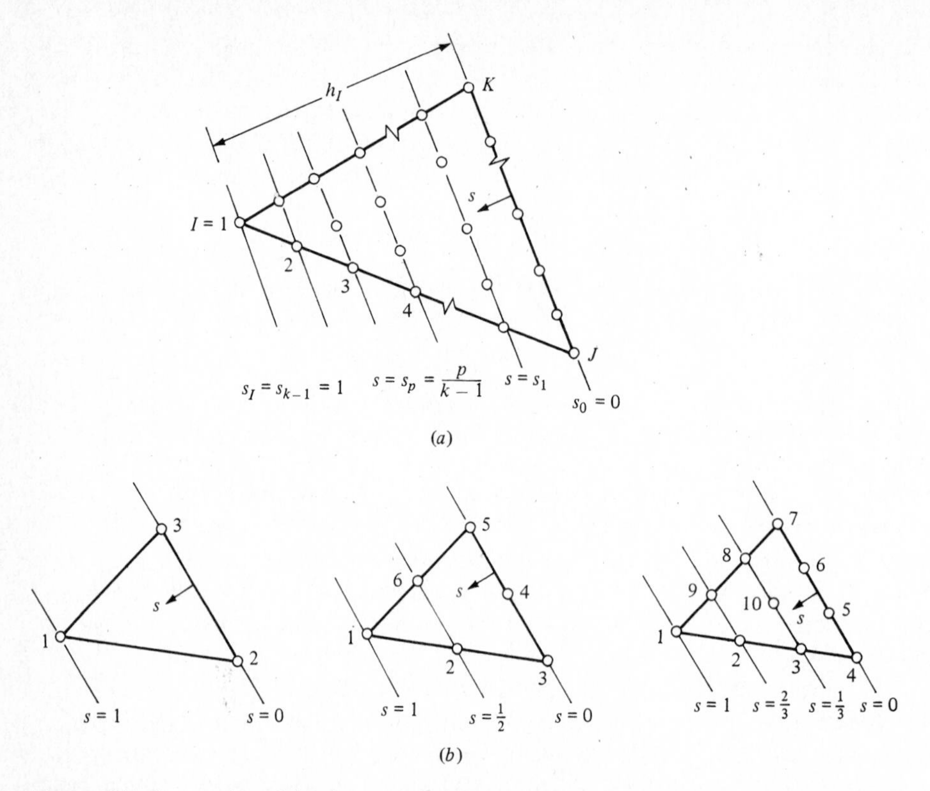

Figure 4.25 Construction of the element interpolation functions of the Lagrange triangular elements. (*a*) An arbitrary $(k - 1)$th-order element. (*b*) Linear, quadratic, and cubic elements.

where x_i and y_i are the coordinates of node i in the (x, y) coordinate system. The interpolation functions satisfy the properties in Eq. (4.26). Now consider a higher-order element with k nodes (equally spaced) per side (see Fig. 4.25*a*). Then the total number of nodes in the element is given by

$$n = \sum_{i=0}^{k-1} (k - i) = k + (k - 1) + \cdots + 1 \tag{4.111}$$

and the degree of the interpolation functions is equal to $k - 1$. For example, for the quadratic element we have $k - 1 = 2$ and $n = 6$. Let the corner (i.e., vertex) nodes be denoted by I, J, and K, and let h_I be the perpendicular distance of the node I from the side connecting nodes J and K. Then the distance s_p to the pth row parallel to side J–K (under the assumption that the nodes are equally spaced along the sides and the rows) is given by

$$s_p = \frac{p}{k - 1} \qquad s_0 = 0 \qquad s_I = s_{k-1} = 1 \tag{4.112}$$

The interpolation function ψ_I should be zero at the nodes on the lines $L_I = 0$, $1/(k - 1), \ldots, p/(k - 1)$ $(p = 0, 1, \ldots, k - 2)$, and ψ_I should be equal to 1 at

$L_I = s_I$. Thus we have the necessary information for constructing the interpolation function ψ_I:

$$\psi_I = \frac{(L_I - s_0)(L_I - s_1)(L_I - s_2)\cdots(L_I - s_{k-2})}{(s_I - s_0)(s_I - s_1)(s_I - s_2)\cdots(s_I - s_{k-2})}$$

$$= \prod_{p=0}^{k-2} \frac{L_I - s_p}{s_{k-1} - s_p} \tag{4.113a}$$

Similar expressions can be derived for nodes located at other than the vertices. In general, ψ_i for node i is given by

$$\psi_i = \prod_{j=1}^{k-1} \frac{f_j}{f_j^i} \tag{4.113b}$$

where f_j are functions of L_1, L_2, and L_3, and f_j^i is the value of f_j at node i. The functions f_j are derived from the equations of $k - 1$ lines which pass through all the nodes except node i. The procedure is illustrated below via an example.

Example 4.6 First consider the triangular element that has two nodes per side (i.e., $k = 2$). This is the linear triangular element with the total number of nodes equal to 3 ($n = 3$). For node 1 (see Fig. 4.25b), we have ($k - 2 = 0$)

$$s_0 = 0 \qquad s_1 = 1$$
$$\psi_1 = \frac{L_1 - s_0}{s_1 - s_0} = L_1 \tag{4.114a}$$

Similarly, for ψ_2 and ψ_3 we get

$$\psi_2 = L_2 \qquad \psi_3 = L_3 \tag{4.114b}$$

Next consider the triangular element with three nodes per side ($k = 3$). The total number of nodes is equal to 6. For node 1, we have

$$s_0 = 0 \qquad s_1 = \tfrac{1}{2} \qquad s_2 = 1$$
$$\psi_1 = \frac{L_1 - s_0}{s_2 - s_0}\frac{L_1 - s_1}{s_2 - s_1} = L_1(2L_1 - 1) \tag{4.115a}$$

The function ψ_2 (see Fig. 4.25b) should vanish at nodes 1, 3, 4, 5, and 6, and should be equal to 1 at node 2. Equivalently, ψ_2 should vanish along the lines connecting nodes 1 and 5, and 3 and 5. These two lines are given in terms of L_1, L_2, and L_3 (note that the subscripts of L refer to the nodes in the three-node triangular element) by $L_2 = 0$ and $L_1 = 0$. Hence,

$$\psi_2 = \frac{L_2 - 0}{\frac{1}{2} - 0}\frac{L_1 - 0}{\frac{1}{2} - 0} = 4L_1L_2 \tag{4.115b}$$

Similarly, we have

$$\psi_3 = L_2(2L_2 - 1) \qquad \psi_4 = 4L_2L_3 \qquad \psi_5 = L_3(2L_3 - 1) \qquad \psi_6 = 4L_1L_3 \tag{4.115c}$$

As a last example, consider the cubic element (i.e., $k - 1 = 3$). For ψ_1 we note that it must vanish along lines $L_1 = 0$, $L_1 = \frac{1}{3}$, and $L_1 = \frac{2}{3}$. Therefore, we have

$$\psi_1 = \frac{L_1 - 0}{1 - 0}\,\frac{L_1 - \frac{1}{3}}{1 - \frac{1}{3}}\,\frac{L_1 - \frac{2}{3}}{1 - \frac{2}{3}} = \frac{L_1}{2}(3L_1 - 1)(3L_1 - 2)$$

Similarly,

$$\psi_2 = \frac{L_2 - 0}{\frac{1}{3} - 0}\,\frac{L_1 - 0}{\frac{2}{3} - 0}\,\frac{L_1 - \frac{1}{3}}{\frac{2}{3} - \frac{1}{3}} = \tfrac{9}{2}L_2L_1(3L_1 - 1)$$

and so on. We have

$$\begin{aligned} \psi_3 &= \tfrac{9}{2}L_1L_2(3L_2 - 1) & \psi_4 &= \tfrac{1}{2}L_2(3L_2 - 1)(3L_2 - 2) \\ \psi_5 &= \tfrac{9}{2}L_2L_3(3L_2 - 1) & \psi_6 &= \tfrac{9}{2}L_2L_3(3L_3 - 1) \\ \psi_7 &= \tfrac{1}{2}L_3(3L_3 - 1)(3L_3 - 2) & \psi_8 &= \tfrac{9}{2}L_3L_1(3L_3 - 1) \\ \psi_9 &= \tfrac{9}{2}L_1L_3(3L_1 - 1) & \psi_{10} &= 27L_1L_2L_3 \end{aligned} \tag{4.116}$$

This completes the example. ■

In closing this section, it should be pointed out that the area coordinates L_i facilitate not only the construction of the interpolation functions for the higher-order elements but also the integration of functions (of L_i) over line paths and areas. The following exact integration formulas prove to be useful:

$$\int_a^b L_1^m L_2^n \, ds = \frac{m!n!}{(m + n + 1)!}(b - a) \tag{4.117a}$$

$$\iint_{\text{area}} L_1^m L_2^n L_3^p \, dA = \frac{m!n!p!}{(m + n + p + 2)!}2A \tag{4.117b}$$

where m, n, and p are arbitrary (positive) integers, A is the area of the domain of integration, and $m!$ denotes the factorial of m. Of course, one should transform the integrals from the x and y coordinates to L_i coordinates using the transformation (4.110) to use Eqs. (4.117).

4.4.3 Rectangular Elements

Analogous to the Lagrange family of triangular elements, the Lagrange family of rectangular elements can be developed from Pascal's triangle or the rectangular array shown in Fig. 4.26. Since a linear rectangular element has four corners (hence, four nodes), the polynomial should have the first four terms 1, x, y, and xy (which form a parallelogram in Pascal's triangle and a rectangle in the array given in Fig. 4.26). In general, a pth-order Lagrange rectangular element has n

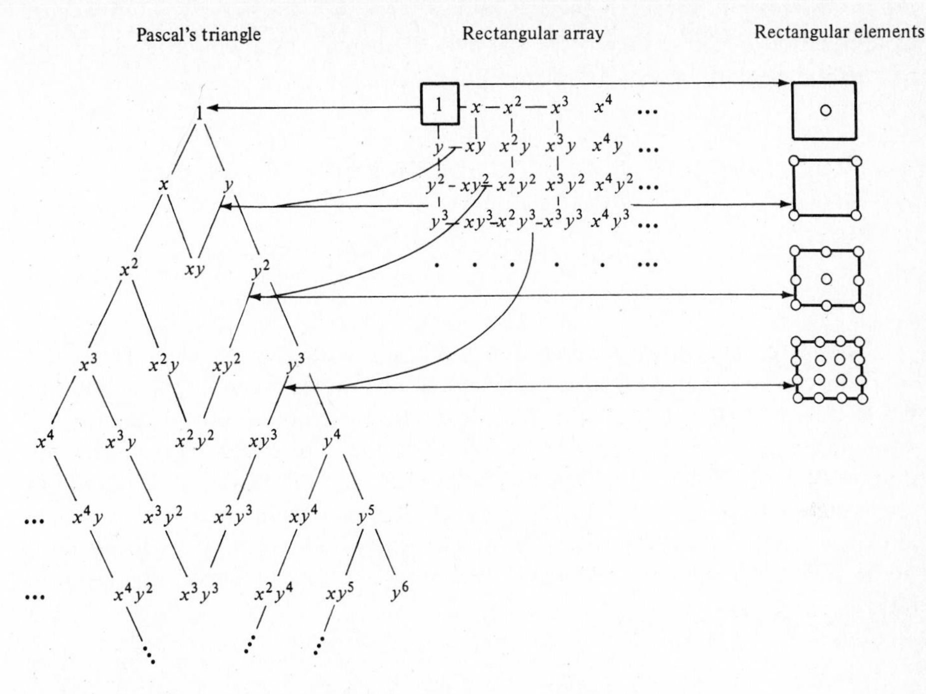

Figure 4.26 Lagrange family of rectangular elements of various order.

nodes, with

$$n = (p + 1)^2 \qquad p = 0, 1, \ldots \tag{4.118}$$

and the associated polynomial contains the terms from the pth parallelogram or the pth rectangle in Fig. 4.26. When $p = 0$, it is understood (as in triangular elements) that the node is at the center of the element (i.e., the variable is a constant on the entire element). The quadratic rectangular element has nine nodes, and the associated polynomial is given by

$$u(x, y) = a_1 + a_2 x + a_3 y + a_4 xy + a_5 x^2 + a_6 y^2 + a_7 x^2 y + a_8 xy^2 + a_9 x^2 y^2 \tag{4.119}$$

The polynomial contains the complete polynomial of the second degree plus the third-degree terms $x^2 y$ and xy^2 and also an $x^2 y^2$ term. Four of the nine nodes are placed at the four corners, four at the midpoints of the sides, and one at the center of the element. The polynomial is uniquely determined by specifying its values at each of the nine nodes. Moreover, along the sides of the element the polynomial is quadratic (with three terms—as can be seen by setting $y = 0$), and is determined by its values at the three nodes on that side. If two rectangular elements share a side, and the polynomial is required to have the same values from both elements at the three nodes of the elements, then u is uniquely defined along the entire side (shared by the two elements). The pth-order Lagrange rectangular element has the pth-degree polynomial

$$u(x, y) = \sum_{i=1}^{n} a_i x^j y^k \qquad j + k \leqslant p + 1 \qquad i, j \leqslant p$$

$$= \sum_{i=1}^{n} u_i \psi_i \tag{4.120}$$

The Lagrange interpolation functions associated with the rectangular elements can be obtained from corresponding one-dimensional Lagrange interpolation functions by taking the so-called tensor product of the x-direction one-dimensional interpolation functions with the y-direction one-dimensional interpolation functions. For the sake of simplicity, let the x and y coordinates be taken along element sides with the origin of the coordinate system at the lower left corner of the rectangle. Then for an element with dimensions a and b along the x and y directions, respectively, the interpolation functions are given as follows:

Linear ($p = 1$)

$$\begin{bmatrix} \psi_1 & \psi_3 \\ \psi_2 & \psi_4 \end{bmatrix} = \begin{Bmatrix} 1 - \dfrac{x}{a} \\ \dfrac{x}{a} \end{Bmatrix} \left\{ 1 - \frac{y}{b} \quad \frac{y}{b} \right\} = \begin{bmatrix} \left(1 - \dfrac{x}{a}\right)\left(1 - \dfrac{y}{b}\right) & \left(1 - \dfrac{x}{a}\right)\dfrac{y}{b} \\ \dfrac{x}{a}\left(1 - \dfrac{y}{b}\right) & \dfrac{x}{a}\dfrac{y}{b} \end{bmatrix} \tag{4.121}$$

Quadratic ($p = 2$)

$$\begin{bmatrix} \psi_1 & \psi_4 & \psi_7 \\ \psi_2 & \psi_5 & \psi_8 \\ \psi_3 & \psi_6 & \psi_9 \end{bmatrix} = \begin{Bmatrix} \dfrac{(x - a/2)(x - a)}{(-a/2)(-a)} \\ \dfrac{x(x - a)}{(a/2)(a/2 - a)} \\ \dfrac{x(x - a/2)}{a(a/2)} \end{Bmatrix} \tag{4.122}$$

$$\times \left\{ \frac{(y - b/2)(y - b)}{b^2/2} \quad \frac{y(y - b)}{-b^2/2} \quad \frac{y(y - b/2)}{b^2/2} \right\}$$

$$\equiv \{f\}\{g\}^T = \begin{bmatrix} f_1 g_1 & f_1 g_2 & f_1 g_3 \\ f_2 g_1 & f_2 g_2 & f_2 g_3 \\ f_3 g_1 & f_3 g_2 & f_3 g_3 \end{bmatrix}$$

where $f_i(x)$ and $g_i(y)$ are the one-dimensional interpolation functions along the

x and y directions, respectively. We get

$$
\begin{aligned}
\psi_1 &= \left(1 - \frac{2x}{a}\right)\left(1 - \frac{x}{a}\right)\left(1 - \frac{2y}{b}\right)\left(1 - \frac{y}{b}\right) \\
\psi_2 &= \frac{4x}{a}\left(1 - \frac{x}{a}\right)\left(1 - \frac{2y}{b}\right)\left(1 - \frac{y}{b}\right) \\
\psi_3 &= \frac{x}{a}\left(\frac{2x}{a} - 1\right)\left(1 - \frac{2y}{b}\right)\left(1 - \frac{y}{b}\right) \\
\psi_4 &= \left(1 - \frac{2x}{a}\right)\left(1 - \frac{x}{a}\right)\frac{4y}{b}\left(1 - \frac{y}{b}\right) \\
\psi_5 &= \frac{4x}{a}\left(1 - \frac{x}{a}\right)\frac{4y}{b}\left(1 - \frac{y}{b}\right) \\
\psi_6 &= \frac{x}{a}\left(\frac{2x}{a} - 1\right)\frac{4y}{b}\left(1 - \frac{y}{b}\right) \\
\psi_7 &= \left(1 - \frac{2x}{a}\right)\left(1 - \frac{x}{a}\right)\frac{y}{b}\left(\frac{2y}{b} - 1\right) \\
\psi_8 &= \frac{4x}{a}\left(1 - \frac{x}{a}\right)\frac{y}{b}\left(\frac{2y}{b} - 1\right) \\
\psi_9 &= \frac{x}{a}\left(\frac{2x}{a} - 1\right)\frac{y}{b}\left(\frac{2y}{b} - 1\right)
\end{aligned}
\tag{4.123}
$$

pth order

$$
\begin{bmatrix}
\psi_1 & \cdots & \psi_{p+2} & \cdots & \psi_k \\
\psi_2 & \ddots & & & \\
\vdots & & & & \vdots \\
\psi_p & & & & \\
\psi_{p+1} & \cdots & \psi_{2p+2} & \cdots & \psi_n
\end{bmatrix}
= \begin{Bmatrix} f_1 \\ f_2 \\ \vdots \\ f_{p+1} \end{Bmatrix} \begin{Bmatrix} g_1 \\ g_2 \\ \vdots \\ g_{p+1} \end{Bmatrix}^T
\tag{4.124}
$$

$$k = (p + 1)p + 1 \qquad n = (p + 1)^2$$

where $f_i(x)$ and $g_i(y)$ are the pth degree interpolants in x and y, respectively. For example, the polynomial (THIS IS PTH ORDER INTERPOLATION FOR ith NODE OF 1-D ELEMENT)

$$
f_i(\xi) = \frac{(\xi - \xi_1)(\xi - \xi_2)\dots(\xi - \xi_{i-1})(\xi - \xi_{i+1})\dots(\xi - \xi_{p+1})}{(\xi_i - \xi_1)(\xi_i - \xi_2)\dots(\xi_i - \xi_{i-1})(\xi_i - \xi_{i+1})\dots(\xi_i - \xi_{p+1})}
\tag{4.125}
$$

where ξ_i is the ξ coordinate of node i, is a pth-degree interpolation polynomial in ξ that vanishes at points $\xi_1, \xi_2, \dots, \xi_{i-1}, \xi_{i+1}, \dots, \xi_{p+1}$.

It is convenient (for numerical integration purposes) to express the interpolation functions in Eq. (4.124) in terms of the natural coordinates ξ and η:

$$
\xi = \frac{2(x - x_1) - a}{a} \qquad \eta = \frac{2(y - y_1) - b}{b}
\tag{4.126}
$$

IN THIS CASE, $x_1 = y_1 = 0$.

where x_1 and y_1 are the global coordinates of node 1. For the coordinate system with origin fixed at node 1, we have $x_1 = y_1 = 0$. In this case, the quadratic interpolation functions in Eqs. (4.123) can be written in terms of ξ and η as

$$\begin{aligned}
\hat{\psi}_1 &= \tfrac{1}{4}(1-\xi)(1-\eta)\xi\eta \\
\hat{\psi}_2 &= -\tfrac{1}{2}(1-\xi^2)(1-\eta)\eta \\
\hat{\psi}_3 &= -\tfrac{1}{4}(1+\xi)(1-\eta)\xi\eta \\
\hat{\psi}_4 &= -\tfrac{1}{2}(1-\xi)(1-\eta^2)\xi \\
\hat{\psi}_5 &= (1-\xi^2)(1-\eta^2) \\
\hat{\psi}_6 &= \tfrac{1}{2}(1+\xi)(1-\eta^2)\xi \\
\hat{\psi}_7 &= -\tfrac{1}{4}(1-\xi)(1+\eta)\xi\eta \\
\hat{\psi}_8 &= \tfrac{1}{2}(1-\xi^2)(1+\eta)\eta \\
\hat{\psi}_9 &= \tfrac{1}{4}(1+\xi)(1+\eta)\xi\eta
\end{aligned} \tag{4.127}$$

It should be cautioned that the subscripts of $\hat{\psi}$ refer to the node numbering used in Fig. 4.27. For any renumbering of the nodes, the subscripts of the interpolation functions should be changed accordingly.

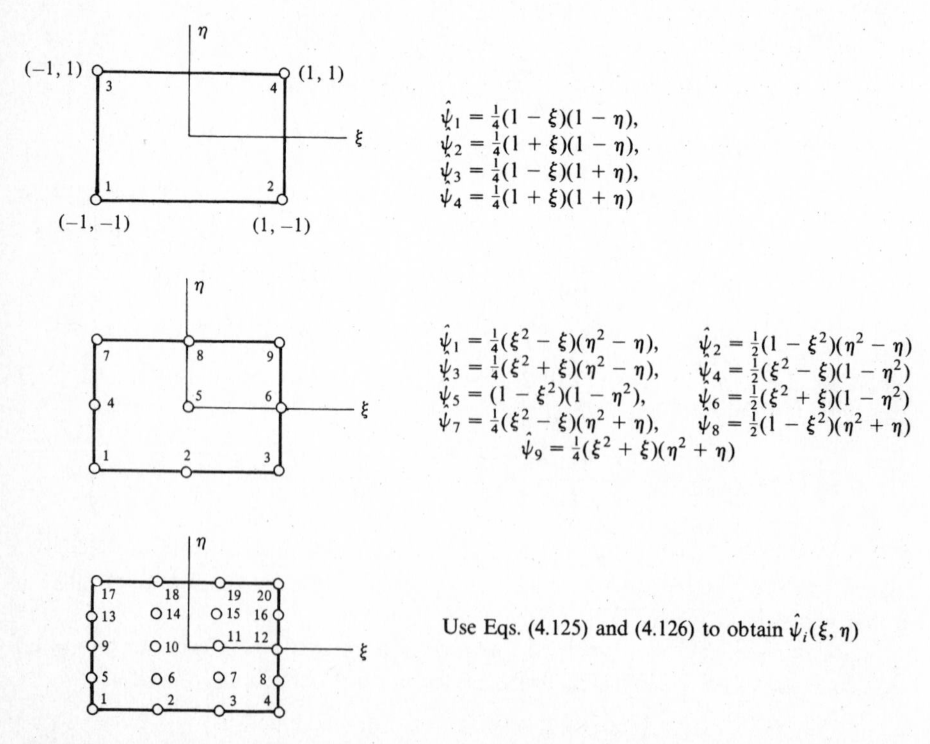

Figure 4.27 Node numbers and interpolation functions for the rectangular elements of the Lagrange family.

4.4.4 The Serendipity Elements

Since the internal nodes of the higher-order elements of the Lagrange family do not contribute to the interelement connectivity, they can be condensed out at the element level so that the size of the element matrices is reduced. Alternatively, one can use the so-called serendipity elements to avoid the internal nodes present in the Lagrange elements. The serendipity elements are those rectangular elements which have no interior nodes. In other words, all the node points are on the boundary of the element. The interpolation functions for serendipity elements cannot be obtained using tensor products of one-dimensional interpolation functions. Instead, an alternative procedure that employs the conditions of the type in Eqs. (4.26) is used. Here we show how to construct the interpolation functions for the eight-node (quadratic) element using the natural coordinates.

The interpolation function for node 1 should take on a value of zero at nodes $2, 3, \ldots, 8$, and a value of unity at node 1. Equivalently, $\hat{\psi}_1$ should vanish on the sides defined by the equations $1 - \xi = 0$, $1 - \eta = 0$, and $1 + \xi + \eta = 0$ (see Fig. 4.28). Therefore $\hat{\psi}_1$ is of the form

$$\hat{\psi}_1(\xi, \eta) = c(1 - \xi)(1 - \eta)(1 + \xi + \eta)$$

where c is a constant which should be determined so as to yield $\hat{\psi}_1(-1, -1) = 1$. We obtain $c = -\frac{1}{4}$, and therefore,

$$\hat{\psi}_1(\xi, \eta) = -\tfrac{1}{4}(1 - \xi)(1 - \eta)(1 + \xi + \eta) \tag{4.128a}$$

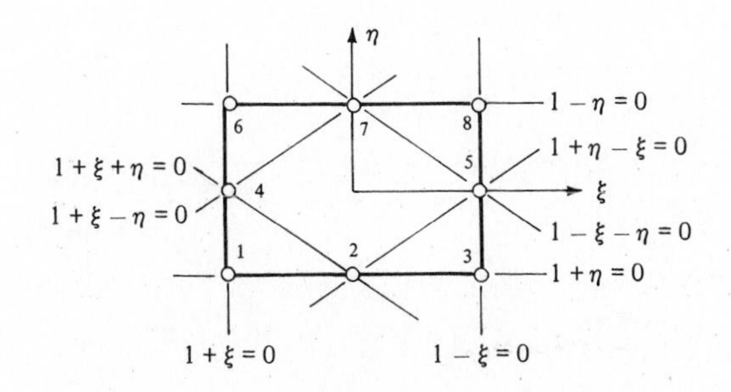

$$\hat{\psi}_1 = \tfrac{1}{4}(1 - \xi)(1 - \eta)(-1 - \xi - \eta)$$
$$\hat{\psi}_2 = \tfrac{1}{2}(1 - \xi^2)(1 - \eta)$$
$$\hat{\psi}_3 = \tfrac{1}{4}(1 + \xi)(1 - \eta)(-1 + \xi - \eta)$$
$$\hat{\psi}_4 = \tfrac{1}{2}(1 - \xi)(1 - \eta^2)$$
$$\hat{\psi}_5 = \tfrac{1}{2}(1 + \xi)(1 - \eta^2)$$
$$\hat{\psi}_6 = \tfrac{1}{4}(1 - \xi)(1 + \eta)(-1 - \xi + \eta)$$
$$\hat{\psi}_7 = \tfrac{1}{2}(1 - \xi^2)(1 + \eta)$$
$$\hat{\psi}_8 = \tfrac{1}{4}(1 + \xi)(1 + \eta)(-1 + \xi + \eta)$$

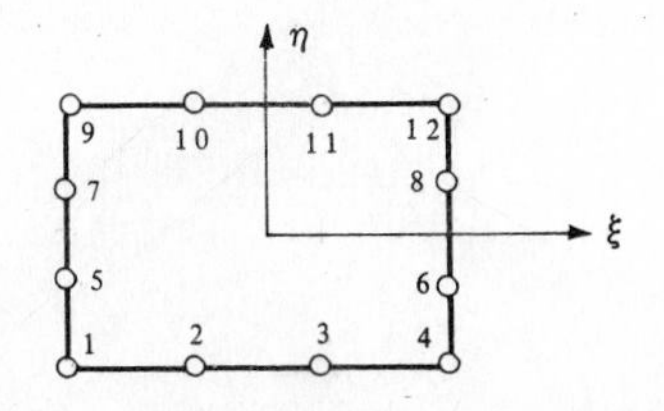

See Eq. (4.130) for the interpolation functions

Figure 4.28 Node numbers and interpolation functions for the serendipity family of elements.

Similarly, we obtain

$$
\begin{aligned}
\hat{\psi}_2 &= \tfrac{1}{2}(1 - \xi^2)(1 - \eta) \\
\hat{\psi}_3 &= \tfrac{1}{4}(1 + \xi)(1 - \eta)(-1 + \xi - \eta) \\
\hat{\psi}_4 &= \tfrac{1}{2}(1 - \xi)(1 - \eta^2) \\
\hat{\psi}_5 &= \tfrac{1}{2}(1 + \xi)(1 - \eta^2) \\
\hat{\psi}_6 &= \tfrac{1}{4}(1 - \xi)(1 + \eta)(-1 - \xi + \eta) \\
\hat{\psi}_7 &= \tfrac{1}{2}(1 - \xi^2)(1 + \eta) \\
\hat{\psi}_8 &= \tfrac{1}{4}(1 + \xi)(1 + \eta)(-1 + \xi + \eta)
\end{aligned}
\tag{4.128b}
$$

Note that all the $\hat{\psi}_i$ for the eight-node element have the form

$$\hat{\psi}_i = c_1 + c_2\xi + c_3\eta + c_4\xi\eta + c_5\xi^2 + c_6\eta^2 + c_7\xi^2\eta + c_8\xi\eta^2 \tag{4.129}$$

Similarly, the interpolation functions for the cubic serendipity element, which has twelve nodes, are given by

$$
\begin{aligned}
\hat{\psi}_1 &= \tfrac{1}{32}(1 - \xi)(1 - \eta)[-10 + 9(\xi^2 + \eta^2)] \\
\hat{\psi}_2 &= \tfrac{9}{32}(1 - \eta)(1 - \xi^2)(1 - 3\xi) \\
\hat{\psi}_3 &= \tfrac{9}{32}(1 - \eta)(1 - \xi^2)(1 + 3\xi) \\
\hat{\psi}_4 &= \tfrac{1}{32}(1 + \xi)(1 - \eta)[-10 + 9(\xi^2 + \eta^2)] \\
\hat{\psi}_5 &= \tfrac{9}{32}(1 - \xi)(1 - \eta^2)(1 - 3\eta) \\
\hat{\psi}_6 &= \tfrac{9}{32}(1 + \xi)(1 - \eta^2)(1 - 3\eta) \\
\hat{\psi}_7 &= \tfrac{9}{32}(1 - \xi)(1 - \eta^2)(1 + 3\eta) \\
\hat{\psi}_8 &= \tfrac{1}{32}(1 + \xi)(1 - \eta^2)(1 + 3\eta) \\
\hat{\psi}_9 &= \tfrac{1}{32}(1 - \xi)(1 + \eta)[-10 + 9(\xi^2 + \eta^2)] \\
\hat{\psi}_{10} &= \tfrac{9}{32}(1 + \eta)(1 - \xi^2)(1 - 3\xi) \\
\hat{\psi}_{11} &= \tfrac{9}{32}(1 + \eta)(1 - \xi^2)(1 + 3\xi) \\
\hat{\psi}_{12} &= \tfrac{1}{32}(1 + \xi)(1 + \eta)[-10 + 9(\xi^2 + \eta^2)]
\end{aligned}
\tag{4.130}
$$

The interpolation functions $\hat{\psi}_i$ for the twelve-node element are of the form

$$\hat{\psi}_i = \text{terms of the form in Eq. (4.129)} + c_9\xi^3 + c_{10}\eta^3 + c_{11}\xi^3\eta + c_{12}\xi\eta^3 \tag{4.131}$$

PROBLEMS

Some of the following problems require use of a computer for the solution of condensed equations. On the other hand, one can use the program FEM2D discussed in Sec. 4.8. The reader is expected to give at least the specified boundary conditions and the force vector for the mesh used in each problem.

4.1 Compute the first four entries of the global force vector for the heat transfer problem in Example 2.4 when q is given by $\hat{q}(y) = a_0 + a_1 y$, $BA = 3h$, node 1 is at B, node 4 is at A, and a_0 and a_1 are constants. Assume that the boundary BA is represented by three linear elements (of equal length).

4.2 Calculate the linear interpolation functions for the triangular elements shown in Fig. P4.2.

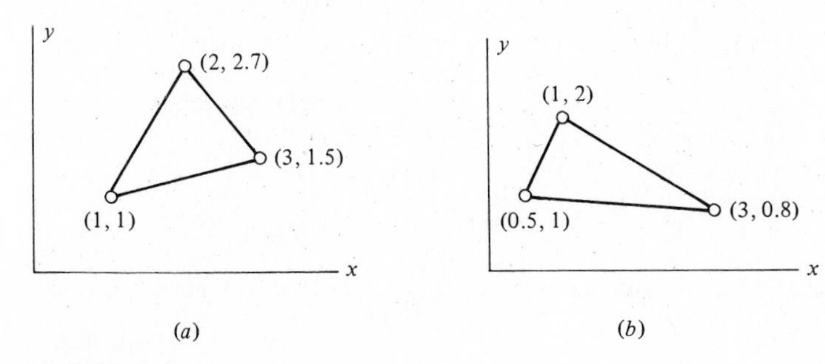

Figure P4.2

4.3 The nodal values of element 13 in the heat conduction problem of Example 4.4 are $T_{10} = 389.79$ K, $T_{11} = 337.19$ K, and $T_{12} = 395.08$ K (see the mesh in Fig. 4.12a). Find the gradient of the temperature in the element. Also determine where the 392 K isotherm intersects the boundary of the element in Fig. P4.3.

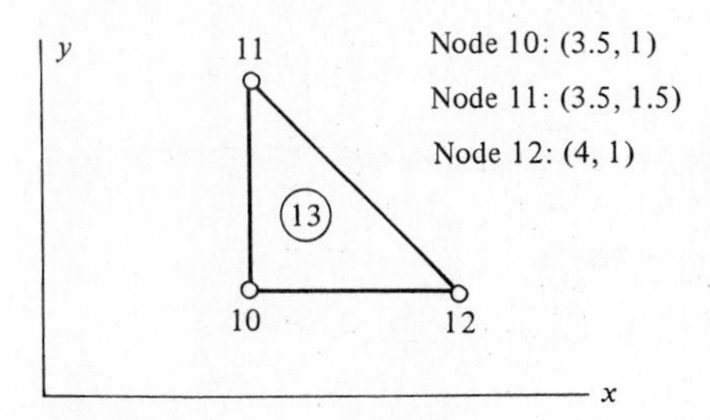

Figure P4.3

4.4 Compute the element matrices

$$K_{ij}^1 = \int_0^a \int_0^b \psi_i \frac{d\psi_j}{d\xi} d\xi \, d\eta \qquad K_{ij}^2 = \int_0^a \int_0^b \psi_i \frac{d\psi_j}{d\eta} d\xi \, d\eta$$

where ψ_i are the linear interpolation functions in Eqs. (4.31).

4.5 Give the assembled coefficient matrix for the finite-element meshes shown in Fig. P4.5. Assume 1 degree of freedom per node, and let $[K^{(e)}]$ denote the element coefficient matrix for the eth element. (Your answer should be in terms of $K_{ij}^{(e)}$.)

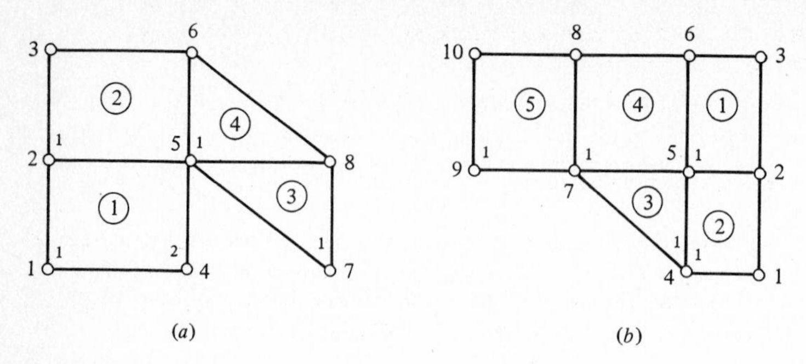

Figure P4.5

4.6 If U_1 and U_4 are specified in the finite-element meshes of Prob. 4.5, modify the assembled set of equations and write the condensed set of equations for the remaining unknown primary degrees of freedom.

4.7 Compute the element matrices of a rectangular element that has two of its sides exposed to ambient temperature (see Fig. P4.7), i.e., experience convection.

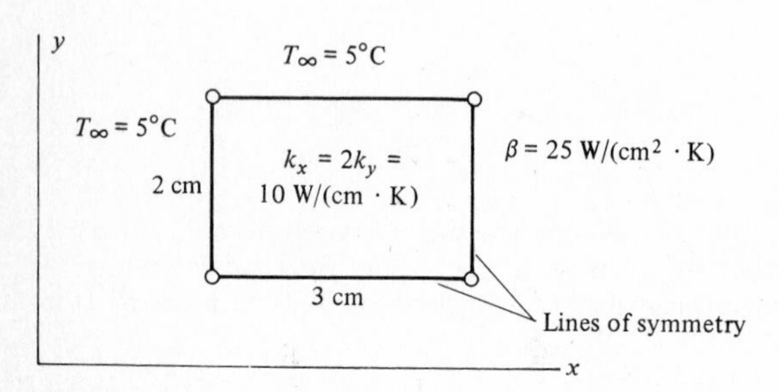

Figure P4.7

4.8 A point source of magnitude 100 W is located inside the triangular element shown in Fig. P4.8. Determine the contribution of the source to the element nodes.

4.9 Repeat Prob. 4.8 for a rectangular element as shown in Fig. P4.9.

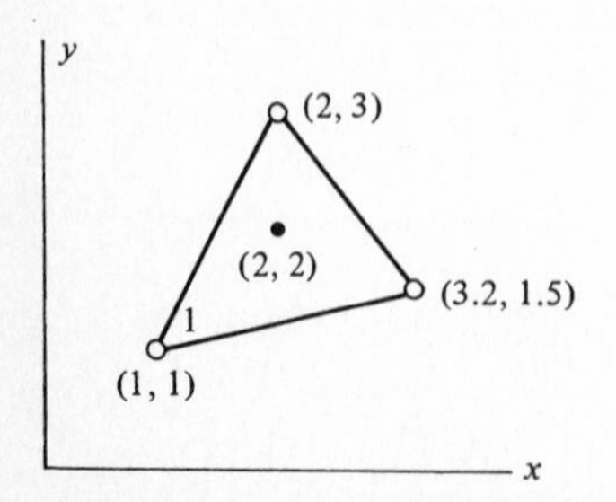

Figure P4.8

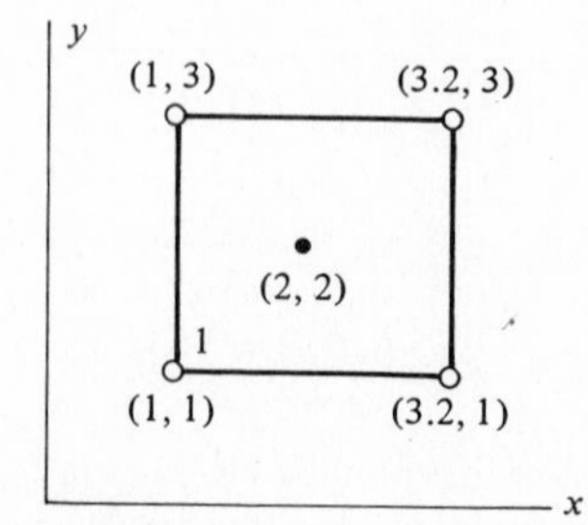

Figure P4.9

4.10 (Area coordinates) Consider an arbitrary triangular element (see Fig. P4.10), and define three coordinates L_1, L_2, and L_3 by

$$L_1 = \frac{A_1}{A}, \qquad L_2 = \frac{A_2}{A}, \qquad L_3 = \frac{A_3}{A}$$

where A_i ($i = 1, 2, 3$) is the area of the triangle formed by nodes j and k and point P, and A is the total area of the triangle. Show that

$$L_i = \psi_i$$

where ψ_i are the linear interpolation functions of the triangular element.

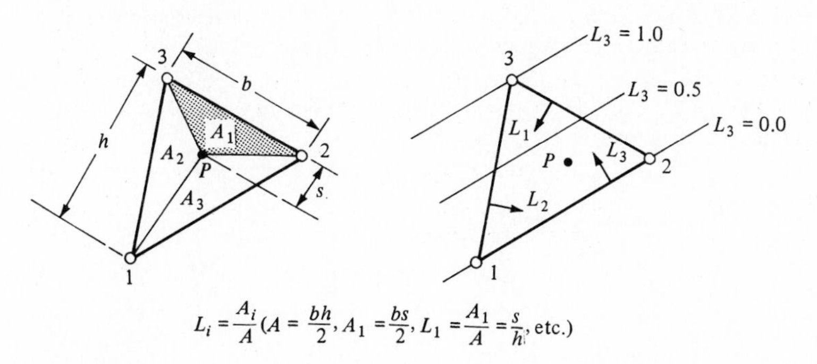

Figure P4.10

4.11 Solve the Laplace equation

$$-\left(\frac{\partial^2 u}{\partial x^2} + \frac{\partial^2 u}{\partial y^2}\right) = 0$$

in a unit square, when $u(0, y) = u(1, y) = u(x, 0) = 0$ and $u(x, 1) = 1.0$. Use the symmetry and (*a*) a mesh of 2×4 triangular elements and (*b*) a mesh of 2×4 rectangular elements (see Fig. P4.11). Compare the finite-element solution with the exact solution

$$u(x, y) = \frac{4}{\pi} \sum_{n=0}^{\infty} \frac{\sin[(2n+1)\pi x] \sinh[(2n+1)\pi y]}{(2n+1)\sinh[(2n+1)\pi]}$$

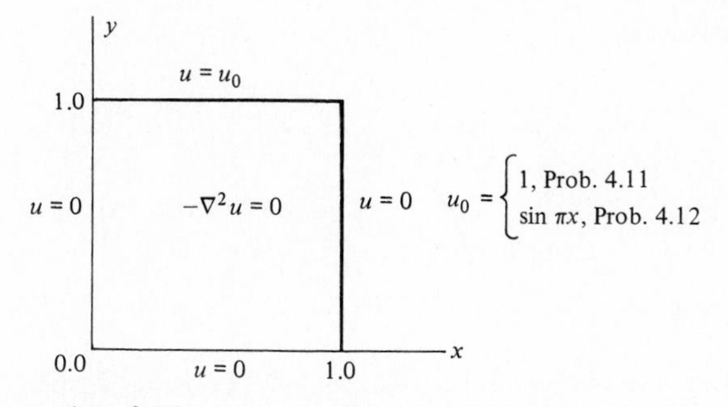

Figure P4.11

4.12 Solve Prob. 4.11 when the boundary condition at $y = 1$ is replaced by

$$u(x, 1) = \sin \pi x$$

Compare the finite-element solution with the exact solution

$$u(x, y) = \frac{\sin \pi x \sinh \pi y}{\sinh \pi}$$

4.13 Perform a convergence study of Prob. 4.11 with linear rectangular elements. Also, investigate the effect of the boundary condition at $x = 0$ and $y = 1$ on the solution (e.g., use $u = 0$, $u = 1$, and $u = 0.5$ at $x = 0$ and $y = 1$ and compare the resulting solutions).

4.14 When a_{ij} in Eq. (4.1) are bilinear functions of x and y, $a_{ij} = a_{ij}^0 + a_{ij}^1 x + a_{ij}^2 y + a_{ij}^3 xy$, determine the explicit form of the coefficient matrix $[K^{(e)}]$ in Eq. (4.14) for the linear triangular element. Express the result in terms of the integrals defined in Eq. (4.36).

4.15 Repeat Prob. 4.14 for the linear rectangular element. Note that the interpolation functions in Eqs. (4.31) are referred to the local coordinate system (ξ, η). Therefore, in evaluating the integrals (4.34), the following transformation (see Fig. P4.15) must be used:

$$x = x_1 + \xi/a \qquad y = y_1 + \eta/b$$

where x_1 and y_1 are the global coordinates of node 1.

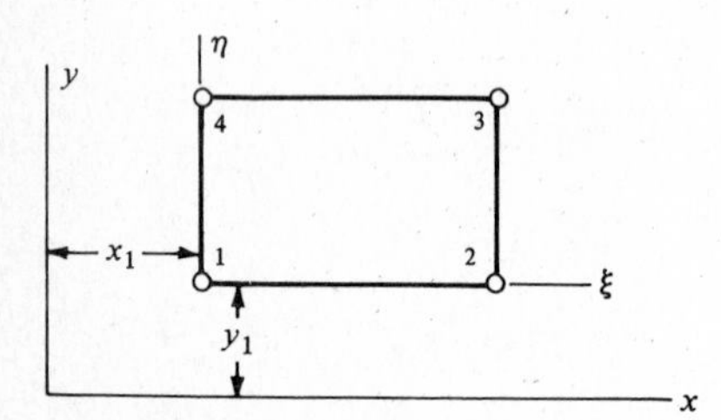

Figure P4.15

4.16 Determine the generalized force vector $\{f^e\}$, where

$$f_i^{(e)} = \int_{\Omega^e} f\psi_i \, dx \, dy$$

when f is a bilinear function of the form

$$f = a_0 + a_1 x + a_2 y$$

for the linear triangular and rectangular elements.

4.17 Repeat Prob. 4.14 for the case

$$a_{ij} = \sum_{k=1}^{3} a_{ij}^k \psi_k \qquad a_{ij}^k = \text{constant}$$

4.18 Repeat Prob. 4.15 for a_{ij} given in Prob. 4.17.

4.19 Repeat Prob. 4.16 for the case

$$f = \sum f_i \psi_i \qquad f_i = \text{constant}$$

4.20 Solve the problem of Couette flow between two parallel flat walls (see Prob. 3.29) using a mesh of 2×5 linear rectangular elements. Take unit length along the x coordinate, $b = 0.5$, $f_0 = -6$, and $u_0 = 1.0$ (see Fig. P4.20). Compare the two-dimensional finite-element solution with the exact solution given in Prob. 3.29.

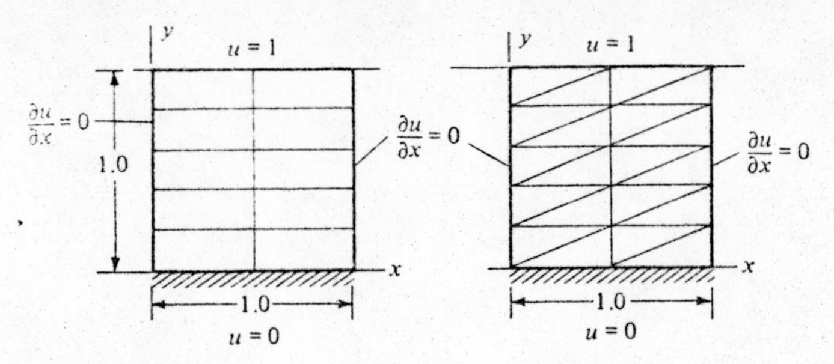

Figure P4.20

Figure P4.21

4.21 Repeat Prob. 4.20 using an equivalent mesh of linear triangular elements (see Fig. P4.21).

4.22 The Prandtl theory of torsion of a cylindrical member leads to

$$-\nabla^2 u = 2G\theta \text{ in } \Omega \qquad u = 0 \text{ on } \Gamma$$

where Ω is the cross section of the cylindrical member being twisted, Γ is the boundary of Ω, G is the shear modulus of the material of the member, θ is the angle of twist, and u is the stress function. Solve the equation for the case in which Ω is (a) an elliptic section and (b) a circular section (see Fig. P4.22) using the linear triangular element. Compare the finite-element solution with the exact solution (for elliptical sections)

$$u = \frac{G\theta a^2 b^2}{a^2 + b^2}\left(1 - \frac{x^2}{a^2} - \frac{y^2}{b^2}\right)$$

4.23 Repeat Prob. 4.22 for the case in which Ω is an equilateral triangle (see Fig. P4.23). The exact solution is given by

$$u = -G\theta\left[\tfrac{1}{2}(x^2 + y^2) - \frac{1}{2a}(x^3 - 3xy^2) - \tfrac{2}{27}a^2\right]$$

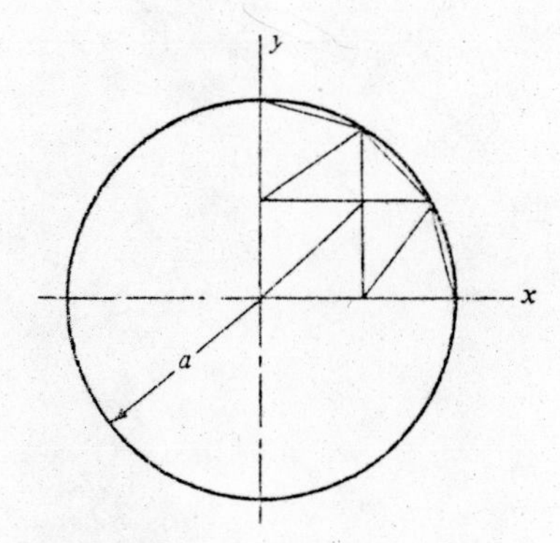

Figure P4.22

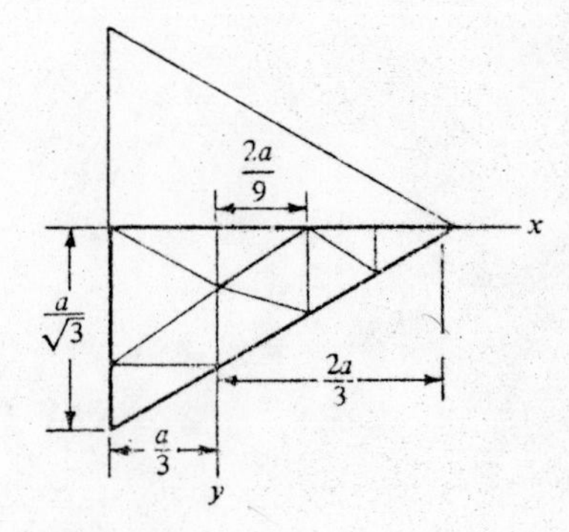

Figure P4.23

4.24 Solve the torsion problem, Prob. 4.22, for the composite cross sections illustrated in Fig. P4.24a and b (only a quadrant is shown). Use the rectangular-element meshes shown.

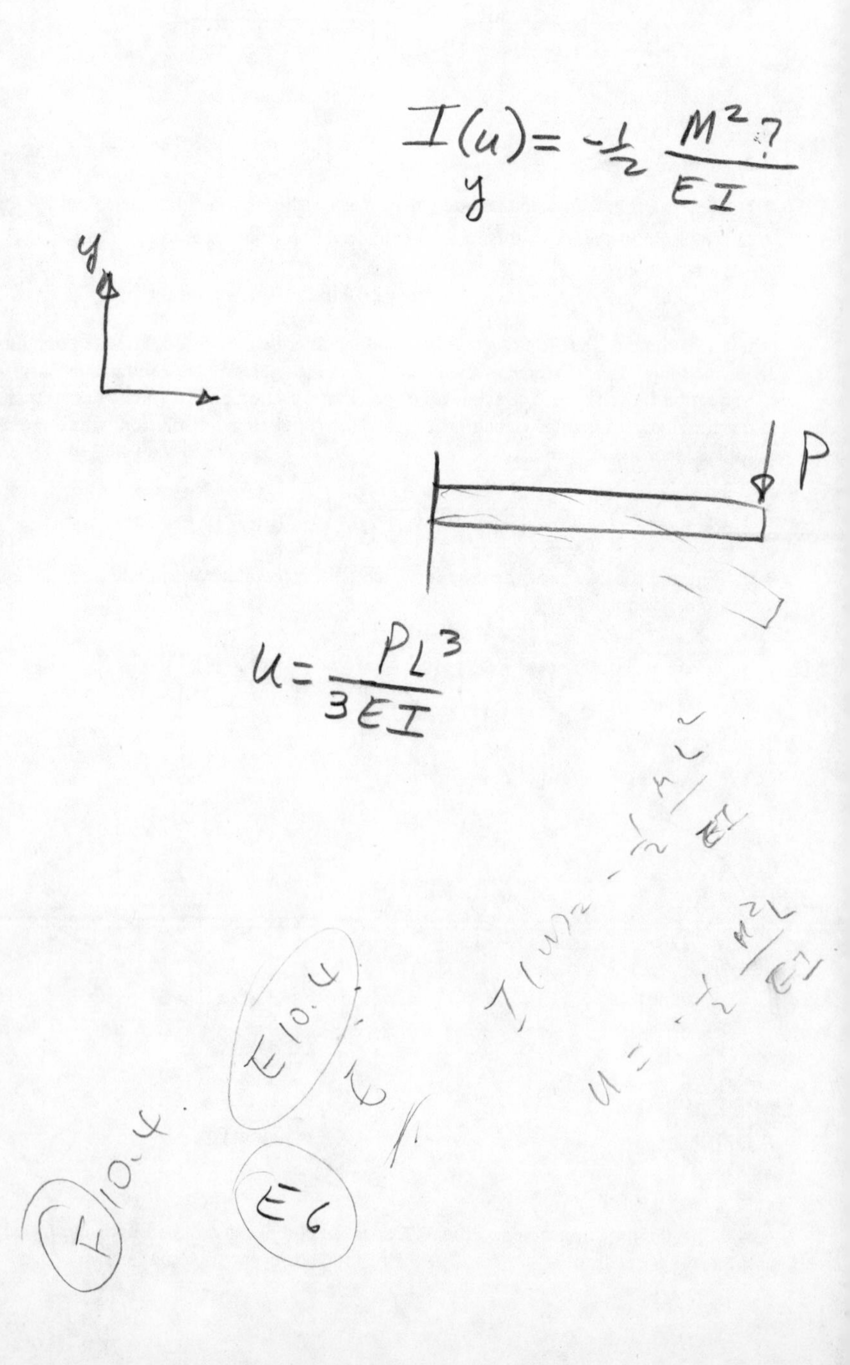

I(u) = -1/2 M²? / EI
y
y
P
u = PL³ / 3EI
I(u) = -1/2 M²L / EI
u = -1/2 M²L / EI
E10.4
E10.4
E6

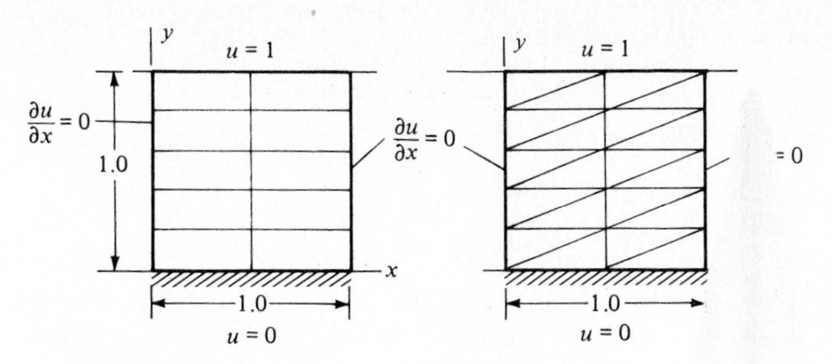

Figure P4.20

Figure P4.21

4.21 Repeat Prob. 4.20 using an equivalent mesh of linear tr[illegible]ents (see Fig. P4.21).

4.22 The Prandtl theory of torsion of a cylindrical member [illegible]

$$-\nabla^2 u = 2G\theta \text{ in } \Omega \qquad u = \text{[illegible]}$$

where Ω is the cross section of the cylindrical member being [illegible] is [illegible]he boundary of Ω, G is the shear modulus of the material of the member, θ is the angle [illegible] s the stress function. Solve the equation for the case in which Ω is (a) an elliptic section [illegible]lar section (see Fig. P4.22) using the linear triangular element. Compare the finite-ele[illegible] with the exact solution (for elliptical sections)

$$u = \frac{G\theta a^2 b^2}{a^2 + b^2}\left(1 - \frac{x^2}{a^2} - \text{[illegible]}\right)$$

4.23 Repeat Prob. 4.22 for the case in which Ω is an equila[illegible] triangle (see Fig. P4.23). The exact solution is given by

$$u = -G\theta\left[\tfrac{1}{2}(x^2 + y^2) - \frac{1}{2a}(x^3 - 3x\text{[illegible]}) - \tfrac{2}{27}a^2\right]$$

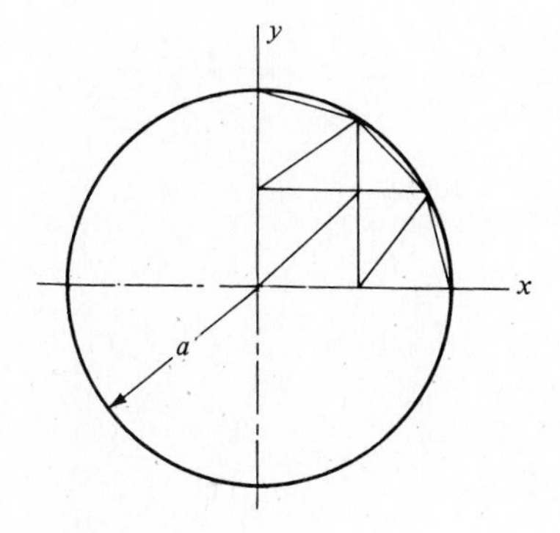

Figure P4.22

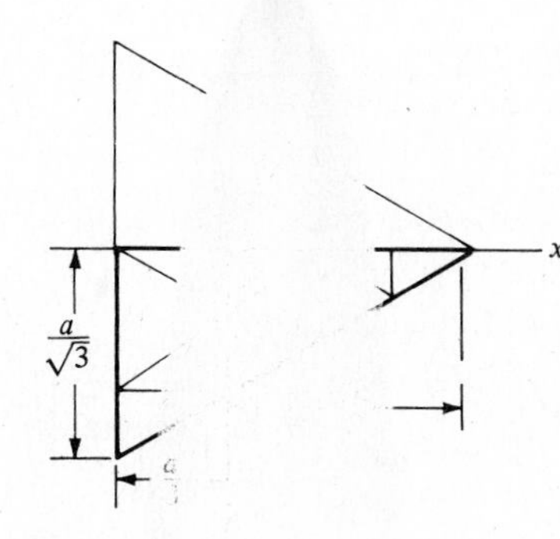

Figure [illegible]

4.24 Solve the torsion problem, Prob. 4.22, for the compo[illegible]llustrated in Fig. P4.24a and b (only a quadrant is shown). Use the rectangular-ele[illegible] shown.

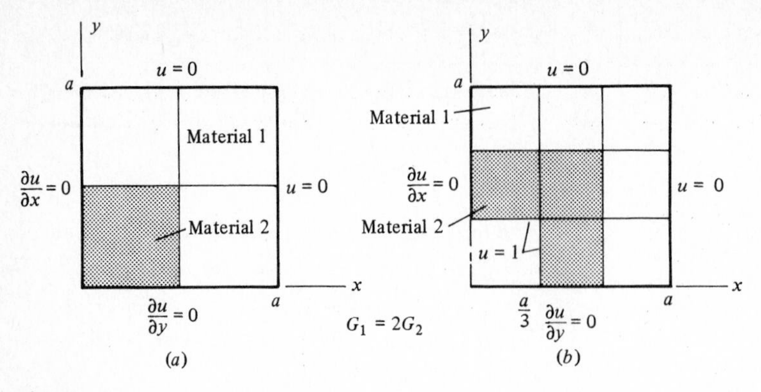

Figure P4.24

4.25 Solve the radial axisymmetric heat transfer problem of Example 3.9 using an equivalent mesh (to that of Example 3.9) of linear triangular elements.

4.26 A series of heating cables have been placed in a conducting medium, as shown in Fig. P4.26. The medium has conductivities of $k_x = 10$ W/(cm · K) and $k_y = 15$ W/(cm · K), the upper surface is exposed to a temperature of -5°C, and the lower surface is bounded by an insulating medium. Assuming that each cable is a point source of 250 W/cm, determine the temperature distribution in the medium. Take the convection coefficient between the medium and the upper surface to be $\beta = 5$ W/(cm^2 · K).

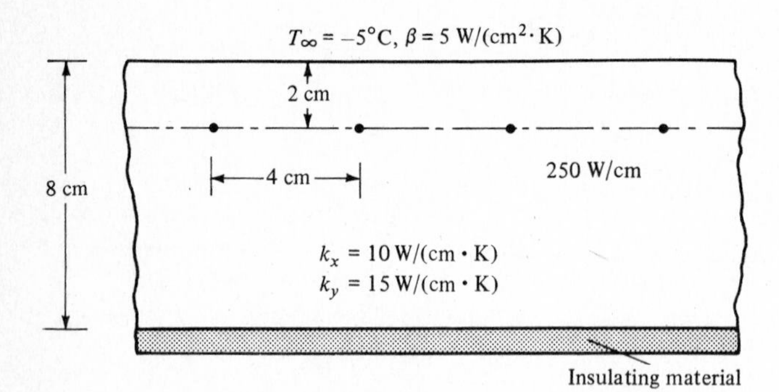

Figure P4.26

4.27 Determine the temperature distribution in the molded asbestos insulation shown in Fig. P4.27. Take $k = 0.1$ Btu/(h · ft · F).

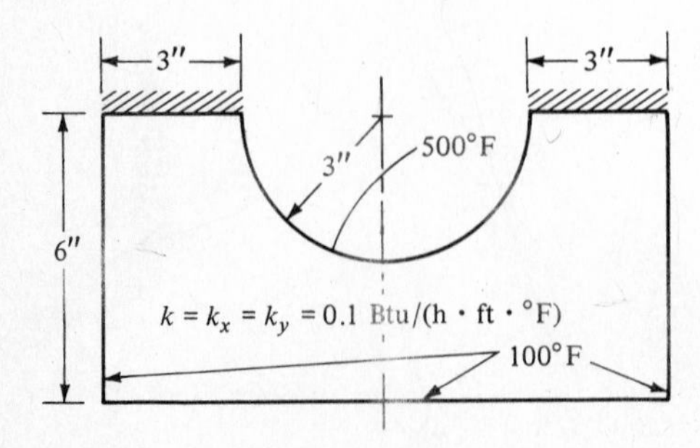

Figure P4.27

4.28 Solve the problem in Example 4.4 when the convective boundary condition on the circular arc is replaced by the essential boundary condition $T = 200°\text{C}$.

4.29 Consider the problem of the flow of groundwater beneath a coffer dam. Solve the problem using the velocity potential formulation. The geometry and boundary conditions are shown in Fig. P4.29.

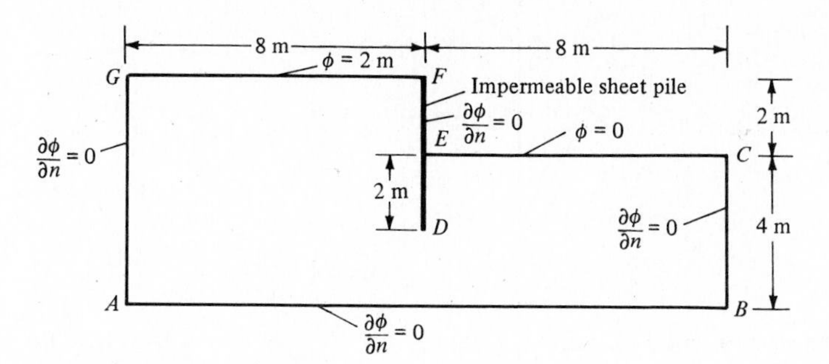

Figure P4.29

4.30 Solve the problem of the flow about an elliptical cylinder using the stream function formulation. The geometry and boundary conditions are shown in Fig. P4.30.

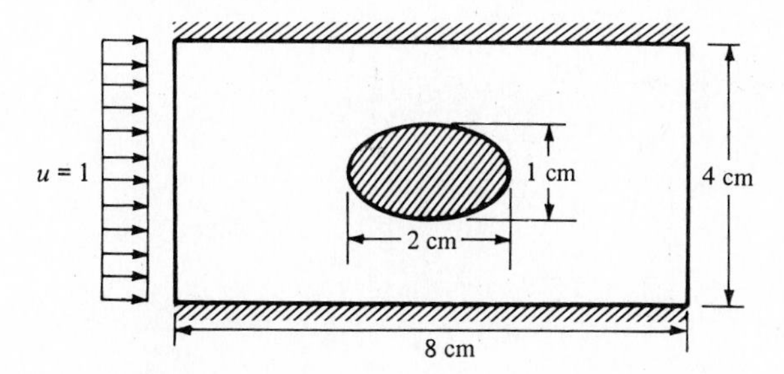

Figure P4.30

4.31 Repeat Prob. 4.30 for the domain given in Fig. P4.31.

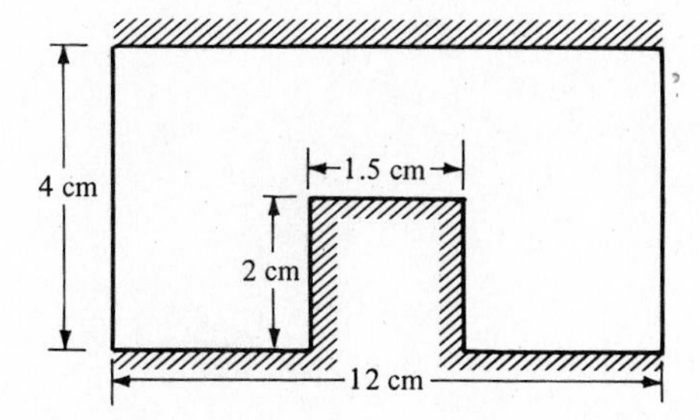

Figure P4.31

4.32 Determine the equipotential lines for the groundwater flow problem of the domain shown in Fig. P4.32.

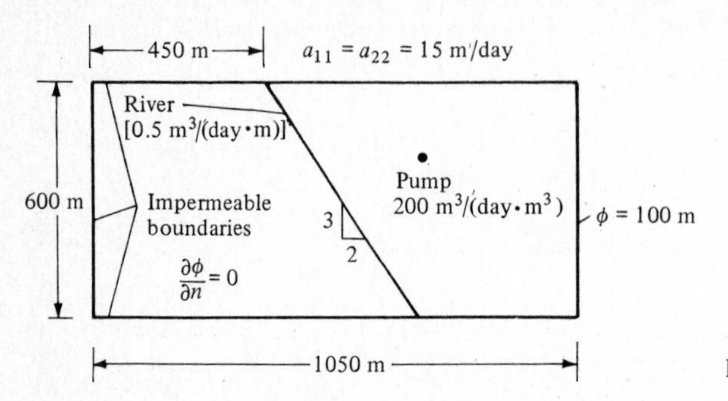

Figure P4.32

4.33 Repeat Prob. 4.32 for the domain shown in Fig. P4.33.

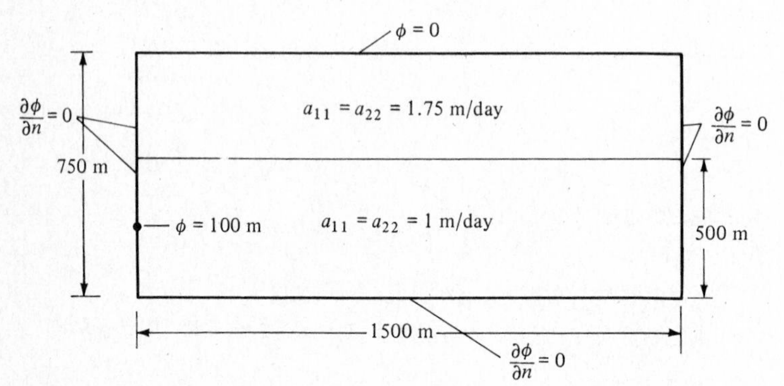

Figure P4.33

4.34 Show that the bilinear interpolation functions for the four-node triangular element given in Fig. P4.34 are of the form

$$\psi_i = a_i + b_i\xi + c_i\eta + d_i\xi\eta \qquad i = 1, 2, 3, 4$$

where

$$a_1 = 1, a_2 = a_3 = a_4 = 0, -b_1 = b_2 = 1/a, b_3 = b_4 = 0, \text{ and}$$

$$c_1 = \frac{6ab - a^2 - 2b^2}{ac(a - 2b)} \qquad c_2 = \frac{2b(a + b)}{ac(a - 2b)} \qquad c_3 = \frac{a + b}{c(a - 2b)} \qquad c_4 = \frac{-9b}{c(a - 2b)}$$

$$d_1 = d_2 = d_3 = -\frac{d_4}{3} = -\frac{3}{c(a - 2b)}$$

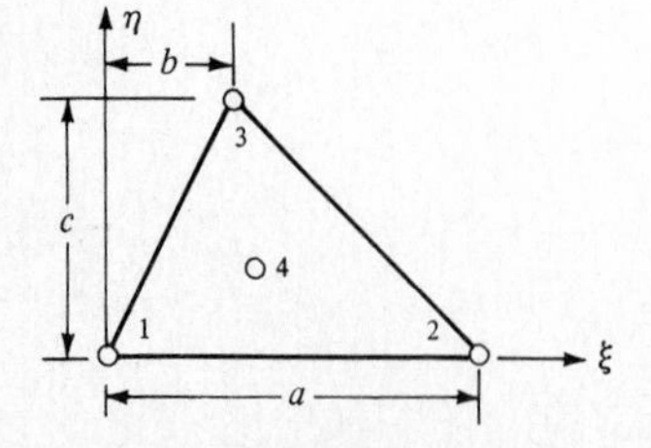

Figure P4.34

4.35 Show that the interpolation functions that involve the term $(x^2 + y^2)$ for the five-node rectangular element shown in Fig. P4.35 are given by

$$\psi_1 = 0.25(-\xi - \eta + \xi\eta) + 0.125(\xi^2 + \eta^2)$$
$$\psi_2 = 0.25(\xi - \eta - \xi\eta) + 0.125(\xi^2 + \eta^2)$$
$$\psi_3 = 0.25(\xi + \eta + \xi\eta) + 0.125(\xi^2 + \eta^2)$$
$$\psi_4 = 0.25(-\xi + \eta - \xi\eta) + 0.125(\xi^2 + \eta^2)$$
$$\psi_5 = 1 - 0.5(\xi^2 + \eta^2)$$

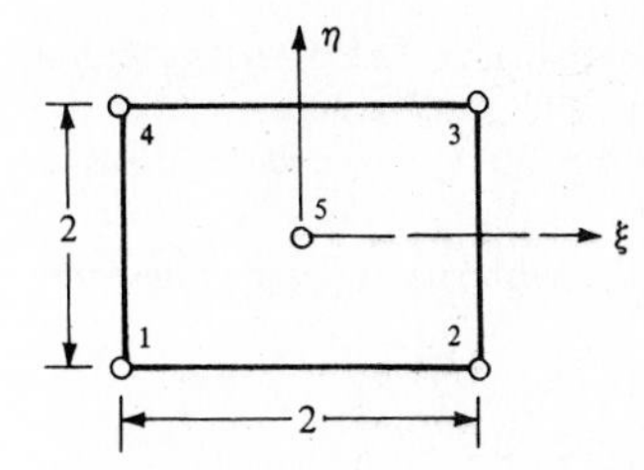

Figure P4.35

4.36 Compute the element matrix

$$K_{ij}^{(e)} = \int_{\Omega^e} \left(\frac{\partial \psi_i}{\partial x} \frac{\partial \psi_j}{\partial x} + \frac{\partial \psi_i}{\partial y} \frac{\partial \psi_j}{\partial y} \right) dx\, dy$$

for the four-node triangular element of Prob. 4.34.

4.37 Repeat Prob. 4.36 for the five-node rectangular element of Prob. 4.35.

4.38 Calculate the interpolation functions for the quadratic triangular element shown in Fig. P4.38.

4.39 Evaluate the integrals

$$\int (\psi_1)^2 \, dA \qquad \text{and} \qquad \int \psi_1 \psi_2 \, dA$$

for the triangular element in Fig. P4.38.

4.40 Determine the interpolation function ψ_{14} for the quintic triangular element in Fig. P4.40.

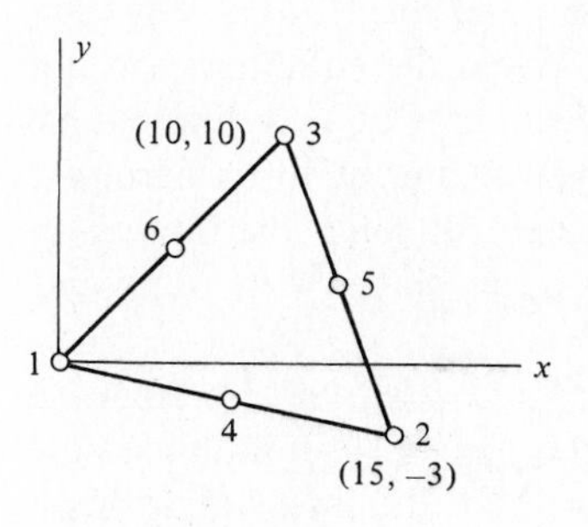

Figure P4.38

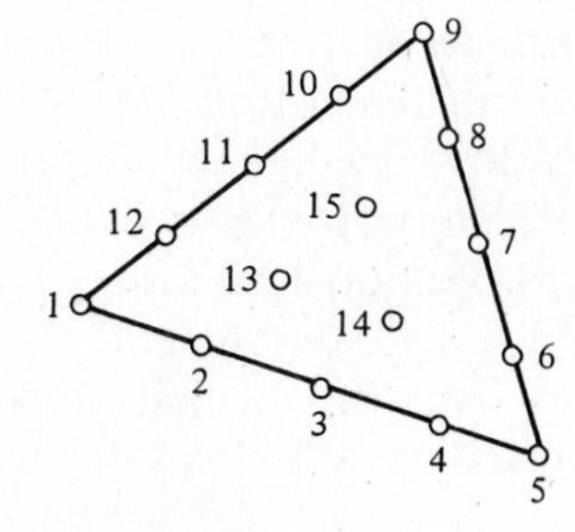

Figure P4.40

4.41 Show that a variable interpolated by the interpolation functions of linear, quadratic, or cubic rectangular elements is continuous along interelement boundaries.

4.42 Drive the interpolation function of a corner node in a cubic serendipity element.

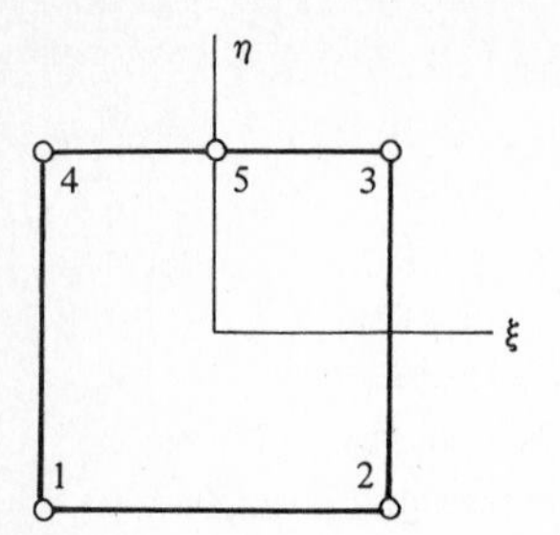

Figure P4.43

4.43 Consider the five-node element shown in Fig. P4.43. Using the basic linear and quadratic interpolations along the coordinate directions ξ and η, derive the interpolation functions for the element. Note that the element can be used as a transition element connecting four-node elements to eight- or nine-node elements.

4.44 (Nodeless variables) Consider the four-node rectangular element with interpolation of the form

$$u = \sum_{i=1}^{4} u_i \psi_i + \sum_{i=1}^{4} c_i \phi_i$$

Determine the form of ψ_i and ϕ_i for the element.

4.5 SECOND-ORDER, MULTIVARIABLE EQUATIONS

4.5.1 Preliminary Comments

In Sec. 4.2, we considered the finite-element analysis of second-order (two-dimensional) problems that involved only one dependent unknown (hence, only one partial differential equation). Here we consider a system (often a pair) of coupled partial differential equations in as many dependent variables as the number of equations. The word "coupled" is used to imply that the same dependent variables appear in more than one equation of the set, and therefore no equation can be solved independent of the other in the set. When the equations are not coupled, the theory presented in Sec. 4.2 applies. Examples of two-dimensional problems in which coupled equations arise (and are of interest to us here) are provided by the plane elastic deformation of a linear elastic solid, the flow of an incompressible viscous fluid, and the bending of elastic plates with transverse shear strains.

The primary objective of this section is twofold: first, to describe the variational formulation and finite-element formulations of the multivariable equations mentioned above; and second, to describe how the linear (Lagrange) elements developed in Sec. 4.2 can be used in the solution of multivariable equations in two dimensions. The treatment of both these subjects proceeds along the same lines as in the one-dimensional problems. (See Sec. 5.2 on alternative formulations which involve two dependent variables and two equations.)

4.5.2 Plane Elasticity

Consider a linear elastic solid Ω of uniform thickness h (say, in the z direction) bounded by two parallel planes (say, by planes $z = -h/2$ and $z = h/2$) and by any closed boundary Γ. If the thickness h is very large, the problem is considered to be a *plane strain problem*, and if the thickness is small compared with the lateral dimensions (x, y in this case), the problem is considered to be a *plane stress problem*. Both of these problems are simplifications of three-dimensional elasticity problems under the following assumptions on loading: the body forces, if any exist, cannot vary in the direction of the body thickness and cannot have components in the z direction; the applied boundary forces cannot have components in the z direction and must be uniformly distributed across the thickness (i.e., constant in the z direction); and no loads can be applied on the parallel planes bounding the top and bottom surfaces.

The assumption that the stresses are zero on the parallel planes implies that for plane stress problems (body is very thin) the stresses in the z direction are negligibly small (see Fig. 4.29),

$$\sigma_z = \tau_{yz} = \tau_{xz} = 0 \tag{4.132}$$

For plane strain problems, where the body is very thick in the z direction, the assumption is that the strains in the z direction are zero (see Fig. 4.30),

$$\varepsilon_z = \gamma_{yz} = \gamma_{xz} = 0 \tag{4.133}$$

An example which illustrates the difference between plane stress and plane strain problems is provided by the bending of a rectangular-cross-section beam. If the beam is narrow, the problem is a plane stress problem, and if the beam is very wide, the problem is a plane strain problem. (The reader is asked to consult appropriate references for details.)

The equations governing the two types of plane elasticity problems discussed above are summarized below.

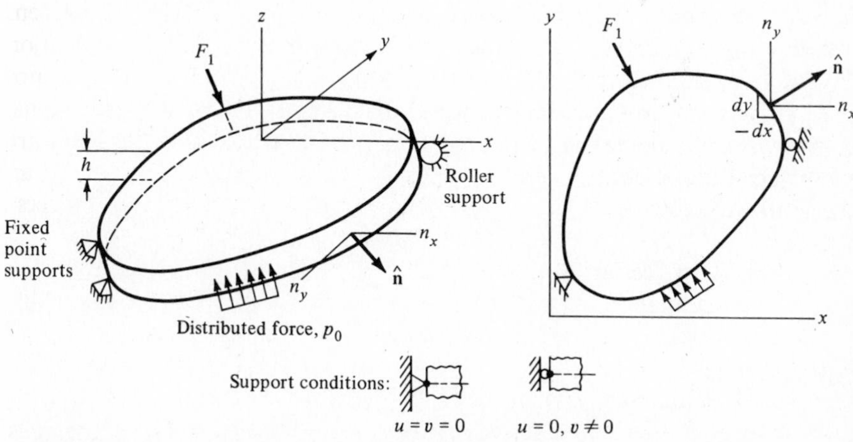

Figure 4.29 Plane stress problems in two-dimensional elasticity.

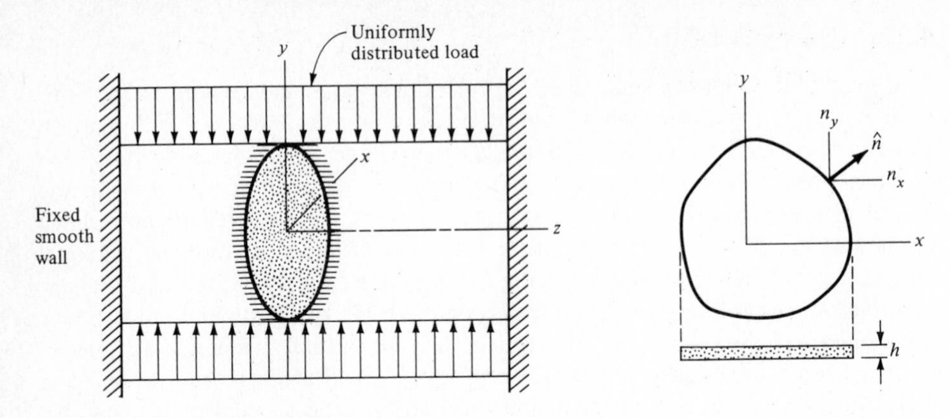

Figure 4.30 Plane strain problems in two-dimensional elasticity.

1. *Equilibrium equations in terms of stresses*

$$\left.\begin{aligned}\frac{\partial \sigma_x}{\partial x} + \frac{\partial \tau_{xy}}{\partial y} + f_x = 0 \\ \frac{\partial \tau_{xy}}{\partial x} + \frac{\partial \sigma_y}{\partial y} + f_y = 0\end{aligned}\right\} \text{ in } \Omega \tag{4.134}$$

where f_x and f_y denote the body forces along the x and y directions, respectively.

2. *Strain-displacement relations*

$$\varepsilon_x = \frac{\partial u}{\partial x} \qquad \varepsilon_y = \frac{\partial v}{\partial y} \qquad \gamma_{xy} = \frac{\partial u}{\partial y} + \frac{\partial v}{\partial x} \tag{4.135}$$

3. *Stress-strain (or constitutive) relations*

$$\begin{aligned}\sigma_x &= C_{11}\varepsilon_x + C_{12}\varepsilon_y \\ \sigma_y &= C_{12}\varepsilon_x + C_{22}\varepsilon_y \\ \tau_{xy} &= C_{33}\gamma_{xy}\end{aligned} \tag{4.136}$$

where $C_{ij}(C_{ji} = C_{ij})$ are the elasticity (material) constants. For an isotropic elastic body, they are given in terms of the modulus of elasticity E and the Poisson's ratio ν by (see Prob. 3.32)

Plane stress

$$C_{11} = C_{22} = \frac{E}{1-\nu^2} \qquad C_{12} = \frac{\nu E}{1-\nu^2} \qquad C_{33} = \frac{E}{2(1+\nu)} \tag{4.137}$$

Plane strain

$$C_{11} = C_{22} = \frac{E(1-\nu)}{(1+\nu)(1-2\nu)} \qquad C_{12} = \frac{E\nu}{(1+\nu)(1-2\nu)} \qquad C_{33} = \frac{E}{2(1+\nu)}$$

4. *Boundary conditions*

Natural

$$\left.\begin{aligned}\sigma_x n_x + \tau_{xy} n_y &= \hat{t}_x \\ \tau_{xy} n_x + \sigma_y n_y &= \hat{t}_y\end{aligned}\right\} \text{ on } \Gamma_1 \tag{4.138}$$

Essential

$$\left.\begin{aligned}u &= \hat{u} \\ v &= \hat{v}\end{aligned}\right\} \text{ on } \Gamma_2 \tag{4.139}$$

where $\hat{\mathbf{n}} = (n_x, n_y)$ denotes the unit normal to the boundary Γ, Γ_1 and Γ_2 are (disjoint) portions of the boundary (Γ_1 and Γ_2 do not overlap except for a small number of discrete points—singular points), $\hat{t}_x$ and $\hat{t}_y$ denote specified boundary (traction) forces, $\hat{u}$ and $\hat{v}$ are specified displacements, and h is the thickness of the plate.

Equations (4.134) through (4.139) can be expressed in terms of only the displacements u and v by substituting Eqs. (4.135) into Eqs. (4.136), and Eqs. (4.136) into Eqs. (4.134) and (4.138). We obtain $c_{ij} = hC_{ij}$

$$\left.\begin{aligned}-\frac{\partial}{\partial x}\left(c_{11}\frac{\partial u}{\partial x} + c_{12}\frac{\partial v}{\partial y}\right) - c_{33}\frac{\partial}{\partial y}\left(\frac{\partial u}{\partial y} + \frac{\partial v}{\partial x}\right) &= f_x \\ -c_{33}\frac{\partial}{\partial x}\left(\frac{\partial u}{\partial y} + \frac{\partial v}{\partial x}\right) - \frac{\partial}{\partial y}\left(c_{12}\frac{\partial u}{\partial x} + c_{22}\frac{\partial v}{\partial y}\right) &= f_y\end{aligned}\right\} \text{ in } \Omega \tag{4.140}$$

$$\left.\begin{aligned}t_x \equiv \left(c_{11}\frac{\partial u}{\partial x} + c_{12}\frac{\partial v}{\partial y}\right)n_x + c_{33}\left(\frac{\partial u}{\partial y} + \frac{\partial v}{\partial x}\right)n_y &= \hat{t}_x \\ t_y \equiv c_{33}\left(\frac{\partial u}{\partial y} + \frac{\partial v}{\partial x}\right)n_x + \left(c_{12}\frac{\partial u}{\partial x} + c_{22}\frac{\partial v}{\partial y}\right)n_y &= \hat{t}_y\end{aligned}\right\} \text{ on } \Gamma_1 \tag{4.141}$$

Here we will present two different ways of constructing the finite-element model of the plane elasticity equations. The first one uses the total potential energy functional expressed in terms of matrices relating displacements to strains and strains relating to stresses. This approach is used in most finite-element texts. The second approach follows a procedure consistent with the previous sections and employs the variational formulation of Eqs. (4.140) and (4.141) to construct the finite-element model. Of course, both methods result in the same final equations.

Matrix formulation of plane elasticity equations First, we rewrite Eqs. (4.134) through (4.136) in matrix form. To this end let

$$\{\varepsilon\}_e = \begin{Bmatrix}\varepsilon_x \\ \varepsilon_y \\ \gamma_{xy}\end{Bmatrix}_e \qquad \{\sigma\}_e = \begin{Bmatrix}\sigma_x \\ \sigma_y \\ \tau_{xy}\end{Bmatrix}_e \qquad \{d\}_e = \begin{Bmatrix}u \\ v\end{Bmatrix}_e \tag{4.142}$$

where the subscript e implies that they are defined on a typical finite element e.

Next, we use the total potential energy functional (not derived here) associated with the equations governing a plane elastic body:

$$\Pi(d) = \tfrac{1}{2}h_e\int_{\Omega^e}\{\varepsilon\}_e^T\{\sigma\}_e dx\,dy - \int_{\Omega^e}\{d\}_e^T\{f\}_e dx\,dy - \oint_{\Gamma^e}\{d\}_e^T\{t\}_e\,ds \tag{4.143}$$

where $\{f\}_e = \{f_x^e, f_y^e\}^T$ is the body force vector, $\{t\}_e = \{t_x^e, t_y^e\}^T$ is the traction vector, and h_e is the thickness of the element. The first term in Eq. (4.143) corresponds to the strain energy, the second term is the work done by the body forces, and the third is the work done by the surface tractions.

The strain-displacement relations (4.135), the stress-strain relations (4.136), and the stress-equilibrium relations (4.134) can be expressed in the form

$$\{\varepsilon\}_e = [D]\{d\}_e \qquad \{\sigma\}_e = [C]_e\{\varepsilon\}_e \qquad -[D]^T\{\sigma\}_e = \{f\}_e \tag{4.144}$$

where $[C] = [C_{ij}]$, and $[D]$ is the matrix of differential operators,

$$[D] = \begin{bmatrix} \dfrac{\partial}{\partial x} & 0 \\ 0 & \dfrac{\partial}{\partial y} \\ \dfrac{\partial}{\partial y} & \dfrac{\partial}{\partial x} \end{bmatrix} \tag{4.145}$$

Let $u^{(e)}$ and $v^{(e)}$ (displacements over element e) be approximated by finite-element interpolation of the form

$$u^{(e)} = \sum_{j=1}^{n} u_j^{(e)}\psi_j^{(e)} \qquad v^{(e)} = \sum_{j=1}^{n} v_j^{(e)}\psi_j^{(e)} \tag{4.146}$$

where $\psi_j^{(e)}$ are appropriate interpolation functions for the variational functional in Eqs. (4.143). We have

$$\{\varepsilon\}_e = [D]\{d\}_e = [D][\Psi]_e\{\Delta\}_e = [B]_e\{\Delta\}_e \tag{4.147}$$

where

$$[B]_e = [D][\Psi]_e \tag{4.148a}$$

and

$$[\Psi]_e = \begin{bmatrix} \psi_1^{(e)} & 0 & \psi_2^{(e)} & 0 & \psi_3^{(e)} & 0 & \psi_4^{(e)} & 0 & \cdots & \psi_n^{(e)} & 0 \\ 0 & \psi_1^{(e)} & 0 & \psi_2^{(e)} & 0 & \psi_3^{(e)} & 0 & \psi_4^{(e)} & \cdots & 0 & \psi_n^{(e)} \end{bmatrix}$$

$$\{\Delta^{(e)}\} = \{u_1^{(e)}, v_1^{(e)}, \ldots, u_n^{(e)}, v_n^{(e)}\}^T \tag{4.148b}$$

Substituting Eqs. (4.144) and (4.146) into Eq. (4.143), and setting the first variation of Π with respect to $\{\Delta\}$ equal to zero (see Sec. 2.3), we obtain

$$[K^{(e)}]\{\Delta^{(e)}\} = \{F^{(e)}\} \tag{4.149}$$

where

$$[K^{(e)}] = h_e \int_{\Omega^e} [B]_e^T [C]_e [B]_e \, dx \, dy$$

$$\{F^{(e)}\} = \int_{\Omega^e} [\Psi]_e^T \{f\}_e dx \, dy - \oint_{\Gamma^e} [\Psi]_e^T \{t\}_e \, ds \tag{4.150}$$

Since the variational functional Eq. (4.143) involves only the first derivatives of u and v with respect to x and y, one can use linear interpolation functions corresponding to the linear triangular element or rectangular element. For the three-node ($n = 3$) linear triangular element (see Fig. 4.31a), we can easily compute $[K^{(e)}]$. We have

$$\psi_{i,x}^{(e)} = \frac{\beta_i^{(e)}}{2A_e} \qquad \psi_{i,y}^{(e)} = \frac{\gamma_i^{(e)}}{2A_e} \qquad A_e = \text{element area} \tag{4.151a}$$

and

$$[B]_e = \frac{1}{2A_e} \begin{bmatrix} \beta_1^{(e)} & 0 & \beta_2^{(e)} & 0 & \beta_3^{(e)} & 0 \\ 0 & \gamma_1^{(e)} & 0 & \gamma_2^{(e)} & 0 & \gamma_3^{(e)} \\ \gamma_1^{(e)} & \beta_1^{(e)} & \gamma_2^{(e)} & \beta_2^{(e)} & \gamma_3^{(e)} & \beta_3^{(e)} \end{bmatrix} \tag{4.151b}$$

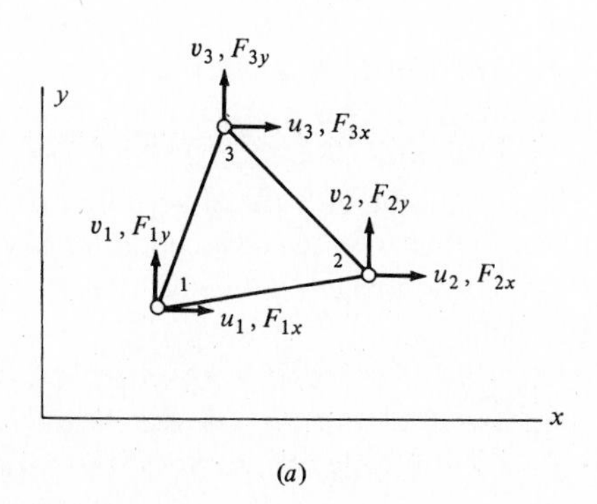

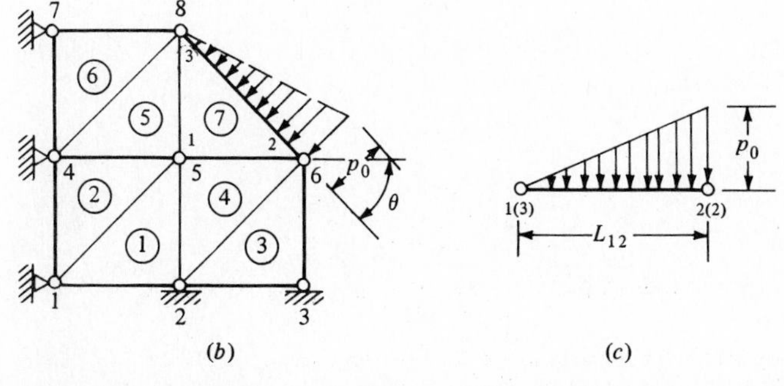

Figure 4.31 Computation of boundary force components for plane elasticity problems. (a) The constant strain triangle (CST). (b) A typical plane elasticity problem. (c) Computation of the boundary forces.

For the case in which f_x and f_y are constant (say, f_0), the first integral in $\{F^{(e)}\}$ is given by

$$\int_{\Omega^e} [\Psi]_e^T \{f\}_e \, dx \, dy = A_e \frac{f_0}{3} \{1, 1, \ldots, 1\}^T \tag{4.152}$$

If $\{t\}$ is specified on the portion Γ_1 of the boundary Γ, it can be computed for those elements which share the boundary Γ_1. For example, the structure shown in Fig. 4.31*b* is subjected to a distributed surface force. Note that the boundary element is a line element with two nodes, and therefore linear interpolation functions for one-dimensional elements can be used to compute the forces at nodes 2 and 3 of element 7, and can be resolved along the global coordinates (see Fig. 4.31*c*).

$$\int_{\Omega^{(7)}} [\Psi]_e^T \{t\}_e \, ds = P_0 L_{12} \{0, 0, \tfrac{1}{6}\cos\theta, \tfrac{1}{6}\sin\theta, \tfrac{1}{3}\cos\theta, \tfrac{1}{3}\sin\theta\} \tag{4.153}$$

Note from Eqs. (4.151*a*) that the linear triangular element yields elementwise constant strains. Therefore, the element is often called the *constant strain triangular* (CST) *element*.

Variational (displacement) formulation of plane elasticity problems Here we present an alternative formulation of plane elasticity problems. The present approach does not assume the knowledge of the quadratic functional (i.e., the total potential energy functional) in Eq. (4.143). We begin with the governing differential equations (4.140) of plane elasticity and construct the associated variational form and the finite-element model. The present approach is precisely in line with the approach we have been using in this book.

The variational form of Eqs. (4.140) over an element Ω^e is given by multiplying the first equation with a test function w_1 and the second one with a test function w_2 and integrating the results by parts [to trade the second differential to w_i, $(i = 1, 2)$]. We have

$$\begin{aligned}
0 &= \int_{\Omega^e} \left\{ \frac{\partial w_1}{\partial x}\left(c_{11}\frac{\partial u}{\partial x} + c_{12}\frac{\partial v}{\partial y}\right) + c_{33}\frac{\partial w_1}{\partial y}\left(\frac{\partial u}{\partial y} + \frac{\partial v}{\partial x}\right) - w_1 f_x \right\} dx\, dy \\
&\quad - \oint_{\Gamma^e} w_1 t_x \, ds \\
0 &= \int_{\Omega^e} \left\{ c_{33}\frac{\partial w_2}{\partial x}\left(\frac{\partial u}{\partial y} + \frac{\partial v}{\partial x}\right) + \frac{\partial w_2}{\partial y}\left(c_{12}\frac{\partial u}{\partial x} + c_{22}\frac{\partial v}{\partial y}\right) - w_2 f_y \right\} dx\, dy \\
&\quad - \oint_{\Gamma^e} w_2 t_y \, ds
\end{aligned} \tag{4.154}$$

where t_x and t_y denote the boundary forces defined in Eqs. (4.141).

Using the finite-element interpolation of the form (4.146) for u and v, and substituting $w_1 = \psi_i$ and $w_2 = \psi_i$ into Eqs. (4.154), we obtain

$$
\begin{aligned}
[K^{11}]\{u\} + [K^{12}]\{v\} &= \{F^1\} \\
[K^{21}]\{u\} + [K^{22}]\{v\} &= \{F^2\}
\end{aligned}
\tag{4.155}
$$

where

$$
\begin{aligned}
K_{ij}^{11} &= \int_{\Omega^e} \left(c_{11} \frac{\partial \psi_i}{\partial x}\frac{\partial \psi_j}{\partial x} + c_{33} \frac{\partial \psi_i}{\partial y}\frac{\partial \psi_j}{\partial y} \right) dx\,dy \\
K_{ij}^{12} = K_{ji}^{21} &= \int_{\Omega^e} \left(c_{12} \frac{\partial \psi_i}{\partial x}\frac{\partial \psi_j}{\partial y} + c_{33} \frac{\partial \psi_i}{\partial y}\frac{\partial \psi_j}{\partial x} \right) dx\,dy \\
K_{ij}^{22} &= \int_{\Omega^e} \left(c_{33} \frac{\partial \psi_i}{\partial x}\frac{\partial \psi_j}{\partial x} + c_{22} \frac{\partial \psi_i}{\partial y}\frac{\partial \psi_j}{\partial y} \right) dx\,dy \\
F_i^1 &= \int_{\Omega^e} \psi_i f_x \,dx\,dy + \oint_{\Gamma^e} \psi_i t_x \,ds \\
F_i^2 &= \int_{\Omega^e} \psi_i f_y \,dx\,dy + \oint_{\Gamma^e} \psi_i t_y \,ds
\end{aligned}
\tag{4.156}
$$

Note that coefficient matrix $[K^{12}]$, for example, corresponds to the coefficient of v in the first equation (i.e., the first superscript corresponds to the equation number and second to the variable number). By expanding expressions (4.150), one can show that they yield the same equations as Eqs. (4.155) for the linear triangular element.

For a linear rectangular element (with sides a and b), the coefficient matrices (4.156) can be computed with the help of Eqs. (4.40). With regard to the boundary conditions, the primary degrees of freedom (i.e., the displacements) or the secondary degrees of freedom (i.e., the forces) can be specified. At any node on the boundary of the domain of an elasticity problem, one has the following four distinct possibilities:

Case 1: u and v are specified (then t_x and t_y are unknown)
Case 2: u and t_y are specified (then t_x and v are unknown)
Case 3: t_x and v are specified (then u and t_y are unknown)
Case 4: t_x and t_y are specified (then u and v are unknown)
(4.157)

In general, only one of the quantities of each of the pairs (u, t_x) and (v, t_y) must be specified at a point on the boundary.

Example 4.7 Consider the linear triangular and rectangular elements shown in Fig. 4.32. We wish to evaluate $[K^{(e)}]$ and $\{F^{(e)}\}$ of Eqs. (4.150) for the element. The element interpolation functions for the triangle are given by

$$
\psi_1 = \tfrac{1}{16}(40 - 3x - 4y) \qquad \psi_2 = \tfrac{1}{16}(-16 + 4x) \qquad \psi_3 = \tfrac{1}{16}(-8 - x + 4y)
\tag{4.158}
$$

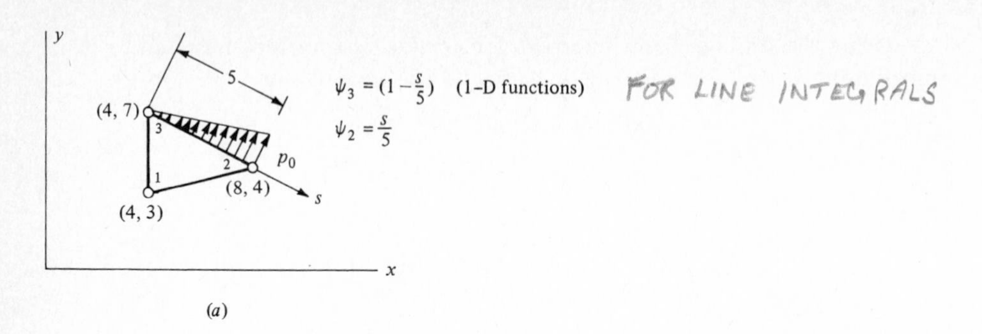

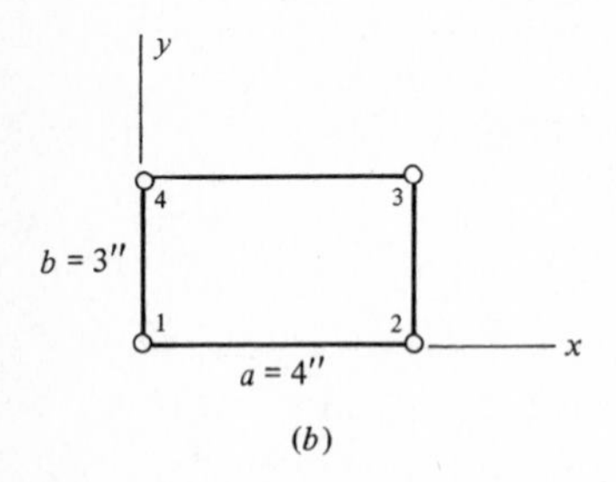

Figure 4.32 Triangular and rectangular elements considered in Example 4.7. (*a*) Triangular element. (*b*) Rectangular element.

The matrix $[B]$ of Eq. (4.148*a*) is given by

$$[B] = \begin{bmatrix} \psi_{1,x} & 0 & \psi_{2,x} & 0 & \psi_{3,x} & 0 \\ 0 & \psi_{1,y} & 0 & \psi_{2,y} & 0 & \psi_{3,y} \\ \psi_{1,y} & \psi_{1,x} & \psi_{2,y} & \psi_{2,x} & \psi_{3,y} & \psi_{3,x} \end{bmatrix}$$

$$= \frac{1}{16} \begin{bmatrix} -3 & 0 & 4 & 0 & -1 & 0 \\ 0 & -4 & 0 & 0 & 0 & 4 \\ -4 & -3 & 0 & 4 & 4 & -1 \end{bmatrix} \tag{4.159}$$

The element stiffness matrix for an orthotropic medium ($c_{13} = c_{23} = 0$) is given by

$$[K] = \frac{1}{8} \begin{bmatrix} \frac{9}{4}c_{11} + 4c_{33} & 3(c_{12} + c_{33}) & -3c_{11} & -4c_{33} & \frac{3}{4}c_{11} - 4c_{33} & c_{33} - 3c_{12} \\ & 4c_{22} + \frac{9}{4}c_{33} & -4c_{12} & -3c_{33} & c_{12} - 3c_{33} & \frac{3}{4}c_{33} - 4c_{22} \\ & & 4c_{11} & 0 & -c_{11} & 4c_{12} \\ \text{Symmetric} & & & 4c_{33} & 4c_{33} & -c_{33} \\ & & & & 4c_{33} + \frac{1}{4}c_{11} & -(c_{12} + c_{33}) \\ & & & & & 4c_{22} + \frac{1}{4}c_{33} \end{bmatrix} \tag{4.160}$$

The force vector $\{F^{(e)}\}$ has contributions only from the second (boundary) integral. Sides 1–2 and 3–1 are interelement boundaries, and side 2–3 is the boundary of the problem on which the normal stress t_n is specified, as shown in

Fig. 4.32*a*. First, we compute the boundary integral

$$f_i = \int_0^5 \left(p_0 \frac{s}{5} \right) \psi_i(s)\, ds \qquad (i = 3, 2) \tag{4.161}$$

$$f_3 = \frac{5p_0}{6} \qquad f_2 = \frac{10p_0}{6}$$

Next, we compute the x and y components of f_i at nodes 2 and 3. We have

$$\hat{F}_{2x} = f_2(0.8) = \tfrac{4}{3}p_0 \qquad \hat{F}_{2y} = f_2(0.6) = p_0$$

$$\hat{F}_{3x} = f_3(0.8) = \tfrac{2}{3}p_0 \qquad \hat{F}_{3y} = f_3(0.6) = \tfrac{1}{2}p_0$$

and

$$\{F^{(e)}\} = \begin{Bmatrix} F_{1x} \\ F_{1y} \\ \hat{F}_{2x} + \tilde{F}_{2x} \\ \hat{F}_{2y} \\ \hat{F}_{3x} + \tilde{F}_{3x} \\ \hat{F}_{3y} \end{Bmatrix} \tag{4.162}$$

where $\tilde{F}_{2x}$ and $\tilde{F}_{3x}$ are internal force components from sides 1–2 and 3–1.

For an isotropic (steel) plate, we take

$$E_1 = E_2 = E = 30 \times 10^6 \text{ lb/in}^2 \qquad \nu_{12} = \nu_{21} = \nu = 0.3 \tag{4.163}$$

so that

$$\begin{aligned} c_{11} &= \frac{E_1 h}{1 - \nu_{12}\nu_{21}} = 32.967h \times 10^6 \qquad c_{12} = c_{11}\nu_{12} = 9.89h \times 10^6 \\ c_{22} &= \frac{E_2 h}{1 - \nu_{12}\nu_{21}} = 32.967h \times 10^6 \qquad c_{33} = G_{12} = 11.538h \times 10^6 \end{aligned} \tag{4.164}$$

The element stiffness matrix (4.160) becomes

$$[K^{(e)}] = 10^6 h \begin{bmatrix} 1.5041 & 0.8036 & -1.2363 & -0.5769 & -0.2679 & -0.2266 \\ & 1.9729 & -0.4945 & -0.4327 & -0.3091 & -1.5402 \\ & & 1.6484 & 0.0 & -0.4121 & 0.4945 \\ \text{Symmetric} & & & 0.5769 & 0.5769 & -0.1442 \\ & & & & 0.6799 & -0.2679 \\ & & & & & 1.6844 \end{bmatrix} \tag{4.165}$$

Now we wish to compute the element stiffness matrix $[K^{(e)}]$ of Eqs. (4.155) for the linear rectangular element shown in Fig. 4.32*b*. We have $a = 4$ in and $b = 3$ in, and c_{ij} are given by Eqs. (4.164). Using Eqs. (4.40) and (4.155), we obtain

$$[K^{(e)}] = 10^7 h \begin{bmatrix} 1.3370 & 0.5357 & -0.5678 & -0.0412 & -0.6685 & -0.5357 & -1.0073 & 0.0412 \\ & 1.7537 & 0.0412 & 0.4441 & -0.5357 & -0.8768 & -0.0412 & -1.3210 \\ & & 1.3370 & -0.5357 & -0.1007 & -0.0412 & -0.6685 & 0.5357 \\ & & & 1.7537 & 0.0412 & -1.3210 & 0.5357 & -0.8768 \\ & \text{Symmetric} & & & 1.3370 & 0.5357 & -0.5678 & -0.0412 \\ & & & & & 1.7537 & 0.0412 & 0.4441 \\ & & & & & & 1.3370 & -0.5357 \\ & & & & & & & 1.7537 \end{bmatrix} \tag{4.166}$$

■

Example 4.8 Consider a thin elastic plate subjected to uniformly distributed edge load, as shown in Fig. 4.33*a*. First we consider a two-element discretization of the plate by triangular elements, and then we perform all the algebra to obtain the nodal displacements.

Element matrices Assuming that the plate is made of an isotropic material, and using the plane stress assumption, we compute the element stiffness matrices for each of the two elements (see Fig. 4.33*b*, *c*).

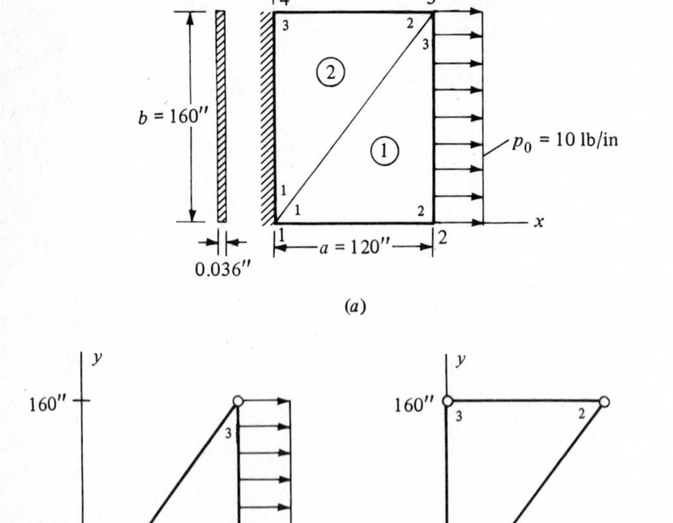

Figure 4.33 Geometry and finite element mesh of a plane elasticity problem by the CST elements. (*a*) A plane elasticity problem. (*b*) Element 1. (*c*) Element 2.

Element 1

$$\begin{aligned} &x_1 = y_1 = y_2 = 0 \qquad x_2 = x_3 = a \qquad y_3 = b \\ &\beta_1 = -b \qquad \beta_2 = b \qquad \beta_3 = 0 \\ &\gamma_1 = 0 \qquad \gamma_2 = -a \qquad \gamma_3 = a \end{aligned} \tag{4.167}$$

$$\{F^{(1)}\} = \{F_{1x}^{(1)}, F_{1y}^{(1)}, F_{2x}^{(1)}, F_{2y}^{(1)}, F_{3x}^{(1)}, F_{3y}^{(1)}\}^T \tag{4.168}$$

$$[K^{(1)}] = C_1 \begin{bmatrix} b^2 & & & & & \\ 0 & \frac{1-\nu}{2}b^2 & & & & \\ -b^2 & \frac{1-\nu}{2}ab & b^2 + \frac{1-\nu}{2}a^2 & & \text{Symmetric} & \\ \nu ab & -\frac{1-\nu}{2}b^2 & -\frac{1+\nu}{2}ab & \frac{1-\nu}{2}b^2 + a^2 & & \\ 0 & -\frac{1-\nu}{2}ab & -\frac{1-\nu}{2}a^2 & \frac{1-\nu}{2}ab & \frac{1-\nu}{2}a^2 & \\ -\nu ab & 0 & \nu ab & -a^2 & 0 & a^2 \end{bmatrix}$$

$$C_1 = \frac{Eh}{2ab(1-\nu^2)} \tag{4.169}$$

Element 2

$$\begin{aligned} &x_1 = y_1 = x_3 = 0 \qquad x_2 = a \qquad y_2 = y_3 = b \\ &\beta_1 = 0 \qquad \beta_2 = b \qquad \beta_3 = -b \\ &\gamma_1 = -a \qquad \gamma_2 = 0 \qquad \gamma_3 = a \end{aligned} \tag{4.170}$$

$$\{F^{(2)}\} = \{F_{1x}^{(2)}, F_{1y}^{(2)}, F_{2x}^{(2)}, F_{2y}^{(2)}, F_{3x}^{(2)}, F_{3y}^{(2)}\}^T \tag{4.171}$$

$$[K^{(2)}] = C_2 \begin{bmatrix} \frac{1-\nu}{2}a^2 & & & & & \\ 0 & a^2 & & & \text{Symmetric} & \\ 0 & -\nu ab & b^2 & & & \\ -\frac{1-\nu}{2}ab & 0 & 0 & \frac{1-\nu}{2}b^2 & & \\ -\frac{1-\nu}{2}a^2 & \nu ab & -b^2 & \frac{1-\nu}{2}ab & b^2 + \frac{1-\nu}{2}a^2 & \\ \frac{1-\nu}{2}ab & -a^2 & \nu ab & -\frac{1-\nu}{2}b^2 & -\frac{1-\nu}{2}ab & a^2 + \frac{1-\nu}{2}b^2 \end{bmatrix}$$

$$C_2 = \frac{Eh}{2ab(1-\nu^2)} \tag{4.172}$$

Clearly (from the inspection of the element nodes of the two elements), $[K^{(2)}]$ can be obtained from $[K^{(1)}]$ by moving the coefficients of nodes 1, 2, and 3 to 2, 3, and 1, respectively. For example, we have

$$K_{\alpha\beta}^{(1)} = K_{pq}^{(2)} \quad \begin{cases} \text{if } \alpha = 2i - 1 \text{ and } \beta = 2k - 1, \text{ then } p = 2j - 1 \text{ and } q = 2l - 1 \\ \text{if } \alpha = 2i \text{ and } \beta = 2k, \text{ then } p = 2j \text{ and } q = 2l \end{cases} \tag{4.173}$$

where i, $k = 1, 2, 3$, and i and j as well as k and l permute in a natural order (i.e., if $i = 1$, then $j = 2$, and if $i = 3$, then $j = 1$, etc.). In particular, we have

$$K_{11}^{(1)} = K_{33}^{(2)} \qquad K_{21}^{(1)} = K_{43}^{(2)}$$

and so on.

Assembly The assembly of element matrices for two-degree-of-freedom (DOF) elements is similar to that described in Sec. 4.2.6. For the finite-element mesh at hand, the correspondence between the global and local nodes and stiffness is given below:

Nodal correspondence		Stiffness correspondence	
Global (DOF)	Local (DOF)	Global	Local
	1 of element 1(1, 2)	K_{11}	$K_{11}^{(1)} + K_{11}^{(2)}$
1(1, 2)		K_{22}	$K_{22}^{(1)} + K_{22}^{(2)}$
	1 of element 2(1, 2)	K_{12}	$K_{12}^{(1)} + K_{12}^{(2)}$
		K_{33}	$K_{33}^{(1)}$
2(3, 4)	2 of element 1(3, 4)	K_{44}	$K_{44}^{(1)}$
		K_{34}	$K_{34}^{(1)}$
	3 of element 1(5, 6)	K_{55}	$K_{55}^{(1)} + K_{33}^{(2)}$
3(5, 6)	2 of element 2(3, 4)	K_{66}	$K_{66}^{(1)} + K_{44}^{(2)}$
		K_{56}	$K_{56}^{(1)} + K_{34}^{(2)}$
		K_{77}	$K_{55}^{(2)}$
4(7, 8)	3 of element 2(5, 6)	K_{88}	$K_{66}^{(2)}$
		K_{78}	$K_{56}^{(2)}$

(4.174)

If two global nodes correspond to two (local) nodes of the same element, then the corresponding stiffness coefficient is nonzero; otherwise it is zero. The assembled matrix is given by

$$[K] = \begin{bmatrix} K_{11}^{(1)} + K_{11}^{(2)} & K_{12}^{(1)} + K_{12}^{(2)} & K_{13}^{(1)} & K_{14}^{(1)} & K_{15}^{(1)} + K_{13}^{(2)} & K_{16}^{(1)} + K_{14}^{(2)} & K_{15}^{(2)} & K_{16}^{(2)} \\ & K_{22}^{(1)} + K_{22}^{(2)} & K_{23}^{(1)} & K_{24}^{(1)} & K_{25}^{(1)} + K_{23}^{(2)} & K_{26}^{(1)} + K_{24}^{(2)} & K_{25}^{(2)} & K_{26}^{(2)} \\ & & K_{33}^{(1)} & K_{34}^{(1)} & K_{35}^{(1)} & K_{36}^{(1)} & 0 & 0 \\ & & & K_{44}^{(1)} & K_{45}^{(1)} & K_{46}^{(1)} & 0 & 0 \\ \text{Symmetric} & & & & K_{55}^{(1)} + K_{33}^{(2)} & K_{56}^{(1)} + K_{34}^{(2)} & K_{35}^{(2)} & K_{36}^{(2)} \\ & & & & & K_{66}^{(1)} + K_{44}^{(2)} & K_{45}^{(2)} & K_{46}^{(2)} \\ & & & & & & K_{55}^{(2)} & K_{56}^{(2)} \\ & & & & & & & K_{66}^{(2)} \end{bmatrix}$$

(4.175)

The assembled force vector is

$$\{F\} = \begin{Bmatrix} F_1^{(1)} + F_1^{(2)} \\ F_2^{(1)} + F_2^{(2)} \\ F_3^{(1)} \\ F_4^{(1)} \\ F_5^{(1)} + F_3^{(2)} \\ F_6^{(1)} + F_4^{(2)} \\ F_5^{(2)} \\ F_6^{(2)} \end{Bmatrix} \tag{4.176}$$

where $F_1 = F_{1x}$, $F_2 = F_{1y}$, $F_3 = F_{2x}$, $F_4 = F_{2y}$, etc.

Boundary conditions The specified (primary) degrees of freedom are

$$U_1 = V_1 = U_4 = V_4 = 0 \tag{4.177}$$

The known (secondary degrees of freedom) forces are already included in $\{F^{(1)}\}$ and $\{F^{(2)}\}$. The equilibrium of forces requires that

$$F_3^{(1)} = \frac{p_0 b}{2} \qquad F_4^{(1)} = 0, \qquad F_5^{(1)} + F_3^{(2)} = \frac{p_0 b}{2} \qquad F_6^{(1)} + F_4^{(2)} = 0 \tag{4.178}$$

Solution of equations The first two rows and columns and the last two rows and columns of the matrix (4.175) can be deleted (since the specified boundary conditions are homogeneous) to obtain the condensed form of the matrix equation

$$\begin{bmatrix} K_{33}^{(1)} & K_{34}^{(1)} & K_{35}^{(1)} & K_{36}^{(1)} \\ K_{43}^{(1)} & K_{44}^{(1)} & K_{45}^{(1)} & K_{46}^{(1)} \\ K_{53}^{(1)} & K_{54}^{(1)} & K_{55}^{(1)} + K_{33}^{(2)} & K_{56}^{(1)} + K_{34}^{(2)} \\ K_{63}^{(1)} & K_{64}^{(1)} & K_{65}^{(1)} + K_{43}^{(2)} & K_{66}^{(1)} + K_{44}^{(2)} \end{bmatrix} \begin{Bmatrix} U_2 \\ V_2 \\ U_3 \\ V_3 \end{Bmatrix} = \begin{Bmatrix} \dfrac{p_0 b}{2} \\ 0 \\ \dfrac{p_0 b}{2} \\ 0 \end{Bmatrix} \tag{4.179}$$

or (using $a = 120$ in, $b = 160$ in, $t = 0.036$ in, $\nu = 0.25$, $E = 30 \times 10^6$ lb/in^2, and $p_0 = 10$ lb/in)

$$10^4 \begin{bmatrix} 93.0 & -36.0 & -16.2 & 14.4 \\ -36.0 & 72.0 & 21.6 & -43.2 \\ -16.2 & 21.6 & 93.0 & 0.0 \\ 14.4 & -43.2 & 0.0 & 72.0 \end{bmatrix} \begin{Bmatrix} U_2 \\ V_2 \\ U_3 \\ V_3 \end{Bmatrix} = \begin{Bmatrix} 800.0 \\ 0.0 \\ 800.0 \\ 0.0 \end{Bmatrix} \tag{4.180}$$

Table 4.6 Finite-element results for a thin plate (plane stress assumption) using various meshes of triangular and rectangular elements and material properties.

Mesh	Material	U_2 $(\times 10^{-4})$	V_2 $(\times 10^{-4})$	U_3 $(\times 10^{-4})$	V_3 $(\times 10^{-4})$
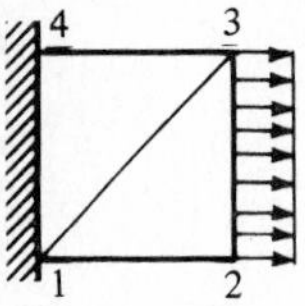	Isotropic: $E = 30 \times 10^6$ lb/in^2 $\nu = 0.25$ $G = E/2(1+\nu)$	11.291 10.853	1.9637 2.3256	10.113 10.853	$-1.08^{\dagger}$ -2.3256
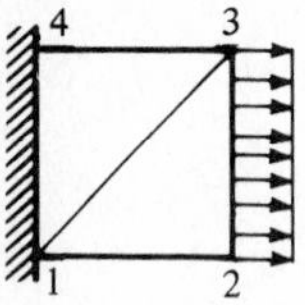	Orthotropic: $E_1 = 31 \times 10^6$ lb/in^2 $E_2 = 2.7 \times 10^6$ lb/in^2 $G_{12} = 0.75 \times 10^6$ lb/in^2 $\nu_{12} = 0.28$	10.767 10.728	1.6662 2.6758	10.650 10.728	-1.579 -2.6758
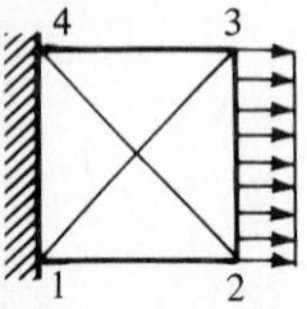	Orthotropic: same as above	10.821 10.778	2.157 2.002	10.821 10.778	-2.157 -2.002

† First row corresponds to triangular elements and the second to one rectangular element.

Inverting the matrix, we obtain

$$\begin{Bmatrix} U_2 \\ V_2 \\ U_3 \\ V_3 \end{Bmatrix} = \frac{10^{-6}}{3} \begin{bmatrix} 4.07 & 2.34 & 0.17 & 0.59 \\ 2.34 & 8.65 & -1.6 & 4.72 \\ 0.17 & -1.6 & 3.63 & -0.99 \\ 0.59 & 4.72 & -0.99 & 6.88 \end{bmatrix} \begin{Bmatrix} 800.0 \\ 0.0 \\ 800.0 \\ 0.0 \end{Bmatrix}$$

$$= 10^{-4} \begin{Bmatrix} 11.28 \\ 1.97 \\ 10.10 \\ -1.09 \end{Bmatrix} \text{in} \tag{4.181}$$

Note that the solution is not symmetric (i.e., $U_2 \neq U_3, V_2 \neq -V_3$) about the horizontal centerline. (The exact solution should be symmetric about the centerline.) This is due to the asymmetry in the finite-element discretization.

Table 4.6 contains the finite-element solution (obtained using a computer) for the displacements at the points (120, 0) and (120, 160) of isotropic plate and orthotropic plates. ■

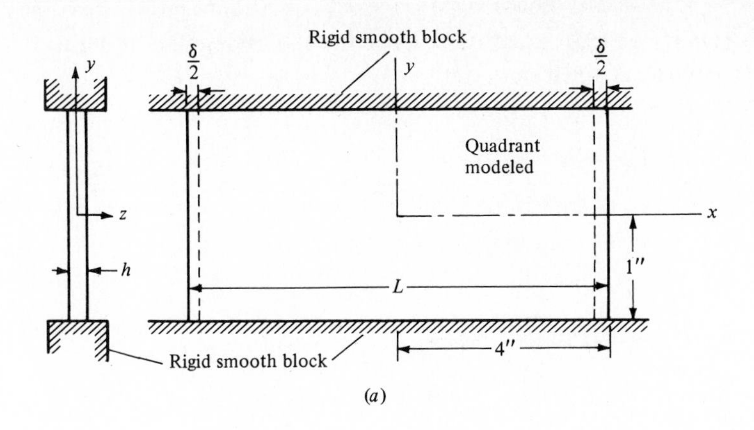

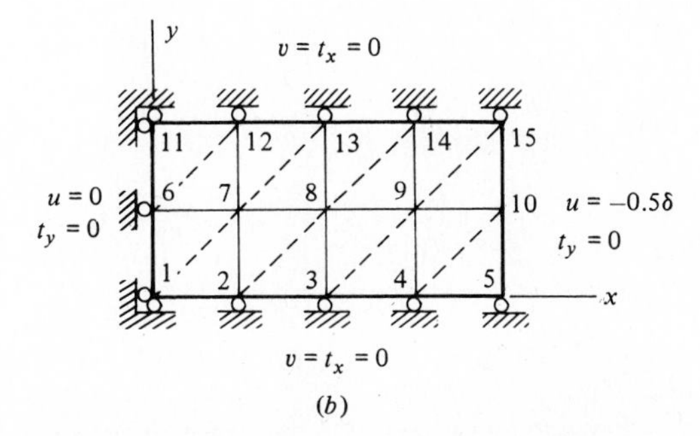

Figure 4.34 Geometry and finite-element mesh of a plane elasticity problem. (*a*) Plane elasticity problem ($\delta = 0.2$). (*b*) Finite-element mesh and boundary conditions for a quadrant of the domain.

Example 4.9 Consider a thin elastic plate held between two smooth rigid walls and subjected to uniform compression δ, as shown in Fig. 4.34*a*. We wish to determine the displacements and stresses in the plate. Due to the biaxial symmetry present in the problem, we model only a quarter of the plate. The boundary conditions associated with the quadrant are shown in Fig. 4.34*b*. Figure 4.34*b* also contains a 4×2 mesh of both triangular and rectangular elements.

We will not discuss the assembly of the element matrices, it being a routine task by now. Let U_i and V_i denote the global degrees of freedom at the global node i. The specified primary degrees of freedom are

$$\begin{aligned} U_1 = V_1 = V_2 = V_3 = V_4 = V_5 = U_6 = U_{11} = 0 \\ V_{11} = V_{12} = V_{13} = V_{14} = V_{15} = 0 \\ U_5 = U_{10} = U_{15} = -0.5\delta = -0.1 \end{aligned} \tag{4.182}$$

The forces associated with the unspecified boundary (primary) degrees of freedom are zero. We have

$$F_{2x} = F_{3x} = F_{4x} = F_{6y} = F_{10y} = F_{12x} = F_{13x} = F_{14x} = 0 \tag{4.183}$$

The solution of the resulting equations (for both the triangular and the rectangular element meshes) is given by

$$\begin{aligned}
U_2 &= U_7 = U_{12} = -0.025 \\
U_3 &= U_8 = U_{13} = -0.050 \\
U_4 &= U_9 = U_{14} = -0.075 \\
U_5 &= U_{10} = U_{15} = -0.100 \\
V_i &= 0 \qquad \text{for all } i
\end{aligned} \tag{4.184}$$

which coincides with the exact solution ($\delta = 0.2$, $L = 8$)

$$u = \frac{x\delta}{L} \qquad v = 0 \tag{4.185}$$

The stresses are (for $\nu = 0.3$)

$$\sigma_x^{(e)} = -0.02747\,E \qquad \sigma_y^{(e)} = -0.008242E \qquad \tau_{xy}^{(e)} = 0 \tag{4.186}$$

for all elements $e = 1, 2, \ldots$. The exact solution for the stresses is given by

$$\sigma_x = -\frac{E\delta}{(1-\nu^2)L} \qquad \sigma_y = -\frac{\nu E\delta}{(1-\nu^2)L} \qquad \tau_{xy} = 0 \tag{4.187}$$

■

4.5.3 Incompressible Fluid Flow

Consider the steady flow of a viscous incompressible fluid in a closed domain Ω. Assume that one of the dimensions (say, along the z direction) of the domain is very long and there is no flow along that direction, and the velocity components in the other two directions do not vary with the z direction. Under these conditions, the flow can be approximated by a two-dimensional model. The governing equations are given by

Conservation of linear momentum:

$$\left.\begin{aligned}
-\frac{\partial}{\partial x}\left(2\mu\frac{\partial u}{\partial x}\right) - \frac{\partial}{\partial y}\left[\mu\left(\frac{\partial u}{\partial y} + \frac{\partial v}{\partial x}\right)\right] + \frac{\partial P}{\partial x} = f_x \\
-\frac{\partial}{\partial x}\left[\mu\left(\frac{\partial u}{\partial y} + \frac{\partial v}{\partial x}\right)\right] - \frac{\partial}{\partial y}\left(2\mu\frac{\partial v}{\partial y}\right) + \frac{\partial P}{\partial y} = f_y
\end{aligned}\right\} \text{ in } \Omega \tag{4.188}$$

Conservation of mass:

$$\frac{\partial u}{\partial x} + \frac{\partial v}{\partial y} = 0 \text{ in } \Omega \tag{4.189}$$

where u and v are the velocity components along the x and y directions, respectively, P is the pressure, f_x and f_y are the body forces (if any) along the x and y directions, and μ is the viscosity of the fluid. The boundary conditions are given by (see Example 2.5)

$$\left.\begin{aligned} t_x &\equiv 2\mu\frac{\partial u}{\partial x}n_x + \mu\left(\frac{\partial u}{\partial y} + \frac{\partial v}{\partial x}\right)n_y - Pn_x = \hat{t}_x \\ t_y &\equiv \mu\left(\frac{\partial u}{\partial y} + \frac{\partial v}{\partial x}\right)n_x + 2\mu\frac{\partial v}{\partial y}n_y - Pn_y = \hat{t}_y \end{aligned}\right\} \text{ on } \Gamma_1 \tag{4.190}$$

$$u = \hat{u} \qquad v = \hat{v} \text{ on } \Gamma_2 \tag{4.191}$$

The same four possibilities as given in cases (4.157) exist for the specification of the boundary conditions in fluid flow problems.

Pressure-velocity finite-element model The variational formulation of Eqs. (4.188) through (4.190) over an element Ω^e is given by

$$0 = \int_{\Omega^e}\left[2\mu\frac{\partial w_1}{\partial x}\frac{\partial u}{\partial x} + \mu\frac{\partial w_1}{\partial y}\left(\frac{\partial u}{\partial y} + \frac{\partial v}{\partial x}\right) - \frac{\partial w_1}{\partial x}P - w_1 f_x\right]dx\,dy - \oint_{\Gamma^e} w_1 t_x\,ds \tag{4.192a}$$

$$0 = \int_{\Omega^e}\left[\mu\frac{\partial w_2}{\partial x}\left(\frac{\partial u}{\partial y} + \frac{\partial v}{\partial x}\right) + 2\mu\frac{\partial w_2}{\partial y}\frac{\partial v}{\partial y} - \frac{\partial w_2}{\partial y}P - w_2 f_y\right]dx\,dy - \oint_{\Gamma^e} w_2 t_y\,ds \tag{4.192b}$$

$$0 = \int_{\Omega^e} w_3\left(\frac{\partial u}{\partial x} + \frac{\partial v}{\partial y}\right)dx\,dy \tag{4.193}$$

where w_1 (variation in u), w_2 (variation in v), and w_3 (variation in P) are appropriate test functions.

Note that terms involving the pressure (in the equations of linear momentum) are also integrated by parts. By trading the differentiation from P onto w_1 and w_2 we gained both physically meaningful natural boundary conditions (which involve P) and the symmetry of the finite-element equations (as we shall see shortly). Although pressure is a primary variable, its specification does not constitute an essential boundary condition because w_3 (variation of P) does not appear in the boundary integrals of the formulation; however, P appears in t_x and t_y. Also note that pressure appears undifferentiated in the momentum equations. Therefore, the weight function w_3 can be a constant (also P), and w_1 and w_2 should be bilinear functions of x and y (so also u and v). In other words, one can use linear interpolation for u and v and a constant for P. Alternatively, a quadratic approximation for u and v as well as a linear approximation for P can be used (see Fig. 4.35). From the governing (momentum) equations it is clear that P is one order less differentiable than both u and v. Hence, the interpolation used for P should be one degree less than that used for u and v (for consistency of approximation).

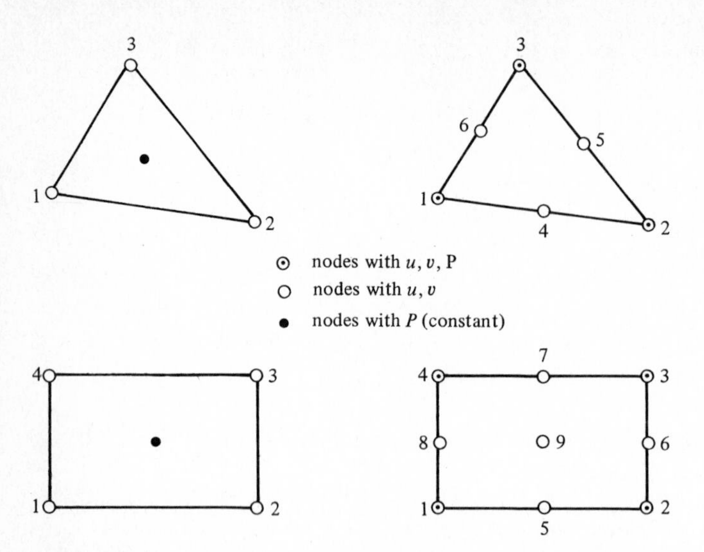

Figure 4.35 Linear and quadratic triangular and rectangular finite elements for the pressure-velocity model of incompressible fluid flow.

Over a typical element Ω_e, we approximate u, v, and P by interpolation of the form

$$u = \sum_{i=1}^{r} u_i\psi_i \qquad v = \sum_{i=1}^{r} v_i\psi_i \qquad P = \sum_{i=1}^{s} P_i\phi_i \tag{4.194}$$

where ψ_i and ϕ_i are interpolation functions of degrees r and s $(r > s)$, respectively. Substituting Eqs. (4.194) into Eqs. (4.192) and (4.193), we obtain

$$\begin{bmatrix} [K^{11}] & [K^{12}] & [K^{13}] \\ & [K^{22}] & [K^{23}] \\ \text{Symmetric} & & [K^{33}] \end{bmatrix} \begin{Bmatrix} \{u\} \\ \{v\} \\ \{P\} \end{Bmatrix} = \begin{Bmatrix} \{F^1\} \\ \{F^2\} \\ \{0\} \end{Bmatrix} \tag{4.195}$$

where

$$[K^{11}] = 2\mu[S^{11}] + \mu[S^{22}] \qquad [K^{12}] = \mu[S^{12}]^T$$

$$[K^{22}] = \mu[S^{11}] + 2\mu[S^{22}] \qquad [K^{33}] = [0] \text{ (zero matrix)}$$

$$S_{ij}^{11} = \int_{\Omega^e} \frac{\partial\psi_i}{\partial x}\frac{\partial\psi_j}{\partial x}\,dx\,dy \qquad S_{ij}^{12} = \int_{\Omega^e} \frac{\partial\psi_i}{\partial x}\frac{\partial\psi_j}{\partial y}\,dx\,dy$$

$$S_{ij}^{22} = \int_{\Omega^e} \frac{\partial\psi_i}{\partial y}\frac{\partial\psi_j}{\partial y}\,dx\,dy \tag{4.196}$$

$$F_i^1 = \int_{\Omega^e} f_x\psi_i\,dx\,dy + \oint_{\Gamma^e} t_x\psi_i\,ds \qquad F_i^2 = \int_{\Omega^e} f_y\psi_i\,dx\,dy + \oint_{\Gamma^e} t_y\psi_i\,ds$$

$$K_{iJ}^{13} = \int_{\Omega^e} \frac{\partial\psi_i}{\partial x}\phi_J\,dx\,dy \qquad K_{iJ}^{23} = \int_{\Omega^e} \frac{\partial\psi_i}{\partial y}\phi_J\,dx\,dy$$

$$i, j = 1, 2, \ldots, r \qquad J = 1, 2, \ldots, s$$

Note that the three matrix equations in Eqs. (4.195) correspond, respectively, to the three governing equations in Eqs. (4.192) and (4.193). The finite-element model in Eq. (4.195) is referred to as the primitive-variable model, the *pressure-velocity model*, or the *mixed model.*

It should be pointed out that the finite-element model in Eq. (4.195) is nonpositive definite because of zeros appearing on the main diagonal. Therefore, equation solvers that employ "pivoting" must be used in the present case. Also, the fact that pressure does not appear as a degree of freedom at every node (see Fig. 4.35) makes the computer implementation of the assembly more complicated. To circumvent this situation, we present an alternative, but approximate, finite-element formulation of Eqs. (4.188) through (4.190).

Penalty finite-element model The penalty function formulation of Eqs. (4.188) and (4.189) involves treating the continuity equation (4.189) as a constraint among the velocity components. To formulate the problem by the penalty method, we first note that the quadratic form corresponding to the variational equations (4.192) and (4.193) is given by

$$I(u, v, P) = \int_{\Omega^e} \left\{ \mu \left[\left(\frac{\partial u}{\partial x} \right)^2 + \left(\frac{\partial v}{\partial y} \right)^2 + \frac{1}{2} \left(\frac{\partial u}{\partial y} + \frac{\partial v}{\partial x} \right)^2 \right] \right.$$
$$\left. \underline{-P\left(\frac{\partial u}{\partial x} + \frac{\partial v}{\partial y} \right)} - f_x u - f_y v \right\} dx\, dy - \oint_{\Gamma^e} (t_x u + t_y v)\, ds \tag{4.197}$$

The reader is asked to verify this result. Now suppose that the velocity components satisfy Eq. (4.189). Then the pressure term in Eq. (4.197) drops out. Let $I_0(u, v)$ denote the quadratic form that results from $I(u, v, P)$ when the underlined term is dropped. Now we can view the problem of solving Eqs. (4.192) and (4.193) as one of minimizing $I_0(u, v)$ subject to the constraint in Eq. (4.189). Then the penalty function method allows us to include the constraint back into the quadratic form in a least-squares approximation sense: Minimize the modified functional

$$I_P(u, v) = I_0(u, v) + \frac{\gamma_e}{2} \int_{\Omega^e} \left(\frac{\partial u}{\partial x} + \frac{\partial v}{\partial y} \right)^2 dx\, dy \tag{4.198}$$

This is equivalent to solving the following variational (or weak) equation (obtained by taking the first variation of I_P and setting it to zero):

$$0 = \int_{\Omega^e} \left\{ 2\mu \left(\frac{\partial w_1}{\partial x} \frac{\partial u}{\partial x} + \frac{\partial w_2}{\partial y} \frac{\partial v}{\partial y} \right) + \mu \left(\frac{\partial w_1}{\partial y} + \frac{\partial w_2}{\partial x} \right) \left(\frac{\partial u}{\partial y} + \frac{\partial v}{\partial x} \right) \right.$$
$$\left. - w_1 f_x - w_2 f_y \right\} dx\, dy - \oint_{\Gamma^e} (w_1 t_x + w_2 t_y)\, ds$$
$$+ \gamma_e \int_{\Omega^e} \left(\frac{\partial w_1}{\partial x} + \frac{\partial w_2}{\partial y} \right) \left(\frac{\partial u}{\partial x} + \frac{\partial v}{\partial y} \right) dx\, dy \tag{4.199}$$

where γ_e denotes the penalty parameter, whose value is preselected, subjected to certain guidelines to be discussed shortly.

The finite-element model associated with the penalty function formulation (4.199) is given by substituting Eqs. (4.194) for u and v:

$$\begin{bmatrix} [\bar{K}^{11}] & [\bar{K}^{12}] \\ [\bar{K}^{12}]^T & [\bar{K}^{22}] \end{bmatrix} \begin{Bmatrix} \{u\} \\ \{v\} \end{Bmatrix} = \begin{Bmatrix} \{F^1\} \\ \{F^2\} \end{Bmatrix} \tag{4.200}$$

where F_i^1 and F_i^2 are as defined in Eqs. (4.196), and

$$\begin{aligned} &[\bar{K}^{11}] = [K^{11}] + \gamma[S^{11}] \qquad [\bar{K}^{12}] = [K^{12}] + \gamma[S^{12}] \\ &[\bar{K}^{22}] = [K^{22}] + \gamma[S^{22}] \end{aligned} \tag{4.201}$$

$K_{ij}^{\alpha\beta}$, $S_{ij}^{\alpha\beta}$ ($\alpha, \beta = 1, 2$) being the element matrices defined in Eqs. (4.196). It is recommended that the value of the penalty parameter be chosen such that

$$\gamma_e = \mu \times 10^6 \leqslant 10^{13} \tag{4.202}$$

The upper limit depends on the computer used. Too large a value for γ will outweigh the other terms in Eqs. (4.201), and the round-off errors in the computation (which depend on the word length in the computer used) could lead to the trivial solution (i.e., $u = v = 0$ everywhere), which certainly satisfies the constraint. To see this, note that the element equation (4.200) (consequently, the assembled equation) is of the form

$$[\bar{K}]\{\Delta\} \equiv [\mu K^1 + \gamma K^2]\{\Delta\} = \{F\} \qquad \{\Delta\} = \begin{Bmatrix} \{u\} \\ \{v\} \end{Bmatrix} \tag{4.203}$$

where $[K^1]$ is the coefficient matrix resulting from the momentum equations (4.188) [equivalently, from the functional in Eq. (4.197)] and $[K^2]$ is the coefficient matrix resulting from the penalty functional (i.e., the last integral) in Eq. (4.199). We have

$$\lim_{\gamma\to\infty} \left[\frac{\mu}{\gamma} K^1 + K^2\right]\{\Delta\} = \lim_{\gamma\to\infty} \left\{\frac{1}{\gamma} F\right\} = 0 \tag{4.204}$$

Thus, for a very large value of γ, one obtains $\{\Delta\} = \{0\}$. On the other hand, for a small value of γ the constraint is not satisfied accurately. Even when γ is sufficiently large [as given in Eq. (4.202)], the penalty terms dominate the nonpenalty terms and yield the trivial solution. To circumvent this, we should make $[K^2]$ singular so that $[K^2]$ cannot be invertible but the total $[K]$ is invertible (because $[K^1]$ is nonsingular). One way to ensure the singularity of $[K^2]$ is to use the so-called reduced integration in the numerical evaluation of the integrals in $[K^2]$. More specifically, when the linear rectangular element is used, one should use the standard 2×2 Gauss quadrature for the evaluation of $[K^1]$ and the 1×1 Gauss quadrature for the evaluation of $[K^2]$ (see Sec. 3.6.4 for the Gauss quadrature; also see Sec. 4.6.2). Table 4.7 contains the explicit form of $S_{ij}^{\alpha\beta}$ ($\alpha, \beta = 0, 1, 2$) computed using exact integration (or the two-point Gauss rule) and two different reduced integration schemes.

Table 4.7 Exact and reduced numerical integration of the matrices in Eqs. (4.196)

Matrix	Exact integration†	Reduced integration: Trapezoidal rule	Reduced integration: One-point Gauss rule
$[S^{00}]$	$\frac{ab}{36}\begin{bmatrix} 4 & 2 & 1 & 2 \\ & 4 & 2 & 1 \\ & & 4 & 2 \\ \text{symm.} & & & 4 \end{bmatrix}$	$\frac{ab}{4}\begin{bmatrix} 1 & 0 & 0 & 0 \\ & 1 & 0 & 0 \\ & & 1 & 0 \\ \text{symm.} & & & 1 \end{bmatrix}$	$\frac{ab}{16}\begin{bmatrix} 1 & 1 & 1 & 1 \\ & 1 & 1 & 1 \\ & & 1 & 1 \\ \text{symm.} & & & 1 \end{bmatrix}$
$[S^{01}]$	$\frac{b}{12}\begin{bmatrix} -2 & 2 & 1 & -1 \\ -2 & 2 & 1 & -1 \\ -1 & 1 & 2 & -2 \\ -1 & 1 & 2 & -2 \end{bmatrix}$	$\frac{b}{4}\begin{bmatrix} -1 & 1 & 0 & 0 \\ -1 & 1 & 0 & 0 \\ 0 & 0 & 1 & -1 \\ 0 & 0 & 1 & -1 \end{bmatrix}$	$\frac{b}{8}\begin{bmatrix} -1 & 1 & 1 & -1 \\ -1 & 1 & 1 & -1 \\ -1 & 1 & 1 & -1 \\ -1 & 1 & 1 & -1 \end{bmatrix}$
$[S^{02}]$	$\frac{a}{12}\begin{bmatrix} -2 & -1 & 1 & 2 \\ -1 & -2 & 2 & 1 \\ -1 & -2 & 2 & 1 \\ -2 & -1 & 1 & 2 \end{bmatrix}$	$\frac{a}{4}\begin{bmatrix} -1 & 0 & 0 & 1 \\ 0 & -1 & 1 & 0 \\ 0 & -1 & 1 & 0 \\ -1 & 0 & 0 & 1 \end{bmatrix}$	$\frac{a}{8}\begin{bmatrix} -1 & -1 & 1 & 1 \\ -1 & -1 & 1 & 1 \\ -1 & -1 & 1 & 1 \\ -1 & -1 & 1 & 1 \end{bmatrix}$
$[S^{11}]$	$\frac{b}{6a}\begin{bmatrix} 2 & -2 & -1 & 1 \\ & 2 & 1 & -1 \\ & & 2 & -2 \\ \text{symm.} & & & 2 \end{bmatrix}$	$\frac{b}{2a}\begin{bmatrix} 1 & -1 & 0 & 0 \\ & 1 & 0 & 0 \\ & & 1 & -1 \\ \text{symm.} & & & 1 \end{bmatrix}$	$\frac{b}{4a}\begin{bmatrix} 1 & -1 & -1 & 1 \\ & 1 & 1 & -1 \\ & & 1 & -1 \\ \text{symm.} & & & 1 \end{bmatrix}$
$[S^{12}]$	$\frac{1}{4}\begin{bmatrix} 1 & 1 & -1 & -1 \\ -1 & -1 & 1 & 1 \\ -1 & -1 & 1 & 1 \\ 1 & 1 & -1 & -1 \end{bmatrix}$	$\frac{1}{4}\begin{bmatrix} 1 & 1 & -1 & -1 \\ -1 & -1 & 1 & 1 \\ -1 & -1 & 1 & 1 \\ 1 & 1 & -1 & -1 \end{bmatrix}$	$\frac{1}{4}\begin{bmatrix} 1 & 1 & -1 & -1 \\ -1 & -1 & 1 & 1 \\ -1 & -1 & 1 & 1 \\ 1 & 1 & -1 & -1 \end{bmatrix}$
$[S^{22}]$	$\frac{a}{6b}\begin{bmatrix} 2 & 1 & -1 & -2 \\ & 2 & -2 & -1 \\ & & 2 & 1 \\ \text{symm.} & & & 2 \end{bmatrix}$	$\frac{a}{2b}\begin{bmatrix} 1 & 0 & 0 & -1 \\ & 1 & -1 & 0 \\ & & 1 & 0 \\ \text{symm.} & & & 1 \end{bmatrix}$	$\frac{a}{4b}\begin{bmatrix} 1 & 1 & -1 & -1 \\ & 1 & -1 & -1 \\ & & 1 & 1 \\ \text{symm.} & & & 1 \end{bmatrix}$

†2-point Gauss rule and the one-third Simpson's rule also give the same result.

The penalty finite-element model allows us to compute the velocities at each node of the finite-element mesh. The pressure, being absent in the model, should be computed by some alternative means. It is well known from the theory of the penalty function method that the Lagrange multiplier (which is equal to the negative of the pressure in the present case) can be computed in each element from the relation

$$-P^e = \gamma_e \operatorname{div} \mathbf{u}^e = \gamma_e \left(\frac{\partial u^e}{\partial x} + \frac{\partial v^e}{\partial y} \right) \tag{4.205}$$

where $\mathbf{u}^e = (u^e, v^e)$ is the velocity vector obtained from the solution of Eq. (4.203). In the finite-element analysis, pressure is computed from the velocity field [which is obtained by solving Eq. (4.200)] in each element. Since the derivatives of the velocities are computed at the Gauss points, the pressure is also computed at the Gauss points. In the case of the linear rectangular element, pressure is computed at the center of each element (because the one-point rule is used for the penalty terms).

Note that the penalty finite-element model (4.200) involves only two primary degrees of freedom per node. The pressure is postcomputed from Eq. (4.205). In the mixed finite-element model, the pressure is obtained as a primary unknown. However, it should be noted that the specification of P does not constitute an essential boundary condition and therefore the pressure can be determined uniquely within a constant. In other words, in the pressure-velocity formulation, P should be specified at least at one point to eliminate the constant state of the pressure. The reader should consult appropriate references at the end of the chapter for additional details.

Example 4.10 Consider the steady, fully developed, laminar flow of an incompressible fluid between two long parallel plates with the top plate moving at a velocity u_0 relative to the bottom plate (see Prob. 3.29). For the fully developed state considered, the velocity field is of the form

$$u = u(y) \qquad v = w = 0 \tag{4.206}$$

Consequently, Eqs. (4.188) take the form

$$\frac{\partial P}{\partial x} = \mu \frac{d^2 u}{dy^2} \qquad \frac{\partial P}{\partial y} = 0 \tag{4.207}$$

and Eq. (4.189) is identically satisfied. Equations (4.207) together with the boundary conditions

$$u(0) = 0 \qquad u(2b) = u_0 \tag{4.208}$$

yield the solution

$$u = \frac{u_0 y}{2b} - \frac{cyb}{\mu}\left(1 - \frac{y}{2b}\right) \qquad c = \frac{dP}{dx} = \text{constant} = \frac{P_1 - P_0}{a}$$

$$\tau_{xy} = \frac{\mu u_0}{2b} - \frac{c}{2}\left(1 - \frac{y}{b}\right) \tag{4.209}$$

where P_0 and P_1 are the values of the pressure at $x = 0$ and $x = a$, respectively. Here we wish to solve the problem using first the pressure-velocity model and then the penalty model. To this end we consider a domain of width a (along the x direction) with the boundary conditions shown in Fig. 4.36a. We take $\mu = 1$, $a = 2$, $b = 3$, $u_0 = 3$, $P_0 = 1$, and $P_1 = 0$ so that $dP/dx = -0.5$.

Pressure-velocity model The eight-node (i.e., serendipity) rectangular element (see Fig. 4.36) is employed with a quadratic approximation of the velocity field and a linear approximation of the pressure:

$$u = \sum_{i=1}^{8} u_i \psi_i \qquad v = \sum_{i=1}^{8} v_i \psi_i \qquad P = \sum_{i=1}^{4} P_i \phi_i \tag{4.210}$$

where ψ_i are the interpolation functions of Eqs. (4.128) (except that the node

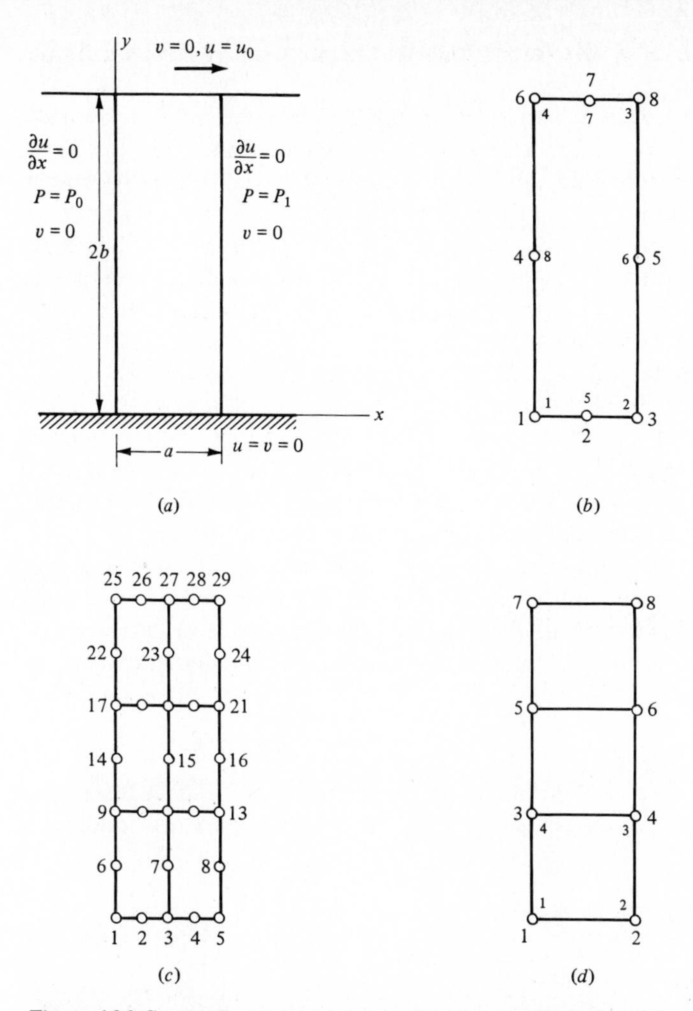

Figure 4.36 Couette flow between two parallel plates. (*a*) Domain and boundary conditions ($b = 3$, $a = 2$, $u_0 = 3$, $P_0 = 1$). (*b*) One-element mesh for the pressure-velocity model. (*c*) 2×3 mesh of eight-node elements for the pressure-velocity model. (*d*) 1×3 mesh of four-node elements for the penalty finite-element model.

numbering in Fig. 4.36*b* is different from that in Fig. 4.28) and ϕ_i are the linear interpolation functions of Eq. (4.121) (also see Fig. 4.27).

The boundary conditions on the velocity field constitute the essential boundary conditions and those on P are included in the natural boundary conditions. Thus we have

$$\begin{aligned} x = 0: \quad & t_x = \left(\mu \frac{\partial u}{\partial x} - P\right) n_x = -P_0(-1) = P_0 \\ x = a: \quad & t_x = \left(\mu \frac{\partial u}{\partial x} - P\right) n_x = -P_1(1) = -P_1 \end{aligned} \tag{4.211}$$

Substituting the values of t_x into the boundary integral in Eq. (4.196), we obtain

$$f_i^{(e)} = \int_{\Gamma^e} t_x \psi_i \, ds = \int_0^h t_x \psi_i(0, y) \, dy \tag{4.212}$$

where $\psi_i(0, y)$ are the quadratic interpolation functions of a one-dimensional element of length h:

$$\psi_1 = \left(1 - \frac{2y}{h}\right)\left(1 - \frac{y}{h}\right) \qquad \psi_2 = \frac{4y}{h}\left(1 - \frac{y}{h}\right) \qquad \psi_3 = -\frac{y}{h}\left(1 - \frac{2y}{h}\right) \tag{4.213}$$

Substituting Eqs. (4.213) into Eqs. (4.212) and performing the integration, we obtain

$$f_1 = t_x\left(\frac{h}{6}\right) \qquad f_2 = t_x\left(\frac{2h}{3}\right) \qquad f_3 = t_x\left(\frac{h}{6}\right) \tag{4.214}$$

Since there are no essential boundary conditions on the pressure, it can be computed only within an arbitrary constant. In the finite-element analysis by the pressure-velocity model, one must specify P at least at one point in order to determine the constant (and hence remove the rigid-body mode in pressure). In the present example, we specify P to be zero at one node of the boundary (say, at node 3):

$$P_3 = 0 \tag{4.215}$$

Now consider a one-element mesh of the domain (see Fig. 4.36*b*). The element (and hence global) matrix is of order 20 × 20. The boundary conditions are

$$\begin{gathered} U_1 = V_1 = U_2 = V_2 = U_3 = V_3 = V_4 = V_5 = V_6 = V_7 = V_8 = 0 \\ U_6 = U_7 = U_8 = 3.0 \end{gathered} \tag{4.216}$$

The solution of the resulting set of equations (which is 5 × 5) is given by

$$U_4 = U_5 = 3.75 \qquad P_1 = 5.0 \qquad P_6 = 17.0 \qquad P_8 = -20.0 \tag{4.217}$$

While the solution for the velocities coincides with the exact solution at the nodes, the pressure is completely erroneous (the true solution is $P = 1.0$ at $x = 0$). This is attributed to the coarse mesh used in the problem (in other words, the pressure is sensitive to the mesh and the boundary conditions).

A 2 × 3 mesh of the eight-node elements (see Fig. 4.36*c*) yields the exact solution for both the velocities and the pressure (when P is specified at node 5).

Penalty finite-element model The boundary conditions for the model are the same as those discussed for the pressure-velocity model (see Fig. 4.36*a*). Use of the element matrices from Table 4.7 in Eq. (4.201) allows us to write the element matrices in explicit form:

$$[\bar{K}^{11}] = \frac{\mu}{6}\left(2\alpha\begin{bmatrix} 2 & -2 & -1 & 1 \\ -2 & 2 & 1 & -1 \\ -1 & 1 & 2 & -2 \\ 1 & -1 & -2 & 2 \end{bmatrix} + \beta\begin{bmatrix} 2 & 1 & -1 & -2 \\ 1 & 2 & -2 & -1 \\ -1 & -2 & 2 & 1 \\ -2 & -1 & 1 & 2 \end{bmatrix} + \frac{\gamma\beta}{4}\begin{bmatrix} 1 & -1 & -1 & 1 \\ -1 & 1 & 1 & -1 \\ -1 & 1 & 1 & -1 \\ 1 & -1 & -1 & 1 \end{bmatrix}\right)$$

$$[\bar{K}^{22}] = \frac{\mu}{6}\left(\alpha\begin{bmatrix} 2 & -2 & -1 & 1 \\ -2 & 2 & 1 & -1 \\ -1 & 1 & 2 & -2 \\ 1 & -1 & -2 & 2 \end{bmatrix} + 2\beta\begin{bmatrix} 2 & 1 & -1 & -2 \\ 1 & 1 & -2 & -1 \\ -1 & -2 & 2 & 1 \\ -2 & -1 & 1 & 2 \end{bmatrix} + \frac{\beta\gamma}{4}\begin{bmatrix} 1 & 1 & -1 & -1 \\ 1 & 1 & -1 & -1 \\ -1 & -1 & 1 & 1 \\ -1 & -1 & 1 & 1 \end{bmatrix}\right)$$

$$\{F^1\} = \begin{Bmatrix} t_x h/2 \\ Q_{x2} \\ Q_{x3} \\ t_x h/2 \end{Bmatrix} \qquad \{F^2\} = \begin{Bmatrix} Q_{y1} \\ Q_{y2} \\ Q_{y3} \\ Q_{y4} \end{Bmatrix} \tag{4.218}$$

where $\alpha = 1/\beta = h_y/h_x$, h_x and h_y being the x and y dimensions of the element. A 1×3 mesh of four-node linear elements is shown in Fig. 4.36*d*.

The penalty finite-element solution for various choices of the penalty parameter γ and meshes is given in Table 4.8. In all cases, the pressure and the stresses are predicted correctly. From the results presented in Table 4.8 it is apparent that the solution converges, for any fixed mesh, with increasing values of the penalty parameter. Also, for any fixed γ, the solution converges as the mesh is refined. Note that for a wide range of the penalty parameter ($\gamma = 10^4$–10^{12}) the solution is virtually unchanged (see Fig. 4.37). This range, generally speaking, depends on the problem, the mesh used, and the word length in the computer used. ■

Example 4.11 Consider the flow of a viscous incompressible material squeezed between two long parallel plates. When the length of the plates is very large compared to both the width of and the distance between the plates, we have a case of plane flow (in the plane formed by the width of and the distance between the plates). Assuming that a state of plane flow exists, we determine the velocity profile of the free surface. An approximate (analytical) solution to this problem is

Table 4.8 Convergence of the finite-element solution with increasing values of the penalty parameter and mesh refinements

		γ								
Mesh	y	1.0	10	10^2	10^3	10^4	10^5–10^{11}	10^{12}	10^{13}	10^{17}
	1.0	—	—	—	—	—	—	—	—	
	2.0	—	—	—	—	—	—	—	—	no
1 × 2	3.0	4.0121	3.8280	3.7597	3.7510	3.7501	3.7500	3.7500	3.7482	inverse
	4.0	—	—	—	—	—	—	—	—	exists
	5.0	—	—	—	—	—	—	—	—	
	6.0	3.0000	3.0000	3.0000	3.0000	3.0000	3.0000	3.0000	3.0000	
	1.0	—	—	—	—	—	—	—	—	—
	2.0	3.2000	3.0541	3.0065	3.0007	3.0001	3.0000	2.9998	2.9967	0.4000
1 × 3	3.0	—	—	—	—	—	—	—	—	—
	4.0	4.2000	4.0541	4.0065	4.0007	4.0001	4.0000	3.9998	3.9960	3.2000
	5.0	—	—	—	—	—	—	—	—	—
	6.0	3.0000	3.0000	3.0000	3.0000	3.0000	3.0000	3.0000	3.0000	3.0000
	1.0	1.9335	1.8096	1.7590	1.7610	1.7501	1.7500	1.7497	1.7455	1.0310
	2.0	3.1649	3.0331	3.0012	3.0000	3.0000	3.0000	2.9995	2.9919	2.0075
1 × 6	3.0	3.9170	3.7978	3.7585	3.7510	3.7501	3.7500	3.7494	3.7404	2.4162
	4.0	4.1649	4.0331	4.0012	4.0000	4.0000	4.0000	3.9994	3.9914	2.6576
	5.0	3.9335	3.8096	3.7590	3.7510	3.7501	3.7500	3.7496	3.7444	2.8420
	6.0	3.0000	3.0000	3.0000	3.0000	3.0000	3.0000	3.0000	3.0000	3.0000

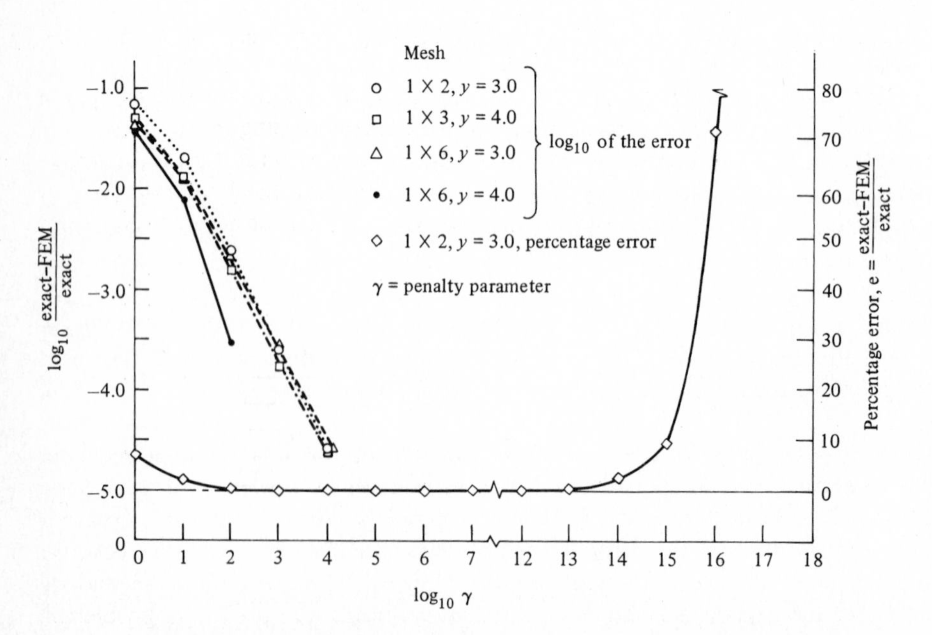

Figure 4.37 Convergence of the penalty finite-element solution with the penalty parameter.

provided by Nadai (1963) (see References for Additional Reading at the end of this chapter):

$$u = \frac{3v_0 x}{2b}\left(1 - \frac{y^2}{b^2}\right) \qquad v = -\frac{v_0 y}{2b}\left(3 - \frac{y^2}{b^2}\right)$$

$$P = \frac{P_0}{a^2}(a^2 + y^2 - x^2) \qquad P_0 = \frac{3\mu v_0 a^2}{2b^3}$$

$$\sigma_x \equiv 2\mu\frac{\partial u}{\partial x} - P = \frac{P_0}{a^2}(x^2 - 3y^2 - a^2 + 2b^2) \tag{4.219}$$

$$\sigma_y \equiv 2\mu\frac{\partial v}{\partial y} - P = \frac{P_0}{a^2}(x^2 + y^2 - a^2 - 2b^2)$$

$$\tau_{xy} \equiv \mu\left(\frac{\partial u}{\partial y} + \frac{\partial v}{\partial x}\right) = \frac{-2P_0 xy}{a^2}$$

Let v_0 be the velocity with which the two plates are moving toward each other (i.e., squeeze out the fluid), and let $2b$ and $2a$ denote, respectively, the distance between and the width of the plates (see Fig. 4.38*a*). Due to the biaxial symmetry present in the problem, it suffices to model only a quadrant of the domain. A 5×3 nonuniform mesh of the nine-node quadratic elements is used in the mixed (i.e., pressure-velocity) model, and a 10×6 mesh of the four-node linear elements is used in the penalty model (see Fig. 4.38*b*). The nonuniform mesh, with smaller elements near the free surface (i.e., at $x = a$), are used to approximate accurately the singularity in the shear stress at the point (a, b). The mesh used for the penalty model has exactly the same number of nodes as in the mesh used for the mixed model.

Plots of the horizontal velocity u as a function of y, at $x = 3$ and $x = 6$, are shown in Fig. 4.39. The finite-element solutions obtained using the mixed and the penalty finite-element models are in good agreement with the approximate solution of Nadai. The pressure at $y = 0.125$ (i.e., near the horizontal axis of symmetry) and the normal stress σ_y at $y = 1.875$ (i.e., near the top plate) are

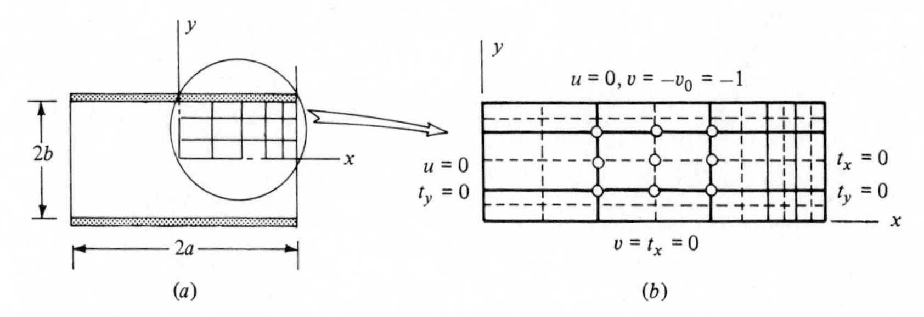

Figure 4.38 Boundary conditions and finite-element meshes for the problem of an incompressible viscous fluid squeezed between plates. (*a*) Geometry of the flow domain. (*b*) Finite-element mesh of a quadrant.

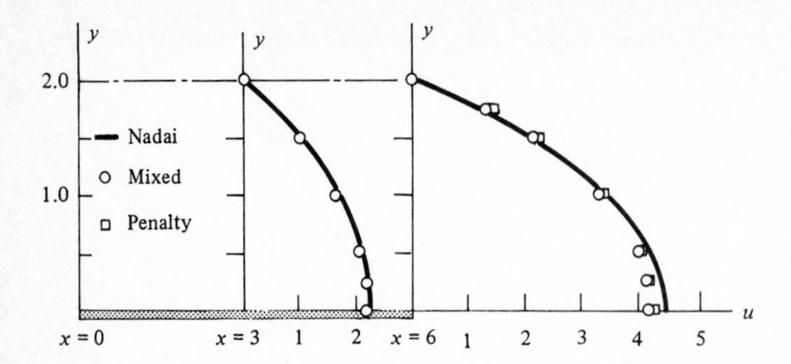

Figure 4.39 Comparison of the finite-element solution u obtained by mixed and penalty models with the approximate solution of Nadai for fluid squeezed between plates ($a = 6$, $b = 2$, $\mu = 1$).

plotted as functions of x in Fig. 4.40. Both P and σ_y are computed at the center of each element of the 10×6 mesh (in both models). In the mixed model, linear interpolation is used to evaluate P at the center of each element of the 10×6 mesh. The finite-element results for P as well as σ_y are in fair agreement with the approximate solution of Nadai.

The finite-element solutions for $u(x, 0.25)$ and $\sigma_y(x, 0.125)$ are compared in Table 4.9 with the analytical solutions of Nadai. The finite-element solution for the displacement is in better agreement with the analytical solution than that for the stress. It should be borne in mind that the analytical solution is not exact. ■

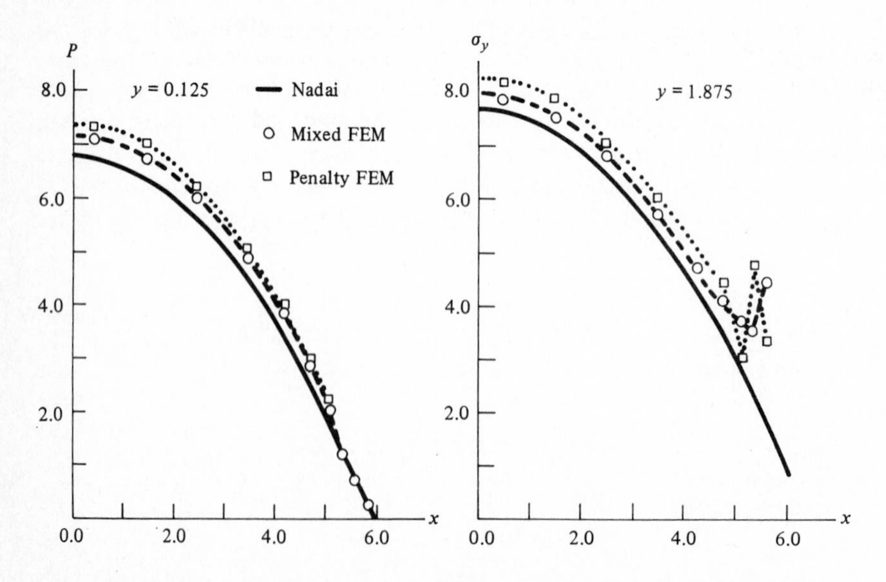

Figure 4.40 Plots of $P(x, 0.125)$ and $\sigma_y(x, 1.875)$ versus the distance along the centerline for the problem of fluid squeezed between parallel plates.

Table 4.9 Comparison of the finite-element solution with the analytical solution of fluid squeezed between two parallel plates

	$u(x, 0.25)$				$-\sigma_y(x, 0.125)$		
x	Mixed	Penalty†	Analytical	x	Mixed	Penalty†	Analytical
0.0	0.0	0.0	0.0	0.5	8.629	8.886	8.200
1.0	0.7378	0.7452	0.7383	1.5	8.261	8.500	7.825
2.0	1.4793	1.4910	1.4766	2.5	7.513	7.766	7.075
3.0	2.2211	2.2376	2.2148	3.5	6.405	6.645	5.950
4.0	2.9744	3.0074	2.9531	4.25	5.406	5.612	4.860
4.5	3.3785	3.3984	3.3223	4.75	4.358	4.578	4.017
5.0	3.7616	3.7996	3.6914	5.12	3.309	3.693	3.323
5.25	3.9269	3.9630	3.8760	5.37	2.270	2.640	2.830
5.50	4.0418	4.1242	4.0605	5.62	1.197	1.546	2.314
5.75	4.1102	4.1986	4.2451	5.87	0.0688	0.0215	1.775
6.00	4.1356	4.2127	4.4297	—	—	—	—

†Penalty parameter $\gamma = 10^9$.

*4.5.4 Bending of Elastic Plates

Here we consider an extension of the beam theory to two dimensions, namely, the bending of plates. The word "plate" is used to indicate bodies that are bounded by two parallel planes whose lateral dimensions are large compared to the separation between these planes. Geometrically, plates are similar to plane elastic bodies considered in Sec. 4.5.2; however, plates are loaded transverse (i.e., perpendicular) to the plane of the plate. Under certain simplifying assumptions, the equations governing the transverse deflection w and the bending slopes S_x and S_y of the midplane of the plate are given by (the so-called shear deformation theory of plates)

$$D_{44}\frac{\partial}{\partial x}\left(S_x + \frac{\partial w}{\partial x}\right) + D_{55}\frac{\partial}{\partial y}\left(S_y + \frac{\partial w}{\partial y}\right) + q = 0 \tag{4.220a}$$

$$\frac{\partial}{\partial x}\left(D_{11}\frac{\partial S_x}{\partial x} + D_{12}\frac{\partial S_y}{\partial y}\right) + D_{33}\frac{\partial}{\partial y}\left(\frac{\partial S_x}{\partial y} + \frac{\partial S_y}{\partial x}\right) - D_{44}\left(S_x + \frac{\partial w}{\partial x}\right) = 0 \tag{4.220b}$$

$$D_{33}\frac{\partial}{\partial x}\left(\frac{\partial S_x}{\partial y} + \frac{\partial S_y}{\partial x}\right) + \frac{\partial}{\partial y}\left(D_{12}\frac{\partial S_x}{\partial x} + D_{22}\frac{\partial S_y}{\partial y}\right) - D_{55}\left(S_y + \frac{\partial w}{\partial y}\right) = 0 \tag{4.220c}$$

Here q is the transversely distributed load and D_{44}, D_{55}, D_{11}, D_{12}, D_{22}, and D_{33}

are orthotropic plate stiffnesses,

$$D_{44} = G_{23}h \qquad D_{55} = G_{13}h \qquad D_{11} = \frac{E_1 h^3}{12(1-\nu_{12}\nu_{21})} \qquad D_{22} = \frac{E_2 D_{11}}{E_1}$$
$$D_{12} = \nu_{12} D_{22} = \nu_{21} D_{11} \qquad D_{33} = \frac{G_{12} h^3}{12} \tag{4.221}$$

where E_1 and E_2 are Young's moduli along the x and y directions, respectively, ν_{12} and ν_{21} $(= \nu_{12}E_2/E_1)$ are Poisson's ratios, G_{12}, G_{13} and G_{23} are the shear moduli in the x–y, x–z, and y–z planes of the material axes of the plate, and h is the thickness of the plate.

The variational formulation of Eqs. (4.220) over a typical element Ω^e is obtained by multiplying each equation by a weight function ϕ_i $(i = 1, 2, 3)$ and integrating the result by parts:

$$\int_{\Omega^e} \phi_1 \{\text{LHS of Eq. } (4.220a)\}\, dx\, dy = 0$$

$$\int_{\Omega^e} \phi_2 \{\text{LHS of Eq. } (4.220b)\}\, dx\, dy = 0$$

$$\int_{\Omega^e} \phi_3 \{\text{LHS of Eq. } (4.220c)\}\, dx\, dy = 0$$

or

$$\int_{\Omega^e} \left\{ D_{44} \frac{\partial \phi_1}{\partial x} \left(S_x + \frac{\partial w}{\partial x} \right) + D_{55} \frac{\partial \phi_1}{\partial y} \left(S_y + \frac{\partial w}{\partial y} \right) \right\} dx\, dy$$
$$= \int_{\Omega^e} \phi_1 q\, dx\, dy + \oint_{\Gamma^e} \phi_1 \left[D_{44} \left(S_x + \frac{\partial w}{\partial x} \right) n_x + D_{55} \left(S_y + \frac{\partial w}{\partial y} \right) n_y \right] ds \tag{4.222a}$$

$$\int_{\Omega^e} \left\{ D_{11} \frac{\partial \phi_2}{\partial x} \frac{\partial S_x}{\partial x} + D_{12} \frac{\partial \phi_2}{\partial x} \frac{\partial S_y}{\partial y} + D_{33} \frac{\partial \phi_2}{\partial y} \left(\frac{\partial S_x}{\partial y} + \frac{\partial S_y}{\partial x} \right) \right.$$
$$\left. + D_{44} \phi_2 \left(S_x + \frac{\partial w}{\partial x} \right) \right\} dx\, dy$$
$$= \oint_{\Gamma^e} \phi_2 \left[\left(D_{11} \frac{\partial S_x}{\partial x} + D_{12} \frac{\partial S_y}{\partial y} \right) n_x + D_{33} \left(\frac{\partial S_x}{\partial y} + \frac{\partial S_y}{\partial x} \right) n_y \right] ds \tag{4.222b}$$

$$\int_{\Omega^e} \left\{ D_{33} \frac{\partial \phi_3}{\partial x} \left(\frac{\partial S_x}{\partial y} + \frac{\partial S_y}{\partial x} \right) + D_{12} \frac{\partial \phi_3}{\partial y} \frac{\partial S_x}{\partial x} + D_{22} \frac{\partial \phi_3}{\partial y} \frac{\partial S_y}{\partial y} \right.$$
$$\left. + D_{55} \phi_3 \left(S_y + \frac{\partial w}{\partial y} \right) \right\} dx\, dy$$
$$= \oint_{\Gamma^e} \phi_3 \left[\left(D_{12} \frac{\partial S_x}{\partial x} + D_{22} \frac{\partial S_y}{\partial y} \right) n_y + D_{33} \left(\frac{\partial S_x}{\partial y} + \frac{\partial S_y}{\partial x} \right) n_x \right] ds \tag{4.222c}$$

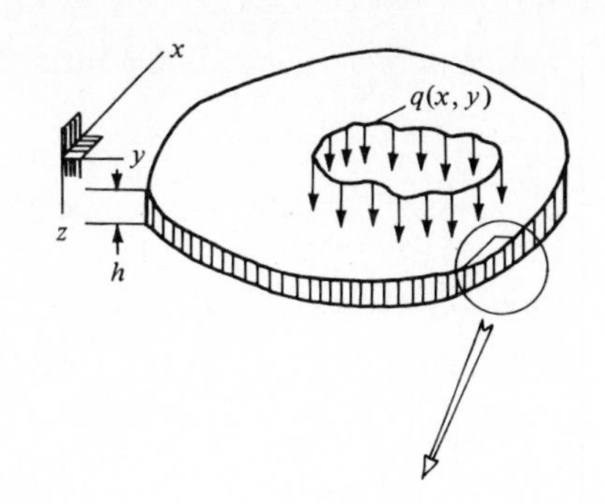

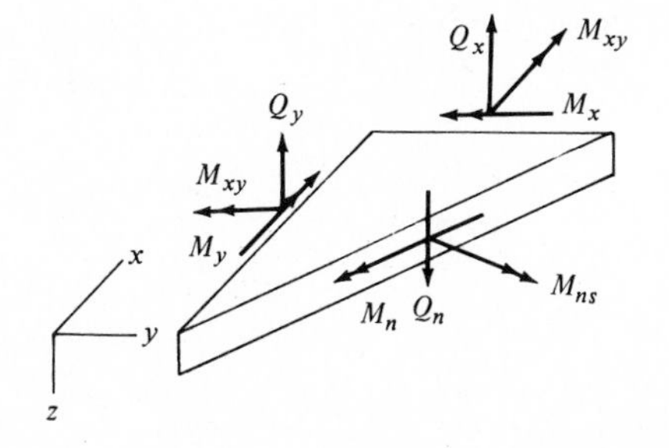

Use the right-hand-rule convention to determine the moment direction (shown with double arrow)

Boundary conditions for plates

Clamped	Simply supported—1	Simply supported—2	Free
$w = 0$	$w = 0$	$w = 0$	$Q_n = 0$
$S_s = 0$	$M_{ns} = 0$	$S_s = 0$	$M_{ns} = 0$
$S_n = 0$	$M_n = 0$	$M_n = 0$	$M_n = 0$

Figure 4.41 Geometry, moment and shear force resultants, and various boundary conditions for a plate element.

The weight functions ϕ_1, ϕ_2, and ϕ_3 can be viewed as the variations δw, δS_x, and δS_y, respectively. From the variational equations (4.222) it is clear that the specifications of the expressions in the square brackets of the boundary terms constitute the natural boundary conditions and the specifications of w, S_x, and S_y constitute the essential boundary conditions. The terms in the natural boundary conditions can be identified with the moment and shear force resultants (see Fig. 4.41):

$$M_x = D_{11}\frac{\partial S_x}{\partial x} + D_{12}\frac{\partial S_y}{\partial y} \qquad M_y = D_{12}\frac{\partial S_x}{\partial x} + D_{22}\frac{\partial S_y}{\partial y}$$

$$M_{xy} = D_{33}\left(\frac{\partial S_x}{\partial y} + \frac{\partial S_y}{\partial x}\right) \tag{4.223}$$

$$Q_x = D_{44}\left(S_x + \frac{\partial w}{\partial x}\right) \qquad Q_y = D_{55}\left(S_y + \frac{\partial w}{\partial y}\right)$$

The finite-element model of Eqs. (4.222) can be derived by assuming interpolation of the form

$$w = \sum_{i=1}^{n} w_i \psi_i \qquad S_x = \sum_{i=1}^{n} S_x^i \psi_i \qquad S_y = \sum_{i=1}^{n} S_y^i \psi_i \tag{4.224}$$

Clearly, ψ_i can be linear, quadratic, and so on. Substituting Eqs. (4.224) and $\phi_\alpha = \psi_i$ ($\alpha = 1, 2, 3$) into Eqs. (4.222), we obtain

$$\begin{bmatrix} [K^{11}] & [K^{12}] & [K^{13}] \\ & [K^{22}] & [K^{23}] \\ \text{Symmetric} & & [K^{33}] \end{bmatrix} \begin{Bmatrix} \{w\} \\ \{S_x\} \\ \{S_y\} \end{Bmatrix} = \begin{Bmatrix} \{F^1\} \\ \{F^2\} \\ \{F^3\} \end{Bmatrix} \tag{4.225}$$

where

$$\begin{aligned}
K_{ij}^{11} &= \int_{\Omega^e} \left(D_{44} \frac{\partial \psi_i}{\partial x} \frac{\partial \psi_j}{\partial x} + D_{55} \frac{\partial \psi_i}{\partial y} \frac{\partial \psi_j}{\partial y} \right) dx\, dy \\
K_{ij}^{12} &= \int_{\Omega^e} D_{44} \frac{\partial \psi_i}{\partial x} \psi_j \, dx\, dy \\
K_{ij}^{13} &= \int_{\Omega^e} D_{55} \frac{\partial \psi_i}{\partial y} \psi_j \, dx\, dy \\
K_{ij}^{22} &= \int_{\Omega^e} \left(D_{11} \frac{\partial \psi_i}{\partial x} \frac{\partial \psi_j}{\partial x} + D_{33} \frac{\partial \psi_i}{\partial y} \frac{\partial \psi_j}{\partial y} + D_{44} \psi_i \psi_j \right) dx\, dy \\
K_{ij}^{23} &= \int_{\Omega^e} \left(D_{12} \frac{\partial \psi_i}{\partial x} \frac{\partial \psi_j}{\partial y} + D_{33} \frac{\partial \psi_i}{\partial y} \frac{\partial \psi_j}{\partial x} \right) dx\, dy \\
K_{ij}^{33} &= \int_{\Omega^e} \left(D_{33} \frac{\partial \psi_i}{\partial x} \frac{\partial \psi_j}{\partial x} + D_{22} \frac{\partial \psi_i}{\partial y} \frac{\partial \psi_j}{\partial y} + D_{55} \psi_i \psi_j \right) dx\, dy \\
F_i^1 &= \int_{\Omega^e} q \psi_i \, dx\, dy + \oint_{\Gamma^e} Q_n \psi_i \, ds \qquad Q_n = Q_x n_x + Q_y n_y \\
F_i^2 &= \oint_{\Gamma^e} M_n \psi_i \, ds \qquad M_n = M_x n_x + M_{xy} n_y \\
F_i^3 &= \oint_{\Gamma^e} M_{ns} \psi_i \, ds \qquad M_{ns} = M_{xy} n_x + M_y n_y
\end{aligned} \tag{4.226}$$

The element stiffness matrix (4.225) is of the order $3n \times 3n$, where n is the number of nodes per element. When the four-node rectangular element is used, then the element stiffness matrix is of the order 12×12, and the coefficients of the matrix can be computed using Eqs. (4.226) and Table 4.7.

It should be noted that the inclusion of the transverse shear strains (i.e., terms involving D_{44} and D_{55}) in the equations presents computational difficulties when the side-to-thickness ratio of the plate is large (i.e., for thin plates). For thin plates, the transverse shear strains are negligible, and consequently the element stiffness matrix becomes stiff (i.e., so-called locking occurs) and yields erroneous results for the generalized displacements (w_i, S_x^i, S_y^i). This phenomenon can be interpreted as one caused by the inclusion of the following slope-deflection relationships:

$$S_x + \frac{\partial w}{\partial x} = 0 \qquad S_y + \frac{\partial w}{\partial y} = 0 \tag{4.227}$$

as constraints into the variational formulation. [Indeed, the present theory of plates is a two-dimensional version of the penalty formulation of beams presented in Eq. (5.30)] To circumvent the computational difficulty (i.e., to avoid the locking), one should evaluate the stiffness coefficients associated with D_{44} and D_{55} using a reduced integration technique. For example, when a four-node rectangular element is used, the 1×1 Gauss rule should be used to evaluate the shear energy terms (i.e., the terms involving D_{44} and D_{55}), and the 2×2 Gauss rule should be used for all other terms. When an eight-node or nine-node rectangular element is used, the 2×2 and 3×3 Gauss rule, respectively, should be used to evaluate the shear and bending terms.

Example 4.12 Consider a simply supported, isotropic, square plate subjected to uniformly distributed transverse load. Due to the biaxial symmetry, we need model only a quadrant of the plate. The essential boundary conditions at simply supported edges ($x = a/2$ and $y = a/2$) are given by

$$\begin{aligned} w = 0 \qquad & S_y = 0 \text{ at } x = a/2 \\ w = 0 \qquad & S_x = 0 \text{ at } y = a/2 \end{aligned} \tag{4.228a}$$

The essential boundary conditions along the symmetry lines ($x = 0$ and $y = 0$) are given by

$$S_x(0, y) = S_y(x, 0) = 0 \tag{4.228b}$$

The natural boundary conditions (which enter the finite-element equations through $\{F^i\}$) in the present case are given by

$$\begin{aligned} Q_n &= 0 \text{ along } x = 0 \text{ and } y = 0 \\ M_x &= 0 \text{ along } y = 0 \text{ and } x = a/2 \\ M_y &= 0 \text{ along } x = 0 \text{ and } y = a/2 \end{aligned} \tag{4.228c}$$

For a linear (four-node) element, the contribution of a uniformly distributed load q_0 is given by

$$\int_0^{h_x} \int_0^{h_y} q_0 \psi_i \, dx \, dy = \frac{h_x h_y}{4} q_0 \tag{4.229}$$

where h_x and h_y are the plane-form dimensions of the element. Except for the above contribution, all other specified forces are zero [because of Eq. (4.228c)].

The effect of reduced integration, thickness, and mesh on the center deflection and stress is investigated, and the results are presented in Table 4.10. Note

Table 4.10 The effect of reduced integration, thickness, and mesh refinement on the center deflection and stress of a simply supported, isotropic ($\nu = 0.25$), square plate ($a = 10$) under uniformly distributed load

		1 × 1 linear		2 × 2 linear		4 × 4 linear		2 × 2 quadratic	
$\frac{a}{h}$	Integration	$\bar{w}$	$\bar{\sigma}_x$	$\bar{w}$	$\bar{\sigma}_x$	$\bar{w}$	$\bar{\sigma}_x$	$\bar{w}$	$\bar{\sigma}_x$
	F	0.964	0.0182	2.474	0.1185	3.883	0.216	4.770	0.2899
10	M	3.950	0.0953	4.712	0.2350	4.773	0.2661	4.799	0.2715
	F	0.270	0.0053	0.957	0.0476	2.363	0.1375	4.570	0.2683
20	M	3.669	0.0954	4.524	0.2350	4.603	0.2660	4.633	0.2715
	F	0.0695	0.0014	0.279	0.0140	0.9443	0.0558	4.505	0.2699
40	M	3.599	0.0953	4.375	0.2349	4.560	0.2661	4.592	0.2714
	F	0.0045	0.0001	0.182	0.0092	0.6515	0.0386	4.496	0.2667
50	M	3.590	0.0953	4.472	0.2350	4.555	0.2660	4.587	0.2714
	F	0.011	0.0002	0.047	0.0024	0.182	0.0108	4.482	0.2664
100	M	3.579	0.0953	4.465	0.235	4.548	0.2661	4.580	0.2715

$\bar{w} = \dfrac{wEh^3 \times 10^2}{q_0 a^4} \qquad \bar{\sigma}_x = \sigma_x(A, A, \pm h/2)\dfrac{h^2}{q_0 a^2},$

$A = \frac{a}{4}$ (1 × 1 linear), $\frac{a}{8}$ (2 × 2 linear), $\frac{a}{16}$ (4 × 4 linear), $0.05283a$ (2 × 2 quadratic).
Classical plate theory solution [from Timoshenko and Woinowsky-Krieger (1959)]: $\bar{w} = 4.580$.

that in general mixed (M) integration gives more accurate results than the full (F) integration. As the mesh is refined or when a higher-order element is used, the effect of reduced integration on the accuracy diminishes (i.e., both full integration and reduced integration give acceptable results). ■

*4.6 TIME-DEPENDENT PROBLEMS

4.6.1 Introduction

This section deals with the finite-element analysis of time-dependent two-dimensional problems. Once again, as in Sec. 3.5, we employ the results of Sec. 2.3.4 to discretize fully semidiscrete finite-element models of time-dependent problems. Since the temporal approximations were already discussed in detail in Sec. 2.3.4, attention is focused here on the semidiscrete finite-element models, and on specific examples from heat transfer, fluid mechanics, and solid mechanics. The examples presented are very simple because they are designed to illustrate the procedure for time-dependent problems; solution of two-dimensional problems with complicated geometry requires the use of isoparametric elements and numerical integration. Section 4.8 is devoted to the discussion of the computer implementation of two-dimensional problems. Numerous subroutines are presented there to aid readers in developing their own programs. Of course, one can use the complete program listed in Appendix II.

4.6.2 Semidiscrete Approximations

Heat transfer (and like) problems Consider the partial differential equation governing the transient heat transfer (and like) problems in a two-dimensional region Ω with total boundary Γ

$$c_1\frac{\partial u}{\partial t} - \frac{\partial}{\partial x}\left(k_1\frac{\partial u}{\partial x}\right) - \frac{\partial}{\partial y}\left(k_2\frac{\partial u}{\partial y}\right) + f = 0 \text{ in } \Omega \qquad 0 < t \leqslant t_0 \tag{4.230}$$

with the boundary conditions

$$\begin{aligned} k_1\frac{\partial u}{\partial x}n_x + k_2\frac{\partial u}{\partial y}n_y + \beta(u - u_\infty) + \hat{q} = 0 \text{ on } \Gamma_1 \qquad t \geqslant 0 \\ u = \hat{u} \text{ on } \Gamma_2 \qquad t \geqslant 0 \end{aligned} \tag{4.231a}$$

and the initial conditions

$$u = u_0 \text{ in } \Omega \qquad t = 0 \tag{4.231b}$$

Here t denotes time, and c_1, k_1, k_2, β, u_∞, $\hat{u}$, u_0, f, and $\hat{q}$ are given functions of position and/or time.

The semidiscrete variational formulation of Eqs. (4.230) and (4.231) over an element $\Omega^{(e)}$ is given by

$$0 = \int_{\Omega^{(e)}} \left(c_1 v \frac{\partial u}{\partial t} + k_1 \frac{\partial v}{\partial x}\frac{\partial u}{\partial x} + k_2 \frac{\partial v}{\partial y}\frac{\partial u}{\partial y} + vf \right) dx\, dy$$

$$+ \oint_{\Gamma^{(e)}} [\beta v(u - u_\infty) + vq]\, ds \tag{4.232}$$

Substituting finite-element interpolation of the form

$$u(x,t) = \sum_{j=1}^{r} u_j(t)\psi_j(x,y) \tag{4.233}$$

and $v = \psi_i$ into Eq. (4.232), we obtain

$$0 = \sum_{j=1}^{r} \left\{ \left(\int_{\Omega^{(e)}} c_1 \psi_i \psi_j \, dx\, dy \right) \frac{du_j}{dt} \right.$$

$$\left. + \left[\int_{\Omega^{(e)}} \left(k_1 \frac{\partial \psi_i}{\partial x}\frac{\partial \psi_j}{\partial x} + k_2 \frac{\partial \psi_i}{\partial y}\frac{\partial \psi_j}{\partial y} \right) dx\, dy + \oint_{\Gamma^{(e)}} \beta \psi_i \psi_j \, ds \right] u_j \right\}$$

$$- \oint_{\Gamma^{(e)}} (\beta u_\infty - q)\psi_i \, ds + \int_{\Omega^{(e)}} \psi_i f \, dx\, dy$$

or, in matrix form,

$$[M^{(e)}]\{\dot{u}\} + [K^{(e)}]\{u\} = \{F^{(e)}\} \tag{4.234}$$

where

$$M_{ij}^{(e)} = \int_{\Omega^{(e)}} c_1 \psi_i \psi_j \, dx\, dy$$

$$K_{ij}^{(e)} = \int_{\Omega^{(e)}} \left(k_1 \frac{\partial \psi_i}{\partial x}\frac{\partial \psi_j}{\partial x} + k_2 \frac{\partial \psi_i}{\partial y}\frac{\partial \psi_j}{\partial y} \right) dx\, dy + \oint_{\Gamma^{(e)}} \beta \psi_i \psi_j \, ds \tag{4.235}$$

$$F_i^{(e)} = -\int_{\Omega^{(e)}} \psi_i f \, dx\, dy + \beta u_\infty \oint_{\Gamma^{(e)}} \psi_i \, ds - \oint_{\Gamma^{(e)}} q\psi_i \, ds$$

Incompressible viscous fluid flow The time-dependent equations associated with Eqs. (4.188) are given by

$$\left.\begin{aligned} c_1 \frac{\partial u}{\partial t} - 2\mu \frac{\partial^2 u}{\partial x^2} - \mu \frac{\partial}{\partial y}\left(\frac{\partial u}{\partial y} + \frac{\partial v}{\partial x} \right) + \frac{\partial P}{\partial x} - f_x = 0 \\ c_1 \frac{\partial v}{\partial t} - \mu \frac{\partial}{\partial x}\left(\frac{\partial u}{\partial y} + \frac{\partial v}{\partial x} \right) - 2\mu \frac{\partial^2 v}{\partial y^2} + \frac{\partial P}{\partial y} - f_y = 0 \\ \frac{\partial u}{\partial x} + \frac{\partial v}{\partial y} = 0 \end{aligned}\right\} \text{in } \Omega \qquad t > 0$$

(4.236a)

with the initial conditions

$$u = u_0 \qquad v = v_0 \text{ in } \Omega \qquad t = 0 \tag{4.236b}$$

The boundary conditions, for $t \geqslant 0$, are given by Eqs. (4.190). Here we give the finite-element models associated with the mixed and penalty function formulations of the equations.

Mixed model The time-derivative terms in Eqs. (4.236) result in a term analogous to that in Eq. (4.234). The element matrix associated with the remaining terms of Eqs. (4.236) is exactly the same as that in Eq. (4.195). We have for an element $\Omega^{(e)}$

$$\begin{bmatrix} [M^{11}] & [0] & [0] \\ [0] & [M^{22}] & [0] \\ [0] & [0] & [0] \end{bmatrix} \begin{Bmatrix} \{\dot{u}\} \\ \{\dot{v}\} \\ \{\dot{P}\} \end{Bmatrix} + \begin{bmatrix} [K^{11}] & [K^{12}] & [K^{13}] \\ [K^{12}]^T & [K^{22}] & [K^{23}] \\ [K^{13}]^T & [K^{23}]^T & [0] \end{bmatrix} \begin{Bmatrix} \{u\} \\ \{v\} \\ \{P\} \end{Bmatrix}$$

$$= \begin{Bmatrix} \{F^1\} \\ \{F^2\} \\ \{0\} \end{Bmatrix} \tag{4.237}$$

where $[K^{\alpha\beta}]$ and $\{F^\alpha\}$ $(\alpha, \beta = 1,2,3)$ are given by Eqs. (4.196), $M_{ij}^{11} = M_{ij}^{22}$ is given by M_{ij} in Eqs. (4.235). Equation (4.237) can be written in the general form

$$[M]\{\dot{\Delta}\} + [K]\{\Delta\} = \{F\} \qquad \{\Delta\} = \begin{Bmatrix} \{u\} \\ \{v\} \\ \{P\} \end{Bmatrix} \tag{4.238}$$

Penalty model The time-dependent model associated with Eqs. (4.236) is given by

$$[M]\{\dot{\Delta}\} + [K]\{\Delta\} = \{F\} \tag{4.239}$$

where $[K]$ and $\{F\}$ are as defined in Eqs. (4.201) and (4.196), and $[M]$ is given by

$$[M] = \begin{bmatrix} [M^{11}] & [0] \\ [0] & [M^{22}] \end{bmatrix}$$

$M_{ij}^{11} = M_{ij}^{22}$ being the coefficient matrices defined in the mixed model above.

Plane elasticity Consider the equations governing the motion of a plane elastic solid [i.e., the dynamic analog of Eqs. (4.140)]:

$$\left.\begin{aligned} c_1 \frac{\partial^2 u}{\partial t^2} - A(u, v) - f_x = 0 \\ c_1 \frac{\partial^2 v}{\partial t^2} - B(u, v) - f_y = 0 \end{aligned}\right\} \text{ in } \Omega \tag{4.240a}$$

with the initial conditions

$$u = u_0 \qquad v = v_0$$
$$\frac{\partial u}{\partial t} = \dot{u}_0 \qquad \frac{\partial v}{\partial t} = \dot{v}_0 \qquad t = 0 \tag{4.240b}$$

where $A(\cdot, \cdot)$ and $B(\cdot, \cdot)$ are the linear (differential) operators on the left-hand side of Eqs. (4.140), and c_1 is a constant which depends on the density ρ ($c_1 = \rho h$) of the body. For a typical element $\Omega^{(e)}$, the semidiscrete finite-element model is given by substituting Eqs. (4.146) into the semidiscrete variational formulation of Eqs. (4.240a):

$$\begin{bmatrix} [M^1] & [0] \\ [0] & [M^2] \end{bmatrix} \begin{Bmatrix} \{\ddot{u}\} \\ \{\ddot{v}\} \end{Bmatrix} + \begin{bmatrix} [K^{11}] & [K^{12}] \\ [K^{12}]^T & [K^{22}] \end{bmatrix} \begin{Bmatrix} \{u\} \\ \{v\} \end{Bmatrix} = \begin{Bmatrix} \{F^1\} \\ \{F^2\} \end{Bmatrix} \tag{4.241}$$

where $M_{ij}^1 = M_{ij}^2$ are the mass matrices defined by M_{ij} in Eqs. (4.235) (with c_1 replaced by ρh), and $K_{ij}^{\alpha\beta}$ and F_i^α ($i, j = 1, 2, \ldots, n$; $\alpha, \beta = 1, 2$) are given by Eqs. (4.156).

Plate Bending The equations governing the motion of elastic plates in bending are given by

$$\left.\begin{aligned} L_1(w, S_x, S_y) + q &= c_1 \frac{\partial^2 w}{\partial t^2} \\ L_2(w, S_x, S_y) &= c_2 \frac{\partial^2 S_x}{\partial t^2} \\ L_3(w, S_x, S_y) &= c_2 \frac{\partial^2 S_y}{\partial t^2} \end{aligned}\right\} \text{ in } \Omega \qquad t \geqslant 0 \tag{4.242}$$

where $L_1(\cdot)$, $L_2(\cdot)$, and $L_3(\cdot)$ are linear differential operators defined by the left-hand sides of Eqs. (4.220), and c_1 and c_2 are constants that depend on the plate properties. Substituting Eqs. (4.224) into the semidiscrete variational formulation (4.242), we obtain

$$\begin{bmatrix} [M^1] & [0] & [0] \\ [0] & [M^2] & [0] \\ [0] & [0] & [M^3] \end{bmatrix} \begin{Bmatrix} \{\ddot{w}\} \\ \{\ddot{S}_x\} \\ \{\ddot{S}_y\} \end{Bmatrix} + [K]\{\Delta\} = \{F\} \tag{4.243}$$

where $\{\Delta\} = \{\{w\} \;\; \{S_x\} \;\; \{S_y\}\}^T$, $c_1 = \rho h$ and $c_2 = \rho h^3/12$, ρ being the density,

$$M_{ij}^1 = \int_{\Omega^{(e)}} c_1 \psi_i \psi_j \, dx \, dy \qquad M_{ij}^2 = M_{ij}^3 = \int_{\Omega^{(e)}} c_2 \psi_i \psi_j \, dx \, dy \tag{4.244}$$

and $[K]$ and $\{F\}$ are the coefficient matrices whose elements are defined by Eqs. (4.226).

4.6.3 Temporal Approximations

Note that the semidiscrete finite-element models (for an element) developed in Sec. 4.6.2 for various physical problems can be classified, according to the order of the time derivatives appearing in the models, into two groups:

$$[M]\{\dot{\Delta}\} + [K]\{\Delta\} = \{F\} \tag{4.245}$$

$$[M]\{\ddot{\Delta}\} + [K]\{\Delta\} = \{F\} \tag{4.246}$$

Equations (4.234), (4.238), and (4.239) are special cases of Eq. (4.245), and Eqs. (4.241) and (4.243) are special cases of Eq. (4.246). Using the developments of Sec. 2.3.4, we derive the fully discretized models of Eqs. (4.245) and (4.246).

The fully discretized element equations associated with both Eqs. (4.245) and (4.246) can be expressed in the form

$$[\hat{K}]\{\Delta\}_{n+1} = \{\hat{F}\} \tag{4.247}$$

where $[\hat{K}]$ and $\{\hat{F}\}$ for Eqs. (4.245) and (4.246) are given, respectively, by

$$\begin{aligned} [\hat{K}] &= [M] + a_4[K] \\ \{\hat{F}\} &= ([M] - a_3[K])\{\Delta\}_n + a_4\{F\}_n + a_3\{F\}_{n+1} \end{aligned} \tag{4.248}$$

and

$$\begin{aligned} [\hat{K}] &= [K] + a_0[M] \\ \{\hat{F}\} &= \{F\}_{n+1} + [M]\left(a_0\{\Delta\}_n + a_1\{\dot{\Delta}\}_n + a_2\{\ddot{\Delta}\}_n\right) \end{aligned} \tag{4.249}$$

Here a_0, a_1, etc. are given by

$$a_0 = \frac{1}{\beta(\Delta t)^2} \qquad a_1 = a_0\,\Delta t \qquad a_2 = \frac{1}{2\beta} - 1$$

$$a_3 = (1 - \alpha)\,\Delta t \qquad a_4 = \alpha\,\Delta t \tag{4.250}$$

Equation (4.247) can be solved for $\{\Delta\}$ at time $t = t_{n+1} \equiv (n + 1)\,\Delta t$ by using known $\{\Delta\}$ at $t = t_n$. At time $t = 0$, the initial conditions of the problem are used to initiate the time-marching scheme. For the second-order (in time) problems, $\{\ddot{\Delta}\}$ at $t = 0$ can be computed from Eq. (4.246).

Example 4.13 We wish to solve the time-dependent heat conduction equation

$$\frac{\partial T}{\partial t} - \nabla^2 T = 1 \text{ in } \Omega = \{(x, y)\colon\ 0 < (x, y) < 1\} \tag{4.251a}$$

subject to the boundary conditions, for $t \geqslant 0$,

$$\begin{aligned} T &= 0 \text{ on } \Gamma_1 = \{\text{lines } x = 1 \text{ and } y = 1\} \\ \frac{\partial T}{\partial n} &= 0 \text{ on } \Gamma_2 = \{\text{lines } x = 0 \text{ and } y = 0\} \end{aligned} \tag{4.251b}$$

and the initial conditions, for (x, y) in Ω,

$$T = 0 \qquad t = 0 \tag{4.251c}$$

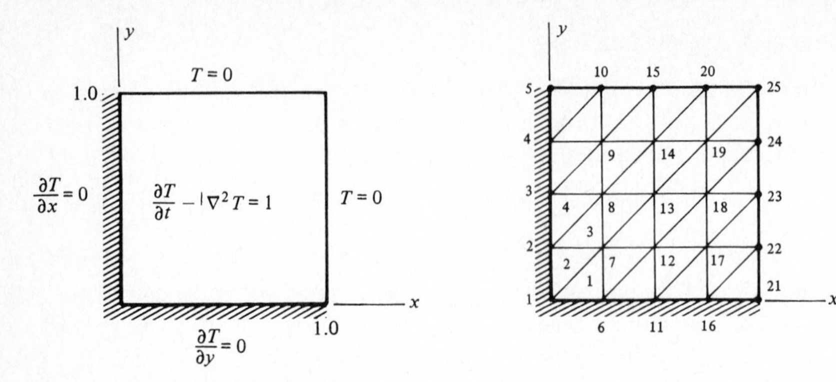

Figure 4.42 Domain, boundary conditions, and finite-element mesh for the transient heat conduction problem of Example 4.13.

We choose a 4×4 mesh of linear triangular elements (see Fig. 4.42) to model the domain and use the Crank-Nicolson method (i.e., $\theta = 0.5$) for the temporal approximation. Since the Crank-Nicolson method is unconditionally stable, one can choose any value of Δt. However, for large values of Δt the solution may not be accurate. To have a reasonable accuracy we compute Δt according to the formula

$$\Delta t \leqslant \frac{2}{\lambda_{\min}} \tag{4.252}$$

where $\lambda_{\min}$ is the minimum eigenvalue of the operator $-\nabla^2$ (see Prob. 4.63). In the present case, we choose $\Delta t = 0.05$.

The element equations are given by Eqs. (4.234), where $c_1 = 1$, $k_1 = k_2 = 1$, $\beta = 0$, and $f = 1$. The boundary conditions of the problem are given by

$$U_5 = U_{10} = U_{15} = U_{20} = U_{21} = U_{22} = U_{23} = U_{24} = U_{25} = 0$$

Beginning with the initial conditions $U_i = 0$ ($i = 1, 2, \ldots, 25$), we solve the assembled set of equations associated with Eqs. (4.247) and (4.248). The temperature $T(x, y, 0)$ versus x for various values of time are shown in Fig. 4.43a. The steady state is reached at time $t = 1.0$. The temperature $T(0, 0, t)$ versus time is shown in Fig. 4.43b, which indicates the evolution of the temperature from zero to the steady state. A comparison of the transient solution at $t = 1.0$ is given in Table 4.11 with the steady-state finite-element, the finite-difference, and the exact solutions. ■

Example 4.14 Consider the Couette flow problem of Example 4.10 with zero initial conditions (i.e., $u_i = v_i = 0$) and boundary conditions in Eq. (4.208). The bilinear element and the same mesh as that in Example 4.10 are used to investigate the transient motion of the fluid. Results obtained by the penalty finite-element model are presented. Figure 4.44 contains plots of the horizontal velocity u versus y for $t = 1.0$, 5.0, 10.0, and 35.0. The flow gradually develops to the fully developed state in 35 units of time (for unit value of the viscosity). ■

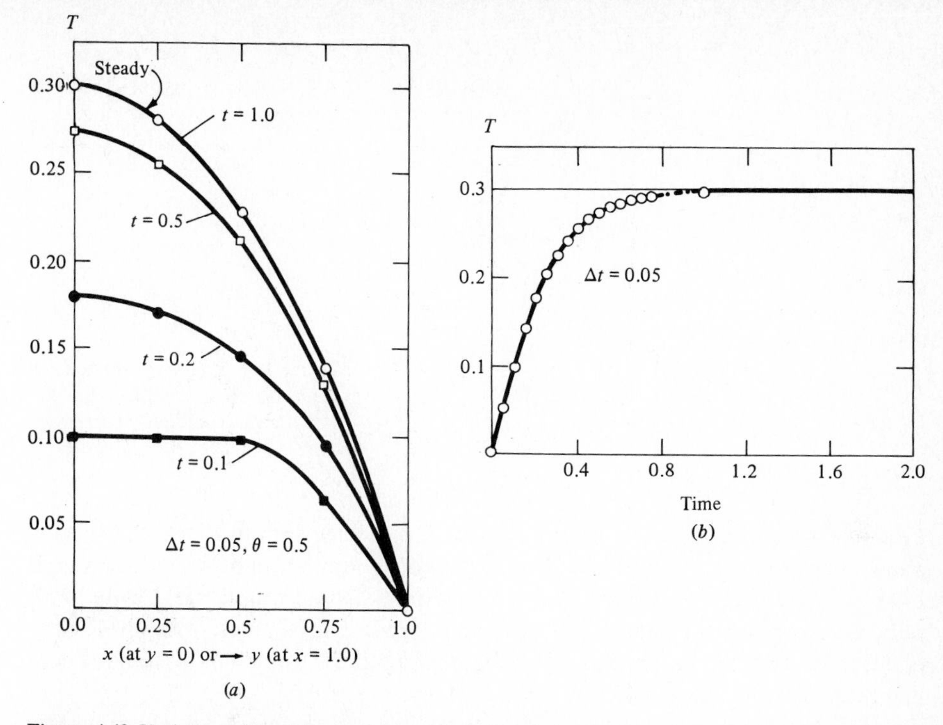

Figure 4.43 Variation of the temperature as a function of position x and time t for the transient heat conduction problem of Example 4.13.

Table 4.11 Comparison of finite-difference and finite-element solutions with the exact solution of the heat conduction problem

Node	Exact solution	Finite-difference solution	Error	Finite-element solution (Steady)	Error	Finite-element solution (Unsteady) at t = 1.0
1	0.2947	0.2911	0.0036	0.3013	−0.0066	0.2993
2	0.2789	0.2755	0.0034	0.2805	−0.0016	0.2786
3	0.2293	0.2266	0.0027	0.2292	0.0001	0.2278
4	0.1397	0.1381	0.0016	0.1392	0.0005	0.1385
5	0.0000	0.0000	0.0000	0.0000	0.0000	0.0000
7	0.2642	0.2609	0.0033	0.2645	−0.0003	0.2628
8	0.2178	0.2151	0.0027	0.2172	0.0006	0.2159
9	0.1333	0.1317	0.0016	0.1327	0.0006	0.1320
10	0.0000	0.0000	0.0000	0.0000	0.0000	0.0000
13	0.1811	0.1787	0.0024	0.1801	0.0010	0.1791
14	0.1127	0.1110	0.0017	0.1117	0.0010	0.1111
15	0.0000	0.0000	0.0000	0.0000	0.0000	0.0000
19	0.0728	0.0711	0.0017	0.0715	0.0013	0.0712
20	0.0000	0.0000	0.0000	0.0000	0.0000	0.0000
25	0.0000	0.0000	0.0000	0.0000	0.0000	0.0000

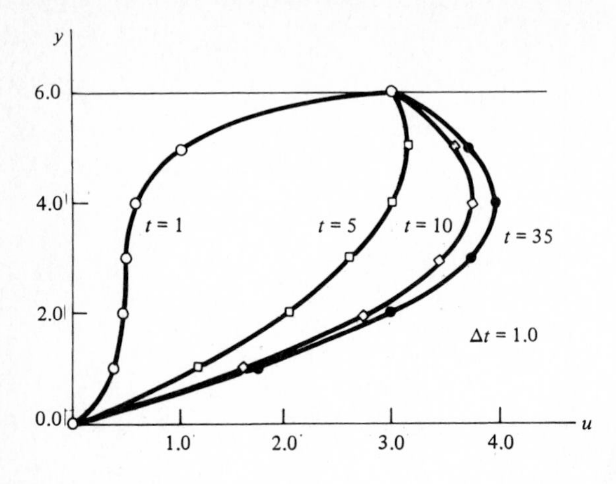

Figure 4.44 Transient solution for the horizontal velocity of the Couette flow problem of Example 4.14.

Example 4.15 Consider an isotropic ($\nu = 0.3$, $\rho = 1.0$), simply supported rectangular plate ($a/b = \sqrt{2}$, $h/b = 0.2$) under suddenly applied uniformly distributed load on a square ($c/b = 0.4$) area at the center. Due to the biaxial symmetry, only one quadrant of the plate can be analyzed. The geometry, the boundary conditions, and the finite-element (nonuniform) mesh of 4×4 nine-node elements are shown in Fig. 4.45.

The time step was selected using the following estimate for conditionally stable time integration schemes for the bending of thick plates:

$$(\Delta t)^2 \leqslant \frac{d^2\left[\rho(1-\nu^2)/E\right]}{2 + (1-\nu)(\pi^2/12)\left[1 + 1.5(d/h)^2\right]} \tag{4.253}$$

where d is the minimum distance between the element node points. Using Eq. (4.253) for the problem at hand, we obtain $\Delta t = 0.03$ (for $d = 0.05$). Since the estimate in Eq. (4.253) is valid for conditionally stable time integration schemes

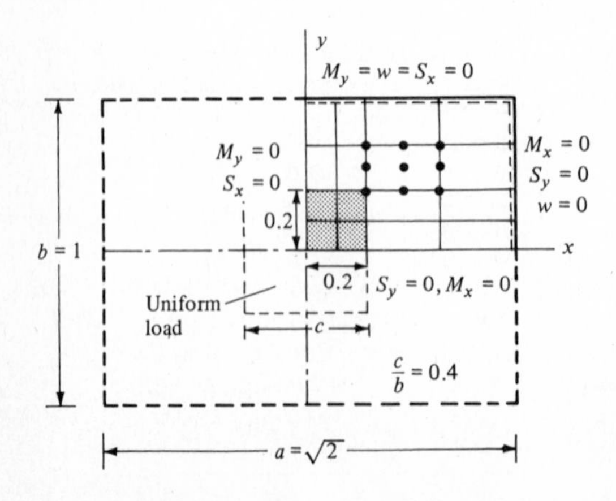

Figure 4.45 Domain, boundary conditions, and finite-element mesh for the bending of a rectangular plate under suddenly applied pulse loading at the central square area.

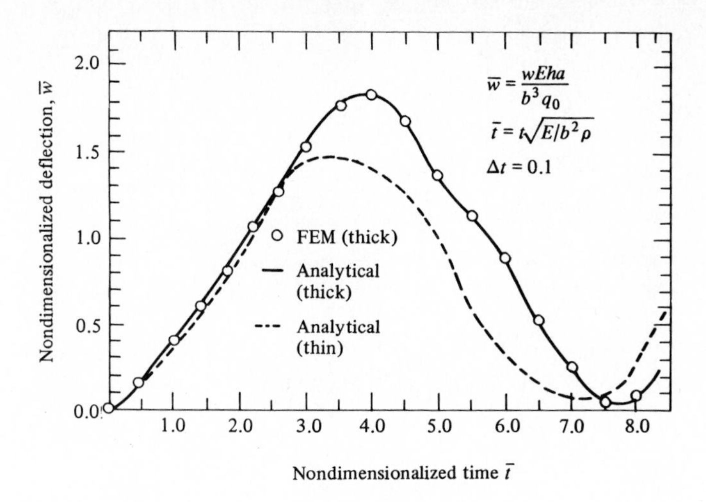

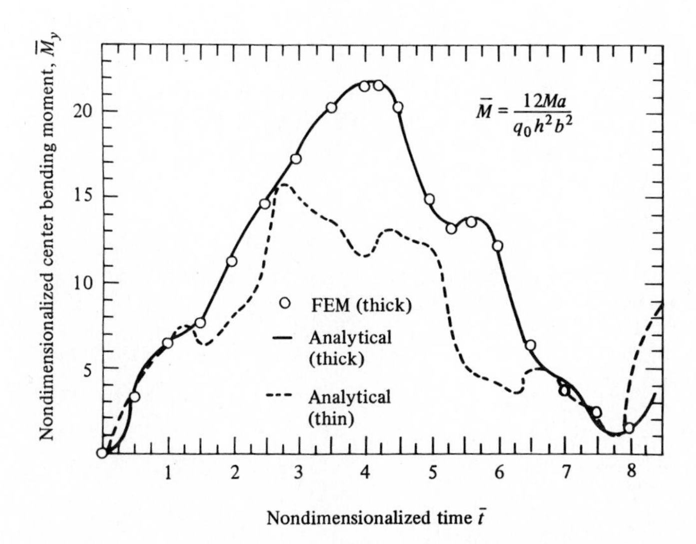

Figure 4.46 Comparison of the finite-element solution with the analytical solution of a simply supported rectangular isotropic plate under suddenly applied pulse loading at the central square area.

and the present scheme is unconditionally stable for $\alpha = 0.5$ and $\beta = 0.25$, we can use a much larger time step, say $\Delta t = 0.1$.

An analytical solution to this problem was obtained by Reismann and Lee (1969) (see References for Additional Reading at the end of this chapter). The finite-element solutions for nondimensionalized transverse deflection and bending moment at the center of the plate are compared in Fig. 4.46 with the analytical solution.The finite-element results are in excellent agreement with the analytical solution. ■

PROBLEMS

4.45 Consider the pair of equations

$$\left.\begin{aligned} \operatorname{grad} u - \frac{1}{k}\mathbf{q} = 0 \\ \operatorname{div} \mathbf{q} + f = 0, \end{aligned}\right\} \text{ in } \Omega$$

where u and $\mathbf{q}$ are the dependent variables and k and f are given functions of position (x, y) in a two-dimensional domain Ω. Derive the finite-element formulation of the equations in the form

$$\begin{bmatrix} [K^{11}] & [K^{12}] & [K^{13}] \\ & [K^{22}] & [K^{23}] \\ \text{Symmetric} & & [K^{33}] \end{bmatrix} \begin{Bmatrix} \{u\} \\ \{q^1\} \\ \{q^2\} \end{Bmatrix} = \begin{Bmatrix} \{F^1\} \\ \{F^2\} \\ \{F^3\} \end{Bmatrix}$$

Caution: Do not eliminate the variable u from the given equations.

4.46 Consider the following equations governing a thin, elastic, isotropic plate:

$$-S(M_1 - \nu M_2) - \frac{\partial^2 w}{\partial x^2} = 0$$

$$-S(M_2 - \nu M_1) - \frac{\partial^2 w}{\partial y^2} = 0$$

$$\frac{2}{(1+\nu)S}\frac{\partial^4 w}{\partial x^2\,\partial y^2} - \frac{\partial^2 M_1}{\partial x^2} - \frac{\partial^2 M_2}{\partial y^2} - q = 0$$

Here M_1 and M_2 are the bending moments, w is the transverse deflection, q is the distributed load, ν is Poisson's ratio, and S is a constant. Give

(a) The variational formulation of the equations

(b) The finite-element formulation (M_1, M_2, and w are the dependent unknowns) in the form

$$\begin{bmatrix} [K^{11}] & [K^{12}] & [K^{13}] \\ & [K^{22}] & [K^{23}] \\ \text{Symmetric} & & [K^{33}] \end{bmatrix} \begin{Bmatrix} \{w\} \\ \{M_1\} \\ \{M_2\} \end{Bmatrix} = \begin{Bmatrix} \{F^1\} \\ \{F^2\} \\ \{F^3\} \end{Bmatrix}$$

4.47 Use the interpolation

$$w = \sum_{j=1}^{4} w_j \psi_j \qquad M_1 = \sum_{j=1}^{2} m_1^j \phi_j^1 \qquad M_2 = \sum_{j=1}^{2} m_2^j \phi_j^2$$

with

$$\psi_1 = (1-x)(1-y) \qquad \psi_2 = x(1-y) \qquad \psi_3 = xy \qquad \psi_4 = (1-x)y$$

$$\phi_1^1 = 1 - x \qquad \phi_2^1 = x \qquad \phi_1^2 = 1 - y \qquad \phi_2^2 = y$$

for a rectangular element with sides a and b to evaluate the matrices $[K^{\alpha\beta}]$, $\alpha, \beta = 1,2,3$, in Prob. 4.46.

4.48 Repeat Prob. 4.47 for the case in which $\phi_i^1 = \phi_i^2 = \psi_i$ $(i = 1,2,3,4)$.

4.49 Compute the contribution of the surface forces to the global force degrees of freedom in the plane elasticity problems given in Fig. P4.49.

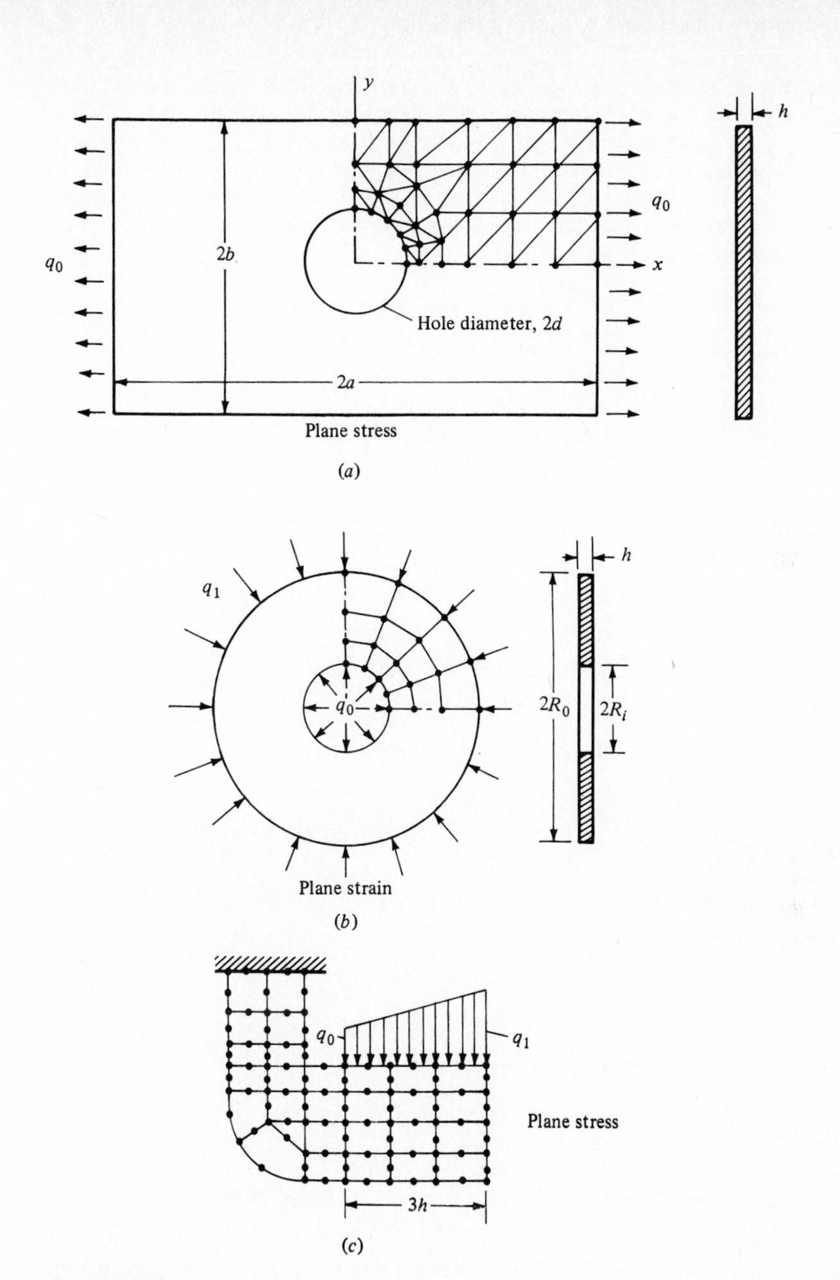
y
h
q_0
q_0
2b
x
Hole diameter, 2d
2a
Plane stress
(a)
h
q_1
q_0
$2R_0$
$2R_i$
Plane strain
(b)
q_0
q_1
Plane stress
3h
(c)

Figure P4.49

4.50 Give the connectivity matrices and the specified primary degrees of freedom for the plane elasticity problems given in Fig. P4.49. Give only the first three rows of the connectivity matrix.

4.51 Consider the cantilevered beam of length 6 cm, height 2 cm, thickness 1 cm, and material properties $E = 3 \times 10^7$ N/cm^2 and $\nu = 0.3$, and subjected to a bending moment of 600 N · cm at the free end, (as shown in P4.51.) Replace the moment by an equivalent pair of in-plane loads at the free end, and model the domain by a 4 × 2 mesh of linear rectangular elements. Identify the specified displacements and global forces.

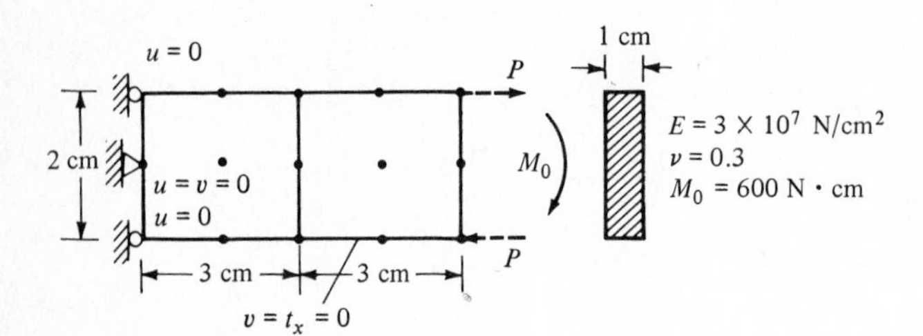

Figure P4.51

4.52 Consider the ("transition") element shown in Fig. P4.52. Define the generalized displacement vector of the element by

$$\{u\} = \{u_1, v_1, \theta_1, u_2, v_2, u_3, v_3\}^T$$

and represent the displacement components u and v by

$$u = \psi_1 u_1 + \psi_2 u_2 + \psi_3 u_3 + \frac{h}{2}\eta\psi_1\theta_1$$

$$v = \psi_1 v_1 + \psi_2 v_2 + \psi_3 v_3$$

where ψ_1 is the interpolation function for the beam, and ψ_2 and ψ_3 are the interpolation functions for nodes 2 and 3:

$$\psi_1 = \tfrac{1}{2}(1 - \xi) \qquad \psi_2 = \tfrac{1}{3}(1 + \xi)(1 - \eta) \qquad \psi_3 = \tfrac{1}{4}(1 + \xi)(1 + \eta)$$

Derive the stiffness matrix for the element using Eq. (4.150).

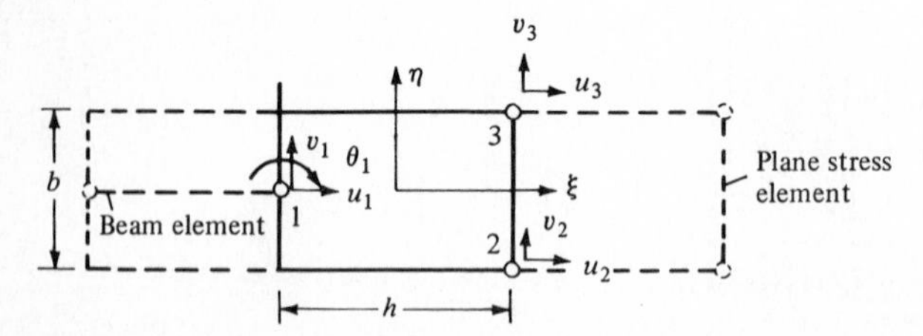

Figure P4.52

4.53 Consider a square, isotropic, elastic body of thickness h shown in Fig. P4.53. Suppose that the displacements are approximated by

$$u(x, y) = (1 - x)yu_1 + x(1 - y)u_2$$

$$v(x, y) = 0$$

Assuming that the body is in a plane state of stress, derive the 2 × 2 stiffness matrix for the unit square

$$[K]\begin{Bmatrix} u_1 \\ u_2 \end{Bmatrix} = \begin{Bmatrix} F_1 \\ F_2 \end{Bmatrix}$$

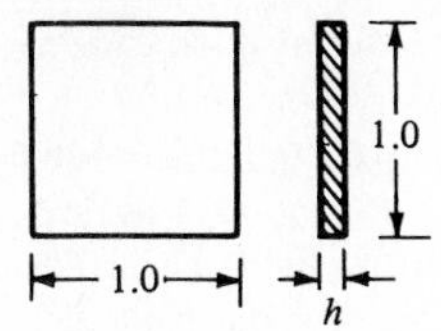

Figure P4.53

4.54 (Natural convection in flow between heated vertical plates) Consider the flow of a viscous incompressible fluid in the presence of a temperature gradient between two stationary long vertical plates. Assuming zero pressure gradient between the plates, we can write $u = u(y)$, $v = 0$, $T = T(y)$, and

$$0 = \rho\beta g(T - T_m) + \mu\frac{d^2u}{dy^2}$$

$$0 = k\frac{d^2T}{dy^2} + \mu\left(\frac{du}{dy}\right)^2$$

where $T_m = \frac{1}{2}(T_0 + T_1)$ is the mean temperature of the two plates, g the gravitational acceleration, ρ the density, β the coefficient of thermal expansion, μ the viscosity, and k the thermal conductivity of the fluid. Give a finite-element formulation of the equations and discuss the solution strategy for the computational scheme.

4.55 The equations given in Prob. P4.54 are a special case of the more general Navier-Stokes equations (which are nonlinear and coupled) for two-dimensional flow:

(a) *Mass continuity*

$$\frac{\partial u}{\partial x} + \frac{\partial v}{\partial y} = 0$$

(b) *Conservation of momentum*

$$\rho\left(u\frac{\partial u}{\partial x} + v\frac{\partial u}{\partial y}\right) = -\frac{\partial P}{\partial x} + \mu\left(\frac{\partial^2 u}{\partial x^2} + \frac{\partial^2 u}{\partial y^2}\right) + \rho g\beta(T - T_m)$$

$$\rho\left(u\frac{\partial v}{\partial x} + v\frac{\partial v}{\partial y}\right) = -\frac{\partial P}{\partial y} + \mu\left(\frac{\partial^2 v}{\partial x^2} + \frac{\partial^2 v}{\partial y^2}\right)$$

(c) *Energy equation*

$$\rho c\left(u\frac{\partial T}{\partial x} + v\frac{\partial T}{\partial y}\right) = k\left(\frac{\partial^2 T}{\partial x^2} + \frac{\partial^2 T}{\partial y^2}\right)$$

where c is the specific heat (the x coordinate is taken vertically downward).

Construct the finite-element model of the equations, and discuss the computational strategy.

4.56 Set the nonlinear terms and the temperature terms in Eq. (b) of Prob. 4.55 to zero (so that the momentum equations are uncoupled from the energy equation) and repeat Prob. 4.55.

4.57 Consider Eqs. (4.188) and (4.189) in cylindrical coordinates (r, ϕ, z). For axisymmetric viscous incompressible flows, we have

$$0 = -\frac{\partial P}{\partial r} + \mu\left(\frac{\partial^2 u}{\partial r^2} + \frac{1}{r}\frac{\partial u}{\partial r} - \frac{u}{r^2} + \frac{\partial^2 u}{\partial z^2}\right)$$

$$0 = -\frac{\partial P}{\partial z} + \mu\left(\frac{\partial^2 w}{\partial r^2} + \frac{1}{r}\frac{\partial w}{\partial r} + \frac{\partial^2 w}{\partial z^2}\right)$$

$$0 = \frac{\partial u}{\partial r} + \frac{u}{r} + \frac{\partial w}{\partial z}$$

Give the finite-element formulation of the equations by (a) the pressure-velocity formulation and (b) the penalty function formulation.

4.58 (Hydrodynamic theory of lubrication) The fluid phenomenon that takes place in oil-lubricated bearings can be analyzed using the equations of a viscous incompressible fluid. The essential features of the motion can be understood by considering an example of a slide block moving on a plane guide surface (see Fig. P4.58). In order to have a nonzero pressure gradient along the guide surface, the slide block should be inclined at a small angle α to the guide surface. For steady-state two-dimensional motion, we assume that the sliding surfaces are very large ($L/h_1 \gg 1$), and that the slide block is at rest while the guide surface is forced to move with a constant velocity U_0. Under these conditions, very large pressure differences will be created in the fluid (i.e., lubricant) occupying the gap between block and guide. Consequently, the sliding surfaces will be separated by the oil film, avoiding metallic contact between the surfaces. For the problem sketched in Fig. P4.58, give the boundary conditions for the domain occupied by the fluid. Assume that the openings are at atmospheric pressure P_0. For a crude finite-element mesh, give the specified primary degrees of freedom and the secondary degrees of freedom for the penalty-finite element model of the problem.

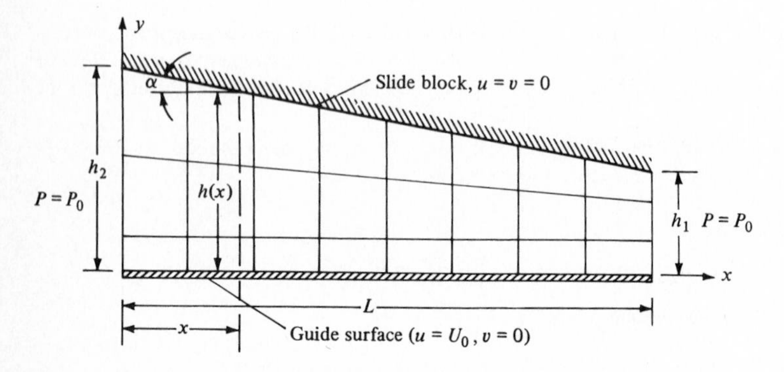

Figure P4.58

4.59 For the viscous flow problems given in Fig. P4.59, give the specified primary and secondary degrees of freedom and their values.

4.60 For the plate bending problems given in Fig. P4.60, give the specified primary and secondary degrees of freedom and their values.

4.61 Compute the element matrices in Eq. (4.225) for a linear triangular element [i.e., ψ_i of Eqs. (4.224) are given by Eqs. (4.25)].

4.62 Give an algebraic form of the elements of the stiffness matrix (in a local coordinate system) for a thin plane elastic body subjected to both in-plane and transverse loads. (*Hint*: combine the element matrices of the plane stress element with the plate bending element—analogous to the construction of a frame element from bar and beam elements.)

4.63 Determine the time step for the transient analysis of the problem

$$\frac{\partial u}{\partial t} - \nabla^2 u = 1 \text{ in } \Omega$$

$$u = 0 \text{ on } \Gamma \text{ for } t \geqslant 0 \qquad u = 0 \text{ in } \Omega \text{ at } t = 0$$

by determining the minimum eigenvalue of the problem

$$-\nabla^2 u = \lambda u \text{ in } \Omega$$
$$u = 0 \text{ on } \Gamma$$

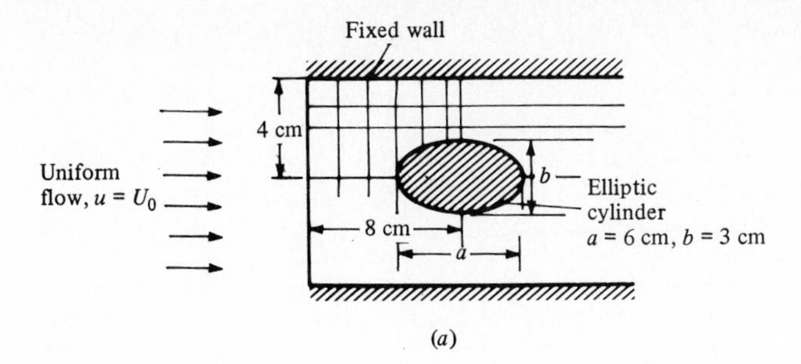

(a)

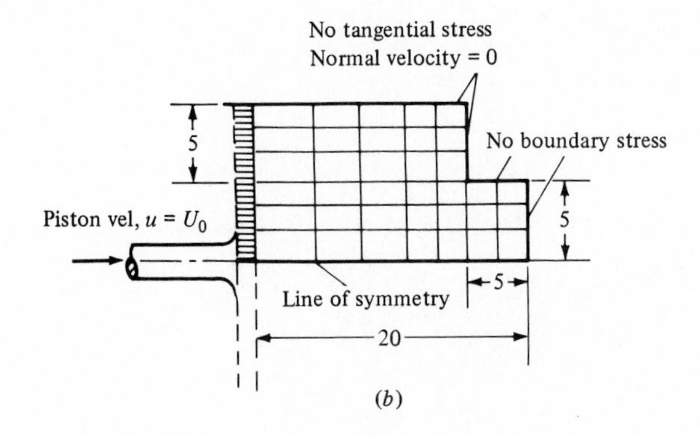

(b)

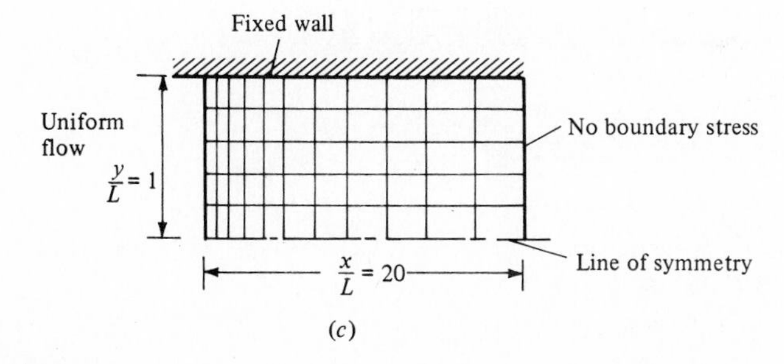

(c)

Figure P4.59

4.64 (Central difference method) Consider the following matrix differential equation in time:

$$[M]\{\ddot{U}\} + [C]\{\dot{U}\} + [K]\{U\} = \{F\}$$

where the superposed dot indicates differentiation with respect to time. Assume

$$\{\ddot{U}\}_n = \frac{1}{(\Delta t)^2}(\{U\}_{n-1} - 2\{U\}_n + \{U\}_{n+1})$$

$$\{\dot{U}\}_n = \frac{1}{2(\Delta t)}(\{U\}_{n+1} - \{U\}_{n-1})$$

and derive the algebraic equations for the solution of $\{U\}_{n+1}$ in the form

$$[A]\{U\}_{n+1} = \{F\}_n - [B]\{U\}_n - [C]\{U\}_{n-1}$$

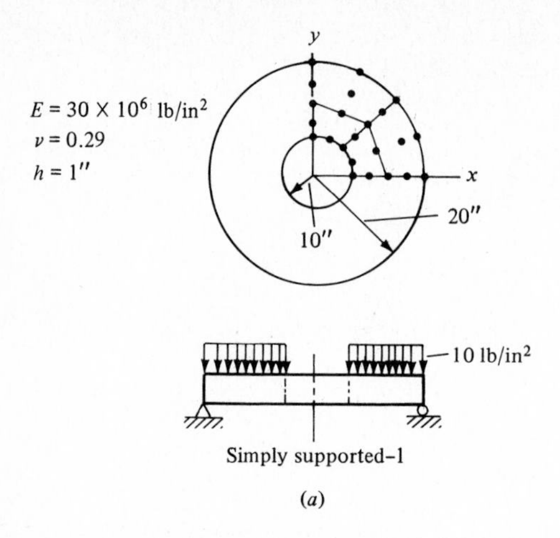

(a)

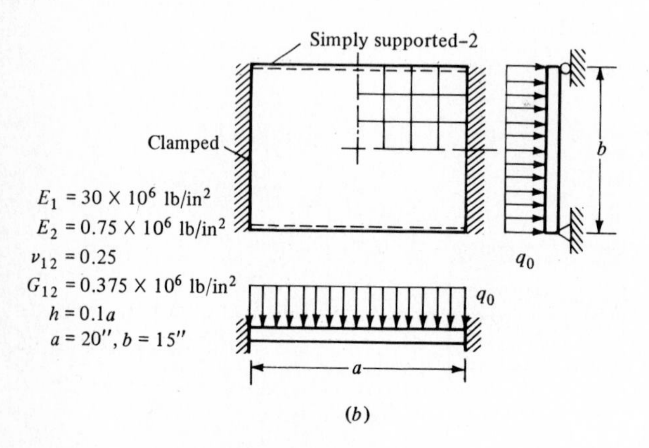

(b)

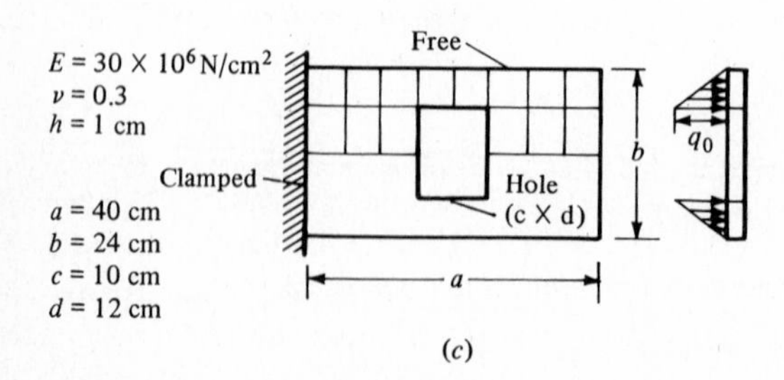

(c)

Figure P4.60

4.65 Consider the first-order differential equation in time

$$a\frac{du}{dt} + bu = f$$

Using linear approximation, $u(t) = u_1\psi_1(t) + u_2\psi_2(t)$, $\psi_1 = 1 - t/\Delta t$, and $\psi_2 = t/\Delta t$, derive the associated algebraic equation and compare with that obtained using the θ family of approximation.

4.66 (Space-time element) Consider the differential equation

$$c\frac{\partial u}{\partial t} - \frac{\partial}{\partial x}\left(a\frac{\partial u}{\partial x}\right) = f \qquad \text{for } 0 < x < L \qquad 0 < t \leqslant T$$

$$u(0, t) = u(L, t) = 0 \qquad \text{for } 0 \leqslant t \leqslant T$$

$$u(x, 0) = u_0(x) \qquad \text{for } 0 < x < L$$

where $c = c(x)$, $a = a(x)$, $f = f(x, t)$, and u_0 are given functions. Consider the rectangular domain defined by

$$\Omega = \{(x, t): \quad 0 < x < L, 0 < t \leqslant T\}$$

A finite-element discretization of Ω by rectangles is a time-space rectangular element (with y replaced by t). Give a finite-element formulation of the equation over a time-space element, and discuss the mathematical/practical limitations of such a formulation. Compute the element matrices for a linear element.

4.7 ISOPARAMETRIC ELEMENTS AND NUMERICAL INTEGRATION

4.7.1 Isoparametric Elements

Recall from Sec. 3.6 that isoparametric elements are those which can be used for the description of both the geometry of the element and the variation of dependent variables:

$$x = \sum_{i=1}^{n} x_i\hat{\psi}_i \qquad y = \sum_{i=1}^{n} y_i\hat{\psi}_i \qquad u = \sum_{i=1}^{n} u_i\hat{\psi}_i \tag{4.254}$$

and so on, where $\hat{\psi}_i = \hat{\psi}_i(\xi, \eta)$ are the element interpolation functions in natural coordinates (ξ, η).

The isoparametric concept is very useful because it facilitates an accurate representation of irregular domains (e.g., domains with curved boundaries). However, the use of curvilinear isoparametric elements makes it difficult to compute the element coefficient matrices and column vectors directly in terms of the global coordinates x and y (which are used to describe the equations of the problem). This difficulty can be overcome by introducing an invertible transformation between a curvilinear element Ω^e, such as that indicated in Fig. 4.47, and a *master* element $\hat{\Omega}$ of simple shape that facilitates numerical integration of element equations. The transformation is accomplished by a coordinate transformation of the form (4.254). Consider, as an example, the master element shown in Fig. 4.47. The coordinates in the master element are chosen to be the natural coordinates (ξ, η) such that $-1 \leqslant (\xi, \eta) \leqslant 1$. This choice is dictated by the limits

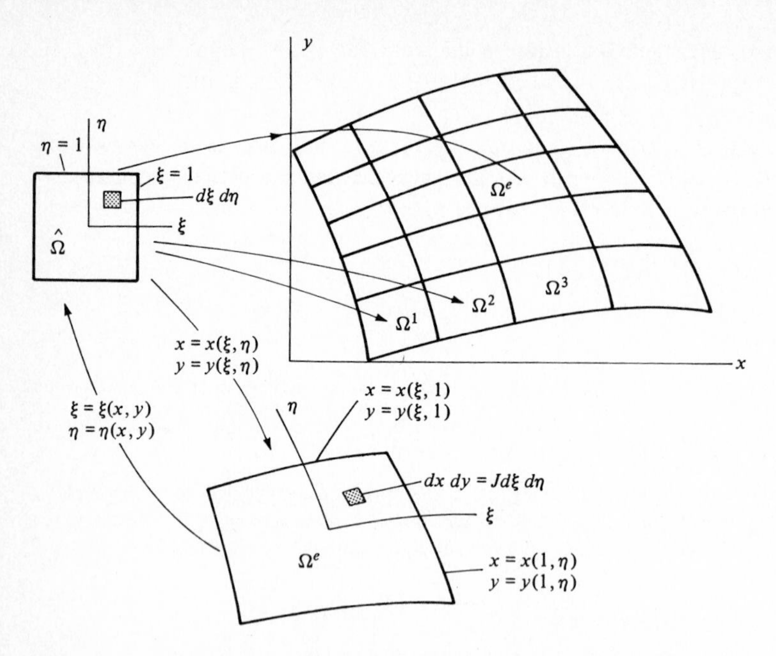

Figure 4.47 Generation of a finite-element mesh from a master element.

of integration in the Gauss quadrature rule. Consider the coordinate transformation

$$x = x(\xi, \eta) \qquad y = y(\xi, \eta) \tag{4.255}$$

of the points (ξ, η) in the master element $\hat{\Omega}$ onto points (x, y) in element Ω^e. For instance, the transformation maps the line $\xi = 1$ in $\hat{\Omega}$ to the curve, defined parametrically by $x = x(1, \eta)$, $y = y(1, \eta)$ in the xy plane. To be more specific, consider a special case of Eqs. (4.255),

$$x = \sum_{i=1}^{4} x_i \hat{\psi}_i(\xi, \eta) \qquad y = \sum_{i=1}^{4} y_i \hat{\psi}_i(\xi, \eta) \tag{4.256}$$

where $\hat{\psi}_i$ are the bilinear interpolation functions in Fig. 4.27 with node numbers 3 and 4 exchanged (also see Fig. 4.48), and (x_i, y_i) are the global coordinates of the ith node of element Ω^e. Now consider the line $\xi = 1$ in $\hat{\Omega}$. We have

$$\begin{aligned} x(1, \eta) &= \sum_{i=1}^{4} x_i \hat{\psi}_i(1, \eta) = x_1 \cdot 0 + \tfrac{1}{2} x_2 (1 - \eta) + \tfrac{1}{2} x_3 (1 + \eta) + x_4 \cdot 0 \\ &= \frac{x_2 + x_3}{2} + \frac{x_3 - x_2}{2} \eta \\ y(1, \eta) &= \sum_{i=1}^{4} y_i \hat{\psi}_i(1, \eta) = \frac{y_2 + y_3}{2} + \frac{y_3 - y_2}{2} \eta \end{aligned}$$

Clearly, x and y are linear functions of η, therefore they define a straight line.

Similarly, the lines $\xi = -1$ and $\eta = \pm 1$ are mapped into straight lines in the element Ω^e. In other words, the master element $\hat{\Omega}$ is transformed, under the linear transformation, into a quadrilateral element in the xy plane. When $\hat{\psi}_i$ are quadratic, then the transformed element in the xy plane will be a curvilinear element. Note that different elements of the finite-element mesh can be generated from a master element by assigning the global coordinates of the elements (see Fig. 4.47). Master elements of different order define different transformations and hence different collections of finite-element meshes. For example, a cubic-order master rectangular element can be used to generate a mesh of cubic curvilinear rectangular elements. Thus, with the help of an appropriate master element, any arbitrary element of a mesh can be generated. However, the transformations of a master element should be such that there exist no spurious gaps between elements and no element overlaps occur. The elements in Figs. 4.27 and 4.28 can be used as master elements.

In order to perform element calculations, we must transform functions of x and y to functions of ξ and η. Using the chain rule of differentiation, we write from Eqs. (4.255)

$$dx = \frac{\partial x}{\partial \xi} d\xi + \frac{\partial x}{\partial \eta} d\eta \qquad dy = \frac{\partial y}{\partial \xi} d\xi + \frac{\partial y}{\partial \eta} d\eta$$

or

$$\begin{Bmatrix} dx \\ dy \end{Bmatrix} = \begin{bmatrix} \dfrac{\partial x}{\partial \xi} & \dfrac{\partial x}{\partial \eta} \\ \dfrac{\partial y}{\partial \xi} & \dfrac{\partial y}{\partial \eta} \end{bmatrix} \begin{Bmatrix} d\xi \\ d\eta \end{Bmatrix} \equiv [J]^T \begin{Bmatrix} d\xi \\ d\eta \end{Bmatrix} \tag{4.257}$$

where $[J]$ is the jacobian matrix of the transformation (4.255). Equation (4.257) represents a linear transformation of line elements $d\xi$ and $d\eta$ in the master element $\hat{\Omega}$ into line elements dx and dy in the xy plane. In order to transform the x and y coordinates into the ξ and η coordinates, the inverse of $[J]$ must exist. A necessary and sufficient condition for Eq. (4.257) to be invertible is that the determinant J (called the jacobian) of the jacobian matrix be nonzero at every point of (ξ, η) in $\hat{\Omega}$:

$$J \equiv \det[J] = \frac{\partial x}{\partial \xi}\frac{\partial y}{\partial \eta} - \frac{\partial x}{\partial \eta}\frac{\partial y}{\partial \xi} \neq 0 \tag{4.258}$$

When $J \neq 0$, we have

$$\begin{Bmatrix} d\xi \\ \hline d\eta \end{Bmatrix} = ([J]^{-1})^T \begin{Bmatrix} dx \\ \hline dy \end{Bmatrix} = \frac{1}{J} \begin{bmatrix} \dfrac{\partial y}{\partial \eta} & -\dfrac{\partial x}{\partial \eta} \\ \hline -\dfrac{\partial y}{\partial \xi} & \dfrac{\partial x}{\partial \xi} \end{bmatrix}^T \begin{Bmatrix} dx \\ \hline dy \end{Bmatrix} \tag{4.259}$$

and

$$\xi = \xi(x, y) \qquad \eta = \eta(x, y) \tag{4.260}$$

From Eq. (4.259) it is clear that the functions $\xi = \xi(x, y)$ and $\eta = \eta(x, y)$ must be continuous, differentiable, and invertible. Moreover, the transformation (4.255) should be algebraically simple so that the jacobian matrix can be easily evaluated. Transformations of the form in Eq. (4.256) satisfy these requirements and the requirement that no spurious gaps between elements or overlapping of elements occur.

Example 4.16 Consider the three-element mesh shown in Fig. 4.48. Let the master element be the four-node square. Elements 1 and 2 have counterclockwise element node numbering consistent with the node numbering in the master element, and element 3 has node numbering opposite to that of the master element. Elements 1

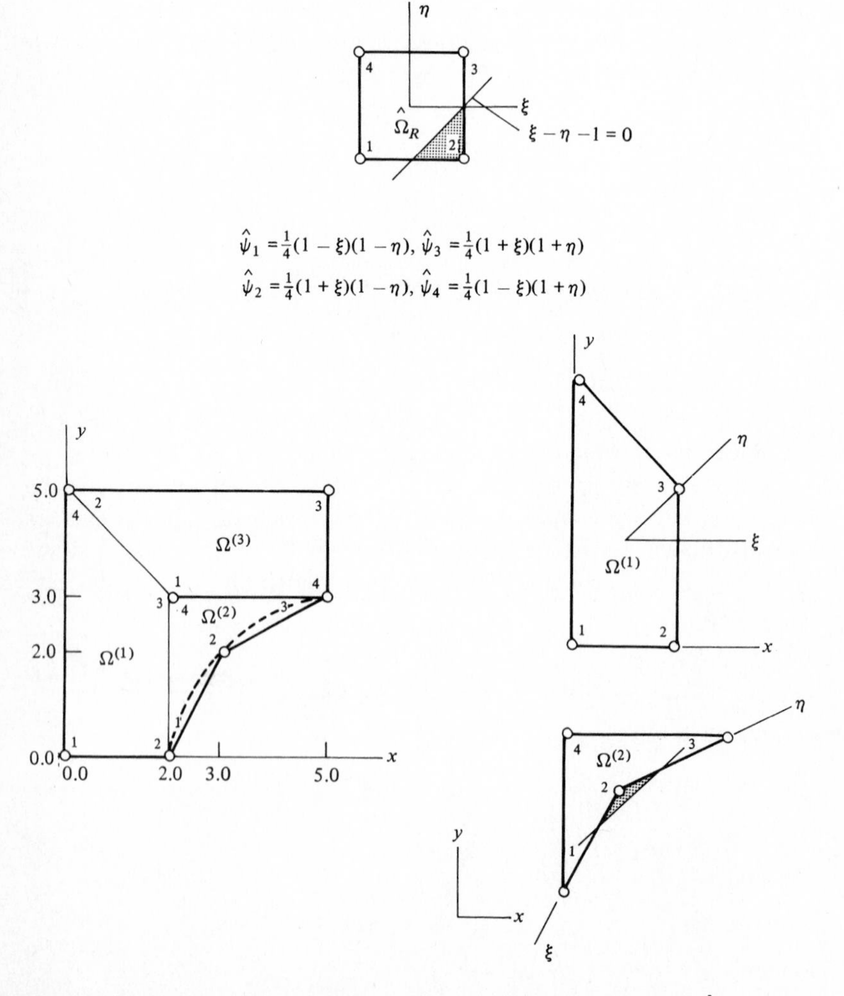

Figure 4.48 Examples of transformations of the master rectangular element $\hat{\Omega}_R$.

and 3 are *convex* domains in the sense that the line segment connecting any two arbitrary points of a convex domain lies entirely in the element. Clearly, element 2 is not convex because, for example, the line segment joining nodes 1 and 3 is not entirely inside the element. In the following paragraphs, we investigate the effect of node numbering and element convexity on the transformations from the master element to each of the three elements.

First, we compute the elements of the jacobian matrix (the interpolation functions are given in Fig. 4.48):

$$\frac{\partial x}{\partial \xi} = \sum_{i=1}^{4} x_i \frac{\partial \hat{\psi}_i}{\partial \xi} = \tfrac{1}{4}[-x_1(1-\eta) + x_2(1-\eta) + x_3(1+\eta) - x_4(1+\eta)]$$

$$\frac{\partial x}{\partial \eta} = \sum_{i=1}^{4} x_i \frac{\partial \hat{\psi}_i}{\partial \eta} = \tfrac{1}{4}[-x_1(1-\xi) - x_2(1+\xi) + x_3(1+\xi) + x_4(1-\xi)]$$

$$(4.261)$$

$$\frac{\partial y}{\partial \xi} = \sum_{i=1}^{4} y_i \frac{\partial \hat{\psi}_i}{\partial \xi} = \tfrac{1}{4}[-y_1(1-\eta) + y_2(1-\eta) + y_3(1+\eta) - y_4(1+\eta)]$$

$$\frac{\partial y}{\partial \eta} = \sum_{i=1}^{4} y_i \frac{\partial \hat{\psi}_i}{\partial \eta} = \tfrac{1}{4}[-y_1(1-\xi) - y_2(1+\xi) + y_3(1+\xi) + y_4(1-\xi)]$$

Next, we evaluate the jacobian for each of the elements.

Element 1 We have $x_1 = x_4 = 0$, $x_2 = x_3 = 2$; $y_1 = y_2 = 0$, $y_3 = 3$, $y_4 = 5$,

$$\begin{aligned} x &= 2\hat{\psi}_2 + 2\hat{\psi}_3 = 1 + \xi \\ y &= 3\hat{\psi}_3 + 5\hat{\psi}_4 = (1+\eta)(2 - \tfrac{1}{2}\xi) \end{aligned} \qquad (4.262a)$$

$$J = \det[J] = \begin{vmatrix} 1 & \dfrac{-(1+\eta)}{2} \\ 0 & 2 - \tfrac{1}{2}\xi \end{vmatrix} = \tfrac{1}{2}(4 - \xi) > 0 \qquad (4.262b)$$

Clearly, the jacobian is linear in ξ, and for all values of ξ in $-1 \leqslant \xi \leqslant 1$, the jacobian is positive. Therefore, the transformation (4.262a) is invertible.

Element 2 Here we have $x_1 = x_4 = 2$, $x_2 = 3$, $x_3 = 5$, $y_1 = 0$, $y_2 = 2$, and $y_3 = y_4 = 3$. The transformation and the jacobian are given by

$$\begin{aligned} x &= 3 + \xi + \frac{\eta}{2} + \frac{\xi\eta}{2} \\ y &= 2 + \frac{\xi}{2} + \eta - \frac{\xi\eta}{2} \end{aligned} \qquad (4.263a)$$

$$J = \begin{vmatrix} 1 + \dfrac{\eta}{2} & \dfrac{1-\eta}{2} \\ \dfrac{1+\xi}{2} & 1 - \dfrac{\xi}{2} \end{vmatrix} = \tfrac{3}{4}(1 + \eta - \xi) \qquad (4.263b)$$

The jacobian is *not* nonzero everywhere in the master element. The jacobian is zero along the line $\xi = 1 + \eta$ shown by the shaded area in the master element. Moreover, the shaded area in the master element is mapped into the shaded area outside element 2. Thus, elements with any interior angle greater than π should not be used in any finite-element mesh.

Element 3 We have $x_1 = 2$, $x_2 = 0$, $x_3 = x_4 = 5$, $y_1 = y_4 = 3$, and $y_2 = y_3 = 5$. The transformation and the jacobian become (note that the nodes are numbered clockwise)

$$x = 3 - \frac{\xi}{2} + 2\eta + \frac{\xi\eta}{2}$$
$$y = 4 + \xi \tag{4.264a}$$

$$J = \begin{vmatrix} -\frac{1}{2}(1-\eta) & 1 \\ 2 + \frac{1}{2}\xi & 0 \end{vmatrix} = -\left(2 + \tfrac{1}{2}\xi\right) < 0 \tag{4.264b}$$

The negative jacobian indicates that a right-hand coordinate system is mapped

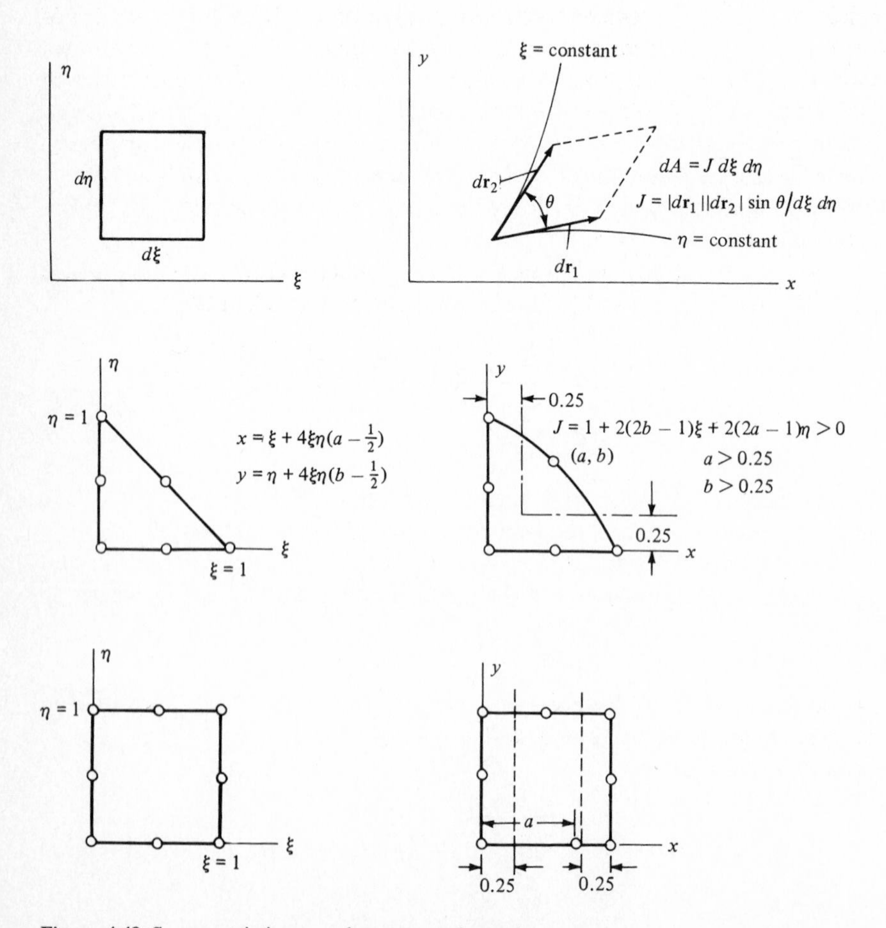

Figure 4.49 Some restrictions on element transformations.

into a left-hand coordinate system. Such coordinate transformations should be avoided. ■

The above example illustrates, for the four-node master element, that non-convex elements are not admissible in finite-element meshes. In general, any interior angle θ (see Fig. 4.49) should not be too small or too large because the jacobian $J = (|d\mathbf{r}_1||d\mathbf{r}_2|\sin\theta)/d\xi\, d\eta$ will be very small. Similar restrictions on elements can be obtained when higher-order master elements are used. Additional restrictions can be derived for higher-order elements. For example, for higher-order elements (both triangular and rectangular) the placing of side (not corner) and interior nodes is restricted. For the eight-node rectangular element, it can be shown that the side nodes should be placed at a distance greater than or equal to a quarter of the length of the side from either corner node (see Fig. 4.49).

4.7.2 Numerical Integration

Numerical evaluation of element coefficient matrices and column vectors, such as those in Eqs. (4.226), is accomplished in much the same way as was described for one-dimensional elements in Sec. 3.6.4. Since the integrals are defined in terms of the global coordinates x and y and the interpolation functions are known in terms of the natural coordinates ξ and η, we should employ the coordinate transformation of the form of Eqs. (4.254) to rewrite the integrals in terms of ξ and η (i.e., transform to the master element). In this section, we discuss the steps involved in the numerical evaluation of element matrices.

First we compute the jacobian matrix $[J]$, which has several uses in the sequel. From Eqs. (4.254) and (4.257), we have for each element Ω^e

$$[J] = \begin{bmatrix} \dfrac{\partial x}{\partial \xi} & \dfrac{\partial y}{\partial \xi} \\ \dfrac{\partial x}{\partial \eta} & \dfrac{\partial y}{\partial \eta} \end{bmatrix} = \begin{bmatrix} \sum_{i=1}^{n} x_i \dfrac{\partial \hat{\psi}_i}{\partial \xi} & \sum_{i=1}^{n} y_i \dfrac{\partial \hat{\psi}_i}{\partial \xi} \\ \sum_{i=1}^{n} x_i \dfrac{\partial \hat{\psi}_i}{\partial \eta} & \sum_{i=1}^{n} y_i \dfrac{\partial \hat{\psi}_i}{\partial \eta} \end{bmatrix} \tag{4.265a}$$

$$= \begin{bmatrix} \dfrac{\partial \hat{\psi}_1}{\partial \xi} & \dfrac{\partial \hat{\psi}_2}{\partial \xi} & \cdots & \dfrac{\partial \hat{\psi}_n}{\partial \xi} \\ \dfrac{\partial \hat{\psi}_1}{\partial \eta} & \dfrac{\partial \hat{\psi}_2}{\partial \eta} & \cdots & \dfrac{\partial \hat{\psi}_n}{\partial \eta} \end{bmatrix} \begin{bmatrix} x_1 & y_1 \\ x_2 & y_2 \\ \vdots & \vdots \\ x_n & y_n \end{bmatrix} \tag{4.265b}$$

Next we derive a relation between the derivatives of the interpolation functions with respect to the x and y, and ξ and η coordinates. Using the chain rule of differentiation, we write

$$\begin{aligned} \frac{\partial \hat{\psi}_i}{\partial x} &= \frac{\partial \hat{\psi}_i}{\partial \xi}\frac{\partial \xi}{\partial x} + \frac{\partial \hat{\psi}_i}{\partial \eta}\frac{\partial \eta}{\partial x} \\ \frac{\partial \hat{\psi}_i}{\partial y} &= \frac{\partial \hat{\psi}_i}{\partial \xi}\frac{\partial \xi}{\partial y} + \frac{\partial \hat{\psi}_i}{\partial \eta}\frac{\partial \eta}{\partial y} \end{aligned} \tag{4.266a}$$

or, in matrix form,

$$\begin{Bmatrix} \dfrac{\partial \hat{\psi}_i}{\partial x} \\[2mm] \dfrac{\partial \hat{\psi}_i}{\partial y} \end{Bmatrix} = \begin{bmatrix} \dfrac{\partial \xi}{\partial x} & \dfrac{\partial \eta}{\partial x} \\[2mm] \dfrac{\partial \xi}{\partial y} & \dfrac{\partial \eta}{\partial y} \end{bmatrix} \begin{Bmatrix} \dfrac{\partial \hat{\psi}_i}{\partial \xi} \\[2mm] \dfrac{\partial \hat{\psi}_i}{\partial \eta} \end{Bmatrix} = [J^*] \begin{Bmatrix} \dfrac{\partial \hat{\psi}_i}{\partial \xi} \\[2mm] \dfrac{\partial \hat{\psi}_i}{\partial \eta} \end{Bmatrix} \tag{4.266b}$$

$$= \begin{Bmatrix} J_{11}^* \dfrac{\partial \hat{\psi}_i}{\partial \xi} + J_{12}^* \dfrac{\partial \hat{\psi}_i}{\partial \eta} \\[2mm] J_{21}^* \dfrac{\partial \hat{\psi}_i}{\partial \xi} + J_{22}^* \dfrac{\partial \hat{\psi}_i}{\partial \eta} \end{Bmatrix} \tag{4.266c}$$

where J_{ij}^* is the element in position (i, j) of the inverse of the jacobian matrix,

$$[J]^{-1} \equiv [J^*] = \begin{bmatrix} J_{11}^* & J_{12}^* \\ J_{21}^* & J_{22}^* \end{bmatrix} \tag{4.267}$$

The elemental area is given by

$$dA \equiv dx\, dy = J\, d\xi\, d\eta \tag{4.268}$$

If the elements have straight sides (i.e., not curved elements), one can still use the corner nodes to define the geometry of the elements. In that case, $\hat{\psi}_i$ in Eqs. (4.265) correspond to linear interpolation functions of the element.

Equations (4.265) and (4.266) provide the necessary relations to transform integral expressions on any element $\Omega^{(e)}$ to an associated master element $\hat{\Omega}$. For instance, consider the following integral expression on an arbitrary element $\Omega^{(e)}$:

$$K_{ij}^{(e)} = \int_{\Omega^{(e)}} \left(a \frac{\partial \hat{\psi}_i}{\partial x} \frac{\partial \hat{\psi}_j}{\partial x} + b \frac{\partial \hat{\psi}_i}{\partial y} \frac{\partial \hat{\psi}_j}{\partial y} + c \hat{\psi}_i \hat{\psi}_j \right) dx\, dy \tag{4.269}$$

where $a = a(x, y)$, $b = b(x, y)$, and $c = c(x, y)$ are functions of x and y. Suppose that the mesh of finite elements is generated by a master element $\hat{\Omega}$. Under the transformation (4.260) [i.e., in view of Eqs. (4.266)], we can write

$$K_{ij}^{(e)} = \int_{\hat{\Omega}} \left[\hat{a} \left(J_{11}^* \frac{\partial \hat{\psi}_i}{\partial \xi} + J_{12}^* \frac{\partial \hat{\psi}_i}{\partial \eta} \right) \left(J_{11}^* \frac{\partial \hat{\psi}_j}{\partial \xi} + J_{12}^* \frac{\partial \hat{\psi}_j}{\partial \eta} \right) \right.$$
$$\left. + \hat{b} \left(J_{21}^* \frac{\partial \hat{\psi}_i}{\partial \xi} + J_{22}^* \frac{\partial \hat{\psi}_i}{\partial \eta} \right) \left(J_{21}^* \frac{\partial \hat{\psi}_j}{\partial \xi} + J_{22}^* \frac{\partial \hat{\psi}_j}{\partial \eta} \right) + \hat{c} \hat{\psi}_i \hat{\psi}_j \right] J\, d\xi\, d\eta$$
$$\equiv \int_{\hat{\Omega}} F(\xi, \eta)\, d\xi\, d\eta \tag{4.270}$$

where J_{ij}^* are the elements of the inverse of the jacobian matrix, and $\hat{a} = a(\xi, \eta)$, and so on. Equations (4.265) through (4.270) are valid for master elements of both rectangular and triangular geometry (see Fig. 4.50).

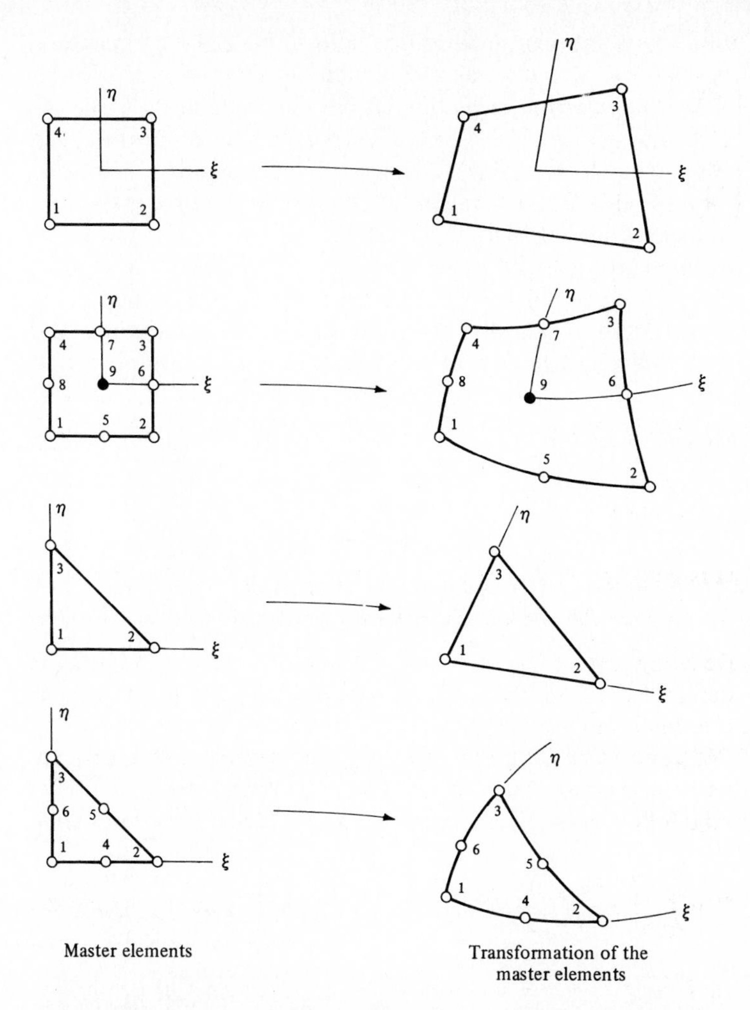

Figure 4.50 Linear and quadratic master elements and their transformations.

Numerical integration over a master rectangular element Quadrature formulas for integrals defined over a rectangular master element $\hat{\Omega}_R$ (such as that shown in Fig. 4.48) can be derived from the one-dimensional quadrature formulas presented in Sec. 3.6.4. We have

$$\int_{\hat{\Omega}_R} F(\xi, \eta)\, d\xi\, d\eta = \int_{-1}^{1} \left(\int_{-1}^{1} F(\xi, \eta)\, d\eta \right) d\xi$$

$$= \int_{-1}^{1} \left(\sum_{J=1}^{N} F(\xi, \eta_J) W_J \right) d\xi$$

$$\simeq \sum_{I=1}^{M} \sum_{J=1}^{N} F(\xi_I, \eta_J) W_I W_J \tag{4.271}$$

where M and N denote the number of quadrature points in the ξ and η directions, (ξ_I, η_J) denote the Gauss points, and W_I and W_J denote the corresponding Gauss weights (see Table 3.8). The selection of the number of Gauss points is based on the same formula as that given in Sec. 3.6.4: a polynomial of degree p is integrated exactly employing $N = (p + 1)/2$ Gauss points; when $p + 1$ is odd, one should pick $N = p/2 + 1$. In most cases, the interpolation functions are of the same degree in both ξ and η, and therefore one has $M = N$. When the integrand is of different degree in ξ and η, the number of Gauss points is selected on the basis of the largest-degree polynomial. Table 4.12 contains information on the selection of the integration order and the location of the Gauss points for linear, quadratic, and cubic Lagrange (and serendipity) quadrilateral elements. The maximum degree of the polynomial refers to the degree of the highest polynomial in ξ or η that is present in the integrands of the element matrices of

Table 4.12 Selection of the integration order and location of the Gauss points for linear, quadratic, and cubic quadrilateral elements (nodes not shown)

Element type	Maximum polynomial degree	Order of integration	Degree of accuracy	Location of integration points in master element
Linear	2	2×2	3	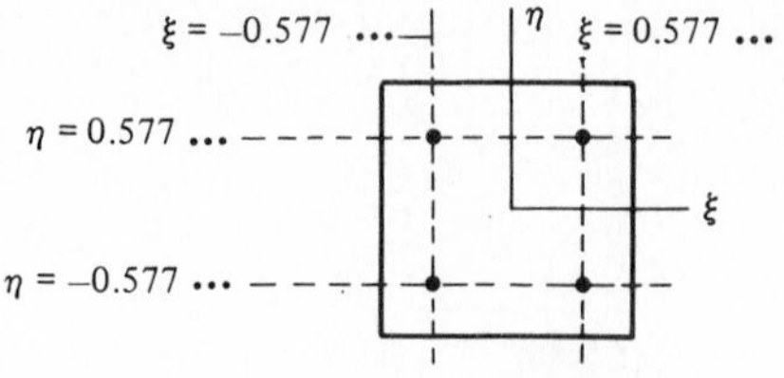 $\xi = -0.577\ldots$, $\xi = 0.577\ldots$; $\eta = 0.577\ldots$, $\eta = -0.577\ldots$
Quadratic	4	3×3	5	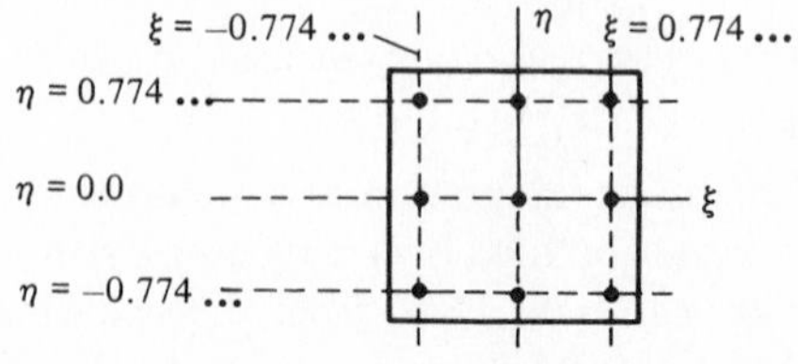 $\xi = -0.774\ldots$, $\xi = 0.774\ldots$; $\eta = 0.774\ldots$, $\eta = 0.0$, $\eta = -0.774\ldots$
Cubic	6	4×4	7	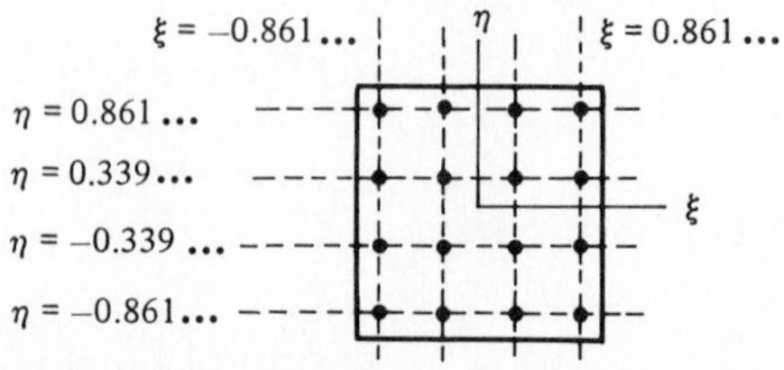 $\xi = -0.861\ldots$, $\xi = 0.861\ldots$; $\eta = 0.861\ldots$, $\eta = 0.339\ldots$, $\eta = -0.339\ldots$, $\eta = -0.861\ldots$

the type

$$M_{ij}^{(e)} = \int_{\hat{\Omega}} J\hat{\psi}_i\hat{\psi}_j \, d\xi \, d\eta$$

$$K_{ij}^{(e)} = \int_{\hat{\Omega}} J\left(\frac{\partial\hat{\psi}_i}{\partial x}\frac{\partial\hat{\psi}_j}{\partial x} + \frac{\partial\hat{\psi}_i}{\partial y}\frac{\partial\hat{\psi}_j}{\partial y}\right) d\xi \, d\eta \tag{4.272}$$

where $\partial\hat{\psi}_i/\partial x$ and $\partial\hat{\psi}_i/\partial y$ are given by Eqs. (4.266). The $N \times N$ Gauss point locations are given by the *tensor product* of one-dimensional Gauss points ξ_i:

$$\begin{Bmatrix} \xi_1 \\ \xi_2 \\ \vdots \\ \xi_N \end{Bmatrix} \{\xi_1, \xi_2, \ldots, \xi_N\} \equiv \begin{bmatrix} (\xi_1, \xi_1) & (\xi_1, \xi_2) \cdots & (\xi_1, \xi_N) \\ (\xi_2, \xi_1) & & \\ \vdots & \ddots & \vdots \\ (\xi_N, \xi_1) & \cdots & (\xi_N, \xi_N) \end{bmatrix} \tag{4.273}$$

The following two examples illustrate the evaluation of the Jacobian and element matrices on rectangular elements.

Example 4.17 Consider the quadrilateral element $\Omega^{(1)}$ shown in Fig. 4.48. We wish to evaluate $\partial\hat{\psi}_i/\partial x$ and $\partial\hat{\psi}_i/\partial y$ at $(\xi, \eta) = (0, 0)$ and $(\frac{1}{2}, \frac{1}{2})$. From Eq. (4.265*b*) we have

$$[J] = \frac{1}{4}\begin{bmatrix} -(1-\eta) & (1-\eta) & (1+\eta) & -(1-\eta) \\ -(1-\xi) & -(1+\xi) & (1+\xi) & (1-\xi) \end{bmatrix}\begin{bmatrix} 0.0 & 0.0 \\ 2.0 & 0.0 \\ 2.0 & 3.0 \\ 0.0 & 5.0 \end{bmatrix}$$

$$= \begin{bmatrix} 1 & -\dfrac{1+\eta}{2} \\ 0 & \dfrac{4-\xi}{2} \end{bmatrix}$$

The inverse of the jacobian matrix is given by

$$[J]^{-1} = \begin{bmatrix} 1 & \dfrac{1+\eta}{4-\xi} \\ 0 & \dfrac{2}{4-\xi} \end{bmatrix}$$

$$J_{11}^* = 1 \qquad J_{21}^* = 0 \qquad J_{12}^* = \frac{1+\eta}{4-\xi} \qquad J_{22}^* = \frac{2}{4-\xi}$$

From Eq. (4.266*c*) we have

$$\frac{\partial\hat{\psi}_i}{\partial x} = \frac{\partial\hat{\psi}_i}{\partial \xi} + \frac{1+\eta}{4-\xi}\frac{\partial\hat{\psi}_i}{\partial\eta} \qquad \frac{\partial\hat{\psi}_i}{\partial y} = \frac{2}{4-\xi}\frac{\partial\hat{\psi}_i}{\partial\eta}$$

where

$$\hat{\psi}_i = \tfrac{1}{4}(1 + \xi\xi_i)(1 + \eta\eta_i)$$
$$\frac{\partial \hat{\psi}_i}{\partial \xi} = \frac{1}{4}\xi_i(1 + \eta\eta_i) \qquad \frac{\partial \hat{\psi}_i}{\partial \eta} = \frac{1}{4}\eta_i(1 + \xi\xi_i) \tag{4.274}$$

(ξ_i, η_i) being the coordinates of the ith node in the master element (see Fig. 4.48):

Node	ξ_i	η_i
1	−1	−1
2	1	−1
3	1	1
4	−1	1

(4.275)

Then we have

$$\frac{\partial \hat{\psi}_i}{\partial x} = \frac{1}{4}\xi_i(1 + \eta\eta_i) + \frac{1}{4}\frac{1 + \eta}{4 - \xi}\eta_i(1 + \xi\xi_i)$$
$$\frac{\partial \hat{\psi}_i}{\partial y} = \frac{1}{4}\frac{2}{4 - \xi}\eta_i(1 + \xi\xi_i)$$

(ξ, η) →	$(0, 0)$	$(\tfrac{1}{2}, \tfrac{1}{2})$
$\dfrac{\partial \hat{\psi}_i}{\partial x}$	$\tfrac{1}{4}\xi_i + \tfrac{1}{16}\eta_i$	$\tfrac{1}{8}\xi_i(2 + \eta_i) + \tfrac{3}{56}\eta_i(2 + \xi_i)$
$\dfrac{\partial \hat{\psi}_i}{\partial y}$	$\tfrac{1}{8}\eta_i$	$\tfrac{1}{14}\eta_i(2 + \xi_i)$

■

Example 4.18 Consider the quadrilateral element in Fig. 4.51*a*. We wish to compute the following element matrices using the Gauss quadrature:

$$S_{ij} = \int_\Omega \hat{\psi}_i \hat{\psi}_j \, dx\, dy$$
$$S_{ij}^{11} = \int_\Omega \frac{\partial \hat{\psi}_i}{\partial x}\frac{\partial \hat{\psi}_j}{\partial x}\, dx\, dy$$
$$S_{ij}^{22} = \int_\Omega \frac{\partial \hat{\psi}_i}{\partial y}\frac{\partial \hat{\psi}_j}{\partial y}\, dx\, dy \tag{4.276}$$
$$S_{ij}^{12} = \int_\Omega \frac{\partial \hat{\psi}_i}{\partial x}\frac{\partial \hat{\psi}_j}{\partial y}\, dx\, dy$$

We have

$$[J] = \frac{1}{4}\begin{bmatrix} 8 - 2\eta & 2\eta \\ -2\xi & 10 + 2\xi \end{bmatrix} \qquad J = (4 - \eta)(5 + \xi) + \xi\eta = 20 + 4\xi - 5\eta \tag{4.277}$$

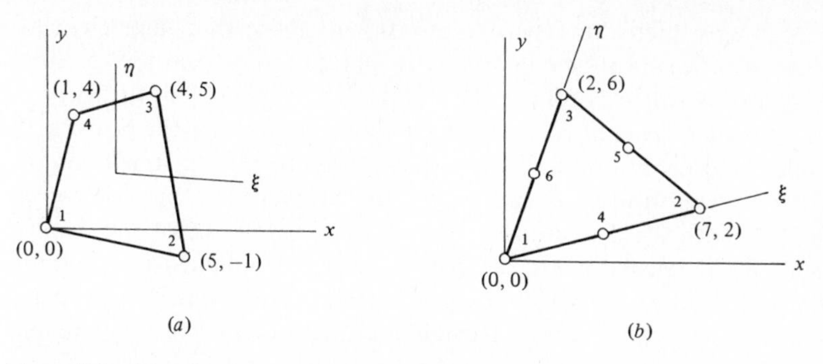

Figure 4.51 Geometry of elements. (*a*) A bilinear quadrilateral element (see Example 4.18). (*b*) A quadratic triangular element (see Example 4.19).

$$S_{ij} = \int_{\Omega} \hat{\psi}_i \hat{\psi}_j \, dx\, dy = \int_{-1}^{1}\int_{-1}^{1} \hat{\psi}_i \hat{\psi}_j J \, d\xi\, d\eta$$

$$S_{ij}^{11} = \int_{\Omega} \frac{\partial \hat{\psi}_i}{\partial x}\frac{\partial \hat{\psi}_j}{\partial x} dx\, dy \tag{4.278}$$

$$= \int_{-1}^{1}\int_{-1}^{1} \left(J_{11}^{*} \frac{\partial \hat{\psi}_i}{\partial \xi} + J_{12}^{*} \frac{\partial \hat{\psi}_i}{\partial \eta} \right)\left(J_{11}^{*} \frac{\partial \hat{\psi}_j}{\partial \xi} + J_{12}^{*} \frac{\partial \hat{\psi}_j}{\partial \eta} \right) J \, d\xi\, d\eta$$

and so on, where $\partial\hat{\psi}_i/\partial\xi$ and $\partial\hat{\psi}_i/\partial\eta$ are given by Eqs. (4.274). Evaluating the integrals in Eq. (4.276), we obtain

$$[S^{11}] = \begin{bmatrix} 0.40995 & -0.36892 & -0.20479 & 0.16376 \\ -0.36892 & 0.34516 & 0.25014 & -0.22639 \\ -0.20479 & 0.25014 & 0.43155 & -0.47690 \\ 0.16376 & -0.22639 & -0.47690 & 0.53953 \end{bmatrix}$$

$$[S^{22}] = \begin{bmatrix} 0.26237 & 0.16389 & -0.13107 & -0.29520 \\ 0.16389 & 0.22090 & -0.23991 & -0.14489 \\ -0.13107 & -0.23991 & 0.27619 & 0.09478 \\ -0.29520 & -0.14489 & 0.09478 & 0.34530 \end{bmatrix}$$

$$[S^{12}] = \begin{bmatrix} 0.24731 & 0.25156 & -0.25297 & -0.24589 \\ -0.24844 & -0.25090 & 0.25172 & 0.24762 \\ -0.25297 & -0.24828 & 0.24671 & 0.25454 \\ 0.25411 & 0.24762 & -0.24546 & -0.25627 \end{bmatrix} \tag{4.279}$$

$$[S] = \begin{bmatrix} 2.27780 & 1.25000 & 0.55556 & 1.00000 \\ 1.25000 & 2.72220 & 1.22220 & 0.55556 \\ 0.55556 & 1.22220 & 2.16670 & 0.97222 \\ 1.00000 & 0.55556 & 0.97222 & 1.72222 \end{bmatrix}$$

■

Numerical integration over a master triangular element In Sec. 4.4 and the preceding pages we discussed quadrilateral isoparametric elements which can be used to model (i.e., represent) very general geometries as well as field variables in a variety of problems. Here we discuss some isoparametric triangular elements that may be convenient (from the computational point of view) and useful. Since quadrilateral elements can be geometrically distorted, it is possible to distort a quadrilateral element to a required triangular element by moving the position of the corner nodes (and the fourth corner in the quadrilateral is merged with one of the neighboring nodes). In practice (i.e., in actual computation), this is achieved by assigning the same global node number to two corner nodes of the quadrilateral element. Thus, master triangular elements can be obtained in a natural way from associated master rectangular elements. Here we discuss the (isoparametric) transformations from a master triangular element to an arbitrary triangular element.

We choose the unit right isosceles triangle (see Fig. 4.52*a*) as the master element. An arbitrary triangular element $\Omega^{(e)}$ can be generated from the master triangular element $\hat{\Omega}_T$ by transformation of the form (4.260). The coordinate lines $\xi = 0$ and $\eta = 0$ in $\hat{\Omega}_T$ correspond to the skew curvilinear coordinate lines 1–3 and 1–2 in $\Omega^{(e)}$.

In Sec. 4.4.2 we introduced the three-node triangular element and illustrated its use in the solution of heat conduction, fluid flow, and plane elasticity problems. There we used explicit integrations to evaluate the integrals appearing in the element equations. Here we discuss the derivation of three-node isopara-

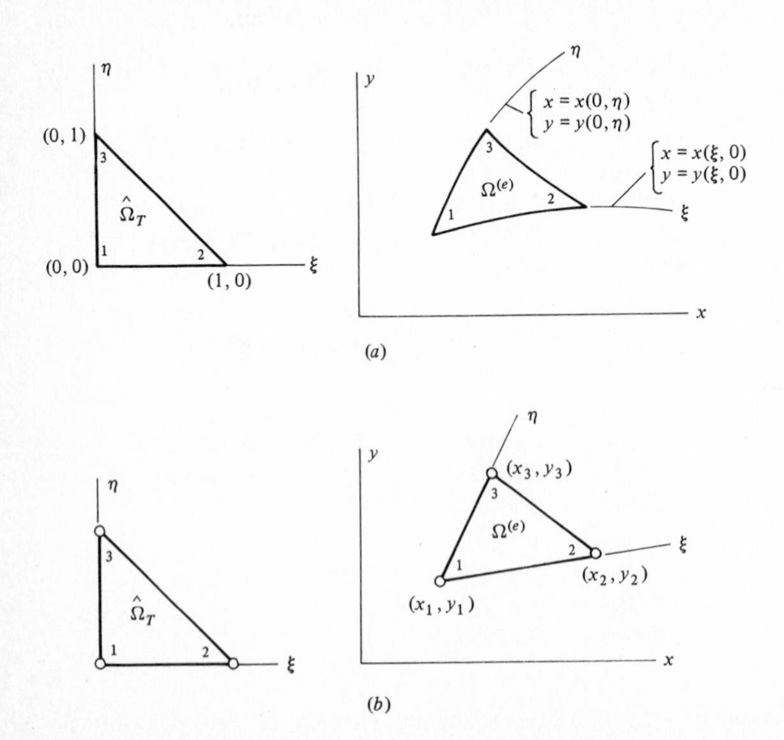

Figure 4.52 Triangular master element and its transformations. (*a*) General transformation. (*b*) Linear transformation of a master element to a triangular element.

metric triangular elements and the evaluation of integrals over such elements. For the three-node triangular element, the transformation (4.260) is taken to be

$$x = \sum_{i=1}^{3} x_i \hat{\psi}_i(\xi, \eta) \qquad y = \sum_{i=1}^{3} y_i \hat{\psi}_i(\xi, \eta) \tag{4.280}$$

where $\hat{\psi}_i(\xi, \eta)$ are the interpolation functions of the master three-node triangular element (see Fig. 4.52*b*)

$$\hat{\psi}_1 = 1 - \xi - \eta \qquad \hat{\psi}_2 = \xi \qquad \hat{\psi}_3 = \eta \tag{4.281}$$

which satisfy the interpolation properties (4.26). The inverse transformation from element $\Omega^{(e)}$ to $\hat{\Omega}_T$ is given by inverting Eqs. (4.280),

$$\begin{aligned} \xi &= \frac{1}{2A}[(x - x_1)(y_3 - y_1) - (y - y_1)(x_3 - x_1)] \\ \eta &= \frac{1}{2A}[(x - x_1)(y_1 - y_2) + (y - y_1)(x_2 - x_1)] \end{aligned} \tag{4.282}$$

where A is the area of $\Omega^{(e)}$.

With the help of Eqs. (4.282), one can show that the interpolation functions in Eqs. (4.25) are equivalent to the $\hat{\psi}_i$ in Eqs. (4.281). Moreover, the area coordinates L_i in Eq. (4.110) are also equivalent to $\hat{\psi}_i$. The interpolation functions for the higher-order master triangular elements can be obtained from the area coordinates, as described in Sec. 4.4.

The jacobian matrix for the linear triangular element is given by

$$[J] = \begin{bmatrix} x_2 - x_1 & y_2 - y_1 \\ x_3 - x_1 & y_3 - y_1 \end{bmatrix} = \begin{bmatrix} \gamma_3 & -\beta_3 \\ -\gamma_2 & \beta_2 \end{bmatrix} \tag{4.283}$$

where β_i and γ_i are the constants defined in Eq. (4.25*b*). The inverse of the jacobian matrix is given by

$$[J]^{-1} = \frac{1}{J}\begin{bmatrix} \beta_2 & \beta_3 \\ \gamma_2 & \gamma_3 \end{bmatrix} \qquad J = \beta_2\gamma_3 - \gamma_2\beta_3 = 2A \tag{4.284}$$

The relations (4.266*c*) take the form

$$\begin{aligned} \frac{\partial \hat{\psi}_1}{\partial x} &= -\frac{\beta_2 + \beta_3}{2A} = \frac{\beta_1}{2A} \\ \frac{\partial \hat{\psi}_1}{\partial y} &= -\frac{\gamma_2 + \gamma_3}{2A} = \frac{\gamma_1}{2A} \\ \frac{\partial \hat{\psi}_2}{\partial x} &= \frac{\beta_2}{2A} \qquad \frac{\partial \hat{\psi}_2}{\partial y} = \frac{\gamma_2}{2A} \\ \frac{\partial \hat{\psi}_3}{\partial x} &= \frac{\beta_3}{2A} \qquad \frac{\partial \hat{\psi}_3}{\partial y} = \frac{\gamma_3}{2A} \end{aligned} \tag{4.285}$$

In a general case, the derivatives of $\hat{\psi}_i$ with respect to the global coordinates can be computed from Eqs. (4.266), which take the form ($L_3 = 1 - L_1 - L_2$)

$$\frac{\partial \hat{\psi}_i}{\partial x} = \frac{\partial \hat{\psi}_i}{\partial L_1}\frac{\partial L_1}{\partial x} + \frac{\partial \hat{\psi}_i}{\partial L_2}\frac{\partial L_2}{\partial x}$$

$$\frac{\partial \hat{\psi}_i}{\partial y} = \frac{\partial \hat{\psi}_i}{\partial L_1}\frac{\partial L_1}{\partial y} + \frac{\partial \hat{\psi}_i}{\partial L_2}\frac{\partial L_2}{\partial y}$$

or

$$\left\{\begin{matrix} \dfrac{\partial \hat{\psi}_i}{\partial x} \\ \dfrac{\partial \hat{\psi}_i}{\partial y} \end{matrix}\right\} = [J]^{-1}\left\{\begin{matrix} \dfrac{\partial \hat{\psi}_i}{\partial L_1} \\ \dfrac{\partial \hat{\psi}_i}{\partial L_2} \end{matrix}\right\} \qquad [J] = \begin{bmatrix} \dfrac{\partial x}{\partial L_1} & \dfrac{\partial y}{\partial L_1} \\ \dfrac{\partial x}{\partial L_2} & \dfrac{\partial y}{\partial L_2} \end{bmatrix} \tag{4.286}$$

In general, integrals on $\hat{\Omega}_T$ have the form ($\hat{G} = GJ$)

$$\int_{\hat{\Omega}_T} G(\xi, \eta)\, d\xi\, d\eta = \int_{\hat{\Omega}_T} \hat{G}(L_1, L_2, L_3)\, dL_1\, dL_2 \tag{4.287}$$

Table 4.13 Quadrature points and weights for triangular elements

Number of integration points	Degree of accuracy and the order of the residual	Location of integration points				
		L_1	L_2	L_3	W	Geometric locations
1	1 $O(h^2)$	$\frac{1}{3}$	$\frac{1}{3}$	$\frac{1}{3}$	1	a 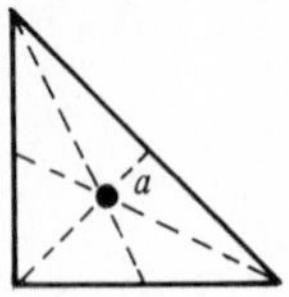
3	2 $O(h^2)$	$\frac{1}{2}$	0	$\frac{1}{2}$	$\frac{1}{3}$	a 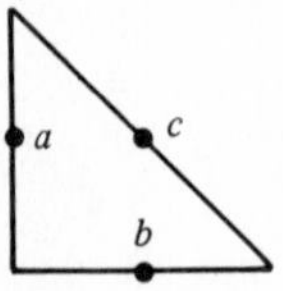
		$\frac{1}{2}$	$\frac{1}{2}$	0	$\frac{1}{3}$	b
		0	$\frac{1}{2}$	$\frac{1}{2}$	$\frac{1}{3}$	c
4	3 $O(h^4)$	$\frac{1}{3}$	$\frac{1}{3}$	$\frac{1}{3}$	$\frac{27}{48}$	a 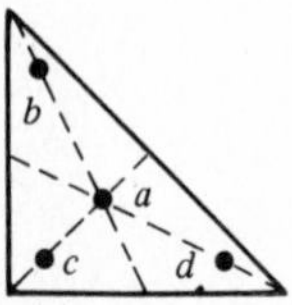
		$\frac{2}{15}$	$\frac{2}{15}$	$\frac{11}{15}$	$\frac{25}{48}$	b
		$\frac{11}{15}$	$\frac{2}{15}$	$\frac{2}{15}$	$\frac{25}{48}$	c
		$\frac{2}{15}$	$\frac{11}{15}$	$\frac{2}{15}$	$\frac{25}{48}$	d

which can be approximated by the quadrature formula ($L_3 = 1 - L_1 - L_2$)

$$\int_{\hat{\Omega}_T} \hat{G}(L_1, L_2, L_3)\, dL_1\, dL_2 \simeq \tfrac{1}{2} \sum_{I=1}^{N} W_I \hat{G}(S_I) \tag{4.288}$$

where W_I and S_I denote the weights and points of the quadrature. Table 4.13 contains information on some quadrature rules for one-point, three-point, and four-point quadrature for triangular elements.

Example 4.19 Consider the quadratic triangular element shown in Fig. 4.51*b*. We wish to calculate $\partial\hat{\psi}_1/\partial x$, $\partial\hat{\psi}_1/\partial y$, $\partial\hat{\psi}_4/\partial x$, and $\partial\hat{\psi}_4/\partial y$ at the point $(x, y) = (2, 4)$ and evaluate the integral of the product $(\partial\hat{\psi}_1/\partial x)(\partial\hat{\psi}_4/\partial x)$.

Since the element has straight edges, its geometry is defined by the interpolation functions of the corner nodes. For the element at hand we have

$$x = \sum_{i=1}^{3} x_i L_i = 7L_2 + 2L_3 = 2 - 2L_1 + 5L_2$$

$$y = \sum_{i=1}^{3} y_i L_i = 2L_2 + 6L_3 = 6 - 6L_1 - 4L_2 \tag{4.289}$$

$$[J] = \begin{bmatrix} -2 & -6 \\ 5 & -4 \end{bmatrix} \qquad [J]^{-1} = \frac{1}{38}\begin{bmatrix} -4 & 6 \\ -5 & -2 \end{bmatrix}$$

$$\begin{Bmatrix} \dfrac{\partial\hat{\psi}_1}{\partial x} \\[2ex] \dfrac{\partial\hat{\psi}_1}{\partial y} \end{Bmatrix} = \frac{1}{38}\begin{bmatrix} -4 & 6 \\[2ex] -5 & -2 \end{bmatrix} \begin{Bmatrix} \dfrac{\partial\hat{\psi}_1}{\partial L_1} \\[2ex] \dfrac{\partial\hat{\psi}_1}{\partial L_2} \end{Bmatrix} = \begin{Bmatrix} \dfrac{6 - 4}{38} \\[2ex] \dfrac{-5 - 2}{38} \end{Bmatrix}$$

$$\begin{Bmatrix} \dfrac{\partial\hat{\psi}_4}{\partial x} \\[2ex] \dfrac{\partial\hat{\psi}_4}{\partial y} \end{Bmatrix} = \frac{1}{38}\begin{bmatrix} -4 & 6 \\[2ex] -5 & -2 \end{bmatrix} \begin{Bmatrix} \dfrac{\partial\hat{\psi}_4}{\partial L_1} \\[2ex] \dfrac{\partial\hat{\psi}_4}{\partial L_2} \end{Bmatrix} = \frac{1}{38}\begin{Bmatrix} -16L_2 + 24L_1 \\[2ex] -20L_2 - 8L_1 \end{Bmatrix}$$

where $\hat{\psi}_1 = L_1$ and $\hat{\psi}_4 = 4L_1L_2$. For the point (2, 4), the area coordinates can be calculated from Eqs. (4.289):

$$2 = 7L_2 + 2L_3$$
$$4 = 2L_2 + 6L_3$$
$$1 = L_1 + L_2 + L_3$$

We obtain

$$L_1 = \frac{5}{19} \qquad L_2 = \frac{2}{19} \qquad L_3 = \frac{12}{19}$$

Evaluating $\partial\hat{\psi}_4/\partial x$ and $\partial\hat{\psi}_4/\partial y$ at the point (2, 4), we obtain

$$\frac{\partial\hat{\psi}_4}{\partial x} = \frac{-16}{(19)^2} + \frac{60}{(19)^2} = \frac{44}{361}$$

$$\frac{\partial\hat{\psi}_4}{\partial y} = \frac{-20}{(19)^2} + \frac{-20}{(19)^2} = \frac{-40}{361}$$

We get

$$\int_{\hat{\Omega}_T} \frac{\partial\hat{\psi}_1}{\partial x}\frac{\partial\hat{\psi}_4}{\partial x}\,dx\,dy = \frac{J}{361}\int_0^1\int_0^{1-L_2}(12L_1 - 8L_2)\,dL_1\,dL_2$$

Since the integrand is linear in both L_1 and L_2, the one-point quadrature is sufficient. The integration point is $(L_1, L_2) = (\frac{1}{3}, \frac{1}{3})$ with a weight of $W = 1$:

$$\frac{38}{361}\int_0^1\int_0^{1-L_2}(12L_1 - 8L_2)\,dL_1 dL_2 = \frac{2}{19}\left(6 \times \frac{1}{3} - \frac{4}{3}\right) = \frac{4}{57}$$

We now verify the result using Eq. (4.117*b*):

$$\frac{1}{361}\int_A (12L_1 - 8L_2)\,dA = \frac{A}{361}\left(\frac{12}{3} - \frac{8}{3}\right) = \frac{4A}{3 \times 361}$$

The area A of the triangle is equal to 19, and therefore we get the same result as above. ■

4.8 COMPUTER IMPLEMENTATION

4.8.1 Introduction

In Sec. 3.7, we discussed some basic ideas surrounding the development of a typical finite-element program. Specific details of various logic units of a finite-element program for one-dimensional problems were given, and the application of program FEM1D was illustrated. Most of the ideas presented there are also valid for two-dimensional problems, including the logic units shown in Fig. 3.20. The imposition of the boundary conditions and the solution of the equations remain the same as in one-dimensional problems. Here we focus attention on the computer implementation of two-dimensional element calculations. The use of a two-dimensional model program FEM2D and the plate analysis program PLATE will also be discussed. The program FEM2D contains the three-node triangular element and four-, eight-, and nine-node isoparametric rectangular elements, and it can be used for the solution of static plane elasticity problems, steady and transient heat conduction/convection problems, and steady and unsteady laminar motion of a viscous incompressible fluid (by the penalty function formulation).

In the preprocessor unit, program MESH is used to generate triangular- and rectangular-element meshes of rectangular domains. The subroutine is not general enough to generate finite-element meshes of arbitrary domains. Of course, one can use any other mesh generation program in place of MESH. The subroutine MESH generates the connectivity matrix (array NOD) and the global coordinates of the nodes (arrays X and Y). When nonrectangular domains are analyzed, the mesh information should be read in.

4.8.2 Element Calculations

In two dimensions the element calculations are more involved due to the following considerations:

1. Various geometric shapes of elements.
2. Single as well as multivariable problems.
3. Reduced-order integrations are considered in certain formulations (plate bending and penalty function formulations).
4. A different number of primary degrees of freedom at different nodes of the element are considered (e.g., see pressure-velocity formulation); consequently, the assembly is more involved.

Element calculations for the linear (three-node) triangular element (see Sec. 4.2) are simplified by the fact that the derivatives of the interpolation functions are constant over the element. When the element is used in the solution of second-order equations, the element coefficient matrix can be evaluated exactly with the aid of Eqs. (4.37) or (4.117). For example, consider the integrals

$$\begin{aligned} K_{ij}^{(e)} &= \int_{\Omega^{(e)}} \left(\frac{\partial \psi_i}{\partial x}\frac{\partial \psi_j}{\partial x} + \frac{\partial \psi_i}{\partial y}\frac{\partial \psi_j}{\partial y} \right) dx\,dy \\ M_{ij}^{(e)} &= \int_{\Omega^{(e)}} \psi_i \psi_j \, dx\,dy \end{aligned} \tag{4.290}$$

over the triangular element $\Omega^{(e)}$. Using Eqs. (4.25) and (4.36), we write

$$\begin{aligned} K_{ij}^{(e)} &= \frac{1}{(2A)^2}(\beta_i\beta_j + \gamma_i\gamma_j)\int_{\Omega^{(e)}} dx\,dy \\ &= \frac{1}{4A}(\beta_i\beta_j + \gamma_i\gamma_j) \\ M_{ij}^{(e)} &= \frac{1}{(2A)^2}\int_{\Omega^{(e)}} (\alpha_i + \beta_i x + \gamma_i y)(\alpha_j + \beta_j x + \gamma_j y)\,dx\,dy \\ &= \frac{1}{(2A)^2}\big[I_{00}\alpha_i\alpha_j + I_{10}(\alpha_i\beta_j + \alpha_j\beta_i) + I_{01}(\alpha_i\gamma_j + \alpha_j\gamma_i) \\ &\qquad + I_{20}\beta_i\beta_j + I_{11}(\beta_i\gamma_j + \beta_j\gamma_i) + I_{02}\gamma_i\gamma_j\big] \end{aligned} \tag{4.291}$$

Computer implementation of the matrices $[K]$ and $[M]$ is straightforward. The description of the arrays and a Fortran form of the expressions for $[K]$ and $[M]$

are given below:

```
C     ..................................................................
C
C      FORTRAN STATEMENTS TO CALCULATE THE ELEMENT MATRICES IN
C       (4.290) AND (4.291) FOR THE LINEAR TRIANGULAR ELEMENT
C
C       NPE.......NODES PER ELEMENT
C       X,Y.......GLOBAL COORDINATES OF THE ELEMENT NODES
C
C       ALPHA...
C       BETA   ...COEFFICIENTS IN THE INTERPOLATION FUNCTIONS
C       GAMA....
C
C       DET.......TWICE THE AREA OF THE ELEMENT
C       AMN.......INTEGRAL OF (X**M)(Y**N) OVER THE TRIANGLE
C       AK(I,J)...COEFFICIENT MATRIX IN EQN.(4.290)
C       AM(I,J)...COEFFICIENT MATRIX IN EQN.(4.291)
C
C       VARIABLES NPE, X(I), AND Y(I) ARE TRANSFERRED FROM MAIN
C
C     ..................................................................
C
      IMPLICIT REAL*8(A-H,O-Z)
      DIMENSION AK(3,3),AM(3,3),ALPHA(3),BETA(3),GAMA(3),
     *          X(3),Y(3)
C
C       DEFINE THE COEFFICIENTS OF THE INTERPOLATION FUNCTIONS
C
       DO 10 I=1,NPE
       J=I+1
       IF (J.GT.NPE) J=J-NPE
       K=J+1
       IF (K.GT.NPE) K=K-NPE
       ALPHA(I)=X(J)*Y(K)-X(K)*Y(J)
       BETA(I)=Y(J)-Y(K)
    10 GAMA(I)=X(K)-X(J)
       DET=X(1)*(Y(2)-Y(3))+X(2)*(Y(3)-Y(1))+X(3)*(Y(1)-Y(2))
C
       XBAR=(X(1)+X(2)+X(3))/3.0
       YBAR=(Y(1)+Y(2)+Y(3))/3.0
       A00=0.5*DET
       A01=A00*YBAR
       A10=A00*XBAR
       A11=A00*(X(1)*Y(1)+X(2)*Y(2)+X(3)*Y(3)+9.0*XBAR*YBAR)/12.0
       A20=A00*(X(1)*X(1)+X(2)*X(2)+X(3)*X(3)+9.0*XBAR*XBAR)/12.0
       A02=A00*(Y(1)*Y(1)+Y(2)*Y(2)+Y(3)*Y(3)+9.0*YBAR*YBAR)/12.0
C
C       COMPUTE THE COEFFICIENTS OF MATRICES IN EQN. (4.290)
C
       DO 20 I=1,NPE
       DO 20 J=1,NPE
    20 AK(I,J)=(AK1*BETA(I)*BETA(J)+AK2*GAMA(I)*GAMA(J))/DET/2.0
C
C       COMPUTE THE COEFFICIENTS OF MATRIX IN EQN. (4.291)
C
       DO 40 I=1,NPE
       DO 40 J=1,NPE
    40 B(I,J)=A00*(ALPHA(I)*ALPHA(J)+
      1       A10*(ALPHA(I)*BETA(J)+ALPHA(J)*BETA(I))+
      2       A11*(BETA(I)*GAMA(J)+BETA(J)*GAMA(I))+
      3       A01*(ALPHA(I)*GAMA(J)+ALPHA(J)*GAMA(I))+
      4       A20*BETA(I)*BETA(J)+A02*GAMA(I)*GAMA(J))/(DET*DET)
C
       RETURN
       END
```

Element calculations for isoparametric rectangular elements (of any order) can be carried out according to the developments presented in Sec. 4.7.2. The principal steps involved are

1. The development of a subroutine, SHAPE, for the evaluation of the interpolation functions and their derivatives with respect to the global coordinates [see Eqs. (4.265) through (4.268)]
2. The numerical integration of the coefficients of the element matrices using the Gauss quadrature [see Eq. (4.271)]
3. The setting up of the element matrices required for the class of problems being solved

The subroutine SHAPE (called in a do-loop on the number of Gauss points) contains the expressions of the interpolation functions for various-order elements and their derivatives with respect to the local (i.e., natural) coordinates. The jacobian matrix is evaluated at the Gauss point using Eq. (4.265*b*). This requires the derivatives of the interpolation functions with respect to natural coordinates and global coordinates of element nodes in a rectangular array form. The derivatives of the interpolation functions with respect to the global coordinates are then computed at the Gauss point by means of relations (4.266).

The coefficients of the element matrices in two-dimensional problems of interest to us here require the evaluation of the element matrices defined in Eqs. (4.276). These matrices involve products of interpolation functions and their derivatives with respect to the global coordinates. Since the integrals are evaluated numerically [see Eq. (4.271)], the integrands must be evaluated at the quadrature points and summed (in each coordinate direction) over the number of integration points (NGP). Thus, the calculation of the integrals in Eqs. (4.276) and the evaluation of the interpolation functions and their derivatives must be carried out inside the do-loops. To fix the ideas, let

SF(I) = interpolation function ψ_i of the ith node of an element

GDSF(K, I) = global derivative with respect to x_k (i.e., derivative with respect to the global coordinate x_k) of interpolation function ψ_i [GDSF(K, I) $= \partial\psi_I/\partial x_k$, $x_1 = x$, $x_2 = y$]

CONST = product of the jacobian (i.e., determinant of $[J]$) with the weights corresponding to the Gauss integration point (ξ_{NI}, η_{NJ}) = DET * WT(NI, NGP) * WT(NJ, NGP)

Then $S_{ij}, S_{ij}^{11}, \ldots$ are given in the Fortran form, by

$$
\begin{aligned}
\text{S(I, J)} &= \text{S(I, J)} + \text{SF(I)} * \text{SF(J)} * \text{CONST} \\
\text{S11(I, J)} &= \text{S11(I, J)} + \text{GDSF(1, I)} * \text{GDSF(1, J)} * \text{CONST} \\
\text{S12(I, J)} &= \text{S12(I, J)} + \text{GDSF(1, I)} * \text{GDSF(2, J)} * \text{CONST} \\
\text{S22(I, J)} &= \text{S22(I, J)} + \text{GDSF(2, I)} * \text{GDSF(2, J)} * \text{CONST}
\end{aligned}
\tag{4.292}
$$

The summed values of S(I, J), S11(I, J), ... represent the numerical values of the integral coefficients in Eqs. (4.276). The Fortran statements listed below summarize the discussion:

```
C     .............................................................
C
C     FORTRAN STATEMENTS TO CALCULATE THE ELEMENT MATRICES IN
C     EQ.(4.276) FOR LINEAR ISOPARAMETRIC RECTANGULAR ELEMENT
C
C      NPE.......NODES PER ELEMENT
C      ELXY......NPE BY 2 ARRAY OF GLOBAL COORDINATES OF ELEMENT
C                NODES: ELXY(I,1)=X(I) AND ELXY(I,2)=Y(I)
C      DET.......DETERMINANT OF THE JACOBIAN
C      SF(I).....ELEMENT INTERPOLATION FUNCTIONS (I=1,NPE)
C      GDSF(1,J).GLOBAL DERIVATIVE OF SF(J) WITH RESPECT TO X
C      GDSF(2,J).GLOBAL DERIVATIVE OF SF(J) WITH RESPECT TO Y
C      WT(I,J)...I-TH GAUSS WEIGHT IN THE J-TH ORDER QUADRATURE
C      GAUS(I,J).I-TH GAUSS POINT IN THE J-TH ORDER QUADRATURE
C      NGP.......NUMBER OF POINTS IN THE QUADRATURE RULE
C      S(I,J)....
C      S11(I,J) .
C      S12(I,J) .COEFFICIENT MATRICES DEFINED IN EQ. (4.276)
C      S22(I,J)..
C
C      VARIABLES NPE,NGP AND ARRAY ELXY ARE TRANSFERRED FROM MAIN
C
C     .............................................................
C
       IMPLICIT REAL*8(A-H,O-Z)
       DIMENSION SF(4),GDSF(4),WT(4,4),GAUSS(4,4),ELXY(4,2),
     1           S(4,4),S11(4,4),S12(4,4),S22(4,4)
C
       DATA GAUS/4*0.0D0,-0.57735027D0,0.57735027D0,2*0.0D0,
      *-0.77459667D0,0.0D0,0.77459667D0,0.0D0,-0.86113631D0,
      *-0.33998104D0,0.33998104D0,0.86113631D0/
C
       DATA WT/2.0D0,3*0.0D0,2*1.0D0,2*0.0D0,0.55555555D0,
      *0.88888888D0,.55555555D0,.0D0,0.34785485D0,2*0.65214515D0,
      *0.34785485D0/
C
C      INITIALIZE THE ARRAYS
C
       DO 20 I = 1, NPE
       DO 20 J = 1, NPE
       S(I,J)=0.0
       S11(I,J) = 0.0
       S12(I,J) = 0.0
    20 S22(I,J) = 0.0
C
C      DO-LOOPS ON NUMERICAL (GAUSS) QUADRATURE BEGIN HERE
C
       DO 60 NI = 1,NGP
       DO 60 NJ = 1,NGP
       XI = GAUSS(NI,NGP)
       ETA = GAUSS(NJ,NGP)
C
C      CALL SUBROUTINE SHAPE TO EVALUATE INTERPOLATION FUNCTIONS
C                     AND THEIR GLOBAL DERIVATIVES
C
       CALL SHAPE(NPE,XI,ETA,SF,GDSF,DET,ELXY)
C
       CONST = DET*WT(NI,NGP)*WT(NJ,NGP)
C
C      COMPUTE VARIOUS COEFFICIENT MATRICES OF EQ. (4.276)
```

```
C
      DO 40 I=1,NPE
      DO 40 J=1,NPE
      S(I,J) = S(I,J)+CONST*SF(I)*SF(J)
      S11(I,J)=S11(I,J)+CONST*GDSF(1,I)*GDSF(1,J)
      S22(I,J)=S22(I,J)+CONST*GDSF(2,I)*GDSF(2,J)
   40 S12(I,J)=S12(I,J)+CONST*GDSF(1,I)*GDSF(2,J)
   60 CONTINUE
C
      RETURN
      END
```

To set up the element coefficient matrices of a given problem, we make use of the element matrices defined above. As an example, consider the problem described by Eq. (4.1). The element coefficient matrix and the column vectors for the problem are given by Eqs. (4.14). The element matrix K_{ij} can be expressed in terms of S_{ij}, $S_{ij}^{11}, \ldots$ by

$$\begin{aligned}\mathrm{STIF(I,J)} = \mathrm{A00*S(I,J)} + \mathrm{A11*S11(I,J)} + \mathrm{A12*S12(I,J)}\\ + \mathrm{A21*S12(J,I)} + \mathrm{A22*S22(I,J)}\end{aligned} \tag{4.293}$$

Of course, one can avoid the introduction of the arrays S, S12,... by defining K_{ij} [= STIF(I, J)] directly in terms of GDSF:

$$\begin{aligned}\mathrm{STIF(I,J)} = \mathrm{STIF(I,J)} + (&\mathrm{A00*SF(I)*SF(J)} + \mathrm{A11*GDSF(1,I)*GDSF(1,J)}\\ &+\mathrm{A12*GDSF(1,I)*GDSF(2,J)}\\ &+\mathrm{A21*GDSF(2,I)*GDSF(1,J)}\\ &+\mathrm{A22*GDSF(2,I)*GDSF(2,J))*CONST}\end{aligned} \tag{4.294}$$

However, it is economical to use [S], [S11],... if they are needed in the definition of the element matrices of several different problems in the same subroutine.

In multivariable problems, the element matrices are themselves defined in terms of submatrices, as was the case in Eqs. (4.155), (4.195), and (4.200). In such cases, the nodal degrees of freedom should be renumbered to reduce the half-bandwidth of the assembled coefficient matrix. For example, consider the element equations (4.155) associated with plane elasticity problems. The element nodal variables Δ_i there are given (say, for a linear rectangular element) by

$$\begin{Bmatrix}\Delta_1\\ \Delta_2\\ \Delta_3\\ \cdot\\ \cdot\\ \cdot\\ \Delta_8\end{Bmatrix} = \begin{Bmatrix}u_1\\ u_2\\ u_3\\ u_4\\ v_1\\ v_2\\ v_3\\ v_4\end{Bmatrix} \tag{4.295a}$$

Thus at any node the difference between the label number of the first degree of freedom and that of the second degree of freedom is 4 (in general, n, where n is the number of nodes per element). This difference contributes to an increase in the half-bandwidth of the assembled coefficient matrix. To remedy this situation, we reorder the nodal degrees of freedom as follows:

$$\begin{Bmatrix} \Delta_1 \\ \Delta_2 \\ \Delta_3 \\ \Delta_4 \\ \cdot \\ \cdot \\ \cdot \\ \Delta_{2n-1} \\ \Delta_{2n} \end{Bmatrix} = \begin{Bmatrix} u_1 \\ v_1 \\ u_2 \\ v_2 \\ \cdot \\ \cdot \\ \cdot \\ u_n \\ v_n \end{Bmatrix} \tag{4.295b}$$

This reordering of the nodal degrees of freedom requires, in order to retain the symmetry present in the equations, renumbering of the individual equations. To illustrate how this can be done, we consider a set of four equations in four unknowns:

$$\begin{aligned} S_{11}^{11}u_1 + S_{12}^{11}u_2 + S_{11}^{12}v_1 + S_{12}^{12}v_2 &= F_1^1 \\ S_{21}^{11}u_1 + S_{22}^{11}u_2 + S_{21}^{12}v_1 + S_{22}^{12}v_2 &= F_2^1 \\ S_{11}^{21}u_1 + S_{12}^{21}u_2 + S_{11}^{22}v_1 + S_{12}^{22}v_2 &= F_1^2 \\ S_{21}^{21}u_1 + S_{22}^{21}u_2 + S_{21}^{22}v_1 + S_{22}^{22}v_2 &= F_2^2 \end{aligned} \tag{4.296a}$$

or, in matrix form,

$$\begin{bmatrix} [S^{11}] & [S^{12}] \\ [S^{21}] & [S^{22}] \end{bmatrix} \begin{Bmatrix} \{u\} \\ \{v\} \end{Bmatrix} = \begin{Bmatrix} \{F^1\} \\ \{F^2\} \end{Bmatrix} \tag{4.296b}$$

Now letting

$$\Delta_1 = u_1 \qquad \Delta_2 = v_1 \qquad \Delta_3 = u_2 \qquad \Delta_4 = v_2 \tag{4.297a}$$

and rearranging Eqs. (4.296), we obtain

$$\begin{aligned} S_{11}^{11}\Delta_1 + S_{11}^{12}\Delta_2 + S_{12}^{11}\Delta_3 + S_{12}^{12}\Delta_4 &= F_1^1 \\ S_{11}^{21}\Delta_1 + S_{11}^{22}\Delta_2 + S_{12}^{21}\Delta_3 + S_{12}^{22}\Delta_4 &= F_1^2 \\ S_{21}^{11}\Delta_1 + S_{21}^{12}\Delta_2 + S_{22}^{11}\Delta_3 + S_{22}^{12}\Delta_4 &= F_2^1 \\ S_{21}^{21}\Delta_1 + S_{21}^{22}\Delta_2 + S_{22}^{21}\Delta_3 + S_{22}^{22}\Delta_4 &= F_2^2 \end{aligned} \tag{4.297b}$$

or, in matrix form,

$$[S]\{\Delta\} = \{F\} \tag{4.297c}$$

where

$$
\begin{aligned}
S_{ij} &= S^{11}_{\alpha\beta} & S_{i,j+1} &= S^{12}_{\alpha\beta} \\
S_{i+1,j} &= S^{21}_{\alpha\beta} & S_{i+1,j+1} &= S^{22}_{\alpha\beta} \\
F_i &= F^1_\alpha & F_{i+1} &= F^2_\alpha
\end{aligned}
\tag{4.298}
$$

$$i = 2\alpha - 1 \qquad j = 2\beta - 1 \qquad \alpha, \beta = 1, 2$$

The above discussion applies to any set of equations of the form of Eqs. (4.296). Computer implementation of Eqs. (4.298) is straightforward. An example of this procedure is given by the following Fortran statements, which correspond to Eq. (4.156):

```
C     ......................................................................
C
C      REARRANGEMENT OF ELEMENT EQUATIONS FOR PROBLEMS WITH
C      SEVERAL DEGREES OF FREEDOM PER NODE (SEE EQS. (4.155))
C
C      NPE..........NODES PER ELEMENT
C      NDF..........NUMBER OF DEGREES OF FREEDOM PER NODE
C      S(I,J),ETC...MATRIX DEFINED IN EQN. (4.292)
C      STIF(I,J)....REARRANGED ELEMENT COEFFICIENT MATRIX
C      C(I,J).......MATRIX OF MATERIAL CONSTANTS
C     ......................................................................
C
       II = 1
       DO 20 I=1,NPE
       JJ = 1
       DO 10 J=1,NPE
       STIF(II,JJ)=C(1,1)*S11(I,J)+C(3,3)*S22(I,J)
       STIF(II+1,JJ+1)=C(3,3)*S11(I,J)+C(2,2)*S22(I,J)
       STIF(II,JJ+1)=C(1,2)*S12(I,J)+C(3,3)*S12(J,I)
       STIF(II+1,JJ)=C(1,2)*S12(J,I)+C(3,3)*S12(I,J)
    10 JJ = NDF*J+1
    20 II = NDF*I+1
```

The variables used in the above Fortran statements have the same meaning as discussed before.

4.8.3 Description of the Computer Program FEM2D

The computer program FEM2D (see Appendix II) can be used to solve the following types of problems:

1. Steady and unsteady solutions of heat transfer problems, including convection type boundary conditions,

$$\frac{\partial u}{\partial t} - \frac{\partial}{\partial x}\left(k_1 \frac{\partial u}{\partial x}\right) - \frac{\partial}{\partial y}\left(k_2 \frac{\partial u}{\partial y}\right) = f$$

2. Steady (i.e., static) solution of plane elasticity problems of Sec. 4.5.2
3. Steady and unsteady solutions of two-dimensional incompressible fluid flows by the penalty function formulation of Sec. 4.5.3 (only by rectangular elements)

The first category of problems is quite general and includes, as special cases, many of the physical problems listed in Table 2.2. The last two categories are specialized to linear elasticity and linear (i.e., Stokes) viscous incompressible fluid flow. Here we describe the meaning of some important variables.

The problem type is specified through the variable ITYPE:

$$\text{ITYPE} = \begin{cases} 2 & \text{Viscous fluid flow (by quadrilateral elements)} \\ 1 & \text{Plane strain elasticity} \\ 0 & \text{Plane stress elasticity} \\ -1 & \text{Heat conduction/convection type problems} \end{cases}$$

The variable ITEM is used to specify steady (ITEM = 0) and unsteady (ITEM = 1) analysis. No unsteady analysis of the plane elasticity problems is included in the program.

The element type (triangular or quadrilateral) is specified by the variable IEL:

$$\text{IEL} = \begin{cases} 0 & \text{Linear triangular element} \\ 1 & \text{Linear quadrilateral element} \\ 2 & \text{Quadratic quadrilateral element} \end{cases}$$

The specification of the number of nodes per element (NPE) further allows us to select either the eight-node or the nine-node quadrilateral (isoparametric) element.

The variable IMESH is used to indicate whether the mesh is to be generated by the subroutine MESH (IMESH = 1) or read into the program (IMESH = 0). The subroutine MESH can be used only to discretize rectangular domains by linear triangular elements, linear rectangular elements, or quadratic (eight- or nine-node) rectangular elements. The local and global node numbering system used in MESH is shown in Fig. 4.53. The connectivity matrix (array NOD) and the global coordinates of the nodes (columns X and Y) are generated in the subroutine MESH.

For heat transfer problems, the variable ICONV is used to indicate the presence (ICONV = 1) or absence (ICONV = 0) of convective boundaries. When convective boundary is involved (i.e., ICONV = 1), the elements whose boundary coincides with the convective boundary will have additional contribution to their coefficient matrices (see Example 4.4). The array IBN is used to store elements that have convective boundaries, and the array INOD is used to store the pairs of element nodes (of elements in array IBN) that are on the convective boundary. If an element has more than one of its sides on the convective boundary, it should be repeated as many times as the number of its sides on the convective boundary.

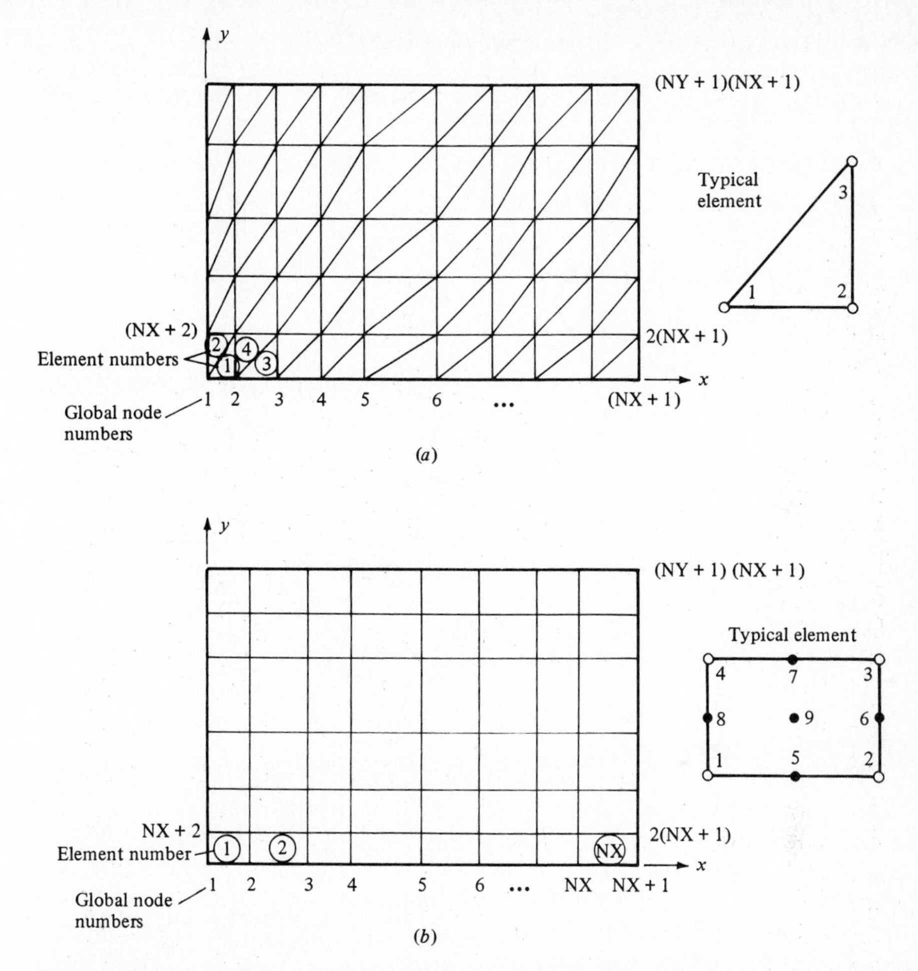

Figure 4.53 Finite-element meshes of triangular and rectangular elements as generated by the program MESH. (*a*) A typical mesh of triangular elements in program MESH. (*b*) A typical mesh of rectangular elements in program MESH (nodes are numbered from left to right and bottom to top).

Linear elements must be used in FEM2D when convective boundary conditions are present in the problem.

A complete description of the input variables of the program FEM2D is given in Table 4.14. In the next section the application of the program FEM2D is illustrated via many examples.

4.8.4 Application of the Computer Program FEM2D

In this section, development of the input data to FEM2D and portions of the output of the program for several example problems is given. The example problems are selected so as to illustrate the capabilities and options available in

Table 4.14 Data input to the computer program FEM2D

COLUMNS	VARIABLE	VARIABLE DESCRIPTION AND NOTES
* PROBLEM DATA CARD 1 (20A4)		ONE CARD PER PROBLEM SOLVED
1-80	TITLE	TITLE OF THE PROBLEM; USED TO LABEL THE PROGRAM OUTPUT
* PROBLEM DATA CARD 2 (16I5)		RIGHT-JUSTIFY EACH ENTRY
1-5	IEL	ELEMENT TYPE: IEL=0, LINEAR TRIANGLE IEL=1, LINEAR QUADRILATERAL IEL=2, QUADRATIC QUADRILATERAL
6-10	NPE	NODES PER ELEMENT: NPE=3 IF IEL =0,NPE=4 IF IEL=1,AND NPE=8 OR 9 IF IEL=2 (8/9-NODE ELEMENT)
11-15	ITYPE	INDICATOR FOR PROBLEM TYPE: ITYPE=-1, HEAT TRANSFER & LIKE ITYPE=0, PLANE STRESS PROBLEMS ITYPE=1, PLANE STRAIN PROBLEMS ITYPE=2, FOR VISCOUS FLOWS
16-20	ICONV	INDICATOR FOR CONVECTION (NEED ONLY WHEN ITYPE = -1): ICONV=1, CONVECTION IS PRESENT ICONV=0, NO CONVECTION
21-25	IMESH	INDICATOR FOR MESH GENERATION: IMESH=1,IF MESH IS TO BE GENERATED IN THE PROGRAM (FOR RECTANGULAR DOMAINS ONLY); IMESH=0 OTHERWISE
26-30	ITEM	INDICATOR FOR TRANSIENT PROBLEM ITEM=1, TRANSIENT ANALYSIS ITEM=0, STATIC ANALYSIS
31-35	NPRNT	INDICATOR FOR PRINT (NPRNT=1) OR NOPRINT (NPRNT=0) OF THE ELEMENT MATIRCES AND VECTORS
* PROBLEM DATA CARD 3 (16I5)		SKIP CARDS 3,4 AND 5 IF IMESH=1
1-5	NEM	NUMBER OF ELEMENTS IN THE MESH
6-10	NNM	NUMBER OF NODES IN THE MESH
* PROBLEM DATA CARD(S) 4 (16I5)		NEM CARDS AND NPE ENTRIES/CARD
1-45	NOD(I,J)	I-TH ROW OF THE CONNECTIVITY MATRIX
* PROBLEM DATA CARDS 5 (8F10.4)		THERE ARE '2*NNM' QUANTITIES, EIGHT QUANTITIES PER CARD
1-80	X(I),Y(I)	X AND Y COORDINATES OF NODE I

* PROBLEM DATA CARD 6 (16I5)		SKIP CARDS 6,7 AND 8 IF IMESH=0
1-5	NX	NUMBER OF SUBDIVISIONS IN THE X DIRECTION
6-10	NY	NUMBER OF SUBDIVISIONS IN THE Y DIRECTION
* PROBLEM DATA CARDS 7&8(8F10.4)		IEL*NX+1 QUANTITIES FOR ARRAY DX AND IEL*NY+1 FOR ARRAY DY
1-80	DX(I)	DISTANCE BETWEEN NODES ALONG THE X-DIRECTION
1-80	DY(I)	DISTANCE BETWEEN NODES ALONG THE Y-DIRECTION
* PROBLEM DATA CARD 9 (8F10.4)		CONSTANTS DEPENDING ON ITYPE: ITYPE=-1: K1,K2,H,TINF, Q ITYPE=0 OR 1:E1,E2,ANU12,G12,T ITYPE=2: AMU, PENALTY PARAMETER
1-10	C1	C1=K1,E1, OR AMU
11-20	C2	C2=K2,E2, OR PENALTY PARAMETER
21-30	C3	C3=H,ANU12, OR ZERO
31-40	C4	C4=TINF,G12, OR ZERO
41-50	C5	C5=Q,T, OR ZERO
* PROBLEM DATA CARD 10 (16I5)		SKIP CARDS 10 AND 11 IF EITHER ITYPE NOT EQUAL TO -1 OR ICONV=0 CONVECTION IS INCLUDED ONLY FOR LINEAR ELEMENTS (IEL=0,1)
1-80	NBE,IBN(I)	NUMBER OF ELEMENTS THAT HAVE SIDES COINCIDING WITH THE CONVECTIVE BOUNDARY (THE ELEMENT SHOULD BE COUNTED AS MANY TIMES AS THE NUMBER OF SIDES ON CONVECTIVE BOUNDARY), AND ARRAY OF ELEMENT NUMBERS WITH CONVECTION
* PROBLEM DATA CARD 11 (16I5)		READ 'NBE' PAIRS OF ENTRIES
1-80	INOD(I,J)	ELEMENT-NODE NUMBERS OF THE SIDE ON CONVECTIVE BOUNDARY
* PROBLEM DATA CARD 12 (16I5)		IF NSDF=0, ENTER ZERO IN COLUMN 5 AND SKIP CARDS 13 AND 14
1-5	NSDF	NUMBER OF SPECIFIED PRIMARY DEGREES OF FREEDOM
* PROBLEM DATA CARDS 13 (16I5)		READ NSDF ENTRIES, 16 PER CARD
1-80	IBDF(I)	SPECIFIED DEGREES OF FREEDOM
* PROBLEM DATA CARDS 14 (8F10.4)		READ NSDF ENTRIES, 8 PER CARD
1-80	VBDF(I)	VALUES OF THE SPECIFIED DEGREES OF FREEDOM IN ARRAY IBDF(I)

Table 4.14 (*Continued*)

* PROBLEM DATA CARD 15 (16I5)		IF NSBF=0, ENTER ZERO IN COLUMN 5 AND SKIP CARDS 16 AND 17
1-5	NSBF	NUMBER OF SPECIFIED NONZERO SECONDARY DEGREES OF FREEDOM
* PROBLEM DATA CARD(S) 16 (16I5)		READ NSBF ENTRIES, 16 PER CARD
1-80	IBSF(I)	SPECIFIED NONZERO SECONDARY VARIABLES
* PROBLEM DATA CARDS 17 (8F10.4)		READ NSBF ENTRIES, 8 PER CARD
1-80	VBSF(I)	VALUES OF SPECIFIED SECONDARY DEGREES OF FREEDOM
* PROBLEM DATA CARD 18 (8F10.4)		SKIP CARDS 18 AND 19 IF ITEM=0
1-10	DT	TIME STEP FOR TRANSIENT ANALYSIS
11-20	THETA	PARAMETER IN TIME-APPROXIMATION
21-30	TO	FINAL TIME VALUE AT WHICH THE TRANSIENT ANALYSIS TERMINATES
* PROBLEM DATA CARDS 19 (8F10.4)		READ 'NEQ' ENTRIES, 8 PER CARD
1-80	GP(I)	INITIAL CONDITIONS ON PRIMARY UNKNOWNS OF THE FEM MODEL

the program FEM2D. A major limitation of the program lies in the mesh generation [i.e., computation of arrays NOD(I, J), X(I), and Y(I)] for nonrectangular domains. For problems involving nonrectangular domains the user is required to input the mesh information (which can be a tedious job if many elements are used). Of course, the program can be modified to accept any other mesh generation subroutine. Other minor limitations (which can be easily overcome by modifying portions of the program) were discussed in the previous section.

Example 4.20 (Solution of the Poisson equation of Example 4.1) We consider the discretization of the domain by linear triangles and rectangles.

Triangular elements We use the 2×2 (i.e., eight element) mesh shown in Fig. 4.54*a*. We have

$$\text{IEL} = 0 \qquad \text{NPE} = 3 \qquad \text{ITYPE} = -1 \qquad \text{ICONV} = 0 \qquad \text{IMESH} = 1$$
$$\text{ITEM} = 0 \qquad \text{NPRNT} = 1$$

Note that we chose to generate the mesh using the subroutine MESH. Therefore, we must specify the number of subdivisions and their lengths (one entry extra) along each direction:

$$\text{NX} = 2 \qquad \text{NY} = 2 \qquad \{\text{DX}\} = \{0.5, 0.5, 0.0\}, \qquad \{\text{DY}\} = \{0.5, 0.5, 0.0\}$$

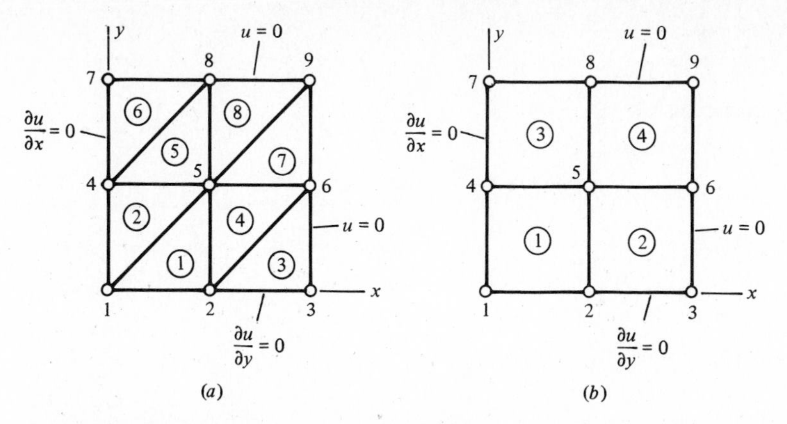

Figure 4.54 Finite-element meshes of linear triangular and rectangular elements used in Example 4.20. (*a*) 2 × 2 mesh of triangles. (*b*) 2 × 2 mesh of rectangles.

The coefficients k_1 and k_2 of the differential equation are unity, and the source term Q has a value of unity:

$$\text{C1}(=\text{K1}) = 1.0 \qquad \text{C2}(=\text{K2}) = 1.0 \qquad \text{C5}(=\text{Q}) = 1.0$$

and C3 (= H) and C4 (= TINF) are zero (no convection). The specified primary degrees of freedom are given by

$$U_3 = U_6 = U_7 = U_8 = U_9 = 0$$

Therefore, we have

$$\text{NSDF} = 5 \qquad \{\text{IBDF}\} = \{3, 6, 7, 8, 9\} \qquad \{\text{VBDF}\} = \{0.0, 0.0, 0.0, 0.0, 0.0\}$$

The value of NSBF is equal to zero because there are no specified nonzero forces in the problem. This completes the data generation for the triangular element case. The printout of the input data and the program output are given in Table 4.15.

Rectangular elements For the 2 × 2 (four element) mesh of rectangular elements (see Fig. 4.54*b*), the data input to the program differs only in the specification of the element type and the number of nodes per element:

$$\text{IEL} = 1 \qquad \text{NPE} = 4$$

The printout of the input data and the program output are given in Table 4.16. ■

Example 4.21 (Flow about a circular cylinder, Example 4.2) Since the domain is not rectangular, we should read in the arrays NOD(I, J), X(I), and Y(I). We consider the 15-element mesh of triangular elements (see Fig. 4.12*a*) used in Example 4.2. We have

$$\text{IEL} = 0 \qquad \text{NPE} = 3 \qquad \text{ITYPE} = -1 \qquad \text{ICONV} = 0 \qquad \text{IMESH} = 0$$
$$\text{ITEM} = 0 \qquad \text{NPRNT} = 0 \qquad \text{NEM} = 15 \qquad \text{NNM} = 14$$
$$\text{C1} = \text{C2} = 1.0 \qquad \text{C3} = \text{C4} = \text{C5} = 0.0$$

Table 4.15 Input and output of the program FEM2D for the triangular-element mesh of the problem in Example 4.20

```
SOLUTION OF THE POISSON EQUATION OF EXAMPLE 4.1 (TRIANGLES)
   0     3    -1     0     1     0     1
   2     2
0.5       0.5
0.5       0.5
1.0       1.0       0.0       0.0       1.0
   5
   3     6     7     8     9
0.0       0.0       0.0       0.0       0.0
   0
..................................................................

  H E A T   T R A N S F E R   T Y P E   P R O B L E M

SOLUTION OF THE POISSON EQUATION OF EXAMPLE 4.1 (TRIANGLES)
         ELEMENT TYPE.............................= 0
         PROBLEM TYPE.............................=-1
         CONVECTION (0:NO, 1:YES).................= 0
         PARAMETERS, C1, C2, C3, C4, AND C5:
              C1 = 0.100D+01
              C2 = 0.100D+01
              C3 = 0.0
              C4 = 0.0
              C5 = 0.100D+01
         ACTUAL NUMBER OF ELEMENTS IN THE MESH....=  8
         NUMBER OF NODES IN THE MESH..............=  9
         TOTAL NUMBER OF EQUATIONS IN THE MODEL...=  9

  COORDINATES OF THE GLOBAL NODES:

  0.0          0.0            0.50000D+00    0.0            0.10000D+01    0.0
  0.0          0.50000D+00    0.50000D+00    0.50000D+00    0.10000D+01    0.50000D+00
  0.0          0.10000D+01    0.50000D+00    0.10000D+01    0.10000D+01    0.10000D+01
```

```
BOOLEAN (CONNECTIVITY) MATRIX NOD(I,J)

          1    1    2    5
          2    1    5    4
          3    2    3    6
          4    2    6    5
          5    4    5    8
          6    4    8    7
          7    5    6    9
          8    5    9    8

  NO. OF SPECIFIED DEGREES OF FREEDOM......=  5
  ARRAY OF THE SPECIFIED DEGREES OF FREEDOM=    3    6    7    8    9
  HALF BAND WIDTH OF GLOBAL STIFFNESS MATRIX =  5
ELEMENT MATRICES:

 0.50000D+00   -0.50000D+00    0.0
-0.50000D+00    0.10000D+01   -0.50000D+00
 0.0           -0.50000D+00    0.50000D+00

SOLUTION VECTOR:

0.31250D+00   0.22917D+00   0.0            0.22917D+00   0.17708D+00   0.0
0.0           0.0           0.0

  THE ANGLE IS DEFINED TO BE THE ARC TANGENT OF Y-COMPONENT DIVIDED BY THE X-COMPONENT

   ELE.NO.  X-COMPONENT      Y-COMPONENT       MAGNITUDE        ANGLE

     1    -0.166667D+00    -0.104167D+00    0.196541D+00    -0.147995D+03
     2    -0.104167D+00    -0.166667D+00    0.196541D+00    -0.122005D+03
     3    -0.458333D+00     0.0             0.458333D+00     0.180000D+03
     4    -0.354167D+00    -0.104167D+00    0.369168D+00    -0.163610D+03
     5    -0.104167D+00    -0.354167D+00    0.369168D+00    -0.106390D+03
     6     0.0             -0.458333D+00    0.458333D+00    -0.900000D+02
     7    -0.354167D+00     0.0             0.354167D+00     0.180000D+03
     8     0.0             -0.354167D+00    0.354167D+00    -0.900000D+02
```

Table 4.16 Input and output of the program FEM2D for the rectangular-element mesh of the problem in Example 4.20

```
SOLUTION OF THE POISSON EQUATION OF EXAMPLE 4.1 (4-NODE RECTANGLES)
    1    4   -1    0    1    0    1
    2    2
0.5       0.5
0.5       0.5
1.0       1.0       0.0       0.0       1.0
    5
    3    6    7    8    9
0.0       0.0       0.0       0.0       0.0
    0

..........................................................................

  H E A T   T R A N S F E R   T Y P E   P R O B L E M

SOLUTION OF THE POISSON EQUATION OF EXAMPLE 4.1 (4-NODE RECTANGLES)
         ELEMENT TYPE.............................= 1
         PROBLEM TYPE.............................=-1
         CONVECTION (0:NO, 1:YES).................= 0
         PARAMETERS, C1, C2, C3, C4, AND C5:
              C1 = 0.100D+01
              C2 = 0.100D+01
              C3 = 0.0
              C4 = 0.0
              C5 = 0.100D+01
         ACTUAL NUMBER OF ELEMENTS IN THE MESH....=  4
         NUMBER OF NODES IN THE MESH..............=  9
         TOTAL NUMBER OF EQUATIONS IN THE MODEL...=  9

  COORDINATES OF THE GLOBAL NODES:

  0.0          0.0           0.50000D+00   0.0           0.10000D+01   0.0
  0.0          0.50000D+00   0.50000D+00   0.50000D+00   0.10000D+01   0.50000D+00
  0.0          0.10000D+01   0.50000D+00   0.10000D+01   0.10000D+01   0.10000D+01
```

```
  BOOLEAN (CONNECTIVITY) MATRIX NOD(I,J)

           1    1    2    5    4
           2    2    3    6    5
           3    4    5    8    7
           4    5    6    9    8

  NO. OF SPECIFIED DEGREES OF FREEDOM......=  5
  ARRAY OF THE SPECIFIED DEGREES OF FREEDOM=    3    6    7    8    9
  HALF BAND WIDTH OF GLOBAL STIFFNESS MATRIX =  5
ELEMENT MATRICES:

 0.66667D+00   -0.16667D+00   -0.33333D+00   -0.16667D+00
-0.16667D+00    0.66667D+00   -0.16667D+00   -0.33333D+00
-0.33333D+00   -0.16667D+00    0.66667D+00   -0.16667D+00
-0.16667D+00   -0.33333D+00   -0.16667D+00    0.66667D+00

SOLUTION VECTOR:

0.31071D+00   0.24107D+00   0.0           0.24107D+00   0.19286D+00   0.0
0.0           0.0           0.0

  THE ANGLE IS DEFINED TO BE THE ARC TANGENT OF Y-COMPONENT DIVIDED BY THE X-COMPONENT

    ELE.NO.  X-COMPONENT      Y-COMPONENT       MAGNITUDE         ANGLE

       1     -0.117857D+00    -0.117857D+00     0.166675D+00    -0.135000D+03
       2     -0.433929D+00    -0.482143D-01     0.436599D+00    -0.173660D+03
       3     -0.482143D-01    -0.433929D+00     0.436599D+00    -0.963402D+02
       4     -0.192857D+00    -0.192857D+00     0.272741D+00    -0.135000D+03
```

The array NOD is given by Eq. (4.77), and X(I) and Y(I) are given by

$$\{X\} = \{0.0, 0.0, 0.0, 1.5, 1.5, 1.5, 3.0, 3.0, 3.0, 3.5, 3.5, 4.0, 4.0, 4.0\}$$
$$\{Y\} = \{0.0, 1.0, 2.0, 0.0, 1.0, 2.0, 0.0, 1.0, 2.0, 1.0, 1.5, 1.0, 1.5, 2.0\}$$

The specified primary degrees of freedom (in the stream function formulation) are given by Eqs. (4.78), and there are no specified nonzero secondary degrees of freedom in the problem. Therefore, we have

$$\text{NSDF} = 10 \qquad \{\text{IBDF}\} = \{1, 2, 3, 4, 6, 7, 9, 10, 12, 14\}$$
$$\{\text{VBDF}\} = \{0.0, 1.0, 2.0, 0.0, 2.0, 0.0, 2.0, 0.0, 0.0, 2.0\} \qquad \text{NSBF} = 0$$

This completes the data generation for the problem. The printout of the input data and the program output are given in Table 4.17.

It should be pointed out that the domain can be discretized by quadrilateral elements. To illustrate this, we consider a finite-element mesh of linear quadrilateral elements (see Fig. 4.55). The mesh closely corresponds to that of triangular elements; except for nodes 7, 8, and 9, all other nodal positions correspond to those in Fig. 4.12. We have

$$\text{IEL} = 1 \qquad \text{NPE} = 4 \qquad \text{NEM} = 8 \qquad \text{NNM} = 15$$

$$[\text{NOD}] = \begin{bmatrix} 1 & 4 & 5 & 2 \\ 2 & 5 & 6 & 3 \\ 4 & 7 & 8 & 5 \\ 5 & 8 & 9 & 6 \\ 7 & 10 & 11 & 8 \\ 8 & 11 & 12 & 9 \\ 11 & 13 & 14 & 12 \\ 12 & 14 & 15 & 9 \end{bmatrix} \qquad \{X\} = \begin{Bmatrix} 0.0 \\ 0.0 \\ 0.0 \\ 1.5 \\ 1.5 \\ 1.5 \\ 2.5 \\ 2.5 \\ 2.5 \\ 3.0 \\ 3.5 \\ 3.5 \\ 4.0 \\ 4.0 \\ 4.0 \end{Bmatrix} \qquad \{Y\} = \begin{Bmatrix} 0.0 \\ 1.0 \\ 2.0 \\ 0.0 \\ 1.0 \\ 2.0 \\ 0.0 \\ 1.0 \\ 2.0 \\ 0.0 \\ 1.0 \\ 1.5 \\ 1.0 \\ 1.5 \\ 2.0 \end{Bmatrix}$$

$$\text{NSDF} = 11 \quad \{\text{IBDF}\} = \{1, 2, 3, 4, 6, 7, 9, 10, 11, 13, 15\}$$
$$\{\text{VBDF}\} = \{0.0, 1.0, 2.0, 0.0, 2.0, 0.0, 2.0, 0.0, 0.0, 0.0, 2.0\} \qquad \text{NSBF} = 0$$

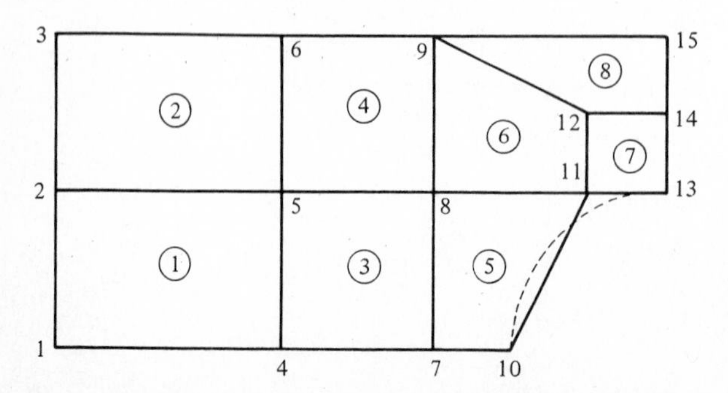

Figure 4.55 Mesh of quadrilateral elements for the flow about a circular cylinder (Example 4.21).

Table 4.17 Input and output for the problem of Example 4.21 (triangular-element mesh)

```
FLOW ABOUT A CIRCULAR CYLINDER BY STREAM FUNCTION FORMULATION (TRIANGLES)
    0     3    -1     0     0     0     0
   15    14
    1     4     5
    1     5     2
    2     5     6
    2     6     3
    4     7     8
    4     8     5
    5     8     9
    5     9     6
    7    10     8
    8    10    11
    8    11     9
   11    14     9
   10    12    11
   12    13    11
   11    13    14
0.0         0.0         0.0         1.0         0.0         2.0         1.5         0.0
1.5         1.0         1.5         2.0         3.0         0.0         3.0         1.0
3.0         2.0         3.5         1.0         3.5         1.5         4.0         1.0
4.0         1.5         4.0         2.0
1.0         1.0         0.0         0.0         0.0
   10
    1     2     3     4     6     7     9    10    12    14
0.0         1.0         2.0         0.0         2.0         0.0         2.0         0.0
0.0         2.0
    0

...........................................................................

   H E A T   T R A N S F E R   T Y P E   P R O B L E M

FLOW ABOUT A CIRCULAR CYLINDER BY STREAM FUNCTION FORMULATION (TRIANGLES)
         ELEMENT TYPE.............................= 0
         PROBLEM TYPE.............................=-1
         CONVECTION (0:NO, 1:YES).................= 0
```

```
      PARAMETERS, C1, C2, C3, C4, AND C5:
           C1 = 0.100D+01
           C2 = 0.100D+01
           C3 = 0.0
           C4 = 0.0
           C5 = 0.0
      ACTUAL NUMBER OF ELEMENTS IN THE MESH....= 15
      NUMBER OF NODES IN THE MESH..............= 14
      TOTAL NUMBER OF EQUATIONS IN THE MODEL...= 14

COORDINATES OF THE GLOBAL NODES (*DELETED IN THE OUTPUT*)
BOOLEAN CONNECTIVITY MATRIX (*DELETED IN THE OUTPUT*)
  NO. OF SPECIFIED DEGREES OF FREEDOM......= 10
  ARRAY OF THE SPECIFIED DEGREES OF FREEDOM=    1    2    3    4    6    7    9   10   12   14

  HALF BAND WIDTH OF GLOBAL STIFFNESS MATRIX =  6

SOLUTION VECTOR:

0.0            0.10000D+01   0.20000D+01   0.0                 0.93913D+00   0.20000D+01
0.0            0.60436D+00   0.20000D+01   0.0                 0.10863D+01   0.0
0.10432D+01    0.20000D+01

  THE ANGLE IS DEFINED TO BE THE ARC TANGENT OF Y-COMPONENT DIVIDED BY THE X-COMPONENT

    ELE.NO.  X-COMPONENT       Y-COMPONENT       MAGNITUDE        ANGLE

      1        0.0             0.939132D+00     0.939132D+00     0.900000D+02
      2      -0.405785D-01     0.100000D+01     0.100082D+01     0.923237D+02
      3      -0.405785D-01     0.106087D+01     0.106164D+01     0.921905D+02
      4        0.0             0.100000D+01     0.100000D+01     0.900000D+02
      5        0.0             0.604360D+00     0.604360D+00     0.900000D+02
      6      -0.223182D+00     0.939132D+00     0.965287D+00     0.103368D+03
      7      -0.223182D+00     0.139564D+01     0.141337D+01     0.990854D+02
      8        0.0             0.106087D+01     0.106087D+01     0.900000D+02
      9      -0.120872D+01     0.604360D+00     0.135139D+01     0.153435D+03
     10      -0.120872D+01     0.217267D+01     0.248627D+01     0.119088D+03
     11      -0.431686D+00     0.139564D+01     0.146088D+01     0.107187D+03
     12        0.0             0.182733D+01     0.182733D+01     0.900000D+02
     13        0.0             0.217267D+01     0.217267D+01     0.900000D+02
     14      -0.863371D-01     0.208634D+01     0.208812D+01     0.923696D+02
     15      -0.863371D-01     0.191366D+01     0.191561D+01     0.925832D+02
```

The input data and the output of the program are given in Table 4.18. Note that the solution obtained using the quadrilateral elements is in agreement with that obtained using triangular elements. ■

Example 4.22 (Convection heat transfer in a plate with a hole, Example 4.4) The data for the problem are the same as those discussed in Example 4.21, except for the following variables:

$$\text{ICONV} = 1 \qquad \text{C1} = \text{C2} = 10.0 \qquad \text{C3} = 15.0 \qquad \text{C4} = 473.0 \qquad \text{C5} = 0.0$$
$$\{\text{VBDF}\} = \{293.0, 293.0, 293.0, 293.0, 293.0, 293.0\} \qquad \text{NSBF} = 0$$

Triangular-element mesh (see Fig. 4.12*a*)

$$\text{NBE} = 2 \qquad \{\text{IBN}\} = \{9, 13\} \qquad \{\text{INOD}\} = \{1, 2;\ 1, 2\}$$
$$\text{NSDF} = 6 \qquad \{\text{IBDF}\} = \{1, 2, 3, 6, 9, 14\}$$

Rectangular-element mesh (see Fig. 4.55)

$$\text{NBE} = 2 \qquad \{\text{IBN}\} = \{5, 7\} \qquad \{\text{INOD}\} = \{2, 3;\ 1, 2\}$$
$$\text{NSDF} = 6 \qquad \{\text{IBDF}\} = \{1, 2, 3, 6, 9, 15\}$$

Table 4.19 contains the input data and output for the triangular-element mesh. ■

Example 4.23 (Plane elasticity problem, Example 4.8) Here we consider both triangular- and rectangular-element meshes (see Fig. 4.56).

Triangular elements We use the 1×1 (two-element) mesh, and specify IMESH = 1 and NX = NY = 1. We have

$$\text{IEL} = 0 \quad \text{NPE} = 3 \quad \text{ITYPE} = 0 \text{ (plane stress)} \quad \text{ICONV} = 0 \quad \text{IMESH} = 1$$
$$\text{C1} = \text{C2} = 3000(10^4) \quad \text{C3} = 0.25 \quad \text{C4} = 1200(10^4) \quad \text{C5} = 0.036$$
$$\text{NSDF} = 4 \qquad \{\text{IBDF}\} = \{1, 2, 5, 6\} \qquad \{\text{VBDF}\} = \{0.0, 0.0, 0.0, 0.0\}$$
$$\text{NSBF} = 2 \qquad \{\text{IBSF}\} = \{3, 7\} \qquad \{\text{VBSF}\} = \{800.0, 800.0\}$$

Linear rectangular element For a 1×1 mesh, the input data are the same as above, except IEL = 1 and NPE = 4.

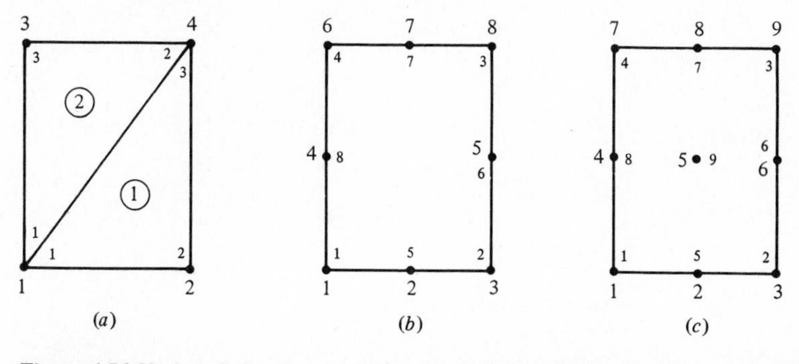

Figure 4.56 Various finite-element meshes for the plane elasticity problem discussed in Example 4.23. (*a*) Linear elements. (*b*) Eight-node element. (*c*) Nine-node element.

Table 4.18 Input and output for the problem in Example 4.21 (quadilateral elements)

```
FLOW ABOUT A CIRCULAR CYLINDER (MESH OF FOUR-NODE QUADRILATERAL ELEMENTS)
   1    4   -1    0    0    0    0
   8   15
   1    4    5    2
   2    5    6    3
   4    7    8    5
   5    8    9    6
   7   10   11    8
   8   11   12    9
  11   13   14   12
  12   14   15    9
0.0       0.0       0.0       1.0       0.0       2.0       1.5       0.0
1.5       1.0       1.5       2.0       2.5       0.0       2.5       1.0
2.5       2.0       3.0       0.0       3.5       1.0       3.5       1.5
4.0       1.0       4.0       1.5       4.0       2.0
1.0       1.0       0.0       0.0       0.0
  11
   1    2    3    4    6    7    9   10   11   13   15
0.0       1.0       2.0       0.0       2.0       0.0       2.0       0.0
0.0       0.0       2.0
   0
.........................................................................

  H E A T   T R A N S F E R   T Y P E   P R O B L E M

FLOW ABOUT A CIRCULAR CYLINDER (MESH OF FOUR-NODE QUADRILATERAL ELEMENTS)
         ELEMENT TYPE.............................= 1
         PROBLEM TYPE.............................=-1
         CONVECTION (0:NO, 1:YES).................= 0
         PARAMETERS, C1, C2, C3, C4, AND C5:
              C1 = 0.100D+01
              C2 = 0.100D+01
              C3 = 0.0
              C4 = 0.0
              C5 = 0.0
```

Table 4.18 (*Continued*)

```
      ACTUAL NUMBER OF ELEMENTS IN THE MESH....=  8
      NUMBER OF NODES IN THE MESH..............= 15
      TOTAL NUMBER OF EQUATIONS IN THE MODEL...= 15

COORDINATES OF THE GLOBAL NODES (*NOT INCLUDED HERE*)
BOOLEAN (CONNECTIVITY) MATRIX NOD(I,J) (*NOT INCLUDED HERE*)
  NO. OF SPECIFIED DEGREES OF FREEDOM......= 11
  ARRAY OF THE SPECIFIED DEGREES OF FREEDOM=    1    2    3    4    6    7    9   10   11   13
          15
  HALF BAND WIDTH OF GLOBAL STIFFNESS MATRIX =  7

SOLUTION VECTOR:

0.0           0.10000D+01   0.20000D+01   0.0           0.98123D+00   0.20000D+01
0.0           0.84356D+00   0.20000D+01   0.0           0.0           0.11260D+01
0.0           0.10437D+01   0.20000D+01

  THE ANGLE IS DEFINED TO BE THE ARC TANGENT OF Y-COMPONENT DIVIDED BY THE X-COMPONENT

    ELE.NO.   X-COMPONENT        Y-COMPONENT        MAGNITUDE          ANGLE

      1       -0.625750D-02      0.990614D+00       0.990634D+00       0.903619D+02
      2       -0.625750D-02      0.100939D+01       0.100941D+01       0.903552D+02
      3       -0.688325D-01      0.912395D+00       0.914988D+00       0.943143D+02
      4       -0.688325D-01      0.108761D+01       0.108978D+01       0.936213D+02
      5       -0.562375D+00      0.562375D+00       0.795318D+00       0.135000D+03
      6       -0.478351D+00      0.152165D+01       0.159507D+01       0.107451D+03
      7       -0.823596D-01      0.216971D+01       0.217127D+01       0.921738D+02
      8       -0.411798D-01      0.178911D+01       0.178958D+01       0.913185D+02
```

Table 4.19 Input and output for the convective heat transfer problem in Example 4.22 (triangular elements)

```
CONVECTION HEAT TRANSFER IN A PLATE WITH HOLE (TRIANGLES)
    0     3    -1     1     0     0     0
   15    14
    1     4     5
    1     5     2
    2     5     6
    2     6     3
    4     7     8
    4     8     5
    5     8     9
    5     9     6
    7    10     8
    8    10    11
    8    11     9
   11    14     9
   10    12    11
   12    13    11
   11    13    14
0.0         0.0         0.0         1.0         0.0         2.0         1.5         0.0
1.5         1.0         1.5         2.0         3.0         0.0         3.0         1.0
3.0         2.0         3.5         1.0         3.5         1.5         4.0         1.0
4.0         1.5         4.0         2.0
10.0        10.0        15.0        473.0       0.0
    2     9    13
    1     2     1     2
    6
    1     2     3     6     9    14
293.0       293.0       293.0       293.0       293.0       293.0
    0
.........................................................................

   H E A T   T R A N S F E R   T Y P E   P R O B L E M

CONVECTION HEAT TRANSFER IN A PLATE WITH HOLE (TRIANGLES)
         ELEMENT TYPE............................= 0
         PROBLEM TYPE............................=-1
         CONVECTION (0:NO, 1:YES)................= 1
```

```
PARAMETERS, C1, C2, C3, C4, AND C5:
     C1 = 0.100D+02
     C2 = 0.100D+02
     C3 = 0.150D+02
     C4 = 0.473D+03
     C5 = 0.0
   ACTUAL NUMBER OF ELEMENTS IN THE MESH....= 15
   NUMBER OF NODES IN THE MESH..............= 14
   TOTAL NUMBER OF EQUATIONS IN THE MODEL...= 14
   CONVECTIVE BOUNDARY DATA: NBE.......=  2
   ARRAY IBN:     9    13
COORDINATES OF THE GLOBAL NODES (*NOT INCLUDED HERE*)
BOOLEAN (CONNECTIVITY) MATRIX NOD(I,J) (*NOT INCLUDED HERE*)

  NO. OF SPECIFIED DEGREES OF FREEDOM......=  6
  ARRAY OF THE SPECIFIED DEGREES OF FREEDOM=    1    2    3    6    9   14
  HALF BAND WIDTH OF GLOBAL STIFFNESS MATRIX =  6

SOLUTION VECTOR:

0.29300D+03   0.29300D+03   0.29300D+03   0.32332D+03   0.31345D+03   0.29300D+03
0.39802D+03   0.35774D+03   0.29300D+03   0.38979D+03   0.33719D+03   0.39508D+03
0.34062D+03   0.29300D+03

  THE ANGLE IS DEFINED TO BE THE ARC TANGENT OF Y-COMPONENT DIVIDED BY THE X-COMPONENT

   ELE.NO.  X-COMPONENT      Y-COMPONENT       MAGNITUDE        ANGLE

     1      0.202111D+03    -0.986267D+02     0.224892D+03    -0.260115D+02
     2      0.136360D+03     0.0              0.136360D+03     0.0
     3      0.136360D+03    -0.204540D+03     0.245827D+03    -0.563099D+02
     4      0.0              0.0              0.0             -0.563099D+02
     5      0.497991D+03    -0.402767D+03     0.640482D+03    -0.389653D+02
     6      0.295231D+03    -0.986267D+02     0.311269D+03    -0.184727D+02
     7      0.295231D+03    -0.647387D+03     0.711528D+03    -0.654853D+02
     8      0.0             -0.204540D+03     0.204540D+03    -0.900000D+02
     9      0.640973D+03    -0.402767D+03     0.757013D+03    -0.321439D+02
    10      0.640973D+03    -0.105188D+04     0.123178D+04    -0.586434D+02
    11      0.236485D+03    -0.647387D+03     0.689228D+03    -0.699332D+02
    12      0.0             -0.883872D+03     0.883872D+03    -0.900000D+02
    13      0.105918D+03    -0.105188D+04     0.105720D+04    -0.842500D+02
    14      0.684804D+02    -0.108931D+04     0.109146D+04    -0.864028D+02
    15      0.684804D+02    -0.952352D+03     0.954811D+03    -0.858871D+02
```

Eight-node rectangular element For a 1×1 mesh (see Fig. 4.56*b*), the changes in the input data are

$$\text{IEL} = 2 \quad \text{NPE} = 8 \quad \text{NSDF} = 6 \quad \{\text{IBDF}\} = \{1, 2, 7, 8, 11, 12\}$$

$$\{\text{VBDF}\} = \{0.0, 0.0, 0.0, 0.0, 0.0, 0.0\} \quad \text{NSBF} = 3$$

$$\{\text{IBSF}\} = \{5, 9, 15\} \quad \{\text{VBSF}\} = \{266.67, 1066.67, 266.67\}$$

Nine-node rectangular element For a 1×1 mesh (see Fig. 4.56*c*), the changes in the input data are

$$\text{IEL} = 2 \quad \text{NPE} = 9 \quad \text{NSDF} = 6 \quad \{\text{IBDF}\} = \{1, 2, 7, 8, 13, 14\}$$

$$\{\text{VBDF}\} = \{0.0, 0.0, 0.0, 0.0, 0.0, 0.0\} \quad \text{NSBF} = 3$$

$$\{\text{IBSF}\} = \{5, 11, 17\} \quad \{\text{VBSF}\} = \{266.67, 1066.67, 266.67\}$$

The input data for all four cases are given in Table 4.20, and the output for each of the cases is given in Table 4.21. ■

Example 4.24 (Viscous fluid squeezed between two parallel plates, Example 4.11) We set up the data for a 10×6 nonuniform mesh of linear rectangular elements and the equivalent 5×3 nonuniform mesh of nine-node elements. Most of the data are exactly the same for both meshes. We have

$$\text{ITYPE} = 2 \quad \text{ICONV} = 0 \quad \text{IMESH} = 1 \quad \text{C1} = 0.0 \quad \text{C2} = 1.0 \times (10^8)$$

and

Linear element

$$\text{IEL} = 1 \quad \text{NPE} = 4 \quad \text{NX} = 10, \quad \text{NY} = 6$$

Nine-node element

$$\text{IEL} = 2 \quad \text{NPE} = 9 \quad \text{NX} = 5 \quad \text{NY} = 3$$

The remaining data are common to both meshes. The coordinates of the subdivisions, the specified boundary degrees of freedom, and the specified values are given in the input data in Tables 4.22 and 4.23. The reader should verify the data on the specified degrees of freedom. Selected parts of the output are also displayed in Tables 4.22 and 4.23. ■

Example 4.25 (Flow of a viscous lubricant in a slider bearing) The slider (or slipper) bearing consists of a short sliding pad moving at a velocity $u = U_0$ relative to a stationary pad inclined at a small angle with respect to the stationary pad, and the small gap between the two pads is filled with a lubricant (see Fig. P4.58). Since the ends of the bearing are generally open, the pressure P_0 there is atmospheric. If the upper pad is parallel to the base plate, the pressure everywhere in the gap must be atmospheric (because dP/dx is a constant for flow between parallel plates), and the bearing cannot support any transverse load. If the upper pad is inclined to the base pad, a pressure distribution (in general, a function of x and y) is set up in the gap. For large values of U_0, the pressure generated can be of sufficient magnitude to support heavy loads normal to the base pad.

Table 4.20 Input data for various meshes of the domain of the plane elasticity problem in Example 4.23

```
PLANE ELASTICITY PROBLEM OF EXAMPLE 4.8 (MESH OF 2 TRIANGULAR ELEMENTS)
    0    3    0    0    1    0    1
    1    1
120.0
160.0
3000.0     3000.0      0.25        1200.0     0.036
    4
    1    2    5    6
0.0        0.0         0.0         0.0
    2
    3    7
800.0      800.0
.........................................................................

PLANE ELASTICITY PROBLEM OF EXAMPLE 4.8 (MESH OF ONE FOUR-NODE ELEMENT)
    1    4    0    0    1    0    1
    1    1
120.0
160.0
3000.0     3000.0      0.25        1200.0     0.036
    4
    1    2    5    6
0.0        0.0         0.0         0.0
    2
    3    7
800.0      800.0
.........................................................................

PLANE ELASTICITY PROBLEM OF EXAMPLE 4.8 (MESH OF ONE EIGHT-NODE ELEMENT)
    2    8    0    0    1    0    0
    1    1
60.0       60.0
80.0       80.0
3000.0     3000.0      0.25        1200.0     0.036
    6
    1    2    7    8   11   12
0.0        0.0         0.0         0.0        0.0        0.0
    3
    5    9   15
266.66667 1066.6667 266.66667
.........................................................................

PLANE ELASTICITY PROBLEM OF EXAMPLE 4.8 (MESH OF ONE NINE-NODE ELEMENT)
    2    9    0    0    1    0    0
    1    1
60.0       60.0
80.0       80.0
3000.0     3000.0      0.25        1200.0     0.036
    6
    1    2    7    8   13   14
0.0        0.0         0.0         0.0        0.0        0.0
    3
    5   11   17
266.66667 1066.6667 266.66667
```

Table 4.21 Output for the plane elasticity problem in Example 4.23

```
   A   P L A N E   E L A S T I C I T Y   O R   F L U I D   F L O W   P R O B L E M

PLANE ELASTICITY PROBLEM OF EXAMPLE 4.8 (*EDITED*)
         ELEMENT TYPE............................= 0
         PROBLEM TYPE............................= 0
         CONVECTION (0:NO, 1:YES)................= 0
         MODULUS OF ELASTICITY, E1...............= 0.300D+04
         MODULUS OF ELASTICITY, E2...............= 0.300D+04
         POISSONS RATIO, ANU12...................= 0.250D+00
         SHEAR MODULUS, G12......................= 0.120D+04
         PLATE THICKNESS, T......................= 0.360D-01
         ACTUAL NUMBER OF ELEMENTS IN THE MESH....=  2
         NUMBER OF NODES IN THE MESH.............=  4
         TOTAL NUMBER OF EQUATIONS IN THE MODEL...=  8

  *EDITED OUTPUT FOR VARIOUS MESHES*:
  -----------------------------------

  SOLUTION VECTOR (FOR TWO TRIANGULAR ELEMENTS):

   1    0.0           0.0                      2    0.11291D+02    0.19637D+01
   3    0.0           0.0                      4    0.10113D+02   -0.10800D+01

    ELE.NO.   STRESS, SXX      STRESS, SYY      STRESS,SXY        PRESSURE

       1     0.106274D+02     0.186156D+01     0.388807D+00
       2     0.970839D+01     0.242710D+01    -0.388807D+00
..................................................................................

  SOLUTION VECTOR (FOR ONE FOUR-NODE ELEMENT):

   1    0.0           0.0                      2    0.10853D+02    0.23256D+01
   3    0.0           0.0                      4    0.10853D+02   -0.23256D+01
```

```
 THE ANGLE IS DEFINED TO BE THE ARC TANGENT OF Y-COMPONENT DIVIDED BY THE X-COMPONENT

  ELE.NO.   STRESS, SXX      STRESS, SYY      STRESS,SXY        PRESSURE

     1      0.100000D+02     0.930233D+00     0.374700D-15

..............................................................................

SOLUTION VECTOR (FOR ONE EIGHT-NODE ELEMENT):

 1    0.0            0.0                           2    0.55212D+01    0.15949D+01
 3    0.11222D+02    0.19144D+01                   4    0.0            0.0
 5    0.10792D+02    0.64632D-14                   6    0.0            0.0
 7    0.55212D+01   -0.15949D+01                   8    0.11222D+02   -0.19144D+01

  ELE.NO.   STRESS, SXX      STRESS, SYY      STRESS,SXY        PRESSURE

     1      0.978584D+01     0.293298D+00     0.528179D-15

..............................................................................

SOLUTION VECTOR (FOR ONE NINE-NODE ELEMENT):

 1    0.0            0.0                           2    0.56874D+01    0.16464D+01
 3    0.10988D+02    0.19156D+01                   4    0.0            0.0
 5    0.52326D+01    0.35043D-14                   6    0.10928D+02    0.10333D-13
 7    0.0            0.0                           8    0.56874D+01   -0.16464D+01
 9    0.10988D+02   -0.19156D+01

  ELE.NO.   STRESS, SXX      STRESS, SYY      STRESS,SXY        PRESSURE

     1      0.989831D+01     0.251914D+00     0.154646D-14
```

Table 4.22 Input and output for the flow problem of Example 4.24 (mesh of four-node elements)

```
FLUID SQUEEZED BETWEEN TWO PARALLEL PLATES (FOUR-NODE QUADRILATERALS)
   1    4    2    0    1    0    0
  10    6
1.0        1.0        1.0        1.0        0.5        0.5        0.25       0.25
0.25       0.25
0.25       0.25       0.5        0.5        0.25       0.25
1.0        1.0
  39
   1    2    4    6    8   10   12   14   16   18   20   22   23   45   67   89
 111  133  134  135  136  137  138  139  140  141  142  143  144  145  146  147
 148  149  150  151  152  153  154
0.0
0.0
0.0        0.0       -1.0        0.0       -1.0        0.0       -1.0        0.0
-1.0       0.0       -1.0        0.0       -1.0        0.0       -1.0        0.0
-1.0       0.0       -1.0        0.0       -1.0        0.0       -1.0
   0
..............................................................................

***EDITED OUTPUT***

FLUID SQUEEZED BETWEEN TWO PARALLEL PLATES (FOUR-NODE QUADRILATERALS)
         ELEMENT TYPE.............................= 1
         PROBLEM TYPE.............................= 2
         CONVECTION (0:NO, 1:YES).................= 0
         VISCOSITY................................= 0.100D+01
         PENALTY PARAMETER........................= 0.100D+09
         ACTUAL NUMBER OF ELEMENTS IN THE MESH....= 60
         NUMBER OF NODES IN THE MESH..............= 77
         TOTAL NUMBER OF EQUATIONS IN THE MODEL...=154
```

SOLUTION VECTOR:

1	0.0	0.0	2	0.75757D+00	0.0
3	0.15135D+01	0.0	4	0.22756D+01	0.0
5	0.30541D+01	0.0	6	0.34648D+01	0.0
7	0.38517D+01	0.0	8	0.40441D+01	0.0
9	0.41712D+01	0.0	10	0.42654D+01	0.0
11	0.42549D+01	0.0	12	0.0	-0.18723D+00
13	0.74517D+00	-0.18845D+00	14	0.14911D+01	-0.18702D+00
15	0.22376D+01	-0.19011D+00	16	0.30074D+01	-0.19694D+00
17	0.33984D+01	-0.20392D+00	18	0.37996D+01	-0.19019D+00
19	0.39630D+01	-0.16554D+00	20	0.41242D+01	-0.12273D+00
21	0.41986D+01	-0.45920D-01	22	0.42127D+01	0.42277D-01
23	0.0	-0.37101D+00	24	0.71032D+00	-0.36855D+00
25	0.14191D+01	-0.37061D+00	26	0.21323D+01	-0.37145D+00
27	0.28526D+01	-0.38812D+00	28	0.32306D+01	-0.39724D+00
29	0.35954D+01	-0.37990D+00	30	0.37883D+01	-0.33212D+00
31	0.39177D+01	-0.24673D+00	32	0.40379D+01	-0.11653D+00
33	0.40360D+01	0.10072D+00	34	0.0	-0.68763D+00
35	0.56777D+00	-0.69097D+00	36	0.11370D+01	-0.68718D+00
37	0.17018D+01	-0.69387D+00	38	0.22655D+01	-0.70773D+00
39	0.25388D+01	-0.72891D+00	40	0.28391D+01	-0.71331D+00
41	0.29831D+01	-0.67262D+00	42	0.31732D+01	-0.54507D+00
43	0.32932D+01	-0.29879D+00	44	0.34093D+01	0.54762D-01
45	0.0	-0.91552D+00	46	0.33168D+00	-0.91281D+00
47	0.66267D+00	-0.91545D+00	48	0.99380D+00	-0.91356D+00
49	0.13051D+01	-0.92558D+00	50	0.14526D+01	-0.93184D+00
51	0.15839D+01	-0.94193D+00	52	0.16782D+01	-0.92065D+00
53	0.17826D+01	-0.88595D+00	54	0.20237D+01	-0.68019D+00
55	0.21793D+01	-0.10719D+00	56	0.0	-0.97396D+00
57	0.17749D+00	-0.98167D+00	58	0.35548D+00	-0.97384D+00
59	0.53189D+00	-0.98206D+00	60	0.69794D+00	-0.97643D+00
61	0.76680D+00	-0.98914D+00	62	0.82738D+00	-0.98056D+00
63	0.84894D+00	-0.99788D+00	64	0.89343D+00	-0.95763D+00
65	0.98981D+00	-0.94599D+00	66	0.15183D+01	-0.52550D+00
67	0.0	-0.10000D+01	68	0.0	-0.10000D+01
69	0.0	-0.10000D+01	70	0.0	-0.10000D+01
71	0.0	-0.10000D+01	72	0.0	-0.10000D+01
73	0.0	-0.10000D+01	74	0.0	-0.10000D+01
75	0.0	-0.10000D+01	76	0.0	-0.10000D+01
77	0.0	-0.10000D+01			

Table 4.23 Input and output of the flow problem in Example 4.24 (nine-node elements)

```
FLUID SQUEEZED BETWEEN TWO PARALLEL PLATES (NINE-NODE QUADRILATERALS)
    2     9     2     0     1     0     0
    5     3
1.0         1.0         1.0         1.0         0.5         0.5         0.25        0.25
0.25        0.25
0.25        0.25        0.5         0.5         0.25        0.25
1.0         1.0
   39
    1     2     4     6     8    10    12    14    16    18    20    22    23    45    67    89
  111   133   134   135   136   137   138   139   140   141   142   143   144   145   146   147
  148   149   150   151   152   153   154
0.0
0.0
0.0         0.0        -1.0         0.0        -1.0         0.0        -1.0         0.0
-1.0        0.0        -1.0         0.0        -1.0         0.0        -1.0         0.0
-1.0        0.0        -1.0         0.0        -1.0         0.0        -1.0
    0

.........................................................................

***EDITED OUTPUT***

FLUID SQUEEZED BETWEEN TWO PARALLEL PLATES (NINE-NODE QUADRILATERALS)
         ELEMENT TYPE.............................= 2
         PROBLEM TYPE.............................= 2
         CONVECTION (0:NO, 1:YES).................= 0
         VISCOSITY................................= 0.100D+01
         PENALTY PARAMETER........................= 0.100D+09
         ACTUAL NUMBER OF ELEMENTS IN THE MESH....= 15
         NUMBER OF NODES IN THE MESH..............= 77
         TOTAL NUMBER OF EQUATIONS IN THE MODEL...=154
```

```
SOLUTION VECTOR:

  1   0.0            0.0                     2   0.75047D+00    0.0
  3   0.14992D+01    0.0                     4   0.22557D+01    0.0
  5   0.30238D+01    0.0                     6   0.34307D+01    0.0
  7   0.38029D+01    0.0                     8   0.39944D+01    0.0
  9   0.41085D+01    0.0                    10   0.42160D+01    0.0
 11   0.41937D+01    0.0                    12   0.0           -0.18591D+00
 13   0.73830D+00   -0.18690D+00            14   0.14777D+01   -0.18599D+00
 15   0.22184D+01   -0.18941D+00            16   0.29796D+01   -0.19502D+00
 17   0.33680D+01   -0.19879D+00            18   0.37572D+01   -0.18294D+00
 19   0.39198D+01   -0.15877D+00            20   0.40721D+01   -0.11729D+00
 21   0.41487D+01   -0.50030D-01            22   0.41719D+01    0.36083D-01
 23   0.0           -0.36935D+00            24   0.70353D+00   -0.36624D+00
 25   0.14052D+01   -0.36947D+00            26   0.21128D+01   -0.37069D+00
 27   0.28250D+01   -0.38578D+00            28   0.32018D+01   -0.38918D+00
 29   0.35561D+01   -0.36773D+00            30   0.37476D+01   -0.31669D+00
 31   0.38626D+01   -0.23743D+00            32   0.39897D+01   -0.11019D+00
 33   0.39621D+01    0.94162D-01            34   0.0           -0.68581D+00
 35   0.56229D+00   -0.68849D+00            36   0.11256D+01   -0.68596D+00
 37   0.16853D+01   -0.69339D+00            38   0.22446D+01   -0.70552D+00
 39   0.25189D+01   -0.72230D+00            40   0.28143D+01   -0.69927D+00
 41   0.29593D+01   -0.64992D+00            42   0.31351D+01   -0.52547D+00
 43   0.32660D+01   -0.29363D+00            44   0.33808D+01    0.63822D-01
 45   0.0           -0.91668D+00            46   0.32824D+00   -0.91282D+00
 47   0.65484D+00   -0.91696D+00            48   0.98238D+00   -0.91476D+00
 49   0.12897D+01   -0.92735D+00            50   0.14366D+01   -0.93279D+00
 51   0.15680D+01   -0.94000D+00            52   0.16680D+01   -0.90892D+00
 53   0.17619D+01   -0.86341D+00            54   0.20059D+01   -0.63400D+00
 55   0.21208D+01   -0.67172D-01            56   0.0           -0.97289D+00
 57   0.17569D+00   -0.97988D+00            58   0.35136D+00   -0.97283D+00
 59   0.52560D+00   -0.98059D+00            60   0.68931D+00   -0.97550D+00
 61   0.75772D+00   -0.98753D+00            62   0.82048D+00   -0.97797D+00
 63   0.85124D+00   -0.98759D+00            64   0.91224D+00   -0.93664D+00
 65   0.10186D+01   -0.89798D+00            66   0.14559D+01   -0.47391D+00
 67   0.0           -0.10000D+01            68   0.0           -0.10000D+01
 69   0.0           -0.10000D+01            70   0.0           -0.10000D+01
 71   0.0           -0.10000D+01            72   0.0           -0.10000D+01
 73   0.0           -0.10000D+01            74   0.0           -0.10000D+01
 75   0.0           -0.10000D+01            76   0.0           -0.10000D+01
 77   0.0           -0.10000D+01
```

Analytical solution Since the width of the gap and the angle of inclination are in general small, it can be assumed with good accuracy that the pressure is not a function of y. Assuming a two-dimensional flow and a small angle of inclination and neglecting the normal stress gradient (in comparison with the shear stress gradient), the equations governing the motion of the lubricant between the pads can be written as

$$\frac{dP}{dx} = \mu\frac{\partial^2 u}{\partial y^2} \qquad u = u(x, y) \qquad v = 0 \tag{4.299a}$$

with the boundary conditions

$$\begin{aligned} y &= 0 \qquad u = U_0 \qquad P = 0 \\ y &= h(x) \qquad u = 0 \qquad P = 0 \end{aligned} \tag{4.299b}$$

where

$$h(x) = h_1 + \frac{h_2 - h_1}{L}x$$

The solution (u, P) of Eqs. (4.299) is given by

$$u = \left(U_0 - \frac{1}{2\mu}h^2\frac{dP}{dx}\frac{y}{h}\right) \qquad P = \frac{6\mu U_0 x(h - h_2)}{h^2(h_1 + h_2)} \tag{4.300}$$

The stresses are given by

$$\tau_{xy} = \mu\frac{\partial u}{\partial y} = \frac{dP}{dx}\left(y - \frac{h}{2}\right) - \mu\frac{U_0}{h} \qquad \sigma_x = \sigma_y = -P \tag{4.301}$$

In our computations, we choose

$$h_2 = 2h_1 = 8 \times 10^{-4} \quad L = 0.36 \quad \mu = 8 \times 10^{-4} \qquad U_0 = 30$$

Finite-element solution First it should be pointed out that the assumption concerning the pressure being not a function of y is not necessary in the finite-element analysis. Here we set up the data using a nonuniform mesh of linear elements to solve the problem. The mesh and the boundary conditions are given in Fig. 4.57. Since the domain is nonrectangular, the mesh information

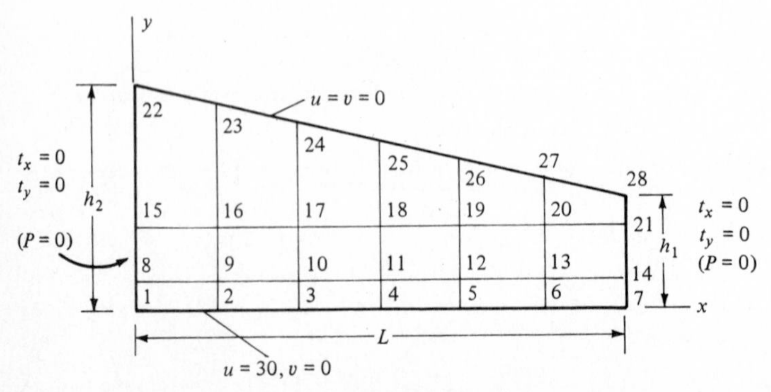

Figure 4.57 Boundary conditions and the finite-element mesh (6 × 3) for the problem in Example 4.25. ($L = 0.36$, $h_1 = 0.0004$, $h_2 = 0.0008$).

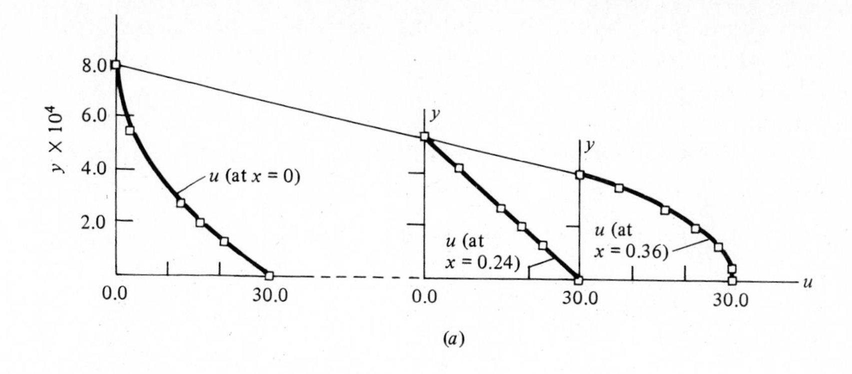

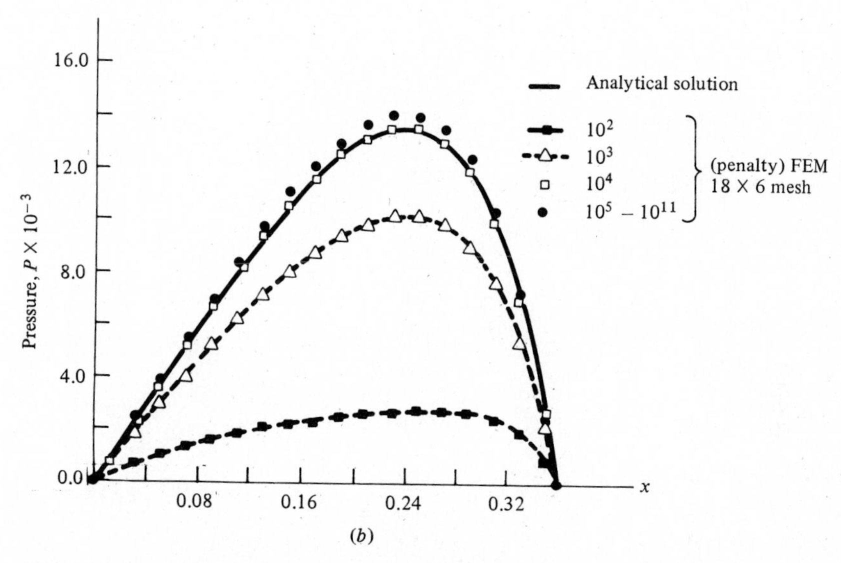

Figure 4.58 Comparison of the finite-element solution with the analytical solution of the problem in Example 4.25. (*a*) Horizontal velocity u versus y. (*b*) Pressure versus x.

should be read in. We have

$$\text{IEL} = 1 \quad \text{NPE} = 4 \quad \text{ITYPE} = 2 \quad \text{IMESH} = 0 \quad \text{NEM} = 18 \quad \text{NNM} = 28$$

The input data and the output of the program are given in Table 4.24 (for the crude mesh). The finite-element solution for the horizontal velocity u and pressure P are not accurate when compared to the analytical solution; a more refined mesh yields a closer solution. Figure 4.58 contains plots of the velocity and pressure distributions (obtained by a mesh of 18×6 linear elements) along the length of the bearing. ■

Example 4.26 (Unsteady solution of the problem in Example 4.13) A 2×2 mesh of either triangles or linear rectangles will have the same number of nodes.

Table 4.24 Input and output of the flow problem in Example 4.25

```
SLIDER BEARING PROBLEM (FOUR-NODE QUADRILATERALS)
    1    4    2    0    0    0    0
   18   28
    1    2    9    8
    2    3   10    9
    3    4   11   10
    4    5   12   11
    5    6   13   12
    6    7   14   13
    8    9   16   15
    9   10   17   16
   10   11   18   17
   11   12   19   18
   12   13   20   19
   13   14   21   20
   15   16   23   22
   16   17   24   23
   17   18   25   24
   18   19   26   25
   19   20   27   26
   20   21   28   27
0.0         0.0         0.06        0.0         0.12        0.0         0.18        0.0
0.24        0.0         0.3         0.0         0.36        0.0         0.0         0.00005
0.06        0.00005     0.12        0.00005     0.18        0.00005     0.24        0.00005
0.3         0.00005     0.36        0.00005     0.0         0.00035     0.06        0.00035
0.12        0.0002833   0.18        0.00025     0.24        0.0002167   0.3         0.0001833
0.36        0.00015     0.0         0.0008      0.06        0.0007333   0.12        0.0006667
0.18        0.0006      0.24        0.0005333   0.3         0.0004667   0.36        0.0004
0.0008      1.0
   28
    2    4    6    8   10   12   14   43   44   45   46   47   48   49   50   51
   52   53   54   55   56    1    3    5    7    9   11   13
0.0         0.0         0.0         0.0         0.0         0.0         0.0         0.0
0.0         0.0         0.0         0.0         0.0         0.0         0.0         0.0
0.0         0.0         0.0         0.0         0.0         30.0        30.0        30.0
30.0        30.0        30.0        30.0
    0

...............................................................................

***EDITED OUTPUT***

SLIDER BEARING PROBLEM (FOUR-NODE QUADRILATERALS)
          ELEMENT TYPE...........................= 1
          PROBLEM TYPE...........................= 2
          CONVECTION (0:NO, 1:YES)...............= 0
```

Table 4.24 (*Continued*)

```
        VISCOSITY................................= 0.800D-03
        PENALTY PARAMETER........................= 0.100D+09
        ACTUAL NUMBER OF ELEMENTS IN THE MESH....= 18
        NUMBER OF NODES IN THE MESH..............= 28
        TOTAL NUMBER OF EQUATIONS IN THE MODEL...= 56

SOLUTION VECTOR:

  1   0.30000D+02   0.0                      2   0.30000D+02   0.0
  3   0.30000D+02   0.0                      4   0.30000D+02   0.0
  5   0.30000D+02   0.0                      6   0.30000D+02   0.0
  7   0.30000D+02   0.0                      8   0.25866D+02  -0.16139D-02
  9   0.25887D+02   0.15969D-02             10   0.26112D+02  -0.17848D-02
 11   0.26431D+02   0.15191D-02             12   0.27110D+02  -0.20853D-02
 13   0.28309D+02   0.10868D-02             14   0.30513D+02  -0.29237D-02
 15   0.71193D+01  -0.52423D-02             16   0.78038D+01   0.17009D-02
 17   0.11343D+02  -0.35837D-03             18   0.14154D+02  -0.37055D-02
 19   0.17624D+02  -0.35013D-02             20   0.22087D+02  -0.72776D-02
 21   0.28045D+02  -0.80164D-02             22   0.0           0.0
 23   0.0           0.0                     24   0.0           0.0
 25   0.0           0.0                     26   0.0           0.0
 27   0.0           0.0                     28   0.0           0.0

   ELE.NO.   STRESS, SXX     STRESS, SYY     STRESS,SXY      PRESSURE

      1     -0.285085D+04   -0.285085D+04   -0.659758D+02   0.285085D+04
      2     -0.847124D+04   -0.847124D+04   -0.640089D+02   0.847124D+04
      3     -0.135567D+05   -0.135567D+05   -0.596546D+02   0.135567D+05
      4     -0.170622D+05   -0.170622D+05   -0.516687D+02   0.170622D+05
      5     -0.170014D+05   -0.170014D+05   -0.366475D+02   0.170014D+05
      6     -0.917713D+04   -0.917718D+04   -0.942770D+01   0.917716D+04
      7     -0.285054D+04   -0.285056D+04   -0.491065D+02   0.285055D+04
      8     -0.847151D+04   -0.847150D+04   -0.492816D+02   0.847150D+04
      9     -0.135565D+05   -0.135565D+05   -0.499350D+02   0.135565D+05
     10     -0.170623D+05   -0.170624D+05   -0.474781D+02   0.170623D+05
     11     -0.170012D+05   -0.170013D+05   -0.418874D+02   0.170013D+05
     12     -0.917708D+04   -0.917727D+04   -0.297975D+02   0.917718D+04
     13     -0.285023D+04   -0.285022D+04   -0.143267D+02   0.285022D+04
     14     -0.847183D+04   -0.847183D+04   -0.199782D+02   0.847183D+04
     15     -0.135562D+05   -0.135562D+05   -0.278125D+02   0.135562D+05
     16     -0.170627D+05   -0.170627D+05   -0.381382D+02   0.170627D+05
     17     -0.170009D+05   -0.170009D+05   -0.529484D+02   0.170009D+05
     18     -0.917769D+04   -0.917760D+04   -0.751884D+02   0.917765D+04
```

Thus we have

$$\text{IEL} = 0 \qquad \text{NPE} = 3 \qquad \text{or} \qquad \text{IEL} = 1 \qquad \text{NPE} = 4$$

and

$$\text{ITYPE} = -1 \qquad \text{ICONV} = 0 \qquad \text{IMESH} = 1 \qquad \text{ITEM} = 1 \qquad \text{NPRNT} = 0$$
$$\text{DT} = 0.05 \qquad \text{ALFA} = 0.5 \qquad \text{T0} = 1.0$$

The input data and selected parts of the computer output for the traingular-element mesh are presented in Table 4.25. ■

Example 4.27 (Unsteady solution of the Couette flow, Example 4.14) We use a mesh of 2×6 linear rectangular elements to solve the problem. We set

$$\text{IEL} = 1 \qquad \text{NPE} = 4 \qquad \text{ITYPE} = 2 \qquad \text{ICONV} = 0$$
$$\text{IMESH} = 1 \qquad \text{ITEM} = 1 \qquad \text{NPRNT} = 0$$
$$\text{DT} = 1.0 \qquad \text{ALFA} = 0.5 \qquad \text{T0} = 40.0$$

The input data and selected parts of the output are presented in Table 4.26. ■

*4.8.5 Description and Application of Computer Program PLATE

The program PLATE is developed using the formulation of the equations of orthotropic plates presented in Secs. 4.5.4 and 4.6.3. From the discussion presented there, it follows that there are three primary degrees of freedom (w, S_x, S_y) per node (i.e., NDF = 3) in the element. The program contains only the four-, eight-, and nine-node quadrilateral isoparametric elements. Except for the subroutines STIFF and STRESS, all other subroutines are the same as those used in FEM2D. The finite-element matrices in Eqs. (4.225) and (4.243) are programmed in subroutine STIFF, and the stresses at the top or bottom of the plate, given by

$$\sigma_x = \left(c_{11}\frac{\partial S_x}{\partial x} + c_{12}\frac{\partial S_y}{\partial y}\right)\frac{h}{2} \qquad \sigma_y = \left(c_{21}\frac{\partial S_x}{\partial x} + c_{22}\frac{\partial S_y}{\partial y}\right)\frac{h}{2}$$

$$\tau_{xy} = c_{33}\left(\frac{\partial S_x}{\partial y} + \frac{\partial S_y}{\partial x}\right)\frac{h}{2} \qquad \tau_{xz} = c_{44}\left(\frac{\partial w}{\partial x} + S_x\right) \qquad \tau_{yz} = c_{55}\left(\frac{\partial w}{\partial y} + S_y\right) \tag{4.302}$$

are computed in the subroutine STRESS.

A description of the input variables is given in Table 4.27. Two examples of application of the computer program PLATE are given below.

Example 4.28 (Bending of a simply supported isotropic plate under uniform load, Example 4.12) The 2×2 mesh of linear elements, the 1×1 mesh of nine-node elements, and the boundary conditions are shown in Fig. 4.59. The data are given

Table 4.25 Input and output for the unsteady analysis of the Poisson equation of Example 4.26 (triangular-element mesh)

```
UNSTEADY SOLUTION OF THE POISSON EQUATION (TRIANGLES)
    0    3   -1    0    1    1    0
    2    2
0.5       0.5
0.5       0.5
1.0       1.0       0.0       0.0       1.0
    5
    3    6    7    8    9
0.0       0.0       0.0       0.0       0.0
    0
0.05      0.5       1.0
0.0
0.0
```

```
UNSTEADY SOLUTION OF THE POISSON EQUATION (TRIANGLES)
         ELEMENT TYPE............................= 0
         PROBLEM TYPE............................=-1
         CONVECTION (0:NO, 1:YES)................= 0
         PARAMETERS, C1, C2, C3, C4, AND C5:
              C1 = 0.100D+01
              C2 = 0.100D+01
              C3 = 0.0
              C4 = 0.0
              C5 = 0.100D+01
         ACTUAL NUMBER OF ELEMENTS IN THE MESH....=  8
         NUMBER OF NODES IN THE MESH..............=  9
         TOTAL NUMBER OF EQUATIONS IN THE MODEL...=  9

***PARTIAL OUTPUT***

  TIME = 0.050
```

Table 4.25 (*Continued*)

```
SOLUTION VECTOR:

0.49549D-01   0.58441D-01   0.0           0.58441D-01   0.53132D-01   0.0
0.0           0.0           0.0
```

ELE.NO.	X-COMPONENT	Y-COMPONENT	MAGNITUDE	ANGLE
1	0.177845D-01	-0.106176D-01	0.207128D-01	-0.308378D+02
2	-0.106176D-01	0.177845D-01	0.207128D-01	0.120838D+03
3	-0.116882D+00	0.0	0.116882D+00	0.180000D+03
4	-0.106264D+00	-0.106176D-01	0.106793D+00	-0.174294D+03
5	-0.106176D-01	-0.106264D+00	0.106793D+00	-0.957058D+02
6	0.0	-0.116882D+00	0.116882D+00	-0.900000D+02
7	-0.106264D+00	0.0	0.106264D+00	0.180000D+03
8	0.0	-0.106264D+00	0.106264D+00	-0.900000D+02

```
TIME =        1.000

SOLUTION VECTOR:

0.31101D+00   0.22820D+00   0.0           0.22820D+00   0.17639D+00   0.0
0.0           0.0           0.0
```

ELE.NO.	X-COMPONENT	Y-COMPONENT	MAGNITUDE	ANGLE
1	-0.165619D+00	-0.103630D+00	0.195368D+00	-0.147965D+03
2	-0.103630D+00	-0.165619D+00	0.195368D+00	-0.122035D+03
3	-0.456401D+00	0.0	0.456401D+00	0.180000D+03
4	-0.352772D+00	-0.103630D+00	0.367678D+00	-0.163629D+03
5	-0.103630D+00	-0.352772D+00	0.367678D+00	-0.106371D+03
6	0.0	-0.456401D+00	0.456401D+00	-0.900000D+02
7	-0.352772D+00	0.0	0.352772D+00	0.180000D+03
8	0.0	-0.352772D+00	0.352772D+00	-0.900000D+02

Table 4.26 Input and output for the unsteady flow problem of Example 4.27

```
UNSTEADY COUETTE FLOW WITH UPPER WALL MOVING (FOUR-NODE QUADRILATERALS)
    1    4    2    0    1    1    0
    2    6
1.0       1.0
1.0       1.0       1.0       1.0       1.0       1.0
1.0       1.0
   22
    1    2    3    4    5    6    8   12   14   18   20   24   26   30   32   36
   38   40   42   37   39   41
0.0
0.0
0.0       0.0       0.0       3.0       3.0       3.0
    5
    7   13   19   25   31
1.0       1.0       1.0       1.0       1.0       1.0
1.0       0.5      40.0
0.0
0.0
0.0

.........................................................................

UNSTEADY COUETTE FLOW WITH UPPER WALL MOVING (FOUR-NODE QUADRILATERALS)
         ELEMENT TYPE...........................= 1
         PROBLEM TYPE...........................= 2
         CONVECTION (0:NO, 1:YES)...............= 0
         VISCOSITY..............................= 0.100D+01
         PENALTY PARAMETER......................= 0.100D+09
         ACTUAL NUMBER OF ELEMENTS IN THE MESH....= 12
         NUMBER OF NODES IN THE MESH.............= 21
         TOTAL NUMBER OF EQUATIONS IN THE MODEL...= 42

**PARTIAL OUTPUT**

  TIME =      1.000
```

Table 4.26 (*Continued*)

```
SOLUTION VECTOR:

  1   0.0           0.0                 2   0.0           0.0
  3   0.0           0.0                 4   0.39659D+00   0.0
  5   0.39659D+00  -0.52406D-08         6   0.39659D+00   0.0
  7   0.48295D+00   0.0                 8   0.48295D+00  -0.18809D-09
  9   0.48295D+00   0.0                10   0.51818D+00   0.0
 11   0.51818D+00   0.89646D-10        12   0.51818D+00   0.0
 13   0.60795D+00   0.0                14   0.60795D+00   0.67119D-09
 15   0.60795D+00   0.0                16   0.10216D+01   0.0
 17   0.10216D+01   0.56341D-08        18   0.10216D+01   0.0
 19   0.30000D+01   0.0                20   0.30000D+01   0.0
 21   0.30000D+01   0.0

TIME =      5.000

SOLUTION VECTOR:

  1   0.0           0.0                 2   0.0           0.0
  3   0.0           0.0                 4   0.11948D+01   0.0
  5   0.11948D+01  -0.55308D-08         6   0.11948D+01   0.0
  7   0.20335D+01   0.0                 8   0.20335D+01   0.42992D-09
  9   0.20335D+01   0.0                10   0.26209D+01   0.0
 11   0.26209D+01  -0.67890D-10        12   0.26209D+01   0.0
 13   0.30447D+01   0.0                14   0.30447D+01  -0.30678D-08
 15   0.30447D+01   0.0                16   0.31748D+01   0.0
 17   0.31748D+01   0.29609D-08        18   0.31748D+01   0.0
 19   0.30000D+01   0.0                20   0.30000D+01   0.0
 21   0.30000D+01   0.0

TIME =     40.000

SOLUTION VECTOR:

  1   0.0           0.0                 2   0.0           0.0
  3   0.0           0.0                 4   0.17500D+01   0.0
  5   0.17500D+01   0.99817D-09         6   0.17500D+01   0.0
  7   0.30000D+01   0.0                 8   0.30000D+01  -0.27725D-09
  9   0.30000D+01   0.0                10   0.37499D+01   0.0
 11   0.37499D+01  -0.14757D-08        12   0.37499D+01   0.0
 13   0.40000D+01   0.0                14   0.40000D+01   0.17192D-08
 15   0.40000D+01   0.0                16   0.37500D+01   0.0
 17   0.37500D+01   0.19195D-08        18   0.37500D+01   0.0
 19   0.30000D+01   0.0                20   0.30000D+01   0.0
 21   0.30000D+01   0.0
```

Table 4.27 Data input to the computer program PLATE

COLUMNS	VARIABLE	VARIABLE DESCRIPTION AND NOTES
* PROBLEM DATA CARD 1 (20A4)		PROBLEM HEADING CARD
1-80	TITLE	THE TITLE OF THE PROBLEM FOR LABELLING THE OUTPUT
* PROBLEM DATA CARD 2 (16I5)		THERE ARE EIGHT ENTRIES ON THE CARD (RIGHT-JUSTIFIED)
1-5	IEL	ELEMENT TYPE: IEL=1, 4-NODE AND IEL=2, EIGHT/NINE-NODE ELEMENT
6-10	NPE	NODES PER ELEMENT: NPE=4 IF IEL=1, AND NPE=8 OR 9 IF IEL=2
11-15	IMESH	INDICATOR FOR MESH GENERATION: MESH=0, MESH INFORMATION IS TO READ IN; IMESH=1, MESH IS TO BE GENERATED (ONLY FOR RECTANGULAR DOMAINS)
16-20	NPRNT	INDICATOR FOR PRINT (NPRNT=1) OR NOPRINT (NPRNT=0) OF THE ELEMENT MATRICES AND VECTORS
21-25	ITEM	INDICATOR FOR TRANSIENT PROBLEM ITEM=0 FOR STATIC ANALYSIS; ITEM=1 FOR TRANSIENT ANALYSIS
26-30	NTIME	TOTAL NUMBER OF TIME STEPS
31-35	NSTP	TIME STEP NUMBER AT WHICH THE LOADING IS TO BE REMOVED
36-40	NOZERO	INDICATOR FOR ZERO (NOZERO=0) OR NONZERO (NOZERO=1) INITIAL CONDITIONS FOR TRANSIENT CASE
* PROBLEM DATA CARD 3 (16I5)		SKIP CARDS 3,4 AND 5 IF IMESH=1
1-5	NEM	NUMBER OF ELEMENTS IN THE MESH
6-10	NNM	NUMBER OF NODES IN THE MESH
* PROBLEM DATA CARD(S) 4 (16I5)		READ 'NEM' CARDS:NPE PER CARD
1-45	NOD(I,J)	I-TH ROW OF THE MATIRX 'NOD'
* PROBLEM DATA CARDS 5 (8F10.4)		INPUT X AND Y COORDINATES; READ '2*NNM' ENTRIES, EIGHT PER CARD
1-80	X(I),Y(I)	GLOBAL COORDINATES OF I-TH NODE
* PROBLEM DATA CARD 6 (16I5)		SKIP CARDS 6,7 AND 8 IF IMESH=0
1-5	NX	NUMBER OF SUBDIVISIONS ALONG THE X-AXIS
6-10	NY	NUMBER OF SUBDIVISIONS ALONG THE Y-AXIS

Table 4.27 (*Continued*)

```
* PROBLEM DATA CARDS 7&8 (8F10.4)   READ 'IEL*NX+1' ENTRIES FOR DX
                                    AND 'IEL*NY+1' ENTRIES FOR DY

   1-80          DX(I)              DISTANCE BETWEEN TWO NODES
                                    ALONG X-COORDINATE DIRECTION

   1-80          DY(I)              DISTANCE BETWEEN TWO NODES
                                    ALONG Y-COORDINATE DIRECTION

* PROBLEM DATA CARD 9 (8E10.3)      RIGHT-JUSTIFY THE ENTRIES

   1-10          E1                 MODULUS ALONG X-DIRECTION

   11-20         E2                 MODULUS ALONG Y-DIRECTION

   21-30         G12                SHEAR MODULUS W.R.TO X,Y-PLANE

   31-40         G13                SHEAR MODULUS W.R.TO X,Z-PLANE

   41-50         G23                SHEAR MODULUS W.R.TO Y,Z-PLANE

   51-60         ANU12              PRINCIPAL POISSON'S RATIO

   61-70         RHO                MATERIAL DENSITY

   71-80         H                  PLATE THICKNESS

* PROBLEM DATA CARD 10 (8F10.4)     FOR LOADS OTHER THAN UNIFORMLY
                                    DISTRIBUTED, ENTER ZERO AND
                                    READ THE FORCES VIA IBSF & VBSF

   1-10          PO                 INTENSITY OF THE UNIFORM LOAD

* PROBLEM DATA CARD 11 (16I5)       **NBDY CANNOT BE A ZERO**

   1-5           NBDY               NUMBER OF SPECIFIED GENERALIZED
                                    DISPLACEMENTS

* PROBLEM DATA CARD(S) 12 (16I5)    READ 'NBDY' ENTRIES,16 PER CARD

   1-80          IBDY(I)            ARRAY OF THE SPECIFIED DEGREES
                                    OF FREEDOM (ORDER: W,SX,SY)

* PROBLEM DATA CARDS 13 (8F10.4)    READ 'NBDY' ENTRIES, 8 PER CARD

   1-80          VBDY(I)            VALUES OF THE SPECIFIED DEGREES
                                    FREEDOM IN THE ARRAY IBDY(I)

* PROBLEM DATA CARD 14 (16I5)       IF NBSF=0, ENTER ZERO IN COL. 5

   1-5           NBSF               NUMBER OF SPECIFIED NONZERO
                                    GENERALIZED FORCES

* PROBLEM DATA CARD(S) 15 (16I5)    ***SKIP IF NBSF=0***

   1-80          IBSF(I)            SPECIFIED NONZERO FORCES

* PROBLEM DATA CARDS 16 (8F10.4)    READ 'NBSF' ENTRIES, 8 PER CARD

   1-80          VBSF(I)            VALUES OF THE SPECIFIED FORCES
```

Table 4.27 (*Continued*)

```
* PROBLEM DATA CARD 17 (8F10.4)       SKIP REMAINING CARDS IF ITEM=0
    1-10          DT                  TIME STEP FOR TRANSIENT PROBLEM
    11-20         ALFA                PARAMETER IN NEWMARK'S METHOD
* PROBLEM DATA CARDS 18 (8F10.4)      SKIP IF NOZERO=0
    1-80          GF0(I)              INITIAL VALUES OF DISPLACEMENTS
* PROBLEM DATA CARDS 19 (8F10.4)      SKIP IF NOZERO=0
    1-80          GF1(I)              INITIAL VALUES OF VELOCITIES
-----------------------------------------------------------------------
```

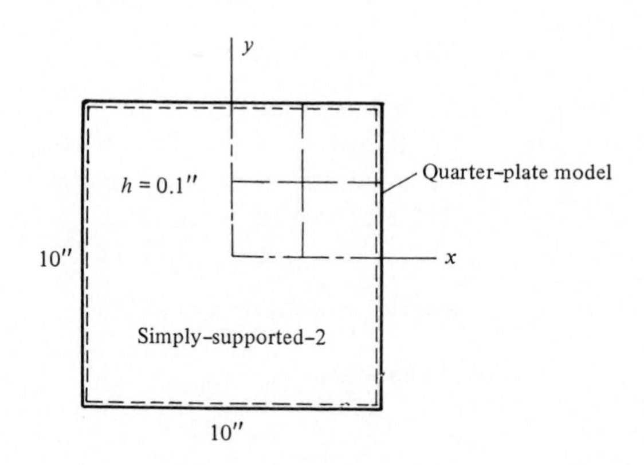

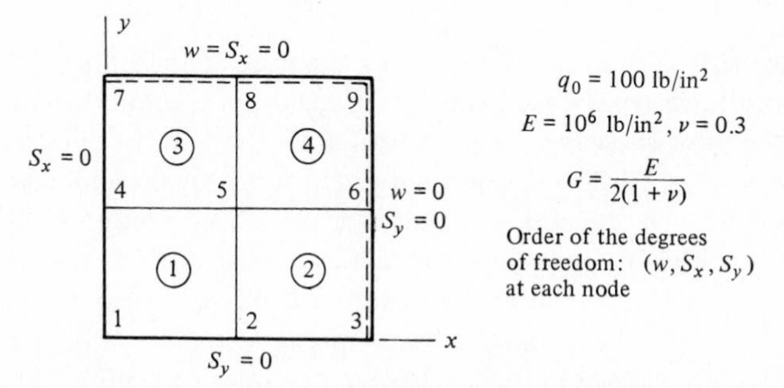

Figure 4.59 Boundary conditions and finite-element mesh for the bending of a simply supported square plate.

by (see Table 4.28)

$$\text{IEL} = 1 \quad \text{NPE} = 4 \ (\text{or IEL} = 2 \quad \text{NPE} = 9) \quad \text{IMESH} = 1$$

$$\text{NPRNT} = 0 \quad \text{ITEM} = 0 \quad \text{NTIME} = 0 \quad \text{NSTP} = 0 \quad \text{NOZERO} = 0$$

$$\text{NX} = \text{NY} = 2 \ (\text{or NX} = \text{NY} = 1) \quad \{\text{DX}\} = \{\text{DY}\} = \{2.5, 2.5\}$$

$$\text{E1} = \text{E2} = 10^6 \quad \text{G12} = \text{G13} = \text{G23} = 0.3846 \times 10^6 \quad \text{ANU12} = 0.3$$

$$\text{RHO} = 0.0 \quad \text{T} = 0.1 \quad \text{P0} = 100.0$$

The specified boundary degrees of freedom (note that there are three degrees of freedom at each node) are given by

$$\{\text{IBDY}\} = \{2, 3, 6, 7, 9, 11, 16, 18, 19, 20, 22, 23, 25, 26, 27\} \quad \text{NBDY} = 15$$

and all values are specified to be zero. Other than the uniformly distributed load, there are no specified nonzero forces in the problem. It should be pointed out that one can include the uniform load through the array IBSF(I) by specifying $P_0 = 0.0$ and computing the contribution of the uniform load to various force degrees of freedom:

$$\{\text{IBSF}\} = \{1, 4, 10, 13\} \quad \text{VBSF} = \{v_0, 2v_0, 2v_0, 4v_0\}$$

$$v_0 = \frac{q_0 h_x h_y}{4} = 156.25$$

Selected parts of the output of the program are given in Table 4.29. The finite-element solution for the center deflection obtained by the linear elements converges from below and that obtained by the quadratic elements converges from above to the classical plate solution:

$$\bar{w} = wEh^3/q_0 a^4$$

Linear elements			Quadratic elements		Classical plate theory
1 × 1	2 × 2	4 × 4	1 × 1	2 × 2	
0.03486	0.04337	0.04416	0.04582	0.04446	0.04434

■

Example 4.29 (Dynamic response of clamped circular plates under suddenly applied uniform transverse loading) Consider a clamped circular plate of radius $R = 100$ in, thickness $h = 20$ in, modulus $E = 100 \text{ lb/in}^2$, Poisson's ratio $\nu = 0.3$, and density $\rho = 10 \ (\text{lb} \cdot \text{s}^2)/\text{in}^4$, subjected to suddenly applied (i.e., $w = S_x = S_y = \dot{w} = \dot{S}_x = \dot{S}_y = 0$ at time $t = 0$) uniform loading of intensity $q_0 = 1 \text{ lb/in}^2$. We solve the problem by modeling one quadrant (because of the biaxial symmetry) of the plate by five nine-node elements (see Fig. 4.60a). The input data are given by

$$\text{IEL} = 2 \quad \text{NPE} = 9 \quad \text{IMESH} = 0 \quad \text{NPRNT} = 0$$

$$\text{ITEM} = 1 \quad \text{NTIME} = 10 \quad \text{NSTP} = 10$$

$$\text{NEM} = 5 \quad \text{NNM} = 29 \quad \text{DT} = 2.5 \quad \text{ALFA} = 0.5$$

Table 4.28 Input data for the plate bending problems in Examples 4.28 and 4.29

```
STATIC BENDING OF A SIMPLY SUPPORTED PLATE UNDER UNIFORM LOAD (2X2L MESH)
   1    4    1    0    0    0    0
   2    2
2.5       2.5
2.5       2.5
    1.0E6     1.0E6  0.3846E6  0.3846E6  0.3846E6    0.30E0     0.0E0     0.1E0
100.0
   15
   2    3    6    7    9   11   16   18   19   20   22   23   25   26   27
0.0
0.0
   0
.................................................................................

STATIC BENDING OF A SIMPLY SUPPORTED PLATE UNDER UNIFORM LOAD (1X1Q9 MESH)
   2    9    1    0    0    0    0
   1    1
2.5       2.5
2.5       2.5
    1.0E6     1.0E6  0.3846E6  0.3846E6  0.3846E6    0.30E0     0.0E0     0.1E0
100.0
   15
   2    3    6    7    9   11   16   18   19   20   22   23   25   26   27
0.0
0.0
   0
.................................................................................

STATIC BENDING OF A SIMPLY SUPPORTED PLATE UNDER UNIFORM LOAD (4X4L MESH)
   1    4    1    0    0    0    0
   4    4
1.25      1.25      1.25      1.25
1.25      1.25      1.25      1.25
    1.0E6     1.0E6  0.3846E6  0.3846E6  0.3846E6    0.30E0     0.0E0     0.1E0
```

Table 4.28 (*Continued*)

```
 100.0
   27
    2     3     6     9    12    13    15    17    28    30    32    43    45    47    58    60
   61    62    64    65    67    68    70    71    73    74    75
0.0
0.0
0.0
0.0
    0

.................................................................................

DYNAMIC BENDING OF A CLAMPED CIRCULAR PLATE UNDER UNIFORM STEP LOAD(5Q9 MESH)
    2     9     0     0     1    10    10     0
    5    29
    1     5     7     9     2     6     8     4     3
    5    15    17     7    10    16    12     6    11
    9     7    17    19     8    12    18    14    13
   15    25    27    17    20    26    22    16    21
   19    17    27    29    18    22    28    24    23
0.0         0.0         16.5        0.0         11.6673     11.6673     0.0         16.5
33.0        0.0         30.488      12.6286     23.3345     23.3345     12.6286     30.488
0.0         33.0        49.5        0.0         45.732      18.9428     35.0        35.0
18.9428     45.732      0.0         49.5        66.0        0.0         60.976      25.2571
46.669      46.669      25.2571     60.976      0.0         66.0        83.0        0.0
76.682      31.7627     58.6898     58.6898     31.7627     76.682      0.0         83.0
100.0       0.0         92.388      38.2683     70.7107     70.7107     38.2683     92.388
0.0         100.0
     1.0E2       1.0E2  0.3845E2  0.3845E2  0.3845E2     0.30E0       1.0E1        2.0E1
1.0
   27
    2     3     6    11    15    26    30    41    45    56    60    71    73    74    75    76
   77    78    79    80    81    82    83    84    85    86    87
0.0
0.0
0.0
0.0
    0
2.5         0.5

.................................................................................
```

Table 4.29 Selected portions of the output for the static bending problem in Example 4.28

```
STATIC BENDING OF A SIMPLY SUPPORTED PLATE UNDER UNIFORM LOAD (2X2L MESH)

    ELEMENT TYPE(1=LINEAR,2=QUADRATIC) = 1       NODES PER ELEMENT= 4
         ACTUAL NUMBER OF ELEMENTS IN THE MESH =  4
         NUMBER OF NODES IN THE MESH =  9
         DEGREES OF FREEDOM = 3

    M A T E R I A L   P R O P E R T I E S:
         MODULUS, E1= 0.10000D+07
         MODULUS, E2= 0.10000D+07
         SHEAR MODULI, G12,G13 AND G23= 0.38460D+06 0.38460D+06 0.38460D+06
         POISSONS RATIO, NU12= 0.30000D+00
         SHEAR CORRECTION COEFFICIENT, K= 0.83333D+00
         MATERIAL DENSITY, RHO= 0.0
         PLATE THICKNESS, H= 0.10000D+00
         LOAD MAGNITUDE, P= 0.10000D+03


    NUMBER OF SPECIFIED DISPLACEMENTS=    15
    SPECIFIED DISPLACEMENTS AND THEIR VALUES FOLLOW:
   2    3    6    7    9   11   16   18   19   20   22   23   25   26   27
  0.0           0.0            0.0            0.0            0.0            0.0            0.0            0.0
  0.0           0.0            0.0            0.0            0.0            0.0            0.0

    NUMBER OF SPECIFIED FORCES=    0
    SPECIFIED FORCE DEGREES OF FREEDOM FOLLOW AND THEIR SPECIFIED VALUES FOLLOW:
   0
  0.0

    BOOLEAN (CONNECTIVITY) MATRIX-NOD(I,J)

   1    1    2    5    4
   2    2    3    6    5
   3    4    5    8    7
   4    5    6    9    8
```

Table 4.29 (*Continued*)

```
  COORDINATES OF THE GLOBAL NODES:

0.0          0.0          0.25000D+01  0.0          0.50000D+01  0.0          0.0          0.25000D+01
0.25000D+01  0.25000D+01  0.50000D+01  0.25000D+01  0.0          0.50000D+01  0.25000D+01  0.50000D+01
0.50000D+01  0.50000D+01

  HALF BAND WIDTH OF GLOBAL STIFFNESS MATRIX =    15


  TRANSVERSE DEFLECTION:

0.43368D+02  0.30838D+02  0.0          0.30838D+02  0.21963D+02  0.0          0.0          0.0
0.0

  BENDING SLOPE, SI-X:

0.0          0.10019D+02  0.14636D+02  0.0          0.70974D+01  0.10460D+02  0.0          0.0
0.0

  BENDING SLOPE, SI-Y:

0.0          0.0          0.0          0.10019D+02  0.70974D+01  0.0          0.14636D+02  0.10460D+02
0.0
    X-COORD     Y-COORD     SIGMAX      SIGMAY      SIGMAXY     SIGMAXZ     SIGMAYZ
   0.1250D+01  0.1250D+01  0.4075D+03  0.4075D+03 -0.3746D+02 -0.6250D+02 -0.6250D+02
   0.3750D+01  0.1250D+01  0.1851D+03  0.1738D+03 -0.9099D+02 -0.2292D+03 -0.2083D+02
   0.1250D+01  0.3750D+01  0.1738D+03  0.1851D+03 -0.9099D+02 -0.2083D+02 -0.2292D+03
   0.3750D+01  0.3750D+01  0.8006D+02  0.8006D+02 -0.2251D+03 -0.1042D+03 -0.1042D+03
..........................................................................................

 STATIC BENDING OF A SIMPLY SUPPORTED PLATE UNDER UNIFORM LOAD (1X1Q9 MESH)

   ELEMENT TYPE(1=LINEAR,2=QUADRATIC) = 2      NODES PER ELEMENT= 9
       ACTUAL NUMBER OF ELEMENTS IN THE MESH =  1
       NUMBER OF NODES IN THE MESH =  9
       DEGREES OF FREEDOM = 3
```

```
  TRANSVERSE DEFLECTION:

0.45821D+02   0.32303D+02   0.0           0.32303D+02   0.22826D+02   0.0           0.0           0.0
0.0

  BENDING SLOPE, SI-X:

0.0           0.99850D+01   0.15019D+02   0.0           0.70177D+01   0.10670D+02   0.0           0.0
0.0

  BENDING SLOPE, SI-Y:

0.0           0.0           0.0           0.99850D+01   0.70177D+01   0.0           0.15019D+02   0.10670D+02
0.0
   X-COORD      Y-COORD      SIGMAX       SIGMAY       SIGMAXY      SIGMAXZ      SIGMAYZ
  0.1057D+01   0.1057D+01   0.4559D+03   0.4559D+03  -0.3329D+02  -0.4675D+02  -0.4675D+02
  0.1057D+01   0.3943D+01   0.1795D+03   0.1998D+03  -0.9042D+02  -0.1354D+02  -0.2365D+03
  0.3943D+01   0.1057D+01   0.1998D+03   0.1795D+03  -0.9042D+02  -0.2365D+03  -0.1354D+02
  0.3943D+01   0.3943D+01   0.8066D+02   0.8066D+02  -0.2556D+03  -0.1125D+03  -0.1125D+03
....................................................................................

STATIC BENDING OF A SIMPLY SUPPORTED PLATE UNDER UNIFORM LOAD (4X4L MESH)

  TRANSVERSE DEFLECTION:

0.44157D+02   0.41011D+02   0.31823D+02   0.17486D+02   0.0           0.41011D+02   0.38100D+02   0.29584D+02
0.16268D+02   0.0           0.31823D+02   0.29584D+02   0.23018D+02   0.12690D+02   0.0           0.17486D+02
0.16268D+02   0.12690D+02   0.70216D+01   0.0           0.0           0.0           0.0           0.0
0.0
```

Table 4.30 Selected portions of the output for the transient analysis of the clamped circular plate problem in Example 4.29

```
DYNAMIC BENDING OF A CLAMPED CIRCULAR PLATE UNDER UNIFORM STEP LOAD(5Q9 MESH)

    ELEMENT TYPE(1=LINEAR,2=QUADRATIC) = 2       NODES PER ELEMENT= 9
         ACTUAL NUMBER OF ELEMENTS IN THE MESH =  5
         NUMBER OF NODES IN THE MESH = 29
         DEGREES OF FREEDOM = 3

    M A T E R I A L   P R O P E R T I E S:
         MODULUS, E1= 0.10000D+03
         MODULUS, E2= 0.10000D+03
         SHEAR MODULI, G12,G13 AND G23= 0.38450D+02 0.38450D+02 0.38450D+02
         POISSONS RATIO, NU12= 0.30000D+00
         SHEAR CORRECTION COEFFICIENT, K= 0.83333D+00
         MATERIAL DENSITY, RHO= 0.10000D+02
         PLATE THICKNESS, H= 0.20000D+02
         LOAD MAGNITUDE, P= 0.10000D+01


    NUMBER OF SPECIFIED DISPLACEMENTS=   27
    SPECIFIED DISPLACEMENTS AND THEIR VALUES FOLLOW:
  2    3    6   11   15   26   30   41   45   56   60   71   73   74   75   76
 77   78   79   80   81   82   83   84   85   86   87
 0.0           0.0           0.0           0.0           0.0           0.0           0.0           0.0
 0.0           0.0           0.0           0.0           0.0           0.0           0.0           0.0
 0.0           0.0           0.0           0.0           0.0           0.0           0.0           0.0
 0.0           0.0           0.0

    NUMBER OF SPECIFIED FORCES=   0
    SPECIFIED FORCE DEGREES OF FREEDOM FOLLOW AND THEIR SPECIFIED VALUES FOLLOW:
  0
 0.0

    BOOLEAN (CONNECTIVITY) MATRIX-NOD(I,J)

  1    1    5    7    9    2    6    8    4    3
  2    5   15   17    7   10   16   12    6   11
  3    9    7   17   19    8   12   18   14   13
  4   15   25   27   17   20   26   22   16   21
  5   19   17   27   29   18   22   28   24   23
```

```
    COORDINATES OF THE GLOBAL NODES:

  0.0            0.0            0.16500D+02    0.0            0.11667D+02    0.11667D+02    0.0            0.16500D+02
  0.33000D+02    0.0            0.30488D+02    0.12629D+02    0.23334D+02    0.23334D+02    0.12629D+02    0.30488D+02
  0.0            0.33000D+02    0.49500D+02    0.0            0.45732D+02    0.18943D+02    0.35000D+02    0.35000D+02
  0.18943D+02    0.45732D+02    0.0            0.49500D+02    0.66000D+02    0.0            0.60976D+02    0.25257D+02
  0.46669D+02    0.46669D+02    0.25257D+02    0.60976D+02    0.0            0.66000D+02    0.83000D+02    0.0
  0.76682D+02    0.31763D+02    0.58690D+02    0.58690D+02    0.31763D+02    0.76682D+02    0.0            0.83000D+02
  0.10000D+03    0.0            0.92388D+02    0.38268D+02    0.70711D+02    0.70711D+02    0.38268D+02    0.92388D+02
  0.0            0.10000D+03

    HALF BAND WIDTH OF GLOBAL STIFFNESS MATRIX =    39


    DT=0.2500D+01      ALFA=0.5000D+00      BETA=0.2500D+00
         TEMPORAL PARAMETERS A0,A1,ETC.:  0.6400D+00  0.8000D+00  0.1600D+01  0.1000D+01  0.1000D+01


    TIME = 0.250D+01


    TRANSVERSE DEFLECTION:

  0.77060D-02    0.77897D-02    0.78680D-02    0.77897D-02    0.75219D-02    0.75974D-02    0.74506D-02    0.75974D-02
  0.75219D-02    0.80174D-02    0.79961D-02    0.80240D-02    0.79961D-02    0.80174D-02    0.64338D-02    0.64725D-02
  0.64261D-02    0.64725D-02    0.64338D-02    0.88882D-02    0.88648D-02    0.88888D-02    0.88648D-02    0.88882D-02
  0.0            0.0            0.0            0.0            0.0

.................................................................................................

    TIME = 0.100D+02


    TRANSVERSE DEFLECTION:

  0.19659D+00    0.19600D+00    0.19349D+00    0.19600D+00    0.19731D+00    0.19785D+00    0.19846D+00    0.19785D+00
  0.19731D+00    0.19262D+00    0.19183D+00    0.19252D+00    0.19183D+00    0.19262D+00    0.18541D+00    0.18788D+00
  0.18554D+00    0.18788D+00    0.18541D+00    0.19929D+00    0.19705D+00    0.19928D+00    0.19705D+00    0.19929D+00
  0.0            0.0            0.0            0.0            0.0
.................................................................................................

    TIME = 0.150D+02
```

```
TRANSVERSE DEFLECTION:

0.46622D+00   0.47675D+00   0.47645D+00   0.47675D+00   0.48188D+00   0.48685D+00   0.48259D+00   0.48685D+00
0.48188D+00   0.46240D+00   0.46035D+00   0.46250D+00   0.46035D+00   0.46240D+00   0.49869D+00   0.50705D+00
0.49860D+00   0.50705D+00   0.49869D+00   0.42642D+00   0.41751D+00   0.42646D+00   0.41751D+00   0.42642D+00
0.0           0.0           0.0           0.0           0.0
.......................................................................

TIME = 0.200D+02

TRANSVERSE DEFLECTION:

0.85851D+00   0.88548D+00   0.89397D+00   0.88548D+00   0.86245D+00   0.87890D+00   0.85258D+00   0.87890D+00
0.86245D+00   0.87583D+00   0.87010D+00   0.87680D+00   0.87010D+00   0.87583D+00   0.95875D+00   0.97620D+00
0.95763D+00   0.97620D+00   0.95875D+00   0.67815D+00   0.65764D+00   0.67825D+00   0.65764D+00   0.67815D+00
0.0           0.0           0.0           0.0           0.0

.......................................................................

TIME = 0.250D+02

TRANSVERSE DEFLECTION:

0.14375D+01   0.14276D+01   0.14104D+01   0.14276D+01   0.13368D+01   0.13596D+01   0.13262D+01   0.13596D+01
0.13368D+01   0.14601D+01   0.14462D+01   0.14592D+01   0.14462D+01   0.14601D+01   0.14928D+01   0.15198D+01
0.14937D+01   0.15198D+01   0.14928D+01   0.92059D+00   0.88705D+00   0.92013D+00   0.88705D+00   0.92059D+00
0.0           0.0           0.0           0.0           0.0
```

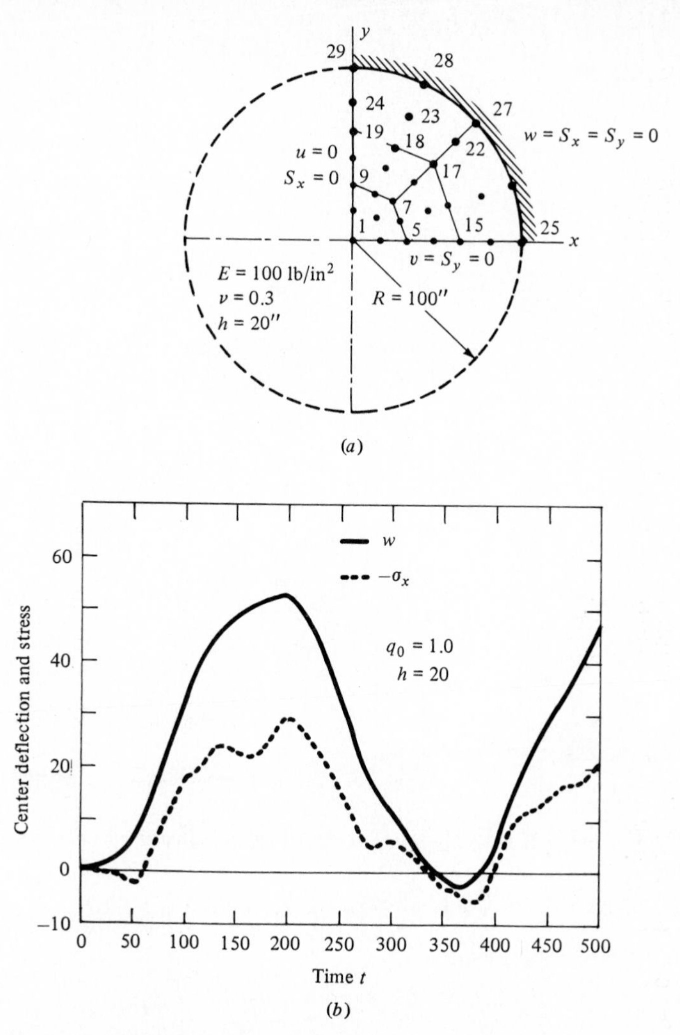

Figure 4.60 Clamped circular plate under suddenly applied uniform load. (*a*) Geometry and boundary conditions. (*b*) Center deflection and stress versus time.

The arrays NOD, X, Y, IBDY, and VBDY can be easily developed using the mesh in Fig. 4.60*a*.

The complete input data and selected portions of the output are listed in Tables 4.28 and 4.30, respectively. Plots of the center deflection and stress versus time are shown in Fig. 4.60*b*. ■

PROBLEMS

4.67 Determine the jacobian matrix and the transformation equations for the elements given in Fig. P4.67.

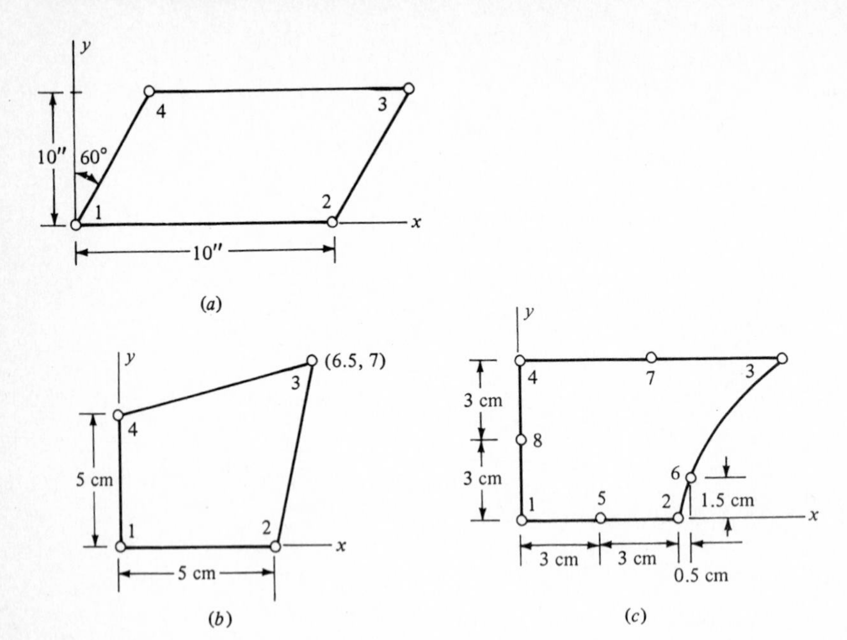

Figure P4.67

4.68 Using the Gauss quadrature, determine the contribution of a constant distributed source to nodal points of the four-node finite element given in Fig. P4.67*b*.

4.69 Show that the side nodes in the eight-node rectangular element of Fig. 4.49 should be located such that $0.25 < a < 0.75$.

4.70 For a 12-node serendipity (cubic) element, as illustrated in Fig. P4.70, show that the determinant J is given by

$$J = \tfrac{9}{2}(2 - 10\xi + 9\xi^2)a + \tfrac{9}{2}(-1 + 8\xi - 9\xi^2)b + \tfrac{1}{2}(2 - 18\xi + 27\xi^2)$$

What can you conclude from the requirement $J > 0$?

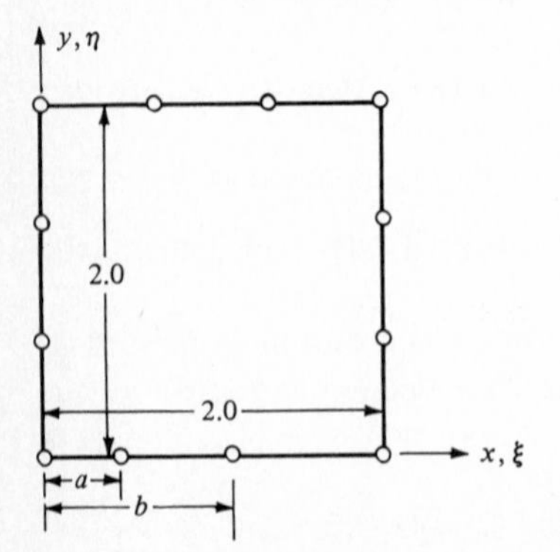

Figure P4.70

4.71 Determine the conditions on the location of node 3 of the quadrilateral element shown in Fig. P4.71. Show that the transformation equations are given by

$$x = \xi + \xi\eta(a - 2) \qquad y = \eta + \xi\eta(b - 2)$$

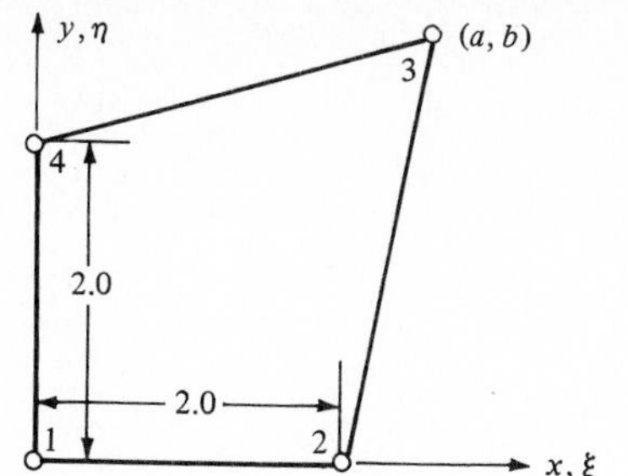

Figure P4.71

4.72 Determine the global derivatives of the interpolation functions for node 3 of the element shown in Fig. P4.67*b*.

For Probs. 4.73 through 4.108, develop the input data required for program FEM2D or program PLATE, and solve them.†

4.73 Prob. 4.11.

4.74 Prob. 4.22.

4.75 Prob. 4.25.

4.76 Prob. 4.26.

4.77 Prob. 4.29.

4.78 (Entrance flow) Flow of a viscous incompressible fluid in a nonrectangular domain shown in Fig. P4.59*b* with $U_0 = 1$.

4.79 Perform the numerical convergence study of the problem in Fig. P4.59*c*, using the meshes 5×5, 7×5, and 11×5, and plot the center axial velocity as a function of x.

4.80 Repeat Prob. 4.78 with an elliptic cylindrical obstruction placed symmetrically (see Fig. P4.59*a*).

4.81 Consider the slider bearing problem discussed in Example 4.25. Take $U_0 = 40$ ft/s, $L = 4$ in $= 0.333$ ft, $\nu = 4 \times 10^{-4}$ ft²/s, $h_2 = 0.008$ in, $h_1 = 0.004$ in, and a nonuniform mesh of 8×6 linear elements to generate the data and the solution. Perform a study of the effect of the penalty parameter (10^3–10^{13}) on the velocity field and the pressure.

4.82 Consider the flow of a viscous incompressible fluid in a square cavity. The flow is induced by the movement of the top wall (or lid) with a constant velocity $u = 1.0$. Using a mesh of 8×8 linear elements, determine the velocity field, and plot the horizontal velocity along the vertical centerline.

4.83 Verify the results of the problem in Example 4.11.

4.84 Consider the flow of a viscous incompressible fluid in a 90° plane tee. Using the symmetry and the mesh shown in Fig. P4.84, determine the velocity distribution at the exit.

4.85 Solve the problem of viscous incompressible flow in the nonrectangular domain shown in Fig. P4.85.

4.86–4.87 Solve the plane elasticity problems in Fig. P4.49*a* and *b*.

4.88–4.92 Analyze the plane elasticity problems illustrated in Figs. P4.88 though P4.92, using $\nu = 0.3$ and the appropriate value of E (for the isotropic case).

4.93 Consider the simply supported plate shown in Fig. 4.45. Use $E_1 = 25 \times 10^6$ lb/in², $E_2 = 10^6$ lb/in², $G_{12} = G_{13} = 0.5 \times 10^6$ lb/in², $G_{23} = 0.2 \times 10^6$ lb/in², and $\nu_{12} = 0.25$ and a 4×4 nonuniform mesh of linear elements to determine the center deflection.

4.94 Repeat Prob. 4.93 when the plate is clamped along vertical sides and simply supported along the horizontal sides.

†It is suggested that the finite-element solutions be compared with other analytical and numerical solutions available in the literature; see References for Additional Reading at the end of this chapter.

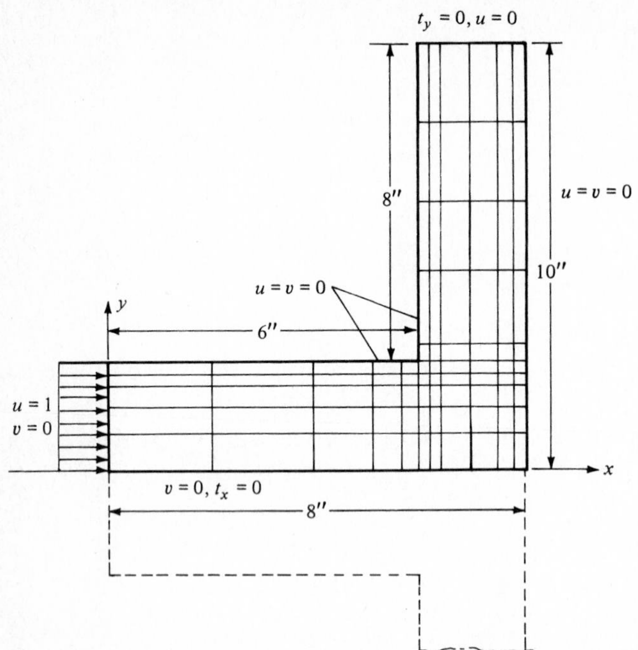

Figure P4.84

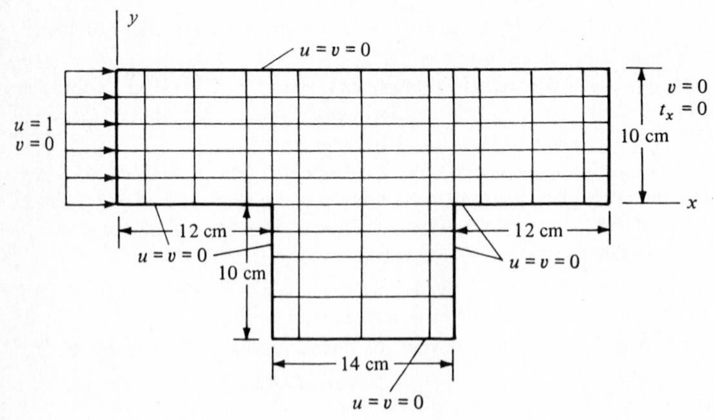

Figure P4.85

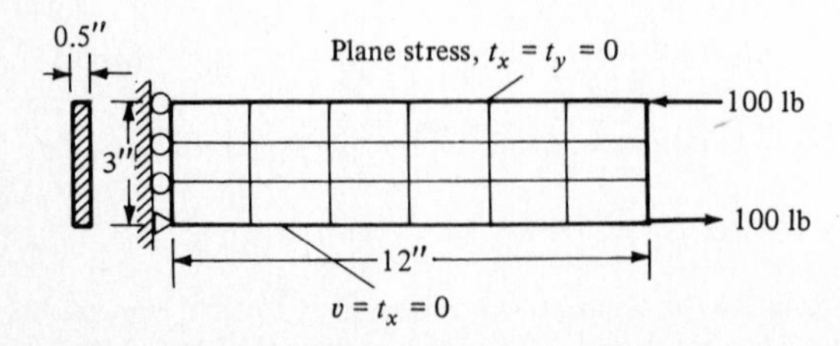

Figure P4.88

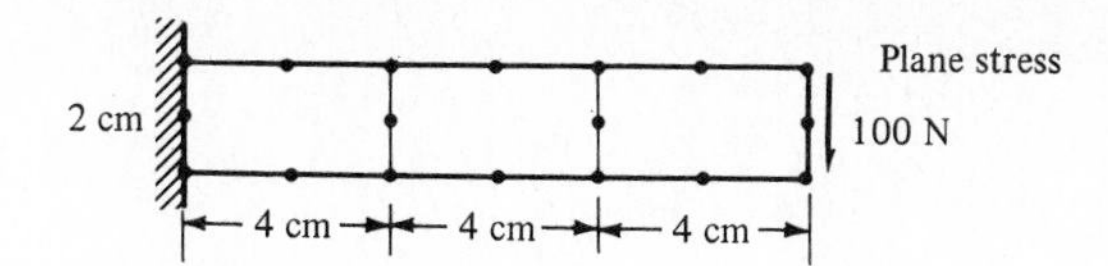

Figure P4.89

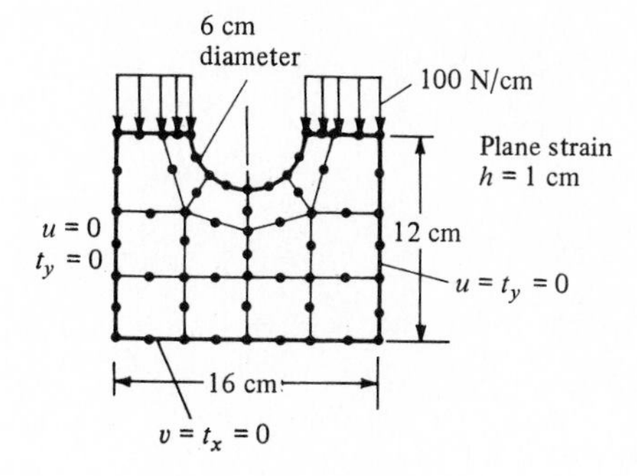

Figure P4.90

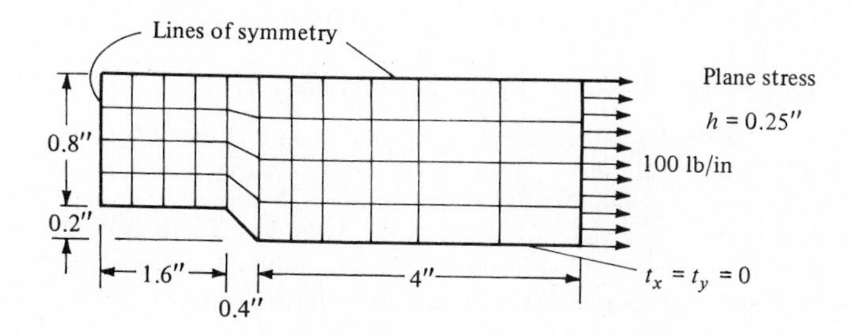

Figure P4.91

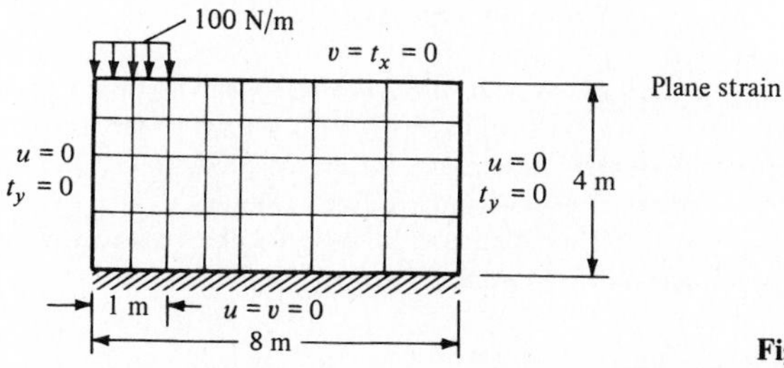

Figure P4.92

4.95 Consider an isotropic annular plate ($E = 30 \times 10^6$ lb/in², $\nu = 0.29$) with the outer edges simply supported, and subjected to uniform loading (see Fig. P4.60a). Determine the deflection at the inner edge.

4.96 Repeat Prob. 4.95 for the case in which the plate is subjected to line loading at the inner edge.

4.97 Repeat Prob. 4.93 for the case in which the plate has a central square hole (of dimension $c = 0.4$) subjected to uniformly distributed load.

In Probs. 4.98 through 4.105, assume zero initial conditions.

4.98 Find the transient solution of Prob. 4.73.

4.99 Find the transient solution of Prob. 4.20.

4.100 Find the transient solution of Prob. 4.82.

4.101 Modify program FEM2D to include the transient analysis of elasticity problems. (*Hint*: Use the logic and formulation used in the program PLATE.)

4.102 Solve the problem of Example 4.12 by the program PLATE.

4.103 Find the transient solution of Prob. 4.95.

4.104 Find the transient solution of Prob. 4.97.

4.105 Find the transient solution of Prob. 4.93 for the case when the load is removed after five time steps.

REFERENCES FOR ADDITIONAL READING

References on fluid mechanics (also see References for Additional Reading in Chap. 3)

Eskinazi, S.: *Principles of Fluid Mechanics*, Allyn and Bacon, Boston (1962).
Nadai, A.: *Theory of Flow and Fracture of Solids*, Vol. II, McGraw-Hill, New York (1963).
Roache, P. J.: *Computational Fluid Dynamics*, Hermosa Publishers, Albuquerque, N.M. (1972).
Schlichting, H.: *Boundary-Layer Theory* (translated by J. Kestin), 7th ed., McGraw-Hill, New York (1979).
Verruijit, A.: *Theory of Groundwater Flow*, Gordon and Breach, New York (1970).

References on heat transfer (also see References for Additional Reading in Chap. 3)

Bruch, J. C., and G. Zyvoloski: "Transient Two-Dimensional Heat Conduction Problems Solved by the Finite Element Method," *International Journal for Numerical Methods in Engineering*, **8**, pp. 481–494 (1974).
Carslaw, H. S., and J. C. Jaeger: *Conduction of Heat in Solids*, Clarendon Press, Oxford (1959).
Compbell, B. C., B. Kaplan, and A. H. Moore: "A Numerical Comparison of the Crandall and Crank-Nicolson Implicit Methods for Solving a Diffusion Equation," *Journal of Heat Transfer, Transactions of the ASME*, **88**, pp. 324–326 (1966).
Donea, J.: "On the Accuracy of Finite Element Solutions to the Transient Heat-Conduction Equation," *International Journal for Numerical Methods in Engineering*, **8**, pp. 103–110 (1974).
Moore, A. H., B. Kaplan, and D. B. Mitchell: "A Comparison of Crandall and Crank-Nicolson Methods for Solving a Transient Heat Conduction Problem," *International Numerical Methods in Engineering*, **9**, pp. 938–943 (1975).
Wilson, E. L., and R. E. Nickell: "Application of the Finite Element Method to Heat Conduction Analysis," *Journal for Nuclear Engineering and Design*, **4**, pp. 276–286 (1966).
Wood, W. L., and R. W. Lewis: "A Comparison of Time Marching Schemes for the Transient Heat Conduction Equation," *International Journal for Numerical Methods in Engineering*, **9**, pp. 679–690 (1975).

References on plane elasticity (also see References for Additional Readings in Chap. 3)

Budynas, R. G.: *Advanced Strength and Applied Stress Analysis*, McGraw-Hill, New York (1977).
Ugural, A. C., and S. K. Fenster, *Advanced Strength and Applied Elasticity*, American Elsevier, New York (1975).
Volterra, E., and J. H. Gaines: *Advanced Strength of Materials*, Prentice-Hall, Englewood Cliffs, N.J. (1971).

References on plate bending

Dym, C. L., and I. H. Shames: *Solid Mechanics: A Variational Approach*, McGraw-Hill, New York (1973).
Jones, R. M.: *Mechanics of Composite Materials*, McGraw-Hill, New York (1975).

Kikuchi, F., and Y. Ando, "Rectangular Finite Element for Plate Bending Analysis Based on Hellinger-Reissner's Variational Principle," *Journal of Nuclear Science and Technology*, **9**, pp. 28–35 (1972).

Reddy, J. N.: "Simple Finite Elements with Relaxed Continuity for Nonlinear Analysis of Plates," *Finite Element Methods in Engineering*, A. P. Kabaila and V. A. Pulmano, eds., The University of New South Wales, Sydney (1979), pp. 265–281.

Reddy, J. N.: "A Penalty-Plate Bending Element for the Analysis of Laminated Anisotropic Composite Plates," *International Journal for Numerical Methods in Engineering*, **15**, pp. 1187–1206 (1980).

Reddy, J. N.: "On the Solutions to Forced Motions of Rectangular Composite Plates," *Journal of Applied Mechanics*, **49**, pp. 403–408 (1982).

Reddy, J. N., and C. S. Tsay: "Mixed Rectangular Finite Elements for Plate Bending," *Proceedings of the Oklahoma Academy of Science*, **57**, pp. 144–148 (1977).

Reismann, H., and Y. Lee: "Forced Motions of Rectangular Plates," in *Developments in Theoretical and Applied Mechanics*, Vol. 4, D. Frederick, ed., Pergamon Press, New York, (1969) pp. 3–18.

Szabo, B. A., and G. C. Lee: "Derivation of Stiffness Matrices for Problems in Plane Elasticity by Galerkin's Method," *International Journal for Numerical Methods in Engineering*, **1**, pp. 301–310 (1969).

Szilard, R.: *Theory and Analysis of Plates*, Prentice-Hall, Englewood Cliffs, N.J. (1974).

Timoshenko, S., and S. Woinowsky-Krieger: *Theory of Plates and Shells*, 2d ed., McGraw-Hill, New York (1959).

Tsay, C. S., and J. N. Reddy: "Bending, Stability, and Free Vibration of Thin Orthotropic Plates by Simplified Mixed Finite Elements," *Journal of Sound and Vibration* (Letter to the Editor), **59**, pp. 307–311 (1978).

References on time approximations [also see Carslaw and Jaeger (1959), Donea (1974), Moore et al. (1975), and Verruijit (1970) above]

Bathe, K. J.: *Finite Element Procedures in Engineering Analysis*, Prentice-Hall, Englewood Cliffs, N.J. (1982).

Chan, S. P., H. L. Cox, and W. A. Benefield, "Transient Analysis of Forced Vibrations of Complex Structural Mechanical Systems," *Journal of the Royal Aeronautical Society*, **66**, pp. 457–460 (1962).

Clough, R. W., and J. Penzien, *Dynamics of Structures*, McGraw-Hill, New York (1975).

Goudreau, G. L., and R. L. Taylor: "Evaluation of Numerical Integration Methods in Elastodynamics," *Journal of Computer Methods in Applied Mechanics and Engineering*, **2**, pp. 69–97 (1973).

Hughes, T. J. R., and W. K. Liu: "Implicit-Explicit Finite Elements in Transient Analysis: Stability Theory, and Implementation and Numerical Examples," *Journal of Applied Mechanics*, **45**, pp. 371–378 (1978).

Johnson, D. E.: "A Proof of the Stability of the Hubolt Method," *Journal of the American Institute for Aeronautics and Astronautics*, **4**, pp. 1450–1451 (1966).

Krieg, R. D.: "Unconditional Stability in Numerical Time Integration Methods," *Journal of Applied Mechanics*, **40**, pp. 417–421 (1973).

Lax, P. D., and R. D. Richtmyer: "Survey of the Stability of Finite Difference Equations," *Communications in Pure and Applied Mathematics*, **9**, pp. 267–293 (1956).

Leech, J. W.: "Stability of Finite-Difference Equations for the Transient Response of a Flat Plate," *Journal of the American Institute for Aeronautics and Astronautics*, **3**, pp. 1772–1773 (1965).

Levy, S., and W. D. Kroll: "Errors Introduced by Finite Space and Time Increments in Dynamic Response Computation," *Journal of Research*, *National Bureau of Standards*, **51**, pp. 57–68 (1953).

Nickell, R. E.: "On the Stability of Approximation Operators in Problems of Structural Dynamics," *International Journal for Solids and Structures*, **7**, pp. 301–319 (1971).

Nickell, R. E.: "Direct Integration Methods in Structural Dynamics," *Journal of the Engineering Mechanics Division, ASCE*, **99** (EM2), (1972).

Tsui, T. Y., and P. Tong: "Stability of Transient Solution of Moderately Thick Plate by Finite Difference Method," *Journal of the American Institute for Aeronautics and Astronautics*, **9**, pp. 2062–2063 (1971).

References on numerical integration (quadrature)

Carnahan, B., H. A. Luther, and J. O. Wilkes: *Applied Numerical Methods*, John Wiley, New York (1969).

Couper, G. R.: "Gaussian Quadrature Formulas for Triangles," *International Journal for Numerical Methods in Engineering*, **7**, pp. 405–408 (1973).

Froberg, C. E.: *Introduction to Numerical Analysis*, Addison-Wesley, Reading, Mass. (1969).

Hammer, P. C., O. P. Marlowe, and A. H. Stroud: "Numerical Integration over Simplexes and Cones," *Mathematics Tables*, *National Research Council* (Washington), Vol. 10 (1956), pp. 130–137.

Irons, B. M.: "Quadrature Rules for Brick-Based Finite Elements," *International Journal for Numerical Methods in Engineering*, **3**, pp. 293–294 (1971).

Loxan, A. N., N. Davids, and A. Levenson: "Table of the Zeros of the Legendre Polynomials of Order 1–16 and the Weight Coefficients for Gauss' Mechanical Quadrature Formula," *Bulletin of the American Mathematical Society*, **48**, pp. 739–743 (1942).

Silvester, P.: "Newton-Cotes Quadrature Formulae for *N*-Dimensional Simplexes," *Proc. 2nd Canadian Congress of Applied Mechanics* (Waterloo, Ont., Canada), (1969).

Stroud, A. H., and D. Secrest: *Gaussian Quadrature Formulas*, Prentice-Hall, Englewood Cliffs, N.J. (1966).

References on solution of finite-element equations

Cantin, G.: "An Equation Solver of Very Large Capacity," *International Journal for Numerical Methods in Engineering*, **3**, pp. 379–388 (1971).

Collins, R. J.: "Bandwidth Reduction by Automatic Renumbering," *International Journal for Numerical Methods in Engineering*, **6**, pp. 345–356 (1973).

Cuthill, E.: "Several Strategies for Reducing the Bandwidth of Matrices," in *Sparse Matrices and Their Applications*, D. J. Rose and R. A. Willoughby, eds., Plenum Press, New York (1972).

Engleman, M. S., R. L. Sani, and P. M. Gresho: "The Implementation of Normal and/or Tangential Boundary Conditions in Finite Element Codes for Incompressible Fluid Flow," *International Journal for Numerical Methods in Fluids*, **2**, pp. 225–238 (1982).

Ergatoudis, J. G., B. M. Irons, and O. C. Zienkiewicz, "Curved, Iso-parametric, Quadrilateral Elements for Finite Element Analysis," *International Journal for Solids and Structures*, **4**, pp. 31–42 (1968).

Faddeeva, V. N.: *Computational Methods of Linear Algebra*, Dover Publications, New York (1959).

Forsythe, G. E., and C. B. Moler: *Computer Solution of Linear Algebraic Systems*, Prentice-Hall, Englewood Cliffs, N. J. (1967), pp. 114–119.

Fried, I.: "A Gradient Computational Procedure for the Solution of Large Problems Arising from the Finite Element Discretization Method," *International Journal for Numerical Methods in Engineering*, **2**, pp. 477–494 (1970).

Hellen, T. K.: "A Frontal Solution for Finite Element Techniques," Central Electricity Generating Board, Report RD/B/N1459, Berkeley Nuclear Laboratories, Berkeley, Gloucestershire (1969).

Irons, B. M.: "A Frontal Solution Program for Finite Element Analysis," *International Journal for Numerical Methods in Engineering*, **2**, pp. 5–32 (1970).

Mondkar, D. P., and G. H. Powell: "Large Capacity Equation Solver for Structural Analysis," *Computers and Structures*, **4**, pp. 699–728 (1974).

Van Norton, R.: "The Solution of Linear Equations by the Gauss-Seidel Method," in *Mathematical Methods for Digital Computers*, vol. 1, A. Ralston and H. S. Wilf, eds., John Wiley, New York (1960).

Wilkinson, J. H.: "The Solution of Ill-Conditioned Linear Equations," in *Mathematical Methods for Digital Computers*, Vol. 2, A. Ralston and H. S. Wilf, eds., John Wiley, New York (1967).

References on space-time finite elements

Kohler, W., and J. Pittr: "Calculation of Transient Temperature Fields with Finite Elements in Space and Time Dimensions," *International Journal for Numerical Methods in Engineering*, **8**, pp. 625–631 (1974).

Oden, J. T.: "A General Theory of Finite Elements II. Applications," *International Journal for Numerical Methods in Engineering*, **1**, pp. 247–259 (1969).

Sobey, R. J.: "Hermitian Space-Time Finite Elements for Estuarine Mass Transport," *International Journal for Numerical Methods in Fluids*, **2**, pp. 277–297 (1982).

Zienkiewicz, O. C., and C. J. Parekh: "Transient Field Problems: Two Dimensional and Three Dimensional Analysis by Isoparametric Finite Elements," *International Journal for Numerical Methods in Engineering*, **2**, pp. 61–71 (1970).

CHAPTER

FIVE

PRELUDE TO ADVANCED TOPICS

5.1 INTRODUCTION

The introduction presented in the preceding chapters is sufficient to provide the essential ideas for the development of finite-element models and the associated computer programs for most linear boundary- and initial-value problems in one and two dimensions. However, there are many additional ideas that deserve some comments. Here we discuss some immediate extensions of the present study to alternative formulations, eigenvalue problems, three-dimensional problems, and nonlinear problems. The discussions are only meant to give some idea of the applicability of the finite-element method to these advanced problems, and therefore the details are not included here. The topics discussed in this chapter are

1. Alternative formulations
2. Eigenvalue problems
3. Nonlinear problems
4. Three-dimensional problems

The reader interested in a detailed treatment of any of these topics is asked to consult finite-element books listed under References for Additional Reading at the end of the chapter.

5.2 ALTERNATIVE FORMULATIONS

The finite-element formulations presented in Chaps. 3 and 4 were based on the weak formulation of governing differential equations. In some cases, the governing equations can be recast in an alternative form which facilitates the use of lower-order interpolations, and yields better accuracies for the secondary variables (which, as the reader knows, are discontinuous at the interelement boundaries). Here we discuss a couple of such formulations and demonstrate the use of the least-squares method in the finite-element modeling of the formulations. More specifically, we will study the following formulations:

1. The least-squares formulation of Eq. (3.1)
2. The mixed formulation of Eq. (3.86)

It should be noted that these specific equations were chosen to illustrate the basic ideas behind the two formulations; the ideas are applicable to other equations, and to two- and three-dimensional problems. For example, the least-squares formulation can be applied to Eq. (3.86), the mixed formulation to Eq. (3.1), and the two formulations to Eq. (4.1). Other methods, such as the penalty function method, can also be used to formulate a given problem.

5.2.1 The Least-Squares Formulation

Consider the problem of finding the variational solution of Eq. (3.1) using the least-squares method. In the finite-element method, the least-squares technique is applied over a typical element. In other words, we minimize the residual in the differential equation (3.1) due to the approximation of u over the element,

$$u = \sum_{j=1}^{n} u_j^{(e)} \psi_j^{(e)} \tag{5.1}$$

where $\psi_j^{(e)}$ $(j = 1, 2, \ldots, n)$ are suitable approximation functions. We have

$$\begin{aligned} 0 &= \frac{\partial}{\partial u_i} \int_{x_e}^{x_{e+1}} \left[-\frac{d}{dx}\left(a \sum_{j=1}^{n} u_j^{(e)} \frac{d\psi_j^{(e)}}{dx} \right) - f \right]^2 dx \\ &= 2\int_{x_e}^{x_{e+1}} \left[-\frac{d}{dx}\left(a \sum_{j=1}^{n} u_j^{(e)} \frac{d\psi_j^{(e)}}{dx} \right) - f \right]\left[-\frac{d}{dx}\left(a \frac{d\psi_i^{(e)}}{dx} \right) \right] dx \end{aligned} \tag{5.2}$$

or

$$\sum_{j=1}^{n} K_{ij}^{(e)} u_j^{(e)} = F_i^{(e)} \tag{5.3}$$

where

$$\begin{aligned} K_{ij}^{(e)} &= \int_{x_e}^{x_{e+1}} \frac{d}{dx}\left(a \frac{d\psi_i^{(e)}}{dx} \right) \frac{d}{dx}\left(a \frac{d\psi_j^{(e)}}{dx} \right) dx \\ F_i^{(e)} &= -\int_{x_e}^{x_{e+1}} f \frac{d}{dx}\left(a \frac{d\psi_i^{(e)}}{dx} \right) dx \end{aligned} \tag{5.4}$$

Before we proceed further with the formulation, we note several undesirable features of the formulation. First, the least-squares method does not include the natural boundary conditions ($a\, du/dx \equiv P$) of the element. Second, an examination of the coefficient matrix $[K^{(e)}]$ in Eqs. (5.4) reveals that $\psi_i^{(e)}$ should be at least quadratic (to yield nonzero coefficients). Third, the coefficient matrix $[K]$ and the column vector $\{F\}$ have lost their physical meaning. For example, in the case of the longitudinal deformation of a bar, the coefficient matrix $[K]$ in Eqs. (5.4) does not correspond to the stiffness matrix, and the column vector $\{F\}$ does not correspond to the force vector. The first two features require the construction of quadratic interpolation functions that satisfy the (essential and natural) boundary conditions of the problem. The last feature makes the application of physical boundary conditions a difficult task. Therefore, we dispense with this procedure and consider the application of the least-squares method to an alternative but equivalent pair of equations.

Equation (3.1) can be written as a pair of first-order equations by introducing an additional variable (the dual variable of the original equation) σ:

$$-\left(\frac{d\sigma}{dx} + f\right) = 0 \qquad a\frac{du}{dx} = \sigma \tag{5.5}$$

In the case of the axial deformation of a bar, the variable σ has the meaning of the tensile force at any point in the bar. Now we proceed to construct the least-squares finite-element approximation of the pair of equations (5.5). Since both u and σ are the primary variables in the present formulation, imposition of the physical boundary conditions becomes simple. Let

$$u = \sum_{j=1}^{n} u_j^{(e)}\phi_j^{(e)} \qquad \sigma = \sum_{j=1}^{m} \sigma_j^{(e)}\psi_j^{(e)} \tag{5.6}$$

In general, the interpolation functions ϕ_i and ψ_i are not the same. For the sake of simplicity, we shall take $\psi_j = \phi_j$. Substituting the approximation (5.6) into Eqs. (5.5) and using the least-squares method on the residuals of Eqs. (5.5), we obtain

$$\begin{aligned} 0 &= \int_{x_e}^{x_{e+1}} \left[\frac{d}{dx}\left(\sum_{j=1}^{m} \sigma_j\psi_j\right) + f\right]\frac{d\psi_i}{dx}\,dx \\ &= \sum_{j=1}^{m} K_{ij}^{(e)}\sigma_j + F_i^{(e)} \end{aligned} \tag{5.7}$$

$$\begin{aligned} 0 &= \int_{x_e}^{x_{e+1}} \left[a^{(e)}\frac{d}{dx}\left(\sum_{j=1}^{n} u_j\phi_j\right) - \sum_{j=1}^{m} \sigma_j\psi_j\right]\frac{d\phi_i}{dx}\,dx \\ &= a^{(e)}\sum_{j=1}^{n} H_{ij}^{(e)}u_j - \sum_{j=1}^{m} G_{ij}^{(e)}\sigma_j \end{aligned} \tag{5.8}$$

where

$$\begin{aligned} K_{ij}^{(e)} &= \int_{x_e}^{x_{e+1}} \frac{d\psi_i}{dx}\frac{d\psi_j}{dx}\,dx \qquad & G_{ij}^{(e)} &= \int_{x_e}^{x_{e+1}} \frac{d\phi_i}{dx}\psi_j\,dx \\ H_{ij}^{(e)} &= \int_{x_e}^{x_{e+1}} \frac{d\phi_i}{dx}\frac{d\phi_j}{dx}\,dx \qquad & F_i^{(e)} &= \int_{x_e}^{x_{e+1}} \frac{d\psi_i}{dx}f\,dx \end{aligned} \tag{5.9}$$

Since the integrals contain only the first derivatives of ψ_i and ϕ_i, we can use the linear interpolation functions for both ψ_i and ϕ_i. We obtain (for f = constant)

$$[K^{(e)}] = \frac{1}{h_e}\begin{bmatrix} 1 & -1 \\ -1 & 1 \end{bmatrix} \qquad [G^{(e)}] = \frac{1}{2}\begin{bmatrix} -1 & -1 \\ 1 & 1 \end{bmatrix}$$
$$\{F^{(e)}\} = f\begin{Bmatrix} 1 \\ -1 \end{Bmatrix} \qquad [H^{(e)}] = [K^{(e)}] \tag{5.10}$$

Consider the finite-element mesh of three equal elements ($h_e = L/3$, $a^{(e)} = a$). We have the following correspondence between the element nodal values and the global nodal values (note that $\sigma_1^{(e)} = -P_1^{(e)}$ and $\sigma_2^{(e)} = P_2^{(e)}$; see Figs. 3.2 and 5.1):

$$\begin{aligned} &u_1^{(1)} = U_1 \quad && u_2^{(1)} = u_1^{(2)} = U_2 \quad && u_2^{(2)} = u_1^{(3)} = U_3 \quad && u_2^{(3)} = U_4 \\ &\sigma_1^{(1)} = R_1 && \sigma_2^{(1)} = \sigma_1^{(2)} = R_2 && \sigma_2^{(2)} = \sigma_1^{(3)} = R_3 && \sigma_2^{(3)} = R_4 \end{aligned} \tag{5.11}$$

where U_i and R_i denote the global values of the displacements and forces, respectively. The assembled equations associated with Eqs. (5.7) and (5.8) are given, respectively, by

$$\frac{1}{h}\begin{bmatrix} 1 & -1 & 0 & 0 \\ -1 & 2 & -1 & 0 \\ 0 & -1 & 2 & -1 \\ 0 & 0 & -1 & 1 \end{bmatrix}\begin{Bmatrix} R_1 \\ R_2 \\ R_3 \\ R_4 \end{Bmatrix} = f\begin{Bmatrix} 1 \\ -1+1 \\ -1+1 \\ -1 \end{Bmatrix} \tag{5.12}$$

$$\frac{a}{h}\begin{bmatrix} 1 & -1 & 0 & 0 \\ -1 & 2 & -1 & 0 \\ 0 & -1 & 2 & -1 \\ 0 & 0 & -1 & 1 \end{bmatrix}\begin{Bmatrix} U_1 \\ U_2 \\ U_3 \\ U_4 \end{Bmatrix} = \frac{1}{2}\begin{bmatrix} -1 & -1 & 0 & 0 \\ 1 & 0 & -1 & 0 \\ 0 & 1 & 0 & -1 \\ 0 & 0 & 1 & 1 \end{bmatrix}\begin{Bmatrix} R_1 \\ R_2 \\ R_3 \\ R_4 \end{Bmatrix} \tag{5.13}$$

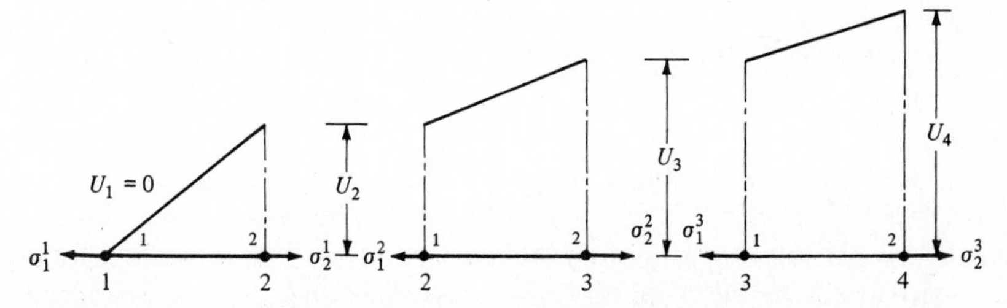

Figure 5.1 Typical finite element and assembly for the alternative formulation of Eq. (3.1).

Imposing the boundary conditions, $U_1 = 0$, and $R_4 = P$, and solving Eqs. (5.12) and (5.13) for the unknowns, we obtain

$$R_1 = P + 3fh \qquad R_2 = P + 2fh \qquad R_3 = P + fh$$
$$U_2 = \frac{5fh^2}{2a} + \frac{Ph}{a} \qquad U_3 = \frac{4fh^2}{a} + \frac{2Ph}{a} \qquad U_4 = \frac{9fh^2}{2a} + \frac{3Ph}{a} \tag{5.14}$$

Clearly, the solution coincides with that in Eqs. (3.39*b*) and (3.40).

5.2.2 The Mixed Formulation

Equation (3.86) can be decomposed into a pair of lower-order equations,

$$\frac{d^2M}{dx^2} + f = 0$$
$$\frac{d^2w}{dx^2} - \frac{M}{b} = 0 \qquad b \neq 0 \tag{5.15}$$

The assumption concerning the function b always holds in practice because neither the modulus of elasticity nor the moment of inertia is zero; in fact, we always have $b > 0$. The boundary conditions in Eq. (3.87) can be rewritten in terms of the dependent variables (w, M) of the problem as

$$w(0) = \frac{dw}{dx}(0) = 0 \qquad M(L) = M_0 \qquad \frac{dM}{dx}(L) = F_0 \tag{5.16}$$

Variational formulation The variational form associated with Eqs. (5.15) over a typical element (a two-node line element) is given by

$$0 = \int_{x_e}^{x_{e+1}} u\left(\frac{d^2M}{dx^2} + f\right) dx$$
$$= -\int_{x_e}^{x_{e+1}} \left(\frac{du}{dx}\frac{dM}{dx} - uf\right) dx - u(x_e)\left(\frac{dM}{dx}\right)\bigg|_{x_e} + u(x_{e+1})\left(\frac{dM}{dx}\right)\bigg|_{x_{e+1}} \tag{5.17a}$$

and

$$0 = \int_{x_e}^{x_{e+1}} v\left(\frac{d^2w}{dx^2} - \frac{M}{b}\right) dx$$
$$= -\int_{x_e}^{x_{e+1}} \left(\frac{dv}{dx}\frac{dw}{dx} + \frac{vM}{b}\right) dx - v(x_e)\left(\frac{dw}{dx}\right)\bigg|_{x_e} + v(x_{e+1})\left(\frac{dw}{dx}\right)\bigg|_{x_{e+1}} \tag{5.17b}$$

where u and v are arbitrary functions and can be viewed as the variations of w and M, respectively. The boundary terms in Eqs. (5.17) indicate that the specification of w and M constitutes the essential boundary conditions, and the specification of the derivatives dw/dx and dM/dx constitutes the natural boundary conditions of the variational problem:

Essential boundary conditions

$$\text{Specify, } w \text{ and } M \tag{5.18a}$$

Natural boundary conditions

$$\text{Specify, } \frac{dw}{dx} \text{ and } \frac{dM}{dx} \tag{5.18b}$$

Note that the natural boundary condition (on the bending moment, M) in the conventional formulation presented in Sec. 3.3 becomes the essential boundary condition in the present formulation.

Finite-element formulation The finite-element model of Eq. (5.17) can be obtained by substituting finite-element approximations of the form (see Fig. 5.2 for the nodal variables)

$$w = \sum_{j=1}^{m} w_j \psi_j \qquad M = \sum_{j=1}^{n} M_j \phi_j \tag{5.19}$$

into the variational formulations (5.17). Note that, technically speaking, w and M can be interpolated (or approximated) by different sets of interpolation functions ψ_j and ϕ_j. The finite-element equations can be conveniently expressed in the form

$$\begin{bmatrix} [K^{11}] & [K^{12}] \\ [K^{12}]^T & [K^{22}] \end{bmatrix} \begin{Bmatrix} \{w\} \\ \{M\} \end{Bmatrix} = \begin{Bmatrix} \{F^1\} \\ \{F^2\} \end{Bmatrix} \tag{5.20a}$$

where the matrix coefficients $K_{ij}^{\alpha\beta}$ of the matrix $[K^{\alpha\beta}]$ ($\alpha, \beta = 1, 2$) are given by

$$K_{ij}^{11} = 0 \qquad K_{ij}^{12} = \int_{x_e}^{x_{e+1}} \frac{d\psi_i}{dx}\frac{d\phi_j}{dx}\,dx = K_{ji}^{21} \qquad K_{ij}^{22} = \int_{x_e}^{x_{e+1}} \frac{1}{b}\phi_i\phi_j\,dx$$

$$F_i^1 = \int_{x_e}^{x_{e+1}} \psi_i f\,dx - \psi_i(x_e)V_1 - \psi_i(x_{e+1})V_2$$

$$F_i^2 = \phi_i(x_e)\theta_1 + \phi_i(x_{e+1})\theta_2 \tag{5.20b}$$

$$V_1 \equiv \left.\frac{dM}{dx}\right|_{x=x_e} \qquad V_2 \equiv -\left.\frac{dM}{dx}\right|_{x_{e+1}} \qquad \theta_1 \equiv -\left.\frac{dw}{dx}\right|_{x_e} \qquad \theta_2 \equiv \left.\frac{dw}{dx}\right|_{x_{e+1}}$$

The quantities V_i and θ_i are the secondary variables associated with w and M at the nodes. It should be pointed out that Eqs. (5.17) are used in an order that gives a symmetric coefficient matrix in Eq. (5.20a).

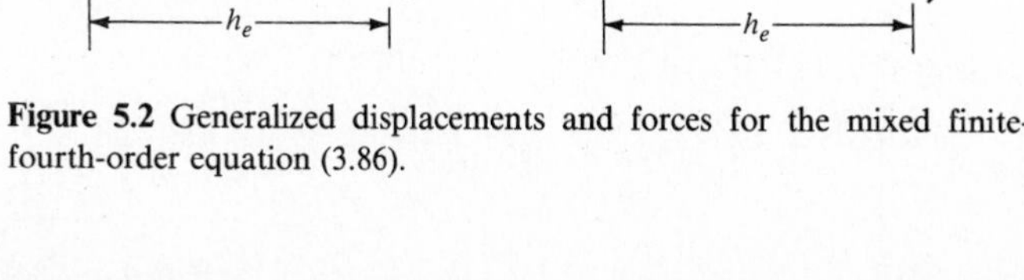

Figure 5.2 Generalized displacements and forces for the mixed finite-element formulation of the fourth-order equation (3.86).

In order to obtain the explicit form of Eq. (5.20), we must choose appropriate functions for ψ_i and ϕ_i. For the sake of simplicity, we take $\psi_i = \phi_i$ and ψ_i to be the linear interpolation functions in Eq. (3.12) or Eq. (3.17*b*). We obtain ($m = n = 2$), for constant b_e and f_e,

$$[K^{12}] = \frac{1}{h_e}\begin{bmatrix} 1 & -1 \\ -1 & 1 \end{bmatrix} \qquad [K^{22}] = \frac{h_e}{6b_e}\begin{bmatrix} 2 & 1 \\ 1 & 2 \end{bmatrix}$$
$$\{F^1\} = \frac{f_e h_e}{2}\begin{Bmatrix} 1 \\ 1 \end{Bmatrix} - \begin{Bmatrix} V_1 \\ V_2 \end{Bmatrix} \qquad \{F^2\} = \begin{Bmatrix} \theta_1 \\ \theta_2 \end{Bmatrix} \tag{5.21}$$

Rearranging the primary variables, element equations (5.20*a*) can be expressed as

$$\frac{1}{h_e}\begin{bmatrix} 0 & 1 & 0 & -1 \\ 1 & 2\alpha_e & -1 & \alpha_e \\ 0 & -1 & 0 & 1 \\ -1 & \alpha_e & 1 & 2\alpha_e \end{bmatrix}\begin{Bmatrix} w_1 \\ M_1 \\ w_2 \\ M_2 \end{Bmatrix} = \frac{f_e h_e}{2}\begin{Bmatrix} 1 \\ 0 \\ 1 \\ 0 \end{Bmatrix} + \begin{Bmatrix} -V_1 \\ \theta_1 \\ -V_2 \\ \theta_2 \end{Bmatrix} \tag{5.22}$$

where $\alpha_e = h_e^2/6b_e$.

Note that the usual assembly procedure, which assumes the interelement continuity of the primary variables, would result in a continuous bending moment field. This feature is not desirable when the problem involves specified concentrated (i.e., point) moments. The following example illustrates this point.

Example 5.1 Consider the clamped, simply supported beam shown in Fig. 5.3. The governing equation is given by Eq. (3.86) with $b = EI$, $a =$ constant, and $f = 0$. The boundary conditions are given by

$$w(0) = 0 \qquad \left(EI\frac{d^2w}{dx^2}\right)\Bigg|_{x=h^-}^{x=h^+} = -M_0 \qquad w(2h) = \theta(2h) = 0 \tag{5.23}$$

We wish to analyze the problem using the mixed formulation discussed above.

For simplicity, we shall use the minimum number of elements that allow us to impose the specified generalized displacements and forces. Thus, we have two elements and three nodes in the mesh. The assembled stiffness matrix is given by

$$\frac{1}{h}\begin{bmatrix} 0 & 1 & 0 & -1 & 0 & 0 \\ 1 & 2\alpha & -1 & \alpha & 0 & 0 \\ 0 & -1 & 0 & 2 & 0 & -1 \\ -1 & \alpha & 2 & 4\alpha & -1 & \alpha \\ 0 & 0 & 0 & -1 & 0 & 1 \\ 0 & 0 & -1 & \alpha & 1 & 2\alpha \end{bmatrix}\begin{Bmatrix} U_1 \\ U_2 \\ U_3 \\ U_4 \\ U_5 \\ U_6 \end{Bmatrix} = \begin{Bmatrix} -V_1^{(1)} \\ \theta_1^{(1)} \\ -V_2^{(1)} - V_1^{(2)} \\ \theta_2^{(1)} + \theta_1^{(2)} \\ -V_2^{(2)} \\ \theta_2^{(2)} \end{Bmatrix} \tag{5.24}$$

Using the boundary conditions

$$U_1 = U_2 = U_5 = 0 \qquad -V_2^{(1)} - V_1^{(2)} = 0 \qquad \theta_2^{(1)} + \theta_1^{(2)} = 0 \qquad \theta_2^{(2)} = 0 \tag{5.25}$$

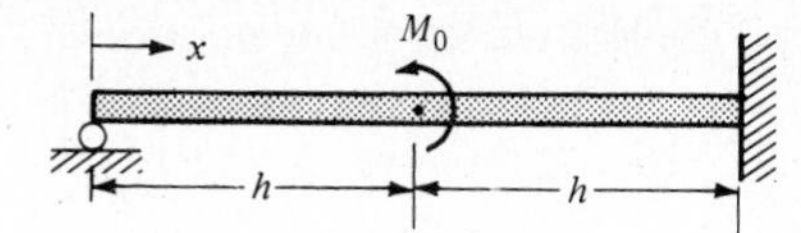

Figure 5.3 Clamped, simply supported beam considered in Example 5.1.

we obtain

$$\frac{1}{h}\begin{bmatrix} 0 & 2 & -1 \\ 2 & 4\alpha & \alpha \\ -1 & \alpha & 2\alpha \end{bmatrix}\begin{Bmatrix} U_3 \\ U_4 \\ U_6 \end{Bmatrix} = \begin{Bmatrix} 0 \\ 0 \\ 0 \end{Bmatrix}$$

If we use the additional boundary condition $U_4 = M_0$, we obtain an inconsistent set of equations. This is the result of the assumption

$$M_2^{(1)} = M_1^{(2)} = U_4 \tag{5.26}$$

used in the assembly. In the present problem, the moment is discontinuous and therefore Eq. (5.26) does not hold. The correct condition is

$$M_2^{(1)} - M_1^{(2)} = M_0 \tag{5.27}$$

Suppose that $U_4 = M_1^{(2)} = M_2^{(1)} - M_0$. Then the equations of Element 2 become

$$\frac{1}{h}\begin{bmatrix} 0 & 1 & 0 & -1 \\ 1 & 2\alpha & -1 & \alpha \\ 0 & -1 & 0 & 1 \\ -1 & \alpha & 1 & 2\alpha \end{bmatrix}\begin{Bmatrix} w_1^{(2)} \\ M_2^{(1)} - M_0 \\ w_2^{(2)} \\ M_2^{(2)} \end{Bmatrix} = \begin{Bmatrix} -V_1^{(2)} \\ \theta_1^{(2)} \\ -V_2^{(2)} \\ \theta_2^{(2)} \end{Bmatrix} \tag{5.28}$$

The assembled equations for the unknowns, after using the boundary condition (5.27), become

$$\frac{1}{h}\begin{bmatrix} 0 & 2 & -1 \\ 2 & 4\alpha & \alpha \\ -1 & \alpha & 2\alpha \end{bmatrix}\begin{Bmatrix} U_3 \\ M_2^{(1)} \\ U_6 \end{Bmatrix} = \frac{M_0}{h}\begin{Bmatrix} 1 \\ 2\alpha \\ \alpha \end{Bmatrix}$$

where the right-hand side comes from the second element when $M_1^{(2)}$ is replaced by $M_1^{(2)} = M_2^{(1)} - M_0$. The solution is given by

$$\begin{gathered} U_3 = -\frac{M_0 h^2}{32EI} \qquad M_2^{(1)} = \frac{9M_0}{16} \qquad U_6 = \frac{M_0}{8} \\ V_1^{(1)} = \frac{9M_0}{16h} \qquad V_2^{(1)} = -\frac{9M_0}{16h} \end{gathered} \tag{5.29}$$

Recalling the sign convention for the bending moments in the mixed element, we note that the solution (5.29) coincides with that given by the conventional model. The reader is asked to verify this statement. ■

The penalty function formulation of the beam bending equation results in a displacement formulation that includes the transverse deflection and the bending slope. The penalty finite-element model can be developed from the modified functional

$$I_p^e(w,\theta) = \int_{x_e}^{x_{e+1}} \left[\frac{b}{2}\left(\frac{d\theta}{dx}\right)^2 + fw \right] dx - Q_1 w(x_e) - Q_2\theta(x_e)$$

$$- Q_3 w(x_{e+1}) - Q_4\theta(x_{e+1}) + \int_{x_e}^{x_{e+1}} \frac{\gamma}{2}\left(\frac{dw}{dx} + \theta\right)^2 dx \quad (5.30)$$

This is left as an exercise to the reader (see Prob. 5.3).

5.3 EIGENVALUE PROBLEMS

In the analysis of natural vibrations (or buckling) of elastic bars, beams, membranes, plane elastic bodies, plates, and so on, we encounter equations of the type

$$Au = \lambda u \text{ in } \Omega \quad (5.31)$$

where A is a linear differential operator, u is the dependent variable of the problem, and λ is a characteristic number (to be determined along with u) that depends on the geometry and the material properties of the problem. All the equilibrium problems considered in the present work can be reduced to the form in Eq. (5.31) by replacing the source term f by λu. (In the case of plane elasticity and plate problems, A denotes a matrix operator of differentials and u denotes the column vector of the dependent variables of the problem.) The solution of Eq. (5.31) by the finite-element method follows a procedure entirely similar to that of the equilibrium equations. The main difference is that the condensed global equations should be solved using an eigenvalue solution procedure.

To gain additional insight into the details, consider the element equations associated with Eq. (5.31). We have

$$[K^{(e)}]\{u^{(e)}\} = \lambda[M^{(e)}]\{u^{(e)}\} + \{Q^{(e)}\} \quad (5.32)$$

where $[K^{(e)}]$ and $\{Q^{(e)}\}$ are the usual coefficient matrices and the column vector of internal forces, respectively, and $[M^{(e)}]$ is the *mass matrix*,

$$M_{ij}^{(e)} = \int_{\Omega^{(e)}} \psi_i^{(e)}\psi_j^{(e)}\, dx\, dy \quad (5.33)$$

The condensed set of global equations is given by

$$[K]\{U\} = \lambda[M]\{U\} \quad (5.34)$$

Note that the order of the matrices is now $(N-r)\times(N-r)$, where N is the total number of primary degrees of freedom in the finite-element mesh and r is the number of specified degrees of freedom. Equation (5.34) can now be solved by any standard eigenvalue-analysis computer package for the eigenvalues λ_i and the associated eigenvector $\{U^{(i)}\}$ $(i = 1, 2, \ldots, N-r)$. In a hand calculation, say, for $N - r \leqslant 3$, the procedure amounts to setting the determinant $|K - \lambda M|$ to zero and determining the roots of the polynomial. The roots are the eigenvalues. For a given eigenvalue λ_i, the components of the eigenvector $\{U^{(i)}\}$ are computed from Eq. (5.34); $N - r - 1$ number of the components can be solved in terms of the remaining component, which is arbitrary. Since one is normally interested in the direction of the vector, rather than in its magnitude, the value of the remaining component can be selected as an arbitrary nonzero constant.

Example 5.2 Consider the differential equation

$$
\begin{gathered}
-\frac{d^2u}{dx^2} = \lambda u \qquad 0 < x < 1 \\
u(0) = u(1) = 0
\end{gathered}
\tag{5.35}
$$

We wish to determine the value or values of λ and u for which the differential equation and the boundary conditions are satisfied. This equation arises, for example, in the transverse natural vibration of a cable fixed at both ends.

The finite-element model and the explicit form of the matrices for a linear element are given by

$$
[K^{(e)}]\{u\} = \lambda[M^{(e)}]\{u\} + \{P^{(e)}\} \tag{5.36a}
$$

$$
\begin{aligned}
K_{ij}^{(e)} &= \int_{x_e}^{x_{e+1}} \frac{d\psi_i}{dx}\frac{d\psi_j}{dx}\,dx \qquad [K^{(e)}] = \frac{1}{h}\begin{bmatrix} 1 & -1 \\ -1 & 1 \end{bmatrix} \\
M_{ij}^{(e)} &= \int_{x_e}^{x_{e+1}} \psi_i\psi_j\,dx \qquad [M^{(e)}] = \frac{h}{6}\begin{bmatrix} 2 & 1 \\ 1 & 2 \end{bmatrix}
\end{aligned}
\tag{5.36b}
$$

For a four-equal-element mesh, the condensed form of the global equations is given by ($h = \frac{1}{4}$)

$$
4\begin{bmatrix} 2 & -1 & 0 \\ -1 & 2 & -1 \\ 0 & -1 & 2 \end{bmatrix}\begin{Bmatrix} U_2 \\ U_3 \\ U_4 \end{Bmatrix} = \frac{\lambda}{24}\begin{bmatrix} 4 & 1 & 0 \\ 1 & 4 & 1 \\ 0 & 1 & 4 \end{bmatrix}\begin{Bmatrix} U_2 \\ U_3 \\ U_4 \end{Bmatrix} \tag{5.37}
$$

The equation has a nontrivial solution ($U_i \neq 0$) only if the following determinant is zero:

$$
\begin{aligned}
0 &= \begin{vmatrix} 8 - \frac{1}{6}\lambda & -4 - \frac{1}{24}\lambda & 0.0 \\ -4 - \frac{1}{24}\lambda & 8 - \frac{1}{6}\lambda & -4 - \frac{1}{24}\lambda \\ 0.0 & -4 - \frac{1}{24}\lambda & 8 - \frac{1}{6}\lambda \end{vmatrix} \\
&= (8 - \tfrac{1}{6}\lambda)\left[(8 - \tfrac{1}{6}\lambda)^2 - (4 + \tfrac{1}{24}\lambda)^2\right] \\
&\quad - (4 + \tfrac{1}{24}\lambda)\left[(4 + \tfrac{1}{24}\lambda)(8 - \tfrac{1}{6}\lambda) - 0\right] \\
&= (8 - \tfrac{1}{6}\lambda)\tfrac{1}{288}(7\lambda^2 - 960\lambda + 9216)
\end{aligned}
\tag{5.38}
$$

The roots of the polynomial are

$$
\lambda_1 = 10.39 \qquad \lambda_2 = 48 \qquad \lambda_3 = 126.76 \tag{5.39a}
$$

The eigenvectors can be obtained from Eq. (5.37). We get

$$
\begin{aligned}
(U_2, U_3, U_4)_1 &= (0.707, 1.0, 0.707) \\
(U_2, U_3, U_4)_2 &= (1.0, 0.0, -1.0) \\
(U_2, U_3, U_4)_3 &= (-0.707, 1.0, -0.707)
\end{aligned}
\tag{5.39b}
$$

The exact solution is given by

$$
\left.\begin{aligned}
\lambda_k &= (k\pi)^2 \\
u_k &= \sin k\pi x
\end{aligned}\right\} \qquad k = 1, 2, \ldots \tag{5.40}
$$

The mass matrix $M_{ij}^{(e)}$ used in Eq. (5.37) is consistent with the variational formulation, and called the *consistent mass matrix*. If it is obtained using an assumption, it is called an *inconsistent mass matrix*. The mass matrix obtained by lumping the total mass of an element equally at the nodes of the element is called the *lumped mass matrix*. In the lumped mass case the mass matrix $[\overline{M}]$ is a diagonal matrix whose elements are given by

$$
\overline{M}_{kk}^{(e)} = \frac{1}{n} \sum_{i,j=1}^{n} \int_{x_e}^{x_{e+1}} \psi_i \psi_j \, dx \qquad k = 1, 2, \ldots, n \tag{5.41}
$$

In the present example, we have

$$
\overline{M}_{11}^{(e)} = \overline{M}_{22}^{(e)} = \frac{h}{2}
$$

This simplifies the eigenvalue analysis. We have

$$
0 = \begin{vmatrix}
8 - \frac{1}{4}\lambda & -4 & 0.0 \\
-4 & 8 - \frac{1}{4}\lambda & -4 \\
0.0 & -4 & 8 - \frac{1}{4}\lambda
\end{vmatrix}
$$

which gives the following eigenvalues and eigenvectors:

$$
\begin{aligned}
\bar{\lambda}_1 = 9.37 \qquad \bar{\lambda}_2 &= 32 \qquad \bar{\lambda}_3 = 54.63 \\
(\overline{U}_2, \overline{U}_3, \overline{U}_4)_1 &= (0.707, 1.0, 0.707) \\
(\overline{U}_2, \overline{U}_3, \overline{U}_4)_2 &= (1.0, 0.0, -1.0) \\
(\overline{U}_2, \overline{U}_3, \overline{U}_4)_3 &= (-0.707, 1.0, -0.707)
\end{aligned}
\tag{5.42}
$$

Note that the eigenvalues $\bar{\lambda}_2$ and $\bar{\lambda}_3$ (higher modes) obtained by using the lumped mass matrix are less accurate than those obtained using the consistent mass matrix. ■

5.4 NONLINEAR PROBLEMS

Many problems of engineering are described by nonlinear differential equations. Under certain simplifying assumptions, these problems can be described by linear differential equations. For example, the equations governing the large-deflection bending of elastic beams are given by

$$
\begin{gathered}
-\frac{d}{dx}\left\{EA\left[\frac{du}{dx}+\frac{1}{2}\left(\frac{dw}{dx}\right)^2\right]\right\}=f \\
\frac{d^2}{dx^2}\left(EI\frac{d^2w}{dx^2}\right)-\frac{d}{dx}\left\{EA\frac{dw}{dx}\left[\frac{du}{dx}+\frac{1}{2}\left(\frac{dw}{dx}\right)^2\right]\right\}+q=0
\end{gathered}
\tag{5.43}
$$

where u is the longitudinal displacement, w is the transverse deflection, E is the modulus of elasticity, A is the area of cross section, f is the axial distributed load and q is the transverse loading. Under the assumption that the slope dw/dx is small compared to unity [i.e., (dw/dx), (du/dx), $(dw/dx)^2 \approx 0$], Eqs. (5.43) become uncoupled and reduce to Eqs. (3.1) and (3.86), respectively. However, when the slope is not too small, we must solve the coupled set of nonlinear equations (5.43).

Another example of a nonlinear problem is provided by the equation governing the flow of a viscous incompressible fluid. When the convective effects are larger than the viscous effects, Eqs. (4.188) should be modified to include the convective terms, that is, the Navier-Stokes equation should be solved:

$$
\begin{gathered}
\rho\left(u\frac{\partial u}{\partial x}+v\frac{\partial u}{\partial y}\right)-\mu\left[2\frac{\partial^2 u}{\partial x^2}+\frac{\partial}{\partial y}\left(\frac{\partial u}{\partial y}+\frac{\partial v}{\partial x}\right)\right]+\frac{\partial P}{\partial x}=f_x \\
\rho\left(u\frac{\partial v}{\partial x}+v\frac{\partial v}{\partial y}\right)-\mu\left[\frac{\partial}{\partial x}\left(\frac{\partial u}{\partial y}+\frac{\partial v}{\partial x}\right)+2\frac{\partial^2 v}{\partial y^2}\right]+\frac{\partial P}{\partial y}=f_y \\
\frac{\partial u}{\partial x}+\frac{\partial v}{\partial y}=0
\end{gathered}
\tag{5.44}
$$

where ρ is the density of the fluid, and all other symbols have the same meaning as described in Sec. 4.5.3.

The finite-element formulation of nonlinear problems proceeds much the same way as the linear problems. The main difference lies in the solution of the finite-element equations. Here we describe some details of the formulation, and comment about the solution procedure using the Navier-Stokes equations (5.44) and the pressure-velocity formulation.

The variational formulation of Eqs. (5.44) over an element is given by

$$
\begin{gathered}
\int_{\Omega^{(e)}}\left\{w_1\rho\left(u\frac{\partial u}{\partial x}+v\frac{\partial u}{\partial y}\right)+[\cdots]\right\}dx\,dy-\oint_{\Gamma^{(e)}}w_1t_x\,ds=0 \\
\int_{\Omega^{(e)}}\left\{w_2\rho\left(u\frac{\partial v}{\partial x}+v\frac{\partial v}{\partial y}\right)+[\cdots]\right\}dx\,dy-\oint_{\Gamma^{(e)}}w_2t_y\,ds=0 \\
\int_{\Omega^{(e)}}w_3\left(\frac{\partial u}{\partial x}+\frac{\partial v}{\partial y}\right)dx\,dy=0
\end{gathered}
\tag{5.45}
$$

where $[\cdots]$ denotes the expression in the square brackets of Eqs. (4.192). The finite-element model of these equations is given by

$$\begin{bmatrix} [\bar{K}^{11}] & [K^{12}] & [K^{13}] \\ [K^{12}]^T & [\bar{K}^{22}] & [K^{23}] \\ [K^{13}]^T & [K^{23}]^T & [0] \end{bmatrix} \begin{Bmatrix} \{u\} \\ \{v\} \\ \{P\} \end{Bmatrix} = \begin{Bmatrix} \{F^1\} \\ \{F^2\} \\ \{0\} \end{Bmatrix} \tag{5.46}$$

where $[K^{\alpha\beta}]$ and $\{F^\alpha\}$ $(\alpha, \beta = 1, 2, 3)$ are as defined in Eqs. (4.196), and

$$\begin{aligned} \bar{K}_{ij}^{11} &= K_{ij}^{11} + \rho \int_{\Omega^{(e)}} \psi_i \left(\bar{u} \frac{\partial \psi_j}{\partial x} + \bar{v} \frac{\partial \psi_j}{\partial y} \right) dx\,dy \\ \bar{K}_{ij}^{22} &= K_{ij}^{22} + \rho \int_{\Omega^{(e)}} \psi_i \left(\bar{u} \frac{\partial \psi_j}{\partial x} + \bar{v} \frac{\partial \psi_j}{\partial y} \right) dx\,dy \\ \bar{u} &= \sum_{i=1}^{r} \bar{u}_i \psi_i \qquad \bar{v} = \sum_{i=1}^{r} \bar{v}_i \psi_i \end{aligned} \tag{5.47}$$

where $\bar{u}$ and $\bar{v}$ are velocity components that are assumed to be known.

Clearly, the element coefficient matrix, and hence the global coefficient matrix, depends on the velocity field, which is not known a priori. Therefore, an iterative solution procedure is required. At the beginning of the first iteration, the velocity field is set to zero and the global equations are solved for the nodal velocities and the pressure [which corresponds to the solution of Eqs. (4.195)]. In the second iteration, the coefficient matrices are evaluated using the velocity field obtained in the first iteration, and the assembled equations are solved again for the nodal velocities. This procedure is repeated until the velocity field obtained at the end of two consecutive iterations differs by a small preassigned number. The convergence criterion can be expressed as

$$\frac{\left[\sum_{i=1}^{N} \left(\left| U_i^{(r)} - U_i^{(r+1)} \right|^2 + \left| V_i^{(r)} - V_i^{(r+1)} \right|^2 \right) \right]^{1/2}}{\left[\sum_{i=1}^{N} \left(\left| U_i^{(r+1)} \right|^2 + \left| V_i^{(r+1)} \right|^2 \right) \right]^{1/2}} \leqslant 10 \text{ percent} \tag{5.48}$$

where $(U_i^{(r)}, V_i^{(r)})$ denote the velocities at node i at the end of iteration r.

There exist other ways of solving the nonlinear equations, such as the Newton-Raphson method, the modified Newton-Raphson method, and others. Interested readers can consult more advanced finite-element texts for details.

5.5 THREE-DIMENSIONAL PROBLEMS

Most of the basic ideas covered in Chaps. 3 and 4 for one- and two-dimensional problems can be extended to three-dimensional problems. Three-dimensional problems are very demanding on storage and computational time. For the sake of completeness, we discuss here the finite-element formulation of the Poisson equation in three dimensions, and describe some of the three-dimensional elements.

Consider the Poisson equation

$$-\frac{\partial}{\partial x}\left(k_1\frac{\partial u}{\partial x}\right)-\frac{\partial}{\partial y}\left(k_2\frac{\partial u}{\partial y}\right)-\frac{\partial}{\partial z}\left(k_3\frac{\partial u}{\partial z}\right)=f \text{ in } \Omega \tag{5.49}$$

$$u=\hat{u} \text{ on } \Gamma_1 \qquad \left(k_1\frac{\partial u}{\partial x}n_x+k_2\frac{\partial u}{\partial y}n_y+k_3\frac{\partial u}{\partial z}n_z\right)=\hat{q} \text{ on } \Gamma_2 \tag{5.50}$$

where $k_i = k_i(x, y, z)$ and $f = f(x, y, z)$ are given functions of position in a three-dimensional domain Ω, and $\hat{u}$ and $\hat{q}$ are specified functions of position on the portions Γ_1 and Γ_2, respectively, of the surface Γ of the domain (see Fig. 5.4). Suppose that the domain Ω is discretized by some three-dimensional elements $\Omega^{(e)}$, such as tetrahedral and prism elements (which are the three-dimensional extensions of the triangular and rectangular elements).

The variational formulation of Eq. (5.49) over an element $\Omega^{(e)}$ is given by

$$\begin{aligned}0&=\int_{\Omega^{(e)}} v\left[-\frac{\partial}{\partial x}\left(k_1\frac{\partial u}{\partial x}\right)-\frac{\partial}{\partial y}\left(k_2\frac{\partial u}{\partial y}\right)-\frac{\partial}{\partial z}\left(k_3\frac{\partial u}{\partial z}\right)-f\right]dx\,dy\,dz\\&=\int_{\Omega^{(e)}}\left[k_1\frac{\partial v}{\partial x}\frac{\partial u}{\partial x}+k_2\frac{\partial v}{\partial y}\frac{\partial u}{\partial y}+k_3\frac{\partial v}{\partial z}\frac{\partial u}{\partial z}-vf\right]dx\,dy\,dz-\oint_{\Gamma^e} vq\,ds\end{aligned} \tag{5.51}$$

where

$$q \equiv k_1\frac{\partial u}{\partial x}n_x+k_2\frac{\partial u}{\partial y}n_y+k_3\frac{\partial u}{\partial z}n_z \tag{5.52}$$

The primary variable is u and the secondary variable is q.

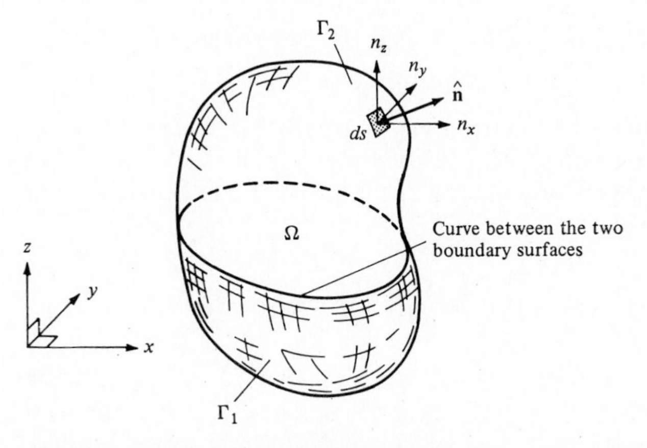

Figure 5.4 Three-dimensional domain showing a surface element ds, unit normal and its components $\hat{\mathbf{n}} = (n_1, n_2)$, and two portions of the boundary Γ.

Assuming finite-element interpolation of the form

$$u = \sum_{j=1}^{n} u_j \psi_j^{(e)}(x, y, z) \tag{5.53}$$

over the element $\Omega^{(e)}$, and substituting $v = \psi_i$ and Eq. (5.53) into Eq. (5.51), we obtain

$$[K^{(e)}]\{u^{(e)}\} = \{f^{(e)}\} + \{Q^{(e)}\} \tag{5.54}$$

where

$$K_{ij}^{(e)} = \int_{\Omega^{(e)}} \left(k_1 \frac{\partial \psi_i}{\partial x} \frac{\partial \psi_j}{\partial x} + k_2 \frac{\partial \psi_i}{\partial y} \frac{\partial \psi_j}{\partial y} + k_3 \frac{\partial \psi_i}{\partial z} \frac{\partial \psi_j}{\partial z} \right) dx\, dy\, dz$$

$$f_i^{(e)} = \int_{\Omega^{(e)}} f \psi_i \, dx\, dy\, dz \qquad Q_i^{(e)} = \oint_{\Gamma^{(e)}} q \psi_i \, ds \tag{5.55}$$

The interpolation functions $\psi_i^{(e)}$ have the same general properties as those for two-dimensional elements:

$$\sum_{i=1}^{n} \psi_i^{(e)}(x, y, z) = 1$$

$$\psi_i^{(e)}(x_j, y_j, z_j) = \delta_{ij} \tag{5.56}$$

The assembly of equations, the imposition of boundary conditions, and the solution of the equations are completely analogous to those described in Chap. 4 for the two-dimensional problems. Here we comment on the geometry of two elements and the element calculations.

The element matrices in Eqs. (5.55) require the use of interpolation functions that are at least linear in x, y, and z. Here we consider two linear elements: the tetrahedral element and the prism (or brick) element. These elements are described by approximations of the form

$$u(x, y, z) = a_0 + a_1 x + a_2 y + a_3 z \qquad \text{four-node tetrahedral element}$$

$$u(x, y, z) = a_0 + a_1 x + a_2 y + a_3 z + a_4 yz + a_5 xz + a_6 xy + a_7 xyz \qquad \text{eight-node prism element} \tag{5.57}$$

The interpolation functions can be determined as described in Chaps. 3 and 4.

If the element matrices are to be evaluated numerically, the isoparametric element concept can be used. The geometry of the elements can be described by the transformation equations

$$x = \sum_{i=1}^{n} x_i \hat{\psi}_i(\xi, \eta, \zeta)$$

$$y = \sum_{i=1}^{n} y_i \hat{\psi}_i(\xi, \eta, \zeta) \tag{5.58}$$

$$z = \sum_{i=1}^{n} z_i \hat{\psi}_i(\xi, \eta, \zeta)$$

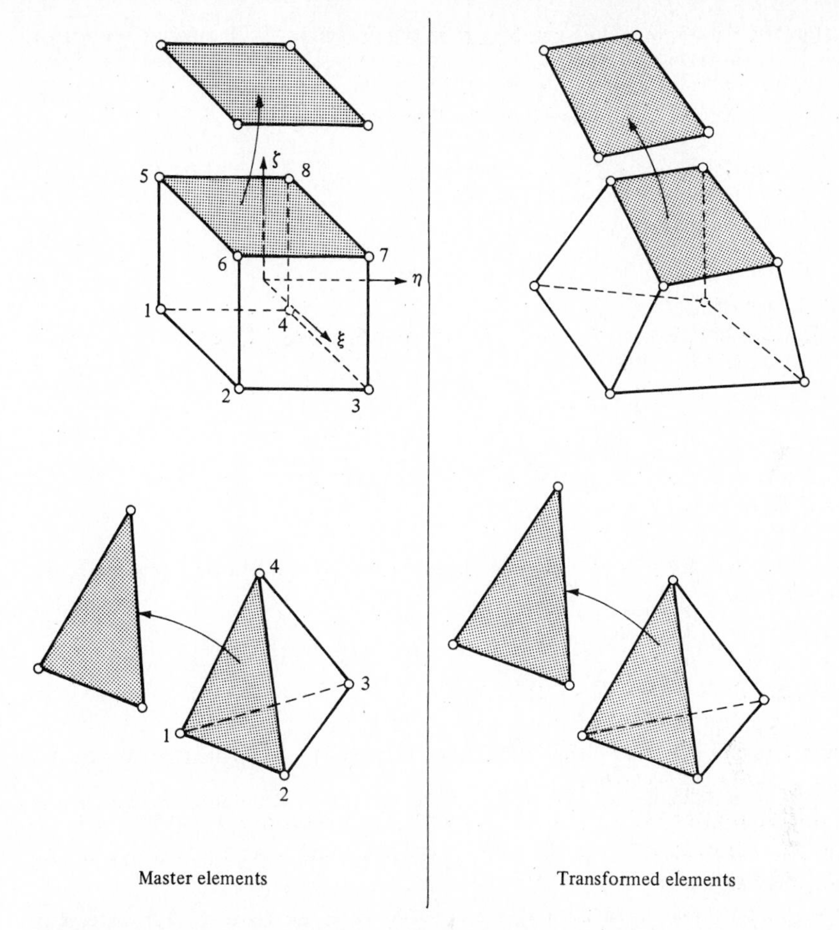

Figure 5.5 Linear three-dimensional elements: tetrahedral element and prism element (whose surfaces are two-dimensional triangular and rectangular elements).

Under these transformations, the master tetrahedral and prism elements transform to arbitrary tetrahedral and hexahedral elements, as shown in Fig. 5.5. The definition of the jacobian matrix and the numerical quadrature rules described in Chap. 4 can be easily extended to the three-dimensional case.

PROBLEMS

5.1 Use the mixed method to formulate Eq. (3.1), and show that the element equations for the linear interpolation case are equivalent to those obtained by the conventional method.

5.2 Use the penalty function method to formulate Eq. (3.1), and derive the element equations in explicit form using the linear interpolation functions.

5.3 Use the quadratic functional in Eq. (5.30) to develop the penalty element, and derive the element matrices using linear interpolation for both transverse deflection w and bending slope θ.

5.4 For a cantilevered beam of length L and flexural rigidity EI, and subjected concentrated moment M_0 at $x = L/2$, obtain the mixed finite-element solution and compare with the exact solution. Do you see any discrepancy in the solutions?

5.5 Solve the eigenvalue problem associated with the differential equation (see Prob. 3.25)

$$-\frac{d^2u}{dx^2} = \lambda u \qquad 0 < x < 1$$

$$u(0) = 0 \qquad u(1) + \frac{du}{dx}(1) = 0$$

Use four linear elements.

5.6 Find the fundamental frequency of a simply supported beam of length L, density ρ, and flexural rigidity EI. Use two elements in the half-beam.

5.7 Find the eigenvalues of the problem

$$-\nabla^2 u = \lambda u \qquad 0 < (x, y) < 1$$

$$u = 0 \text{ on all edges}$$

using the four-element mesh given in Fig. 4.8.

5.8 The quadratic functional associated with the coupled nonlinear differential equations (5.43) over an element is given by

$$I(u, w) = \int_{x_e}^{x_{e+1}} \frac{1}{2}\left[EA\left(\frac{du}{dx}\right)^2 + EA\frac{du}{dx}\left(\frac{dw}{dx}\right)^2 + EI\left(\frac{d^2w}{dx^2}\right)^2\right]$$

$$\times dx + \int_{x_e}^{x_{e+1}} (fu + qw)\, dx + \text{boundary terms}$$

Using the linear interpolation for u and the cubic interpolation for w, derive the element coefficient matrices in algebraic form.

5.9 Evaluate the element matrices in Prob. 5.8 by assuming the nonlinear parts in the element coefficients to be constant.

5.10 Give the finite-element formulation of the following nonlinear equation over an element (x_e, x_{e+1}):

$$-\frac{d}{dx}\left(u\frac{du}{dx}\right) + 1 = 0 \qquad 0 < x < 1$$

$$\frac{du}{dx}(0) = 0 \qquad u(1) = \sqrt{2}$$

5.11 Derive the interpolation functions ψ_1, ψ_5, and ψ_8 for the eight-node prism element using the alternative procedure described in Sec. 4.4.3.

5.12 Give the finite-element formulation of the heat conduction equation in three dimensions with convective boundary condition.

5.13 Give the penalty finite-element model of the three-dimensional flow equations.

5.14 Discuss the three-dimensional formulation of Eq. (4.230) using a space-time element. Assume finite-element approximation of the form (see Prob. 4.66)

$$u(x, y, t) = \sum_{j=1}^{n} u_j \psi_j(x, y, t)$$

5.15 Evaluate the force vector components $f_i^{(e)}$ and $Q_i^{(e)}$ over a master prism element when f is a constant, f_0, and q is a constant, q_0, in Eqs. (5.55).

REFERENCES FOR ADDITIONAL READING

Bathe, K. J.: *Finite Element Procedures in Engineering Analysis*, Prentice-Hall, Englewood Cliffs, N.J. (1982).

Cook, R. D.: *Concepts and Applications of Finite Element Analysis*, 2d ed., John Wiley, New York (1981).

Desai, C. S., and J. F. Abel: *Introduction to the Finite Element Method*, Van Nostrand Reinhold, New York (1972).

Gallagher, R. H.: *Finite Element Analysis Fundamentals*, Prentice-Hall, Englewood Cliffs, N.J. (1975).

Oden, J. T.: *Finite Elements of Nonlinear Continua*, McGraw-Hill, New York (1972)

Zienkiewicz, O. C.: *The Finite Element Method*, 3d ed., McGraw-Hill, New York (1977).

APPENDIX

I

COMPUTER PROGRAM FEM1D

```
C                   P R O G R A M    F E M 1 D                              FEM00010
C      (AN IN-CORE FINITE-ELEMENT ANALYSIS COMPUTER PROGRAM)                FEM00020
C                                                                           FEM00030
C     ......................................................................FEM00040
C     .                                                                    .FEM00050
C     . FINITE-ELEMENT ANALYSIS OF SECOND- AND FOURTH-ORDER EQUATIONS     .FEM00060
C     .                    IN ONE DIMENSION                                .FEM00070
C     .                                                                    .FEM00080
C     ......................................................................FEM00090
C     .                                                                    .FEM00100
C     .     D E S C R I P T I O N   O F   T H E   V A R I A B L E S      .FEM00110
C     .                                                                    .FEM00120
C     .                                                                    .FEM00130
C     . AL.......LENGTH OF THE DOMAIN                                      .FEM00140
C     .                                                                    .FEM00150
C     . AX0,AX1,                                                           .FEM00160
C     . BX0,BX1,..COEFFICIENTS IN THE DIFFERENTIAL EQUATION(SEE STIFF).FEM00170
C     . CX0,CX1                                                            .FEM00180
C     .                                                                    .FEM00190
C     . ALFA,BETA,                                                         .FEM00200
C     . A0,A1,....PARAMETERS IN THE TIME APPROXIMATION SCHEMES             .FEM00210
C     . A3,A4                                                              .FEM00220
C     .                                                                    .FEM00230
C     . COEF(I,J).ARRAY FOR STORING AX0,AX1,...,F0,F1:                     .FEM00240
C     .           AX0=COEF(I,1),AX1=COEF(I,2), ETC. FOR I-TH ELEMENT      .FEM00250
C     .                                                                    .FEM00260
C     . DT........TIME INCREMENT FOR TIME-DEPENDENT PROBLEMS               .FEM00270
C     . EC(I,J)...ARRAY OF GLOBAL COORDINATES OF NODE J OF ELEMENT I      .FEM00280
C     . ELX.......SAME AS EC(I,J) BUT USED TO TRANSFER TO 'STIF'          .FEM00290
C     . ELSTIF....ELEMENT COEFFICIENT (STIFFNESS) MATRIX                   .FEM00300
C     . ELF.......ELEMENT FORCE VECTOR                                     .FEM00310
C     . GF,GF0....COLUMN VECTORS FOR DISPLACEMENTS AND THEIR TIME          .FEM00320
C     . GF1,GF2   DERIVATIVES (GF1=D(GF)/DT, GF2=D(GF1)/DT)                .FEM00330
C     . ICONT.....INDICATOR FOR CONTINUITY OF THE DATA (0, NO;1, YES)     .FEM00340
C     .                                                                    .FEM00350
C     . IELEM.....INDICATOR FOR THE TYPE OF FINITE ELEMENT:                .FEM00360
C     .           IELEM=0, 2-NODE ELEMENT FOR FOURTH-ORDER PROBLEMS       .FEM00370
C     .           IELEM=1, 2-NODE ELEMENT FOR SECOND-ORDER PROBLEMS       .FEM00380
C     .           IELEM=2, 3-NODE ELEMENT FOR SECOND-ORDER PROBLEMS       .FEM00390
C     .           IELEM=3, 4-NODE ELEMENT FOR SECOND-ORDER PROBLEMS       .FEM00400
C     .                                                                    .FEM00410
C     . ITEM......INDICATOR FOR TRANSIENT (TIME-DEPENDENT) ANALYSIS       .FEM00420
C     .           ITEM=0, STATIC (NOT TIME-DEPENDENT) ANALYSIS            .FEM00430
C     .           ITEM=1, FIRST-ORDER TIME DERIVATIVES INVOLVED           .FEM00440
C     .           ITEM=2, SECOND-ORDER TIME DERIVATIVES INVOLVED          .FEM00450
```

```
C       .                                                             .FEM00460
C       . GSTIF.....GLOBAL (THAT IS, ASSEMBLED) COEFFICIENT MATRIX       .FEM00470
C       .           IN UPPER-HALF BANDED FORM (OF ORDER NEQ BY NHBW)     .FEM00480
C       . GF........COLUMN VECTOR OF GLOBAL FORCES BEFORE GOING INTO     .FEM00490
C       .           THE SUBROUTINE 'SOLVE', AND CONTAINS THE SOLUTION    .FEM00500
C       .           WHEN COMES OUT OF THE SUBROUTINE 'SOLVE'             .FEM00510
C       . NBDY......NUMBER OF SPECIFIED PRIMARY DEGREES OF FREEDOM       .FEM00520
C       . IBDY......COLUMN OF SPECIFIED PRIMARY DEGREES OF FREEDOM       .FEM00530
C       . VBDY......COLUMN OF SPECIFIED VALUES OF THE ENTRIES IN IBDY    .FEM00540
C       . NBF.......NUMBER OF SPECIFIED SECONDARY DEGREES OF FREEDOM     .FEM00550
C       . IBF.......COLUMN OF SPECIFIED SECONDARY DEGREES OF FREEDOM     .FEM00560
C       . VBF.......COLUMN OF SPECIFIED VALUES OF THE ENTRIES IN IBF     .FEM00570
C       . NRMAX.....ROW-DIMENSION OF 'GSTIF' IN DIMENSION STATEMENT      .FEM00580
C       . NCMAX.....COLUMN-DIMENSION OF 'GSTIF' IN DIMENSION STATEMENT   .FEM00590
C       . NPE.......NUMBER OF NODES PER ELEMENT                          .FEM00600
C       . NDF.......NUMBER OF DEGREES OF FREEDOM PER NODE                .FEM00610
C       . NEM.......NUMBER OF ELEMENTS IN THE MESH                       .FEM00620
C       . NEQ.......TOTAL DEGREES OF FREEDOM IN THE FINITE ELEMENT MODEL.FEM00630
C       . NHBW......HALF BAND WIDTH OF THE GLOBAL COEFFICIENT MATRIX     .FEM00640
C       . NGP.......NUMBER OF GAUSS POINTS USED IN THE INTEGRATION       .FEM00650
C       . NN .......NUMBER OF TOTAL DEGREES OF FREEDOM IN THE ELEMENT    .FEM00660
C       . NNM.......NUMBER OF NODES IN THE FINITE ELEMENT MESH           .FEM00670
C       . NOD(I,J)..GLOBAL NODE NUMBER CORRESPONDING TO THE J-TH NODE    .FEM00680
C       .           OF THE I-TH ELEMENT (CONNECTIVITY MATRIX)            .FEM00690
C       . NPRNT.....INDICATOR FOR PRINT (NPRNT=1) OR NOPRINT (NPRNT=0)   .FEM00700
C       .           OF ELEMENT MATRICES AND GLOBAL MATRICES              .FEM00710
C       . NTIME.....NUMBER OF TIME STEPS (IN THE TRANSIENT ANALYSIS)     .FEM00720
C       . X.........VECTOR OF THE GLOBAL COORDINATES OF GLOBAL NODES     .FEM00730
C       .                                                                .FEM00740
C       . NOTE: DIMENSION STATEMENTS IN THE MAIN PROGRAM SHOULD BE       .FEM00750
C       .       MODIFIED TO MEET THE REQUIREMENTS OF THE PROBLEM         .FEM00760
C       ..................................................................FEM00770
C                                                                        FEM00780
C                                                                        FEM00790
      IMPLICIT REAL*8(A-H,O-Z)                                           FEM00800
      DIMENSION  GSTIF(55,4),GF(55),GF0(55),GF1(55),GF2(55),IBDY(9),     FEM00810
     *           VBDY(9),X(55),IBF(10),VBF(10),TITLE(20),NOD(55,4),      FEM00820
     *           EC(55,4),COEF(55,8)                                     FEM00830
      COMMON/STF/ELSTIF(4,4),ELF(4),ELX(4),W0(4),W1(4),W2(4)             FEM00840
      COMMON/STF1/A0,A1,A2,A3,A4,AX0,AX1,BX0,BX1,CX0,CX1,C0,F0,F1        FEM00850
      DATA NRMAX,NCMAX/55,4/                                             FEM00860
C                                                                        FEM00870
C       ............................................                     FEM00880
C       .                                          .                     FEM00890
C       .      P R E P R O C E S S O R   U N I T   .                     FEM00900
C       .                                          .                     FEM00910
C       ............................................                     FEM00920
C                                                                        FEM00930
C       R E A D  I N  T H E  P R O B L E M  D A T A  H E R E             FEM00940
C                                                                        FEM00950
      READ 300, TITLE                                                    FEM00960
      READ 310, NPRNT,IELEM,NPE,NDF,NEM,NBDY,NBF,ICONT,ITEM,NTIME        FEM00970
      READ 320, AL,X0                                                    FEM00980
      IF(ICONT.EQ.1)READ 320, AX0,AX1,BX0,BX1,CX0,CX1,F0,F1              FEM00990
      IF(ITEM .GT. 0)READ 320, ALFA,BETA,DT,C0                           FEM01000
      READ 310, (IBDY(I),I=1,NBDY)                                       FEM01010
      READ 320, (VBDY(I),I=1,NBDY)                                       FEM01020
      IF(NBF .EQ. 0)GOTO 5                                               FEM01030
      READ 310, (IBF(I),I=1,NBF)                                         FEM01040
      READ 320, (VBF(I),I=1,NBF)                                         FEM01050
C                                                                        FEM01060
    5 NT=0                                                               FEM01070
      T=0.0                                                              FEM01080
      NHBW = NPE*NDF                                                     FEM01090
      PRINT 330                                                          FEM01100
      IF(IELEM .EQ. 0)PRINT 340                                          FEM01110
      IF(IELEM .NE. 0)PRINT 350                                          FEM01120
      PRINT 300, TITLE                                                   FEM01130
      IF(ITEM .NE. 0)PRINT 370                                           FEM01140
      IF(ITEM .GT. 0)PRINT 480, ALFA,BETA,DT,C0                          FEM01150
      NNM = NEM*(NPE-1)+1                                                FEM01160
      NEQ = NNM*NDF                                                      FEM01170
      NN = NPE*NDF                                                       FEM01180
      DO 10 I=1,NPE                                                      FEM01190
   10 NOD(1,I)=I                                                         FEM01200
      DO 20 N=2,NEM                                                      FEM01210
```

```
      DO 20 I=1,NPE                                                     FEM01220
   20 NOD(N,I) = NOD(N-1,I)+NPE-1                                       FEM01230
      PRINT 390,NEM,NNM,NDF                                             FEM01240
      IF(ICONT .EQ. 0)GOTO 40                                           FEM01250
C                                                                       FEM01260
C     GLOBAL COORDINATES FOR A UNIFORM MESH ONLY                        FEM01270
C                                                                       FEM01280
      DX = AL/(NNM-1)                                                   FEM01290
      DO 30 I=1,NNM                                                     FEM01300
   30 X(I)=DX*(I-1)+X0                                                  FEM01310
      PRINT 380,(X(I),I=1,NNM)                                          FEM01320
      PRINT 380, AX0,AX1,BX0,BX1,CX0,CX1,F0,F1                          FEM01330
   40 IF (ITEM .EQ. 0)GOTO 60                                           FEM01340
C                                                                       FEM01350
C     INPUT THE INITIAL CONDITIONS HERE                                 FEM01360
C                                                                       FEM01370
      READ 320, (GF0(I),I=1,NEQ)                                        FEM01380
      READ 320, (GF1(I),I=1,NEQ)                                        FEM01390
      READ 320, (GF(I),I=1,NEQ)                                         FEM01400
C                                                                       FEM01410
C     COMPUTE PARAMETERS FOR THE TIME-APPROXIMATION SCHEMES             FEM01420
C                                                                       FEM01430
      A0=1.0/BETA/(DT*DT)                                               FEM01440
      A1=A0*DT                                                          FEM01450
      A2=0.5/BETA-1.0                                                   FEM01460
      A3=(1.0-ALFA)*DT                                                  FEM01470
      A4=ALFA*DT                                                        FEM01480
C                                                                       FEM01490
C     ..............................................                    FEM01500
C     .                                            .                    FEM01510
C     .          P R O C E S S O R   U N I T       .                    FEM01520
C     .                                            .                    FEM01530
C     ..............................................                    FEM01540
C                                                                       FEM01550
C     DO-LOOP ON TIME BEGINS HERE.  FOR ITEM=2, THE INITIAL CONDITIONS  FEM01560
C     ON SECOND DERIVATIVE (GF2(I)) ARE COMPUTED IN THE PROGRAM         FEM01570
C                                                                       FEM01580
      IF(ITEM.EQ.2)GOTO 60                                              FEM01590
   50 T=T+DT                                                            FEM01600
      PRINT 420, T                                                      FEM01610
C                                                                       FEM01620
C     INITIALIZE GLOBAL MATRICES                                        FEM01630
C                                                                       FEM01640
   60 DO 70 I=1,NEQ                                                     FEM01650
      IF(NT.EQ.1)GF2(I)=GF(I)                                           FEM01660
      GF(I)=0.0                                                         FEM01670
      DO 70 J=1,NHBW                                                    FEM01680
   70 GSTIF(I,J)=0.0                                                    FEM01690
C                                                                       FEM01700
C     DO-LOOP ON NUMBER OF ELEMENTS (FOR ELEMENT CALCULATIONS) BEGINS   FEM01710
C                                                                       FEM01720
      DO 150 N = 1, NEM                                                 FEM01730
      IF(ICONT .EQ. 1)GOTO 90                                           FEM01740
      IF(NT.GT.0)GOTO 80                                                FEM01750
C                                                                       FEM01760
C     R E A D I N   E L E M E N T   D A T A   H E R E                   FEM01770
C                                                                       FEM01780
      READ 320, (EC(N,I),I=1,NPE)                                       FEM01790
      READ 320, (COEF(N,I),I=1,8)                                       FEM01800
      PRINT 430, N,(EC(N,I),I=1,NPE)                                    FEM01810
      PRINT 380, (COEF(N,I),I=1,8)                                      FEM01820
   80 AX0=COEF(N,1)                                                     FEM01830
      AX1=COEF(N,2)                                                     FEM01840
      BX0=COEF(N,3)                                                     FEM01850
      BX1=COEF(N,4)                                                     FEM01860
      CX0=COEF(N,5)                                                     FEM01870
      CX1=COEF(N,6)                                                     FEM01880
      F0=COEF(N,7)                                                      FEM01890
      F1=COEF(N,8)                                                      FEM01900
   90 L=0                                                               FEM01910
C                                                                       FEM01920
C     TRANSFER THE GLOBAL DATA TO THE ELEMENT                           FEM01930
C                                                                       FEM01940
      DO 100 I=1,NPE                                                    FEM01950
      NI=NOD(N,I)                                                       FEM01960
```

```
      IF(ICONT.EQ.1)ELX(I)=X(NI)                                        FEM01970
      IF(ICONT.EQ.0)ELX(I)=EC(N,I)                                      FEM01980
      IF(ITEM .EQ. 0)GOTO 100                                           FEM01990
      LI=(NI-1)*NDF                                                     FEM02000
      DO 95 J=1,NDF                                                     FEM02010
      LI=LI+1                                                           FEM02020
      L=L+1                                                             FEM02030
      W0(L)=GF0(LI)                                                     FEM02040
      W1(L)=GF1(LI)                                                     FEM02050
   95 W2(L)=GF2(LI)                                                     FEM02060
  100 CONTINUE                                                          FEM02070
C                                                                       FEM02080
      CALL STIFF (IELEM,NN,NPE,T,ITEM)                                  FEM02090
      IF(T .GT. 0.0)GOTO 120                                            FEM02100
      IF(NPRNT .EQ. 0)GOTO 120                                          FEM02110
      IF(N.GT.1 .OR. NT.GT.1)GOTO 120                                   FEM02120
      PRINT 440                                                         FEM02130
      DO 110 I=1,NN                                                     FEM02140
  110 PRINT 380,(ELSTIF(I,J),J=1,NN)                                    FEM02150
      PRINT 380, (ELF(I),I=1,NN)                                        FEM02160
  120 CONTINUE                                                          FEM02170
C                                                                       FEM02180
C     ASSEMBLE ELEMENT MATRICES INTO BANDED GLOBAL MATRIX               FEM02190
C                                                                       FEM02200
      DO 140 I = 1, NPE                                                 FEM02210
      NR = (NOD(N,I) - 1)*NDF                                           FEM02220
      DO 140 II = 1, NDF                                                FEM02230
      NR = NR + 1                                                       FEM02240
      L = (I-1)*NDF + II                                                FEM02250
      GF(NR) = GF(NR) + ELF(L)                                          FEM02260
      DO 140 J = 1, NPE                                                 FEM02270
      NCL = (NOD(N,J)-1)*NDF                                            FEM02280
      DO 140 JJ = 1, NDF                                                FEM02290
      M = (J-1)*NDF + JJ                                                FEM02300
      NC = NCL-NR+JJ+1                                                  FEM02310
      IF(NC)140,140,130                                                 FEM02320
  130 GSTIF(NR,NC) = GSTIF(NR,NC) + ELSTIF(L,M)                         FEM02330
  140 CONTINUE                                                          FEM02340
  150 CONTINUE                                                          FEM02350
      IF(NT.GT.1)GOTO 160                                               FEM02360
      PRINT 400, (IBDY(I),I=1,NBDY)                                     FEM02370
      PRINT 410, (VBDY(I),I=1,NBDY)                                     FEM02380
      PRINT 500, (IBF(I),I=1,NBF)                                       FEM02390
      PRINT 510, (VBF(I),I=1,NBF)                                       FEM02400
C                                                                       FEM02410
C     IMPOSE SPECIFIED NONZERO VALUES OF THE SECONDARY VARIABLES        FEM02420
C                                                                       FEM02430
  160 IF(NBF .EQ. 0)GOTO 180                                            FEM02440
      DO 170 NF=1, NBF                                                  FEM02450
      NB=IBF(NF)                                                        FEM02460
  170 GF(NB)=GF(NB)+VBF(NF)                                             FEM02470
  180 IF(T .EQ. 0.0)GOTO 200                                            FEM02480
      IF(NPRNT .LE. 1 .OR. NT .GT. 1)GOTO 200                           FEM02490
      PRINT 490                                                         FEM02500
      DO 190 I=1,NEQ                                                    FEM02510
  190 PRINT 380,(GSTIF(I,J),J=1,NHBW)                                   FEM02520
      PRINT 380, (GF(I),I=1,NEQ)                                        FEM02530
C                                                                       FEM02540
C     IMPOSE SPECIFIED VALUES OF THE PRIMARY VARIABLES                  FEM02550
C                                                                       FEM02560
  200 CALL BNDY(NRMAX,NCMAX,NEQ,NHBW,GSTIF,GF,NBDY,IBDY,VBDY)           FEM02570
      IRES = 0                                                          FEM02580
C                                                                       FEM02590
C     SOLVE THE BANDED, SYMMETRIC EQUATIONS                             FEM02600
C                                                                       FEM02610
      CALL SOLVE(NRMAX,NCMAX,NEQ,NHBW,GSTIF,GF,IRES)                    FEM02620
C                                                                       FEM02630
C     SOLUTION IS RETURNED IN GF                                        FEM02640
C                                                                       FEM02650
      IF(ITEM.NE.0)GOTO 210                                             FEM02660
      PRINT 450                                                         FEM02670
      PRINT 380,(GF(NI),NI=1,NEQ)                                       FEM02680
      GOTO 230                                                          FEM02690
  210 NT=NT+1                                                           FEM02700
      IF(T.EQ.0.0)GOTO 50                                               FEM02710
```

```
C                                                                       FEM02720
C     COMPUTE AND PRINT CURRENT VALUES OF GF0, GF1, AND GF2             FEM02730
C                                                                       FEM02740
      DO 220 I=1,NEQ                                                    FEM02750
      GF0(I)=A0*(GF(I)-GF0(I))-A1*GF1(I)-A2*GF2(I)                      FEM02760
      GF1(I)=GF1(I)+A3*GF2(I)+A4*GF0(I)                                 FEM02770
      GF2(I)=GF0(I)                                                     FEM02780
  220 GF0(I)=GF(I)                                                      FEM02790
      PRINT 450                                                         FEM02800
      PRINT 380, (GF0(I),I=1,NEQ)                                       FEM02810
      PRINT 380, (GF1(I),I=1,NEQ)                                       FEM02820
      IF(IELEM.EQ.0)PRINT 380, (GF2(I),I=1,NEQ)                         FEM02830
  230 CONTINUE                                                          FEM02840
C     ...............................................                   FEM02850
C     .                                             .                   FEM02860
C     .     P O S T P R O C E S S O R    U N I T    .                   FEM02870
C     .                                             .                   FEM02880
C     ...............................................                   FEM02890
C                                                                       FEM02900
C     COMPUTE THE DERIVATIVES OF THE SOLUTION                           FEM02910
C                                                                       FEM02920
      PRINT 460                                                         FEM02930
      DO 250 N=1,NEM                                                    FEM02940
      L=0                                                               FEM02950
      DO 240 I=1,NPE                                                    FEM02960
      NI=NOD(N,I)                                                       FEM02970
      IF(ICONT.EQ.1)ELX(I)=X(NI)                                        FEM02980
      IF(ICONT.EQ.0)ELX(I)=EC(N,I)                                      FEM02990
      LI=(NI-1)*NDF                                                     FEM03000
      DO 240 J=1,NDF                                                    FEM03010
      LI=LI+1                                                           FEM03020
      L=L+1                                                             FEM03030
  240 W0(L)=GF(LI)                                                      FEM03040
  250 CALL STRESS(NPE,NDF,IELEM,W0,ELX)                                 FEM03050
      PRINT 470                                                         FEM03060
      IF(ITEM.EQ.0)GOTO 270                                             FEM03070
      IF(NT.GT.NTIME)GOTO 270                                           FEM03080
  260 GOTO 50                                                           FEM03090
  270 CONTINUE                                                          FEM03100
C                                                                       FEM03110
C                   F O R M A T S                                       FEM03120
C                                                                       FEM03130
  300 FORMAT(20A4)                                                      FEM03140
  310 FORMAT(16I5)                                                      FEM03150
  320 FORMAT(8F10.4)                                                    FEM03160
  330 FORMAT(1H1)                                                       FEM03170
  340 FORMAT(5X,'*** F O U R T H - O R D E R   P R O B L E M S ***',/)  FEM03180
  350 FORMAT(5X,'*** S E C O N D - O R D E R   P R O B L E M S ***',/)  FEM03190
  370 FORMAT(/,10X,'TIME-DEPENDENT (TRANSIENT) ANALYSIS ',/)            FEM03200
  380 FORMAT(10X,6E15.5)                                                FEM03210
  390 FORMAT(10X,'NO. OF ELEMENTS IN THE MESH ......... =',I4,/         FEM03220
     *10X,'NO. OF NODES IN THE MESH ........... =',I4,/,                FEM03230
     *10X,'NO. OF DEG. OF FREEDOM PER NODE ......=',I4,/,10X,'FEM MESH  FEM03240
     *AND COEFFICIENTS (AX0,AX1, ETC.) FOLLOW :',/)                     FEM03250
  400 FORMAT(10X,'SPECIFIED DEGREES OF FREEDOM:',//,10X,10I5,/)         FEM03260
  410 FORMAT(/,10X,'VALUES OF SPECIFIED DEGREES OF FREEDOM:',//,10X,    FEM03270
     *6E10.3,/,10X,6E10.3)                                              FEM03280
  420 FORMAT(/,10X,'TIME = ',E13.5,/)                                   FEM03290
  430 FORMAT(/,10X,I5,8E13.4)                                           FEM03300
  440 FORMAT(/,10X,'ELEMENT MATRICES :',/)                              FEM03310
  450 FORMAT(/,8X,'S O L U T I O N : ',/)                               FEM03320
  460 FORMAT(/,15X,'  X  ',7X,' DISPL. ',4X,'1ST DERIV.',4X,'2ND DERIV.FEM03330
     *',/)                                                              FEM03340
  470 FORMAT(120('.'))                                                  FEM03350
  480 FORMAT(/,10X,'ALFA =',E13.4,2X,'BETA =',E13.4,2X,'TIME INCREMENT =FEM03360
     *',E13.4,2X,'C0 =',E11.4,/)                                        FEM03370
  490 FORMAT(/,10X,'GLOBAL MATRICES :',/)                               FEM03380
  500 FORMAT(/,10X,'SPECIFIED NONZERO SECONDARY DEGREES OF FREEDOM :',  FEM03390
     *//,10X,10I5)                                                      FEM03400
  510 FORMAT(/,10X,'VALUES OF THE SPECIFIED SECONDARY DEGREES OF FREEDOMFEM03410
     * :',//,10X,6E10.3)                                                FEM03420
      STOP                                                              FEM03430
      END                                                               FEM03440
      SUBROUTINE STIFF (IELEM,NN,NPE,T,ITEM)                            STF00010
C                                                                       STF00020
```

```
C     ---------------------------------------------------------------   STF00030
C      MODEL EQUATION USED IN THE PROGRAM:                              STF00040
C                                                                       STF00050
C           -(AX.U')' + (BX.U'')'' + CX.U = F + CO(U* + U**)            STF00060
C                                                                       STF00070
C      HERE SYMBOLS ' AND * DENOTE DIFFERENTIATIONS WITH RESPECT TO X   STF00080
C      AND T (TIME), RESPECTIVELY, AND AX,BX,CX,CO, AND F ARE GIVEN     STF00090
C      FUNCTIONS (DATA) OF THE INDEPENDENT VARIABLES, X AND/OR T.       STF00100
C                                                                       STF00110
C        X.........GLOBAL COORDINATE                                    STF00120
C        XI .......LOCAL COORDINATE                                     STF00130
C        H.........ELEMENT LENGTH                                       STF00140
C        SF........ELEMENT INTERPOLATION FUNCTIONS                      STF00150
C        GDSF......SECOND DERIVATIVE OF SF W.R.T. LOCAL COORDINATE      STF00160
C        GDDSF.....SECOND DERIVATIVE OF SF W.R.T. GLOBAL COORDINATE     STF00170
C        GJ........JACOBIAN OF THE TRANSFORMATION                       STF00180
C        GAUSS.....4 BY 4 MATRIX OF GAUSS POINTS: COLUMNS CORRESPOND    STF00190
C                  TO GAUSS POINTS FOR NGP=COLUMN NUMBER                STF00200
C        WT........GAUSS WEIGHTS CORRESPONDING TO THE GAUSS POINTS      STF00210
C        A,B,C,G...ELEMENT MATRICES NEEDED TO COMPUTE ELSTIF            STF00220
C                                                                       STF00230
C     ---------------------------------------------------------------   STF00240
C                                                                       STF00250
      IMPLICIT REAL*8(A-H,O-Z)                                          STF00260
      COMMON/STF/ELSTIF(4,4),ELF(4),ELX(4),WO(4),W1(4),W2(4)            STF00270
      COMMON/STF1/A0,A1,A2,A3,A4,AX0,AX1,BX0,BX1,CX0,CX1,C0,F0,F1       STF00280
      COMMON/SHP/SF(4),GDSF(4),GDDSF(4),GJ                              STF00290
      DIMENSION GAUSS(4,4),WT(4,4),A(4,4),B(4,4),C(4,4),G(4,4)          STF00300
      DATA GAUSS/4*0.0D0,-.57735027D0,.57735027D0,2*0.0D0,-.77459667D0, STF00310
     *0.0D0,.77459667D0,0.0D0,-.86113631D0,                             STF00320
     *-.33998104D0,.33998104D0,.86113631D0/                             STF00330
      DATA WT/2.0D0,3*0.0D0,2*1.0D0,2*0.0D0,.55555555D0,.88888888D0,    STF00340
     *.55555555D0,0.0D0,.34785485D0,2*.65214515D0,.34785485D0/          STF00350
C                                                                       STF00360
      NGP = NPE                                                         STF00370
      IF (IELEM .EQ. 0)NGP=4                                            STF00380
      H = ELX(NPE)-ELX(1)                                               STF00390
      DO 10 I=1,NN                                                      STF00400
   10 ELF(I) = 0.0                                                      STF00410
      NET = NPE                                                         STF00420
      IF (IELEM .EQ. 0)NET = 4                                          STF00430
      DO 20 I = 1,NET                                                   STF00440
      DO 20 J = 1,NET                                                   STF00450
      A(I,J) = 0.0                                                      STF00460
      B(I,J) = 0.0                                                      STF00470
      C(I,J) = 0.0                                                      STF00480
   20 G(I,J) = 0.0                                                      STF00490
C                                                                       STF00500
C     DO-LOOP ON NUMBER OF GAUSS POINTS BEGINS HERE                     STF00510
C                                                                       STF00520
      DO 70 NI=1,NGP                                                    STF00530
      XI = GAUSS(NI,NGP)                                                STF00540
C                                                                       STF00550
C     CALL SUBROUTINE SHAPE TO EVALUATE THE INTERPOLATION FUNCTIONS AND STF00560
C           THEIR GLOBAL DERIVATIVES AT THE GAUSS POINT XI              STF00570
C                                                                       STF00580
      CALL SHAPE(XI,H,NPE,NET,IELEM)                                    STF00590
      CONST = GJ*WT(NI,NGP)                                             STF00600
      X = 0.5*H*(1.0+XI)+ELX(1)                                         STF00610
C                                                                       STF00620
C....... DEFINE THE COEFFICIENTS OF THE DIFFERENTIAL EQUATIONS ......   STF00630
C          (ONLY UPTO LINEAR VARIATION IS ACCOUNTED HERE)               STF00640
C                                                                       STF00650
      AX=AX0+AX1*X                                                      STF00660
      BX=BX0+BX1*X                                                      STF00670
      CX=CX0+CX1*X                                                      STF00680
      F=F0+F1*X                                                         STF00690
C                                                                       STF00700
C     COMPUTE BASIC COEFFICIENT MATRICES THAT ARE NEEDED TO GENERATE    STF00710
C                 ELEMENT MATRICES FOR VARIOUS PROBLEMS                 STF00720
C                                                                       STF00730
      DO 60 I = 1,NET                                                   STF00740
      ELF(I) = ELF(I) + CONST*SF(I)*F                                   STF00750
      DO 60 J = 1,NET                                                   STF00760
      IF(ITEM.EQ.0)GOTO 50                                              STF00770
```

```
      G(I,J) = G(I,J) + CONST*SF(I)*SF(J)*CO                            STF00780
   50 C(I,J) = C(I,J) + CONST*SF(I)*SF(J)*CX                            STF00790
      A(I,J) = A(I,J) + CONST*GDSF(I)*GDSF(J)*AX                        STF00800
   60 B(I,J) = B(I,J) + CONST*GDDSF(I)*GDDSF(J)*BX                      STF00810
   70 CONTINUE                                                          STF00820
      IF(ITEM .NE. 0)GOTO 120                                           STF00830
C                                                                       STF00840
C     ELEMENT MATRICES FOR STATIC ANALYSIS                              STF00850
C                                                                       STF00860
      IF (IELEM .NE. 0)GOTO 90                                          STF00870
      DO 80 I=1,4                                                       STF00880
      DO 80 J=1,4                                                       STF00890
   80 ELSTIF(I,J) = B(I,J)+ C(I,J)                                      STF00900
      RETURN                                                            STF00910
C                                                                       STF00920
   90 CONTINUE                                                          STF00930
      DO 100 I = 1, NPE                                                 STF00940
      DO 100 J=1, NPE                                                   STF00950
  100 ELSTIF(I,J) = A(I,J) + B(I,J) + C(I,J)                            STF00960
      RETURN                                                            STF00970
C                                                                       STF00980
C     ELEMENT MATRICES FOR DYNAMIC ANALYSIS                             STF00990
C     ONLY CONSTANT SOURCE TERM IS ASSUMED HERE                         STF01000
C                                                                       STF01010
  120 IF (IELEM.EQ.0)GOTO 190                                           STF01020
C                                                                       STF01030
C     TIME APPROXIMATION SCHEME FOR FIRST-ORDER EQUATIONS (IN TIME)     STF01040
C                                                                       STF01050
      IF (ITEM .EQ. 2)GOTO 150                                          STF01060
      DO 140 I=1,NPE                                                    STF01070
      SUM=0.0                                                           STF01080
      DO 130 J=1,NPE                                                    STF01090
      SUM=SUM+(G(I,J)-A3*A(I,J))*WO(J)                                  STF01100
  130 ELSTIF(I,J)=G(I,J)+A4*A(I,J)                                      STF01110
  140 ELF(I)=(A3+A4)*ELF(I)+SUM                                         STF01120
      RETURN                                                            STF01130
C                                                                       STF01140
C     TIME APPROXIMATION SCHEME FOR SECOND-ORDER EQUATIONS (IN TIME)    STF01150
C                                                                       STF01160
  150 IF(T.GT.0.0)GOTO 170                                              STF01170
      DO 160 I=1,NPE                                                    STF01180
      ELF(I)=0.0                                                        STF01190
      DO 160 J=1,NPE                                                    STF01200
      ELF(I)=ELF(I)-A(I,J)*WO(J)                                        STF01210
  160 ELSTIF(I,J)=G(I,J)                                                STF01220
      RETURN                                                            STF01230
  170 DO 180 I=1,NPE                                                    STF01240
      DO 180 J=1,NPE                                                    STF01250
      ELF(I)=ELF(I)+G(I,J)*(AO*WO(J)+A1*W1(J)+A2*W2(J))                 STF01260
  180 ELSTIF(I,J)=A(I,J)+AO*G(I,J)                                      STF01270
      RETURN                                                            STF01280
C                                                                       STF01290
C     COMPUTE GF2 AT TIME T=0                                           STF01300
C                                                                       STF01310
  190 IF(T .GT. 0.0)GOTO 210                                            STF01320
      DO 200 I=1,4                                                      STF01330
      ELF(I)=0.0                                                        STF01340
      DO 200 J=1,4                                                      STF01350
      ELF(I)=ELF(I)-B(I,J)*WO(J)                                        STF01360
  200 ELSTIF(I,J)=G(I,J)                                                STF01370
      RETURN                                                            STF01380
C                                                                       STF01390
  210 DO 230 I=1,4                                                      STF01400
      DO 220 J=1,4                                                      STF01410
      ELF(I)=ELF(I)+G(I,J)*(AO*WO(J)+A1*W1(J)+A2*W2(J))                 STF01420
  220 ELSTIF(I,J)=B(I,J)+AO*G(I,J)                                      STF01430
  230 CONTINUE                                                          STF01440
      RETURN                                                            STF01450
      END                                                               STF01460
      SUBROUTINE SHAPE (XI,H,NPE,NET,IELEM)                             SHP00010
C                                                                       SHP00020
C     ..................................................................SHP00030
C     EVALUATION OF THE INTERPOLATION FUNCTIONS AND THEIR GLOBAL DERI-  SHP00040
C                    VATIVES AT THE GAUSS POINTS                        SHP00050
C                                                                       SHP00060
```

```
C         X.........GLOBAL COORDINATE                                              SHP00070
C         XI .......LOCAL COORDINATE                                               SHP00080
C         H.........ELEMENT LENGTH                                                 SHP00090
C         SF........ELEMENT INTERPOLATION FUNCTIONS                                SHP00100
C         DSF.......FIRST DERIVATIVE OF SF W.R.T. LOCAL COORDINATE, XI             SHP00110
C         GDSF......FIRST DERIVATIVE OF SF W.R.T. GLOBAL COORDINATE, X             SHP00120
C         DDSF......SECOND DERIVATIVE OF SF W.R.T. LOCAL COORDINATE, XI            SHP00130
C         GDDSF.....SECOND DERIVATIVE OF SF W.R.T. GLOBAL COORDINATE, XI           SHP00140
C         GJ........JACOBIAN OF THE TRANSFORMATION                                 SHP00150
C.........................................................................         SHP00160
C                                                                                  SHP00170
      IMPLICIT REAL*8 (A-H,O-Z)                                                    SHP00180
      COMMON/SHP/SF(4),GDSF(4),GDDSF(4),GJ                                         SHP00190
      DIMENSION DSF(4),DDSF(4)                                                     SHP00200
      IF (IELEM .NE. 0)GOTO 20                                                     SHP00210
C                                                                                  SHP00220
C     HERMITE INTERPOLATION FUNCTIONS FOR THE TWO-NODE ELEMENT FOR THE             SHP00230
C        FOURTH-ORDER DIFFERENTIAL EQUATION (SEE SUBROUTINE STIFF)                 SHP00240
C                                                                                  SHP00250
      SF(1) = 0.25*(2.0-3.0*XI+XI**3)                                              SHP00260
      SF(2) =   -H*(1.0-XI)*(1.0-XI*XI)/8.0                                        SHP00270
      SF(3) = 0.25*(2.0+3.0*XI-XI**3)                                              SHP00280
      SF(4) =    H*(1.0+XI)*(1.0-XI*XI)/8.0                                        SHP00290
      DSF(1) = -0.75*(1.0-XI*XI)                                                   SHP00300
      DSF(2) =  H*(1.0+2.0*XI-3.0*XI*XI)/8.0                                       SHP00310
      DSF(3) =  0.75*(1.0-XI*XI)                                                   SHP00320
      DSF(4) =  H*(1.0-2.0*XI-3.0*XI*XI)/8.0                                       SHP00330
      DDSF(1) = 1.5*XI                                                             SHP00340
      DDSF(2) = 0.25*H*(1.0-3.0*XI)                                                SHP00350
      DDSF(3) = -1.5*XI                                                            SHP00360
      DDSF(4) =-0.25*(1.0+3.0*XI)*H                                                SHP00370
      GOTO 80                                                                      SHP00380
C                                                                                  SHP00390
   20 IF(IELEM-2)30,40,50                                                          SHP00400
C                                                                                  SHP00410
C     INTERPOLATION FUNCTIONS  FOR TWO, THREE, AND FOUR NODE ELEMENTS              SHP00420
C       FOR SECOND-ORDER DIFFERENTIAL EQUATION (SEE COMMENTS ABOVE)                SHP00430
C                                                                                  SHP00440
C     LINEAR (LAGRANGE)INTERPOLATION FUNCTIONS                                     SHP00450
C                                                                                  SHP00460
   30 SF(1) = 0.5*(1.0-XI)                                                         SHP00470
      SF(2) = 0.5*(1.0+XI)                                                         SHP00480
      DSF(1) = -0.5                                                                SHP00490
      DSF(2) = 0.5                                                                 SHP00500
      GOTO 60                                                                      SHP00510
C                                                                                  SHP00520
C     QUADRATIC (LAGRANGE) INTERPOLATION FUNCTIONS                                 SHP00530
C                                                                                  SHP00540
   40 SF(1) = -0.5*XI*(1.0-XI)                                                     SHP00550
      SF(2) = 1.0-XI*XI                                                            SHP00560
      SF(3) = 0.5*XI*(1.0+XI)                                                      SHP00570
      DSF(1) = -0.5*(1.0-2.0*XI)                                                   SHP00580
      DSF(2) = -2.0*XI                                                             SHP00590
      DSF(3) = 0.5*(1.0+2.0*XI)                                                    SHP00600
      GOTO 60                                                                      SHP00610
C                                                                                  SHP00620
C     CUBIC (LAGRANGE) INTERPOLATION FUNCTIONS                                     SHP00630
C                                                                                  SHP00640
   50 SF(1) = 0.0625*(1.0-XI)*(9.0*XI*XI-1.)                                       SHP00650
      SF(2) = 0.5625*(1.0-XI*XI)*(1.0-3.0*XI)                                      SHP00660
      SF(3) = 0.5625*(1.0-XI*XI)*(1.0+3.0*XI)                                      SHP00670
      SF(4) = 0.0625*(9.0*XI*XI-1.0)*(1.0+XI)                                      SHP00680
      DSF(1) = 0.0625*(1.0+18.0*XI-27.0*XI*XI)                                     SHP00690
      DSF(2) = 0.5625*(-3.0-2.0*XI+9.0*XI*XI)                                      SHP00700
      DSF(3) = 0.5625*(3.0-2.0*XI-9.0*XI*XI)                                       SHP00710
      DSF(4) = 0.0625*(18.0*XI+27.0*XI*XI-1.0)                                     SHP00720
   60 DO 70 I=1,NPE                                                                SHP00730
   70 DDSF(I)=0.0                                                                  SHP00740
C                                                                                  SHP00750
C     COMPUTE THE DERIVATIVES OF SF(I) WITH RESPECT TO X                           SHP00760
C                                                                                  SHP00770
   80 GJ =  H*0.5                                                                  SHP00780
      DO 90 I = 1,NET                                                              SHP00790
      GDSF(I) = DSF(I)/GJ                                                          SHP00800
   90 GDDSF(I) = DDSF(I)/GJ/GJ                                                     SHP00810
```

```
      RETURN                                                            SHP00820
      END                                                               SHP00830
      SUBROUTINE STRESS(NPE,NDF,IELEM,W0,ELX)                           STR00010
C                                                                       STR00020
C     ------------------------------------------------------------------STR00030
C        X.........GLOBAL COORDINATE                                    STR00040
C        XI .......LOCAL COORDINATE                                     STR00050
C        SF........ELEMENT INTERPOLATION FUNCTIONS                      STR00060
C        GDSF......FIRST DERIVATIVE OF SF W.R.T. GLOBAL COORDINATE      STR00070
C        GDDSF.....SECOND DERIVATIVE OF SF W.R.T. GLOBAL COORDINATE     STR00080
C        W0........COLUMN OF GENERALIZED DISPLACEMENTS                  STR00090
C        W.........INTERPOLATED GENERALIZED DISPLACEMENT                STR00100
C        DW........FIRST DERIVATIVE OF W: DW/DX                         STR00110
C        DDW.......SECOND DERIVATIVE OF W: D(DW)/DX                     STR00120
C                                                                       STR00130
C        NOTE: W,  DW, AND DDW ARE COMPUTED AT NINE POINTS OF EACH      STR00140
C               ELEMENT (DW AND DDW ARE NOT EXPECTED TO BE ACCURATE     STR00150
C               AT THE NODAL POINTS OF THE ELEMENT)                     STR00160
C     ------------------------------------------------------------------STR00170
C                                                                       STR00180
      IMPLICIT REAL*8 (A-H,O-Z)                                         STR00190
      COMMON/SHP/SF(4),GDSF(4),GDDSF(4),GJ                              STR00200
      DIMENSION   GAUSS(9),W0(4),ELX(4)                                 STR00210
      DATA GAUSS/-1.0D0,-0.75D0,-0.50D0,-0.25D0,0.0D0,0.25D0,0.50D0,    STR00220
     *           0.75D0,1.0D0/                                          STR00230
C                                                                       STR00240
      NET=NPE                                                           STR00250
      IF(IELEM .EQ. 0)NET=4                                             STR00260
      H = ELX(NPE)-ELX(1)                                               STR00270
      DO 70 NI=1,9                                                      STR00280
      XI = GAUSS(NI)                                                    STR00290
      CALL SHAPE(XI,H,NPE,NET,IELEM)                                    STR00300
      X = 0.5*H*(1.0+XI)+ELX(1)                                         STR00310
      W=0.0                                                             STR00320
      DW=0.0                                                            STR00330
      DDW=0.0                                                           STR00340
      DO 65 I=1,NET                                                     STR00350
      W = W + SF(I)*W0(I)                                               STR00360
      DW=DW+GDSF(I)*W0(I)                                               STR00370
      IF(IELEM .NE. 0)GOTO 65                                           STR00380
      DDW=DDW+GDDSF(I)*W0(I)                                            STR00390
   65 CONTINUE                                                          STR00400
      IF(IELEM.EQ.0) PRINT 10,X,W,DW,DDW                                STR00410
      IF(IELEM.GT.0) PRINT 10, X,W,DW                                   STR00420
   70 CONTINUE                                                          STR00430
   10 FORMAT (10X,8E13.5)                                               STR00440
      RETURN                                                            STR00450
      END                                                               STR00460
      SUBROUTINE BNDY(NRMAX,NCMAX,NEQ,NHBW,S,SL,NBDY,IBDY,VBDY)         BND00010
C                                                                       BND00020
C     ..................................................................BND00030
C     SUBROUTINE USED TO IMPOSE BOUNDARY CONDITIONS ON BANDED EQUATIONS BND00040
C     ..................................................................BND00050
C                                                                       BND00060
      IMPLICIT REAL*8 (A-H,O-Z)                                         BND00070
      DIMENSION S(NRMAX,NCMAX),SL(NRMAX)                                BND00080
      DIMENSION IBDY(NBDY),VBDY(NBDY)                                   BND00090
      DO 300 NB = 1, NBDY                                               BND00100
      IE = IBDY(NB)                                                     BND00110
      SVAL = VBDY(NB)                                                   BND00120
      IT=NHBW-1                                                         BND00130
      I=IE-NHBW                                                         BND00140
      DO 100 II=1,IT                                                    BND00150
      I=I+1                                                             BND00160
      IF (I .LT. 1)  GO TO 100                                          BND00170
      J=IE-I+1                                                          BND00180
      SL(I)=SL(I)-S(I,J)*SVAL                                           BND00190
      S(I,J)=0.0                                                        BND00200
  100 CONTINUE                                                          BND00210
      S(IE,1)=1.0                                                       BND00220
      SL(IE)=SVAL                                                       BND00230
      I=IE                                                              BND00240
      DO 200 II=2,NHBW                                                  BND00250
      I=I+1                                                             BND00260
      IF (I .GT. NEQ)  GO TO 200                                        BND00270
```

```
      SL(I)=SL(I)-S(IE,II)*SVAL                                     BND00280
      S(IE,II)=0.0                                                  BND00290
  200 CONTINUE                                                      BND00300
  300 CONTINUE                                                      BND00310
      RETURN                                                        BND00320
      END                                                           BND00330
      SUBROUTINE SOLVE(NRM,NCM,NEQNS,NBW,BAND,RHS,IRES)             SLV00010
C     .........................................................     SLV00020
C     SOLVING A BANDED SYMMETRIC SYSTEM OF EQUATIONS                SLV00030
C     IN RESOLVING, IRES .GT. 0, LHS ELIMINATION IS SKIPPED         SLV00040
C     .........................................................     SLV00050
C                                                                   SLV00060
      IMPLICIT REAL*8 (A-H,O-Z)                                     SLV00070
      DIMENSION BAND(NRM,NCM),RHS(NRM)                              SLV00080
      MEQNS=NEQNS-1                                                 SLV00090
      IF (IRES .GT. 0)  GO TO 90                                    SLV00100
      DO 500 NPIV=1,MEQNS                                           SLV00110
      NPIVOT=NPIV+1                                                 SLV00120
      LSTSUB=NPIV+NBW-1                                             SLV00130
      IF(LSTSUB.GT.NEQNS) LSTSUB=NEQNS                              SLV00140
      DO 400 NROW=NPIVOT,LSTSUB                                     SLV00150
C     INVERT ROWS AND COLUMNS FOR ROW FACTOR                        SLV00160
      NCOL=NROW-NPIV+1                                              SLV00170
      FACTOR=BAND(NPIV,NCOL)/BAND(NPIV,1)                           SLV00180
      DO 200 NCOL=NROW,LSTSUB                                       SLV00190
      ICOL=NCOL-NROW+1                                              SLV00200
      JCOL=NCOL-NPIV+1                                              SLV00210
  200 BAND(NROW,ICOL)=BAND(NROW,ICOL)-FACTOR*BAND(NPIV,JCOL)        SLV00220
  400 RHS(NROW)=RHS(NROW)-FACTOR*RHS(NPIV)                          SLV00230
  500 CONTINUE                                                      SLV00240
      GO TO 101                                                     SLV00250
   90 DO 100 NPIV=1,MEQNS                                           SLV00260
      NPIVOT=NPIV+1                                                 SLV00270
      LSTSUB=NPIV+NBW-1                                             SLV00280
      IF(LSTSUB.GT.NEQNS) LSTSUB=NEQNS                              SLV00290
      DO 110 NROW=NPIVOT,LSTSUB                                     SLV00300
      NCOL=NROW-NPIV+1                                              SLV00310
      FACTOR=BAND(NPIV,NCOL)/BAND(NPIV,1)                           SLV00320
  110 RHS(NROW)=RHS(NROW)-FACTOR*RHS(NPIV)                          SLV00330
 100  CONTINUE                                                      SLV00340
C     BACK SUBSTITUTION                                             SLV00350
  101 DO 800 IJK=2,NEQNS                                            SLV00360
      NPIV=NEQNS-IJK+2                                              SLV00370
      RHS(NPIV)=RHS(NPIV)/BAND(NPIV,1)                              SLV00380
      LSTSUB=NPIV-NBW+1                                             SLV00390
      IF(LSTSUB.LT.1) LSTSUB=1                                      SLV00400
      NPIVOT=NPIV-1                                                 SLV00410
      DO 700 JKI=LSTSUB,NPIVOT                                      SLV00420
      NROW=NPIVOT-JKI+LSTSUB                                        SLV00430
      NCOL=NPIV-NROW+1                                              SLV00440
      FACTOR=BAND(NROW,NCOL)                                        SLV00450
  700 RHS(NROW)=RHS(NROW)-FACTOR*RHS(NPIV)                          SLV00460
  800 CONTINUE                                                      SLV00470
      RHS(1)=RHS(1)/BAND(1,1)                                       SLV00480
      RETURN                                                        SLV00490
      END                                                           SLV00500
```

APPENDIX

II

COMPUTER PROGRAM FEM2D

```
C                      P R O G R A M   F E M 2 D                    FEM00010
C         (AN IN-CORE FINITE-ELEMENT ANALYSIS COMPUTER PROGRAM)      FEM00020
C                                                                    FEM00030
C   .................................................................FEM00040
C   .                                                               .FEM00050
C   . FINITE-ELEMENT ANALYSIS OF TWO-DIMENSIONAL PROBLEMS, INCLUDING .FEM00060
C   .         HEAT TRANSFER, FLUID FLOW, AND PLANE ELASTICITY       .FEM00070
C   .                                                               .FEM00080
C   .                                                               .FEM00090
C   .   THE PROGRAM ILLUSTRATES THE USE OF THE THREE-NODE (LINEAR)  .FEM00100
C   .   TRIANGULAR ELEMENT AND THE LINEAR AND QUADRATIC ISOPARAMETRIC .FEM00110
C   .   QUADRILATERAL ELEMENTS FOR THE SOLUTION OF HEAT CONDUCTION/ .FEM00120
C   .   CONVECTION TYPE PROBLEMS, PLANE ELASTICITY PROBLEMS, AND    .FEM00130
C   .   VISCOUS INCOMPRESSIBLE FLUID FLOW PROBLEMS BY THE PENALTY   .FEM00140
C   .   FUNCTION FORMULATION.  OTHER SECOND-ORDER PROBLEMS CAN BE   .FEM00150
C   .   SOLVED BY MODIFYING APPROPRIATE STATEMENTS IN SUBROUTINE STIFF.FEM00160
C   .                                                               .FEM00170
C   .................................................................FEM00180
C                                                                    FEM00190
C          D E S C R I P T I O N   O F   T H E   V A R I A B L E S   FEM00200
C                                                                    FEM00210
C      AMU.......VISCOSITY IN THE PENALTY FORMULATION OF STOKES FLOW  FEM00220
C      ANU12.....POISSON'S RATIO FOR ELASTICITY PROBLEMS (ANU12=C3)   FEM00230
C      ANU21.....POISSON'S RATIO FOR ORTHOTROPIC MEDIUM=ANU12*(E2/E1) FEM00240
C      B(I,J)....MASS MATRIX FOR TRANSIENT ANALYSIS                   FEM00250
C      C(I,J)....MATERIAL STIFFNESS COEFFICIENTS OF THE ORTHOTROPIC   FEM00260
C                ELASTICITY (COMPUTED IN THE PROGRAM).                FEM00270
C                                                                    FEM00280
C      C1,C2,....MATERIAL CONSTANTS AND SOURCE TERM AS DEFINED BELOW  FEM00290
C                HEAT TRANSFER TYPE PROBLEMS:                         FEM00300
C                     C1=K1, C2=K2, C3=H, C4=TINF, C5=Q               FEM00310
C                ELASTICITY PROBLEMS:                                 FEM00320
C                     C1=E1, C2=E2, C3=ANU12, C4=G12, C5=T            FEM00330
C                INCOMPRESSIBLE FLUID FLOW PROBLEMS:                  FEM00340
C                     C1=AMU, C2=PENALTY PARAMETER                    FEM00350
C                                                                    FEM00360
C      DT........TIME STEP FOR TRANSIENT ANALYSIS                     FEM00370
C      E1,E2.....ORTHOTROPIC MODULI (E1=C1,E2=C2)                     FEM00380
C      ELSTIF....ELEMENT COEFFICIENT (OR 'STIFFNESS') MATRIX          FEM00390
C      ELXY......GLOBAL COORDINATES OF ELEMENT NODES:                 FEM00400
C                ELXY(I,1)=XE(I);  ELXY(I,2)=YE(I), I=1,NPE.          FEM00410
C      F(I)......ELEMENT 'FORCE' VECTOR (SHOULD BE MODIFIED FOR NON-  FEM00420
C                CONSTANT SOURCE TERM(S) IN THE PROBLEM).             FEM00430
```

```
C        G12.......SHEAR MODULUS FOR ELASTICITY PROBLEMS (G12=C4)          FEM00440
C        GF(I).....ASSEMBLED 'FORCE' VECTOR. ALSO USED TO RETURN THE       FEM00450
C                  SOLUTION FROM SUBROUTINE 'SOLVE'.                       FEM00460
C        GP(I).....ARRAY OF INITIAL VALUES (ALSO ARRAY FOR THE PREVIOUS    FEM00470
C                  TIME-STEP SOLUTION) USED IN TRANSIENT ANALYSIS.         FEM00480
C        GSTIF.....GLOBAL COEFFICIENT (OR 'STIFFNESS') MATRIX. NOTE        FEM00490
C                  THAT IT IS ASSEMBLED AND STORED IN A BANDED FORM.       FEM00500
C        H.........FILM COEFFICIENT IN HEAT TRANSFER PROBLEMS (H=C3)       FEM00510
C        IBDF(I)...ARRAY OF SPECIFIED BOUNDARY DEGREES OF FREEDOM          FEM00520
C        IBN(I)....ARRAY OF BOUNDARY ELEMENTS FOR CONVECTIVE B.C.          FEM00530
C        IBSF(I)...ARRAY OF SPECIFIED NONZERO FORCES                       FEM00540
C                                                                          FEM00550
C        ICONV.....INDICATOR FOR CONVECTION BOUNDARY CONDITION:            FEM00560
C                        ICONV=1, CONVECTION BOUNDARY IS PRESENT           FEM00570
C                        ICONV=0, NO CONVECTION.                           FEM00580
C                                                                          FEM00590
C        IEL.......INDICATOR FOR THE TYPE OF ELEMENT:                      FEM00600
C                        IEL=0, THREE-NODE TRIANGULAR ELEMENT              FEM00610
C                        IEL=1, FOUR-NODE QUADRILATERAL ELEMENT            FEM00620
C                        IEL=2, EIGHT/NINE-NODE QUADRILATERAL ELEMENT.     FEM00630
C                                                                          FEM00640
C        IMESH.....INDICATOR FOR MESH GENERATION:                          FEM00650
C                        IMESH=0, MESH INFORMATION IS TO BE READIN         FEM00660
C                        IMESH=1, MESH IS GENERATED IN THE PROGRAM         FEM00670
C                                                                          FEM00680
C        INOD(I,J).THE J-TH NODE OF THE I-TH ELEMENT FOR THE ELEMENTS      FEM00690
C                  (ON THE CONVECTIVE BOUNDARY)IN ARRAY IBN(N).            FEM00700
C                                                                          FEM00710
C        ITEM......INDICATOR FOR TRANSIENT ANALYSIS                        FEM00720
C                                                                          FEM00730
C        ITYPE.....INDICATOR FOR THE TYPE PROBLEM BEING SOLVED:            FEM00740
C                        ITYPE=2, VISCOUS INCOMPRESSIBLE FLOW (IEL=1)      FEM00750
C                        ITYPE=1, PLANE STRAIN ELASTICITY                  FEM00760
C                        ITYPE=0, PLANE STRESS ELASTICITY                  FEM00770
C                        ITYPE=-1, STEADY HEAT TRANSFER                    FEM00780
C                                                                          FEM00790
C        K1,K2.....THERMAL CONDUCTIVITIES FOR HEAT TRANSFER PROBLEMS       FEM00800
C                  (K1=C1, K2=C2).                                         FEM00810
C        NBE.......NUMBER OF BOUNDARY ELEMENTS FOR CONVECTION B.C.         FEM00820
C        NCMAX.....COLUMN DIMENSION OF GSTIF IN THE DIMENSION STATEMENT    FEM00830
C        NDF.......NUMBER OF DEGREES OF FREEDOM PER NODE:                  FEM00840
C                  NDF=1, FOR HEAT TRANSFER TYPE PROBLEMS                  FEM00850
C                  NDF=2, FOR ELASTICITY AND INCOMPRESSIBLE FLUID FLOW.    FEM00860
C        NEQ.......TOTAL NUMBER OF EQUATIONS IN THE PROBLEM (=NNM*NDF)     FEM00870
C        NEM.......NUMBER OF ELEMENTS IN THE MESH                          FEM00880
C        NHBW......HALF BAND WIDTH OF THE COEFFICIENT MATRIX, GSTIF        FEM00890
C        NN........TOTAL DEGREES OF FREEDOM PER ELEMENT (=NPE*NDF)         FEM00900
C        NNM.......NUMBER OF NODES IN THE FINITE-ELEMENT MESH              FEM00910
C        NOD(I,J)..GLOBAL NODE NUMBER CORRESPONDING TO THE J-TH NODE OF    FEM00920
C                  ELEMENT I (BOOLEAN CONNECTIVITY MATRIX).                FEM00930
C        NPE.......NUMBER OF NODES PER ELEMENT                             FEM00940
C        NRMAX.....ROW DIMENSION OF GSTIF IN THE DIMENSION STATEMENT       FEM00950
C        NSBF......NUMBER OF SPECIFIED NONZERO BOUNDARY 'FORCES'           FEM00960
C        NSDF......NUMBER OF SPECIFIED PRIMARY DEGREES OF FREEDOM          FEM00970
C        NX,NY.....NUMBER OF ELEMENTS IN X AND Y DIRECTIONS, RESPEC.       FEM00980
C        Q.........VALUE OF THE CONSTANT SOURCE TERM (Q=C5)                FEM00990
C        T.........THICKNESS OF THE PLATE IN ELASTICITY PROBLEMS (T=C5)    FEM01000
C        TO........MAXIMUM LIMIT ON TIME IN THE TRANSIENT ANALYSIS         FEM01010
C        TINF......AMBIENT TEMPERATURE IN HEAT TRANSFER PROB. (TINF=C4)    FEM01020
C                                                                          FEM01030
C        THETA.....PARAMETER IN TIME APPROXIMATION:                        FEM01040
C                        THETA=0, FORWARD-DIFFERENCE SCHEME                FEM01050
C                        THETA=0.5, THE CRANK-NICOLSON SCHEME              FEM01060
C                        THETA=2/3, THE GALERKIN SCHEME                    FEM01070
C                        THETA=1, BACKWARD-DIFFERENCE SCHEME               FEM01080
C                                                                          FEM01090
C        VBDF(I)...ARRAY OF THE VALUES CORRESPONDING TO THE SPECIFIED      FEM01100
C                  DEGREES OF FREEDOM IN ARRAY IBDF(I).                    FEM01110
C        VBSF(I)...ARRAY OF THE VALUES CORRESPONDING TO THE SPECIFIED      FEM01120
C                  NONZERO FORCES IN ARRAY IBSF(I).                        FEM01130
C        W(I,J)....VALUE OF THE I-TH DEGREE OF FREEDOM AT THE J-TH NODE    FEM01140
C                  OF THE ELEMENT.                                         FEM01150
C        X(I)......X-COORDINATE OF GLOBAL NODE I (I=1,NNM)                 FEM01160
C        Y(I)......Y-COORDINATE OF GLOBAL NODE I (I=1,NNM)                 FEM01170
C                                                                          FEM01180
```

```
C     .................................................................FEM01190
C                                                                      FEM01200
C        D I M E N S I O N S   F O R   V A R I O U S   A R R A Y S     FEM01210
C                                                                      FEM01220
C      ARRAYS ELSTIF,X,Y,F,B,C,BETA,AND GAMA ARE OF FIXED DIMENSION.   FEM01230
C      DIMENSIONS FOR THE OTHER ARRAYS SHOULD BE AS INDICATED BELOW:   FEM01240
C                                                                      FEM01250
C          GSTIF(NRMAX,NCMAX); NRMAX.GE.NEQ AND NCMAX.GE.NHBW          FEM01260
C                   GF(NEQ), GP(NEQ), X(NNM), Y(NNM)                   FEM01270
C        NOD(NEM,9),INODE(NBE,2),IBDF(NSDF),VBDF(NSDF),IBN(NBE)        FEM01280
C                                                                      FEM01290
C      NOTE: THE VALUES OF NCMAX AND NRMAX IN THE DATA STATEMENT MUST  FEM01300
C             MATCH THOSE IN THE DIMENSION STATEMENT FOR GSTIF         FEM01310
C     .................................................................FEM01320
C                                                                      FEM01330
      IMPLICIT REAL*8(A-H,O-Z)                                         FEM01340
      DIMENSION GSTIF(450,50),GF(450),GP(450),W(2,9),TITLE(20),        FEM01350
     1          IBDF(100),VBDF(100),IBSF(50),VBSF(50),IBN(20),INOD(20,2)FEM01360
      COMMON/MSH/NOD(200,9),X(225),Y(225),DX(15),DY(15)                FEM01370
      COMMON/STF/ELSTIF(18,18),ELXY(9,2),C(3,3),F(18),WO(18),AO,A1     FEM01380
      DATA NRMAX,NCMAX/450,50/                                         FEM01390
C                                                                      FEM01400
C     ...............................................................  FEM01410
C     .                                                             .  FEM01420
C     .                 P R E P R O C E S S O R   U N I T           .  FEM01430
C     .                                                             .  FEM01440
C     ...............................................................  FEM01450
C                                                                      FEM01460
C     R E A D I N   T H E   I N P U T   D A T A   H E R E              FEM01470
C                                                                      FEM01480
      READ 400,TITLE                                                   FEM01490
      READ 520, IEL,NPE,ITYPE,ICONV,IMESH,ITEM,NPRNT                   FEM01500
      IF(ITYPE.EQ.2 .AND. IEL.EQ.0)STOP                                FEM01510
      IF(ITYPE.GE.0)PRINT 550                                          FEM01520
      IF(ITYPE.LT.0)PRINT 570                                          FEM01530
      PRINT 400, TITLE                                                 FEM01540
      IF(IMESH.EQ.1)GOTO 30                                            FEM01550
      READ 520, NEM,NNM                                                FEM01560
      DO 20 N=1,NEM                                                    FEM01570
   20 READ 520, (NOD(N,I),I=1,NPE)                                     FEM01580
      READ 580, (X(I),Y(I),I=1,NNM)                                    FEM01590
      GOTO 40                                                          FEM01600
   30 READ 520, NX,NY                                                  FEM01610
      NXX1=IEL*NX+1                                                    FEM01620
      NYY1=IEL*NY+1                                                    FEM01630
      IF(IEL.EQ.0)NXX1=NX+1                                            FEM01640
      IF(IEL.EQ.0)NYY1=NY+1                                            FEM01650
      READ 580, (DX(I),I=1,NXX1)                                       FEM01660
      READ 580, (DY(I),I=1,NYY1)                                       FEM01670
      CALL MESH(IEL,NX,NY,NPE,NNM,NEM)                                 FEM01680
   40 READ 580, C1,C2,C3,C4,C5                                         FEM01690
      IF(ITYPE.GE.0 .OR. ICONV.EQ.0)GOTO 50                            FEM01700
      READ 520,NBE,(IBN(I),I=1,NBE)                                    FEM01710
      READ 520,((INOD(I,J),J=1,2),I=1,NBE)                             FEM01720
   50 READ 520, NSDF                                                   FEM01730
      IF(NSDF.EQ.0)GOTO 60                                             FEM01740
      READ 520, (IBDF(I),I=1,NSDF)                                     FEM01750
      READ 580, (VBDF(I),I=1,NSDF)                                     FEM01760
   60 READ 520, NSBF                                                   FEM01770
      IF(NSBF.EQ.0)GOTO 70                                             FEM01780
      READ 520, (IBSF(I),I=1,NSBF)                                     FEM01790
      READ 580, (VBSF(I),I=1,NSBF)                                     FEM01800
   70 IF(ITEM.EQ.1)READ 580, DT,THETA,TO                               FEM01810
      IF(ITEM.EQ.1)READ 580, (GP(I),I=1,NEQ)                           FEM01820
C                                                                      FEM01830
C     E N D   O F   T H E   D A T A   I N P U T                        FEM01840
C                                                                      FEM01850
      IF (ITYPE.LT.0) NDF=1                                            FEM01860
      IF (ITYPE.GE.0) NDF=2                                            FEM01870
      NEQ=NNM*NDF                                                      FEM01880
      NN=NPE*NDF                                                       FEM01890
      PRINT 660, IEL,ITYPE,ICONV                                       FEM01900
      IF(ITYPE.EQ.2)C2=C2*1.0D8                                        FEM01910
      IF(ITYPE.EQ.2)PRINT 650, C1,C2                                   FEM01920
      IF(ITYPE.EQ.0 .OR. ITYPE.EQ.1)PRINT 530, C1,C2,C3,C4,C5          FEM01930
```

```
      IF(ITYPE.LT.0)PRINT 510, C1,C2,C3,C4,C5                          FEM01940
      PRINT 450, NEM,NNM,NEQ                                            FEM01950
      IF(ICONV.EQ.1)PRINT 470, NBE, (IBN(I),I=1,NBE)                    FEM01960
      PRINT 560                                                         FEM01970
      PRINT 480, (X(I),Y(I),I=1,NNM)                                    FEM01980
      PRINT 500                                                         FEM01990
      DO 75 I=1,NEM                                                     FEM02000
   75 PRINT 440, I,(NOD(I,J),J=1,NPE)                                   FEM02010
      IF(NSDF.GT.0)PRINT 460, NSDF,(IBDF(I),I=1,NSDF)                   FEM02020
      IF(NSBF.EQ.0)GOTO 80                                              FEM02030
      PRINT 640, NSBF,(IBSF(I),I=1,NSBF)                                FEM02040
      PRINT 600                                                         FEM02050
      PRINT 480, (VBSF(I),I=1,NSBF)                                     FEM02060
   80 IF(ITYPE.LT.0 .OR. ITYPE.EQ.2)GOTO 100                            FEM02070
C                                                                       FEM02080
C     ............................................................     FEM02090
C     .                                                          .     FEM02100
C     .                 P R O C E S S O R   U N I T              .     FEM02110
C     .                                                          .     FEM02120
C     ............................................................     FEM02130
C                                                                       FEM02140
C     COMPUTE THE MATERIAL-CONSTANTS FOR PLANE ELASTICITY PROBLEMS      FEM02150
C                                                                       FEM02160
      ANU21=C3*C2/C1                                                    FEM02170
      DENOM=1.0-C3*ANU21                                                FEM02180
C     PLANE STRESS CASE                                                 FEM02190
      C(1,1)=C1*C5/DENOM                                                FEM02200
      C(1,2)=ANU21*C(1,1)                                               FEM02210
      C(2,2)=C2*C(1,1)/C1                                               FEM02220
      IF(ITYPE.EQ.0)GOTO 90                                             FEM02230
C     PLANE STRAIN CASE                                                 FEM02240
      S0=(1.0-3.0*C3*ANU21-C3*ANU21*(C3+ANU21))                         FEM02250
      C(1,1)=C1*C5*DENOM/S0                                             FEM02260
      C(1,2)=C1*C5*ANU21*(1.0+C3)/S0                                    FEM02270
      C(2,2)=C2*C(1,1)/C1                                               FEM02280
   90 C(3,3)=C4*C5                                                      FEM02290
      C(1,3)=0.0                                                        FEM02300
      C(2,3)=0.0                                                        FEM02310
      C(2,1)=C(1,2)                                                     FEM02320
      C(3,1)=C(1,3)                                                     FEM02330
      C(3,2)=C(2,3)                                                     FEM02340
C                                                                       FEM02350
C     COMPUTE THE HALF BAND WIDTH                                       FEM02360
C                                                                       FEM02370
  100 NHBW=0                                                            FEM02380
      DO 110 N=1,NEM                                                    FEM02390
      DO 110 I=1,NPE                                                    FEM02400
      DO 110 J=1,NPE                                                    FEM02410
      NW=(IABS(NOD(N,I)-NOD(N,J))+1)*NDF                                FEM02420
  110 IF (NHBW.LT.NW) NHBW=NW                                           FEM02430
      PRINT 590, NHBW                                                   FEM02440
      IF(ITEM.NE.1)GOTO 125                                             FEM02450
      TIME=0.0                                                          FEM02460
      PRINT 420, DT,THETA,TO                                            FEM02470
      A0=(1.0-THETA)*DT                                                 FEM02480
      A1=THETA*DT                                                       FEM02490
  120 TIME=TIME+DT                                                      FEM02500
      IF(TIME.GT.TO)STOP                                                FEM02510
C                                                                       FEM02520
C     INITIALIZE THE GLOBAL STIFFNESS MATRIX AND FORCE VECTOR           FEM02530
C                                                                       FEM02540
  125 DO 130 I=1,NEQ                                                    FEM02550
      GF(I)=0.0                                                         FEM02560
      DO 130 J=1,NHBW                                                   FEM02570
  130 GSTIF(I,J)=0.0                                                    FEM02580
C                                                                       FEM02590
C     DO-LOOP ON THE NUMBER OF ELEMENTS TO CALCULATE THE ELEMENT        FEM02600
C     MATRICES, AND ASSEMBLY OF THE ELEMENT MATRICES BEGINS HERE        FEM02610
C                                                                       FEM02620
  140 DO 300 N=1,NEM                                                    FEM02630
      DO 150 I=1,NPE                                                    FEM02640
      NI=NOD(N,I)                                                       FEM02650
      ELXY(I,1)=X(NI)                                                   FEM02660
      ELXY(I,2)=Y(NI)                                                   FEM02670
      IF(ITEM.EQ.0)GOTO 150                                             FEM02680
```

```
      L=NI*NDF-1                                                        FEM02690
      K=I*NDF-1                                                         FEM02700
      IF(NDF.EQ.1)K=I                                                   FEM02710
      IF(NDF.EQ.1)L=NI                                                  FEM02720
      WO(K)=GP(L)                                                       FEM02730
      WO(K+1)=GP(L+1)                                                   FEM02740
  150 CONTINUE                                                          FEM02750
      IF(IEL.GT.0)CALL STIFFQ(NPE,NN,IEL,ITYPE,ITEM,C1,C2,C5)           FEM02760
      IF(IEL.EQ.0)CALL STIFFT(NPE,NN,IEL,ITYPE,ITEM,C1,C2,C5)           FEM02770
      IF(N.NE.NPRNT)GOTO 170                                            FEM02780
      PRINT 680                                                         FEM02790
      DO 160 I=1,NN                                                     FEM02800
  160 PRINT 670, (ELSTIF(I,J),J=1,NN)                                   FEM02810
  170 IF (ICONV.EQ.0) GO TO 200                                         FEM02820
C                                                                       FEM02830
C     ADDITION OF THE CONVECTIVE TERMS TO THE ELEMENT MATRIX            FEM02840
C                                                                       FEM02850
      DO 180 M=1,NBE                                                    FEM02860
      IF (IBN(M).NE.N) GO TO 180                                        FEM02870
      M1=INOD(M,1)                                                      FEM02880
      M2=INOD(M,2)                                                      FEM02890
      NM1=NOD(N,M1)                                                     FEM02900
      NM2=NOD(N,M2)                                                     FEM02910
      DL=DSQRT((X(NM2)-X(NM1))**2+(Y(NM2)-Y(NM1))**2)                   FEM02920
      BL=C3*DL                                                          FEM02930
      ELSTIF(M1,M1)=ELSTIF(M1,M1)+BL/3.0                                FEM02940
      ELSTIF(M2,M2)=ELSTIF(M2,M2)+BL/3.0                                FEM02950
      ELSTIF(M1,M2)=ELSTIF(M1,M2)+BL/6.0                                FEM02960
      ELSTIF(M2,M1)=ELSTIF(M1,M2)                                       FEM02970
      F(M1)=F(M1)+0.5*BL*C4                                             FEM02980
      F(M2)=F(M2)+0.5*BL*C4                                             FEM02990
  180 CONTINUE                                                          FEM03000
C                                                                       FEM03010
C     ASSEMBLE ELEMENT MATRICES TO OBTAIN GLOBAL MATRIX                 FEM03020
C                                                                       FEM03030
  200 DO 280 I=1,NPE                                                    FEM03040
      NR=(NOD(N,I)-1)*NDF                                               FEM03050
      DO 280 II=1,NDF                                                   FEM03060
      NR=NR+1                                                           FEM03070
      L=(I-1)*NDF+II                                                    FEM03080
      GF(NR)=GF(NR)+F(L)                                                FEM03090
      DO 260 J=1,NPE                                                    FEM03100
      NCL=(NOD(N,J)-1)*NDF                                              FEM03110
      DO 260 JJ=1,NDF                                                   FEM03120
      M=(J-1)*NDF+JJ                                                    FEM03130
      NC=NCL+JJ+1-NR                                                    FEM03140
      IF (NC) 260,260,250                                               FEM03150
  250 GSTIF(NR,NC)=GSTIF(NR,NC)+ELSTIF(L,M)                             FEM03160
  260 CONTINUE                                                          FEM03170
  280 CONTINUE                                                          FEM03180
  300 CONTINUE                                                          FEM03190
C                                                                       FEM03200
C     ASSEMBLED MATRIX EQUATIONS ARE NOW READY FOR IMPLEMENTATION OF    FEM03210
C       THE BOUNDARY CONDITIONS ON PRIMARY AND SECONDARY VARIABLES      FEM03220
C                                                                       FEM03230
      IRES=0                                                            FEM03240
      IF (NSBF.EQ.0) GO TO 320                                          FEM03250
      DO 310 I=1,NSBF                                                   FEM03260
      II=IBSF(I)                                                        FEM03270
  310 GF(II)=VBSF(I)+GF(II)                                             FEM03280
  320 IF(NPRNT.LE.1)GOTO 340                                            FEM03290
      PRINT 700                                                         FEM03300
      DO 330 I=1,NEQ                                                    FEM03310
  330 PRINT 480, (GSTIF(I,J),J=1,NHBW)                                  FEM03320
      PRINT 710                                                         FEM03330
      PRINT 480, (GF(I),I=1,NEQ)                                        FEM03340
  340 IF (NSDF.EQ.0) GO TO 350                                          FEM03350
      CALL BNDY(NRMAX,NCMAX,NEQ,NHBW,GSTIF,GF,NSDF,IBDF,VBDF)           FEM03360
C                                                                       FEM03370
C     CALL SUBROUTINE 'SOLVE' TO SOLVE THE SYSTEM OF EQUATIONS FOR THE  FEM03380
C     PRIMARY DEGREES OF FREEDOM (THE SOLUTION IS RETURNED IN ARRAY GF) FEM03390
C                                                                       FEM03400
  350 CALL SOLVE (NRMAX,NCMAX,NEQ,NHBW,GSTIF,GF,IRES)                   FEM03410
C                                                                       FEM03420
```

```
C     ...............................................................   FEM03430
C     .                                                             .   FEM03440
C     .                 P O S T P R O C E S S O R   U N I T         .   FEM03450
C     .                                                             .   FEM03460
C     ...............................................................   FEM03470
C                                                                       FEM03480
      IF(ITEM.EQ.0)GOTO 370                                             FEM03490
      DO 360 I=1,NEQ                                                    FEM03500
  360 GP(I)=GF(I)                                                       FEM03510
      PRINT 430, TIME                                                   FEM03520
  370 PRINT 540                                                         FEM03530
      IF(ITYPE.GE.0)PRINT 485,(I,GF(I+I-1),GF(I+I),I=1,NNM)             FEM03540
      IF(ITYPE.LT.0)PRINT 480,(GF(I),I=1,NNM)                           FEM03550
      PRINT 690                                                         FEM03560
      IF(ITYPE.GE.0)PRINT 620                                           FEM03570
      IF(ITYPE.LT.0)PRINT 410                                           FEM03580
      DO 390 N=1,NEM                                                    FEM03590
      DO 380 I=1,NPE                                                    FEM03600
      NI=NOD(N,I)                                                       FEM03610
      L=NI*NDF-1                                                        FEM03620
      IF(NDF.EQ.1)L=NI                                                  FEM03630
      W(1,I)=GF(L)                                                      FEM03640
      IF(NDF.EQ.2)W(2,I)=GF(L+1)                                        FEM03650
      ELXY(I,1)=X(NI)                                                   FEM03660
  380 ELXY(I,2)=Y(NI)                                                   FEM03670
  390 CALL STRESS(N,NPE,ELXY,C,W,C1,C2,ITYPE,IEL)                       FEM03680
      IF (ITEM.EQ.1)GOTO 120                                            FEM03690
      STOP                                                              FEM03700
C                                                                       FEM03710
C                          F  O  R  M  A  T  S                          FEM03720
C                                                                       FEM03730
  400 FORMAT(20A4)                                                      FEM03740
  410 FORMAT ( 7X,'ELE.NO.',2X,'X-COMPONENT',6X,'Y-COMPONENT',7X,'MAGNITFEM03750
     1UDE',9X,'ANGLE',/)                                                FEM03760
  420 FORMAT (10X,'THETA=',E10.3,2X,'TIME STEP=',E10.3,2X,'MAX. TIME='  FEM03770
     1,E10.3)                                                           FEM03780
  430 FORMAT (/,3X,'TIME =',F10.3,/)                                    FEM03790
  440 FORMAT (10X,20I5)                                                 FEM03800
  450 FORMAT (10X,'ACTUAL NUMBER OF ELEMENTS IN THE MESH....=',I3,/,    FEM03810
     *10X,'NUMBER OF NODES IN THE MESH.............=',I3,/,10X, 'TOTAL FEM03820
     2NUMBER OF EQUATIONS IN THE MODEL...=',I3,/)                       FEM03830
  460 FORMAT (/,5X,'NO. OF SPECIFIED DEGREES OF FREEDOM......=',I3,/,   FEM03840
     *5X,'ARRAY OF THE SPECIFIED DEGREES OF FREEDOM=',10I5,/,10X,       FEM03850
     *20I5,/,10X,20I5,/,10X,20I5,/,10X,20I5,/)                          FEM03860
  470 FORMAT (/,10X,'CONVECTIVE BOUNDARY DATA: NBE.......=',I3,/,10X,   FEM03870
     *'ARRAY IBN:',10I5)                                                FEM03880
  480 FORMAT (6(2X,E12.5))                                              FEM03890
  485 FORMAT (2(I6,2E15.5,10X))                                         FEM03900
  500 FORMAT (/,5X,'BOOLEAN (CONNECTIVITY) MATRIX NOD(I,J) ',/)         FEM03910
  510 FORMAT (10X,'PARAMETERS, C1, C2, C3, C4, AND C5:',/,15X,'C1 =',   FEM03920
     *E10.3,/,15X,'C2 =',E10.3,/,15X,'C3 =',E10.3,/,15X,'C4 =',E10.3,/, FEM03930
     *15X,'C5 =',E10.3)                                                 FEM03940
  520 FORMAT (16I5)                                                     FEM03950
  530 FORMAT (10X,'MODULUS OF ELASTICITY, E1...............=',E10.3,/,  FEM03960
     *10X,'MODULUS OF ELASTICITY, E2...............=',E10.3,/,10X,      FEM03970
     *'POISSONS RATIO, ANU12...................=',E10.3,/,10X,          FEM03980
     *'SHEAR MODULUS, G12......................=',E10.3,/,10X,          FEM03990
     *'PLATE THICKNESS, T......................=',E10.3)                FEM04000
  540 FORMAT (/,3X,'SOLUTION VECTOR:',/)                                FEM04010
  550 FORMAT (5X,'A   P L A N E   E L A S T I C I T Y   O R   F L U I D FEM04020
     1 F L O W   P R O B L E M',/)                                      FEM04030
  560 FORMAT (3X,'COORDINATES OF THE GLOBAL NODES: ',/)                 FEM04040
  570 FORMAT (3X,'H E A T   T R A N S F E R   T Y P E   P R O B L E M',/)FEM04050
  580 FORMAT (8F10.4)                                                   FEM04060
  590 FORMAT (5X,'HALF BAND WIDTH OF GLOBAL STIFFNESS MATRIX =',I3)     FEM04070
  600 FORMAT(/,5X,'VALUES OF THE SPECIFIED FORCES:',/)                  FEM04080
  620 FORMAT (/,6X,'ELE.NO.',3X,'STRESS, SXX',6X,'STRESS, SYY',6X,'STRESFEM04090
     *S,SXY',8X,'PRESSURE',/)                                           FEM04100
  640 FORMAT(/,5X,'NO. OF SPECIFIED FORCES =',I5,/,5X,'SPECIFIED FORCE  FEM04110
     1DEGREES OF FREEDOM :',15I5,/)                                     FEM04120
  650 FORMAT(10X,'VISCOSITY...............................=',E10.3,/,  FEM04130
     *10X,'PENALTY PARAMETER.......................=',E10.3)            FEM04140
  660 FORMAT(10X,'ELEMENT TYPE............................=',I2,/,     FEM04150
     *10X,'PROBLEM TYPE............................=',I2,/,10X,        FEM04160
```

```
     *'CONVECTION (0:NO, 1:YES).................=',I2)                  FEM04170
 670 FORMAT(5E15.5)                                                     FEM04180
 680 FORMAT(3X,'ELEMENT MATRICES:',/)                                   FEM04190
 690 FORMAT(/,5X,'THE ANGLE IS DEFINED TO BE THE ARC TANGENT OF Y-COMPOFEM04200
    1NENT DIVIDED BY THE X-COMPONENT',/)                                FEM04210
 700 FORMAT(3X,'GLOBAL STIFFNESS MATRIX:',/)                            FEM04220
 710 FORMAT(/,3X,'GLOBAL FORCE VECTOR:',/)                              FEM04230
     END                                                                FEM04240
     SUBROUTINE STIFFT(NPE,NN,IEL,ITYPE,ITEM,AK1,AK2,Q)                 STF00010
C                                                                       STF00020
C    .................................................................  STF00030
C                                                                       STF00040
C    THE SUBROUTINE COMPUTES THE ELEMENT COEFFICIENT MATRICES  FOR      STF00050
C    HEAT TRANSFER TYPE AND PLANE ELASTICITY PROBLEMS WHEN THE LINEAR   STF00060
C    TRIANGULAR ELEMENT IS USED.  CONSTANT SOURCE TERM IS USED FOR      STF00070
C    THE HEAT TRANSFER TYPE PROBLEMS AND ZERO FORCE TERMS ARE ASSUMED   STF00080
C    FOR PLANE ELASICITY PROBLEMS IN COMPUTING THE 'FORCE' VECTOR.      STF00090
C    .................................................................  STF00100
C                                                                       STF00110
     IMPLICIT REAL*8(A-H,O-Z)                                           STF00120
     COMMON/STF/ELSTIF(18,18),ELXY(9,2),C(3,3),F(18),W0(18),A0,A1       STF00130
     DIMENSION B(3,6),BT(6,3),STR(3,6),ALPHA(3),BETA(3),GAMA(3)         STF00140
    1          ,X(3),Y(3)                                               STF00150
C                                                                       STF00160
C    DEFINE THE COEFFICIENTS OF THE INTERPOLATION FUNCTIONS             STF00170
C                                                                       STF00180
     DO 10 I=1,NPE                                                      STF00190
     X(I)=ELXY(I,1)                                                     STF00200
     Y(I)=ELXY(I,2)                                                     STF00210
     DO 10 J=1,NN                                                       STF00220
  10 B(I,J)=0.0                                                         STF00230
     DO 20 I=1,NPE                                                      STF00240
     J=I+1                                                              STF00250
     IF (J.GT.NPE) J=J-NPE                                              STF00260
     K=J+1                                                              STF00270
     IF (K.GT.NPE) K=K-NPE                                              STF00280
     ALPHA(I)=X(J)*Y(K)-X(K)*Y(J)                                       STF00290
     BETA(I)=Y(J)-Y(K)                                                  STF00300
  20 GAMA(I)=X(K)-X(J)                                                  STF00310
     DET=X(1)*(Y(2)-Y(3))+X(2)*(Y(3)-Y(1))+X(3)*(Y(1)-Y(2))             STF00320
     IF(ITYPE)100,40,40                                                 STF00330
C                                                                       STF00340
C    COMPUTE THE ELEMENT STIFFNESS MATRIX FOR PLANE ELASTICITY PROBLEM  STF00350
C                         K =CONST* BT  C  B                            STF00360
C                  ZERO BODY FORCES ARE ASSUMED: F(I)=0.0               STF00370
C                                                                       STF00380
  40 DO 60 I=1,NPE                                                      STF00390
     J=2*I-1                                                            STF00400
     B(1,J)=BETA(I)/DET                                                 STF00410
     B(3,J)=GAMA(I)/DET                                                 STF00420
     B(2,J+1)=GAMA(I)/DET                                               STF00430
  60 B(3,J+1)=BETA(I)/DET                                               STF00440
     DO 70 I=1,3                                                        STF00450
     DO 70 J=1,NN                                                       STF00460
     BT(J,I)=0.5*DET*B(I,J)                                             STF00470
     STR(I,J)=0.0                                                       STF00480
     DO 70 K=1,3                                                        STF00490
  70 STR(I,J)=STR(I,J)+C(I,K)*B(K,J)                                    STF00500
     DO 80 I=1,NN                                                       STF00510
     F(I)=0.0                                                           STF00520
     DO 80 J=1,NN                                                       STF00530
     ELSTIF(I,J)=0.0                                                    STF00540
     DO 80 K=1,3                                                        STF00550
  80 ELSTIF(I,J)=ELSTIF(I,J)+BT(I,K)*STR(K,J)                           STF00560
     RETURN                                                             STF00570
C                                                                       STF00580
C    COMPUTE THE COEFFICIENT MATRIX FOR HEAT TRANSFER TYPE PROBLEMS     STF00590
C                                                                       STF00600
 100 XBAR=(X(1)+X(2)+X(3))/3.0                                          STF00610
     YBAR=(Y(1)+Y(2)+Y(3))/3.0                                          STF00620
     A00=0.5*DET                                                        STF00630
     A01=A00*YBAR                                                       STF00640
     A10=A00*XBAR                                                       STF00650
     A11=A00*(X(1)*Y(1)+X(2)*Y(2)+X(3)*Y(3)+9.0*XBAR*YBAR)/12.0         STF00660
     A20=A00*(X(1)*X(1)+X(2)*X(2)+X(3)*X(3)+9.0*XBAR*XBAR)/12.0         STF00670
```

```
      A02=A00*(Y(1)*Y(1)+Y(2)*Y(2)+Y(3)*Y(3)+9.0*YBAR*YBAR)/12.0        STF00680
C     ONLY CONSTANT SOURCE TERM IS ASSUMED                              STF00690
      DO 120 I=1,NPE                                                    STF00700
      F(I)=0.5*Q*(ALPHA(I)+BETA(I)*XBAR+GAMA(I)*YBAR)                   STF00710
      DO 120 J=1,NPE                                                    STF00720
  120 ELSTIF(I,J)=(AK1*BETA(I)*BETA(J)+AK2*GAMA(I)*GAMA(J))/DET/2.0     STF00730
      IF(ITEM.EQ.0)RETURN                                               STF00740
C                                                                       STF00750
C     DEFINE THE MASS MATRIX FOR TRANSIENT PROBLEMS                     STF00760
C                                                                       STF00770
      DO 140 I=1,NPE                                                    STF00780
      DO 140 J=1,NPE                                                    STF00790
  140 B(I,J)=(ALPHA(I)*ALPHA(J)*A00+(ALPHA(I)*BETA(J)+ALPHA(J)*BETA(I))*STF00800
     1A10+(BETA(I)*GAMA(J)+GAMA(I)*BETA(J))*A11+(ALPHA(I)*GAMA(J)+GAMA(ISTF00810
     2)*ALPHA(J))*A01+BETA(I)*BETA(J)*A20+GAMA(I)*GAMA(J)*A02)/(DET*DET)STF00820
      DO 160 I=1,NPE                                                    STF00830
      SUM=0.0                                                           STF00840
      DO 150 J=1,NPE                                                    STF00850
      SUM=SUM+(B(I,J)-A0*ELSTIF(I,J))*W0(J)                             STF00860
  150 ELSTIF(I,J)=B(I,J)+A1*ELSTIF(I,J)                                 STF00870
  160 F(I)=(A0+A1)*F(I)+SUM                                             STF00880
      RETURN                                                            STF00890
      END                                                               STF00900
      SUBROUTINE STIFFQ(NPE,NN,IEL,ITYPE,ITEM,AK1,AK2,Q)                STQ00010
C                                                                       STQ00020
C     ..........................................................        STQ00030
C     STIFFNESS MATRIX FOR ISOPARAMETRIC QUADRILATERAL ELEMENTS         STQ00040
C     ..........................................................        STQ00050
C                                                                       STQ00060
      IMPLICIT REAL*8(A-H,O-Z)                                          STQ00070
      COMMON/STF/ELSTIF(18,18),ELXY(9,2),C(3,3),F(18),W0(18),A0,A1      STQ00080
      DIMENSION SF(9),GDSF(2,9),GAUSS(4,4),WT(4,4),SS(18,18),S(9,9)     STQ00090
     *          ,SXY(9,9),SX(9,9),SY(9,9)                               STQ00100
C                                                                       STQ00110
      DATA GAUSS/4*0.0D0,-.57735027D0,.57735027D0,2*0.0D0,-.77459667D0, STQ00120
     20.0D0,.77459667D0,0.0D0,-.86113631D0,                             STQ00130
     3-.33998104D0,.33998104D0,.86113631D0/                             STQ00140
C                                                                       STQ00150
      DATA WT/2.0D0,3*0.0D0,2*1.0D0,2*0.0D0,.55555555D0,.88888888D0,    STQ00160
     2.55555555D0,0.0D0,.34785485D0,2*.65214515D0,.34785485D0/          STQ00170
C                                                                       STQ00180
      NDF = NN/NPE                                                      STQ00190
      NGP = IEL+1                                                       STQ00200
C                                                                       STQ00210
C     INITIALIZE THE ARRAYS                                             STQ00220
C                                                                       STQ00230
      DO 20 I = 1, NPE                                                  STQ00240
      DO 20 J = 1, NPE                                                  STQ00250
      S(I,J)=0.0                                                        STQ00260
      SX(I,J) = 0.0                                                     STQ00270
      SY(I,J) = 0.0                                                     STQ00280
   20 SXY(I,J) = 0.0                                                    STQ00290
      DO 30 I = 1,NN                                                    STQ00300
      F(I) = 0.0                                                        STQ00310
      DO 30 J = 1,NN                                                    STQ00320
   30 SS(I,J) = 0.0                                                     STQ00330
C                                                                       STQ00340
C     DO-LOOPS ON NUMERICAL (GAUSS) QUADRATURE BEGIN HERE               STQ00350
C                                                                       STQ00360
      DO 100 NI = 1,NGP                                                 STQ00370
      DO 100 NJ = 1,NGP                                                 STQ00380
      XI = GAUSS(NI,NGP)                                                STQ00390
      ETA = GAUSS(NJ,NGP)                                               STQ00400
      CALL SHAPE(NPE,XI,ETA,SF,GDSF,DET,ELXY)                           STQ00410
      CONST = DET*WT(NI,NGP)*WT(NJ,NGP)                                 STQ00420
      DO 80 I=1,NPE                                                     STQ00430
      IF(ITYPE.GE.0)GOTO 60                                             STQ00440
C     CONSTANT SOURCE TERM IS ASSUMED ONLY FOR HEAT TRANSFER TYPE PROB. STQ00450
      F(I) = F(I) + Q*SF(I)*CONST                                       STQ00460
   60 DO 80 J=1,NPE                                                     STQ00470
      S(I,J) = S(I,J)+CONST*SF(I)*SF(J)                                 STQ00480
      SX(I,J)=SX(I,J)+CONST*GDSF(1,I)*GDSF(1,J)                         STQ00490
      SY(I,J)=SY(I,J)+CONST*GDSF(2,I)*GDSF(2,J)                         STQ00500
      SXY(I,J)=SXY(I,J)+CONST* GDSF(1,I)*GDSF(2,J)                      STQ00510
   80 CONTINUE                                                          STQ00520
```

```
  100 CONTINUE                                                            STQ00530
C                                                                         STQ00540
C     ELEMENT CACULATIONS FOR HEAT TRANSFER TYPE PROBLEMS                 STQ00550
C                                                                         STQ00560
      IF(ITYPE .GE. 0)GOTO 140                                            STQ00570
      DO 110 I=1,NPE                                                      STQ00580
      DO 110 J=1,NPE                                                      STQ00590
  110 ELSTIF(I,J)=AK1*SX(I,J)+AK2*SY(I,J)                                 STQ00600
      IF(ITEM.EQ.0)RETURN                                                 STQ00610
      DO 130 I=1,NPE                                                      STQ00620
      SUM=0.0                                                             STQ00630
      DO 120 J=1,NPE                                                      STQ00640
      SUM=SUM+(S(I,J)-A0*ELSTIF(I,J))*W0(J)                               STQ00650
  120 ELSTIF(I,J)=S(I,J)+A1*ELSTIF(I,J)                                   STQ00660
  130 F(I)=(A0+A1)*F(I)+SUM                                               STQ00670
      RETURN                                                              STQ00680
  140 IF(ITYPE.EQ.2)GOTO 230                                              STQ00690
C                                                                         STQ00700
C     ELEMENT CALCULATIONS FOR PLANE ELASTICITY PROBLEMS                  STQ00710
C             NO BODY FORCES ARE ACCOUNTED FOR                            STQ00720
C                                                                         STQ00730
      II = 1                                                              STQ00740
      DO 200 I=1,NPE                                                      STQ00750
      JJ = 1                                                              STQ00760
      DO 150 J=1,NPE                                                      STQ00770
      ELSTIF(II,JJ)=C(1,1)*SX(I,J)+C(3,3)*SY(I,J)                         STQ00780
      ELSTIF(II+1,JJ+1)=C(3,3)*SX(I,J)+C(2,2)*SY(I,J)                     STQ00790
      ELSTIF(II,JJ+1)=C(1,2)*SXY(I,J)+C(3,3)*SXY(J,I)                     STQ00800
      ELSTIF(II+1,JJ)=C(1,2)*SXY(J,I)+C(3,3)*SXY(I,J)                     STQ00810
  150 JJ = NDF*J+1                                                        STQ00820
  200 II = NDF*I+1                                                        STQ00830
      RETURN                                                              STQ00840
C                                                                         STQ00850
C     ELEMENT CALCULATIONS FOR INCOMPRESSIBLE FLUID FLOW                  STQ00860
C           USE FULL INTEGRATION ON VISCOUS TERMS                         STQ00870
C           NO BODY FORCES ARE ACCOUNTED FOR HERE                         STQ00880
C                                                                         STQ00890
  230 II = 1                                                              STQ00900
      DO 250 I=1,NPE                                                      STQ00910
      JJ = 1                                                              STQ00920
      DO 240 J =1,NPE                                                     STQ00930
      ELSTIF(II,JJ) = AK1*(2.0*SX(I,J)+SY(I,J))                           STQ00940
      ELSTIF(II+1,JJ) = AK1*SXY(I,J)                                      STQ00950
      ELSTIF(II,JJ+1) = AK1*SXY(J,I)                                      STQ00960
      ELSTIF(II+1,JJ+1) = AK1*(SX(I,J)+2.0*SY(I,J))                       STQ00970
      IF(ITEM.EQ.0)GOTO 240                                               STQ00980
      SS(II,JJ)=S(I,J)                                                    STQ00990
      SS(II+1,JJ+1)=S(I,J)                                                STQ01000
  240 JJ=NDF*J+1                                                          STQ01010
  250 II=NDF*I+1                                                          STQ01020
C                                                                         STQ01030
C     USE REDUCED INTEGRATION ON PENALTY TERMS                            STQ01040
C                                                                         STQ01050
      DO 300 NI=1,IEL                                                     STQ01060
      DO 300 NJ=1,IEL                                                     STQ01070
      XI = GAUSS(NI,IEL)                                                  STQ01080
      ETA = GAUSS(NJ,IEL)                                                 STQ01090
      CALL SHAPE(NPE,XI,ETA,SF,GDSF,DET,ELXY)                             STQ01100
      CONST=DET*WT(NI,IEL)*WT(NJ,IEL)                                     STQ01110
      II=1                                                                STQ01120
      DO 280 I=1,NPE                                                      STQ01130
      JJ = 1                                                              STQ01140
      DO 260 J=1,NPE                                                      STQ01150
      ELSTIF(II,JJ)=ELSTIF(II,JJ)+AK2*GDSF(1,I)*GDSF(1,J)*CONST           STQ01160
      ELSTIF(II+1,JJ)=ELSTIF(II+1,JJ)+AK2*GDSF(2,I)*GDSF(1,J)*CONST       STQ01170
      ELSTIF(II,JJ+1)=ELSTIF(II,JJ+1)+AK2*GDSF(1,I)*GDSF(2,J)*CONST       STQ01180
      ELSTIF(II+1,JJ+1)=ELSTIF(II+1,JJ+1)+AK2*GDSF(2,I)*GDSF(2,J)*CONST   STQ01190
  260 JJ=NDF*J+1                                                          STQ01200
  280 II=NDF*I+1                                                          STQ01210
  300 CONTINUE                                                            STQ01220
      IF(ITEM.EQ.0)RETURN                                                 STQ01230
      DO 320 I=1,NN                                                       STQ01240
      DO 320 J=1,NN                                                       STQ01250
      F(I)=F(I)+(SS(I,J)-A0*ELSTIF(I,J))*W0(J)                            STQ01260
  320 ELSTIF(I,J)=SS(I,J)+A1*ELSTIF(I,J)                                  STQ01270
```

```
      RETURN                                                            STQ01280
      END                                                               STQ01290
      SUBROUTINE SHAPE(NPE,XI,ETA,SF,GDSF,DET,ELXY)                     SHP00010
C                                                                       SHP00020
C     ..................................................................SHP00030
C     THE SUBROUTINE EVALUATES THE INTERPOLATION FUNCTIONS (SF(I)) AND  SHP00040
C     ITS DERIVATIVES WITH RESPECT TO NATURAL COORDINATES (DSF(I,J)),   SHP00050
C     AND THE DERIVATIVES OF SF(I) WITH RESPECT TO GLOBAL COORDINATES   SHP00060
C     FOR FOUR, EIGHT, AND NINE NODE RECTANGULAR ISOPARAMETRIC ELEMENTS.SHP00070
C                                                                       SHP00080
C     SF(I)........INTERPOLATION FUNCTION FOR NODE I OF THE ELEMENT     SHP00090
C     DSF(I,J).....DERIVATIVE OF SF(J) WITH RESPECT TO XI IF I=1 AND    SHP00100
C                  WITH RESPECT TO ETA IF I=2.                          SHP00110
C     GDSF(I,J)....DERIVATIVE OF SF(J) WITH RESPECT TO X IF I=1 AND     SHP00120
C                  WITH RESPECT TO Y IF I=2.                            SHP00130
C     XNODE(I,J)...J-TH (J=1,2) COORDINATE OF NODE I OF THE ELEMENT     SHP00140
C     NP(I)........ARRAY OF ELEMENT NODES (USED FOR DEFINING SF AND DSF)SHP00150
C     GJ(I,J)......JACOBIAN MATRIX                                      SHP00160
C     GIINV(I,J)...INVERSE OF THE JACOBIAN MATRIX                       SHP00170
C     ..................................................................SHP00180
C                                                                       SHP00190
      IMPLICIT REAL*8 (A-H,O-Z)                                         SHP00200
      DIMENSION ELXY(9,2),XNODE(9,2),NP(9),DSF(2,9),GJ(2,2),GJINV(2,2), SHP00210
     1          SF(9),GDSF(2,9)                                         SHP00220
      DATA XNODE/-1.0D0,2*1.0D0,-1.0D0,0.0D0,1.0D0,0.0D0,-1.0D0,0.0D0,  SHP00230
     2           2*-1.0D0,2*1.0D0,-1.0D0,0.0D0,1.0D0,2*0.0D0/           SHP00240
      DATA NP/1,2,3,4,5,7,6,8,9/                                        SHP00250
C                                                                       SHP00260
      FNC(A,B) = A*B                                                    SHP00270
      IF (NPE-8) 60,10,80                                               SHP00280
C                                                                       SHP00290
C     QUADRATIC INTERPOLATION FUNCTIONS (FOR THE EIGHT-NODE ELEMENT)    SHP00300
C                                                                       SHP00310
   10 DO 40 I = 1, NPE                                                  SHP00320
      NI = NP(I)                                                        SHP00330
      XP = XNODE(NI,1)                                                  SHP00340
      YP = XNODE(NI,2)                                                  SHP00350
      XI0 = 1.0+XI*XP                                                   SHP00360
      ETA0 = 1.0+ETA*YP                                                 SHP00370
      XI1 = 1.0-XI*XI                                                   SHP00380
      ETA1 = 1.0-ETA*ETA                                                SHP00390
      IF(I.GT.4) GO TO 20                                               SHP00400
      SF(NI) = 0.25*FNC(XI0,ETA0)*(XI*XP+ETA*YP-1.0)                    SHP00410
      DSF(1,NI) = 0.25*FNC(ETA0,XP)*(2.0*XI*XP+ETA*YP)                  SHP00420
      DSF(2,NI) = 0.25*FNC(XI0,YP)*(2.0*ETA*YP+XI*XP)                   SHP00430
      GO TO 40                                                          SHP00440
   20 IF(I.GT.6) GO TO 30                                               SHP00450
      SF(NI) = 0.5*FNC(XI1,ETA0)                                        SHP00460
      DSF(1,NI) = -FNC(XI,ETA0)                                         SHP00470
      DSF(2,NI) = 0.5*FNC(YP,XI1)                                       SHP00480
      GO TO 40                                                          SHP00490
   30 SF(NI) = 0.5*FNC(ETA1,XI0)                                        SHP00500
      DSF(1,NI) = 0.5*FNC(XP,ETA1)                                      SHP00510
      DSF(2,NI) = -FNC(ETA,XI0)                                         SHP00520
   40 CONTINUE                                                          SHP00530
      GO TO 130                                                         SHP00540
C                                                                       SHP00550
C     LINEAR INTERPOLATION FUNCTIONS (FOR FOUR-NODE ELEMENT)            SHP00560
C                                                                       SHP00570
   60 DO 70 I = 1, NPE                                                  SHP00580
      XP=XNODE(I,1)                                                     SHP00590
      YP=XNODE(I,2)                                                     SHP00600
      XI0=1.0+XI*XP                                                     SHP00610
      ETA0=1.0+ETA*YP                                                   SHP00620
      SF(I)=0.25*FNC(XI0,ETA0)                                          SHP00630
      DSF(1,I)=0.25*FNC(XP,ETA0)                                        SHP00640
   70 DSF(2,I)=0.25*FNC(YP,XI0)                                         SHP00650
      GO TO 130                                                         SHP00660
C                                                                       SHP00670
C     QUDRATIC INTERPOLATION FUNCTIONS (FOR THE NINE-NODE ELEMENT)      SHP00680
C                                                                       SHP00690
   80 DO 120 I=1,NPE                                                    SHP00700
      NI = NP(I)                                                        SHP00710
      XP = XNODE(NI,1)                                                  SHP00720
      YP = XNODE(NI,2)                                                  SHP00730
```

```
      XI0 = 1.0+XI*XP                                                   SHP00740
      ETA0 = 1.0+ETA*YP                                                 SHP00750
      XI1 = 1.0-XI*XI                                                   SHP00760
      ETA1 = 1.0-ETA*ETA                                                SHP00770
      XI2 = XP*XI                                                       SHP00780
      ETA2 = YP*ETA                                                     SHP00790
      IF(I .GT. 4)GOTO 90                                               SHP00800
      SF(NI) = 0.25*FNC(XI0,ETA0)*XI2*ETA2                              SHP00810
      DSF(1,NI)=0.25*XP*FNC(ETA2,ETA0)*(1.0+2.0*XI2)                    SHP00820
      DSF(2,NI)=0.25*YP*FNC(XI2,XI0)*(1.0+2.0*ETA2)                     SHP00830
      GO TO 120                                                         SHP00840
   90 IF(I .GT. 6)GO TO 100                                             SHP00850
      SF(NI) = 0.5*FNC(XI1,ETA0)*ETA2                                   SHP00860
      DSF(1,NI) = -XI*FNC(ETA2,ETA0)                                    SHP00870
      DSF(2,NI) = 0.5*FNC(XI1,YP)*(1.0+2.0*ETA2)                        SHP00880
      GO TO 120                                                         SHP00890
  100 IF(I .GT. 8)GO TO 110                                             SHP00900
      SF(NI) = 0.5*FNC(ETA1,XI0)*XI2                                    SHP00910
      DSF(2,NI) = -ETA*FNC(XI2,XI0)                                     SHP00920
      DSF(1,NI) = 0.5*FNC(ETA1,XP)*(1.0+2.0*XI2)                        SHP00930
      GO TO 120                                                         SHP00940
  110 SF(NI) = FNC(XI1,ETA1)                                            SHP00950
      DSF(1,NI) = -2.0*XI*ETA1                                          SHP00960
      DSF(2,NI) = -2.0*ETA*XI1                                          SHP00970
  120 CONTINUE                                                          SHP00980
  130 DO 140 I = 1,2                                                    SHP00990
      DO 140 J = 1,2                                                    SHP01000
      GJ(I,J) = 0.0                                                     SHP01010
      DO 140 K = 1,NPE                                                  SHP01020
  140 GJ(I,J) = GJ(I,J) + DSF(I,K)*ELXY(K,J)                            SHP01030
      DET = GJ(1,1)*GJ(2,2)-GJ(1,2)*GJ(2,1)                             SHP01040
      GJINV(1,1) = GJ(2,2)/DET                                          SHP01050
      GJINV(2,2) = GJ(1,1)/DET                                          SHP01060
      GJINV(1,2) = -GJ(1,2)/DET                                         SHP01070
      GJINV(2,1) = -GJ(2,1)/DET                                         SHP01080
      DO 150 I = 1,2                                                    SHP01090
      DO 150 J = 1,NPE                                                  SHP01100
      GDSF(I,J) = 0.0                                                   SHP01110
      DO 150 K = 1, 2                                                   SHP01120
  150 GDSF(I,J) = GDSF(I,J) + GJINV(I,K)*DSF(K,J)                       SHP01130
      RETURN                                                            SHP01140
      END                                                               SHP01150
      SUBROUTINE STRESS(N,NPE,ELXY,C,W,AK1,AK2,ITYPE,IEL)               STR00010
C                                                                       STR00020
C     ....................................................              STR00030
C     SUBROUTINE TO COMPUTE THE GRADIENT OF SOLUTION AND                STR00040
C     STRESSES FOR TRIANGULAR AND QUADRILATERAL ELEMENTS                STR00050
C     ....................................................              STR00060
C                                                                       STR00070
      IMPLICIT REAL*8 (A-H,O-Z)                                         STR00080
      DIMENSION SF(9),GDSF(2,9),C(3,3),X(3),Y(3),GAMA(3),BETA(3)        STR00090
     1          ,ELXY(9,2),W(2,9)                                       STR00100
      CONST=180.0/3.14159265                                            STR00110
      UX=0.0                                                            STR00120
      UY=0.0                                                            STR00130
      VX=0.0                                                            STR00140
      VY=0.0                                                            STR00150
      IF(IEL.GT.0)GOTO 100                                              STR00160
C     ...................................................               STR00170
C     STRESS/GRADIENT COMPUTATION FOR TRIANGULAR ELEMENTS               STR00180
C     ...................................................               STR00190
      DO 10 I=1,NPE                                                     STR00200
      X(I)=ELXY(I,1)                                                    STR00210
   10 Y(I)=ELXY(I,2)                                                    STR00220
      DO 20 I=1,NPE                                                     STR00230
      J=I+1                                                             STR00240
      IF(J.GT.NPE)J=J-NPE                                               STR00250
      K=J+1                                                             STR00260
      IF(K.GT.NPE)K=K-NPE                                               STR00270
      BETA(I)=Y(J)-Y(K)                                                 STR00280
   20 GAMA(I)=X(K)-X(J)                                                 STR00290
      DET=X(1)*(Y(2)-Y(3))+X(2)*(Y(3)-Y(1))+X(3)*(Y(1)-Y(2))            STR00300
      IF(ITYPE)30,70,70                                                 STR00310
```

```
C                                                                       STR00320
   30 DO 40 I=1,NPE                                                     STR00330
      UX=UX+W(1,I)*BETA(I)*AK1/DET                                      STR00340
   40 UY=UY+W(1,I)*GAMA(I)*AK2/DET                                      STR00350
      VALU=DSQRT(UX**2+UY**2)                                           STR00360
      IF(UX.EQ.0.0)GOTO 50                                              STR00370
      ANGLE=DATAN2(UY,UX)*CONST                                         STR00380
      GOTO 60                                                           STR00390
   50 IF(UY.LT.0.0)ANGLE=-90.0                                          STR00400
      IF(UY.GT.0.0)ANGLE=90.0                                           STR00410
   60 PRINT 4, N,UX,UY,VALU,ANGLE                                       STR00420
      RETURN                                                            STR00430
C                                                                       STR00440
C     CALCULATION OF STRESSES FOR PLANE ELASTICITY PROBLEMS             STR00450
C                                                                       STR00460
   70 DO 80 I=1,NPE                                                     STR00470
      UX=UX+BETA(I)*W(1,I)/DET                                          STR00480
      VY=VY+GAMA(I)*W(2,I)/DET                                          STR00490
      UY=UY+GAMA(I)*W(1,I)/DET                                          STR00500
   80 VX=VX+BETA(I)*W(2,I)/DET                                          STR00510
      SX=C(1,1)*UX+C(1,2)*UY                                            STR00520
      SY=C(1,2)*UX+C(2,2)*UY                                            STR00530
      SXY=C(3,3)*(UY+VX)                                                STR00540
      PRINT 4, N,SX,SY,SXY                                              STR00550
      RETURN                                                            STR00560
C     ..................................................                STR00570
C     CACULATION OF STRESSES/GRADIENT OF SOLUTION AT THE                STR00580
C             CENTER OF THE QUADRILATERAL ELEMENTS                      STR00590
C     ..................................................                STR00600
C                                                                       STR00610
  100 XI = 0.0                                                          STR00620
      ETA = 0.0                                                         STR00630
      CALL SHAPE(NPE,XI,ETA,SF,GDSF,DET,ELXY)                           STR00640
      DO 110 I=1,NPE                                                    STR00650
      UX=UX + W(1,I)*GDSF(1,I)                                          STR00660
      UY=UY + W(1,I)*GDSF(2,I)                                          STR00670
      VX=VX + W(2,I)*GDSF(1,I)                                          STR00680
  110 VY=VY + W(2,I)*GDSF(2,I)                                          STR00690
      IF(ITYPE.LT.0)GOTO 120                                            STR00700
      IF(ITYPE.EQ.2)GOTO 150                                            STR00710
C                                                                       STR00720
C     STRESS COMPUTATION FOR PLANE ELASTICITY PROBLEMS                  STR00730
C                                                                       STR00740
      SX=C(1,1)*UX+C(1,2)*VY                                            STR00750
      SY=C(1,2)*UX+C(2,2)*VY                                            STR00760
      SXY= C(3,3)*(UY+VX)                                               STR00770
      PRINT 4,N,SX,SY,SXY                                               STR00780
      RETURN                                                            STR00790
C                                                                       STR00800
C     COMPUTATION OF THE GRADIENT OF THE SOLUTION IN HEAT-CONDUTION     STR00810
C            TYPE PROBLEMS (ONE DEGREE-OF-FREEDOM PROBLEMS)             STR00820
C                                                                       STR00830
  120 SX=AK1*UX                                                         STR00840
      SY=AK2*UY                                                         STR00850
      VALU=DSQRT(SX**2+SY**2)                                           STR00860
      IF(SX.EQ.0.0)GOTO 130                                             STR00870
      ANGLE=DATAN2(SY,SX)*CONST                                         STR00880
      GOTO 140                                                          STR00890
  130 IF(SY.GT.0.0)ANGLE=90.0                                           STR00900
      IF(SY.LT.0.0)ANGLE=-90.0                                          STR00910
  140 PRINT 4,N,SX,SY,VALU,ANGLE                                        STR00920
      RETURN                                                            STR00930
C                                                                       STR00940
C     STRESS COMPUTATION FOR THE STOKES FLOW (PENALTY METHOD)           STR00950
C                                                                       STR00960
  150 P=-AK2*(UX+VY)                                                    STR00970
      SX=2.0*AK1*UX-P                                                   STR00980
      SY=2.0*AK1*VY-P                                                   STR00990
      SXY=AK1*(UY+VX)                                                   STR01000
      P= -AK2*(UX+VY)                                                   STR01010
      PRINT 4, N,SX,SY,SXY,P                                            STR01020
    4 FORMAT(5X,I5,4(4X,E13.6))                                         STR01030
      RETURN                                                            STR01040
```

```
      END                                                               STR01050
      SUBROUTINE MESH(IEL,NX,NY,NPE,NNM,NEM)                            MSH00010
C                                                                       MSH00020
C     ................................................................. MSH00030
C      THE SUBROUTINE GENERATES ARRAY NOD(I,J), COORDINATES X(I),Y(I),  MSH00040
C      AND MESH INFORMATION (NNM,NEM,NPE) FOR RECTANGULAR DOMAINS. THE  MSH00050
C      DOMAIN IS DIVIDED INTO LINEAR TRIANGULAR ELEMENTS OR QUADRILA-   MSH00060
C      TERAL ELEMENTS (NX BY NY NONUNIFORM MESH IN GENERAL).            MSH00070
C     ................................................................. MSH00080
C                                                                       MSH00090
      IMPLICIT REAL*8 (A-H,O-Z)                                         MSH00100
      COMMON/MSH/NOD(200,9),X(225),Y(225),DX(15),DY(15)                 MSH00110
      IF(IEL.GT.0)GOTO 100                                              MSH00120
C                                                                       MSH00130
C     MESH OF TRIANGULAR ELEMENTS                                       MSH00140
C                                                                       MSH00150
      NEM = 2*NX*NY                                                     MSH00160
      NX1=NX+1                                                          MSH00170
      NY1=NY+1                                                          MSH00180
      NXX1=2*NX                                                         MSH00190
      NYY1=2*NY                                                         MSH00200
      NNM=NX1*NY1                                                       MSH00210
      NOD(1,1)=1                                                        MSH00220
      NOD(1,2)=2                                                        MSH00230
      NOD(1,3)=NX1+2                                                    MSH00240
      NOD(2,1)=1                                                        MSH00250
      NOD(2,2)=NX1+2                                                    MSH00260
      NOD(2,3)=NX1+1                                                    MSH00270
      K=3                                                               MSH00280
      DO 60 IY=1,NY                                                     MSH00290
      L=IY*NXX1                                                         MSH00300
      M=(IY-1)*NXX1                                                     MSH00310
      IF(NX.EQ.1)GOTO 40                                                MSH00320
      DO 30 N=K,L,2                                                     MSH00330
      DO 20 I=1,NPE                                                     MSH00340
      NOD(N,I)=NOD(N-2,I)+1                                             MSH00350
   20 NOD(N+1,I)=NOD(N-1,I)+1                                           MSH00360
   30 CONTINUE                                                          MSH00370
   40 IF(NY.EQ.1)GOTO 60                                                MSH00380
      DO 50 I=1,NPE                                                     MSH00390
      NOD(L+1,I)=NOD(M+1,I)+NX1                                         MSH00400
   50 NOD(L+2,I)=NOD(M+2,I)+NX1                                         MSH00410
   60 K=L+3                                                             MSH00420
   70 L = 0                                                             MSH00430
      YC = 0.0                                                          MSH00440
      DO 90 J=1,NY1                                                     MSH00450
      XC = 0.0                                                          MSH00460
      DO 80 I=1,NX1                                                     MSH00470
      L = L + 1                                                         MSH00480
      X(L) = XC                                                         MSH00490
      Y(L) = YC                                                         MSH00500
   80 XC = XC + DX(I)                                                   MSH00510
   90 YC = YC + DY(J)                                                   MSH00520
      RETURN                                                            MSH00530
C                                                                       MSH00540
C     MESH OF QUADRILATERAL ELEMENTS WITH FOUR,EIGHT, OR NINE NODES     MSH00550
C                                                                       MSH00560
  100 NEX1=NX+1                                                         MSH00570
      NEY1=NY+1                                                         MSH00580
      NXX = IEL*NX                                                      MSH00590
      NYY = IEL*NY                                                      MSH00600
      NXX1 = NXX + 1                                                    MSH00610
      NYY1 = NYY + 1                                                    MSH00620
      NEM=NX*NY                                                         MSH00630
      NNM = NXX1*NYY1 - (IEL-1)*NX*NY                                   MSH00640
      IF (NPE .EQ. 9)NNM=NXX1*NYY1                                      MSH00650
      KO = 0                                                            MSH00660
      IF (NPE .EQ. 9) KO=1                                              MSH00670
C                                                                       MSH00680
C     GENERATE THE ARRAY NOD(I,J)                                       MSH00690
C                                                                       MSH00700
      NOD(1,1) = 1                                                      MSH00710
      NOD(1,2) = IEL+1                                                  MSH00720
      NOD(1,3) = NXX1+(IEL-1)*NEX1+IEL+1                                MSH00730
      IF (NPE .EQ. 9)NOD(1,3)=4*NX+5                                    MSH00740
```

```
      NOD(1,4) = NOD(1,3) - IEL                                        MSH00750
      IF(NPE .EQ. 4)GO TO 200                                          MSH00760
      NOD(1,5) = 2                                                     MSH00770
      NOD(1,6) = NXX1 + (NPE-6)                                        MSH00780
      NOD(1,7) = NOD(1,3) - 1                                          MSH00790
      NOD(1,8) = NXX1+1                                                MSH00800
      IF (NPE .EQ. 9)NOD(1,9)=NXX1+2                                   MSH00810
  200 IF(NY .EQ. 1)GOTO 230                                            MSH00820
      M = 1                                                            MSH00830
      DO 220 N = 2,NY                                                  MSH00840
      L = (N-1)*NX + 1                                                 MSH00850
      DO 210 I = 1,NPE                                                 MSH00860
  210 NOD(L,I) = NOD(M,I)+NXX1+(IEL-1)*NEX1+KO*NX                      MSH00870
  220 M=L                                                              MSH00880
  230 IF(NX .EQ .1)GO TO 270                                           MSH00890
      DO 260 NI = 2,NX                                                 MSH00900
      DO 240 I = 1,NPE                                                 MSH00910
      K1 = IEL                                                         MSH00920
      IF(I .EQ. 6 .OR. I .EQ. 8)K1=1+KO                                MSH00930
  240 NOD(NI,I) = NOD(NI-1,I)+K1                                       MSH00940
      M = NI                                                           MSH00950
      DO 260 NJ = 2,NY                                                 MSH00960
      L = (NJ-1)*NX+NI                                                 MSH00970
      DO 250 J = 1,NPE                                                 MSH00980
  250 NOD(L,J) = NOD(M,J)+NXX1+(IEL-1)*NEX1+KO*NX                      MSH00990
  260 M = L                                                            MSH01000
C                                                                      MSH01010
C     GENERATE THE COORDINATES X(I) AND Y(I)                           MSH01020
C                                                                      MSH01030
  270 YC=0.0                                                           MSH01040
      IF (NPE .EQ. 9) GOTO 310                                         MSH01050
      DO 300 NI = 1, NEY1                                              MSH01060
      I = (NXX1+(IEL-1)*NEX1)*(NI-1)+1                                 MSH01070
      J = (NI-1)*IEL+1                                                 MSH01080
      X(I) = 0.0                                                       MSH01090
      Y(I) = YC                                                        MSH01100
      DO 280 NJ = 1,NXX                                                MSH01110
      I=I+1                                                            MSH01120
      X(I) = X(I-1)+DX(NJ)                                             MSH01130
  280 Y(I) = YC                                                        MSH01140
      IF(NI.GT.NY .OR. IEL.EQ.1)GO TO 300                              MSH01150
      J = J+1                                                          MSH01160
      YC = YC+DY(J-1)                                                  MSH01170
      I = I+1                                                          MSH01180
      X(I) = 0.0                                                       MSH01190
      Y(I) = YC                                                        MSH01200
      DO 290 II = 1, NX                                                MSH01210
      K = 2*II-1                                                       MSH01220
      I = I+1                                                          MSH01230
      X(I) = X(I-1)+DX(K)+DX(K+1)                                      MSH01240
  290 Y(I) = YC                                                        MSH01250
  300 YC = YC+DY(J)                                                    MSH01260
      RETURN                                                           MSH01270
C                                                                      MSH01280
  310 DO 330 NI=1,NYY1                                                 MSH01290
      I=NXX1*(NI-1)                                                    MSH01300
      XC=0.0                                                           MSH01310
      DO 320 NJ=1,NXX1                                                 MSH01320
      I = I+1                                                          MSH01330
      X(I) = XC                                                        MSH01340
      Y(I) = YC                                                        MSH01350
  320 XC = XC + DX(NJ)                                                 MSH01360
  330 YC = YC + DY(NI)                                                 MSH01370
      RETURN                                                           MSH01380
      END                                                              MSH01390
```

APPENDIX

III

COMPUTER PROGRAM PLATE

```
C             C O M P U T E R   P R O G R A M   P L A T E                PLT00010
C    (STATIC AND TRANSIENT ANALYSIS OF BENDING OF ORTHOTROPIC PLATES)    PLT00020
C                                                                        PLT00030
C    ................................................................    PLT00040
C    .                                                              .    PLT00050
C    .      D E S C R I P T I O  O F  T H E  V A R I A B L E S      .    PLT00060
C    .                                                              .    PLT00070
C    .     A1,...,A4.PARAMETERS IN THE TIME-APPROXIMATION SCHEME    .    PLT00080
C    .     AK........SHEAR CORRECTION COEFFICIENT                   .    PLT00090
C    .     ALFA......PARAMETER IN THE NEWMARK SCHEME                .    PLT00100
C    .     BETA......PARAMETER IN THE NEWMARK SCHEME                .    PLT00110
C    .     C.........COEFFICIENT OF THE TIME DERIVATIVE TERM        .    PLT00120
C    .     D(I,J)....MATRIX OF MATERIAL COEFFICIENTS                .    PLT00130
C    .     D44,D55...MATERIAL COEFFICIENTS (SHEAR)                  .    PLT00140
C    .     DT........TIME INCREMENT IN THE TRANSIENT ANALYSIS       .    PLT00150
C    .     E1,E2.....MODULI ALONG X AND Y DIRECTIONS OF THE PLATE   .    PLT00160
C    .     ELP(I)....ELEMENT FORCE VECTOR                           .    PLT00170
C    .     ELXY(I,J) J-TH COORDINATE OF ELEMENT NODE I (J=1,2)      .    PLT00180
C    .     GF(I).....GLOBAL FORCE VECTOR; SOLUTION VECTOR FROM 'SOLVE' . PLT00190
C    .     GF0(I)....SOLUTION VECTOR AT CURRENT TIME                .    PLT00200
C    .     GF1(I)....FIRST TIME DERIVATIVE OF THE SOLUTION          .    PLT00210
C    .     GF2(I)....SECOND TIME DERIVATIVE OF THE SOLUTION         .    PLT00220
C    .     GSTIF.....GLOBAL STIFNESS MATRIX (IN BANDED FORM)        .    PLT00230
C    .     H.........THICKNESS OF THE PLATE                         .    PLT00240
C    .     IBDY(I)...ARRAY OF SPECIFIED GLOBAL DISPLACEMENTS        .    PLT00250
C    .     IBSF(I)...ARRAY OF SPECIFIED NONZERO GLOBAL FORCES       .    PLT00260
C    .                                                              .    PLT00270
C    .     IEL.......INDICATOR FOR THE ELEMENT TYPE:                .    PLT00280
C    .                  IEL=1, 4-NODE ELEMENT                       .    PLT00290
C    .                  IEL=2, 8- OR 9-NDE ELEMENT                  .    PLT00300
C    .                                                              .    PLT00310
C    .     ITEM......INDICATOR FOR TRANSIENT ANALYSIS (1, YES; 0,NO)  .  PLT00320
C    .     NCMAX.....VALUE OF THE COLUMN-DIMENSION OF GSTIF         .    PLT00330
C    .     NRMAX.....VALUE OF THE ROW-DIMENSION OF GSTIF            .    PLT00340
C    .     NOZERO....INDICATOR FOR ZERO(NOZERO=0) OR NONZERO(NOZERO=1) . PLT00350
C    .                   INITIAL CONDITIONS FOR TRANSIENT ANALYSIS  .    PLT00360
C    .     NSTP......TIME STEP AT WHICH THE LOAD IS REMOVED FROM THE  .  PLT00370
C    .               PLATE (IN THE TRANSIENT ANALYSIS)              .    PLT00380
C    .     NTIME.....NUMBER OF TIME STEPS IN THE TRANSIENT ANALYSIS  .   PLT00390
C    .     NOD(I,J)..CONNECTIVITY MATRIX                            .    PLT00400
C    .     NBDY......TOTAL NUMBER OF SPECIFIED DEGREES OF FREEDOM   .    PLT00410
C    .     NBSF......TOTAL NUMBER OF SPECIFIED NONZERO FORCES       .    PLT00420
```

```
C   .     PO.......INTENSITY OF THE DISTRIBUTED OR POINT LOAD         .  PLT00430
C   .     STIF.....ELEMENT STIFFNESS MATRIX                           .  PLT00440
C   .     T........TIME VARIABLE IN THE TRANSIENT ANALYSIS            .  PLT00450
C   .     VBDY.....VALUES OF THE DISPLACEMENTS IN THE ARRAY IBDY      .  PLT00460
C   .     VBSF.....VALUES OF THE SPECIFIED FORCES IN THE ARRAY IBSF   .  PLT00470
C   .     W0,W1,W2  ARRAYS CORRESPONDING TO GF0,GF1,GF2 IN AN ELEMENT .  PLT00480
C   .     X,Y......ARRAYS OF X AND Y-COORDINATES OF GLOBAL NODES      .  PLT00490
C   .                                                                 .  PLT00500
C   .                                                                 .  PLT00510
C   ...................................................................  PLT00520
C                                                                        PLT00530
      IMPLICIT REAL*8(A-H,O-Z)                                           PLT00530
      DIMENSION GSTIF(243,63),GF(243),GF0(243),GF1(243),GF2(243),        PLT00540
     1          VBDY(85),IBDY(85),VBSF(20),IBSF(25),TITLE(20)            PLT00550
      COMMON/STF/ELXY(9,2),STIF(45,45),ELP(45),W0(45),W1(45),W2(45),     PLT00560
     1             D(3,3),A0,A1,A2,A3,A4,D44,D55,C1,C2                   PLT00570
      COMMON/MSH/NOD(200,9),X(225),Y(225),DX(15),DY(15)                  PLT00580
C                                                                        PLT00590
      DATA NDF,NRMAX,NCMAX/3,243,63/                                     PLT00600
C                                                                        PLT00610
C     ..................................................................PLT00620
C     .               P R E P R O C E S S O R   U N I T                 .PLT00630
C     ..................................................................PLT00640
C                                                                        PLT00650
      READ 260, TITLE                                                    PLT00660
      READ 270, IEL,NPE,IMESH,NPRNT,ITEM,NTIME,NSTP,NOZERO               PLT00670
      IF(IMESH.EQ.1)GOTO 20                                              PLT00680
      READ 270, NEM,NNM                                                  PLT00690
      DO 10 I=1,NEM                                                      PLT00700
   10 READ 270, (NOD(I,J),J=1,NPE)                                       PLT00710
      READ 280, (X(I),Y(I),I=1,NNM)                                      PLT00720
      GOTO 30                                                            PLT00730
   20 READ 270, NX,NY                                                    PLT00740
      NX1=IEL*NX+1                                                       PLT00750
      NY1=IEL*NY+1                                                       PLT00760
      READ 280, (DX(I),I=1,NX1)                                          PLT00770
      READ 280, (DY(I),I=1,NY1)                                          PLT00780
      CALL MESH(IEL,NX,NY,NPE,NNM,NEM)                                   PLT00790
   30 READ 290, E1,E2,G12,G13,G23,ANU12,RHO,H                            PLT00800
      READ 280, PO                                                       PLT00810
      READ 270, NBDY                                                     PLT00820
      READ 270, (IBDY(I),I=1,NBDY)                                       PLT00830
      READ 280, (VBDY(I),I=1,NBDY)                                       PLT00840
      READ 270, NBSF                                                     PLT00850
      IF(NBSF.EQ.0)GOTO 35                                               PLT00860
      READ 270, (IBSF(I),I=1,NBSF)                                       PLT00870
      READ 280, (VBSF(I),I=1,NBSF)                                       PLT00880
   35 IF(ITEM.EQ.0)GOTO 40                                               PLT00890
      READ 280, DT,ALFA                                                  PLT00900
      IF(NOZERO.EQ.0)GOTO 36                                             PLT00910
      READ 280, (GF0(I),I=1,NEQ)                                         PLT00920
      READ 280, (GF1(I),I=1,NEQ)                                         PLT00930
   36 BETA=0.25*(0.5+ALFA)**2                                            PLT00940
      DT2=DT*DT                                                          PLT00950
      A0=1.0/BETA/DT2                                                    PLT00960
      A2=1.0/BETA/DT                                                     PLT00970
      A1=ALFA*A2                                                         PLT00980
      A3=0.5/BETA-1.0                                                    PLT00990
      A4=ALFA/BETA-1.0                                                   PLT01000
      IF(NOZERO.EQ.1)GOTO 40                                             PLT01010
      DO 38 I=1,NEQ                                                      PLT01020
      GF0(I)=0.0                                                         PLT01030
      GF1(I)=0.0                                                         PLT01040
   38 GF2(I)=0.0                                                         PLT01050
C                                                                        PLT01060
C     E N D   O F   D A T A I N P U T   T O   T H E   P R O G R A M      PLT01070
C                                                                        PLT01080
C     ..................................................................PLT01090
C     .                 P R O C E S S O R   U N I T                     .PLT01100
C     ..................................................................PLT01110
C                                                                        PLT01120
   40 NEQ=NNM*NDF                                                        PLT01130
      NN=NPE*NDF                                                         PLT01140
      AK=5.0/6.0                                                         PLT01150
      ANU21=ANU12*E2/E1                                                  PLT01160
      DENOM=(1.0-ANU12*ANU21)                                            PLT01170
```

```
      TOP=(H**3)/12.0                                                   PLT01180
      C1=RHO*H                                                          PLT01190
      C2=RHO*TOP                                                        PLT01200
      D(1,1)=E1*TOP/DENOM                                               PLT01210
      D(1,2)=ANU12*E2*TOP/DENOM                                         PLT01220
      D(1,3)=0.0                                                        PLT01230
      D(2,2)=D(1,1)*E2/E1                                               PLT01240
      D(2,3)=0.0                                                        PLT01250
      D(3,3)=G12*TOP                                                    PLT01260
      D44=G13*H*AK                                                      PLT01270
      D55=G23*H*AK                                                      PLT01280
      DO 50 I=1,3                                                       PLT01290
      DO 50 J=I,3                                                       PLT01300
   50 D(J,I)=D(I,J)                                                     PLT01310
C                                                                       PLT01320
C     PRINT THE DATA INPUT AND THE MESH INFORMATION                     PLT01330
C                                                                       PLT01340
      PRINT 260, TITLE                                                  PLT01350
      PRINT 310, IEL,NPE                                                PLT01360
      PRINT 320, NEM,NNM,NDF                                            PLT01370
      PRINT 330, E1,E2,G12,G13,G23,ANU12,AK,RHO,H,PO                    PLT01380
      PRINT 340, NBDY                                                   PLT01390
      PRINT 270, (IBDY(I),I=1,NBDY)                                     PLT01400
      PRINT 300, (VBDY(I),I=1,NBDY)                                     PLT01410
      PRINT 350, NBSF                                                   PLT01420
      PRINT 270, (IBSF(I),I=1,NBSF)                                     PLT01430
      PRINT 300, (VBSF(I),I=1,NBSF)                                     PLT01440
      PRINT 360                                                         PLT01450
      DO 60 I=1,NEM                                                     PLT01460
   60 PRINT 270, I,(NOD(I,J),J=1,NPE)                                   PLT01470
      PRINT 370                                                         PLT01480
      PRINT 300, (X(I),Y(I),I=1,NNM)                                    PLT01490
C                                                                       PLT01500
C     COMPUTE THE HALF BAND WIDTH                                       PLT01510
C                                                                       PLT01520
      NHBW=0                                                            PLT01530
      DO 70 N=1,NEM                                                     PLT01540
      DO 70 I=1,NPE                                                     PLT01550
      DO 70 J=1,NPE                                                     PLT01560
      NW=(IABS(NOD(N,I)-NOD(N,J))+1)*NDF                                PLT01570
   70 IF (NHBW.LT.NW) NHBW=NW                                           PLT01580
      PRINT 400, NHBW                                                   PLT01590
C                                                                       PLT01600
C     DO-LOOP ON NUMBER OF TIME STEPS BEGINS HERE                       PLT01610
C                                                                       PLT01620
      T = 0.0                                                           PLT01630
      IF(ITEM.EQ.1)PRINT 460, DT,ALFA,BETA, A0,A1,A2,A3,A4              PLT01640
      DO 220 NT=1,NTIME                                                 PLT01650
      IF(ITEM.EQ.1 .AND. NT.GE.NSTP)PO=0.0                              PLT01660
C                                                                       PLT01670
C     INITIALIZE THE GLOBAL STIFFNESS MATRIX AND FORCE VECTOR           PLT01680
C                                                                       PLT01690
      DO 80 I=1,NEQ                                                     PLT01700
      GF(I)=0.0                                                         PLT01710
      DO 80 J=1,NHBW                                                    PLT01720
   80 GSTIF(I,J)=0.0                                                    PLT01730
      DO 130 N=1,NEM                                                    PLT01740
      L=0                                                               PLT01750
      DO 90 I=1,NPE                                                     PLT01760
      NI=NOD(N,I)                                                       PLT01770
      ELXY(I,1)=X(NI)                                                   PLT01780
      ELXY(I,2)=Y(NI)                                                   PLT01790
      LI=(NI-1)*NDF                                                     PLT01800
      DO 90 J=1,NDF                                                     PLT01810
      LI=LI+1                                                           PLT01820
      L=L+1                                                             PLT01830
      W0(L)=GF0(LI)                                                     PLT01840
      W1(L)=GF1(LI)                                                     PLT01850
   90 W2(L)=GF2(LI)                                                     PLT01860
      CALL STIFF(IEL,NPE,NN,PO,ITEM,NT,NOZERO)                          PLT01870
      IF (NPRNT.EQ.0) GO TO 110                                         PLT01880
      IF (N.GT.1) GO TO 110                                             PLT01890
      PRINT 380                                                         PLT01900
      DO 100 I=1,NN                                                     PLT01910
  100 PRINT 300, (STIF(I,J),J=1,NN)                                     PLT01920
```

```
      PRINT 410                                                          PLT01930
      PRINT 300, (ELP(I),I=1,NN)                                         PLT01940
      PRINT 410                                                          PLT01950
  110 CONTINUE                                                           PLT01960
C                                                                        PLT01970
C     ASSEMBLE ELEMENT STIFFNESS MATRICES TO GET GLOBAL STIFFNESS MATRIXPLT01980
C                                                                        PLT01990
      DO 130 I=1,NPE                                                     PLT02000
      NR=(NOD(N,I)-1)*NDF                                                PLT02010
      DO 130 II=1,NDF                                                    PLT02020
      NR=NR+1                                                            PLT02030
      L=(I-1)*NDF+II                                                     PLT02040
      GF(NR) = GF(NR) + ELP(L)                                           PLT02050
      DO 130 J=1,NPE                                                     PLT02060
      NCL=(NOD(N,J)-1)*NDF                                               PLT02070
      DO 130 JJ=1,NDF                                                    PLT02080
      M=(J-1)*NDF+JJ                                                     PLT02090
      NC=NCL+JJ-NR+1                                                     PLT02100
      IF (NC) 130,130,120                                                PLT02110
  120 GSTIF(NR,NC)=GSTIF(NR,NC)+STIF(L,M)                                PLT02120
  130 CONTINUE                                                           PLT02130
C                                                                        PLT02140
C     THE GLOBAL SYSTEM EQUATIONS ARE NOW READY FOR IMPLEMENTING THE     PLT02150
C           THE FORCE AND DISPLACEMENT BOUNDARY CONDITIONS               PLT02160
C                                                                        PLT02170
      IF(NBSF.EQ.0)GOTO 145                                              PLT02180
      IF(NOZERO.EQ.1 .AND. ITEM.EQ.1)GOTO 145                            PLT02190
      DO 140 I=1,NBSF                                                    PLT02200
      NB=IBSF(I)                                                         PLT02210
  140 GF(NB)=GF(NB)+VBSF(I)                                              PLT02220
  145 CALL BNDY(NRMAX,NCMAX,NEQ,NHBW,GSTIF,GF,NBDY,IBDY,VBDY)            PLT02230
C                                                                        PLT02240
C     CALL SUBROUTINE 'SOLVE' TO SOLVE THE SYSTEM OF EQUATIONS           PLT02250
C                   THE SOLUTION IS RETURNED IN GF                       PLT02260
C                                                                        PLT02270
      CALL SOLVE (NRMAX,NCMAX,NEQ,NHBW,GSTIF,GF,0)                       PLT02280
      IF(ITEM.EQ.0)GOTO 180                                              PLT02290
      IF(NOZERO.EQ.0)GOTO 160                                            PLT02300
      IF(NT.GT.1)GOTO 160                                                PLT02310
C                                                                        PLT02320
C     CALCULATE THE SECOND TIME DERIVATIVE WHEN INITIAL CONDITIONS       PLT02330
C                        ARE NON ZERO                                    PLT02340
C                                                                        PLT02350
      DO 150 I=1,NEQ                                                     PLT02360
  150 GF2(I)=GF(I)                                                       PLT02370
      GOTO 220                                                           PLT02380
C                                                                        PLT02390
C     CALCULATE NEW VELOCITIES AND ACCELERATIONS                         PLT02400
C                                                                        PLT02410
  160 T=T+DT                                                             PLT02420
      DO 170 I=1,NEQ                                                     PLT02430
      GF0(I)=A0*(GF(I)-GF0(I))-A2*GF1(I)-A3*GF2(I)                       PLT02440
      GF1(I)=GF1(I)+DT*(1.0-ALFA)*GF2(I)+DT*ALFA*GF0(I)                  PLT02450
      GF2(I)=GF0(I)                                                      PLT02460
  170 GF0(I)=GF(I)                                                       PLT02470
      PRINT 470, T                                                       PLT02480
  180 PRINT 420                                                          PLT02490
      PRINT 300,(GF(I),I=1,NEQ,NDF)                                      PLT02500
      PRINT 430                                                          PLT02510
      PRINT 300,(GF(I),I=2,NEQ,NDF)                                      PLT02520
      PRINT 440                                                          PLT02530
      PRINT 300,(GF(I),I=3,NEQ,NDF)                                      PLT02540
C                                                                        PLT02550
C     ..................................................................PLT02560
C     .                 P O S T P R O C E S S O R   U N I T            .PLT02570
C     ..................................................................PLT02580
C                                                                        PLT02590
C     COMPUTE BENDING STRESSES (AT THE GAUSS POINTS)                     PLT02600
C                                                                        PLT02610
      PRINT 450                                                          PLT02620
      DO 200 N=1,NEM                                                     PLT02630
      L=0                                                                PLT02640
      DO 190 I=1,NPE                                                     PLT02650
      NI=NOD(N,I)                                                        PLT02660
      ELXY(I,1)=X(NI)                                                    PLT02670
```

```
      ELXY(I,2)=Y(NI)                                                   PLT02680
      LI=(NI-1)*NDF                                                     PLT02690
      DO 190 J=1,NDF                                                    PLT02700
      LI=LI+1                                                           PLT02710
      L=L+1                                                             PLT02720
  190 WO(L)=GF(LI)                                                      PLT02730
  200 CALL STRESS (NPE,NDF,IEL,ELXY,WO,D,D44,D55,H)                     PLT02740
      IF(ITEM.EQ.0)GOTO 230                                             PLT02750
  220 PRINT 390                                                         PLT02760
  230 STOP                                                              PLT02770
C                                                                       PLT02780
C                      F   O   R   M   A   T   S                        PLT02790
C                                                                       PLT02800
  260 FORMAT (20A4)                                                     PLT02810
  270 FORMAT (16I5)                                                     PLT02820
  280 FORMAT (8F10.4)                                                   PLT02830
  290 FORMAT (8E10.3)                                                   PLT02840
  300 FORMAT (8(2X,E12.5))                                              PLT02850
  310 FORMAT (/,5X,'ELEMENT TYPE(1=LINEAR,2=QUADRATIC) =',I2,5X,' NODES PLT02860
     1PER ELEMENT=',I2)                                                 PLT02870
  320 FORMAT (10X,'ACTUAL NUMBER OF ELEMENTS IN THE MESH =',I3,/,10X,'NUPLT02880
     1MBER OF NODES IN THE MESH =',I3,/,10X,'DEGREES OF FREEDOM =',I2,/)PLT02890
  330 FORMAT(5X,'M A T E R I A L   P R O P E R T I E S:',/,10X,'MODULUS, PLT02900
     1E1=',E12.5,/,10X,'MODULUS, E2=',E12.5,/,10X,'SHEAR MODULI, G12,G13PLT02910
     2 AND G23=',3E12.5,/,10X,'POISSONS RATIO, NU12=',E12.5,/,10X,'SHEARPLT02920
     3 CORRECTION COEFFICIENT, K=',E12.5,/,10X,'MATERIAL DENSITY, RHO=',PLT02930
     4E12.5,/,10X,'PLATE THICKNESS, H=',E12.5,/,10X,'LOAD MAGNITUDE, P='PLT02940
     5 ,E12.5,/)                                                        PLT02950
  340 FORMAT (/,5X,'NUMBER OF SPECIFIED DISPLACEMENTS=',I5,/,5X,'SPECIFIPLT02960
     1ED DISPLACEMENTS AND THEIR VALUES FOLLOW:')                       PLT02970
  350 FORMAT (/,5X,'NUMBER OF SPECIFIED FORCES=',I4,/,5X,'SPECIFIED FORCPLT02980
     1E DEGREES OF FREEDOM FOLLOW AND THEIR SPECIFIED VALUES FOLLOW:')  PLT02990
  360 FORMAT (/,5X,'BOOLEAN (CONNECTIVITY) MATRIX-NOD(I,J) ',/)         PLT03000
  370 FORMAT (/,5X,'COORDINATES OF THE GLOBAL NODES:',/)                PLT03010
  380 FORMAT (/,5X,'ELEMENT STIFFNESS AND FORCE MATRICES:',/)           PLT03020
  390 FORMAT (120('.'),//)                                              PLT03030
  400 FORMAT (/,5X,'HALF BAND WIDTH OF GLOBAL STIFFNESS MATRIX = ',I5,/)PLT03040
  410 FORMAT (//)                                                       PLT03050
  420 FORMAT (/,5X,'TRANSVERSE DEFLECTION:',/)                          PLT03060
  430 FORMAT (/,5X,'BENDING SLOPE, SI-X:',/)                            PLT03070
  440 FORMAT (/,5X,'BENDING SLOPE, SI-Y:',/)                            PLT03080
  450 FORMAT(8X,'X-COORD',5X,'Y-COORD',5X,'SIGMAX ',5X,'SIGMAY ',5X,    PLT03090
     1'SIGMAXY',5X,'SIGMAXZ',5X,'SIGMAYZ')                              PLT03100
  460 FORMAT(/,5X,'DT=',E10.4,5X,'ALFA=',E10.4,5X,'BETA=',E10.4,/,10X,  PLT03110
     1'TEMPORAL PARAMETERS A0,A1,ETC.:',5E12.4,/)                       PLT03120
  470 FORMAT (/,5X,'TIME =',E10.3,/)                                    PLT03130
      END                                                               PLT03140
      SUBROUTINE STIFF(IEL,NPE,NN,PO,ITEM,NT,NOZERO)                    STF00010
C                                                                       STF00020
C     ..................................................................STF00030
C     THE PROGRAM IS WRITTEN FOR ORTHOTROPIC PLATES.  THE ELEMENT IS    STF00040
C     BASED ON A SHEAR-DEFORMABLE THEORY.  HERE THE FOUR-, EIGHT- OR    STF00050
C     NINE-NODE ISOPARAMETRIC ELEMENT WITH THREE DEGREES OF FREEDOM     STF00060
C     (W,SX,SY) PER NODE CAN BE USED BY SPECIFYING THE ELEMENT TYPE.    STF00070
C     ..................................................................STF00080
C                                                                       STF00090
      IMPLICIT REAL*8(A-H,O-Z)                                          STF00100
      COMMON/STF/ELXY(9,2),STIF(45,45),ELP(45),WO(45),W1(45),W2(45),    STF00110
     1           D(3,3),A0,A1,A2,A3,A4,D44,D55,C1,C2                    STF00120
      COMMON/SHP/SF(9),GDSF(2,9)                                        STF00130
      DIMENSION GAUSS(4,4),WT(4,4),H(45,45)                             STF00140
C                                                                       STF00150
      DATA GAUSS/4*0.0D0,-.57735027D0,.57735027D0,2*0.0D0,-.77459667D0,0STF00160
     1.0D0,.77459667D0,0.0D0,-.86113631D0,-.33998104D0,.33998104D0,.8611STF00170
     23631D0/                                                           STF00180
C                                                                       STF00190
      DATA WT/2.0D0,3*0.0D0,2*1.0D0,2*0.0D0,.55555555D0,.88888888D0,.555STF00200
     155555D0,0.0D0,.34785485D0,2*.65214515D0,.34785485D0/              STF00210
C                                                                       STF00220
      NGP=IEL+1                                                         STF00230
      LGP=IEL                                                           STF00240
      NDF=NN/NPE                                                        STF00250
C                                                                       STF00260
C     INITIALIZE THE ELEMENT MATRICES AND FORCE VECTOR                  STF00270
C                                                                       STF00280
```

```
      DO 10 I=1,NN                                                      STF00290
      ELP(I)=0.0                                                        STF00300
      DO 10 J=1,NN                                                      STF00310
      H(I,J) = 0.0                                                      STF00320
   10 STIF(I,J)=0.0                                                     STF00330
C                                                                       STF00340
C     GAUSS QUADRATURE (FULL INTEGRATION) ON BENDING TERMS BEGINS HERE  STF00350
C                                                                       STF00360
      DO 80 NI=1,NGP                                                    STF00370
      DO 80 NJ=1,NGP                                                    STF00380
      XI=GAUSS(NI,NGP)                                                  STF00390
      ETA=GAUSS(NJ,NGP)                                                 STF00400
      CALL SHAPE (NPE,XI,ETA,ELXY,DET)                                  STF00410
      CNST=DET*WT(NI,NGP)*WT(NJ,NGP)                                    STF00420
      DO 30 I=1,NPE                                                     STF00430
      L=(I-1)*NDF+1                                                     STF00440
      ELP(L)=ELP(L)+CNST*SF(I)*PO                                       STF00450
   30 CONTINUE                                                          STF00460
C                                                                       STF00470
C     DEFINE STIFFNESS MATRIX 'STIF' AND MASS MATRIX 'H' COEFFICIENTS   STF00480
C                                                                       STF00490
      II=1                                                              STF00500
      DO 70 I=1,NPE                                                     STF00510
      JJ=1                                                              STF00520
      DO 60 J=1,NPE                                                     STF00530
       STIF(II+1,JJ+1)=STIF(II+1,JJ+1)+(D(1,1)*GDSF(1,I)*GDSF(1,J)      STF00540
     1                           +D(3,3)*GDSF(2,I)*GDSF(2,J))*CNST      STF00550
       STIF(II+1,JJ+2)=STIF(II+1,JJ+2)+(D(1,2)*GDSF(1,I)*GDSF(2,J)      STF00560
     1                           +D(3,3)*GDSF(2,I)*GDSF(1,J))*CNST      STF00570
       STIF(II+2,JJ+1)=STIF(II+2,JJ+1)+(D(1,2)*GDSF(2,I)*GDSF(1,J)      STF00580
     1                           +D(3,3)*GDSF(1,I)*GDSF(2,J))*CNST      STF00590
       STIF(II+2,JJ+2)=STIF(II+2,JJ+2)+(D(3,3)*GDSF(1,I)*GDSF(1,J)      STF00600
     1                           +D(2,2)*GDSF(2,I)*GDSF(2,J))*CNST      STF00610
       IF(ITEM.EQ.0)GOTO 60                                             STF00620
       H(II,JJ)=H(II,JJ)           + C1*SF(I)*SF(J)*CNST                STF00630
       H(II+1,JJ+1)=H(II+1,JJ+1) + C2*SF(I)*SF(J)*CNST                  STF00640
       H(II+2,JJ+2)=H(II+2,JJ+2) + C2*SF(I)*SF(J)*CNST                  STF00650
   60 JJ=NDF*J+1                                                        STF00660
   70 II=NDF*I+1                                                        STF00670
   80 CONTINUE                                                          STF00680
C                                                                       STF00690
C     GAUSS QUADRATURE (REDUCED INTEGRATION) ON SHEAR TERMS BEGINS HERE STF00700
C                                                                       STF00710
      DO 110 NI=1,LGP                                                   STF00720
      DO 110 NJ=1,LGP                                                   STF00730
      XI=GAUSS(NI,LGP)                                                  STF00740
      ETA=GAUSS(NJ,LGP)                                                 STF00750
      CALL SHAPE (NPE,XI,ETA,ELXY,DET)                                  STF00760
      CNST=DET*WT(NI,LGP)*WT(NJ,LGP)                                    STF00770
      II=1                                                              STF00780
      DO 100 I=1,NPE                                                    STF00790
      JJ=1                                                              STF00800
      DO 90 J=1,NPE                                                     STF00810
       STIF(II,JJ)  = STIF(II,JJ)+(D44*GDSF(1,I)*GDSF(1,J)+             STF00820
     1                        D55*GDSF(2,I)*GDSF(2,J))*CNST             STF00830
       STIF(II,JJ+1)=STIF(II,JJ+1)+D44*GDSF(1,I)*SF(J)*CNST             STF00840
       STIF(II+1,JJ)=STIF(II+1,JJ)+D44*SF(I)*GDSF(1,J)*CNST             STF00850
       STIF(II,JJ+2)=STIF(II,JJ+2)+D55*GDSF(2,I)*SF(J)*CNST             STF00860
       STIF(II+2,JJ)=STIF(II+2,JJ)+D55*SF(I)*GDSF(2,J)*CNST             STF00870
       STIF(II+1,JJ+1)=STIF(II+1,JJ+1)+D44*SF(I)*SF(J)*CNST             STF00880
       STIF(II+2,JJ+2)=STIF(II+2,JJ+2)+D55*SF(I)*SF(J)*CNST             STF00890
   90 JJ=NDF*J+1                                                        STF00900
  100 II=NDF*I+1                                                        STF00910
  110 CONTINUE                                                          STF00920
      IF(ITEM.EQ.0)RETURN                                               STF00930
C                                                                       STF00940
C     ELEMENT CALCULATIONS FOR TRANSIENT ANALYSIS BEGIN HERE            STF00950
C                                                                       STF00960
      IF(NOZERO.EQ.0 .OR. NT.GT.1)GOTO 130                              STF00970
      DO 120 I=1,NN                                                     STF00980
      ELP(I)=0.0                                                        STF00990
      DO 120 J=1,NN                                                     STF01000
      ELP(I)=ELP(I)-STIF(I,J)*WO(J)                                     STF01010
  120 STIF(I,J)=H(I,J)                                                  STF01020
      RETURN                                                            STF01030
```

```
  130 DO 140 I=1,NN                                                     STF01040
      DO 140 J=1,NN                                                     STF01050
      ELP(I)=ELP(I)+H(I,J)*(A0*W0(J)+A2*W1(J)+A3*W2(J))                 STF01060
  140 STIF(I,J)=STIF(I,J)+A0*H(I,J)                                     STF01070
      RETURN                                                            STF01080
      END                                                               STF01090
      SUBROUTINE STRESS (NPE,NDF,IEL,ELXY,W,D,D44,D55,H)                STR00010
C                                                                       STR00020
C     ................................................................. STR00030
C     THE PROGRAM EVALUATES THE BENDING STRESSES AT THE GAUSS POINTS    STR00040
C     USING THE REDUCED INTEGRATION.  THE STRESSES ARE AT THE TOP (OR   STR00050
C     NEGATIVE OF THOSE AT THE BOTTOM) OF THE PLATE.                    STR00060
C     ................................................................. STR00070
C                                                                       STR00080
      IMPLICIT REAL*8 (A-H,O-Z)                                         STR00090
      COMMON/SHP/SF(9),GDSF(2,9)                                        STR00100
      DIMENSION GAUSS(4,4),ELXY(9,2),W(45),D(3,3)                       STR00110
      DATA GAUSS/4*0.0D0,-.57735027D0,.57735027D0,2*0.0D0,-.77459667D0,0STR00120
     1.0D0,.77459667D0,0.0D0,-.86113631D0,-.33998104D0,.33998104D0,.8611STR00130
     23631D0/                                                           STR00140
C                                                                       STR00150
      NGP=IEL                                                           STR00160
      DO 40 NI=1,NGP                                                    STR00170
      DO 40 NJ=1,NGP                                                    STR00180
      XI=GAUSS(NI,NGP)                                                  STR00190
      ETA=GAUSS(NJ,NGP)                                                 STR00200
      CALL SHAPE (NPE,XI,ETA,ELXY,DET)                                  STR00210
      SIX=0.0                                                           STR00220
      SIY=0.0                                                           STR00230
      DWX=0.0                                                           STR00240
      DWY=0.0                                                           STR00250
      DSXY=0.0                                                          STR00260
      DSYX=0.0                                                          STR00270
      DSXX=0.0                                                          STR00280
      DSYY=0.0                                                          STR00290
      X=0.0                                                             STR00300
      Y=0.0                                                             STR00310
      DO 20 I=1,NPE                                                     STR00320
      L=(I-1)*NDF+1                                                     STR00330
      X=X+SF(I)*ELXY(I,1)                                               STR00340
      Y=Y+SF(I)*ELXY(I,2)                                               STR00350
   10 DWX=DWX+GDSF(1,I)*W(L)                                            STR00360
      DWY=DWY+GDSF(2,I)*W(L)                                            STR00370
      SIX=SIX+SF(I)*W(L+1)                                              STR00380
      SIY=SIY+SF(I)*W(L+2)                                              STR00390
      DSXX=DSXX+GDSF(1,I)*W(L+1)                                        STR00400
      DSXY=DSXY+GDSF(2,I)*W(L+1)                                        STR00410
      DSYX=DSYX+GDSF(1,I)*W(L+2)                                        STR00420
   20 DSYY=DSYY+GDSF(2,I)*W(L+2)                                        STR00430
      SGMAX=(D(1,1)*DSXX+D(1,2)*DSYY)*6.0/(H**2)                        STR00440
      SGMAY=(D(1,2)*DSXX+D(2,2)*DSYY)*6.0/(H**2)                        STR00450
      SGMXY=D(3,3)*(DSXY+DSYX)*6.0/(H**2)                               STR00460
      SGMXZ=D44*(DWX+SIX)/H                                             STR00470
      SGMYZ=D55*(DWY+SIY)/H                                             STR00480
      PRINT 50, X,Y,SGMAX,SGMAY,SGMXY,SGMXZ,SGMYZ                       STR00490
   40 CONTINUE                                                          STR00500
      RETURN                                                            STR00510
   50 FORMAT (5X,10E12.4)                                               STR00520
      END                                                               STR00530
```

```
.....................................................................
. SUBROUTINES 'BNDY' AND 'SOLVE' ARE THE SAME AS IN PROGRAM FEM1D.
. SUBROUTINES 'MESH' AND 'SHAPE' ARE THE SAME AS IN PROGRAM FEM2D.
.....................................................................
```

ANSWERS TO SELECTED PROBLEMS

Chapter 2

The weak forms of Probs. 2.1 and 2.3 are given in the form $B(v, u) = l(v)$, where v is the test function and u is the dependent variable. Whenever B is symmetric, the functional is given by Eq. (2.33).

2.1 $$B(v,u) = \int_0^1 a\frac{dv}{dx}\frac{du}{dx}\,dx + hv(1)u(1) \qquad \text{(symmetric)}$$

$$l(v) = -\int_0^1 vf\,dx + v(1)(q + hu_\infty)$$

2.3 $$B(v,u) = \int_0^1 u\frac{dv}{dx}\frac{du}{dx}\,dx \qquad (\textit{not}\ \text{symmetric})$$

$$l(v) = -\int_0^1 vf\,dx$$

2.5 $$0 = \int_0^L \left\{ a\frac{d\xi}{dx}\left[\frac{du}{dx} + \frac{1}{2}\left(\frac{dv}{dx}\right)^2\right] + \xi P \right\} dx$$

$$0 = \int_0^L \left\{ b\frac{d^2\zeta}{dx^2}\frac{d^2v}{dx^2} + a\frac{d\zeta}{dx}\frac{dv}{dx}\left[\frac{du}{dx} + \frac{1}{2}\left(\frac{dv}{dx}\right)^2\right] + \zeta f \right\} dx - \frac{d\zeta}{dx}(L)M_0$$

$$I(u,v) = \int_0^L \left\{ \frac{1}{2}\left[a\left(\frac{du}{dx}\right)^2 + b\left(\frac{d^2v}{dx^2}\right)^2 + a\frac{du}{dx}\left(\frac{dv}{dx}\right)^2 + \frac{a}{4}\left(\frac{dv}{dx}\right)^4\right] + uP + vf \right\} dx - \frac{dv}{dx}(L)M_0$$

2.7 $$0 = \int_\Omega \left[\xi \left(u \frac{\partial u}{\partial x} + v \frac{\partial u}{\partial y} \right) - \frac{1}{\rho} \frac{\partial \xi}{\partial x} P + \nu \left(\frac{\partial \xi}{\partial x} \frac{\partial u}{\partial x} + \frac{\partial \xi}{\partial y} \frac{\partial u}{\partial y} \right) \right] dx\, dy - \int_{\Gamma_2} \xi \hat{t}_x \, ds$$

$$0 = \int_\Omega \left[\zeta \left(u \frac{\partial v}{\partial x} + v \frac{\partial v}{\partial y} \right) - \frac{1}{\rho} \frac{\partial \zeta}{\partial y} P + \nu \left(\frac{\partial \zeta}{\partial x} \frac{\partial v}{\partial x} + \frac{\partial \zeta}{\partial y} \frac{\partial v}{\partial y} \right) \right] dx\, dy - \int_{\Gamma_2} \zeta \hat{t}_y \, ds$$

$$0 = \int_\Omega \lambda \left(\frac{\partial u}{\partial x} + \frac{\partial v}{\partial y} \right) dx\, dy$$

where ξ, ζ, and λ are test functions and $\hat{t}_x$ and $\hat{t}_y$ are the specified values of the secondary variables.

2.9 $$0 = \int_0^1 \left(a \frac{\partial v}{\partial x} \frac{\partial u}{\partial x} + b\, v \frac{\partial u}{\partial t} - vf \right) dx - v(1) T_0$$
subject to the conditions $u(0, t) = 0$ and $u(x, 0) = u_0$.

2.11 $$0 = \int_\Omega \left(k \nabla v \cdot \nabla u + \rho v \frac{\partial u}{\partial t} - vf \right) dx\, dy - \int_{\Gamma_2} v \hat{t} \, ds$$
subject to the conditions $u = 0$ on Γ_1 for any t, and $u = u_0$ for $t = 0$ and any (x, y) in Ω.

2.12 $$b_{ij} = \frac{ij}{i+j-1} - \frac{ij+i+j}{i+j} + \frac{1-ij}{i+j+1} + \frac{(i+1)(j+1)}{i+j+2}$$

$$l_i = \frac{1}{(1+i)(2+i)} \qquad c_1 = \frac{55}{131} \qquad c_2 = -\frac{20}{131}$$

2.13 $$\phi_0 = \sin \frac{\pi x}{2} \qquad \phi_1 = \sin \pi x \qquad \phi_2 = \sin 2\pi x$$

2.15 (*a*) $$w = \sum_{i=1}^{N} c_i x^i (L - x)$$

$$b_{ij} = EIijL^{i+j-1} \left[\frac{(i-1)(j-1)}{i+j-3} - \frac{2(ij-1)}{i+j-2} + \frac{(i+1)(j+1)}{i+j-1} \right]$$

To compute l_i, use mathematical tables to evaluate the integral

$$\int x^m \sin ax \, dx = -\frac{1}{a} x^m \cos ax + \frac{m}{a} \int x^{m-1} \cos ax \, dx$$

(*b*) $$w = \sum_{i=1}^{N} c_i \sin \frac{i \pi x}{L}$$

$$b_{ii} = EI(i)^4 \left(\frac{\pi}{L} \right)^4 \frac{L}{2} \qquad b_{ij} = 0 \qquad \text{for } i \neq j$$

$$l_1 = -\frac{L}{2} f_0 \qquad l_i = 0 \qquad i \neq 1$$

$$c_1 = -\frac{f_0 L^4}{EI \pi^4} \qquad c_i = 0 \text{ if } i \neq 1$$

The Ritz solution coincides with the exact solution,

$$w_0 = -\frac{f_0 L^4}{EI \pi^4} \sin \frac{\pi x}{L}$$

2.17 For the eigenvalue problem we set $u(0) = u(1) = 0$.

$$u = \sum_{i=1}^{N} c_i \phi_i \qquad \phi_i = x^i(1 - x)$$

$$N = 2: \frac{1}{60}\begin{bmatrix} 30 & 17 \\ 17 & 14 \end{bmatrix}\begin{Bmatrix} c_1 \\ c_2 \end{Bmatrix} = \frac{\lambda}{420}\begin{bmatrix} 14 & 7 \\ 7 & 4 \end{bmatrix}\begin{Bmatrix} c_1 \\ c_2 \end{Bmatrix}$$

$$\lambda_1 = 14.42 \qquad \mathbf{e}_1 = (1.000, -0.4474)$$

$$\lambda_2 = 63.58 \qquad \mathbf{e}_2 = (-0.4794, 1.000)$$

2.19 $u = c_1 x^2 \qquad v = x + d_1(1 - 2x + x^2)$

$$\int_0^1 \left(\frac{d^2 v}{dx^2} + g\right)(1 - 2x + x^2)\, dx = 0 \qquad d_1 = -\frac{g}{2}$$

$$\int_0^1 \left(\frac{d^2 u}{dx^2} - \frac{v}{a} - f\right) x^2\, dx = 0 \qquad c_1 = \frac{3}{8a} - \frac{g}{40a} + \frac{f}{2}$$

The exact solutions are

$$u_0 = \left(\frac{f}{2} + \frac{g}{4a}\right)x^2 - \left(\frac{1+g}{a}\right)x^3 + \frac{gx^4}{24a}$$

$$v_0 = -\frac{g}{2} + (1 + g)x - \frac{g}{2}x^2$$

2.20 (a) $\phi_0 = 1 - x \qquad \phi_1 = x(1 - x) \qquad c_1^2 + c_1 - 2 = 0$ which gives $(c_1)_1 = 1$ and $(c_1)_2 = -2$. Choose $c_1 = 1$ on the basis of the criterion that $\int_0^1 E\, dx$ is a minimum.

2.21 $u_1 = c_{11} \sin \pi x \sin \pi y \qquad c_{11} = \dfrac{8}{\pi^4}$

2.23 $w_N = \displaystyle\sum_{i=1}^{N}\sum_{j=1}^{N} c_{ij} \sin\frac{i\pi x}{a} \sin\frac{j\pi y}{a}$

$$\sum_{i=1}^{N}\sum_{j=1}^{N}\left[\frac{D\pi^4}{a^4}(i^2 + j^2)^2 - N_{cr}\pi^2\left(\frac{j}{a}\right)^2\right] c_{ij} = 0$$

$$N_{cr} = \frac{4\pi^2 D}{a^2}$$

2.25 $u_3 = c_1(x - x^2) + c_2(x^2 - x^3) + c_3(x^3 - x^4)$

Solving the eigenvalue problem on a calculator, we obtain

$$\lambda_1 = 9.968 \qquad \lambda_2 = 32.00 \qquad \lambda_3 = 51.37$$

The exact solution is $\lambda_n = (n\pi)^2$.

2.26 (a) See the text by Reddy and Rasmussen (1982), pp. 406–412.

2.27 (a, b) $\phi_0 = 0$, $\phi_1 = 2(1 + x) - x^2$, so that the boundary conditions are satisfied.

$$c_1 = \frac{35}{108} e^{-(20/108)t}[2(1 + x) - x^2]$$

2.29 $\phi_0 = 0 \qquad \phi_1 = x(1 - x) \qquad \phi_2 = x^2(1 - x)$

This choice gives the same solution as that in Prob. 2.28:

$$c_1 = \cos 10.95445t \qquad c_2 = 0 \qquad u(x, t) = (x - x^2)\cos 10.95445t$$

Chapter 3

3.1 $\psi_1(\bar{x}) = 1 - \dfrac{\bar{x}}{h_e} \qquad \psi_2(\bar{x}) = \dfrac{\bar{x}}{h_e}$ (origin at node 1)

3.2 $\psi_1(\bar{x}) = 1 - 3\dfrac{\bar{x}}{h_e} + 2\left(\dfrac{\bar{x}}{h_e}\right)^2 \qquad \psi_2(\bar{x}) = 4\dfrac{\bar{x}}{h_e}\left(1 - \dfrac{\bar{x}}{h_e}\right)$

$\psi_3(\bar{x}) = -\dfrac{\bar{x}}{h_e}\left(1 - 2\dfrac{\bar{x}}{h_e}\right)$ (origin at node 1, and node 2 is in the middle)

3.3 $K_{ij}^e = \displaystyle\int_{x_A}^{x_B}\left(a\frac{d\psi_i}{dx}\frac{d\psi_j}{dx} + b\psi_i\frac{d\psi_j}{dx} + c\psi_i\psi_j\right)dx$

$F_i^e = \displaystyle\int_{x_A}^{x_B} f\psi_i\,dx + P_i$ where x_A and x_B are endpoints of the element

3.4 For $L = 1$, $a = 1$, $f = x$, and $P = 0$ the following element coefficient matrices, column vectors, and solution are obtained:

Quadratic elements (2):

$$[K^{(1)}] = [K^{(2)}] = \frac{1}{3}\begin{bmatrix} 14 & -16 & 2 \\ -16 & 32 & -16 \\ 2 & -16 & 14 \end{bmatrix} \qquad \{f^{(1)}\} = \begin{Bmatrix} 0 \\ \frac{1}{12} \\ \frac{1}{24} \end{Bmatrix}$$

$U_1 = 0.0,\ \ U_2 = 0.1224,\ \ U_3 = 0.22917,\ \ U_4 = 0.30469,\ \ U_5 = 0.3333.$

Linear elements (4): $e = 1, 2, 3, 4$

$$[K^{(e)}] = 4\begin{bmatrix} 1.0 & -1.0 \\ -1.0 & 1.0 \end{bmatrix} \qquad \{F^{(1)}\} = \frac{1}{96}\begin{Bmatrix} 1 \\ 2 \end{Bmatrix}$$

$U_1 = 0.0,\ \ U_2 = 0.1224,\ \ U_3 = 0.22917,\ \ U_4 = 0.30469,\ U_5 = 0.3333.$

3.5 $[K^{(1)}] = \dfrac{1}{18}\begin{bmatrix} 52 & -55 \\ -55 & 52 \end{bmatrix}$

(*a*) $U_1 = 0.0,\ \ U_2 = -0.029992,\ \ U_3 = -0.04257,\ \ U_4 = 0.0;\ \ -P_1^{(1)} = 0.08998.$

(*b*) $U_1 = 0.0,\ \ U_2 = 0.4134,\ \ U_3 = 0.79584,\ \ U_4 = 1.142;\ \ -P_1^{(1)} = -1.240.$

(*c*) $U_1 = 0.3451,\ \ U_2 = 0.0,\ \ U_3 = 0.4382,\ \ U_4 = 0.8791;\ \ -P_1^{(1)} = 1.035.$

3.6 $u_1^{(1)} = U_1 = 0,\ \ u_2^{(1)} = u_1^{(2)} = u_1^{(3)} = U_2,\ \ u_2^{(2)} = U_3 = 0,\ \ u_2^{(3)} = U_4 = 0.$

$P_2^{(1)} + P_1^{(2)} + P_1^{(3)} = 100$ kips

$$[K^{(e)}] = \frac{A_e E_e}{h_e}\begin{bmatrix} 1 & -1 \\ -1 & 1 \end{bmatrix}$$

where A_e = area of cross-section, h_e = length, E_e = Young's modulus.

Assembled matrix:

$$[K] = \begin{bmatrix} K_{11}^{(1)} & K_{12}^{(1)} & 0 & 0 \\ K_{21}^{(1)} & K_{22}^{(1)} + K_{11}^{(2)} + K_{11}^{(3)} & K_{12}^{(2)} & K_{12}^{(3)} \\ 0 & K_{21}^{(2)} & K_{22}^{(2)} & 0 \\ 0 & K_{21}^{(3)} & 0 & K_{22}^{(3)} \end{bmatrix}$$

Solution:

$$\left(K_{22}^{(1)} + K_{11}^{(2)} + K_{11}^{(3)}\right)U_2 = 100, \quad P_1^{(1)} = K_{12}^{(1)}U_2, \quad P_2^{(2)} = K_{21}^{(2)}U_2, \quad P_2^{(3)} = K_{21}^{(3)}U_2.$$

3.8 The equations of a typical element are

$$\frac{1}{R_e}\begin{bmatrix} 1 & -1 \\ -1 & 1 \end{bmatrix}\begin{Bmatrix} P_1^{(e)} \\ P_2^{(e)} \end{Bmatrix} = \begin{Bmatrix} Q_1^{(e)} \\ Q_2^{(e)} \end{Bmatrix}$$

Condensed set of equations:

$$\frac{1}{12a}\begin{bmatrix} 8 & -6 & 0 \\ -6 & 16 & -10 \\ 0 & -10 & 16 \end{bmatrix}\begin{Bmatrix} P_1 \\ P_2 \\ P_3 \end{Bmatrix} = \begin{Bmatrix} Q \\ 0 \\ 0 \end{Bmatrix}$$

Solution:

$$P_1 = \frac{39}{14}Qa \qquad P_2 = \frac{1}{7}Qa \qquad P_3 = \frac{20}{7}Qa$$

3.10 $U_1 = 50°\text{C}$, $U_2 = 27.58°\text{C}$, $U_3 = 5.159°\text{C}$; $P_1^{(1)} = 4.484\ \text{watts/cm}^2$.

3.12 See Example 3.2.

3.14 $U_1 = 0.0$, $U_2 = 15\alpha$, $U_3 = 46.67\alpha$, $U_4 = 550\alpha$, $\alpha = (\pi \times 10^6)/4$.

3.15 $K_{ij}^{(e)} = \int_{r_A}^{r_B} kr\frac{d\psi_i}{dr}\frac{d\psi_j}{dr}\,dr + [\beta r\psi_i\psi_j]_{r_A}^{r_B}$

$$F_i^{(e)} = \int_{r_A}^{r_B} rf\psi_i\,dr + [r\psi_i\beta(u_\infty - q)]_{r_A}^{r_B}$$

3.16 $U_1 = 45.322$, $U_2 = 47.232$, $U_3 = 47.869$, $U_4 = 48.505$, $U_5 = 49.142$, $U_6 = 49.663$, $U_7 = 50.0$. The exact solution is given by

$$u(r) = \frac{Q}{2\pi k}\log\left(\frac{r}{r_0}\right) + u_0 \qquad (Q = 150, k = 25, u_0 = 50, r_0 = L = 200)$$

3.17 $\Phi_1 = 1.3862$, $\Phi_2 = 1.119$, $\Phi_3 = 0.8108$, $\Phi_4 = 0.44595$, $\Phi_5 = 0.0$. Note that $P_1^{(1)} = 2u_0$. The exact solution is $(A_0 = 1)$

$$\phi_0 = -2u_0 L\log\left(2 - \frac{x}{L}\right)$$

3.19 The perimeter, $c = 2\sqrt{(0.125)^2 + (3)^2}/3$ per unit width (perpendicular to the plane of the paper), and the area is $A = (0.25x/3)\ \text{in}^2$.

$$T_1(\text{tip}) = 166.188°\text{F} \qquad T_2 = 191.1°\text{F} \qquad T_3 = 218.87°\text{F}$$

3.21 $K_{ij}^{11} = 0 \qquad K_{ij}^{12} = \int_{x_A}^{x_B}\frac{d\psi_i}{dx}\frac{d\psi_j}{dx}\,dx \qquad F_i^1 = \int_{x_A}^{x_B} f\psi_i\,dx + Q_1\psi_i(x_A) + Q_2\psi_i(x_B)$

$$K_{ij}^{21} = K_{ji}^{12} \qquad K_{ij}^{22} = \frac{1}{EI}\int_{x_A}^{x_B}\psi_i\psi_j\,dx \qquad F_i^2 = P_1\psi_i(x_A) + P_2\psi_i(x_B)$$

$$Q_1 = -\left.\frac{dM}{dx}\right|_{x=x_A} \qquad Q_2 = \left.\frac{dM}{dx}\right|_{x=x_B}$$

$$P_1 = -\left.\frac{dw}{dx}\right|_{x=x_A} \qquad P_2 = \left.\frac{dw}{dx}\right|_{x=x_B}$$

3.22 $w_1 = 0 \qquad w_2 = -\frac{27}{12}\frac{f_0h^4}{EI} \qquad w_3 = -\frac{19}{6}\frac{f_0h^4}{EI} \qquad M_1 = 0 \qquad M_2 = \frac{3}{2}f_0h^2 \qquad M_3 = 2f_0h^2$

3.23 $\lambda_1 = 10.387 \qquad \lambda_2 = 48.0 \qquad \lambda_3 = 126.756$. Also see Example 5.2.

3.25 $\lambda_1 = 4.49 \qquad \lambda_2 = 36.648$

3.26 $[M^{(e)}]\{\dot{u}\} + [K^{(e)}]\{u\} = \{F^{(e)}\}$ where $M_{ij}^{(e)} = \int_{x_e}^{x_{e+1}} \psi_i \psi_j \, dx$

and other matrices and vectors are defined in Eq. (3.73).

3.27 $$\frac{h}{6}\begin{bmatrix} 2 & 1 & 0 \\ 1 & 4 & 1 \\ 0 & 1 & 2 \end{bmatrix}\begin{Bmatrix} \dot{U}_1 \\ \dot{U}_2 \\ \dot{U}_3 \end{Bmatrix} + \frac{1}{h}\begin{bmatrix} 1 & -1 & 0 \\ -1 & 2 & -1 \\ 0 & -1 & 1 \end{bmatrix}\begin{Bmatrix} U_1 \\ U_2 \\ U_3 \end{Bmatrix} = \begin{Bmatrix} P_1^{(1)} \\ P_2^{(1)} + P_1^{(2)} \\ P_2^{(2)} \end{Bmatrix}$$

$$P_2^{(1)} + P_1^{(2)} = 0, \quad U_1 = 0, \quad P_2^{(2)} \equiv \left(a\frac{du}{dx}\right)\bigg|_{x=L} = -\beta(U_3 - u_\infty), \quad h = \frac{L}{2}.$$

The condensed equations become

$$\frac{h}{6}\begin{bmatrix} 4 & 1 \\ 1 & 2 \end{bmatrix}\begin{Bmatrix} \dot{U}_2 \\ \dot{U}_3 \end{Bmatrix} + \frac{1}{h}\begin{bmatrix} 2 & -1 \\ -1 & 1+\beta h \end{bmatrix}\begin{Bmatrix} U_2 \\ U_3 \end{Bmatrix} = \begin{Bmatrix} 0 \\ \beta u_\infty \end{Bmatrix}$$

Using the Laplace transforms, one can obtain the solution of these equations. For example, $U_2(t)$ is given by

$$U_2(t) = -\frac{5}{28}\left[\frac{d}{ab} + \frac{a^2 + ca + d}{a(a-b)}e^{-at} + \frac{b^2 + cb + d}{b(b-a)}e^{-bt}\right]$$

where

$$c = \frac{24}{5h^2}[10 + \beta h(5 - u_\infty)] \qquad d = -\frac{144}{5h^3}$$

and a and b are the roots of the equation

$$x^2 + \frac{12}{7h^2}(5 + 2\beta h)x + \frac{36}{7h^4}(1 + 2\beta h) = 0$$

3.28 *3-element (uniform) mesh:* $U_1 = -0.5f_0$, $U_2 = -0.4444f_0$, $U_3 = -0.2777f_0$, $U_4 = 0$.
4-element (uniform) mesh: $U_1 = -0.5f_0$, $U_2 = -0.46875f_0$, $U_3 = -0.375f_0$, $U_4 = -0.2187f_0$, $U_5 = 0.0$.

3.29 $U_1 = 0.0$, $U_2 = 1.84$, $U_3 = 2.015$, $U_4 = 1.76$, $U_5 = 1.0$.

3.31 $T_1 = 400.0°\text{F}$, $T_2 = 270.27°\text{F}$, $T_3 = 166.49°\text{F}$, $T_4 = 80.0°\text{F}$.

3.33 $$\sum_{j=1}^{n} K_{ij}\sigma_j + f_i = 0 \qquad \sum_{j=1}^{n} K_{ji}u_j - \sum_{j=1}^{n} G_{ij}\sigma_j = 0$$

where

$$K_{ij} = \int_{x_e}^{x_{e+1}} \frac{d\psi_i}{dx}\psi_j \, dx \qquad G_{ij} = \int_{x_e}^{x_{e+1}} \frac{1}{a}\psi_i\psi_j \, dx \qquad F_i = \int_{x_e}^{x_{e+1}} f\psi_i \, dx + P_i$$

$$P_1^{(e)} = -\sigma(x_e) \qquad P_2^{(e)} = \sigma(x_{e+1})$$

3.34 See Sec. 5.2.

3.35 (*a*) $$[K] = \frac{2EI}{h^3}\begin{bmatrix} 6 & -3h & -6 & -3h & 0 & 0 \\ & 2h^2 & 3h & h^2 & 0 & 0 \\ & & 12 & 0 & -6 & -3h \\ & \text{symm.} & & 4h^2 & 3h & h^2 \\ & & & & 6 & 3h \\ & & & & & 2h^2 \end{bmatrix}$$

$$\{F\} = -\frac{f_0 h}{12}\begin{Bmatrix} 6 \\ -h \\ 6 \\ h \\ 0 \\ 0 \end{Bmatrix} + \begin{Bmatrix} Q_1^{(1)} \\ Q_2^{(1)} \\ Q_3^{(1)} + Q_1^{(2)} \\ Q_4^{(1)} + Q_2^{(2)} \\ Q_3^{(2)} \\ Q_4^{(2)} \end{Bmatrix}$$

(*b*) $U_1 = U_2 = U_5 = U_6 = 0$

$$Q_3^{(1)} + Q_1^{(2)} = 0 \qquad Q_4^{(1)} + Q_2^{(2)} = 0$$

(*c*) $\dfrac{2EI}{h^3}\begin{bmatrix} 12 & 0 \\ 0 & 4h^2 \end{bmatrix}\begin{Bmatrix} U_3 \\ U_4 \end{Bmatrix} = -\dfrac{f_0 h}{12}\begin{Bmatrix} 6 \\ h \end{Bmatrix}$

or

$$U_3 = -\frac{f_0 h^4}{48EI} \qquad U_4 = -\frac{f_0 h^3}{96EI}$$

$$Q_1^{(1)} = \frac{f_0 h}{2} - (6U_3 + 3hU_4)\frac{2EI}{h^3} = \frac{13}{16}f_0 h$$

$$Q_2^{(1)} = -\frac{f_0 h^2}{12} + \left(3hU_3 + h^2 U_4\right)\frac{2EI}{h^3} = -\frac{11}{48}f_0 h^2$$

$$Q_3^{(2)} = \frac{2EI}{h^3}(-6U_3 + 3hU_4) = \frac{3}{16}f_0 h$$

$$Q_4^{(2)} = \frac{2EI}{h^3}\left(-3hU_3 + h^2 U_4\right) = \frac{5}{48}f_0 h^2$$

$$M^c = EI\left.\frac{d^2 w}{dx^2}\right|_{x=0.6h} = EI\sum_{i=1}^{4} U_i \left.\frac{d^2\phi_i}{dx^2}\right|_{x=0.6h}$$

$$= EI\left.\left(U_3\frac{d^2\phi_3}{dx^2} + U_4\frac{d^2\phi_4}{dx^2}\right)\right|_{x=0.6h} = \frac{f_0 h^2}{24}$$

3.36 $U_4 = 0.0651$, $U_5 = -0.071615$ m, $U_6 = 0.16927$; $M^c = -20{,}833$ N·m

3.37 $[K^{(1)}] = \dfrac{E}{h^3}\log 2\begin{bmatrix} 324 & -180h & -324 & -144h \\ & 100h^2 & 180h & 80h^2 \\ \text{symm.} & & 324 & 144h \\ & & & 64h^2 \end{bmatrix}$

$$-\frac{E}{h^3}\begin{bmatrix} 216 & -120h & -216 & -96h \\ & 64h^2 & 120h & 54h^2 \\ \text{symm.} & & 216 & 96h \\ & & & 42h^2 \end{bmatrix}$$

$[K^{(2)}]$ is given by Eq. (3.102) with $b = EI_2$. $U_1 = U_2 = 0$. $Q_3^{(1)} + Q_1^{(2)} = 0$, $Q_4^{(1)} + Q_2^{(2)} = 0$, $Q_3^{(2)} = -F_0$, $Q_4^{(2)} = 0$.

$$M^c = \frac{E}{1.5}\left.\left(\frac{d^2 w}{dx^2}\right)\right|_{x=h/2}$$

3.38 $U_3 = -0.1215$ ft, $U_4 = 0.0315$, $U_5 = -0.3321$ ft, $U_6 = 0.036$; $M^c = -525$ ft · lb.

3.39 $U_3 = -0.056588$ in, $U_4 = 0.005305$, $U_5 = -0.27447$ in, $U_6 = 0.015365$, $U_7 = -0.80813$ in, $U_8 = 0.032342$; $M^c = -265.26$ lb · in.

3.41 The components of force vector due to the distributed load are given by

$$\{f^{(1)}\}^T = \{-346.67, 426.67, -186.67, -320.0\}$$

$$\bar{U}_3 = -0.74695, \quad \bar{U}_4 = 0.1634, \quad \bar{U}_5 = -1.8714, \quad \bar{U}_6 = 0.1994,$$

where $\bar{U}_i = U_i(EI \times 10^{-6})$;
$M^c = -22613$ ft · lb.

3.43 $U_2 = 0.000051$, $U_3 = -0.00013$ m, $U_4 = -0.0000188$, $U_5 = -0.000039$ m, $U_6 = -0.000036$, $U_8 = 0.000033$; $M^c = 115.94$ N · m.

3.44 $\bar{U}_2 = 0.12187$, $\bar{U}_3 = -0.4566$, $\bar{U}_4 = 0.0302$, $\bar{U}_5 = -0.32986$, $\bar{U}_6 = -0.06979$, $\bar{U}_8 = 0.0010417$, $\bar{U}_9 = -0.39583$, $\bar{U}_{10} = 0.1052$, where $\bar{U}_i = U_i(EI \times 10^{-5})$; $M^c = 1833.33$ lb · ft.

3.45 The components of force vector due to the distributed load are given by

$$\{f^{(1)}\}^T = \{-375.0, 416.67, -875.0, -625.0\}$$

$\bar{U}_2 = 0.24826$, $\bar{U}_3 = -0.99537$, $\bar{U}_4 = 0.111115$, $\bar{U}_5 = -0.9838$, $\bar{U}_6 = -0.11806$, $\bar{U}_8 = -0.23611$, where $\bar{U}_i = U_i(EI \times 10^{-5})$; $M^c = 4583.3$ N · m.

3.47 $U_1 = \dfrac{3f_0L^4}{8(3EI + kL^3)} \qquad Q_1^{(1)} = \text{spring force} = -kU_1$

3.48 $$U_3 = \frac{-f_0h^4\left[2.5 + (11/12600)(kh^4/EI)\right]}{12EI + \left[784/(420)^2\right]\left[kh^4/EI\right] + 455(k^2h^8/EI)}$$

Note that as $k \to 0$, we get $U_3 = -(5f_0L^4/384)$ $(h = L/2)$.

3.49 (*b*) $$K_{ij}^{(e)} = \int_{r_1}^{r_2}\left[D_{11}\frac{d^2\phi_i}{dr^2}\frac{d^2\phi_j}{dr^2} + D_{22}\frac{1}{r^2}\frac{d\phi_i}{dr}\frac{d\phi_j}{dr}\right]r\,dr$$

$$f_i^{(e)} = \int_{r_1}^{r_2} rf\phi_i\,dr$$

3.50 The variational formulation is given by

$$0 = \int_{r_i}^{r_0}\left[-r\frac{dv}{dr}\frac{dM_r}{dr} + v(M_r - M_\theta) + rvf\right]dr - Q_1v(r_i) - Q_2v(r_0)$$

$$0 = \int_{r_i}^{r_0}\left[-\frac{d}{dr}(r\xi)\frac{dw}{dr} + \xi(\bar{D}_{12}M_\theta - \bar{D}_{22}M_r)r\right]dr - \theta_1\xi(r_i) - \theta_2\xi(r_0)$$

$$0 = \int_{r_i}^{r_0}\left[\zeta\frac{dw}{dr} + \zeta(\bar{D}_{12}M_r - \bar{D}_{11}M_\theta)r\right]dr$$

where v, ξ, and ζ are the test functions corresponding to w, M_r, and M_θ, and

$$Q_1 = r\left.\frac{dM_r}{dr}\right|_{r=r_i} \qquad Q_2 = -r\left.\frac{dM_r}{dr}\right|_{r=r_0} \qquad \theta_1 = \left.\frac{dw}{dr}\right|_{r=r_i} \qquad \theta_2 = -\left.\frac{dw}{dr}\right|_{r=r_0}$$

3.51 The two-element (in half length) is given by $U_1 = w(0) = 0.171455 \times 10^{-3}$, $U_2 = 0$, $U_3 = 0.9627 \times 10^{-4}$, $U_4 = 0.255 \times 10^{-3}$, $U_5 = U_6 = 0.0$. The exact solution is given by

$$w(r) = \frac{f_0}{64D}\left(r_0^2 - r^2\right)^2 \qquad D = \frac{Eh^3}{12(1-\nu^2)}$$

where r_0 = radius of the plate and h = thickness of the plate.

3.52 Let $w(\bar{x}) = \alpha_0 + \alpha_1 \bar{x} + \alpha_2 \bar{x}^2 + \alpha_3 \bar{x}^3 + \alpha_4 \bar{x}^4 + \alpha_5 \bar{x}^5$ where $\bar{x}$ is the local coordinate with the origin at node 1.

$$\begin{Bmatrix} w_1 \\ \theta_1 \\ \kappa_1 \\ w_2 \\ \theta_2 \\ \kappa_2 \end{Bmatrix} = \begin{bmatrix} 1 & 0 & 0 & 0 & 0 & 0 \\ 0 & 1 & 0 & 0 & 0 & 0 \\ 0 & 0 & 2 & 0 & 0 & 0 \\ 1 & L & L^2 & L^3 & L^4 & L^5 \\ 0 & 1 & 2L & 3L^2 & 4L^3 & 5L^4 \\ 0 & 0 & 2 & 6L & 12L^2 & 20L^3 \end{bmatrix} \begin{Bmatrix} \alpha_0 \\ \alpha_1 \\ \alpha_2 \\ \alpha_3 \\ \alpha_4 \\ \alpha_5 \end{Bmatrix}$$

$$\begin{Bmatrix} \alpha_0 \\ \alpha_1 \\ \alpha_2 \\ \alpha_3 \\ \alpha_4 \\ \alpha_5 \end{Bmatrix} = \frac{1}{2L^5} \begin{bmatrix} 2L^5 & 0 & 0 & 0 & 0 & 0 \\ 0 & 2L^5 & 0 & 0 & 0 & 0 \\ 0 & 0 & L^5 & 0 & 0 & 0 \\ -20L^2 & -12L^3 & -3L^4 & -20L^2 & 8L^3 & L^4 \\ 30L & 16L^2 & 3L^3 & 30L & -14L^2 & -2L^3 \\ -12 & -6L & -L^2 & -12 & 6L & L^2 \end{bmatrix} \begin{Bmatrix} w_1 \\ \theta_1 \\ \kappa_1 \\ w_2 \\ \theta_2 \\ \kappa_2 \end{Bmatrix}$$

$$[K] = \frac{EI}{70L^3} \begin{bmatrix} 1200 & 600L & 30L^2 & -1200 & 600L & -30L^2 \\ & 384L^2 & 22L^3 & -600L & 216L^2 & -8L^3 \\ & & 6L^4 & -30L^2 & 8L^3 & L^4 \\ & & & 1200 & -600L & 30L^2 \\ & \text{symmetric} & & & 384L^2 & -22L^3 \\ & & & & & 6L^4 \end{bmatrix}$$

3.53 (*a*) We denote the vertical portion of the frame as element 1. Then the element matrices are given by

$$[K^{(1)}] = \frac{2EI}{h^3} \begin{bmatrix} 6 & & & & & \\ 0 & \mu & & & & \text{symm.} \\ -3h & 0 & 2h^2 & & & \\ -6 & 0 & 3h & 6 & & \\ 0 & -\mu & 0 & 0 & \mu & \\ -3h & 0 & h^2 & 3h & 0 & 2h^2 \end{bmatrix}$$

$[K^{(2)}]$ is the same as in Eq. (3.129), with $b = EI$.

$$\{F^{(1)}\} = \begin{Bmatrix} -F_1^{(1)} \\ P_1^{(1)} \\ F_2^{(1)} \\ -F_3^{(1)} \\ P_2^{(1)} \\ F_4^{(1)} \end{Bmatrix} \qquad \{F^{(2)}\} = \begin{Bmatrix} P_1^{(2)} \\ F_1^{(2)} \\ F_2^{(2)} \\ P_2^{(2)} \\ F_3^{(2)} \\ F_4^{(2)} \end{Bmatrix}$$

(*b*) Assembled equations:

$$[K] = \frac{2EI}{h^3} \begin{bmatrix} 6 & 0 & 3h & -6 & 0 & 3h & & & \\ & \mu & 0 & 0 & -\mu & 0 & & \mathbf{0} & \\ & & 2h^2 & -3h & 0 & 2h^2 & & & \\ & & & 6+\mu & 0 & -3h & -\mu & 0 & 0 \\ & & & & 6+\mu & -3h & 0 & -6 & -3h \\ & \text{symmetric} & & & & 4h^2 & 0 & 3h & h^2 \\ & & & & & & \mu & 0 & 0 \\ & & & & & & & 6 & 3h \\ & & & & & & & & 2h^2 \end{bmatrix} \begin{Bmatrix} U_1 \\ U_2 \\ U_3 \\ U_4 \\ U_5 \\ U_6 \\ U_7 \\ U_8 \\ U_9 \end{Bmatrix}$$

$$\{F\} = \begin{Bmatrix} -F_1^{(1)} \\ P_1^{(1)} \\ F_2^{(1)} \\ -F_3^{(1)} + P_1^{(2)} \\ P_2^{(1)} + F_1^{(2)} \\ F_4^{(1)} + F_2^{(2)} \\ P_2^{(2)} \\ F_3^{(2)} \\ F_4^{(2)} \end{Bmatrix}$$

(*c*) The boundary conditions are given by $-F_3^{(1)} + P_1^{(2)} = 0$, $P_2^{(1)} + F_1^{(2)} = 0$, $F_4^{(1)} + F_2^{(2)} = 0$, $P_2^{(2)} = P_0$, $F_3^{(2)} = -F_0$, $F_4^{(2)} = 0$; $U_1 = U_2 = U_3 = 0$.
Then the condensed equations are given by the 6×6 matrix equations associated with the last six rows and columns.

3.56 (*a*) $[K^{(1)}]$ is the same as in Eq. (3.130), with $\sin\alpha = \frac{3}{5}$ and $\cos\alpha = \frac{4}{5}$. $[K^{(2)}]$ is the same as in Eq. (3.129).

(*c*) The condensed equations are given by

$$\begin{bmatrix} \frac{16\alpha + 54}{25} c_1 + \beta c_2 & \frac{12(\alpha - 6)}{25} c_1 & -1080 c_1 \\ \frac{12(\alpha - 6)}{25} c_1 & \frac{9\alpha + 96}{25} c_1 + 6 c_2 & 1440 c_1 - 1080 c_2 \\ -1080 c_1 & 1440 c_1 - 1080 c_2 & 2(600)^2 c_1 + 2(360)^2 c_2 \end{bmatrix} \begin{Bmatrix} U_4 \\ U_5 \\ U_6 \end{Bmatrix} = \begin{Bmatrix} 0 \\ -100 \\ 0 \end{Bmatrix}$$

and

$$\alpha = \frac{A}{I}\frac{(600)^2}{2} \qquad \beta = \frac{A}{I}\frac{(360)^2}{2} \qquad c_1 = \frac{2EI}{(600)^3} \qquad c_2 = \frac{2EI}{(360)^3}$$

For $EI = 10^7$ and $EA = 10^4$, the solution of the equations is given by $U_4 = 2.597$ in, $U_5 = -14.62$ in, $U_6 = -0.0255$ rad.

3.57 Condensed set of equations

$$\frac{EI}{h^3}\begin{bmatrix} \frac{4h^2}{\sqrt{2}} & & & & \\ \frac{-3h}{\sqrt{2}} & \frac{\mu + 6}{2\sqrt{2}} + \mu & & \text{symmetric} & \\ \frac{3h}{\sqrt{2}} & \frac{\mu - 6}{2\sqrt{2}} & \frac{\mu + 6}{2\sqrt{2}} + 12 & & \\ \frac{2h^2}{\sqrt{2}} & -\frac{3h}{\sqrt{2}} & \frac{3h}{\sqrt{2}} + 6h & \frac{4h^2}{\sqrt{2}} + 4h^2 & \\ 0 & 0 & 6h & 2h^2 & 4h^2 \end{bmatrix} \begin{Bmatrix} U_3 \\ U_4 \\ U_5 \\ U_6 \\ U_9 \end{Bmatrix} = \begin{Bmatrix} 0 \\ 0 \\ -P \\ 0 \\ 0 \end{Bmatrix}$$

where $\mu = (A/I)h^2$.

3.59 Follow the same procedure as that used in Example 3.5.

3.60 $\dfrac{dx}{d\xi} = h(1 - 2\alpha\xi)$

$$\frac{du}{d\xi} = \frac{du}{dx}\frac{dx}{d\xi} = h(1 - 2\alpha\xi)\frac{du}{dx}$$

$$\frac{d^2u}{d\xi^2} = h^2(1 - 2\alpha\xi)^2\frac{d^2u}{dx^2} - 2\alpha h\frac{du}{dx}\cdots$$

$$u_h(\xi) = \sum_{i=1}^{3} u_i\psi_i(\xi) \qquad \psi_i = \text{quadratic interpolation functions in natural coordinates}$$

By Taylor's theorem,

$$|u - u_h| \leqslant c(\xi)\max\left|\frac{d^3u}{d\xi^3}\right| = c(\xi)0(h^2)$$

In view of Eq. (a), the energy norm becomes

$$\|u - u_h\|_1 \leqslant c\frac{\alpha}{1 - 2\alpha}h$$

3.61 $[\hat{K}]\{u\}_{n+1} = [\hat{M}]\{u\}_n$

$$[\hat{K}] = [M] + \Delta t\,\theta[K]$$

$$[\hat{M}] = [M] - (1 - \theta)\,\Delta t[K]$$

Solve the eigenvalue problem:

$$|[\hat{M}] - \lambda[\hat{K}]| = 0$$

Set $\lambda_{\min} = 0$ for stable scheme ($\lambda_{\min} \geqslant -1$ for stable oscillations), and solve for $\alpha = \Delta t/h$. We get $\alpha = -\sqrt{2} + \frac{5}{3}$. The scheme is stable with oscillations for all values of α.

3.62 Axial displacement ($u \times 10^3$) at $t = 0.08\ \mu$s (at selected points)

x	0.0	300	350	360	370	380	390	400
u	0.0	-1.135×10^{-3}	-1.905×10^{-2}	-0.178	0.03	0.381	0.821	1.348
x	410	420	430	440	450	460	480	500
u	1.80	2.346	2.806	3.338	3.815	4.331	5.326	6.32

3.63 The exact solution is given by

$$u(x, t) = 1 - x - \frac{2}{\pi}\sum_{n=1}^{\infty}\frac{\sin n\pi x\, e^{-n^2\pi^2 t}}{n}$$

The finite-element solution at $t = 0.5$ is given below.

	x						
Mesh	0.125	0.25	0.375	0.5	0.625	0.75	0.875
$4L(\Delta t = 0.05)$	0.8735	0.7470	0.6212	0.4954	0.3712	0.2470	0.1235
$2Q(\Delta t = 0.05)$	0.8729	0.7463	0.6201	0.4944	0.3701	0.2463	0.1229
$8L(\Delta t = 0.025)$	0.8732	0.7465	0.6206	0.4951	0.3706	0.2466	0.1232
$4Q(\Delta t = 0.025)$	0.8731	0.7462	0.6204	0.4947	0.3704	0.2463	0.1231
$4Q(\Delta t = 0.05)$	0.8758	0.7406	0.6214	0.4942	0.3697	0.2461	0.1230
Exact solution	0.8733	0.7468	0.6208	0.4954	0.3708	0.2468	0.1233

3.64 Transverse deflection ($-w \times 10^3$) at the midspan:

$t(\mu s)$		1	5	10	20	40	50
Exact		0.00053	0.0287	0.1441	0.4984	0.8861	1.024
FEM	5-element	0.00012	0.0091	0.0686	0.3640	0.8596	1.0123
	10-element	0.00025	0.0173	0.1672	0.4868	—	—

3.65 Transverse deflection ($-w \times 10^6$) obtained by using eight linear elements in the full beam:

$t(\mu s)$	1	5	10	20	25	30	40	50
$w\left(\frac{L}{4}, t\right)$	0.0971	7.319	55.55	302.83	444.81	575.82	799.55	975.56
$w\left(\frac{L}{2}, t\right)$	−0.0044	−0.332	−2.466	−11.81	−14.90	−14.79	−2.6039	25.613

3.67 Tip deflection ($-w \times 10^3$) of the cantilevered beam:

$t(\mu s)$	5	10	20	25	30	35	40	45	50
w	0.0371	0.2817	1.5355	2.2534	2.9093	3.4880	3.9779	4.3723	4.6695

3.68 Center deflection ($-w \times 10^3$) of the clamped beam:

$t(\mu s)$	5	10	20	25	30	35	40	45	50
w	0.0091	0.0686	0.3640	0.5218	0.6559	0.7671	0.8596	0.9396	1.0124

3.70 $a_0 M_{33} - \lambda(K_{33} + a_0 M_{33}) = 0$ gives

$$\lambda = \frac{1}{(1260/156)\alpha^2 + 1} \qquad \text{which is always greater than zero}$$

3.71 $\psi_1(\xi) = \frac{1}{2s}(1 - \xi)(-1 + s - \xi)$

$$\psi_2(\xi) = \frac{1}{s(2 - s)}(1 - \xi^2)$$

$$\psi_3(\xi) = \frac{1}{2s}(1 + \xi)(1 + s - \xi)$$

3.72 $\phi_1(\xi) = \frac{1}{4}(2 - 3\xi + \xi^3) \qquad \phi_2(\xi) = -\frac{h}{8}(1 - \xi)(1 - \xi^2) \qquad \phi_3(\xi) = \frac{1}{4}(2 + 3\xi - \xi^3)$

$$\phi_4(\xi) = \frac{h}{8}(1 + \xi)(1 - \xi^2)$$

3.74 $\psi_1(\xi) = \sin\dfrac{\pi(\xi+\xi^2)}{4} \qquad \psi_2(\xi) = \sin\dfrac{\pi(1-\xi^2)}{2} \qquad \psi_3(\xi) = \sin\dfrac{\pi(\xi^2-\xi)}{4}$

3.75 See Sec. 3.4.

3.76 The two-point integration gives the exact values.

$$K_{12} = -\frac{1}{h}\left[1 + \frac{1}{2}(x_e + x_{e+1})\right] \qquad G_{12} = \frac{h}{6}\left(1 + x_e + \frac{h}{2}\right)$$

3.78 $n = 1$: $K_{11} = 0.0, \quad G_{11} = \dfrac{h}{4}$

$n = 2$: $K_{11} = \dfrac{12}{h^3}, \quad G_{11} = 0.398148h$

$n = 3$: $K_{11} = \dfrac{12}{h^3}, \quad G_{11} = 0.37h$

$n = 4$: $K_{11} = \dfrac{12}{h^3}, \quad G_{11} = 0.371429h\left(= \dfrac{13h}{35}\right) \quad$ (exact)

3.79 $K_{11} = \dfrac{120}{7h^3}, \quad G_{11} = \dfrac{181h}{462} \quad$ (exact for five-point quadrature)

3.81 Both one-element and two-element meshes give the exact solution at the nodes.

3.82 (*a*)
$$\begin{bmatrix} 9.0 & -5.5 & 0.0 & 0.0 \\ -5.5 & 12.0 & -6.5 & 0.0 \\ 0.0 & -6.5 & 14.0 & -7.5 \\ 0.0 & 0.0 & -7.5 & 8.5 \end{bmatrix} \begin{Bmatrix} U_2 \\ U_3 \\ U_4 \\ U_5 \end{Bmatrix} = \begin{Bmatrix} 0.50 \\ 0.75 \\ 1.00 \\ \beta + 0.5833 \end{Bmatrix}$$

3.91 See Prob. 3.81 and Example 3.10.

3.92 The data input to FEM1D is as follows (for two-element mesh of quadratic elements):
- NPRNT = 1, IELEM = 2, NPE = 3, NDF = 1, NEM = 2, NBDY = 2, NBF = 0, ICONT = 1, ITEM = 0, NTIME = 0
- AL = 1.0, X0 = 0.0
- AX0 = 1.0, AX1 = 1.0, BX0 = BX1 = CX0 = CX1 = 0.0
  F0 = 1.0, F1 = 4.0
- IBDY(1) = 1, IBDY(2) = 5
- VBDY(1) = 0.0, VBDY(2) = 0.0

3.94
- NPRNT = 1, IELEM = 1, NPE = 2, NDF = 1, NEM = 6, NBDY = 1, NBF = 1, ICONT = 0, ITEM = 0, NTIME = 0
- AL = 200.0, X0 = 0.0
- IBDY(1) = 7
- VBDY(1) = 50.0
- IBF(1) = 1
- VBF(1) = $-150/2\pi$
- EC(1,1) = 0.0, EC(1,2) = 10.0
- COEF(1,1) = 0.0, COEF(1,2) = 157.08, all other COEF are zero
- EC(2,1) = 10.0 EC(2,2) = 20.0
- COEF(2,1) = 0.0, COEF(2,2) = 157.08, all other COEF are zero
- EC(3,1) = 20.0, EC(3,2) = 40.0
- COEF(3,1) = 0.0, COEF(3,2) = 157.08, all other COEF are zero

- EC(4,1) = 40.0, EC(4,2) = 80.0
- COEF(4,1) = 0.0, COEF(4,2) = 157.08, all other COEF are zero
- EC(5,1) = 80.0, EC(5,2) = 140.0
- COEF(5,1) = 0.0, COEF(5,2) = 157.08, all other COEF are zero
- EC(6,1) = 140.0, EC(6,2) = 200.0
- COEF(6,1) = 0.0, COEF(6,2) = 157.08, all other COEF are zero.

See Prob. 3.16 for the answer.

For Probs. **3.95** to **3.110** see answers to the corresponding problems in the earlier problem set.

Chapter 4

4.1 $F_1 = f_1^{(1)} = \dfrac{a_0 h}{2} + \dfrac{a_1 h^2}{6} \qquad F_2 = f_2^{(1)} + f_1^{(2)} = a_0 h + a_1 h^2$

$$F_3 = f_2^{(2)} + f_1^{(3)} = a_0 h + 2a_1 h^2 \qquad F_4 = f_2^{(3)} = \frac{a_0 h}{2} + \frac{4}{3} a_1 h^2$$

4.2 (a) $\psi_1 = c(5.1 - 1.2x - y) \qquad \psi_2 = c(-0.7 + 1.7x - y) \qquad \psi_3 = c(-1.5 - 0.5x + 2y)$

$c = 1/2.9$

4.3 $\text{grad}\, T = \dfrac{1}{2A} \sum_{i=1}^{3} (\beta_i T_i \hat{\mathbf{e}}_1 + \gamma_i T_i \hat{\mathbf{e}}_2) = 10.76\hat{\mathbf{e}}_1 - 105.02\hat{\mathbf{e}}_2$

where $\hat{\mathbf{e}}_1$ and $\hat{\mathbf{e}}_2$ are unit basis vectors along the x and y coordinates. The 392 K isotherm intersects at points (3.71375, 1.0) and (3.9734, 1.0266).

4.4 $[K^1] = [S^{01}]$ and $[K^2] = [S^{02}]$, where $[S^{01}]$ and $[S^{02}]$ are given in Table 4.7

4.5 (b) The assembled matrix is of order 10 × 10. Some of the coefficients of the assembled matrix are given by

$$K_{11} = K_{22}^{(2)} \qquad K_{22} = K_{33}^{(2)} + K_{22}^{(1)} \qquad K_{55} = K_{11}^{(1)} + K_{44}^{(2)} + K_{22}^{(3)} + K_{22}^{(4)}$$

$$K_{57} = K_{23}^{(3)} + K_{21}^{(4)} \qquad K_{15} = K_{24}^{(2)} \qquad K_{13} = 0, \ldots$$

4.6 (a)

$$\begin{bmatrix} K_{22} & & & & & \\ K_{32} & K_{33} & & \text{symmetric} & & \\ K_{52} & K_{53} & K_{55} & & & \\ K_{62} & K_{63} & K_{65} & K_{66} & & \\ K_{72} & K_{73} & K_{75} & K_{76} & K_{77} & \\ K_{82} & K_{83} & K_{85} & K_{86} & K_{87} & K_{88} \end{bmatrix} \begin{Bmatrix} U_2 \\ U_3 \\ U_5 \\ U_6 \\ U_7 \\ U_8 \end{Bmatrix} = \begin{Bmatrix} F_2 - K_{21}U_1 - K_{24}U_4 \\ F_3 - K_{31}U_1 - K_{34}U_4 \\ F_5 - K_{51}U_1 - K_{54}U_4 \\ F_6 - K_{61}U_1 - K_{64}U_4 \\ F_7 - K_{71}U_1 - K_{74}U_4 \\ F_8 - K_{81}U_1 - K_{84}U_4 \end{Bmatrix}$$

4.7 Matrices in Eq. (4.96) for the element given here are

$$[K] = 10[S^{11}] + 5[S^{22}] \qquad a = 3, \quad b = 2$$

$$[H] = \frac{25}{6} \begin{bmatrix} 4 & 0 & 0 & 2 \\ 0 & 0 & 0 & 0 \\ 0 & 0 & 6 & 3 \\ 2 & 0 & 3 & 10 \end{bmatrix}$$

$$\{P\} = \frac{25}{2}(273 + 5) \begin{Bmatrix} 2 \\ 0 \\ 3 \\ 5 \end{Bmatrix}$$

where $[S^{11}]$ and $[S^{22}]$ are given by Eq. (4.40).

4.8 $f_1 = 30.77W, \quad f_2 = 25.64W, \quad f_3 = 43.59W.$

4.9 $f_1 = f_4 = 27.27W, \quad f_2 = f_3 = 22.73W.$

4.11

Node	5	6	8	9	11	12
Exact	0.06796	0.09541	0.18203	0.25	0.43203	0.54053
FEM*						
Triangles	0.07143	0.09821	0.18750	0.25	0.42857	0.52679
Rectangles	0.07154	0.10092	0.19286	0.27857	0.50703	0.58479

*$u(0, 1) = 1$.

See also Prob. 4.73.

4.12

Node	5	6	8	9	11	12
Exact	0.05318	0.07522	0.14090	0.19927	0.32010	0.45269
FEM						
Triangles	0.05834	0.08252	0.15087	0.21337	0.33178	0.46923
Rectangles	0.04763	0.06736	0.12994	0.18376	0.30683	0.43391

4.13 Comparison of $u(0.5, 0.5)$ obtained by various rectangular meshes.

1 × 2	2 × 4	4 × 8	8 × 16	Exact
0.375	0.27857	0.25597	0.25144	0.2500

4.14 $K_{ij} = \dfrac{1}{4A^2} \sum_{k=0}^{3} \left(\sum_{i,j=1}^{2} a_{ij}^k [S^{ijk}] + a_{00}^k [S^k] \right)$

where $S_{mn}^{ijk} = \int_{\Omega^e} x_k \dfrac{\partial \psi_m}{\partial x_i} \dfrac{\partial \psi_n}{\partial x_j} \, dx_1 \, dx_2 \qquad S_{mn}^k = \int_{\Omega^e} x_k \psi_m \psi_n \, dx_1 \, dx_2 \qquad x_0 = 1 \qquad x_1 = x$

$x_2 = y \qquad x_3 = xy \qquad \dfrac{\partial \psi_m}{\partial x_i} = \dfrac{1}{2A}(\beta_m \delta_{i1} + \gamma_m \delta_{i2}) \qquad \delta_{ij} = \text{Kronecker delta}$

4.16 *Triangular element*

$$f_i = \frac{1}{2A}\left[a_0 I_{00} \alpha_i + I_{10}(a_1 \alpha_i + a_0 \beta_i) + I_{01}(a_2 \alpha_i + a_0 \gamma_i) \right.$$

$$\left. + I_{11}(a_2 \beta_i + a_1 \gamma_i) + a_1 \beta_i I_{20} + a_2 \gamma_i I_{02} \right]$$

4.17 Same as Prob. 4.14, with a_{ij}^0, a_{ij}^1, a_{ij}^2, and a_{ij}^3 given by $a_{ij}^0 = a_{ij}^k \alpha_k / 2A$, $a_{ij}^1 = a_{ij}^k \beta_k / 2A$, $a_{ij}^2 = a_{ij}^k \gamma_k / 2A$, and $a_{ij}^3 = 0$, where summation on the repeated k is implied.

4.19 Same as Prob. 4.16 with a_0, a_1, and a_2 given by $a_0 = f_k \alpha_k / 2A$, $a_1 = f_k \beta_k / 2A$, $a_2 = f_k \gamma_k / 2A$, and summation on k is implied ($f_k \alpha_k = f_1 \alpha_1 + f_2 \alpha_2 + f_3 \alpha_3$).

4.20 Both two-dimensional and one-dimensional analyses, with uniform meshes, give the exact solution. The values at the nodes along any vertical line are given by (0.0, 1.16, 1.84, 2.04, 1.76, 1.0).

4.22 For $2G\theta = 10$, the finite-element and exact solutions are compared in the following table (see also Prob. 4.74).

Elliptic cross section ($a = 1, b = 3/2$)				Circular cross section ($a = b = 1$)			
x	y	Exact	FEM	x	y	Exact	FEM
0.0	0.0	3.4615	3.6753	0.0	0.0	2.5000	2.6744
0.5	0.0	3.0769	3.0903	0.5	0.0	1.875	1.8411
1.0	0.0	1.9230	1.8093	0.0	0.5	1.875	1.8411
0.0	0.5528	2.4037	2.3531	0.5	0.5	1.25	1.1848
0.5	0.5528	2.0191	1.9878				
1.0	0.5528	0.8652	0.8436				
0.0	0.7454	1.5382	1.5000				
0.5	0.7454	1.1536	1.1095				

4.23 For $a = 1$ and $2G\theta = 10$, the solution is tabulated.

(x, y)	(0.0, 0.19245)	(0.0, 0.0)	(0.222, 0.0)	(0.444, 0.0)
Exact	0.2777	0.3704	0.2743	0.0960
FEM	0.2780	0.3256	0.2897	0.0774

4.24 The solution of this problem requires minor modifications of the program FEM2D. For $a = 0.5$ and $G_1\theta = 2G_2\theta = 5$, the solution for part (a) is given by $u(0,0) = 1.5536$, $u(0.5, 0.0) = 1.2054$, $u(0.5, 0.5) = 0.96429$.

4.25 For a sector of angle 5° and 12 linear triangular elements (6 intervals along the radial direction), the solution is given by 400.0, 331.01, 270.13, 215.66, 166.38, 121.39, 80.00. The solution agrees with that given in table 3.9 (see also Prob. 4.75).

4.26 A 6 × 16 nonuniform mesh of linear rectangular elements is used. The element lengths DX(I) and DY(I) are

DX = (0.2, 0.2, 0.4, 0.4, 0.4, 0.4)

DY = (1.2, 1.0, 0.8, 0.8, 0.6, 0.4, 0.4, 0.4, 0.2, 0.2, 0.2, 0.2, 0.6, 0.6, 0.2, 0.2)

The origin of the coordinate system is at the lower left corner, and the source 125W (per half the domain) is located at node 71. The computed temperatures are: $T_{64} = 293.38$ K, $T_{65} = 292.38$ K, $T_{66} = 390.46$ K, $T_{71} = 299.87$ K, $T_{72} = 292.04$ K, $T_{73} = 290.29$ K, $T_{78} = 292.54$ K, $T_{79} = 291.55$ K, $T_{80} = 289.63$ K, $T_{113} = 280.89$ K, $T_{114} = 280.87$ K, $T_{115} = 280.81$ K.

4.28 For the mesh shown in Fig. 4.12, the temperatures of the interior nodes are $T_4 = 345.74$ K, $T_5 = 329.18$ K, $T_8 = 409.51$ K, $T_{11} = 373.93$ K, $T_{13} = 378.46$ K.

4.29 The origin of the coordinate system is taken at the lower left corner of the domain. The domain is divided into linear rectangular elements with the following increments:

DX = {2.0, 2.0, 2.0, 1.0, 0.6, 0.4, 0.4, 0.6, 1.0, 2.0, 2.0, 2.0}

DY = {1.6, 0.4, 0.4, 0.4, 0.4, 0.4, 0.4, 0.4, 0.6, 1.0}

The mesh consists of 102 elements and 130 nodes; nodes (70, 71, 72, 73, 74) have the same geometric coordinates as nodes (78, 79, 80, 81, 82). The two sets form the left and right side of the sheet pile. The values of the velocity potential along the sheet pile are given by $\phi_{68} = 0.97185$, $\phi_{69} = 1.0415$, $\phi_{70} = 1.2867$, $\phi_{71} = 1.4113$, $\phi_{72} = 1.5115$, $\phi_{73} = 1.5973$, $\phi_{74} = 1.6741$, $\phi_{78} = 0.9215$, $\phi_{79} = 0.8727$, $\phi_{80} = 0.7071$, $\phi_{81} = 0.4077$, $\phi_{82} = 0.0$.

4.37 $[K] = \frac{ab}{24}\begin{bmatrix} 5 & 0 & -1 & 0 & -4 \\ 0 & 5 & 0 & -1 & -4 \\ -1 & 0 & 5 & 0 & -4 \\ 0 & -1 & 0 & 5 & -4 \\ -4 & -4 & -4 & -4 & 16 \end{bmatrix}$

4.38 $\psi_1 = L_1(2L_1 - 1),\ \psi_2 = L_2(2L_2 - 1),\ \psi_3 = L_3(2L_3 - 1),$

$\psi_4 = 4L_1L_2,\quad \psi_5 = 4L_2L_3,\quad \psi_6 = 4L_1L_3.$

4.39 3 and $-1/2$.

4.40 $\psi_{14} = 32L_1L_2L_3(4L_2 - 1)$

4.43 $\psi_1 = \frac{1}{4}(1-\eta)(1-\xi),\ \psi_2 = \frac{1}{4}(1+\xi)(1-\eta),\ \psi_3 = \frac{1}{4}(1+\xi)(1+\eta)\xi,$

$\psi_4 = \frac{1}{4}(1-\xi)(1+\eta)\xi,\quad \psi_5 = \frac{1}{2}(1-\xi^2)(1+\eta).$

4.44 ψ_i, $i = 1, 2, 3, 4$ are the linear interpolation functions of the four-node rectangular element, and ϕ_i are given by $\phi_1 = (\xi^2 - 1),\ \ \phi_2 = (\eta^2 - 1),\ \ \phi_3 = (\xi^2 - 1)\eta,\ \ \phi_4 = \xi(\eta^2 - 1).$

4.45 *Variational form:*

$$\int_{\Omega^e} (\text{grad}\, w \cdot \mathbf{q} - wf)\, dx\, dy - \oint_{\Gamma^e} w q_n\, ds = 0$$

$$\int_{\Omega^e} \mathbf{v} \cdot \left(\text{grad}\, \mathbf{u} - \frac{1}{k}\mathbf{q}\right) dx\, dy = 0 \text{ (2 eqns, one for } v_1 \text{ and one for } v_2\text{)}$$

where $\mathbf{v} = (v_1, v_2)$ and w are test functions (or, variations in $\mathbf{q}$ and u, respectively), and $q_n = \hat{\mathbf{n}} \cdot \mathbf{q}$

Finite-element matrices: For the case when u_1, q_1, q_2 are interpolated by same ψ_i, we have

$$K_{ij}^{11} = 0 \qquad K_{ij}^{12} = \int_{\Omega^e} \frac{\partial \psi_i}{\partial x}\psi_j\, dx\, dy \qquad K_{ij}^{13} = \int_{\Omega^e} \frac{\partial \psi_i}{\partial y}\psi_j\, dx\, dy \qquad F_i^1 = \oint_{\Gamma^e} \psi_i q_n\, ds,$$

$$K_{ij}^{21} = K_{ji}^{12} \qquad K_{ij}^{22} = -\int_{\Omega^e} \frac{1}{k}\psi_i\psi_j\, dx\, dy \qquad K_{ij}^{23} = 0 \qquad F_i^2 = 0 \qquad K_{ij}^{31} = K_{ji}^{13} \qquad K_{ij}^{32} = 0$$

$$K_{ij}^{33} = K_{ij}^{22} \qquad F_i^3 = 0.$$

4.46 (*a*) *Variational formulation:*

$$0 = \int_{\Omega^e} \left[\frac{\partial v_1}{\partial x}\frac{\partial w}{\partial x} - v_1(D_{22}M_1 - D_{12}M_2)\right] dx\, dy - \oint_{\Gamma^e} v_1 t_1\, ds$$

$$0 = \int_{\Omega^e} \left[\frac{\partial v_2}{\partial y}\frac{\partial w}{\partial y} - v_2(D_{11}M_2 - D_{12}M_1)\right] dx\, dy - \oint_{\Gamma^e} v_2 t_2\, ds$$

$$0 = \int_{\Omega^e} \left(\frac{\partial u}{\partial x}\frac{\partial M_1}{\partial x} + \frac{\partial u}{\partial y}\frac{\partial M_2}{\partial y} + 2D_{33}\frac{\partial^2 u}{\partial x\, \partial y}\frac{\partial^2 w}{\partial x\, \partial y} - uq\right) dx\, dy$$

$$-\oint_{\Gamma^e} \left[t_3 u + t_4\left(n_x \frac{\partial u}{\partial y} + n_y \frac{\partial u}{\partial x}\right)\right] ds$$

where v_1, v_2, and u are the test functions associated with M_1, M_2 and w, respectively, and

$$t_1 = \frac{\partial w}{\partial x}n_x \qquad t_2 = \frac{\partial w}{\partial y}n_y \qquad t_3 = \frac{\partial M_1}{\partial x}n_x + \frac{\partial M_2}{\partial y}n_y - D_{33}\left(\frac{\partial^3 w}{\partial x^2\, \partial y}n_y + \frac{\partial^3 w}{\partial x\, \partial y^2}n_x\right)$$

$$t_4 = 2\frac{\partial^2 w}{\partial x\, \partial y} \qquad D_{11} = S \qquad D_{22} = D_{11} \qquad D_{12} = \nu D_{11} \qquad 2D_{33} = \frac{2}{(1+\nu)S}$$

4.47 The element matrices are given by

$$[K^{11}] = \frac{2D_{33}}{ab}\begin{bmatrix} 1 & -1 & 1 & -1 \\ -1 & 1 & -1 & 1 \\ 1 & -1 & 1 & -1 \\ -1 & 1 & -1 & 1 \end{bmatrix} \qquad [K^{12}] = \frac{b}{2a}\begin{bmatrix} 1 & -1 \\ -1 & 1 \\ -1 & 1 \\ 1 & -1 \end{bmatrix} = [K^{21}]^T$$

$$[K^{13}] = [K^{31}]^T = \frac{a}{2b}\begin{bmatrix} 1 & -1 \\ 1 & -1 \\ -1 & 1 \\ -1 & 1 \end{bmatrix} \qquad [K^{22}] = -\frac{D_{22}ab}{6}\begin{bmatrix} 2 & 1 \\ 1 & 2 \end{bmatrix}$$

$$[K^{23}] = [K^{32}]^T = \frac{D_{12}ab}{4}\begin{bmatrix} 1 & 1 \\ 1 & 1 \end{bmatrix} \qquad [K^{33}] = \frac{D_{11}}{D_{22}}[K^{22}]$$

4.48 $[K^{11}]$ = same as in Prob. 4.47 $\quad [K^{12}] = [K^{21}]^T = [S^{11}] \quad [K^{13}] = [K^{31}]^T = [S^{22}]$
$[K^{22}] = -D_{22}[S] \quad [K^{23}] = [K^{32}]^T = D_{12}[S] \quad [K^{33}] = -D_{11}[S]$
where $[S^{11}]$, $[S^{22}]$, and $[S]$ are given in Table 4.7.

4.51 $P = M_0/2$, $U_1 = 0$, $U_6 = 0$, $V_6 = 0$, $U_{11} = 0$, $F_{x5} = -P$, $F_{x15} = P$, F_{x1}, F_{x6}, F_{y6}, and F_{x11} are unknown, and all other forces are zero.

4.53 $[K] = \dfrac{(\alpha+\beta)h}{6}\begin{bmatrix} 2 & -1 \\ -1 & 2 \end{bmatrix} \qquad \alpha = \dfrac{E}{1-\nu^2} \qquad \beta = \dfrac{E}{2(1+\nu)}$

4.54

$$\mu[K]\{u\} - \rho\beta g[G]\{T\} = \{F^1\} \qquad \text{(i)}$$

$$k[K]\{T\} = \{F^2\} \qquad \text{(ii)}$$

$$K_{ij} = \int_0^h \frac{d\psi_i}{dy}\frac{d\psi_j}{dy}\,dy \qquad G_{ij} = \int_0^h \psi_i\psi_j\,dy$$

$$F_i^1 = -\int_0^h \rho\beta g\psi_i\,dy + P_i \qquad P_1 = -\mu\left(\frac{du}{dy}\right)_0 \qquad P_2 = \mu\left(\frac{du}{dy}\right)_h$$

$$F_i^2 = \int_0^h \mu\left(\frac{du}{dy}\right)^2 \psi_i\,dy + Q_i \qquad Q_1 = -k\left(\frac{dT}{dy}\right)_0 \qquad Q_2 = k\left(\frac{dT}{dy}\right)_h$$

Solution strategy: Solve the assembled equations corresponding to Eq. (ii) for T, subject to boundary conditions and initial values of $u = 0$. Use the temperatures thus obtained in the assembled equations associated with Eq. (i) and solve for u. Then resolve Eq. (ii) with the updated F_i^2 (because of the newly computed u). Iterate the procedure until u and T obtained in two consecutive iterations differ by, say, 1 percent.

4.56 The finite-element model is the same as that in Eq. (4.195), with F_i^1 replaced by

$$F_i^1 = \oint_{\Gamma^e} t_x\psi_i\,ds + \rho g\beta\int_{\Omega^e} T\psi_i\,dx\,dy$$

In addition, we have the finite-element model associated with the energy equation.

$$[K]\{T\} = \{Q\}$$

$$K_{ij} = \int_{\Omega^e} k\left(\frac{\partial\psi_i}{\partial x}\frac{\partial\psi_j}{\partial x} + \frac{\partial\psi_i}{\partial y}\frac{\partial\psi_j}{\partial y}\right) dx\,dy$$

$$Q_i = \oint_{\Gamma^e} k\frac{\partial T}{\partial n}\psi_i\,ds$$

4.58 $U_1 = U_2 = \cdots = U_9 = U_0$, $V_1 = V_2 = \cdots = V_9 = 0$; $U_{28} = U_{29} = \cdots = U_{36} = V_{28} = V_{29} = \cdots = V_{36} = 0.0$; $F_{10x} = F_{18x} = F_{19x} = F_{27x} = P_0$, $F_{10y} = F_{18y} = F_{19y} = F_{27y} = 0.0$.

4.59 In general, both velocity components are zero on fixed walls, and shear stress is zero along the line of symmetry (see the discussion in the text). Nodes on the inlet have zero vertical velocities and specified u_0 horizontal velocities (see Prob. 4.79).

4.60 (*a*) At nodes on the boundary: $w = 0.0$; at nodes along the x axis: $S_y = 0.0$; at nodes along the y axis: $S_x = 0.0$. See also Prob. 4.95.

(*b*) At nodes on the top (simply supported) boundary: $w = S_x = 0.0$; at nodes on the right (clamped) boundary: $w = S_x = S_y = 0.0$; at nodes along the axes of symmetry: see part (*a*).

(*c*) At nodes along the clamped boundary: $w = S_x = S_y = 0.0$.

4.61 $K_{ij}^{11} = \frac{1}{4A}(D_{44}\beta_i\beta_j + D_{55}\gamma_i\gamma_j) \quad K_{ij}^{12} = \frac{D_{44}}{4A^2}\beta_i(I_{00}\alpha_j + I_{10}\beta_j + I_{01}\gamma_j)$

$$K_{ij}^{13} = \frac{D_{55}}{4A^2}\gamma_i(I_{00}\alpha_j + I_{10}\beta_j + I_{01}\gamma_j)$$

$$K_{ij}^{22} = \frac{1}{4A}\left(D_{11}\beta_i\beta_j + D_{33}\gamma_i\gamma_j\right) + \frac{D_{44}}{4A^2}\Big[I_{00}\alpha_i\alpha_j + I_{10}\left(\alpha_i\beta_j + \alpha_j\beta_i\right)$$

$$+I_{01}\left(\alpha_i\gamma_j + \alpha_j\gamma_i\right) + I_{20}\beta_i\beta_j + I_{11}\left(\beta_i\gamma_j + \beta_j\gamma_i\right) + I_{02}\gamma_i\gamma_j\Big] \ldots$$

4.63 Minimum eigenvalues:

1×1 mesh of rectangles (in a quadrant): $\lambda_1 = 24.0$
2×2 mesh of rectangles: $\lambda_1 = 20.77$
1×1 mesh of triangles: $\lambda_1 = 24.0$
2×2 mesh of rectangles: $\lambda_1 = 21.66$

4.65 $\left(a + \frac{2}{3}\Delta t\, b\right)u_{n+1} = \left(a - \frac{1}{3}\Delta t\, b\right)u_n + \Delta t\left(\frac{1}{3}f_n + \frac{2}{3}f_{n+1}\right)$

where $u_{n+1} = u(t_{n+1})$. The present scheme is a special case of the θ family of approximation for $\theta = \frac{2}{3}$.

4.66 $[K]\{u\} = \{F\} \qquad K_{ij} = \int_0^{\Delta t}\int_{x_e}^{x_{e+1}}\left(c\psi_i\frac{\partial\psi_j}{\partial t} + a\frac{\partial\psi_i}{\partial x}\frac{\partial\psi_j}{\partial x}\right)dx\,dt$

$$F_i = \int_0^{\Delta t}\left(a\frac{\partial u}{\partial x}\psi_i\right)\Bigg|_{x=x_e}^{x=x_{e+1}} dt + \int_{x_e}^{x_{e+1}}\int_0^{\Delta t} f\psi_i\,dx\,dt$$

For the case in which a and c are constant and ψ_1 are the linear interpolation functions of a (time-space) rectangular element, we have

$$[K] = \frac{h}{12}\begin{bmatrix} -2c + 4a\alpha & -c - 4a\alpha & c - 2a\alpha & 2(c + a\alpha) \\ -c - 4a\alpha & -2c + 4a\alpha & 2(c + a\alpha) & c - 2a\alpha \\ -c - 2a\alpha & -2c - 2a\alpha & 2c + 4a\alpha & c - 4a\alpha \\ 2(c + a\alpha) & -c - 2a\alpha & -c - 4a\alpha & c + 4a\alpha \end{bmatrix}$$

where $\alpha = \Delta t/h^2$.

4.67 (*b*) $x = \frac{1}{4}(1 + \xi)(11.5 + 1.5\eta), \quad y = \frac{1}{4}(1 + \eta)(12 + 2\xi), \quad J = \frac{1}{8}(67.5 + 7.5\eta + 10\xi) > 0.$

4.68 $f_0(7.7083, 8.5417, 9.1667, 8.3333)$.

4.71 $J = 1 + \eta(a - 2) + \xi(b - 2), J > 0$ gives $a > \frac{3}{2}, b > \frac{3}{2}$, and $a + b > \frac{7}{2}$. Note that $\xi = \eta = 2$ at node 3 of the master element.

4.72 $\frac{\partial\hat{\psi}_3}{\partial x} = \frac{5(1 + \eta)}{8J} \qquad \frac{\partial\hat{\psi}_3}{\partial y} = \frac{5(1 + \xi)}{8J} \qquad J = \frac{135 + 15\eta + 20\xi}{16}$

4.73–4.105 Input data and edited computer output from program FEM2D for some of these problems are given below.

```
PROBLEM 4.73: DATA AND PARTIAL SOLUTION TO PROBLEM 4.11 (TRIANGULAR ELEMENTS)
    0    3   -1    0    1    0    0
    2    4
 0.25      0.25
 0.25      0.25      0.25      0.25
 1.0       1.0       0.0       0.0       0.0
    9
    1    2    3    4    7   10   13   14   15
 0.0       0.0       0.0       0.0       0.0       0.0       1.0       1.0
 1.0
    0

SOLUTION VECTOR (VALUES OF U AT THE NODES):

   0.0            0.0            0.0            0.0            0.71429D-01   0.98214D-01
   0.0            0.18750D+00    0.25000D+00    0.0            0.42857D+00   0.52679D+00
   0.10000D+01    0.10000D+01    0.10000D+01
```

....

```
PROBLEM 4.73: DATA AND PARTIAL SOLUTION TO PROBLEM 4.11 (RECTANGULAR ELEMENTS)
    1    4   -1    0    1    0    0
    2    4
 0.25      0.25
 0.25      0.25      0.25      0.25
 1.0       1.0       0.0       0.0       0.0
    9
    1    2    3    4    7   10   13   14   15
 0.0       0.0       0.0       0.0       0.0       0.0       1.0       1.0
 1.0
    0

SOLUTION VECTOR (VALUES OF U AT THE NODES):

   0.0            0.0            0.0            0.0            0.71544D-01   0.10092D+00
   0.0            0.19286D+00    0.27857D+00    0.0            0.50703D+00   0.58479D+00
   0.10000D+01  , 0.10000D+01    0.10000D+01
```

....

PROBLEM 4.74: DATA AND PARTIAL OUTPUT FOR PROBLEM 4.22 (ELLIPTIC SECTION)

```
   0    3   -1     0      0     0     0
  14   13
   1    2    6
   1    6    5
   2    3    7
   2    7    6
   3    4    8
   3    8    7
   5    6   10
   5   10    9
   6    7   11
   6   11   10
   7    8   11
   9   10   12
   9   12   13
  10   11   12
 0.0      0.0       0.5       0.0       1.0       0.0       1.5       0.0
 0.0      0.5528    0.50      0.5528    1.0       0.5528    1.25      0.5528
 0.0      0.7454    0.5       0.7454    1.0       0.7454    0.5       0.9428
 0.0      1.0
 1.0      1.0       0.0       0.0      10.0
   5
   4    8   11    12    13
 0.0
   0
```

SOLUTION VECTOR (STRESS FUNCTION VALUES AT THE NODES):

```
0.36753D+01   0.30903D+01   0.18093D+01   0.0           0.23531D+01   0.19878D+01
0.84365D+00   0.0           0.15000D+01   0.11095D+01   0.0           0.0
0.0
```

ELE.NO.	X-COMPONENT (-TAUYZ)	Y-COMPONENT (TAUXZ)	MAGNITUDE	ANGLE
1	-0.116995D+01	-0.199444D+01	0.231227D+01	-0.120396D+03
2	-0.730596D+00	-0.239184D+01	0.250093D+01	-0.106985D+03
3	-0.256206D+01	-0.174685D+01	0.310091D+01	-0.145713D+03
4	-0.228832D+01	-0.199444D+01	0.303549D+01	-0.138925D+03
5	-0.361862D+01	-0.163650D+01	0.397147D+01	-0.155665D+03
6	-0.337461D+01	-0.174685D+01	0.379993D+01	-0.152632D+03
7	-0.730596D+00	-0.456032D+01	0.461847D+01	-0.991018D+02
8	-0.781053D+00	-0.442933D+01	0.449767D+01	-0.100001D+03
9	-0.228832D+01	-0.438033D+01	0.494203D+01	-0.117583D+03
10	-0.221898D+01	-0.456032D+01	0.507153D+01	-0.115947D+03
11	-0.337461D+01	-0.438033D+01	0.552949D+01	-0.127611D+03
12	-0.781053D+00	-0.562053D+01	0.567454D+01	-0.979114D+02
13	-0.674007D+00	-0.589167D+01	0.593010D+01	-0.965262D+02
14	-0.221898D+01	-0.562053D+01	0.604270D+01	-0.111544D+03

....

```
PROBLEM 4.75:DATA AND PARTIAL SOLUTION FOR PROBLEM 4.25(TRIANGLES;5 DEG. SECTOR)
    0    3   -1    0    0    0    0
   12   14
    1    2    9
    1    9    8
    2    3   10
    2   10    9
    3    4   11
    3   11   10
    4    5   12
    4   12   11
    5    6   13
    5   13   12
    6    7   14
    6   14   13
 1.75      0.0       2.0       0.0       2.25      0.0       2.5       0.0
 2.75      0.0       3.0       0.0       3.25      0.0       1.74334   0.1525225
 1.99239   0.1743    2.24144   0.1961    2.49047   0.217889  2.739535  0.2397
 2.9886    0.26147   3.2376    2.283256
 1.0       1.0       0.0       0.0       0.0
    4
    1    8    7   14
 400.0     400.0     80.0      80.0
    0

SOLUTION VECTOR (TEMPERATURE, IN DEG  F, AT THE NODES):

   0.40000D+03   0.32328D+03   0.25559D+03   0.19503D+03   0.14026D+03   0.90357D+02
   0.80000D+02   0.40000D+03   0.32328D+03   0.25559D+03   0.19502D+03   0.14020D+03
   0.90037D+02   0.80000D+02

..........................................................................................
```

```
PROBLEM 4.76: DATA AND PARTIAL SOLUTION OF PROBLEM 4.26 (RECTANGLES)
    1    4   -1    1    1    0    0
    6   16
  0.2       0.2       0.4       0.4       0.4       0.4
  1.2       1.0       0.8       0.8       0.6       0.4       0.4       0.4
  0.2       0.2       0.2       0.2       0.6       0.6       0.2       0.2
  0.0
 10.0      15.0       5.0     268.0       0.0
    6   91   92   93   94   95   96
    4    3    4    3    4    3    4    3    4    3    4    3
    0
    1
   71
125.0

SOLUTION (TEMPERATURE, IN DEG K, AT THE NODES; SOURCE 125W IS AT NODE 71):

   0.28883D+03   0.28883D+03   0.28883D+03   0.28883D+03   0.28883D+03   0.28883D+03
   0.28883D+03   0.28884D+03   0.28884D+03   0.28884D+03   0.28883D+03   0.28883D+03
   0.28883D+03   0.28883D+03   0.28885D+03   0.28885D+03   0.28885D+03   0.28884D+03
   0.28883D+03   0.28882D+03   0.28881D+03   0.28889D+03   0.28889D+03   0.28888D+03
   0.28885D+03   0.28881D+03   0.28878D+03   0.28877D+03   0.28902D+03   0.28901D+03
   0.28898D+03   0.28889D+03   0.28877D+03   0.28869D+03   0.28865D+03   0.28926D+03
   0.28923D+03   0.28917D+03   0.28894D+03   0.28869D+03   0.28851D+03   0.28844D+03
   0.28958D+03   0.28954D+03   0.28940D+03   0.28899D+03   0.28857D+03   0.28829D+03
   0.28820D+03   0.29021D+03   0.29010D+03   0.28981D+03   0.28900D+03   0.28834D+03
   0.28795D+03   0.28782D+03   0.29168D+03   0.29128D+03   0.29038D+03   0.28882D+03
   0.28792D+03   0.28744D+03   0.28728D+03   0.29338D+03   0.29238D+03   0.29046D+03
   0.28860D+03   0.28761D+03   0.28710D+03   0.28694D+03   0.29987D+03   0.29204D+03
   0.29029D+03   0.28824D+03   0.28724D+03   0.28671D+03   0.28655D+03   0.29254D+03
   0.29155D+03   0.28963D+03   0.28777D+03   0.28678D+03   0.28627D+03   0.28610D+03
   0.28999D+03   0.28960D+03   0.28873D+03   0.28717D+03   0.28625D+03   0.28576D+03
   0.28561D+03   0.28566D+03   0.28561D+03   0.28545D+03   0.28487D+03   0.28430D+03
   0.28395D+03   0.28383D+03   0.28267D+03   0.28264D+03   0.28256D+03   0.28230D+03
   0.28200D+03   0.28178D+03   0.28170D+03   0.28176D+03   0.28174D+03   0.28167D+03
   0.28145D+03   0.28119D+03   0.28100D+03   0.28093D+03   0.28089D+03   0.28087D+03
   0.28081D+03   0.28061D+03   0.28037D+03   0.28019D+03   0.28013D+03

.............................................................................
```

```
PROBLEM 4.79: CONVERGENCE STUDY OF THE VISCOUS FLOW PROBLEM IN FIG. P4.59C
    1     4     2     0     1     0     0
    5     5
 2.0          4.0          4.0          4.0          6.0
 0.2          0.2          0.2          0.2          0.2
 1.0          1.0
   27
    1     2     4     6     8    10    12    13    14    25    26    37    38    49    50    61
   62    63    64    65    66    67    68    69    70    71    72
 1.0          0.0          0.0          0.0          0.0          0.0          0.0          1.0
 0.0          1.0          0.0          1.0          0.0          1.0          0.0          0.0
 0.0          0.0          0.0          0.0          0.0          0.0          0.0          0.0
 0.0          0.0          0.0
    0

SOLUTION (VELOCITIES AT SELECTED NODES; PENALTY PARAMETER=1.0D8):
SOLUTION FOR 5X5 MESH:

     1    0.10000D+01    0.0                          2    0.12932D+01    0.0
     3    0.14302D+01    0.0                          4    0.13140D+01    0.0
     5    0.13922D+01    0.0                          6    0.13407D+01    0.0
     7    0.10000D+01    0.0                          8    0.12585D+01   -0.55169D-01
    11    0.13317D+01    0.25894D-01                 12    0.12908D+01   -0.22813D-01
    13    0.10000D+01    0.0                         14    0.11476D+01   -0.95780D-01
    17    0.11511D+01    0.51809D-01                 18    0.11407D+01   -0.47018D-01
    19    0.10000D+01    0.0                         20    0.93127D+00   -0.10367D+00
    23    0.85558D+00    0.68971D-01                 24    0.88642D+00   -0.64859D-01
    25    0.10000D+01    0.0                         26    0.51602D+00   -0.48398D-01
    29    0.46544D+00    0.36681D-01                 30    0.51170D+00   -0.35139D-01
```

```
SOLUTION FOR 7X5 MESH (INPUT DATA NOT GIVEN HERE):

   1    0.10000D+01    0.0                     2    0.12380D+01    0.0
   3    0.14309D+01    0.0                     4    0.13410D+01    0.0
   5    0.13760D+01    0.0                     6    0.13530D+01    0.0
   7    0.13749D+01    0.0                     8    0.13528D+01    0.0
   9    0.10000D+01    0.0                    10    0.12109D+01   -0.89791D-01
  15    0.13182D+01   -0.40566D-02            16    0.13004D+01    0.60522D-02
  17    0.10000D+01    0.0                    18    0.11034D+01   -0.15267D+00
  23    0.11490D+01   -0.71132D-02            24    0.11421D+01    0.10342D-01
  25    0.10000D+01    0.0                    26    0.94666D+00   -0.16269D+00
  31    0.86719D+00    0.51296D-02            32    0.87802D+00   -0.20979D-02
  33    0.10000D+01    0.0                    34    0.61995D+00   -0.76011D-01
  39    0.47819D+00    0.10176D-01            40    0.50309D+00   -0.89307D-02

SOLUTION FOR 11X5 MESH (INPUT DATA NOT GIVEN HERE):

   1    0.10000D+01    0.0                     2    0.10649D+01    0.0
   3    0.11892D+01    0.0                     4    0.13411D+01    0.0
   5    0.13731D+01    0.0                     6    0.13594D+01    0.0
   7    0.13674D+01    0.0                     8    0.13586D+01    0.0
   9    0.13695D+01    0.0                    10    0.13567D+01    0.0
  11    0.13729D+01    0.0                    12    0.13546D+01    0.0
  13    0.10000D+01    0.0                    14    0.10604D+01   -0.10029D+00
  23    0.13162D+01   -0.87214D-02            24    0.13021D+01    0.10343D-01
  25    0.10000D+01    0.0                    26    0.10558D+01   -0.19324D+00
  35    0.11470D+01   -0.14692D-01            36    0.11439D+01    0.17175D-01
  37    0.10000D+01    0.0                    38    0.10204D+01   -0.25420D+00
  47    0.86770D+00   -0.16533D-01            48    0.87762D+00    0.18675D-01
  49    0.10000D+01    0.0                    50    0.83090D+00   -0.13528D+00
  59    0.48260D+00   -0.10151D-01            60    0.49906D+00    0.10974D-01

...............................................................................
```

```
PROBLEM 4.82: FLOW IN WALL-DRIVEN SQUARE CAVITY (BY THE PENALTY FEM MODEL)
   1    4    2    0    1    0    0
   8    8
 0.1     0.125     0.125     0.15      0.15      0.125     0.125     0.1
 0.0
 0.15    0.125     0.125     0.125     0.125     0.125     0.125     0.1
 0.0
 1.0     100.0
  64
   1    2    3    4    5    6    7    8    9   10   11   12   13   14   15   16
  17   18   19   20   35   36   37   38   53   54   55   56   71   72   73   74
  89   90   91   92  107  108  109  110  125  126  127  128  143  144  146  148
 150  152  154  156  158  160  162  145  147  149  151  153  155  157  159  161
**** INCLUDE SEVEN ROWS OF ZEROS ****
 1.0     1.0       1.0       1.0       1.0       1.0       1.0       0.0
   0
```

SOLUTION (VELOCITIES AT SELECTED NODES FOR TWO SETS OF BOUNDARY CONDITIONS):

NODE	U=0 AT BOTH SINGULAR POINTS U73 = U81 = 0.0		U=1 AT BOTH SINGULAR POINTS U73 = U81 = 1.0	
	U	V	U	V
14	-0.64176D-01	0.26762D-07	-0.69202D-01	0.12581D-07
23	-0.16789D+00	0.67272D-08	-0.11107D+00	0.58670D-08
32	-0.13663D+00	0.94054D-08	-0.14636D+00	-0.12640D-07
41	-0.29892D+00	-0.26532D-07	-0.16206D+00	-0.71277D-07
50	-0.12927D+00	-0.86178D-07	-0.95458D-01	-0.10898D-06
59	-0.39691D-01	-0.12529D-06	0.11133D+00	-0.13989D-06
68	0.49221D+00	-0.82541D-07	0.53305D+00	-0.84628D-07

```
PROBLEM 4.87: ANNULAR DISK WITH INTERNAL PRESSURE,QO=100 (FIG.P4.49B)
    1    4    1    0    0    0    0
   12   20
    1    2    6    5
    2    3    7    6
    3    4    8    7
    5    6   10    9
    6    7   11   10
    7    8   12   11
    9   10   14   13
   10   11   15   14
   11   12   16   15
   13   14   18   17
   14   15   19   18
   15   16   20   19
 2.0        0.0        4.0        0.0        6.0        0.0        8.0        0.0
 1.848      0.765      3.696      1.531      5.543      2.296      7.391      3.061
 1.414      1.414      2.828      2.828      4.243      4.243      5.657      5.657
 0.765      1.848      1.531      3.696      2.296      5.543      3.061      7.391
 0.0        2.0        0.0        4.0        0.0        6.0        0.0        8.0
30000000. 30000000. 0.3        11538462. 1.0
    8
    2    4    6    8   33   35   37   39
 0.0
    8
    1    9   10   17   18   25   26   34
 39.27      72.56      30.056     55.536     55.536     30.056     72.56      39.27

SOLUTION (DISPLACEMENTS IN X AND Y DIRECTIONS AT THE FIRST FOUR NODES):

     1    0.86387D-05    0.0                    2    0.47413D-05    0.0
     3    0.35437D-05    0.0                    4    0.30489D-05    0.0

      --------------------------------------------------------------
      ELE.NO.    STRESS, SXX       STRESS, SYY       STRESS,SXY
      --------------------------------------------------------------
         1     -0.364341D+02     0.526484D+02     -0.184509D+02
         2     -0.858335D+01     0.218416D+02     -0.630355D+01
         3     -0.120962D+01     0.141050D+02     -0.317155D+01
      --------------------------------------------------------------
.............................................................................
```

```
PROBLEM 4.88: PLANE ELASTICITY PROBLEM OF FIGURE P4.88
   1    4    0    0    1    0    0
   6    3
 2.0       2.0       2.0       2.0       2.0       2.0       2.0
 1.0       1.0       1.0       1.0
30000000. 30000000. 0.3       11538462. 0.5
  11
   2    4    6    8   10   12   14    1   15   29   43
 0.0
 0.0
   2
  13   55
 100.0    -100.0


SOLUTION (DISPLACEMENTS AT SELECTED NODES):

    7    0.22089D-04    0.0                        8    0.0            0.17038D-07
   14    0.60676D-05    0.54905D-05               15    0.0            0.25494D-07
   21   -0.40879D-05    0.95755D-05               22    0.0            0.27810D-07
   27   -0.61619D-05   -0.38926D-05               28   -0.26049D-04    0.13109D-04

      ---------------------------------------------------------
      ELE.NO.   STRESS, SXX      STRESS, SYY      STRESS,SXY
      ---------------------------------------------------------
         1     -0.139286D+00    0.322345D+00     0.663894D-01
         6      0.759464D+02    0.401529D+02    -0.381496D+02
         7      0.879431D-02    0.196623D+00     0.152073D+00
        12      0.422825D+01    0.228827D+02    -0.165697D+02
        13      0.130491D+00    0.290535D-01     0.938342D-01
        18     -0.801746D+02    0.608983D+01    -0.303968D+02
      ---------------------------------------------------------

..........................................................................
```

```
PROBLEM 4.89: CANTILEVERED BEAM BY PLANE STRESS ELEMENTS (FIG.P.4.89)
   2    8    0    0    1    0    0
   3    1
  2.0     2.0        2.0       2.0        2.0        2.0
  1.0     1.0
20000000. 20000000. 0.3         7692308. 0.5
   6
   1    2   15   16   23   24
 0.0      0.0        0.0       0.0        0.0        0.0
   1
  36
 -100.0

SOLUTION (DISPLACEMENTS AT SELECTED NODES):

    5   -0.94189D-03   -0.44053D-02                7   -0.10592D-02   -0.85146D-02
   11    0.36346D-06   -0.85206D-02               18    0.10667D-02   -0.85340D-02

      ---------------------------------------------------------
      ELE.NO.   STRESS, SXX       STRESS, SYY      STRESS,SXY
      ---------------------------------------------------------
          1    -0.726235D-01    0.182124D+00    -0.221517D+02
          2     0.132818D+01    0.539333D+00    -0.230334D+02
          3     0.327726D+01    0.113591D+02    -0.227092D+02
      ---------------------------------------------------------

.....................................................................................
```

```
PROBLEM 4.93: BENDING OF A SIMPLY SUPPORTED PLATE UNDER UNIFORM LOAD (FIG.4.45)
   1    4    1    0    0    0    0    0
   4    4
 0.1       0.1        0.2535     0.2535     0.2535
 0.1       0.1        0.15       0.15       0.15
   25.0E6      1.0E6      0.5E6      0.5E6      0.2E6     0.25E0      0.0E0 0.20000E0
 0.00
   27
    2    3    6    9   12   13   15   17   28   30   32   43   45   47   58   60
   61   62   64   65   67   68   70   71   73   74   75
 0.0
 0.0
 0.0
 0.0
    9
    1    4    7   16   19   22   31   34   37
 0.025     0.05       0.025      0.05       0.1        0.05       0.025      0.05
 0.025

TRANSVERSE DEFLECTION AT THE NODES (LOAD, PO=10 PSI):

   0.16335D-04  0.15639D-04  0.13687D-04  0.69013D-05  0.0          0.15198D-04  0.14621D-04  0.12760D-04
   0.64698D-05  0.0          0.11816D-04  0.11374D-04  0.10115D-04  0.53652D-05  0.0          0.55145D-05
   0.53698D-05  0.49607D-05  0.29166D-05  0.0          0.0          0.0          0.0          0.0
   0.0
```

```
PROBLEM 4.95: STATIC BENDING OF A SIMPLY SUPPORTED CIRCULAR PLATE UNDER UDL
   2    8    0    0    0    0    0    0
   4   21
   1    3   11    9    2    7   10    6
   3    5   13   11    4    8   12    7
   9   11   19   17   10   15   18   14
  11   13   21   19   12   16   20   15
5.0      0.0      6.25     0.0      7.5      0.0      8.75     0.0
10.0     0.0      4.6194   1.9134   6.9291   2.8701   9.2388   3.8268
3.5355   3.5355   4.4194   4.4194   5.3033   5.3033   6.1872   6.1872
7.071    7.071    1.9134   4.6194   2.8701   6.9291   3.8268   9.2388
0.0      5.0      0.0      6.25     0.0      7.5      0.0      8.75
0.0      10.0
  30.0E6    30.0E6 11.6279E6 11.6279E6 11.6279E6    0.29E0    0.0E0    1.0E0
10.0
  15
   3    6    9   12   13   15   22   37   46   61   50   53   56   59   62
0.0
0.0
   0
```

```
TRANSVERSE DEFLECTION AT THE NODES (LOAD PO=10 PSI):

  0.22831D-02  0.16990D-02  0.11296D-02  0.56082D-03  0.0       0.22809D-02  0.11257D-02  0.0
  0.22831D-02  0.16990D-02  0.11296D-02  0.56079D-03  0.0       0.22809D-02  0.11257D-02  0.0
  0.22831D-02  0.16990D-02  0.11296D-02  0.56082D-03  0.0
```

```
PROBLEM 4.98: TRANSIENT ANALYSIS OF THE EQUATION IN PROBLEM 4.11 OR 4.73
   0    3   -1    0    1    1    0
   2    4
0.25      0.25
0.25      0.25      0.25      0.25
1.0       1.0       0.0       0.0       0.0
   9
   1    2    3    4    7    10    13    14    15
0.0       0.0       0.0       0.0       0.0       0.0       1.0       1.0
1.0
   0
0.05      0.5       1.0
0.0
0.0

SOLUTION (U AT THE NODES OF THE TRIANGULAR-ELEMENT MESH):

TIME= 0.050

   0.0             0.0             0.0             0.0             0.24682D-02     0.62607D-02
   0.0             0.19734D-01     0.36582D-01     0.0             0.14397D+00     0.19816D+00
   0.10000D+01     0.10000D+01     0.10000D+01

TIME= 0.500

   0.0             0.0             0.0             0.0             0.71851D-01     0.97631D-01
   0.0             0.18633D+00     0.25127D+00     0.0             0.42988D+00     0.52564D+00
   0.10000D+01     0.10000D+01     0.10000D+01

TIME= 0.750

   0.0             0.0             0.0             0.0             0.71311D-01     0.98480D-01
   0.0             0.18780D+00     0.24956D+00     0.0             0.42828D+00     0.52710D+00
   0.10000D+01     0.10000D+01     0.10000D+01
```

```
TIME= 1.000

   0.0            0.0            0.0            0.0            0.71465D-01    0.98129D-01
   0.0            0.18742D+00    0.25013D+00    0.0            0.42865D+00    0.52670D+00
   0.10000D+01    0.10000D+01    0.10000D+01

SOLUTION (U AT THE NODES OF THE RECTANGULAR-ELEMENT MESH; INPUT DATA NOT GIVEN):

TIME= 0.050

   0.0            0.0            0.0            0.0            0.20104D-02    0.24122D-02
   0.0            0.14181D-01    0.25287D-01    0.0            0.15876D+00    0.16143D+00
   0.10000D+01    0.10000D+01    0.10000D+01

TIME= 0.500

   0.0            0.0            0.0            0.0            0.71890D-01    0.10086D+00
   0.0            0.19217D+00    0.27890D+00    0.0            0.50774D+00    0.58421D+00
   0.10000D+01    0.10000D+01    0.10000D+01

TIME= 0.750

   0.0            0.0            0.0            0.0            0.71481D-01    0.10099D+00
   0.0            0.19297D+00    0.27845D+00    0.0            0.50693D+00    0.58491D+00
   0.10000D+01    0.10000D+01    0.10000D+01

TIME= 1.000

   0.0            0.0            0.0            0.0            0.71557D-01    0.10090D+00
   0.0            0.19284D+00    0.27860D+00    0.0            0.50704D+00    0.58477D+00
   0.10000D+01    0.10000D+01    0.10000D+01

...............................................................................
```

```
PROBLEM 4.100: TRANSIENT ANALYSIS OF THE WALL-DRIVEN SQUARE CAVITY (PROB.4.82)
   1     4     2     0     1     1     0
   8     8
0.1       0.125     0.125     0.15      0.15      0.125     0.125     0.1
0.0
0.15      0.125     0.125     0.125     0.125     0.125     0.125     0.1
0.0
1.0      100.0      0.0       0.0       0.0
  64
   1     2     3     4     5     6     7     8     9    10    11    12    13    14    15    16
  17    18    19    20    35    36    37    38    53    54    55    56    71    72    73    74
  89    90    91    92   125   126   127   128   143   144   145   146   148   150   152   154
 156   158   160   161   162   107   108   109   110   147   149   151   153   155   157   159
**** SEVEN ROWS OF ZEROS GO HERE ****
0.0       1.0       1.0       1.0       1.0       1.0       1.0       1.0
   0
0.05      0.5       1.00
**** TWENTY TWO ROWS OF ZEROS GO HERE ****
```

VELOCITIES AT SELECTED NODES FOR VARIOUS TIMES:

	NODE	U	V	NODE	U	V
T=.05	14	-0.34104D-01	-0.53545D-08	50	-0.11693D+00	0.22379D-07
	23	-0.11353D+00	0.10310D-07	59	-0.13027D+00	0.14882D-07
	32	-0.66084D-01	0.22431D-08	68	0.33218D+00	-0.20727D-08
	41	-0.23463D+00	-0.78096D-08	77	0.10000D+01	0.0
T=.5	14	-0.64281D-01	-0.31532D-07	50	-0.12825D+00	-0.49570D-07
	23	-0.16820D+00	-0.24267D-07	59	-0.45167D-01	0.17425D-07
	32	-0.13697D+00	0.46672D-07	68	0.49755D+00	0.63478D-07
	41	-0.29850D+00	0.46453D-07	77	0.10000D+01	0.0
T=1.	14	-0.64209D-01	0.67872D-09	50	-0.12856D+00	0.10767D-07
	23	-0.16787D+00	-0.50488D-08	59	-0.40853D-01	-0.72040D-07
	32	-0.13663D+00	-0.71011D-07	68	0.49145D+00	0.23391D-07
	41	-0.29925D+00	0.28405D-07	77	0.10000D+01	0.0

```
PROBLEM 4.103:DYNAMIC ANALYSIS OF A SIMPLY SUPPORTED CIRCULAR PLATE UNDER UDL
   2     8     0     0     1    10    10     0
   4    21
   1     3    11     9     2     7    10     6
   3     5    13    11     4     8    12     7
   9    11    19    17    10    15    18    14
  11    13    21    19    12    16    20    15
5.0       0.0       6.25      0.0       7.5       0.0       8.75      0.0
10.0      0.0       4.6194    1.9134    6.9291    2.8701    9.2388    3.8268
3.5355    3.5355    4.4194    4.4194    5.3033    5.3033    6.1872    6.1872
7.071     7.071     1.9134    4.6194    2.8701    6.9291    3.8268    9.2388
0.0       5.0       0.0       6.25      0.0       7.5       0.0       8.75
0.0       10.0
   30.0E6     30.0E6 11.6279E6 11.6279E6 11.6279E6     0.29E0      1.0E0      1.0E0
10.0
  15
   3     6     9    12    15    13    22    37    46    61    50    53    56    59    62
0.0
0.0
   0
0.1       0.5


TRANSVERSE DEFLECTION AT THE NODES (LOAD PO=10 PSI):

TIME = 0.100D+00

   0.21585D-02   0.16065D-02   0.10684D-02   0.53058D-03   0.0              0.21564D-02   0.10647D-02   0.0
   0.21585D-02   0.16065D-02   0.10684D-02   0.53054D-03   0.0              0.21564D-02   0.10647D-02   0.0
   0.21585D-02   0.16065D-02   0.10684D-02   0.53058D-03   0.0
```

```
TIME = 0.500D+00

  0.18268D-02    0.13604D-02    0.90543D-03    0.45000D-03    0.0            0.18250D-02    0.90231D-03    0.0
  0.18268D-02  , 0.13604D-02    0.90542D-03    0.44997D-03    0.0            0.18250D-02    0.90231D-03    0.0
  0.18268D-02    0.13604D-02    0.90543D-03    0.45000D-03    0.0

TIME = 0.100D+01

 -0.39423D-03   -0.29245D-03   -0.19357D-03   -0.95684D-04    0.0           -0.39385D-03   -0.19291D-03    0.0
 -0.39424D-03   -0.29245D-03   -0.19357D-03   -0.95678D-04    0.0           -0.39385D-03   -0.19291D-03    0.0
 -0.39423D-03   -0.29245D-03   -0.19357D-03   -0.95684D-04    0.0
```

Chapter 5

5.1 The finite-element equations associated with Eqs. (5.5) are given by

$$\frac{1}{2}\begin{bmatrix} 1 & -1 \\ -1 & 1 \end{bmatrix}\begin{Bmatrix} \sigma_1^e \\ \sigma_2^e \end{Bmatrix} = \frac{f_e h_e}{2}\begin{Bmatrix} 1 \\ 1 \end{Bmatrix} + \begin{Bmatrix} P_1^e \\ P_2^e \end{Bmatrix}$$

$$\frac{1}{2}\begin{bmatrix} -1 & 1 \\ -1 & 1 \end{bmatrix}\begin{Bmatrix} u_1^e \\ u_2^e \end{Bmatrix} = \frac{h_e}{6a_3}\begin{bmatrix} 2 & 1 \\ 1 & 2 \end{bmatrix}\begin{Bmatrix} \sigma_1^e \\ \sigma_2^e \end{Bmatrix}$$

5.2 When the second equation in Eq. (5.5) is used as a constraint, the following finite-element equations can be derived over an element:

$$[K^e]\{u^e\} - [G^e]\{\sigma^e\} = \{F^e\} \qquad [G^e]^T\{u^e\} - [H^e]\{\sigma^e\} = \{0\}$$

$$K_{ij}^e = \int_{x_e}^{x_{e+1}} (a_e + \gamma_e)\frac{d\psi_i^e}{dx}\frac{d\psi_j^e}{dx}\,dx \qquad G_{ij}^e = \int_{x_e}^{x_{e+1}} \frac{\gamma_e}{a_e}\frac{d\psi_i^e}{dx}\phi_j^e\,dx$$

$$H_{ij}^e = \int_{x_e}^{x_{e+1}} \frac{\gamma_e}{a_e^2}\phi_i^e\phi_j^e\,dx \qquad F_i^e = \int_{x_e}^{x_{e+1}} f\psi_i^e\,dx + P_i^e$$

where ψ_i^e and ϕ_i^e are the interpolation functions used to approximate u and σ, respectively.

5.3 The finite-element mode associated with the penalty functional in Eq. (5.30) is given by

$$\begin{bmatrix} [K^{11}] & [K^{12}] \\ [K^{12}]^T & [K^{22}] \end{bmatrix}\begin{Bmatrix} \{w\} \\ \{\theta\} \end{Bmatrix} = \begin{Bmatrix} \{F^1\} \\ \{F^2\} \end{Bmatrix}$$

where

$$K_{ij}^{11} = \int_{x_e}^{x_{e+1}} \gamma\frac{d\psi_i}{dx}\frac{d\psi_j}{dx}\,dx \qquad K_{ij}^{12} = \int_{x_e}^{x_{e+1}} \gamma\frac{d\psi_i}{dx}\phi_j\,dx = K_{ji}^{21}$$

$$K_{ij}^{22} = \int_{x_e}^{x_{e+1}} \left[b\frac{d\phi_i}{dx}\frac{d\phi_j}{dx}\,dx + \gamma\phi_i\phi_j\right]dx \qquad F_i^1 = -\int_{x_e}^{x_{e+1}} \psi_i f\,dx + Q_1\psi_i(x_e) + Q_3\psi_i(x_{e+1})$$

$$F_i^2 = Q_2\phi_i(x_e) + Q_4\phi_i(x_{e+1})$$

5.5 For four-element mesh, the eigenvalues are given by $\lambda_1 = 4.2048$, $\lambda_2 = 27.336$, $\lambda_3 = 85.776$, and $\lambda_4 = 177.6$. The exact solution is given by the roots of the equation $\lambda + \tan\lambda = 0$.

5.6 For the crude meshes of triangular elements and rectangular elements shown in Figs. 4.8 and 4.9, the first eigenvalue is given by $\lambda_{11} = 5.415$ (triangles), $\lambda_{11} = 5.193$ (rectangles). The exact solution is given by $\lambda_{mn} = [(m/a)^2 + (n/a)^2]\pi^2$ $(a = 2)$.

5.8
$$\begin{bmatrix} [K^{11}] & [K^{12}] \\ [K^{21}] & [K^{22}] \end{bmatrix}\begin{Bmatrix} \{\Delta^1\} \\ \{\Delta^2\} \end{Bmatrix} = \begin{Bmatrix} \{F^1\} \\ \{F^2\} \end{Bmatrix} \qquad \{\Delta^1\} = \begin{Bmatrix} u_1 \\ u_2 \end{Bmatrix} \qquad \{\Delta^2\} = \begin{Bmatrix} w_1 \\ \theta_1 \\ w_2 \\ \theta_2 \end{Bmatrix}$$

$$K_{ij}^{11} = \int_{x_e}^{x_{e+1}} EA\frac{d\psi_i}{dx}\frac{d\psi_j}{dx}\,dx \qquad K_{iJ}^{12} = \frac{1}{2}\int_{x_e}^{x_{e+1}} EA\left(\frac{dw}{dx}\right)\frac{d\psi_i}{dx}\frac{d\phi_J}{dx}\,dx$$

$$K_{Ij}^{21} = \int_{x_e}^{x_{e+1}} EA\left(\frac{dw}{dx}\right)\frac{d\phi_I}{dx}\frac{d\psi_j}{dx}\,dx \qquad K_{IJ}^{22} = \int_{x_e}^{x_{e+1}} EI\frac{d^2\phi_I}{dx^2}\frac{d^2\phi_J}{dx^2}\,dx$$

$$F_i^1 = \int_{x_e}^{x_{e+1}} f\psi_i\,dx + P_i \qquad F_I^2 = \int_{x_e}^{x_{e+1}} q\phi_I\,dx + Q_I$$

for $i, j = 1, 2$ and $I, J = 1, 2, 3, 4$. Here f is the axially distributed force and q is the transversely distributed force.

5.9 $[K^{11}] = \frac{EA}{h}\begin{bmatrix} 1 & -1 \\ -1 & 1 \end{bmatrix} \qquad [K^{12}] = \frac{1}{2}[K^{21}] = \frac{EAN}{h}\begin{bmatrix} 1 & 0 & -1 & 0 \\ -1 & 0 & 1 & 0 \end{bmatrix}$

$[K^{22}] =$ given by Eq. (3.102) $\qquad N = \frac{1}{2}\left(\frac{dw}{dx}\right)$

5.10 $[K(\bar{u})]\{u\} = \{F\} \qquad K_{ij} = \int_{x_e}^{x_{e+1}} \bar{u}\frac{d\psi_i}{dx}\frac{d\psi_j}{dx}dx \qquad F_i = -\int_{x_e}^{x_{e+1}} \psi_i\, dx + P_i$

where $\bar{u}$ is the solution from the previous iteration ($\bar{u} = 0$ at the start of the iteration).

5.11 $\psi_1 = \frac{1}{8}(1-\xi)(1-\eta)(1-\zeta) \qquad \psi_5 = \frac{1}{8}(1-\xi)(1-\eta)(1+\zeta)$

$\psi_8 = \frac{1}{8}(1-\xi)(1+\eta)(1+\zeta)$

5.12 $[K^e + H^e]\{T^e\} = \{F^e\}$

$$K_{ij}^e = \int_{\Omega^e}\left(k_x\frac{\partial\psi_i}{\partial x}\frac{\partial\psi_j}{\partial x} + k_y\frac{\partial\psi_i}{\partial y}\frac{\partial\psi_j}{\partial y} + k_z\frac{\partial\psi_i}{\partial z}\frac{\partial\psi_j}{\partial z}\right)dx\,dy\,dz$$

$$H_{ij}^e = \beta\int_{S^e}\psi_i\psi_j\,ds \qquad F_i^e = \int_{\Omega^e} q\psi_i\,dx\,dy\,dz + \beta T_\infty\int_{S^e}\psi_i\,ds$$

5.13
$$\begin{bmatrix} [K^{11}] & [K^{12}] & [K^{13}] \\ [K^{12}]^T & [K^{22}] & [K^{23}] \\ [K^{13}]^T & [K^{23}]^T & [K^{33}] \end{bmatrix}\begin{Bmatrix} \{u\} \\ \{v\} \\ \{w\} \end{Bmatrix} = \begin{Bmatrix} \{F^1\} \\ \{F^2\} \\ \{F^3\} \end{Bmatrix}$$

where the coefficient matrices K_{ij}^{IJ} and F_i^I ($i, j = 1, 2, \ldots, n$; $I, J = 1, 2, 3$) are given by

$$K_{ij}^{11} = \int_{\Omega^e}\left[\mu\left(2\frac{\partial\psi_i}{\partial x}\frac{\partial\psi_j}{\partial x} + \frac{\partial\psi_i}{\partial y}\frac{\partial\psi_j}{\partial y} + \frac{\partial\psi_i}{\partial z}\frac{\partial\psi_j}{\partial z}\right) + \gamma\frac{\partial\psi_i}{\partial x}\frac{\partial\psi_j}{\partial x}\right]dx\,dy\,dz$$

$$K_{ij}^{12} = \int_{\Omega^e}\left(\mu\frac{\partial\psi_i}{\partial y}\frac{\partial\psi_j}{\partial x} + \gamma\frac{\partial\psi_i}{\partial x}\frac{\partial\psi_j}{\partial y}\right)dx\,dy\,dz = K_{ji}^{21}$$

$$K_{ij}^{13} = \int_{\Omega^e}\left(\mu\frac{\partial\psi_i}{\partial z}\frac{\partial\psi_j}{\partial x} + \gamma\frac{\partial\psi_i}{\partial x}\frac{\partial\psi_j}{\partial z}\right)dx\,dy\,dz = K_{ji}^{31}$$

$$K_{ij}^{22} = \int_{\Omega^e}\left[\mu\left(\frac{\partial\psi_i}{\partial x}\frac{\partial\psi_j}{\partial x} + 2\frac{\partial\psi_i}{\partial y}\frac{\partial\psi_j}{\partial y} + \frac{\partial\psi_i}{\partial z}\frac{\partial\psi_j}{\partial z}\right) + \gamma\frac{\partial\psi_i}{\partial y}\frac{\partial\psi_j}{\partial y}\right]dx\,dy\,dz$$

$$K_{ij}^{23} = \int_{\Omega^e}\left(\mu\frac{\partial\psi_i}{\partial z}\frac{\partial\psi_j}{\partial y} + \gamma\frac{\partial\psi_i}{\partial y}\frac{\partial\psi_j}{\partial z}\right)dx\,dy\,dz = K_{ji}^{32}$$

$$K_{ij}^{33} = \int_{\Omega^e}\left[\mu\left(\frac{\partial\psi_i}{\partial x}\frac{\partial\psi_j}{\partial x} + \frac{\partial\psi_i}{\partial y}\frac{\partial\psi_j}{\partial y} + 2\frac{\partial\psi_i}{\partial z}\frac{\partial\psi_j}{\partial z}\right) + \gamma\frac{\partial\psi_i}{\partial z}\frac{\partial\psi_j}{\partial z}\right]dx\,dy\,dz$$

$$F_i^1 = \int_{\Omega^e} f_x\psi_i\,dx\,dy\,dz + \oint_{\Gamma^e} t_x\psi_i\,ds$$

$$F_i^2 = \int_{\Omega^e} f_y\psi_i\,dx\,dy\,dz + \oint_{\Gamma^e} t_y\psi_i\,ds$$

$$F_i^3 = \int_{\Omega^e} f_z\psi_i\,dx\,dy\,dz + \oint_{\Gamma^e} t_z\psi_i\,ds$$

5.15 $f_i^e = 1 \qquad Q_i^e = 1 \qquad$ for $i = 1, 2, \ldots, 8$

INDEX